AF530730

Rainwater Management (Theory and Practice)

Dr. M.L. Jat, obtained his Ph.D. degree from the MPUAT in 2002 in the field of soil water engineering. He has 30 years teaching, extension and research experience in the field of dryland agriculture, groundwater, rainwater manage-ment, resource characterization and drought predication and management. He has published five books and more than 100 research papers and has actively participated in many national and international scientific conferences and seminar. He is the member of several organizations and societies. He is presently working as Professor (S.W.E.) in the All India Coordinated Research Project on Dryland Agriculture at Dryland Farming Research Station, Arjia, Bhilwara (Rajasthan), MPUAT, Udaipur.

Dr. Om Prakash Gill, obtained his Ph.D. from RAU, Bikaner in 1992. He took over the additional charge as the Vice-Chancellor, MPUAT, Udaipur 2012 to 2015. He has 38 years experience in teaching, research, extension, administration and shouldered many additional responsibilities. He has contributed significantly in specialized areas like Integrated Farming System; Agromet Advisory Services; Climate Change; Organic Farming; Micro Irrigation and Weed Management. The technologies developed by him to achieve sustainable increase in productivity and minimization of pre and post harvest losses due to pests and weeds cover crops like groundnut, onion, radish, isabgol, pearl millet, wheat, mustard, chickpea, etc. He has published 42 research papers in different international and national journals, more than 300 popular articles in different periodicals and publications, three edited books and delivered more than 120 radio and TV talks.

Dr. B.S. Deora, obtained his Ph.D. (SWC Engg.) MPUAT, Udaipur presently working, Professor at Centre for Natural Resources Management, S.D. Agricultural University, S.K. Nagar, Gujarat (India). He for the last 32 years is actively involved in teaching, research & development, extension and techno-administration. He gave 9 scientific recommendations, published 86 international/ national publications/articles and reports; 5 book chapters, 2 books deliveral 10 T.V. and 4 Radio programmes to his credit. He also has been a member of the Governing Board, the Academic Councils, various selection committees and panel of examiners of several organizations/universities; and *Chairman/Member Secretary/Member* of more than 140 scientific/administrative committees.

Dr. Vivek Kumar, is presently working as Assistant Professor in the Department of Agriculture Engeneering & Food Technology, Sardar Vallab Bhai Patel University of Agriculture and Technology, Modipuram, Meerut. He obtained his Ph.D. (Agricultural Engeneering) from Sardar Vallab Bhai Patel University of Agriculture and Technology, Meerut. He has 12 years experience of teaching UG & PG classes. He guided more than six students of M. Tech and one Ph. D. student as Chairman of advisory Committee. He also worked in International Collobration with IRRI, Philipines and CIMMYT Mexico and has International exposure also. He has published two books, four Technical Mannuals and more than 30 research papers in International and national repute Journals.

Rainwater Management

Theory and Practice

Authors

M. L. Jat

O. P. Gill

B. S. Deora

Vivek Kumar

Published by:

Scientific Publishers (India)
5 A, New Pali Road, P.O. Box 91
Jodhpur - 342 001 (India)

Branch Office
Scientific Publishers (India)
4806/24, Ansari Road, Daryaganj
New Delhi - 110 002 (India)

E-mail: info@scientificpub.com
Website: www.scientificpub.com

ISBN: 978-93-86347-23-7

Printed in India

FOREWORD

The mainstay of economy of India's is agriculture, which is currently exhibiting trends of increasing un-sustainability. Further, rainfed farming will remain the main stay for the livelihood support of millions of small and marginal farmers across the country even after realizing the complete irrigation potential. It faces major challenges imposed by climate change in further expansion of irrigated area in the country. Impacts of climate change in agriculture are witnessed all over the world, but country like India is more vulnerable in view of the high population pressure on agriculture and natural resources with poor coping mechanisms. Though, much of the enormous water resources in the country remain unutilized. Thus, unplanned and unscientific planning of rainwater quite often causes disruptions of hydrologic and environmental regimes at micro and regional level are common to our experiences, such action are eventually lead to break drawn of economic fabrics and social setup. The efficient utilization of rainwater through better management is the core issue if the cropping intensity and production are to be enhanced for livelihood security of dryland farmers. The rainwater management can be implemented efficiency and economically as a viable alternative to on-farm irrigation, drinking and multiple uses, considering the fact that any land anywhere can be used to manage rainwater. This considered being an ideal solution to water scarcity regimes.

The scope of rainwater management is wide and varied. It covers diversified areas such as understanding of dryland engineering, dryland farming and its concept, weather forecasting, rainwater management for drought proofing and mitigation, risk management under drought situation, water harvesting either *in-situ* or *ex-situ* and their efficient recycling, reduction of water losses, design of water harvesting structures and irrigation systems and the planning operation drainage of recycling system, waste water use in agricultural lands and crop planning, watershed management for better utilization of rainwater. This book deals with all these aspects. It amply serves a widely felt need for a suitable and comprehensive book on rainwater management with emphasis on water harvesting either *in-situ* or *ex-situ*, the analysis of rainfall and water balance, efficient crop planning to enhance productivity, recycling of waste water, watershed management, planning operation of recycling system.

I very much appreciate the efforts made by Drs. M.L. Jat, O.P. Gill & B.S. Deora for bringing out this book on "Rainwater Management" which I hope will be useful for students, teachers and researchers, policy makers and other having interest in the areas related to rainwater management and dryland agriculture.

Prof. U.S. Sharma
Vice-Chancellor
MPUAT, Udaipur (Raj.)

PREFACE

This book is intended as professional textbook for students of agricultural engineering, students of agriculture soil & water engineering. It will also serve as a text for students at the undergraduate and postgraduate levels in civil engineering. In addition to the students, the book will be a valuable reference to professional engineers and agricultural scientists working in the field of rainwater management, dryland farming and irrigation water management.

In addition, the book will also be a valuable reference for the candidates appearing in competitive examinations for the Agricultural Research Service and for the examinations leading to the Associate Membership of the Institution of Engineers (India). Professional training institutions in Soil Conservation and Water Management, Agricultural Polytechnics, Krishi Vigyan Kendras and Rural Institutes and similar other organizations having a programme in water management would find this book useful.

Though written primarily to serve as a text/reference for students of agricultural and engineering institutions and practicing engineers and technologists in developmental organizations in India, it is hoped that the book will be valuable to similar groups in the developing countries.

Rainwater Management, Theory and Practice covers all the areas of rainwater management, precipitation and its analysis, drought indices, weather forecasting, water harvesting, design of water harvesting structure, water losses and control, water lifting devices, irrigation scheduling, planning operation of recycling system, water recycling system related to agriculture. It is a comprehensive treatment of rainwater resources, water balance, drought, measurement of water and field layout, soil-plant-water relationships, water conveyance and control, water application methods and water recycling in salty land, drainage of agricultural lands.

The details regarding design and layout of irrigation practices in the field, including step-by-step procedures, and necessary equipment, are included wherever applicable. Solved examples have been included to emphasize the design principles and facilitate the understanding of the subject matter in each chapters. Tables, figures and articles are numbered chapter-wise.

A list of important references is given for each chapter. A set of problems for home work is provided for answers at the end of each chapter. Examples and illustrations are provided wherever applicable, make easy the understanding of the subject. The entire book has been written using MKS of weights and measures. Conversion tables of units from one system to another are given in Appendix.

The author is deeply indebted to many societies, dignitaries and organizations for the supply of useful material for inclusion in the manuscript and the assistance at various levels for the preparation of the manuscript.

I express my sincere gratitude to Dr. S. K. Sharma, Zonal Director Research, ARS, MPUAT, Udaipur, Dr. J. K. Balyan, Assistant Professor (Agronomy), Dryland Farming Research Station, Arjia, Bhilwara and Dr. C. M. Yadav, Assistant Professor, Krishi Vigyan Kendra, Bhilwara for extending us valuable help from time to time during preparation of manuscript and also be highly acknowledge direct-indirect cooperation to our all well wishers for their.

The authors would welcome suggestions from the readers to improve the text.

- Authors

LIST OF TABLES

LIST OF FIGURES

CONTENTS

Chapter 1

INTRODUCTION

1.1 Introduction

Rainwater management is essentially the process of managing every drop of rain for productive use *viz.*, crop production, drinking water for human and livestock, industrial purposes and also for artificial recharge of groundwater etc. without deteoriating land and groundwater depletion. Watershed is the basis of rainwater management in the water scarcity areas or dryland ecosystem.

The success of rainfed agriculture, industrial development and drinking water for human and livestock depends on the efficient utilization of rainwater. The core strategies of rainwater conservation were based on improving the water availability to the crops, increasing the groundwater recharge and harvesting rainwater for providing life saving irrigation to the crops during the dryspell period. The approaches were on building in-situ soil moisture reserves to tide over the recurring dryspells, preventing subsequent loss of profile stored soil moisture, permitting safe disposal of runoff, its collection above or below ground and tactical recycling of harvested stored water. The aim of rainwater management is to enhance crop yield per drop. Applicability of techniques so developed depends upon the soil characteristics and rainfall.

1.2 Water Resources

India receives annually about 4000 billion m^3 water through precipitation both during the south-west and north-west monsoons. Out of this, about 1280 billion m^3 is lost as evaporation, about 850 billion m^3 seeps into the land and the remaining 1870 billion m^3 constitutes the average surface water flow. Theoretically, this water is adequate to provide 1 metre depth on the entire cultivated area in the country. However, only about 50% surface water can be put to beneficial use because of topographical and other constraints. Out of the country's utilizable water resource of about 1100 billion m^3, surface water and groundwater account for about 60 and 40%, respectively. As the main rainy season concedes with hot temperatures and high evapotranspiration demand, up to 50% precipitation is lost through evapotranspiration. Additional 30% runs off, because tropical, rainfall has tendency to raplenish in a few large storms interspersed by discrete dry periods. Estimates reveal that areas receiving up to 1000 mm annual precipitation hold a potential to add 63 million m^3 water equivalent through runoff (Singh *et al.*, 2000).

The country has been divided into 20 river basins – comprising 12 major river basins having a catchment area >20,000 km^2 and 8 composite river basins (Sharma and Paul, 1999). Other water resources such as reservoirs, tanks, ponds and lakes over about 7 million ha surface area. Because of the uneven precipitation, the availability of renewable fresh water varies in different river basins. In the semi-arid region, the peninsular rivers are rainfed and therefore fluctuate in flow volume. The coastal rivers are short in length with limited catchment areas. The rivers and tributaries of these river systems are intermittent and ephemeral in nature. The Himalayan Rivers, flowing through semi-arid region, are snowed and perennial. Floods and droughts affect on an average about 9 million ha $year^{-1}$ and about one-third of the country, respectively.

In recent years, there has been marked trend towards development of groundwater for irrigation. With the advent of electrical and diesel pumps, groundwater extraction has tremendously increased to exploitation level in many parts of the country. Specifically after the introduction of high yielding varieties of crops in the mid sixties, rapid progress in the agricultural sector has been seam with the development of groundwater based irrigation system through dug wells, bore wells, tube wells and dug-cum-bore wells. However, due to overexploitation leading to groundwater mining and limited recharge, the groundwater is getting depleted at a faster rate. This depletion is particularly marked in most of the semi-arid region in the States of Andhra Pradesh, Karnataka, Madhya Pradesh. Maharashtra, Orissa, Rajasthan and Tamilnadu Intrusion of seawater into fresh water aquifers due to overexploitation of groundwater adds to the problem in coastal areas. The Central Groundwater Authority set up under the Environment (Protection) Act, 1986 regulates the sinking of rube wells in the country, particularly where the level of groundwater decline is severe. However, the only control for regulating the groundwater exploitation is through indirect measures being adopted by the financial institutions by the way of prescribing for interspacing criterion between two consecutive wells and insisting for technical clearance for the proposed programme from the competent authority. Assessing the stage of groundwater development in the particular block, i.e., white, grey or dark, based on the stage of groundwater development as ratio of replenishable groundwater, also regulates the groundwater development schemes. The sanction of groundwater development schemes in grey or dark areas require more intensive studies through micro-level surveys and evaluation for detained assessment of the groundwater potential.

1.3 Need for Rainwater Management

Water is a major component of the environment and the human beings occupy the centre-stage in it. Water is needed for agricultural production, for drinking and other purposes by the people and animals and for municipal and commercial use. Water is an essential input for all industries. Energy production and navigation are dependent on water. This all-pervasive needs for water has brought into focus the fragility of environment associated with the fulfillment of these requirements and the need to guard against it.

The status of water in the environment is unique. Water is use only distributed and a scarce economic resource. The challenge in meeting water demand is to balance the needs for development with environmental health for sustainable growth.

1.3.1 Availability

It is hard to believe that the earth on which water seems to be the dominant element should ever face its shortage. But the reality is that 97.3 per cent of its reserve on earth is saline water. Only 2.7 per cent is fresh water. Of that nearly 70 per cent is frozen in the ice caps of Antarctica and Greenland and most of the balance is present either as soil moisture or lies in deep underground aquifers as ground water not accessible to human use. As a result, less than 1 per cent of all the water on earth is accessible for direct human use. This water is found in lakes, rivers, reservoirs and those underground sources that are shallow enough to be tapped at an affordable cost. Only this tiny fraction of the planet's water is renewed by nature's water cycle.

1.3.2 Distribution

Water resources are very unevenly distributed. At one extreme are the deserts with almost no rainfall. On the other hand the humid regions can receive several metres of rainfall in a year. Most of the flow is in a limited number of rivers. Even in parts of the world with large river flows, there is large variation in the availability of water. Most of the annual water flow may come as floods following heavy rains or the melting of snow. Unless stored in reservoirs, the water flows back to the sea, sometimes causing even seasonal flooding. Latter in the year, the same areas may experience drought.

Water requirement for agricultural production and domestic needs like drinking, cooking, sanitation and municipal services on the one hand and for trade commerce and industries on the other constitutes the bulk of its total consumption. The requirement is closely related to population, demand for food, production of non-food agricultural and industrial items, output and for improvement in the quality of life and for preserving ecology of the nation. Fresh water is a renewable but a finite and vulnerable resource, essential to sustain life, development and the environment.

Water scarcities caused by population, environmental change and degradation and the unequal distribution of water resources. Water is getting scarce due to rising population, rapid industrialization and growing industrial demands.

The total storage built up from projects completed upto 1995 was about 174 km^3, while another 76 km^3 will be available from projects now under construction. Small tanks provide about 3 km^3, giving a total of 253 km^3 on completion of on going projects. There are identified future projects for storage of 132 km^3. If these are taken up and commissioned, the total storage would become to 385 km^3. It can be shown that this is the bare minimum live storage needed to balance seasonal flows in an average year. Without availability of this much storage, the assumption of 690 km^3 of utilizable surface water would not be valid. For comparison for about the same annual runoff USA has storage of 700 km^3.

The world's population in 2003 is estimated at 6.45 billion. Nearly 170 million people are being added to it annually and will increase upto 10 billion by the year 2025. The high growth ratio of population is posing a serious threat to the future prospects of human well being. If the population continues to grow at this rate, it will perhaps be too large to be supported by the limited resources of the earth in a few decades.

Water scarcities are likely to occur sooner in regions where its per capita availability is already very low following the high population growth. The demand gets more and more serious if its per capita demand keeps growing up following changes in its consumption pattern. People dump wastes, untreated sewage and chemical discharges which pollute the sources of water like the rivers, lakes, ponds and the underground reserves. The ever-increasing demand for water has resulted in two billion people craving for this basic resource worldwide.

1.3.3 Projection of water demand

National Commission for Integrated Water Resources Development (1999) estimated the future water demand exhaustive in India. The projected population in the year 2050 is taken as 1581 million for high and 1346 m for low estimates. For meeting adequate nutritional requirement, the requirement of food grains comes to 449 million tonnes for 'high' and 385 million tonnes for 'low' estimates of different sectors is shown in Table 1.1, based on the yields per ha are projected as 4.0 tonnes for irrigated and 1.5 tonnes for un-irrigated areas.

Table 1.1: Projected water demand for the year 2050

Sector	High (km^2)	Low (km^2)
Irrigation	68.4	64.54
Domestic	9.40	9.25
Industrial	6.86	8.32
Power	5.93	6.47
Inland navigation	63.13	1.5
Environmental release	1.10	2.05
Evaporation losses	6.44	7.8

Continued unchecked pollution through industrial wastes, untreated sewage and agrochemicals adds to the multidimensional negative consequences. The share of agriculture for good quality water is expected to decrease from 90% in 1990 to 71% in 2025. Though India receives annual precipitation of 119 cm (highest per unit area in any country of this size), assured water availability is greatly jeopardized by highly erratic and uncertain rainfall in terms of quantity, time and space. It is increasingly realized that achievement of the much-emphasized food security and sustainable agriculture will not be possible without most efficient and economic water management.

Agriculture will have to use more of the marginal waters and to produce more with less water. Thus, water shortage is going to be the most serious constraint to economic development and sustainable agricultural production in future the world over. If appropriate measures are not taken, we may have to even face disastrous water famines, conflicts and wars. The United Nations is seized of this impending danger. The recent interstate disputes for sharing river basin waters in India give enough signal of the imminent danger arising from the water shortage crisis. The overall per capita water availability in India is already approximately half of the world's average.

1.3.4 Indian scenario

India has 2.45 per cent of the world's land resources and 4 per cent of its fresh water resources which is estimated as 1953 km^3 (109 m^3 or BCM). Due to uneven distribution and limitation of suitable storage sites the un-utilizable resource is estimated as 690 km^3 surface and 396 km^3 groundwater equal to a total of 1086 km^3. Twelve per cent areas of the country receive an average rainfall of less than 610 millimeters (mm) annually. Only 8 percent areas receive more than 2500 mm of water. Cherrapunji in the eastern part of Meghalaya receives 11,000 mm of rainfall while western Rajasthan receives only 100 mm. The variability of the rainfall from month to month and year to year for the same place is very high. With high coefficient of variation ranging from 60-70 per cent and frequent occurrence of drought of floods (Singh *et al.*, 2000). Drought occurs once in 3 to 5 years either due to deficit rainfall in seasonal rainfall or inadequate water availability due to dryspells (Ramakrishnan, *et al.*, 1999). Even low rainfall areas, especially in some parts of Gujarat and Rajasthan, are prone to occasional high intensity storms.

Rajasthan is the driest state in the country and is having per capital water availability below 1000 cum/year, since 1990 (Sukla and Vohra, 2003).

More than 90 per cent of the annual runoff in peninsular rivers and over 80 per cent of the annual runoff in the Himalayan rivers occurs between June and September. Many of the small rivers totally dry up during the summer. The depletion of forests has further aggravated the problem.

Rajasthan has only 1 per cent of the country's water resources. Even those who live in areas of high rainfall often face drought. For instance, Cherrapunji in Meghalaya receives the highest level of mean rainfall recorded in the world and yet, because of heavy seasonal rainfall and the nature of topography, much of the runoff cannot be retained. This region now suffers from excessive flooding for 3 or 4 months and frequent droughts during the rest of the year.

India is the second largest country after China in terms of population. The Indians numbered about 108.8 crore in 2003, accounting for nearly 17 per cent of the world's population. India has roughly about 4 percent of the world's fresh water resources to support 17 per cent of the total world population. It is evident that a large number of Indians are facing acute water shortage. The number of those with no access to safe water supply and sanitation is unacceptably high. Nearly 50 per cent of the total sown area will remain rainfed even after harnessing all available water resources for creating the envisaged ultimate irrigation potential of 113.5 m ha.

Resource illiteracy makes hardly 30-40 per cent of water available for plant uptake. The water delivery and distribution systems are faulty not matching with the crop needs at different growth stages, thereby resulting in low water production efficiency (20-30 per cent). Excessive and irrational pumping of groundwater not being regulated in conformity with the annual recharge rate of aquifers in many parts of India endowed with good quality water, has led to severe water shortage particularly in the intensively cultivated areas and drought prone area. Rainwater management is essential for food and livelihood security for providing food security for millions of people. Based on the minimum per unit water consumption figure and the production figure, the water requirement for all the 17 categories of industries was assessed as 15 BCM per year in 1977. The demand for small scale industries was assessed as 7 BCM per year. The water demand in future for industries estimated as 70 BCM and 81 km^3 for the years 2025 and 2050, respectively.

Food grain production has touched an all time high of 198.2 million tonnes in 1996-97. The per capita availability of food grains has increased from 525 to 580 g per day between 1980 and 1996. Advances being made in the fields of bio-technology raise hope for increasing agricultural production further by virtue of its tremendous potential. However, a trend of decline is observed in so far as consumption is concerned. Per capita cereal consumption declined by about 1 kg per month and 600 g per month in rural and urban India, respectively between the period from 1987-88 to 1993-94. Even a marginal decline in per capita consumption of most varieties of pulses has been observed in the corresponding period (National Sample Survey Report No. 400, 402.

Growth rate of per capita food grain demand = Growth rate of (Food grain - Population). Therefore, growth rate of food grain demand is found out to be 3.67% per annum. The average net food grain availability in the period 1900-96 has been around 156.47 million tonnes. The net food grain demand in 2020 works out to be around 371.62 million tonnes. Assuming that the whole of this amount would he produced indigenously, the gross production of food grains should be around 42 million tonnes in 2020. The figure, although looks astounding, is in fact not so, because our food grains demand 25 years ago was around 85 million tonnes which is half the amount demanded now (i.e., 169.4 million tonnes in 1996). Hence, we can find a rationale for the figure that we have arrived at for 2020. However, considering different population growth rates in the coming years, we arrive at different levels of food grain demand in 2020 (Table 1.2), subject to assumption (iii) given earlier.

Table 1.2: Net and gross food grain availability under different population scenario

Population growth rate (% per annum)	Net food grain availability in 2020 10^6 tonnes	Gross food grain production in 2020 (for self sufficiency) 10^6 tonnes
1.0	332.45	380
1.5	352.33	406
1.75	362.28	414
2.0	371.62	425

Calculation based on data available from Economic Survey, 1996-97
(Source: Sawaminathan, 1998)

While good monsoon makes direct contribution in achieving higher production both in the un-irrigated areas and the irrigated commands due to proper filling of reservoirs and effective groundwater recharging poor monsoon can spell disaster by suppressing the production potential (Fig. 1.1). Creation of more live storage capacities and the promotion of traditional water harvesting methods can, to a large extent, mitigate the hazards of uncertain monsoon both in irrigated and rainfed areas

Water being a vital component of the life support system and a key input in agriculture, human population has centred largely around the rivers, seas, tanks, ponds, lakes and other water bodies as also in areas with otherwise good water availability. Trends in water consumption indicate that demand for water for household and industrial uses in developing countries could double as a proportion of total water demand in the next 25 years. Scope for water supply expansion will at the same time be limited because development of irrigation and urban water supplies is becoming increasingly expensive, and often involves high costs in terms of environmental degradation and human resettlement. Without fundamental reform of water management, the rapid growth in urban water demand will require large transfers of water from irrigated agriculture, thereby threatening food security. Hence, water supply and demand should be managed in an integrated fashion, considering simultaneously all uses and sources.

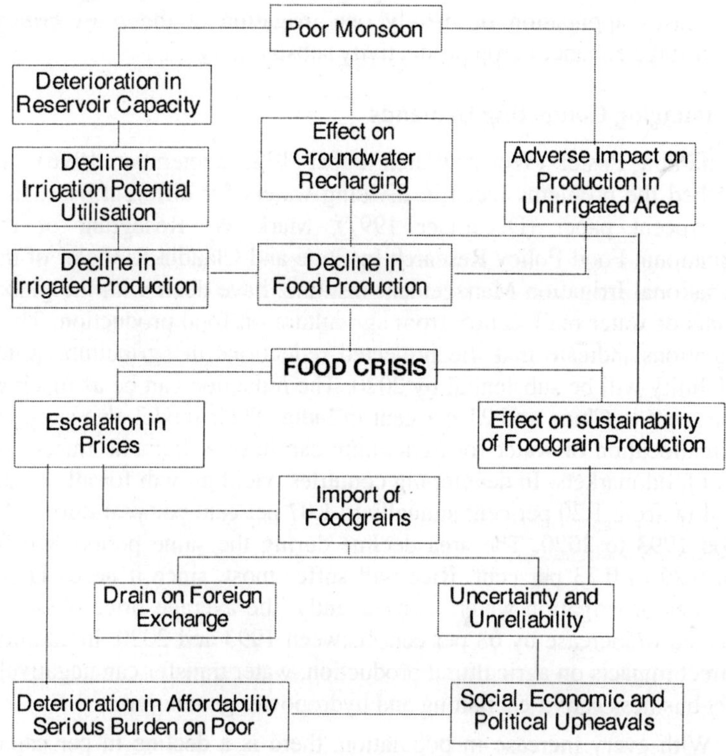

Fig. 1.1: Crisis due to poor monsoon (*Source*: Sawaminathan, 1998)

In order to meet the demands of the fast swelling population huge investments were made by several countries for creating water resources, thereby bringing roughly 255 m ha under irrigation at the global level with a target of 300 m ha by 2000 AD. About one-sixth (17%) of the world's cropped land is irrigated, but this area produces approximately one-third (33%) of the food supply. Irrigation enhances the production, capacity in otherwise harsh environments and reduces the need for horizontal expansion of rainfed agriculture. However, many civilizations flourished and perished depending upon how water was used, as in the wake of mismanagement of rainwater including surface and groundwater.

In 1995, the National Academy of Agricultural Sciences in India brought out the Agricultural Scientists perceptions on National Water Policy, but not much headway has been made in this respect so far. An increase of even 10 to 15% in irrigation efficiency, can result in enormous water saving and in bringing more areas under irrigation. In many cases application of merely one irrigation at the most critical growth stage enhances crop productivity substantially.

1.4 Managing Competing Demands

Sandra Postel (World Watch Paper 132, September 1996) has described the problems faced in dividing waters for different uses. In a more recent paper (December 1997), Mark W. Rosegrant of the International Food Policy Research Institute and Claudia Rinngler of the International Irrigation Management Institute, have dealt with the global impacts of water reallocation from agriculture on food production. Their calculations indicate that the projected reductions in agriculture water availability will be substantial by 2020. The reduction can be as much as 24 per cent in China and 21 per cent in India. Their model also suggests that reallocation of water for agriculture can have a dramatic impact on global food markets. In developing countries, yield growth for all cereals will slow from 1.20 per cent annually to 1.07 per cent per year during the period 1993 to 2020. The area decline during the same period will be from 0.29 to 0.23 per cent. Rice will suffer most, since it needs larger quantities of irrigation water. Consequently, the average price of rice is projected to increase by 68 per cent between 1993 and 2020. In addition to direct impacts on agricultural production, water transfer can negatively affect business activities, fishing and hydropower generation.

With every increase in population, there is a decline in per capita availability of water, a situation which can be altered only by new technologies such as solar desalination of sea water. For example, water availability per capita in India was over 5000 cubic meters per annum in 1950. It now stands at hardly more than 2000 cubic meters per capita. By year 2025, per capita availability is projected at only 1500 cubic maters (Fig. 1.2). Such quantitative data alone are not adequate to get a real picture of the water availability status. Pollution affects water quality both in rivers and groundwater. Also there are gross inequalities between basins and geographic regions. Water markets, if they are organized in a non-exploitative manner, can help to meet the minimum household and agricultural needs.

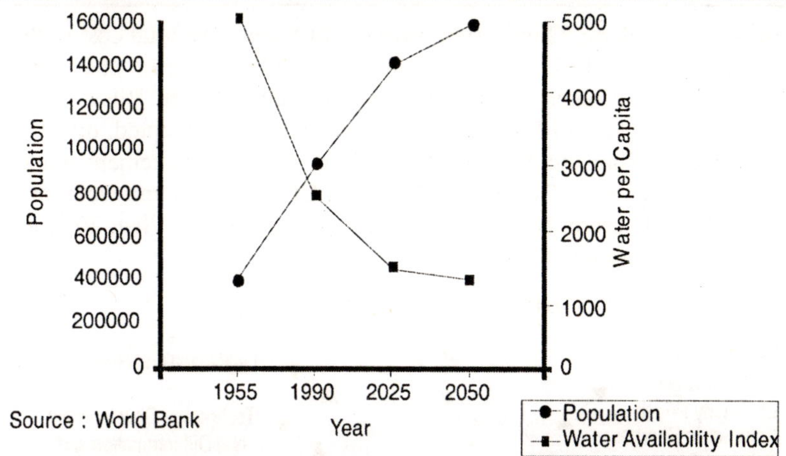

Fig. 1.2: Population growth and water availability index in India (cum) (*Source:* Sawaminathan, 1998)

An institutional structure for dealing with the multiple dimensions of water management at the country level is proposed in Fig. 1.3. The term National Water Trust is used instead of the commonly used term National Water Authority, to emphasize that water management should be carried out in the trusteeship mode. We must consider ourselves as trustees of water sources and not as owners. This will help to understand that management of water should not only be in the interests of the present generation but also of the generations yet to be born. Thus, the guiding principle for the work of the National Water Trust will be intra- and inter-generated equity.

The Centre is actively involved in and preventing further pollution of the rivers as can be seen from the activities of the National River Conservation Authority which is implementing the Action plans for de-polluting the Ganga, the Yamuna and the Gomati rivers. The Lake Conservation Plan at the national level aims at conserving various urban water bodies. The main component of the Ganga Action Plan is the diversion of sewage flowing into the river to other locations for treatment and its conversion into a source of energy. As against a target of creating infrastructure capable of intercepting, diverting and treating 873 million litres per day (mild) of domestic waste, infrastructure capable of treating 835 mld of municipal sewage has been created. Out of the 261 schemes sanctioned, 258 have been completed.

The Ganga Action Plan Phase II has been merged with the National River Conservation Plan (NRCP). The expanded NRCP now covers 141 towns located along 22 inter-state rivers in 14 States. The total cost of the schemes is Rs.20 13 crore. Under this Action Plan pollution abatement works are being taken up in 29 towns in Uttar Pradesh, Bihar, West Bengal. About 618 mld of sewage is targeted to be intercepted, diverted and treated. Under the Yamuna Action Plan pollution abatement works are being taken up in 21 towns in Haryana, Uttar Pradesh and Delhi at a cost of Rs. 496 crore. Similarly, under the Gomati Action Plan, 269 mld of sewage is proposed to be treated.

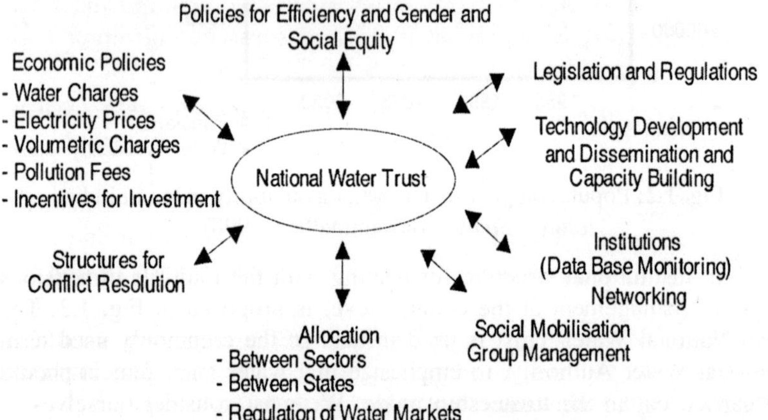

Fig. 1.3: National system for the conservation and sustainable and equitable use of water

The people by and large have also to come forward in water conservation, and supplement the governmental efforts. They have to develop a proper civic sense to do nothing that would spoil the element so vital for our survival.

1.5 Basic Principle of Rainwater Management

There are following basic principles of rainwater management:

i. Caputuring rainwater in a variable environment (water harvesting *viz.*, in-situ and ex-situ),

ii. Controlling water losses from soil surface and water surface,

iii. Efficient recycling of harvested rainwater,

iv. Selection of crops and varieties under variable environment,

v. Improvement in industrial plant which reduces water requirements and also reuse of waste water, and

vi. Controlling population growth rate.

Capturing water in a variable environment

Semi-arid regions are characterized by low erratic rainfall, poor nutrient soils and high temperatures and pose serious limitation in crop productivity, especially when water supply is inadequate. This can be manifested through poor development of crop yield structures and ultimately low yields. Semi-arid areas have at least one entirely rainless month/year and the amount off rainfall ranges from 500-1000 mm/annum in most areas. This means that conditions of water deficit, water stress or drought are common in these areas. In cases of extreme drought stress, crops yield poorly or not at all if drought stress during reproductive growth is severe and persistent (Barrow, 1999).

Recurrent drought is common phenomenon in these areas. Changes in severity of drought contribute to dryland degradation and desertification. This process has been evident in the Sahel region (West Africa) where rainfall levels have declined by 20-40 per cent in recent decades accompanied by severe land degradation. Similarly in India year 2002 was a wide spread severe drought year, during which rainfall was declined by 60%.

The approach summarized for rainwater management should be preventive, progressive, corrective as well as curative. The important initiatives for the rainwater management are as follows:

i. Conserving soil and water,
ii. Improving the ability of land to hold water and productivity,
iii. Rainwater harvesting and recharging,
iv. Growing greenery – trees, crops and grasses,
v. Optimal utilization of marginal lands, and
vi. Development of rural manpower and energy management systems.

1.6 National Water Policy

The first National Water Policy was adopted in September, 1987. It was reviewed and updated in 2002 and later in 2013 to adapt to new challenges. The present policy lays emphasize the on conservaton and efficient use, Ecological needs of river, Adaptation strategies in view of climate change, Water auditing be developed to ensure efficient use of water Community participation through Water Users Association planning and management practices and preparation annual water balances. According to Vora and Sharma (2015) the salient features of National Water Policy (2012) are;

(i) To ensure access to a minimum quantity of potato water for essential health and hygiene to all the citizens, available within easy reach of household

(ii) To curtail subsidy to agricultural electricity users

(iii) Setting up of Water Regulatory Authority

(iv) To keep aside a portion of the river flow to meet and the ecological needs and to ensure that the low high flow releases correspond in time closely to the natural flow regime.

(v) To give statutory powers to Water Users Associations to maintain the distribution system

(vi) Project benefited families to bear part of the cost of resettlement & rehabilitation of project affected families

(vii) To remove the large disparity between stipulations for water supply in urban areas and in rural areas

(viii) To support a National Water Framework Law

Integrated Water Resources Management Plan

The IWRM is an ongoing plan undertaken the Ministry in to promote the coordinated development and management of land, water and related resources in order to maximize the resultant economic and social welfare in an equitable manner without compromising the sustainability of vital ecosystems. The Plan will have following benefits

* Integration of Surface water - Groundwater, all sources of water, all sectoral demands, quantity and quality,

- Availability of Water at Different nodal points at Base line scenario
- Different demands at Base line scenario
- Water Balance at basin scale
- Future scenario depending upon the national challenges e.g. environmental sustainability, population growth, domestic need, agricultural need, power needs, water and food security, Will help in resolving interstate issues
- Will help in implementation of various awards of tribunals

The plan envisages following activities

- Identification and finalization of basins and obtain concurrence of co-basin State Governments.

- Appointment of Consultants, preparation and finalization of ToR for the study, estimates
- Formulation of multi-disciplinary committee comprising of officials from CWC, CGWB, CEA, NIH, IMD, CPCB, NRSC, NWDA, Departments of State Governments, Premier technical/ research institutes located within basin or its vicinity etc
- Preliminary identification of issues in the basin through consultation process study of interstate agreement, Tribunal awards, policies/acts etc
- Inventorization of required data, collection of data (Including spatial data) (30 years)
- Digitization and compilation of data, analysis, validation and formating of available data, filling up of data gaps, identification and elimination of erroneous data
- Preparation of schematic details, selection of appropriate models, and modelling of components and input of data in the model in the desired format
- Calibration and validation of model with observed data, analysis of results and formulation of baseline scenario
- Scenario planning in consultation with stakeholders including Climate Change scenarios
- Modelling of various scenarios, analysis of results, comparison between various scenarios, drawing conclusions, and report preparation
- Preparation of evidence-based Master Plan in consultation with all the States, prioritization of actions and preparation of details action plan.

1.6.1 Environment acts

Article 47 of Indian Constitution provides for protection an improvement of health. Article 48(A) is directed towards protection and improvement of environment and protection of forest and wildlife. Article 51(A) states that it is the duty of every citizen to protect and improve natural environment. Following the UN Conference on Human Environment (Stockholm, 1972), a constitutional amendment (42) was inserted in 1976 in the relevant provisions for environment protection in Constitution in Part IV- Directive Principles and Part IV A-Fundamental Duties. In order to ensure sustainable development of water resources, Government of India have also enacted various Acts and Legislations.

Prominent among these is the Environment (Protection) Act, 1986 through which the Government has acquired wide powers for protecting the environment. Some other Acts related to Water and Environment are Water (Prevention and Control) of Pollution Act, 1974 (amended in 1988), Water (Prevention and Control) of Pollution (Cess) Act. 1977 (amended in 1991), Forest Conservation Act, 1980 and Environmental Impact Assessment Notification of 1994 (amended in 1997) and the Ministry of Environment & Forest's Notification of Jan. 1997, constituting the Central Ground Water Authority (CGWA). The Water (Prevention and Control of Pollution) Act, 1974 seeks to maintain or restore "wholesomeness of Water" and the Central and State Pollution Control Boards have been established under this Act. According to the Water Cess Act, 1997, both Central and State Governments have to provide funds to Boards for implementing this act. The Forest Conservation Act, 1980 provides for compensatory afforestation to make up for the diversion of forest land to non-forest use.

The Environment (Protection), 1986 was enacted in 1986 for the protection and improvement of human environment. The EJA Notification of 1994 has made the environmental clearance mandatory for all the new projects as well as for expansion/ modernization of existing projects covering 29 disciplines including hydropower, major irrigation and flood control projects. Amended Notification of 1997 has made it mandatory to hold environmental public hearing before according environmental clearance to the listed category of projects.

Various pollutants are discharged in the surface waters including rivers and streams, which affect their quality and as the concentration increases problem of pollution is severely arises in water bodies, which are not suitable for drinking and irrigation purposes and their capacity of regeneration is essential. Dumping of soil and toxic wastes on land may lead to leaching of hazardous chemicals into groundwater thus, pollutes such water resources. In the arid region Gujarat, facing water shortages and frequent drought, a number of chemical and other industries have led to depletion of groundwater resource.

1.6.2 Policy issues

Under the umbrella of Ministry of Water Resources, The National River Conservation Directorate (NRCD), Central Groundwater Authority (CGWA, Central Water Commission (CWC), Central Pollution Control Board (CPCB), Command Area Development Authority (CADA) and State Level Institutions of these organizations are responsible for water resources development in the country.

National Water Policy, 1987

The Ministry of Water Resources, Govt. of India prepared a National Water Policy in September 1987. The Policy is a broad statement covering the following aspects -

In the planning and operation systems, water allocation priorities have been given to - Drinking water, Irrigation, Hydro-power, Navigation and Industrial and other uses.

Water resource development projects should be planned and developed as multi-purpose projects. The projects should provide for irrigation, flood mitigation, hydroelectric power generation, navigation, pisciculture and recreation wherever possible. Drinking water would be the prime consideration. Recycling and reuse of water should be an integral part of water resource development. Resource planning has to be for a hydrological unit such as a drainage basin as a whole or a sub-basin. Water zoning for use and economic development would be undertaken.

1.6.3 Prevention & Control of Pollution Act, 1974 as amended in 1988

This act deals with issues for the prevention, control and abatement of water pollution in the country. Water polluting industries are required, under the act, to obtain consent from Pollution Control Boards set up in every State for the purpose of taking appropriate measures to prevent and control pollution from effluents.

Despite providing the enabling environment for taking corrective action where necessary for resource conservation through the promulgation of a number of laws on environmental issues, their implementation remains tardy. This is a major drawback in protection of the environment and the natural resources of the country. Pollution of land and water remains one of the major problems of concern for the deterioration of environmental quality.

1.6.4 Water quality assessment authority

Recently, Government of India has constituted "Water Quality Assessment Authority" (WQAA) vide MOEF/No. J-15011/8/ 2000-NRCD Dated 29th May, 2001 under the Chairmanship of Secretary, MoEF, exercising the powers under the Environment (Protection) Act 1786. This authority shall exercise the powers and functions under the said Act for the following functions -

- To direct various agencies to standardize methods for water quality monitoring;
- To ensure quality of data generation of utilization thereof;
- To take measures so as to ensure proper treatment of waste water with a view to restoring the water quality of the river/water bodies to meet the designated-best-uses;
- To take up research and development activities in the area of water quality management;
- To promote recycling/reuse of treated sewage/trade effluent for irrigation in development to agriculture;
- To draw action plans for quality improvement in water bodies;
- To monitor and review/assess implementation of schemes launched/to be launched to that effect;
- To draw schemes(s) for imposition of restriction. in water abstraction and discharge of treated sewage/trade effluent on land, rivers and other water bodies with a view to mitigating crisis of water quality;
- To maintain minimum discharge for sustenance of aquatic life forms in riverine system;
- To promote rain water harvesting;
- To utilize self assimilation capacities at the critical river stretches;
- To minimize cost of effluent treatment;
- To provide information to Pollution Control Authorities to facilitate allocation of waste load;
- To review the status of quality of National Water Resources (both surface water and ground water);
- To identify "hot spots" for taking necessary actions for improvement in water quality;
- To interact with the authorities/committees relating to be constituted under the provisions of the said Act for matters relating to management of water resources:
- To constitute/setup state level Water Quality Review Committees (WQRCs) to coordinate the works to be assigned to such Committees, and;
- To deal with any environmental issues concerning surface and ground water quality referred to it by Central or the State

Government relating to the respective areas, for maintaining and/or restoration to sustain designated best uses.

MOWR is providing secretarial assistance to this multi-disciplinary high level authority created recently. The impact of functioning of the high level WQAA would take considerable time, in view of the alarming situation of water pollution in our river bodies.

1.6.5 Environmental and forest clearance of projects

The National Committee on Environmental Planning and Co-ordination was constituted in 1972 by the Government of India for detailed appraisal of the various environmental aspects of the Projects. A separate Department of Environment was created by the Government of India with the responsibility of providing constant vigil on the environmental requirements of development projects. It was serving as the nodal agency for environmental protection and eco-system development in coordinating role. It had the responsibility to earn out environmental appraisal of development projects. It also had the direct administrative responsibility for pollution monitoring and regulation and conservation of critical eco-systems. Initially all major and medium river valley projects were examined in respect of their technical feasibility and economic viability, before their acceptance by the Planning Commission in the developmental plans of various States. Clearance from environmental angle of all projects including major irrigation, multipurpose and flood control projects was later on made mandatory by enacting the Environmental Protection Act.

At present, all new development projects and/or proposals for expansion of existing projects in the public sector are referred to MoEF for environmental appraisal. The Environmental Impact Assessment (EIA) of the projects referred is carried out by the Impact Assessment Division (IAD) of the Department. The IAD, which has a multi-disciplinary base, carries out EIA of projects with the help of inter-ministerial appraisal committee, comprising of experts from related disciplines. The Feasibility Reports along with Questionnaires and Environmental Management Plans (EMP) furnished by the IAD and the project proponents are advised to plug the information gaps, if any. The project is then considered b\ the Expert Environmental Appraisal Committee (EAC). Site visits are undertaken by the sub-committees constituted as and when necessary to supplement the information submitted by the project authorities to collect the relevant information for EIA.

1.7 Guidelines for Management and Assessment

Central Water Commission has published "The Guidelines for sustainable Water Resources Development and Management (Sept. 1992)'' to meet the Indian requirement for data collection and analysis, impacts assessment and management plans. Together with EIA, for water resources projects, Environmental Management Plan (EMP) is also essential to ensure sustainable development and to limit the stress/load on the system within its carrying capacity. Special attention has to be given for the aspect of rehabilitation and resettlement compensatory afforestation and catchment area treatment (of directly drainage areas) in preparation of EMP. In the Environmental Assessment following types of information is essential –

- A summary of positive and negative environmental impacts, and
- An analysis of the economic implications of the impacts that can be quantified, as well as qualitative statements on unquantifiable impacts.

Apart from above, Environmental Management Directorate has published three documents for reference -

i. Guidelines for Environmental Monitoring of Water Resources Projects, January, 1996, 2nd Edition, CWC Publication No. 35/97, March 1998,

ii. Pamphlet "Hydropower and Environment", March, 2000, and

iii. River Valley Projects and Environment: Concerns and Management, CWC Publication No. 6 1/2000, February, 2001.

1.7.1 Environmental monitoring

Government of India has made it obligatory to get environmental clearance for all major irrigation, multipurpose and flood control projects from the Union Ministry of Environment and Forests (MoEF). The project is considered by the Environmental Appraisal Committee (EAC) about the acceptance of the project from the environmental angle stipulating certain environmental safeguards to be implemented along with the construction of the project. Accordingly, three tier inter-disciplinary Environmental Monitoring Committees for River Valley Projects at National, States' and Projects levels (NEMCRVP, SEMCs & PEMCs) were constituted, as per the decision of Union Ministry of Water Resources (MoWR), 1990. Such committees have been constituted for overseeing and ensuring the implementation of environmental safeguards, stipulated by MoEF in granting the environmental and forest clearances of multi-purpose river valley, irrigation, flood control and hydropower projects. NEMCRVP is presently monitoring 85 river valley projects, geographically well distributed all over the country. Representatives of the Planning Commission and Union Ministries of Agriculture and Cooperation, Environment and Forests, Tribal Development, Water Resources and CWC regularly visit these projects and participate in the meeting of NEMCRVP.

1.7.2 Complexities of environmental impacts

Following complexities should be kept in mind while discussing, assessing and managing environmental impacts of water resources projects -

i. It is as important to identify the contribution of favourable environmental effects of a project, as it is to identify the negative ones and the constraints they may impose. Sometimes the same physical component may produce both benefits as well as costs. For example, the creation of a dam may block a fish migration route. Thereby reducing fish stock and catch, at the same time, it may create a potential lake fishery as well as recreational benefits.

ii. Most of the environmental impacts in water resources projects are interconnected. People displaced from the inundated area of the reservoir, or those whose movement is facilitated by the reservoir and dam construction activities, may move upstream into the watershed. Use of forests and cultivation by them may create additional erosion, leading to increased sedimentation in the present and future reservoirs thereby reducing storage capacity. Sediment, in turn, affects water quality, and may reduce the capacity for power generation, and provision of water supplied and other intended benefits for irrigation and water supply users.

iii. In general, the characteristics of water releases from the dam affect such factors as downstream river flow, stream-bed configuration, water quality and associated cultivation. The dam can also affect conditions elsewhere in the river basin through changes in subsurface water levels, resettlement of displaced people and land use patterns.

iv. As a major investment, a dam may have impact on regional development and induce growth of new population centers and industrial activity. The cumulative effects of such projects may be substantially different from the effects of individual projects. Consequently, consideration must also be given to other projects, those under development in the basin, or those induced as a result of a particular project. It is to be noted that the combined adverse impacts of large number of small and medium projects in a river basin may be many times more serious compared to a large projects.

v. It is to be highlighted that poverty and low productivity levels in dryland areas themselves have tremendous social costs. Acute scarcity of water in these areas add on to the social and economic problem of the inhabitants.

vi. Well planned transfer of surface water can be a great boon in providing drinking water, removing or reducing water-borne diseases, raising agricultural products of the region and improving ecology.

vii. The design of development has to augment the sustaining capability of the environment and provide for a more active interaction between mail technology and available land eco-resources.

viii. Water is essential to solve the social and economic problems of the inhabitants. Well planned out transfer of surface water may be a boon by way of provision of drinking water, reduction in water borne diseases, increase in agricultural productivity of the region, availability of power to reduce deforestation and variety of benefits and improved ecology.

ix. Water resources project may really not induce significant adverse environmental impacts (for example Ramganga Project). Whether these exist or not, must be determined through the environmental impact assessment (EIA) at different stages of the project: considering both the positive and negative socio-environmental impacts.

x. The effect of "induced growth" in an area, once it starts to get industrialized, is not easy to predict quantitatively; but this has a greater socioeconomic effect in the long run through multiplier chain concept of benefits of river valley projects than many other items.

1.8. Assessment and Management of Environmental Impacts

Water quality monitoring

Monitoring the quality of rivers and other water bodies is the most essential first step in restoring water quality. CPCB has a nationwide responsibility of water quality monitoring and management of rivers. The water quality monitoring by CWC was started with the limited objective of classification of water for irrigation and other related uses, but presently, it also includes monitoring the rate of silt flows, chemical indices like sodium absorption ratio, sodium percentage, residual sodium carbonate and hardness number as well as other pollution parameters. Ground Water quality is being monitored by the CGWB.

1.8.1 Minimum flows in the river

It is necessary that certain allocation of water should be made for environment conservation. As of now there are no established methods, to assess the requirement of minimum flow in the river from environmental considerations point of view. To begin with, perhaps at least 50 percent of the lean period flow before the structure was built (average of 1-2 months); over and above the downstream committed use may be allowed to pass on to dams downstream of hydraulic structures. Low level outlets may be provided in the dams to pass required minimum flows downstream. Maintenance of minimum flow in river has also to be considered as a water use since it restricts the quantity of water that can be diverted for other uses. Necessity for maintaining minimum flow may arise out of the need to maintain water quality, river regime, maintenance of river eco-system or other public necessities such as bathing etc. To assess the requirement of minimum flow in the river to maintain water quality, NCIWRDP considered the bio-chemical oxygen demand (BOD) in river like Ganga (between Rishikesh and Patna), Yamuna (D/s of Okhla). It has been observed that water quality has deteriorated badly and as much as 3 BCM of fresh water may have to be released in Yamuna to bring the water quality to safe limit of 3 mg/litre of BOD. Considering inadequacy of data about water quality of all the rivers, NCIWRDP made a total provision of water demand of 5 BCM, 10 BCM and 20 BCM for ecology in the years 2010, 2025 and 2050, respectively.

1.8.2 Pollution control

The question of sustainability is not only a matter of adequate water availability but also one of maintaining the water quality. Hence, a war should be declared on the increasing menace of water pollution. Effective measures should be taken not only to stop the on-going deterioration of both surface and ground water quality hut also to improve it further. For this purpose, special attention needs to be paid to control urban wastewater and industrial effluents which arc the prime causes of water pollution. Pollution from other sources would also need to be controlled. Both the pro and anti-dam lobby should join hands together to fight the common enemy. "Polluter pays" or "Polluter gets punished". The polluting agency should be made responsible to meet the costs of cleaning the polluted waters. Actually, recycling and reuse of industrial and municipal wastewater after treatment are highly desirable to reduce pollution of water bodies.

1.8.3 Use of bio-technology in water pollution control

In fact, use of bio-technology is the prime method for treatment of polluted water, when pollutants are biodegradable. Domestic sewage which contains mostly the biodegradable pollutants is being treated by microorganisms since the old times. Difficulty lies due to many modern vulnerable industrial wastes, which are generally not degradable by conventional methods. New researches in the field of bio-technology are opening the possibility of treatment of such types of industrial wastes in cost effective manner. At the global level, use of biotechnology has already become popular. In India biotechnology is still not on the fast track and we should increase our share in global market.

1.8.4 Improvement of city water fronts

From time immemorial, human civilizations have thrived along water fronts. Majority of the population' depends on the streams, rivers and lakes for fresh water supplies Unfortunately, the same streams, rivers and lakes have also been converted by public into convenient, inexpensive and apparently inexhaustible dumping areas for human and animal wastes and residues from industries. Sustaining the environment along water bodies calls for monitoring and management of the water quality. There is also an urgent need for development and proper maintenance of the city water fronts and lakes in an environmentally sustainable manner. CWC had taken lead in this direction by organizing an outstanding National Symposium on Protection of Environment of City Water Fronts in December, 1990. State Governments and municipal agencies should lay attention in this respect.

1.9 National Commission for Integrated Water Source Development Plan

The Ministry of Water Resources has set up a National Commission for Integrated Water Sources Development Plan both for surface and groundwater with the following objectives –

i. To prepare an integrated plan for development of water resources for drinking, irrigation, industrial, flood control, and other uses.

ii. To suggest modalities for transfer of surplus water to water deficit basins by interlinking river for achieving the above objectives.

iii. To identify important ongoing projects as well as new projects which should be completed on priority basis in a phased manner.

iv. To identify a technological and interdisciplinary research plan for the water sector with a view to maximizing benefits.

v. To suggest the strategies for generation of physical and financial resources for the water sector.

1.9.1 Any other issues related to integrated water resource development

The Commission has since submitted its report. The Ministry has constituted a Committee for implementing the recommendations of the Commission.

1. Management of surface and groundwater resources

Currently, barely 29% of the total rainfall is conserved. Further, it is estimated that about 10% of the surface water is unutilized. Though water is a precious and scarce resource, its application and use efficiencies have been low. The full irrigation potential of the country has been revised recently from 113.5 to 139.5 million ha, comprising 58.5 million ha from major and medium schemes, 15 million ha from major irrigation schemes and 66 million ha from groundwater exploitation. Present rate of creation of irrigation potential of about 2.5 million ha year^{-1} needs to be stepped up considerably to meet the targets of increasing demand for food, It is estimated that even after achieving the full irrigation potential, nearly 75 million ha of the cultivated area will remain rainfed. In addition, only 30% of the annual sustainable groundwater potential is utilized. The country also loses a large volume of water due to faulty irrigation practices. Most irrigation projects are operating at low water use efficiency in the range of 30-40% only, thereby losing 60-70% irrigation water in conveyance (ICAR, 1999). In another estimate, about 45% irrigation water seeps through unlined field channels and 15% is in excess of what is being required (TERI: GREEN Report 2047, 1997). Large areas in Gujarat, Haryana, Punjab, Rajasthan and Uttar Pradesh degraded due to faulty irrigation practices.

Severe problems of water logging have been observed in the Indira Gandhi Nahar Pariyojana, Rajasthan and 'in the State of Haryana where excess irrigation to soil having gypsum-rich barriers at shallow depth dominate. In the coastal alluvial plains of Kutch and Saurashtra regions in Gujarat, overexploitation of potable groundwater along the sea has resulted in saline water ingress into the freshwater aquifers. In many instances, although the water is available in wells constructed in the village land, people are denied access to its use due to social conflicts and social order. Experience shows that water will continue to be conserved less and wasted more until (i) people, who are its primary users, remain alienated from the schemes on development and management of water resources, (ii) it is delivered free or it is priced so low that even the receipts are unable to neutralize the system maintenance costs, (iii) efficient conveyance and application systems are enforced, (iv) irrigation suitability of land remains defined and area under inefficient crops un-rationalized, and (v) greater commitment for research on irrigation water, environment and society are made.

Integrated water resource development planning should take into account –

- Water resources *viz.*, rainfall, surface water, soil moisture storage and groundwater,
- Watershed-scale to river basin-scale should receive priority, and
- Sector-based demand to be integrated in a dynamic way in order to allow the variability of demand during the next 50 years when the human population and demand for water are likely to be stabilized. This should also cater to prioritise within the river basin from one watershed to the other.

2. Artificial recharge of groundwater

The process of artificially increasing the rate and volume of water entry into the groundwater reserves is known as artificial recharge. The increasing demand for groundwater as a source of fresh water has brought into focus the need for artificial recharge. The artificial recharge of groundwater is undertaken to -

- Conserve and dispose surface runoff,
- Augment the groundwater availability, and
- Reduce or prevent seawater intrusion in coastal aquifers.

Flooding is the most common surface method of groundwater recharge when the aquifer is easily accessible from the ground surface. For low transmissivity soil formations, ditches and furrows at smaller intervals are preferred. In case of confined aquifers, injection wells with appropriate filtering systems are the most suitable.

3. Policy issues

In order to meet the country's water needs, the following issues need to be taken up –

- Identify the water scarcity areas – both natural as well as artificial scarcity,
- Apply measures for conservation of water in the irrigation projects,
- Integrate cost effective and traditional technologies of rainwater harvesting both in the rural as well as in the urban areas (rooftop rainwater harvesting),
- Address problems of falling water table and groundwater pollution, and
- Rationalize water and energy pricing and subsidies for the rural sector.

4. Strengthening water resources database

The National Water Policy mentions the urgency of strengthening database on different aspects of water resources. There have been some attempts to improve the data-gathering network. However, the situation is far from satisfactory. At most places the data are still gathered manually and at infrequent intervals. The requirement of using modern tools such as remote sensing, GIS, GPS, etc. for such purposes need not to be over-emphasized. Also, documentation of data and its availability to the users needs improvement and formulation of clear guidelines.

5. Issue of pricing of water

Water use efficiency in India compares unfavourably with not only the developed world but also with several developing countries. The industry also uses water equally wastefully. Supplying electric power at grossly subsidized rates has encouraged farmers to sink more and more tube wells and to pump increasingly larger quantities of water. Efforts to regulate the withdrawal of water have not been successful; imposing credit restrictions have not worked because of rich farmers using diesel-powered pumps where water table is comparatively high (TERI GREEN Report 2047, 1997). Abnormally lower pricing of water leads to its misuse pricing also needs to be studied in terms of man-hour lost in a day for simply collecting water both in the rural and urban areas. These costs need to be estimated in the overall pricing of water.

Tube wells are mainly affordable by the rich farmers. The sinking of a large number of tube wells resulted in the steady decline of water table in the water deficit States. The increasing are number of tube wells, subsidy and cheaper energy have led to overexploitation of groundwater leading to serious problems of soil salinization and water logging thereby causing decline in productivity. The poor and marginal farmers, and landless are affected the most during droughts.

Inter-State sharing f river water has been contentious and involves conflicts. However, the co-basin States are applying of agreements of sharing of water in the basin through mutual management in state cases. The tribunals constituted under Inter-State Water Disputes Act have also given awards on sharing of Inter-State river waters. Mega hydel projects have been successful in India. Large dams such as Bhakra, Nagarjuna Sagar, Hirakund and Mettur helped in development of the people. These projects have contributed in making India self-sufficient in food grain production, besides providing drinking water and hydro-electric energy to the growing population. The Sardar Samand project has been built in the drought-affected areas of Gujarat.

6. Water transfer

The distribution of water is highly skewed. There are areas with seasonal water surplus, where as there is water deficit in other areas during the same period. The technical feasibility of inter-basin transfer of water has been investigated. Garland Canal was proposed long ago to convey surplus water of north India to balance deficit in peninsular India. A perspective modular plan, which can be implemented in stages over a period of time, is required.

7. Maintenance of minimal flows

Water management should also include environmental issues such as minimal flow of water throughout the year at various stages of the river / rivulet and also indication of maximum flow, to achieve recharge of groundwater, to be managed by underground and surface storage. Minimal flow of water also helps its regeneration after mixing with domestic effluents.

8. Maintenance of water quality

Prevention of pollution addressed in the Policy on Abatement of Pollution, 1992 requires to be given special attention for safeguarding the quality of available water. Three sectors need to be looked at simultaneo-usly, these eventually affect the river as water quality *viz.*, (i) dome-stic sanitation specially in urban areas, (ii) industrial effluents-treatment and encouraging recycle and reuse, and (iii) agricultural pollution due to excessive use of chemical fertilizers and pesticides.

Shift in emphasis from supply to demand management. In general creation of water supply is expensive than demand management through promotion of water conservation practices, rationalization of prices and involvement of local communities. It is also possible to augment supplies through reuse of wastewater. Policies for promotion of wastewater use need to be framed.

9. Reallocation of water

Reallocation of existing resources to more efficient use is a mean of avoiding high expenditure on new infrastructure. Policy makers are faced with two options. Reallocation to more efficient use through adminis-trative procedures or through introduction of a system of property rights, water pricing and tradable water entitlement allowing the market forces to operate.

10. Climate change

The global warming will lead to changes in rainfall, evaporation and rise in sea water level. At many places this would introduce changes in water supply and demand patterns. In the light of expected changes it is imperative that the status of water resources be reassessed. Policy initiatives to formulate appropriate water development and user plan are called for.

11. Institutional constraints

The efforts to develop and manage water resources on sustainable basis still remain the responsibility of Central and State Governments. Further, this responsibility is spread amongst several institutions with overlapping jurisdiction. The irrigation sector, which uses most of the water, is a State subject, while most of the major rivers are interstate rivers. The concept of river basin planning is emerging but when the interests of partner States are in conflict, it takes long time to settle these disputes and in the process the water resources development suffers. Decentralized water management policy would be more effective.

References

CWC. 2004. Bhagirath. Vol. LI(1): 3-33. Central Water Commission, Sewa Bhawan, R.K. Puram, New Delhi.

ICAR. 1999. "Fifty Years of Natural Resource Management Research". (ed. G.B. Singh and B.R. Sharma). Indian Council of Agricultural Research, New Delhi, India.

Jat, M.L. and Bhakar, S.R. 2009. Groundwater Hydrology (Theory and Practice): Agro Tech Publishing House, Udaipur, pp. 548.

Katyal, J.C. 1997. "Research and Development in Rainfed Agriculture: Retrospective and Perspective." In: Productivity of Land and Water. New Age International Publisher, New Delhi, India, pp. 256-270.

NBSSLUP. 2001. "Classification of Dryland Regions in India". National Bureau of Soil Survey and Land Use Planning, Nagpur, India.

Ramakrishna, Y.S., G.G.S.N. Rao, B.V. Ramana Rao and P. Vijay Kumar. 1999. Resource Management Research (ed. G.B. Singh and B.R. Sharma). Indian Council of Agricultural Research, New Delhi, India, pp. 505-536.

Sharma, B.R. and D.K. Paul. 1999. "Water Resources of India". In: Fifty Years of Natural Resource Management Research (ed. G.B. Singh and B.R. Sharma). Indian Council of Agricultural Research, New Delhi, India, pp. 31-48.

Shukla and Vohra. 2003. Development of Surface Water Resource in Rajasthan. *Indian Journal of Water Resources Society*, 23(3): 51-52.

Singh, H.P., B. Venkateswarlu, K.P.R. Vittal and K. Ramchandran. 2000. "Management of Rainfed Agro-ecosystem". In: Natural Resource Management for Agricultural Production in India. *Indian Society of Soil Science*, New Delhi, India, pp. 669-774.

Singh, R.P. 1986. "Farm Ponds". Central Research Institute for Dryland Agriculture, Hyderabad, India.

Swaminathan, M.S. 1998. Drinking Water Security. Proceeding of National Seminar on "Water Management for Sustainable Agriculture", Problem and Prospects for 21st Century, IARI, New Delhi, April, 15-17: 15-25.

TERI: GREEN Report 2047. 1997. "Looking Back to Think Ahead". Tata Energy Research Institute, New Delhi, India.

Vora, K. and Sharma, P.K. 2016. Climate change and sustainable development of Water resources in India. Indian J. Water resources 36 (2) : 4-10.

EXERCISES

1.1. Discuss the water resources in India.

1.2. Why rainwater management is essential in India?

1.3. How the competing demands is managed in India?

1.4. Explain the principles of rainwater management.

1.5. How the water is captured in a variable environment?

1.6. What is the National Water Policy in India?

1.7. Explain the guidelines for management and assessment for sustainable water resources development in India.

1.8. What is the pollution control?

1.9. Discuss the use of bio-technology in water pollution control.

1.10. What is the integrated water resource development plan?

■■■■

Chapter 2

PRECIPITATION

2.1 Introduction

Precipitation refers to all forms of water derived from atmospheric vapours and deposited on the earth surface. It occurs due to the condensation of atmospheric water vapour. The different forms of precipitation are described in subsequent section (2.3). Rainfall is the predominant form of precipitation in India and is responsible for causing stream flow in majority of Indian rivers. However, appreciable snowfall is taking place at high altitudes in the Himalayan Mountain and rivers in Northern India receive heavy snow-melt during summer months. The precipitation is seen to be changeable considerably with time and space.

For occurrence of precipitation is different forms, the following three conditions should be satisfied: (a) water vapour must become saturated usually through cooling, (b) the water vapour must change its phase to liquid and/or solid, and (c) the drops or crystals of water must grow in size so that they may fall. Sometimes clouds may form as a result of conditions (a) and (b) but no precipitation may occur due to third condition not getting satisfied.

2.2 Hydrological Cycle

Hydrologic cycle is the water transfer cycle which occurs continuously in nature. Hydrology deals with the occurrence and movement of water above, on and below the surface of earth. There is a continuous transformation from one form of water to another or of movement from one location of storage to another, however, the total amount of water above, on and below the surface remains constant. Thus, hydrologic cycle is a closed system having no beginning and end.

The hydrologic cycle (Fig. 2.1) consists of four phases as given below:

(a) Precipitation,

(b) Combination of evaporation and evapotranspiration,

(c) Stream or river flow, and

(d) Groundwater.

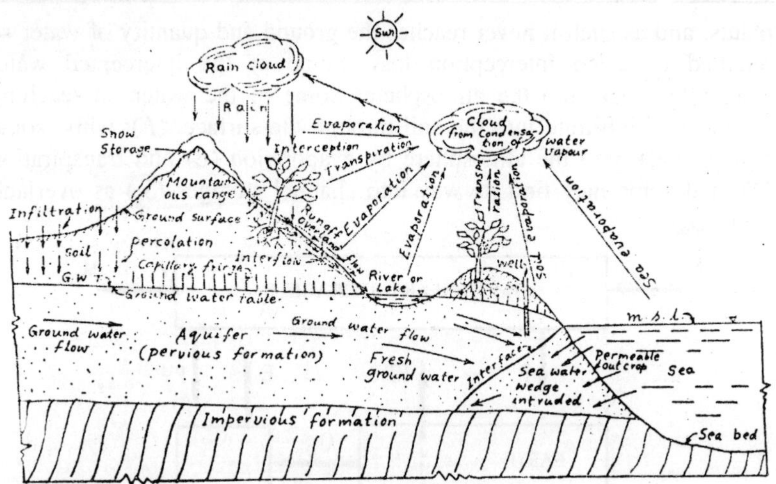

Fig. 2.1: The hydrologic cycle

The world has one-third land and two-third ocean. Evaporation from the surfaces of ponds, lakes, reservoirs, oceans, etc. and transpiration from surface vegetation, *i.e.* from plant leaves of cropped land and forests, etc. take place. These vapours rise to the sky and are condensed at higher altitudes by condensation nuclei and from clouds, resulting in droplet growth. The clouds melt and sometimes burst resulting in precipitation of different forms like rain, snow, hail, sleet, mist, dew and frost. A part of this precipitation flows over the land called runoff and a part infiltrates into the soil which builds up the ground water. The surface runoff joins the streams and the water is stored in reservoirs. A portion of surface runoff and ground water flows back to ocean. Again evaporation starts from the surfaces of lakes, reservoirs and ocean and the cycle repeats. Of these four phases, runoff phase is important for agriculturist and farmers for in-situ moisture conservation and water harvesting and recycling it to crop during dryspell period to enhance productivity.

The energy to run this cycle is provided by solar radiation. A qualitative description of such a cycle is given in Fig. 2.2.

It is evident from the Fig. 2.2 that the rectangle shows various forms of water storage in different segments of hydrological cycle and arrows indicate the transfer of water from one form of storage to another in this cycle. Atmospheric moisture (S_A) moves towards the earth's surface in the form of precipitation, *(P)* (the various forms of which are rain, hail, sleet, snow, etc.). A portion of this is intercepted by buildings, trees and plants, and as such it never reaches the ground and quantity of water so retained is called interception loss. Some of the intercepted water evaporates back into the atmosphere. Some of the water on reaching ground will infiltrate into the soil through the surface, (*F*); while some will go back into the atmosphere by evaporation (*E*) and transpiration (*T*), and some may find its way into channel storages (S_c) as overland flow (Q_o).

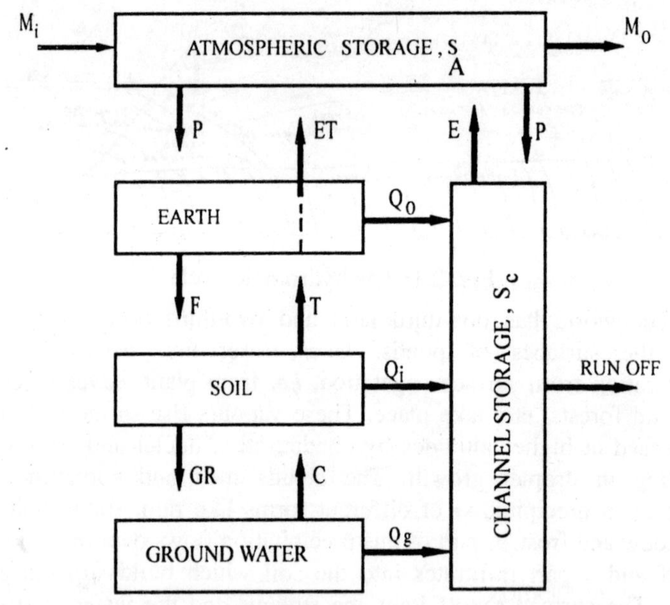

Fig. 2.2: Hydrological cycle – qualitative description

There may, also be an interlow (Q_i) through the soil which under favourable conditions (*e.g.*, proper gradient) shall find its way into the channel network. Again the groundwater storage may also be depleted if there is a groundwater flow (Q_g), into the channel network which mostly happens during dry weather, and in the dry weather flow in stream channel is sustained by (Q_g) which is also known as base flow.

2.2.1 Runoff process

Runoff process is one of the aspects of hydrological cycle and is shown schematically in Fig. 2.3.

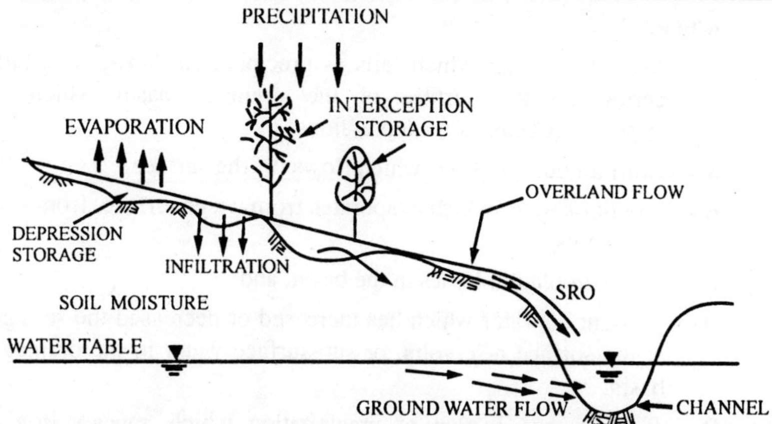

Fig. 2.3: Schematic representation of runoff process

2.2.2 Conceptional model of hydrological cycle

Fig. 2.2 does not indicates quantitative analysis required for an engineering study or mathematical modelling and one has, therefore, to make reasonably/plausible assumptions. Therefore, one such idealization or conceptual model is shown in Fig. 2.4.

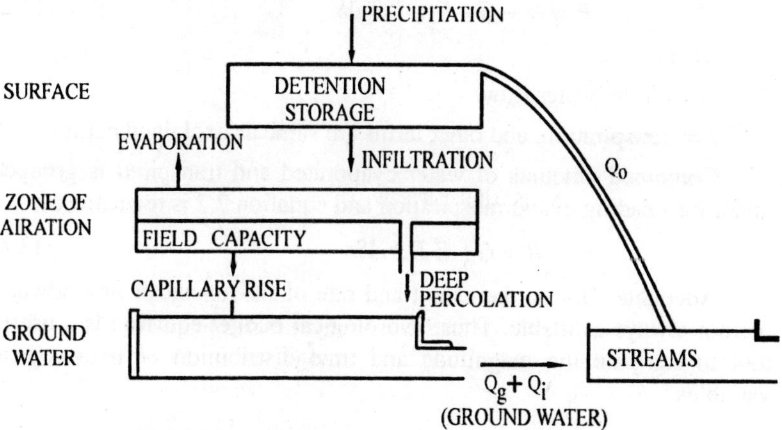

Fig. 2.4: Simplified watershed model

2.2.3 Hydrological equation

The hydrological cycle for a drainage basin or water sheet may be expressed by the following relationship:

$$P = R + E + D \pm \Delta S \tag{2.1}$$

where,

P = amount of water which fails as precipitation during a certain period on the surface of the drainage basin which is impervious and has no depressions,

R = runoff amount of water which flows off the surface,

E = amount of water which evaporates from water surface, from soil and plants,

D = Deep percolation losses in the basin, and

ΔS = amount of water which has increased or decreased the storage at the surface reservoirs, or sub-surface water in the drainage basin.

Runoff R is that portion of precipitation which appears in the streams after satisfying the needs of interception storage, depression storage evapotranspiration and is not absorbed by the deep strata in the drainage basin. It is very necessary that all the elements of this equation should be referred to the same period of time. Qualitatively, hydrological cycle does not change over the last 50 years and perhaps there is no possibility of any fundamental change in the future. The hydrological budget for a region can be expressed as,

$$P = R + G + E + T \pm \Delta S \quad \text{....(2.2)}$$

where,

G = groundwater flow,

T = transpiration, and other terms are same as explained earlier.

Combined amounts of water evaporated and transpired is grouped under the heading evapotranspiration and equation 2.2 is rewritten as,

$$P = R + G + ET \pm \Delta S \quad \text{....(2.3)}$$

Adequate data on the extent and rate of movement of groundwater are not always available. Thus, hydrological budget equation is a useful tool to estimate the magnitude and time distribution of hydrological variables.

Example 2.1: A drainage basin is having an area of 12,000 km^2 received 90 cm of rainfall in 2010. The resulting runoff was measured in a river draining the area and its annual rate of flow was 200 cubic metres per second (Cumecs). Determine evapo-transpiration making suitable assumptions, wherever necessary from the area.

Solution:

Step I. On using equation (2.3) as,

$$ET = P - R - G - \Delta S$$

In this equation *P* and *R* are known already, *G* and *ΔS* are to be known or assumed,

Step II. Assume that phreatic and topographic divide are coincident, thus, *G=0,*

Step Ill. Further assume that the ground water volume has not changed during the year 2010. Thus, assume $\Delta S = 0$,

Step IV. Convert runoff, *R* into cm depth per year

$$R = \frac{200 \times 60 \times 60 \times 24 \times 365}{12000 \times 10^6} \times 100 = 52.56 \, cm/year$$

Step V. $ET = 90 - 52.56 = 37.44$ cm/year

Ans. ET = 37.44 cm/year

2.2.4 Hydrological cycle in India

In India monsoons occur mostly during the four months from June to September. During these four months, there are two distinct currents emanating from Bay of Bengal and Arabian Sea. Bay of Bengal currents carry moisture amounting to 340 million hectare metre (Mham) and Arabian Sea currents 770 Mham. Out of these, only 25-30% of the moisture available during monsoons precipitates in the form of rainfall. During the rest of the year, the moisture which contributes to rainfall including snowfall is of the order of 100 Mham. Rainfall over the country varies from place to place and time to time. The average annual precipitation (considering monsoon and non-monsoon period) is 400 Mham.

According to Parthasarthy and Dhar (1975), the south west monsoon (June-September) rainfall contributes about 75% of average annual rainfall.

The approximate distribution of average annual water resources in million hectare metres is shown in Fig. 2.5. The figures given in the round bracket and square bracket respectively, indicate values in the year 1974 and 2025 AD (When the full development of water resources is expected to take place). The figures given in these brackets should be taken as indicative rather than definitive. In order to have some rough idea of extent of contribution of various sources to and from surface and groundwater with a view to know the total utilizable resources, the numerical figures arrived at by National Commission on Agriculture have been adopted. It is seen that total utilizable water resources is 38 Mham.

It is evident from the figures for the year 1974 that basic surface water resources derived from precipitation (rainfall and snowfall) total upto 115 Mham. Out of which snowfall contribution is about 10% of the total. Catchments lying outside the country contribute another 20 Mham. The portion of precipitation that goes to become part of ground water is of the order of 50 Mham. Thus, the total annual basic water resources of the country are 185 Mham. However, there is an interchange at various stages between surface water and ground water besides some recirculation (Fig. 2.5). When water resources are fully harnessed/ developed, total ground water increases to 85 Mham and the surface water to 185 Mham. It may he noted that all the above mentioned surface and not groundwater resources are utilizable.

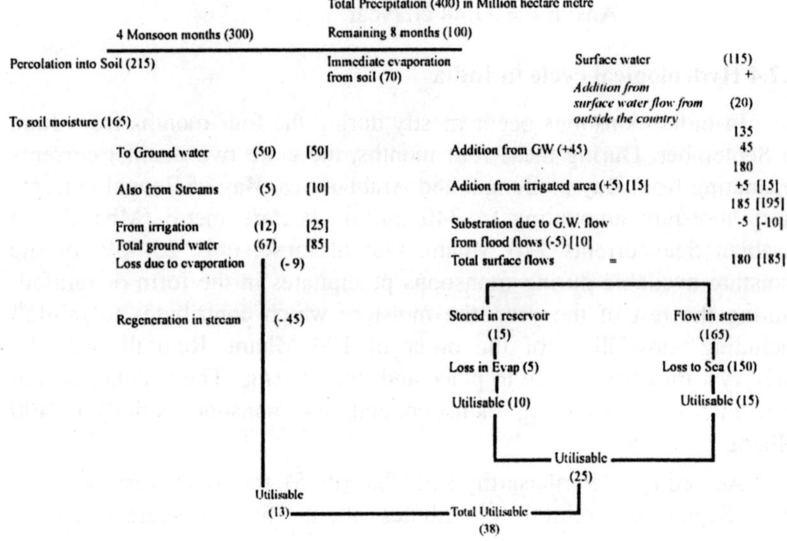

Fig. 2.5: Distribution of average annual water resources of India

2.3 Different Forms of Precipitation

There are following different forms of precipitation:

Drizzle- It seems to a light steady rain in fine drops (0.5 mm) and intensity <1 mm/hr.

Rain - It refers the condensed water vapour of the atmosphere falling in drops (> 0.5 mm, maximum size-6 mm) from the clouds.

Glaze - It refers freezing of drizzle or rain when they come in contact with cold objects.

Sleet - It is defined as frozen rain drops while falling through air at subfreezing temperature.

Snow - It is defined as ice crystals resulting from sublimation (*i.e.* water vapour condenses to ice).

Snow flakes - It is defined as ice crystals fused together.

Hail- It is defined as small lumps of ice (>5 mm dia) formed by alternate freezing and melting, when they are carried up and down in highly turbulent air currents.

Dew -It is defined as moisture condensed from the atmosphere in small drops upon cool surfaces.

Frost- It is defined as a feathery deposit of ice formed on the ground or on the surface of exposed objects by dew or water vapour that has frozen.

Fog- It is defined as a thin cloud of varying size formed at the surface of the earth by condensation of atmospheric vapour.

Mist- It is defined as a very thin fog.

2.4. Types of Precipitation

The precipitation occurs when drops or crystals of water or ice combine together in large drops or crystals and move towards the earth resulting in rain, hail, sleet and snow. This type of condensation may result from dynamic cooling of mixture of two air masses at different temperatures. This is the first form of precipitation in the form of clouds. The precipitation may be classified into following categories:

(i) ***Convectional precipitation*** - This type of precipitation occurs due to local whirling thunder storms and is typical of the tropics. The air close to the warm earth gets heated and rises due to its low density, cools adiabatically to form a cauliflower shaped cloud which finally bursts into a thunder storm. When accompanied by destructive winds, they are called 'tornador'.

(ii) ***Frontal precipitation*** - When two air masses of contrasting temperatures and densities clash with each other then condensation and precipitation occur at the surface of contact (Fig 2.6).

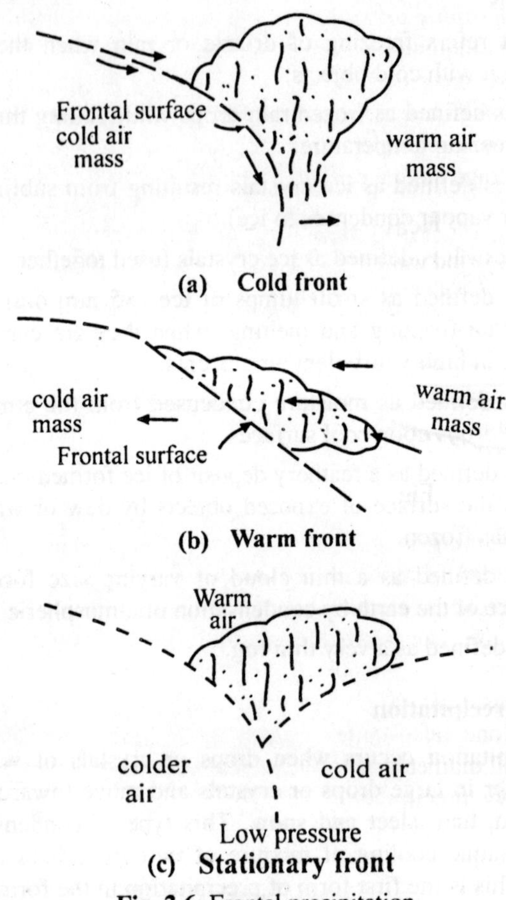

Fig. 2.6: Frontal precipitation

This surface of contact is called a 'front' or 'frontal surface'. If a cold air mass drives out a warm air mass it is called a 'cold front' and if a warm air mass replaces the retreating cold air mass and it is known as 'warm front'. On the other hand, if the two air masses are drawn simultaneously towards a low pressure area, the front developed is stationary and is called a 'stationary front'. Cold front causes intense precipitation on comparatively small areas, while the precipitation due to warm front is less intense but is spread over a comparatively larger area. Cold fronts move faster than warm fronts and usually overtake them and the frontal surfaces of cold and warm air sliding against each other. This phenomenon is called 'occlusion' and the resulting frontal surface is called an 'occluded front'.

(iii) ***Orographic precipitation*** – The mechanical lifting of moist air over mountain barriers, causes heavy precipitation on the windward side (Fig. 2.7). Cherrapunji in the Himalayan range and Agumbe in the Western Ghats of South India get very heavy orographic precipitation of 1250 cm and 900 cm (average annual rainfall), respectively.

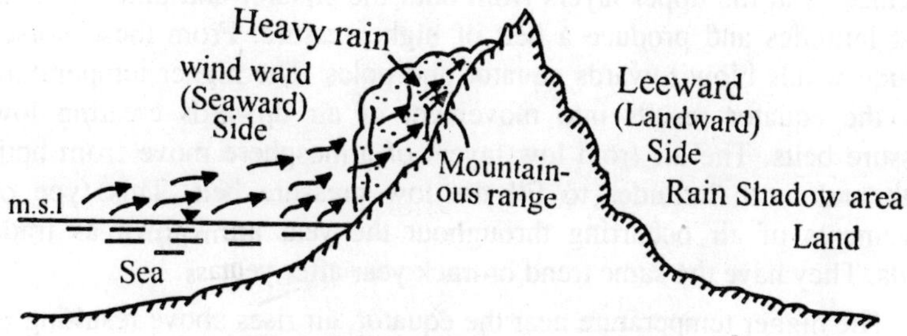

Fig. 2.7: Orographic precipitation

(iv) ***Cyclonic precipitation*** - This type of precipitation occurs due to lifting of moist air into a low pressure belt or due to pressure differences created by the unequal heating of the earth's surface. The winds blow spirally inward counter clockwise in northern hemisphere and clockwise in the southern hemisphere. There are two main types of cyclones-tropical cyclone (also called hurricane or typhoon) of comparatively small diameter of 300-1500 km causing high wind velocity and heavy precipitation, and the extra-tropical cyclone of large diameter up to 3000 km causing wide spread frontal type precipitation.

2.5 Monsoons

The word monsoon has originated from the Arabic term which means season. Arabs, used word with reference tot winds only, but in later years, it was used to mean rainfall. The course of trade winds changes during an year due to local factors like topography, oceans etc. These trade winds with the change in direction are known as monsoon winds. The direction of monsoon winds changes with season.

High and low pressure areas are developed due to differences in the amount of solar radiation received on earth and differences in absorption of heat by water and earth and rotation of the earth. Solar radiation received near the equator is more as compared to poles. Thus, temperatures are high near the equator resulting in rising up of the air and decreasing of pressure in the lower layers of the atmosphere. The low-pressure areas near the equator are called doldrums.

The air in the polar region is swung to the temperate zones by the rotation of the earth. The movement of air from the polar region pressure decreases at polar region and is known as polar calms. The latitude belt between 30 to 35° of both south and north are called horse-latitudes. The air currents at the upper layers from both the equator and poles meet at horst latitudes and produce a belt of high pressure. From these horse-latitude winds blow towards equator and poles. The higher temperature near the equator results into movement of air upwards creating low pressure belts. The air from low layers of atmosphere move from both south and north latitudes to fill the low pressure belt. This type of movements of air occurring throughout the year are known as trade winds. They have the same trend or track year after year.

The higher temperature near the equator, air rises above resulting in high pressure area in the upper layers of the atmosphere which force the air to move south-wards or north-wards from the equator. These winds are called trade winds but occur at higher layers of the atmosphere.

Types of monsoon in India

Following types of monsoon occurs in India:

(i) South-west monsoon, and

(ii) North-east monsoon

2.5.1 South-west monsoon

India is situated in north-east trade wind zone and these trade winds continue throughout the year. However, regular monsoon system is developed due to physiographic features. Central Asia and arid zones of North India get heated up resulting in low pressure areas during summer (Fig. 2.8).

The presence of vast Indian Ocean on the southern side develops high pressure area. Actually the trade winds have to start from north of India towards equator. Because of the presence of Indian Ocean in the south and dry areas in north (*i.e.* due to local factors the direction of the trade winds is changed. Thus, south-east trade winds initiated from the south of equator. When they across the Indian Ocean, they absorb large amounts of moisture. While crossing the equator, they are caught suddenly in the air circulation over India and deflected as south-west winds. These moist winds reach India with violent thunder and lignite. The sudden advent of the violent rain-bearing winds is known as the burst break of the monsoon. Generally, south-west monsoon reach South India (Kerala) around first of June every year. These rain-bearing winds are very strong blowing having an average speed of 30 km/hr. In about one month's time, they overrun almost the entire country. These monsoon winds cover India as two branches, one is Arabian sea branch and the other is Bay of Bengal branch. The Bay of Bengal branch moves up to Assam, then deflects towards north-westwards due to obstruction of Himalayas. The Arabian Sea branch moves northwards. Both these branches occasionally clash near Delhi.

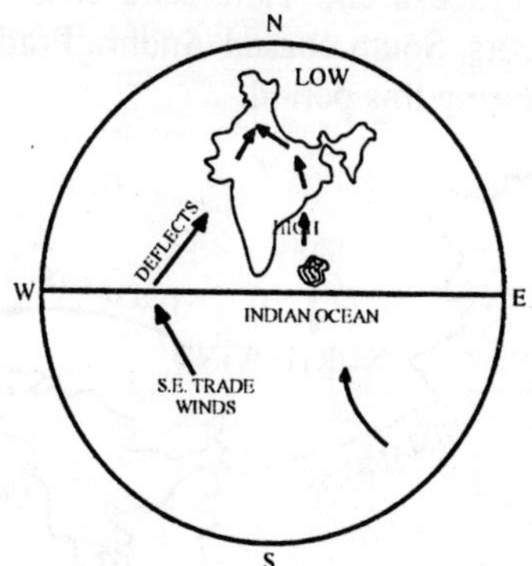

Fig. 2.8: Schematic diagram of south-west monsoon

The south-west monsoon period (June-September) is called grand period of rainfall in India. Bulk of the rainfall is received during this season in almost every part of India with sole exception of east coast of Tamilnadu. The orographic influence (presencc of mountains) on the distribution of rainfall is also especially pronounced during this season. For instance, the windward side of western ghats receives around 2,500 mm of rainfall while the leeward side receives only 500 mm. The amount of rainfall received in north-western India is less than that received in north-eastern India. The monsoon, even in the wettest part, is not continuous. The downpours are interspersed with rainless periods. This phenomenon is closely connected with the cyclonic depressions originating at the head of the Bay of Bengal and their crossing over into the mainland. The intensity and frequency of these depressions are responsible for the overall strength or weakness of the monsoon in a given year.

2.5.2 North-east monsoon

It occurs in wither because off movement of sun towards south of equator resulting in high-pressure area. Low-pressure areas are developed over the Indian Ocean due to high temperature. Thus, air moves from high-pressure areas in China and Russia to low-pressure areas in the Indian Ocean. These winds are cool and dry, while they reach India, they are obstructed by the Himalayas and deflected to east (Fig. 2.9). These north-east winds subsequently deflect to south-west they become warm and very little rainfall occurs due to these winds. On moving further south-westerly across the Bay of Bengal, the air masses pick up moisture and are warmed. As they strike the cool land surface of south coastal Andhra Pradesh and Tamilnadu coast, the air masses are cooled and rainfall occurs. South coastal Andhra Pradesh and Tamilnadu get sufficient rainfall during this period.

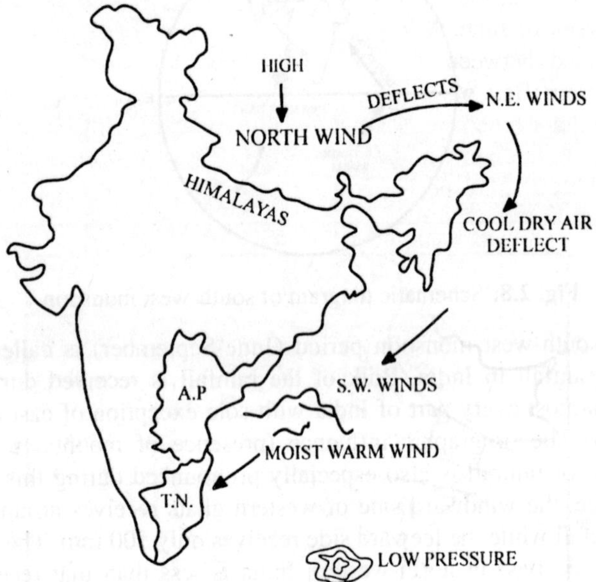

Fig. 2.9: Schematic diagram of north-east monsoon

Seasons

There are two major seasons *viz.*, (i) south-west monsoon (June to September) and (ii) winter season (December to February) in the Indian sub-continent from the point of view of climate. However, there are two transitional periods in between these (i) post-monsoon transition

(October to November) and summer transition (March to May). The important characteristics of these four seasons are discussed as follows.

South-west monsoon

During this period, over 75 per cent of annual rainfall is received over major portion of India, while, the highest rainfall occurs in the month of July. The South-west monsoon originates from Indian ocean and indicates its appearance in the southern part of Kerala by the end of May. The monsoon winds advance across the country in two branches-Arabian sea branch and Bay of Bengal branch. Arabian sea branch sets in at the extreme southern part of Kerala in the first week of June, moves northwards over Karnataka, Maharashtra and Gujarat and reaches Delhi about the last week of June. Bay of Bengal branch sets in at Assam also in the first week of June, first covers the north-eastern regions of India and turns westward to enter into Bihar and U.P. and reaches Delhi also in the last week of June. A low pressure region known as monsoon trough is developed between these two branches. The low pressure region extends from Bay of Bengal to Rajasthan and the rainfall pattern is generally determined by its position. The onset of monsoon, south-west monsoon rainfall and withdrawal of monsoon is shown in Fig. 2.10, Fig. 2.11 and Fig. 2.12, respectively.

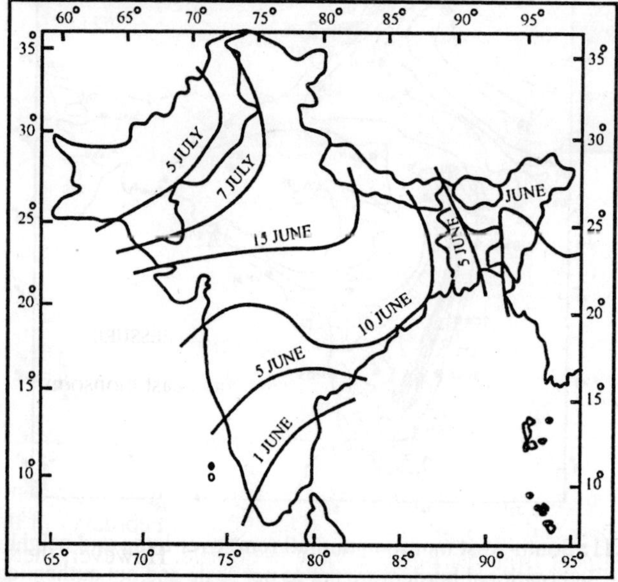

Fig. 2.10: Onset of monsoon

Post-monsoon (October-November)

As the south-west monsoon recedes, low pressure areas develop in the Bay of Bengal and the available moisture is picked up by the north-easterly winds. When this mass of air strikes the east-coast of southern peninsula, rainfall occurs in southern part of India. Severe tropical cyclones also occur during this time in the Bay of Bengal and Arabian sea. When these cyclones strike east and west coast they cause severe damage to life and property.

Winter season (December-February)

Disturbances of extra tropical originated and travel towards Pakistan and Afghanistan in the middle of December or so. It results into moderate to heavy rainfall and snow in the Himalayas and Jammu and Kashmir. During the period December to February, some light rainfall also occurs in the northern plains and low pressure area developed in Bay of Bengal causes rainfall in the southern parts of India (Tamil Nadu).

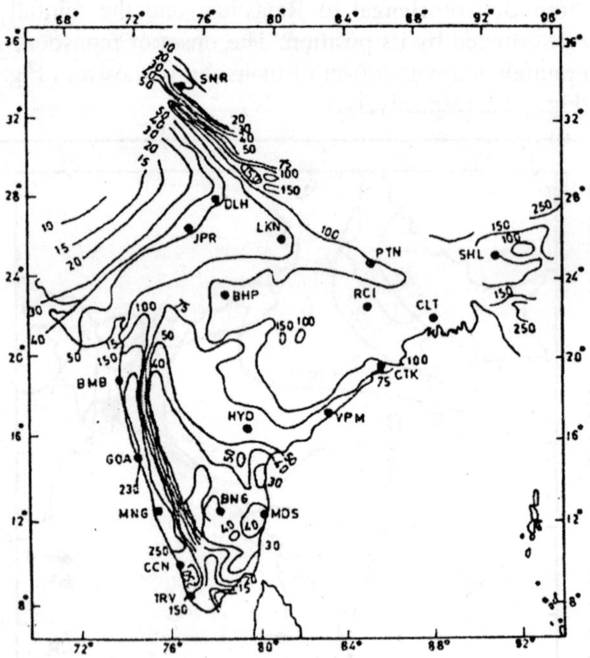

Fig. 2.11: South-west monsoon rainfall (cm) over India and neighbourhood (*Note*: International borders are not as per scale and not authenticated by G.S.I.)

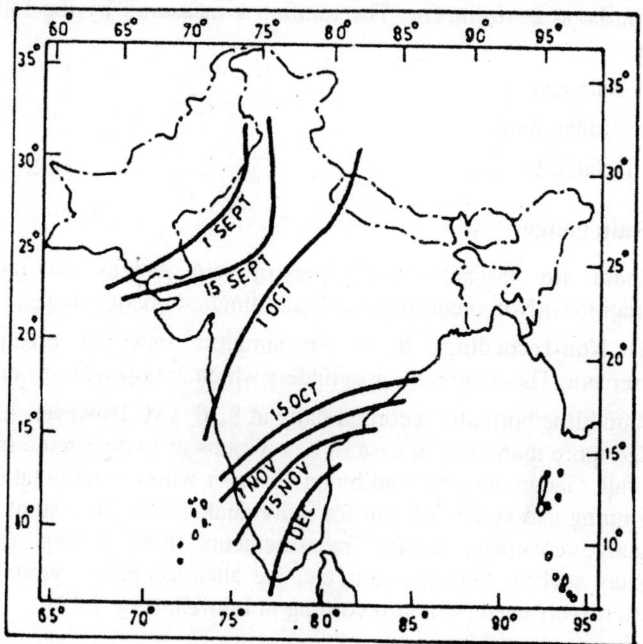

Fig. 2.12: Withdrawal of monsoon (Note: International borders are not as per scale and not authenticated by G.S.I.)

Summer transition (March-May)

Very little rainfall may likely to occur during this period in India. However, convective spells cause thunderstorms in West Bengal, Assam and Kerala. Cyclonic activity also mostly occurs on the east coast.

2.6 Measurement of Point Precipitation

The aim of precipitation measurement is to determine the true representative value of precipitation in terms of the region. Rain is measured as vertical depth of water (mm) which accumulates on a level surface such that none of it flows away or is lost through percolation and evaporation. While, snow, sleet and hail are measured in terms of equivalent water depth. The representative precipitation values are those values which would have occurred at a select location in the absence of a gauge. However, this is not possible because of the uncertainty arising due to material non-homogeneity of precipitation and error associated with the gauges. The measurement of precipitation is influenced by the neighbourhood environment. The rainfall is measured by the following devices:

(i) Rain gauge,

(ii) Radar, and

(iii) Satellite.

2.6.1 Rain gauge

There are basically two types of rain gauges for recording precipitation (i) Non-recording, (ii) Recording/automatic gauges.

(i) Non-recording: It is the simplest form of precipitation measurement. These are hollow cylinders which are open at one end.

Rainfall is normally recorded daily at 8.30 AM. However, it can be recorded more than once in case of heavy rains or to determine intensity of rainfall. Gauge must be read by an observer who normally takes, time of beginning and ending of rain for individual storms. In case of limited rain gauge coverage, rainfall measurements from various types of containers such as buckets, cans etc. are also accepted. Symon Gauge was extensively used for non-recording of rainfall. (Fig. 2.13)

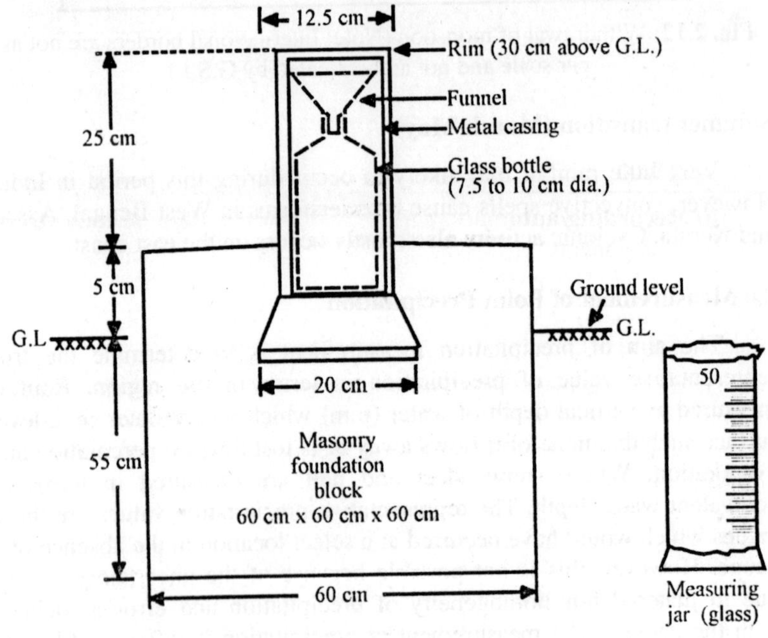

Fig. 2.13: Non-recording Symon gauge

This gauge is modified as per IS: 5225-1969 and is shown in Fig. 2.14. There are four such types of rain gauges as given in Table 2.1.

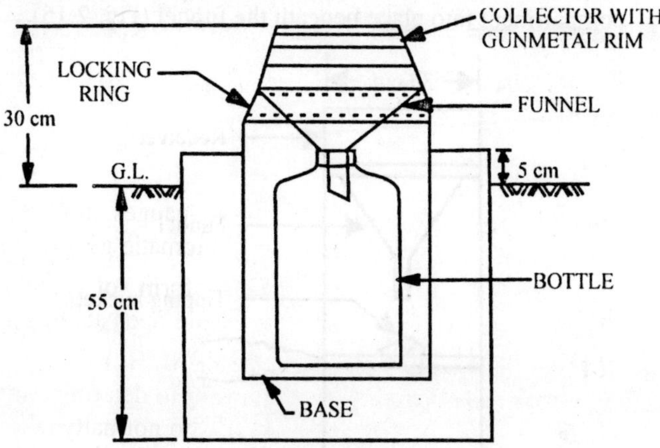

Fig. 2.14: Standard rain gauge (ISI)

Table 2.1: Nominal measuring capacities and combinations

Nominal measuring capacity (mm rainfall)	Collector size (cm)	Base	Polythene bottle capacity (litre)
100	200	Small	2
200	200	Small	4
400	100	Small	4
1000	100	Large	10

(ii) Recording/automatic gauges: This type of rain gauge consists of three components *viz.*, ponents-collector/receiver, a funnel, and a measuring tube. The collector is deep and its rim has a knife edge. In order to minimize the splash of rain drops and to guide the water into the measuring tube a 45 degree funnel is used.

These rain gauges are automatic which record cumulative rainfall continuously with respect to time in the form of chart. Recording gauges are essential to determine short-duration precipitation. These are of following types:

(a) Tipping bucket (b) Weighing gauge and (c) Float gauge.

(a) Tipping bucket gauge: In this type of gauge rainwater is caught in the collector and is funneled into two balanced buckets, which tip back and forth as they are filled in turn by the rainfall. 0.25 mm or 0.01 inch of rainfall will fill one compartment and over balance the bucket so that it tips, empties itself into a reservoir, and moves the second compartment into place beneath the funnel (Fig. 2.15).

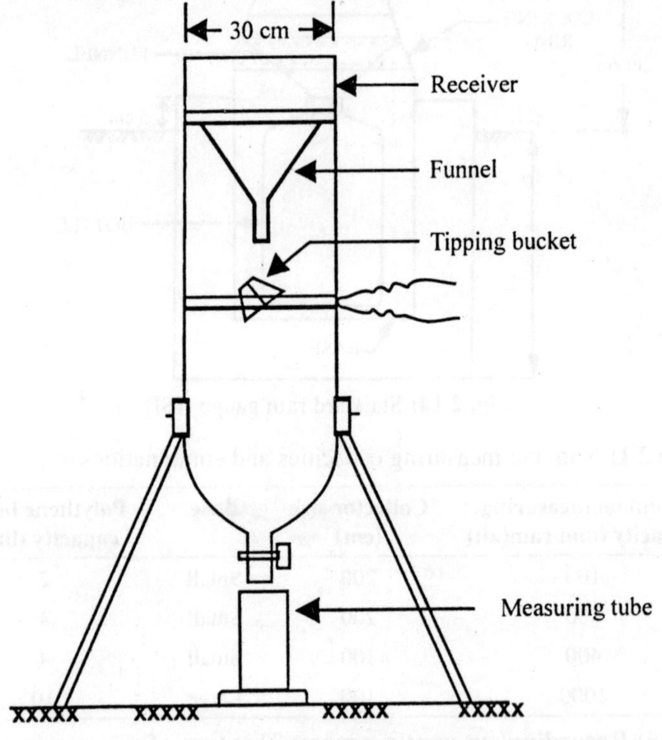

Fig. 2.15: Tipping bucket gauge

An electric circuit gets actuated when the bucket is tipped. This causes the pen to put a mark on a recorder or on graph. If the bucket tipping is designed for a particular intensity, due to inertia the buckets will tip either too soon or too late for other intensities. Again if the intensity of rainfall is very high, the buckets will tip so quickly that the jogs in the record will tend to overlap and make the correct reading of record very difficult and sometime impossible. Such gauges require frequent servicing. Unless heating arrangements are provided such gauges can not be used to measure snowfall.

(b) **Weighing type gauge:** In this type of gauges, the receiver rests on a weighing scale of the spring type (Fig. 2.16).

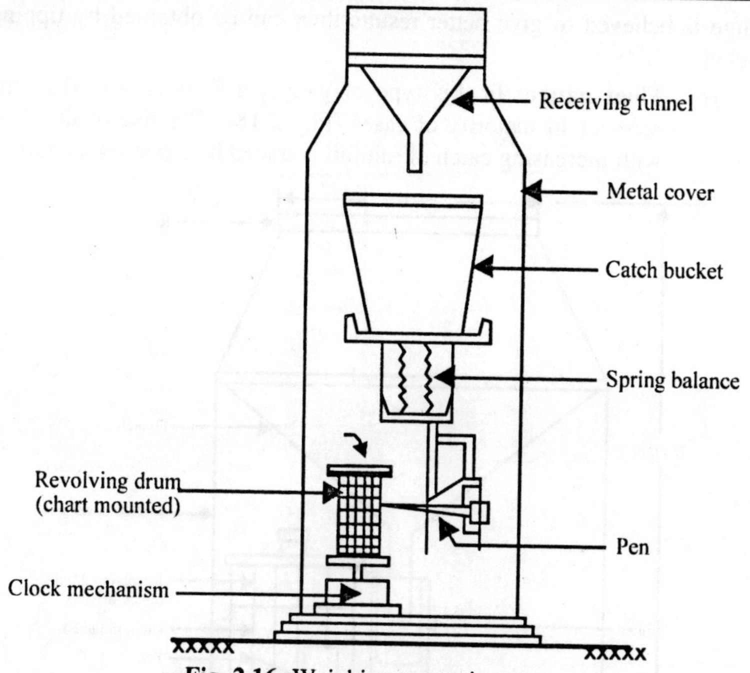

Fig. 2.16: Weighing type rain gauge

The spring is compressed as precipitation accumulates in the receiver, due to which a pen is actuated which draws graph in the form of a mass diagram (Fig. 2.17).

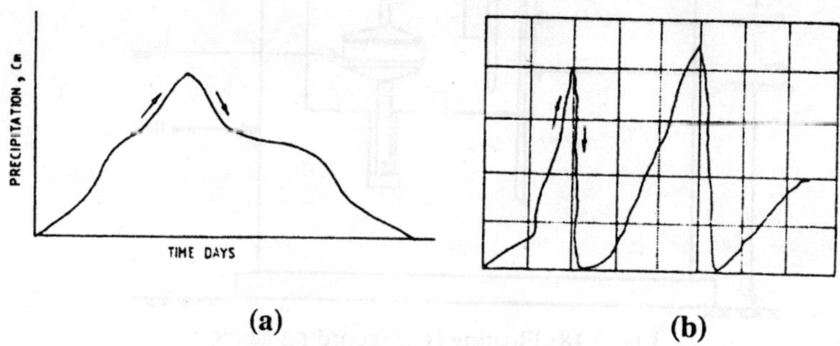

(a) (b)

Fig. 2.17: Mass curve of rainfall

The slope of the curve with respect to horizontal axis gives the intensity of rainfall. Hence in this type of gauge duration of rainfall, total rainfall and its intensity can be determined simultaneously. This type of gauge is suitable both for rainfall and snow measurement. This type of gauge is believed to give better results than can be obtained by tipping bucket.

(c) Float gauge: In this type of gauge, a float is placed in the receiver in majority of cases (Fig. 2.18). The rise of the float with increasing catch of rainfall is traced by a pen on a chart.

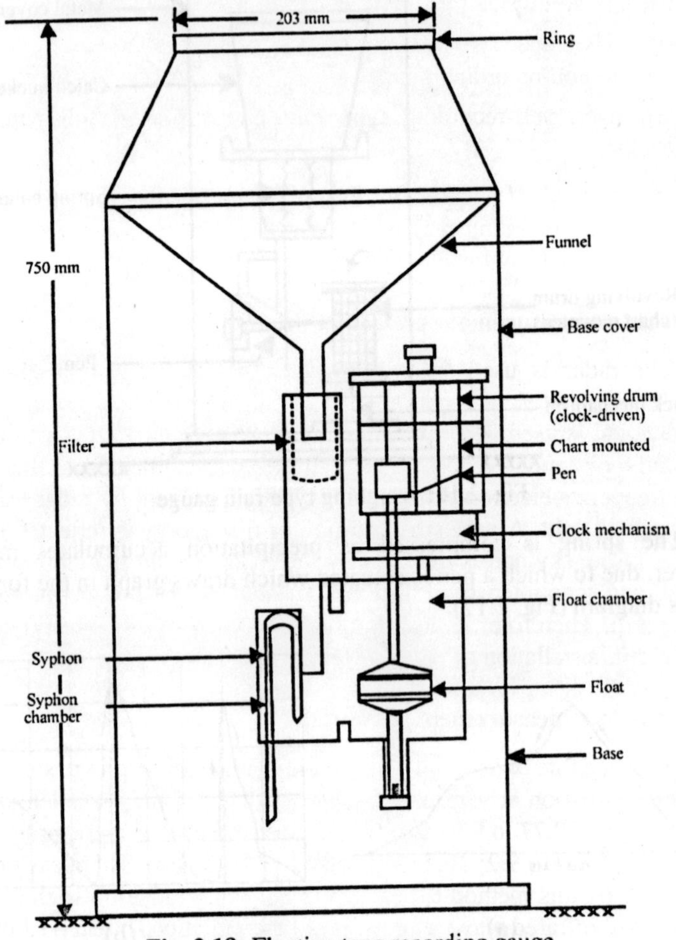

Fig. 2.18: Floating type recording gauge

The record produced in this way is also in the form of mass diagram as shown in Fig. 2.17. There are chances of floats being damaged if the rainfall catch freezes. These float gauges work like floating water stage recorders. Syphonic arrangement is provided for emptying the float chamber quickly as soon as chamber gets filled up, thus obviating the necessity of providing a big chamber. Float type recording gauge is also called as *Syphon Rain Gauge*. It may be of self-emptying type.

The self recording gauges are having following merits:

(i) Recording gauges are the only means by which data can be collected from inaccessible areas,

(ii) They are free from error due to human judgement,

(iii) The capacity of these gauges is relatively larger as compared to non-recording gauges.

However, self-recording type rain gauge having following two drawbacks:

(i) Initial cost is high, and

(ii) The recording may be interrupted because of electrical and/or mechanical faults.

2.6.2 Radar measurement of precipitation

The radar is used for measurement of rainfall and it gives the rainfall within 10% accuracy of the amount recorded by a rain gauge. This method is used in inaccessible areas, where rain gauges can not be installed, continuous record of rainfall can be obtained from the radar. High frequency electromagnetic waves which are sent by radar travel at the speed of light. A very small portion of this energy is reflected by the objects in the sky and is detected by the radar. By proper calibration of rainfall and echo intensity it becomes feasible to measure the intensity of the rainfall. Therefore, its use is limited. Because of very heavy cost is involved in installation of radar system.

2.6.3 Satellite measurement of rainfall

The satellite works on the principle that atmosphere selectively transmits radiation at various wavelengths. The visible wave lengths are of the order of 0.77 to 9.91 μm and thermal infra red wave lengths are in the spans of 8.0 to 9.2; 10.2 to 12.4 and 17 to 22 μm. The measurement of rainfall by this method depend upon detecting radiation in the visible and thermal infrared wave lengths. From the radiation so received images arc produced by the satellites. These images are then made available in various forms for computer processing and the data obtained there from are made available for use by the Indian Meteorological Department.

2.6.4 Errors in precipitation measurements

Any obstruction placed in fluid (air) flow will cause turbulence. Rain gauges, therefore, generate turbulence in the air flow at the place where they are situated and there is an under-catch of rainfall. Similarly turbulence provided by immediate environment *e.g.* trees, buildings, etc will affect the catch. The turbulence is directly a function of wind strength and the area. Since the wind speed increases with height above the ground it is better to keep the receiver top as low as possible. Gauges should be located away from trees, buildings, etc. by at least 4 times their height. Sometime wind shields are used to reduce the effect of wind turbulence.

Losses due to evaporation may introduce substantial errors in case of exposure of gauges to the free air for long periods of time. Oil can be used to reduce evaporation losses.

Errors made in reading the graduations on the scale are usually random and compensating. At times such errors may be large. Other observational errors are usually small but cumulative. Instrument errors on the other hand are quite large and are also cumulative *e.g.* when measuring stick is put in the measuring tube, it increases the reading by 1 per cent. If the gauge is inclined by 10° and rain is falling vertically, loss in catch may be to the extent of 1.5 per cent.

Inaccurate measurements of rainfall may result due to frictional problems in the weighing mechanism of weighing type gauges and in the float guides of float type gauges.

2.7 Analysis of Precipitation

For the proper design and operation of water harvesting structures, *e.g.* size of spillways in a dam, flood protection works, for crop planning to make best use of rain, knowledge of maximum flows in a stream is essential. Most of the time such information is not available or is scanty. But precipitation records for long periods of time are usually available at locations near the project site. Study of rainfall data, and correlating it with run off data is feasible and can help in extending run off records.

Variation in precipitation

Generally, precipitation varied as follows:

1. Precipitation tends to be heavier near the coastlines,
2. Direction of prevailing winds from the ocean also affects the precipitation,

3. Mean annual rainfall being heavier near the shore,
4. Precipitation will go on decreasing slowly towards the interior. However, if the prevailing winds are from land to the ocean then the mean annual rainfall shall not be excessive. Since mean annual temperature affects rainfall as these affects evaporation and wind movement,
5. Precipitation has been generally found to increase with the increase in the altitude, and
6. Precipitation also varies with time.

Therefore, due to the above reasons true long-term mean rainfall cannot be estimated when the limited rainfall records are available at a station. Alexander Binnie found a periodic variation of rainfall and made the following observations:

1. A record of five-year length is likely to be nearly 15 per cent in error,
2. A record of 10-year length is probably 8.2 per cent of the true mean,
3. One 20-year in length should be within 3.3 per cent of the correct value, and so on,
4. Records of 30 or 40 year length in all probability give the true long-term mean rainfall with an average error of about 2 per cent.

However, rainfall records of 30-40 years length is considered good enough for engineering or practical purposes and agro- meteorology to draw any conclusion.

Isohyets

It is the graph in which lines join points of equal precipitation and it is contour map of rainfall. A typical isohyetal map of 2-hour rainfall occurring at various rain gauge stations resulting from a simple thunderstorm is shown in Fig. 2.19.

2.8 Precipitation Gauge Network

Precipitation characteristics *viz.*, intensity, duration, etc. vary with storm, topography and general climate. The precipitation gauge network is essential to have complete knowledge of the precipitation character over areas. The size of such a network (or density) depends upon the economics, purpose of observation, and other such factors. ISI recommended the following minimum densities of precipitation network for general hydro-meteorological purposes:

1. One station per 520 km^2 for plains area.
2. One station per 260-360 km^2 for regions of average elevation.
3. One station per 130 km^2 for predominantly hilly areas with heavy rainfall.

Out of these about 10 per cent of rain gauge stations should be equipped with self recording rain gauges for measurement of rainfall intensities.

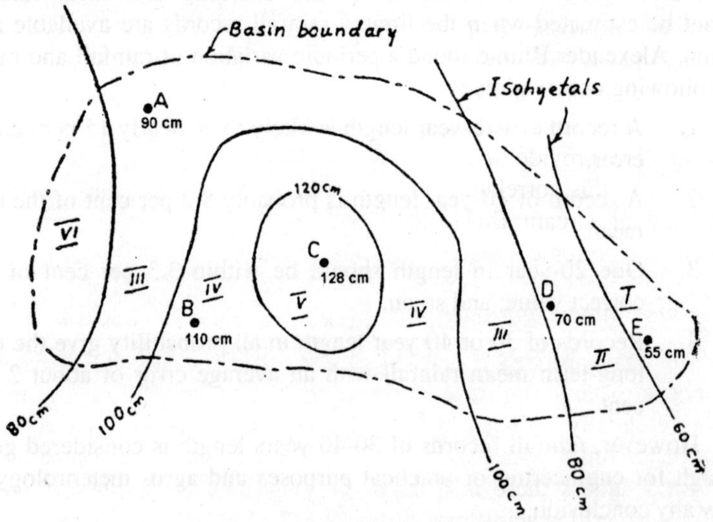

Fig. 2.19: Typical Isohyetal map

Networks may be of different types: National network, mountain network and watershed network. National networks are designed to serve a highly variable need and are extensive in nature. The design of network depends upon economics, accessibility, etc. Watershed networks recommended for projects to come up or are already being executed are sparsely populated, for accessibility to many locations in these area. While, mountain network is recommended if the topography is heterogeneous as in mountaneous areas rainfall pattern is also heterogeneous. Gauge spacing in such cases shall have to be closer as compared to the gauge spacing in the plains.

The density of network primarily is determined by the objective to be fulfilled and the variability of the precipitation. It has been noted that variability of precipitation is much less when the duration of rainfall and the area over which it occurs is large compared to when duration is short and area is small. Watershed network will provide data which may be used as indices or as absolute values and are helpful in water balance studies.

2.8.1 Network design

There are many approaches to watershed network design. Some of these are described below:

Index approach

This approach is based on correlation between the number of gauges and location s(rainfall and distance). Therefore, at least one gauge should be placed in each homogenous contributing area, in order to forecast the results from present experimentation. The network design should also provide indices for sub-watersheds. The accuracy to be achieved in this correlation procedure should be commensurate with the accuracy of stream flow measurements or other measurements with which meteorological records are being compared with.

Absolute estimate approach

In this approach a large number of rain gauges are provided in watershed even to the limit of saturation. The number of rain gauge can be substantially reduced without loss of accuracy within the desired confidence limits with passage of time. Uniform grid network should be followed, unless there is a strong meteorological non-homogeneity in which case grid spacing may be shortened. This approach is suitable for experimental basins.

Transposition approach

When a new project is taken up at a site, there is a remote possibility of existence of a dense network. But if in the neighbouring area, a dense network exists and that area is meteorologically homogenous to the basin under consideration; then it is possible to design the network on the basis of experience of the dense network. The error is determined with the less dense network by comparing with dense network. The error is determined according to McGuiness (1963) on using a less dense network by the flowing relationship:

$$E = 0.03\, P^{0.54}\, G^{0.24} \qquad \text{....(2.4)}$$

where,

E = the absolute difference (inches);

P = the rainfall in inches for the true network, and

G = the gauging network density in square miles per gauge for the reduced network.

2.8.2 Optimum number of rain gauges

The optimal number of rain gauge stations (n) can be determined by the following relationship:

$$n = \left(\frac{C_v}{E}\right)^2 \qquad \text{....(2.5)}$$

where,

E = the percentage error, usually taken as 10 per cent, and

C_v = coefficient of variation of the rainfall values at the existing n stations expressed in percent.

Let us consider that $P_1, P_2, P_3, \ldots P_i \ldots P_n \ldots$ be the precipitation amounts at n rain gauge stations, then coefficient of variation of rainfall is determined as,

$$C_v = \left(\frac{\sigma_{n-1}}{\mu}\right) \qquad \text{....(2.6)}$$

where,

μ = the mean precipitation = $\frac{1}{n}\sum_{i=1}^{n} P_i$, and σ_{n-1} = the standard deviation

The standard deviation is determined by the following relationship:

$$(\sigma_{n-1})^2 = \frac{\sum_{i=1}^{n} (P_i)^2 - \frac{1}{n}\left(\sum_{i=1}^{n} P_i\right)^2}{n-1} \qquad \text{....(2.7)}$$

The additional rain gauges are required to be installed as a result of above noted procedure, according to the spatial distribution of existing rain gauge stations and variability of rainfall over the drainage basin.

Example 2.2: Determine the additional number of rain gauges for optimum design for a drainage basin so as to limit the percentage error within 10 per cent. At present, the drainage basin is provided with 5 rain gauges. The average rainfall (cm) at these stations is: 95, 85, 70, 55, 40.

Solution:

Step I: Compute the mean rainfall, $\mu = \frac{1}{n}\sum_{1}^{n} P_i$

i.e. $$\mu = \frac{95 + 85 + 70 + 55 + 40}{5}$$

$$= \frac{345}{5} = 69\,cm$$

Step II: Compute the sum of squares of rainfall at the stations

$$i.e. \sum_{1}^{n} P_i^2 = 95^2 + 85^2 + 70^2 + 55^2 + 40^2$$

$$= 9025 + 7225 + 4900 + 3025 + 1600$$

$$= 25775$$

Step III: Compute the factor $\frac{1}{n}\left(\sum_{i=1}^{n} Pi\right)^2$

$$\frac{1}{n}\left(\sum_{i=1}^{n} Pi\right)^2 = \frac{345^2}{5} = 23855$$

Step IV: Compute square of standard deviation $(\sigma_n - 1)^2$

$$(\sigma_{n-1})^2 = \frac{\sum_{i=1}^{n} (P_i)^2 - \frac{1}{n}\left[\sum_{i=1}^{n} P_i\right]^2}{n-1} = \frac{25775 - 23805}{5 - 1}$$

$$= 394 \therefore \sigma_{n-1} = 19.85$$

or

En-1 = 19.85

Step V: Compute the coefficient of variation (C_v)

$$C_v = \left(\frac{\sigma_{n-1}}{\mu}\right).100$$

$$= \frac{19.85}{69} = 100 = = 28.76$$

Step VI: Optimum number of rain gauge, N

$$N = \left(\frac{C_v}{E}\right)^2 = (28.76/10)^2 = 8.27 = 8$$

The additional rain gauge required = 8 – 5 = 3.

2.8.3 Estimation of missing data of precipitation

The estimation of missing precipitation at a rain gauge station may be made from the observed precipitation values at about three other nearby stations. Of course, these stations should be almost evenly spaced around the station (*A*). *A* simple arithmetic average would provide the estimate if the normal period precipitation at the nearby stations used far reconstruction of records is within 10 per cent of that for station, *i.e.* the one with the missing record. Otherwise, the ratios of storm rainfall to average annual rainfall at each index station should be equated *i.e.* $P_1/N_1 = P_2/N_2 = P_3/N_3 = P_4/N_4$, which shall lead to the establishment of following relationship:

$$P_A = \frac{1}{3}\left[\frac{N_A P_1}{N_1} + \frac{N_A P_2}{N_2} + \frac{N_A P_3}{N_3}\right] \quad ....(2.8)$$

where,

P_A = the missing precipitation depth for station (A),

P_1, P_2, P_3 = precipitation values at stations 1, 2 and 3 respectively during the period under study,

N_1 N_2, N_3, and A = long-time precipitation values at stations 1,2,3 and *A* stations, respectively.

This method of estimating the missing precipitation known as 'Normal Ratio Method' would work well if the precipitation regime of nearby stations is the same as that of the station at which the record is missing.

Missing precipitation at a station (*A*), can also be estimated by a method known as *National Weather Service Method.* In this method a set of coordinate axes passing through the location of station are drawn. The distance of the nearest surrounding stations (say *B, C, D, E*, etc.) from point *A* are determined (Fig. 2.20) as $d^2 = X^2 + Y^2$.

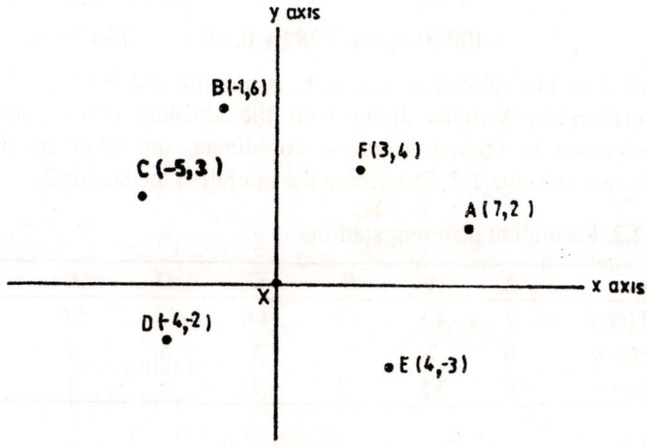

Fig. 2.20: Location of stations on co-ordinate axes

The missing precipitation at station (*A*), is computed by weighted average of precipitations at the stations *B, C, D,* and *E*. Weights would naturally be proportional to the reciprocal of squares of distances d^2 between station *A* and other nearby stations *B, C,* and *D i.e.* $W = 1/d^2$. The missing rainfall at station (*A*), can then be estimated as by the following relationship:

$$P_A = \frac{\Sigma P.W}{\Sigma W} \quad \text{....(2.9)}$$

Example 2.3: Rain gauge station *X* was in operation for some days of the month during which the storm occurred. The respective storm totals at the three neighbouring stations *A, B* and *C* were 107, 90 and 120 mm. The normal annual precipitation values at stations *A, B, C* and *X* are 980, 1100, 890 and 1200 mm, respectively. Determine the missing precipitation for station *X*.

Solution:

The mean precipitation at station × is determined:

$$X = \frac{1}{3}\left[\frac{P_1 N_A}{N_1} + \frac{P_2 N_A}{N_2} + \frac{P_3 N_A}{N_3}\right]$$

$$= \frac{1}{3}\left[\frac{107}{980}1200 + \frac{900}{1100}1200 + \frac{120}{890}1200\right]$$

$$= 400\ (0.109 + 0.082 + 0.135) \quad = 130.3 \text{ mm}$$

Example 2.4: The rainfall at station *X*, is missing and the rainfall at the surrounding stations along with the absolute coordinates with reference to Station *A* whose coordinates are taken as (0,0) is shown in Table 2.2. Determine the rainfall at the station *X*.

Table 2.2. Rainfall at different stations

Station	***X***	***A***	***B***	***C***	***D***	***E***	***F***
Rainfall (cm)	2	4.5	5.0	4.6	5.2	5.0	5.5
Coordinate *X*	0	7	2	5	5	4	3
Coordinate *Y*	0	3.5	10	3	5	3	4

Solution:

Computations have been shown in the tabular form (Table 2.3) and the coordinates of stations shown in Fig. 2.21.

$$Px = \frac{\Sigma P.W}{\Sigma W} = \frac{1025.2}{203.28} = 5.04 \text{ cm.}$$

Table 2.3: Computation for missing rainfall

Station	*Rainfall P. (cm)*	*X*	*Y*	$d^2 = X^2 + Y^2$	$W = \frac{1}{d_2} \times 10^{-3}$	*(P×W)*
X	-	0	0	0	0	0
A	4.5	7	2	53	18.87	84.92
B	5.0	-2	6	40	25.00	125.00
C	4.6	-5	3	34	29.41	135.28
D	5.2	-4	-2	20	50.00	260.00
E	5.0	4	-3	25	40.00	200.00
F	5.5	3	4	25	40.00	220.00
Total					203.28	1025.20

Ans. Rainfall at the station X = 5.04 cm

Double mass curve technique is usually used to test the homogeneity of the precipitation and runoff data. Since the past relation to the present day conditions, the data are usually assigned in severe condition.

2.8.4 Consistency of records

The method of testing homogeneity or consistency of records by double mass curve is explained by the following example.

Example 2.5: The annual rainfall for stations *A, B, C, D* and *E* is given in Table 2.4. Using double mass analysis establish the consistency of the record at station *B*.

Solution:

Step I: Cumulative rainfall of the gauge at each of the stations *A, B, C, D* and *E* is computed and is shown in Col. 8, 9, 10, 11, 12 of the Table-Part B. Cumulative mean values are also worked out as shown in Col. 13.

Table 2.4: Double mass analysis for example 2.5

	Part A – Data Part Annual rainfall for stations (cm)						**B – Computations Cumulative annual rainfall for stations (cm)**					
	A	**B**	**C**	**D**	**E**	**Mean**	**A**	**B**	**C**	**D**	**E**	**Mean**
(1)	(2)	(3)	(4)	(5)	(6)	(7)	(8)	(9)	(10)	(11)	(12)	(13)
1990	58.28	48.46	66.53	60.09	60.45	58.76	58.28	48.46	66.56	60.09	60.45	58.76
1989	47.08	40.53	40.83	43.75	43.66	43.17	105.36	88.99	107.06	103.84	104.11	101.87
1988	60.02	63.06	60.78	49.54	64.16	59.51	165.38	152.05	167.84	153.38	168.27	161.38
1987	63.19	52.81	63.31	52.18	59.50	58.20	228.57	204.86	231.15	205.56	227.77	219.58
1986	58.48	49.45	54.95	64.29	64.35	58.35	287.05	254.81	286.10	269.85	292.12	277.92
1985	58.25	56.02	49.73	56.05	48.17	53.64	345.30	310.83	335.83	325.90	340.29	331.42
1984	41.45	50.14	44.96	39.78	54.38	46.14	386.75	360.97	380.79	365.68	397.67	377.77
1983	57.01	61.12	39.03	52.78	48.34	51.65	443.76	422.09	419.82	418.38	443.01	429.41
1982	57.66	52.83	45.88	41.20	59.41	51.41	501.42	474.92	465.70	459.58	502.42	480.81
1981	50.61	37.55	41.58	34.15	38.53	40.48	552.03	512.47	507.28	493.73	540.95	521.29
1980	61.10	50.90	60.77	49.04	58.72	56.11	613.13	563.37	568.05	542.77	599.67	577.40
1979	70.54	66.67	79.18	46.50	74.16	67.41	683.67	630.04	647.23	589.27	673.83	644.70

1978	40.07	57.32	46.63	46.13	30.18	44.07	723.74	687.36	693.86	635.40	704.01	688.87
1977	52.56	43.99	41.26	50.70	37.57	45.22	776.30	731.35	735.12	686.10	741.58	733.29
1976	53.35	61.29	51.36	53.34	42.56	52.38	829.65	792.64	786.48	739.44	784.14	786.47
1975	50.25	48.92	52.06	39.20	35.66	45.22	879.90	841.56	838.54	778.64	819.80	831.69

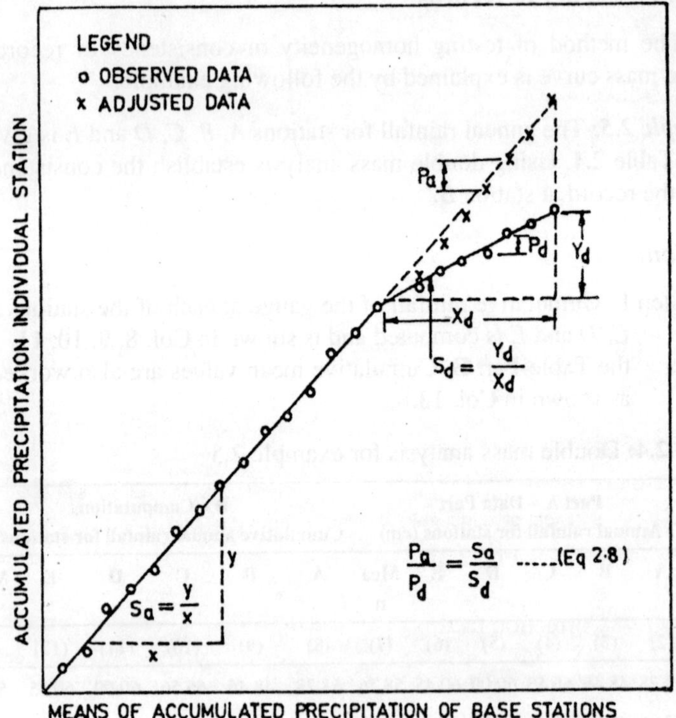

Fig. 2.21: Double mass curve

Step II: The cumulative rainfall values of gauge at stations *B* are plotted against the combined cumulative rainfall values of surrounding stations *A, B, C* and *D* (Fig. 2.22).

Step III: Since the plot is not a straight line for stations B and has a break against the year 1987, the record has become inconsistent with effect from the year 1987. There is no inconsistency is record for station *A*.

Step IV: Work out the slopes of the two portions of the line *OAB*

Line *OA* Slope = 0.83

Line *AB* Slope = 1.05

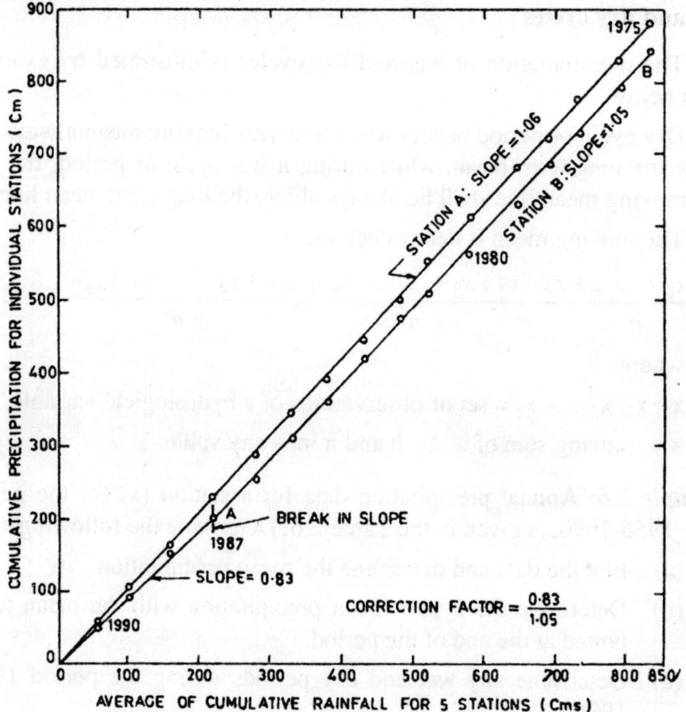

Fig. 2.22: Double mass analysis

Step V: Correction factor = $\frac{0.83}{1.05} = 0.79$

Step VI: Values of the earliest period of records are adjusted to the most recent period and are given in Table 2.5.

Table 2.5: Adjusted values at station B

Year	Original value (cm)	Adjusted value (cm)	Year	Original value (cm)	Adjusted value (cm)
1986	49.45	46.20	1980	50.90	40.21
1985	56.02	44.26	1979	66.67	52.67
1984	50.14	39.61	1978	57.32	45.2.8
1983	61.12	48.28	1977	43.99	34.75
1982	52.83	41.74	1976	61.29	48.42
1981	37.55	29.66	1975	48.92	38.65

Wet and dry cycles

The determination of wet and dry cycles is illustrated by example given below:

Dry cycle or period occurs when five year moving mean always lies below the long term mean, while during a wet cycle or period, the five year moving mean line shall lie always above the long term mean line.

The moving mean is determined as,

$$\frac{x_1 + x_2 + \dots\dots + x_n}{n}, \frac{x_2 + x_3 + \dots\dots + x_{n-1}}{n}, \frac{x_3 + x_4 + \dots\dots + x_{n+2}}{n} \quad (2.10)$$

where,

$x_1, x_2, x_3 \dots\dots x_n$ = set of observations of a hydrological variable,

n = moving sum of .... n and n may any value.

Example 2.6: Annual precipitation data for a station (x) for the period 1956-1990, is given in the Table 2.6. Determine the following:

(a) Plot the data and determine the mean precipitation.

(b) Determine the 5-year mean precipitation with the mean to be potted at the end of the period.

(c) Determine any wet and dry periods during the period 1956-1990.

Solution:

(a) Plot is shown in Fig. 2.23; 35-year mean = 36.16 cm.

Table 2.6. Annual precipitation for station (x)

Year	Annual precipitation (cm)	Year	Annual precipitation (cm)	Year	Annual precipitation (cm)
1956	40.30	1968	18.10	1980	52.38
1957	30.52	1969	40.24	1981	37.46
1958	50.40	1970	22.36	1982	39.35
1959	50.63	1971	38.20	1983	24.68
1960	35.02	1972	55.60	1984	36.57
1961	45.20	1973	50.42	1985	25.52
1962	17.50	1974	25.38	1986	25.30
1963	30.33	1975	50.48	1987	35.12
1964	30.50	1976	40.22	1988	45.68

1965	22.64	1977	62.52	1989	40.27
1966	22.50	1978	30.18	1990	35.00
1967	27.12	1979	30.10	-	-

(b) Five-year mean has been determined and is shown in Table 2.7.

(c) 5-year moving means have been plotted in Fig. 2.23.

Dry Cycles: 1963-1972

1984-1990

Wet cycles: 1972-1984

5-year mean has been plotted at the end of the period, rather than at the mid-point of the period so that it reflects the natural Storage in the hydrological system in a more exact manner.

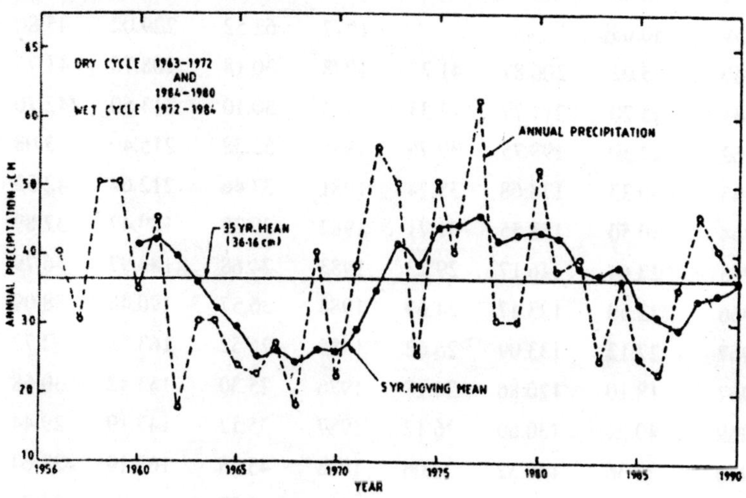

Fig. 2.23: Dry and wet cycles

2.9 Precipitation Analysis and its Presentation

The through knowledge in respect of precipitation is required for various purposes such as relation between precipitation and runoff on long-term basis, short term basis, and for individual storm period. The precipitation data collected for a station with the help of recording and non-recording gauges are usually analysed in the following ways.

(a) Determination of physical distribution of precipitation *viz.*, as time distributing (Depth-time-relation) and areal distribution

(Depth- area relation) or a combination of these (Depth-area-duration or time-area-depth relation).

(b) Determination of statistical distribution like frequencies analysis of precipitation data.

Table 2.7: Computation for 5-year mean

Year	Annual precipi-tation (cm)	5-year total	5-year mean (cm)	Year	Annual precipi-tation (cm)	5-year total	5-year mean (cm)
1956	40.30	-	-	1974	25.38	191.96	38.39
1957	30.32	-	-	1975	50.48	220.08	44.02
1958	50.40	-	-	1976	40.22	222.10	44.42
1959	50.63	-	-	1977	62.52	229.02	45.80
1960	35.02	206.87	41.37	1978	30.18	208.78	41.75
1961	45.20	211.77	42.35	1979	30.10	213.50	42.70
1962	17.50	198.75	39.75	1980	52.38	215.40	43.08
1963	30.33	178.68	35.74	1981	37.46	212.64	42.53
1964	30.50	158.55	31.71	1982	39.25	189.47	37.89
1965	22.64	146.17	29.23	1983	24.68	183.97	36.79
1966	22.50	123.47	24.69	1984	36.57	190.44	38.09
1967	27.12	133.09	26.62	1985	25.52	163.58	32.72
1967	18.10	120.86	24.17	1986	25.30	153.42	30.68
1969	40.24	130.60	26.12	1987	35.12	147.19	29.44
1970	22.36	130.32	26.06	1988	45.68	168.19	33.64
1971	38.20	146.02	29.20	1989	40.27	171.89	34.38
1972	55.60	174.50	34.90	1990	35.00	181.37	36.27
1973	50.42	206.82	41.36	-	-	-	-

Depth-time graph

It is a graph for showing time-distribution of rainfall (Fig. 2.24). It is evident that storm patterns are found to vary regionally and seasonally. The depth of rainfall for selected intervals of time may be determined which is not provided by non-recording gauges. The intervals of time usually vary from 5 minutes to an hour. Intensity of rainfall for each interval of time can be determined by dividing depth of precipitation by the interval during which that much depth of precipitation occurred. When graphs are plotted in such a way that having time interval divisions as one hour and rainfall in cm on vertical axis gives automatically the intensity of rainfall (cm/hr.). Such curves are helpful in studying long-duration storms which is essential for hydro-meteorological studies.

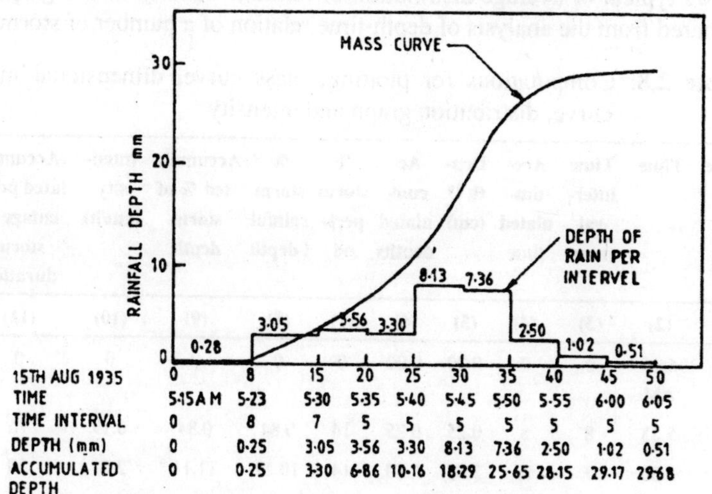

15TH AUG 1935	0	8	15	20	25	30	35	40	45	50
TIME	5·15 AM	5·23	5·30	5·35	5·40	5·45	5·50	5·55	6·00	6·05
TIME INTERVAL	0	8	7	5	5	5	5	5	5	5
DEPTH (mm)	0	0·25	3·05	3·56	3·30	8·13	7·36	2·50	1·02	0·51
ACCUMULATED DEPTH	0	0·25	3·30	6·86	10·16	18·29	25·65	28·15	29·17	29·68

Fig. 2.24: Depth-time curve and mass curve

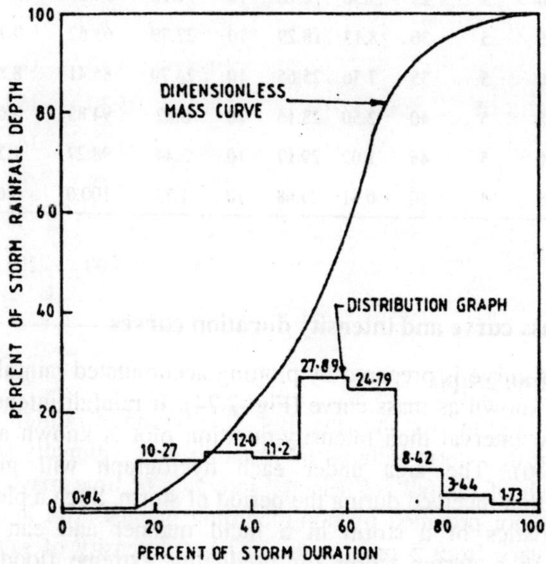

Fig. 2.25: Percentage distribution graph and dimensionless mass curve

2.9.1 Distribution graph

This graph is prepared from the depth-time relationship (Fig. 2.25). Percentage distribution graph gives the relationship between percentage of storm rainfall depth and percentage of storm duration. This graph shows typical or average distribution of rainfall. Usually such a graph is prepared from the analysis of depth-time relation of a number of storms.

Table 2.8: Computations for plotting, mass curve, dimensional mass curve, distribution graph and intensity

Date	Time	Time inter-val (mm)	Acc-um-ulated time	Dep-th D (cm)	Ac-cum-ulated depths	% storm peri-od	% storm rainfal l depth	Accumula -ted % of storm depth	Inten-sity (cm/h)	Accumu-lated perc-entage of storm duration
(1)	(2)	(3)	(4)	(5)	(6)	(7)	(8)	(9)	(10)	(11)
15th Aug. 1935	5.15 AM	0	0	0.00	0.00	0	0	0	0	0
	5.23	8	8	0.25	0.25	16	0.84	0.84	0.19	16
	5.30	7	15	3.05	3.30	14	10.27	11.11	2.61	30
	5.35	5	20	3.56	6.86	10	12.00	23.11	4.27	40
	5.40	5	25	3.30	10.16	10	11.12	34.23	3.96	50
	5.45	5	30	8.13	18.29	10	27.39	61.62	9.76	60
	5.50	5	35	7.36	25.65	10	24.79	86.41	8.83	70
	5.55	5	40	2.50	28.15	10	8.42	94.83	3.00	80
	6.00	5	45	1.02	29.17	10	3.44	98.27	1.22	90
	6.05	5	50	0.51	29.68	10	1.73	100.0	0.61	100

2.9.2 Mass curve and intensity duration curves

This curve is prepared by plotting accumulated rainfall versus time and it is known as mass curve (Fig. 2.24). If rainfall intensity is plotted with time interval then intensity-duration plot is known as *hyetograph* (Fig. 2.26). The area under each hyetograph will give the total precipitation received during the period of storm. Such a plot exhibits the characteristics of a storm in a lucid manner and can be used for developing a design storm for predicting extreme floods of a given frequency.

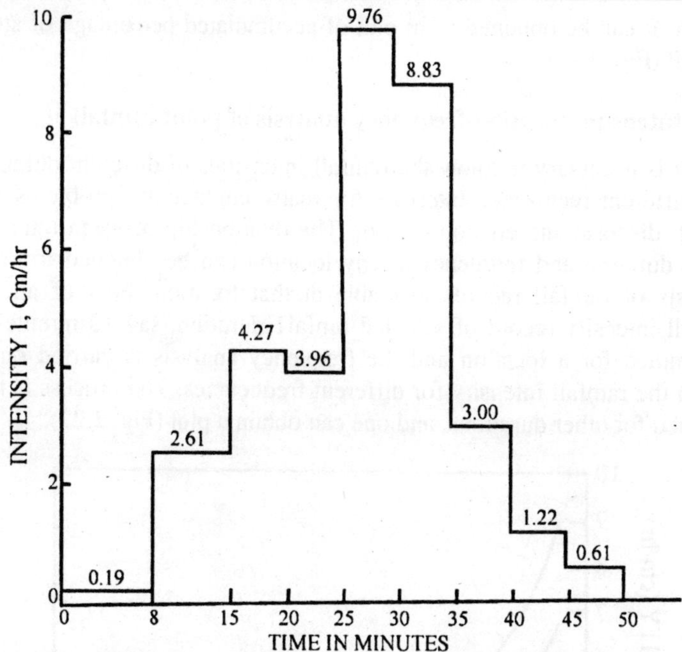

Fig. 2.26: Intensity-time graph (Hyetograph)

Mass curve and intensity-duration plot (Fig. 2.24 and 2.26) are closely related to each other. The ordinate of mass curve at any time is equivalent to the area under the intensity duration plot from the beginning of rainfall to that time (Fig. 2.24). This may be checked as follows.

Area under intensity time graph

$$= \left(0.19 \times \frac{8}{60} \times 10\right) + \left(2.61 \times \frac{7}{60} \times 10\right)$$

$$= 3.30 \text{ mm (at } t = 15 \text{ minutes)}$$

Ordinate of mass curve at 15 minutes = 3.30 mm.

If a line is drawn through the horizontal steps of time-intensity plot in such a way that the areas above and below the line are same then one can obtain a continuous function of intensity-vs-time.

2.9.3 Dimensionless mass curve

It is the graph when plotted accumulated storm rainfall percentage versus storm duration percentage for a large number of storms in a region. It can be obtained from plot of accumulated percentage of storm rainfall (Fig. 2.25).

2.9.4 Intensity-duration-frequency analysis of point rainfall

It is necessary to know the rainfall intensities of different durations and different recurrence intervals for many engineering problems *viz.*, runoff, disposal and erosion control. The relationship between intensity, storm duration and frequency at any location can be obtained from the analysis of rainfall records available at that location. First of all, the rainfall intensity record of selected ranfall duration, say 10 minutes, is determined for a location and the frequency analysis is carried out to obtain the rainfall intensity for different frequencies. The process is then repeated for other durations, and one can obtain a plot (Fig. 2.27).

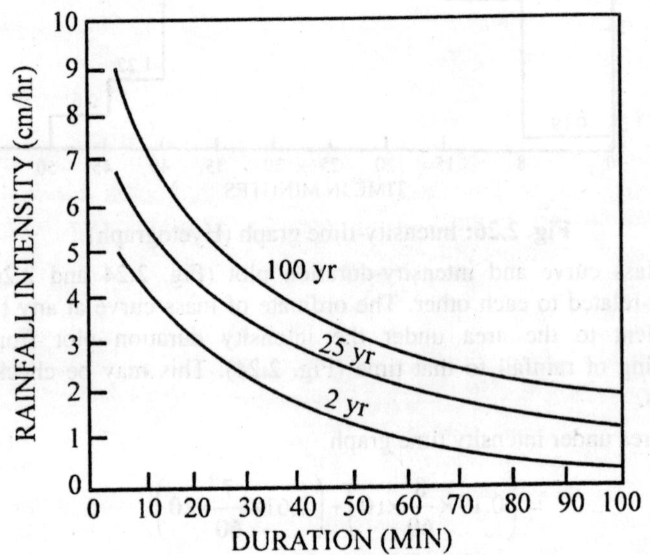

Fig. 2.27: Intensity-duration-frequency curve

The intensity-duration-frequency curve can be represented by the following expression:

$$t=\frac{KT^b}{(D+a)^n} \qquad \text{....(2.11)}$$

where,

i = the intensity of rainfall, (cm/h),

T = the recurrence interval,

D = the duration of rainfall, and

k, b, a, n = constants and do not have a fixed value but will vary from one catchment to another.

For number of locations in a region these constants can he determined and latter on can be used for any engineering design. Alternatively to this, we may use is pluvial maps (map showing lines connecting equal depths of rainfall).

Theoretical probability distribution curves

In collecting hydrological data, two types of measurements are recorded] routine and extremes. In design problems, extreme values are extensively used *e.g.* greatest 24-h rainfall in a particular year. Treatment of extreme is again divided into two series made up of annual extremes and partial duration series. A statistical analysis can only be applied to annual series or any weeks or months during that period assuming as annual series.

If the distribution of annual extremes in precipitation is skewed then theoretical normal probability distribution should not be fitted. However, skewed data can be normalized by using logarithmic transformation or other transformation.

2.9.5 Station-year method

Station-year method is used to determine the recurrence interval for the larger rains which otherwise could not he better estimated by usual frequency analysis methods. In frequency analysis, for reliable estimation there should be enough number of samples of an event. Now the largest rain in the set of observations at a particular gauge station covering period of say 50 years shall have a recurrence interval, T equal to 50 and there is only one 50-year sample. If during the next 50 years, four rains larger than this one were to occur, then its recurrence interval would be 20. On the other hand if no rain occurred during the next 50 years, then recurrence interval would be 100 years. Hence the frequency analysis can only provide an approximate value of frequency of largest rain and the correct value could with equal probability, be smaller or greater than the computed value. This limitation of frequency analysis also applies to other rains having large recurrence intervals.

The procedure for determination of recurrence interval of intense rainfall by station-year method is illustrated in the example given below.

Example 2.7: The maximum summer, 24-hour rains at 20 rain gauge stations within 100 km from station X is given in Table 2.9 and these are rearranged in descending order. The total duration of records from 50 rain gauge stations was 1200 years. Determine the frequency of larger rains on using station-year method.

Table 2.9. 24 hour rain tora and computation recurrence intervals

Order number (m)	Station, *X* *t* = 50 years		Station, *Y* *t* = 1200 years	
	24 h rainfall (cm)	Recurrence interval *T*	24 h rainfall (cm)	Recurrence interval *T*
(1)	(2)	(3)	(4)	(5)
1	16.81	50.0	25.95	1200
2	16.02	25.0	22.78	600
3	14.43	16.7	20.70	400
4	12.52	12.5	20.22	300
5	12.11	10.0	20.19	240
6	11.92	8.3	19.54	200
7	9.83	7.1	18.90	171
8	9.54	6.2	18.09	150
9	9.45	5.6	17.91	133
10	9.32	5.0	17.64	120
12	9.21	4.2	17.44	100
14	8.92	3.6	17.40	86
16	8.61	3.1	16.80	75
18	8.46	2.8	16.26	67
20	7.71	2.5	15.90	60

Solution:

Step I: Determine the recurrence interval of rains using the relation, $T = Hm$ as shown in columns (3) and (5) of Table 2.9 for station *X* and station year.

Step II: Then make a plot as shown in Fig. 2.28. Line *AB* represents plot for *X* station and *CD* is the plot for station year.

Step III: Extend the line *AB* to E guided by the line *CD*.

Step IV: Line *BE* can then be used to determine recurrence intervals of rains larger than 16.81 cm.

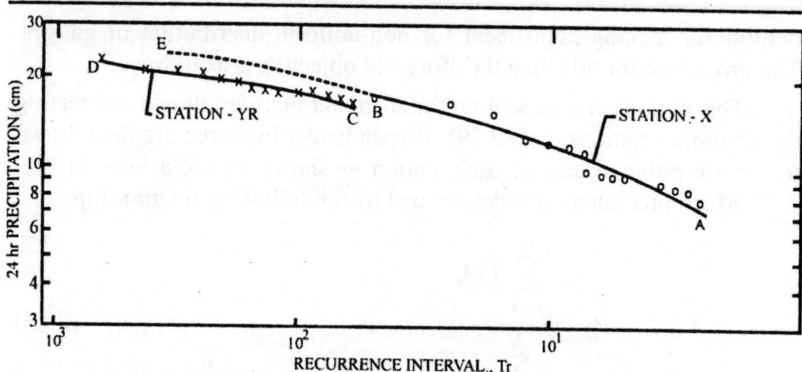

Fig. 2.28. Theoretical probability distribution curve

2.9.6 Mean precipitation over a basin

The precipitation data recorded at any rain gauge stations refers to the point precipitation. However, it can not be representative of the precipitation over a large area except in its immediate neighbourhood. For most hydrological purpose areal distribution of precipitation is required. The most common technique used to determine the areal distribution of precipitation is described below:

Arithmetic average method

This method implies simply determining the average of all observed values of precipitation records at the rain gauge stations. This method shall give correct average value provided that the rain gauge stations are uniformly distributed over the drainage basin. The arithmetic average is determined by the following relationship:

$$P_{av} = \left(\sum_{i=1}^{N} Pi \right) \div N \qquad ....(2.12)$$

where,

P_i = the precipitation depth at station (i), and

N = the total number of rain gauge stations.

Thiessen polygon method

In this method, weights are assigned to precipitation for each rain gauge station in direct proportion to the area it represents. This method is helpful for making adjustment for non-uniform distribution of gauges. The procedure for fulfilling the aforesaid objective is as follows.

The stations are marked on the map and lines are drawn connecting the adjoining stations (Fig. 2.29). Perpendicular bisectors are then drawn to form a polygon around each station as shown by solid lines in Fig. 2.29. Mean precipitation is determined by the following relationship:

$$P_{av} = \frac{\sum_{i=1}^{N} PiA_i}{\sum_{i=1}^{N} Ai} \tag{2.13}$$

where,

A_i = the area assigned to station (*i*), and

P_i = precipitation for its station.

The precipitation in this method is assumed to vary linearly between stations. Thus, not possible to account for topographic or it is other characteristics. Further if any station is added or subtracted from the network then polygons have to be redrawn all over the basin. The drawing of these polygons do not require much skill. This method gives better mean precipitation as compared to arithmetic average method as it takes care of non-uniform distribution of stations and also omits consideration of those stations which have no influence.

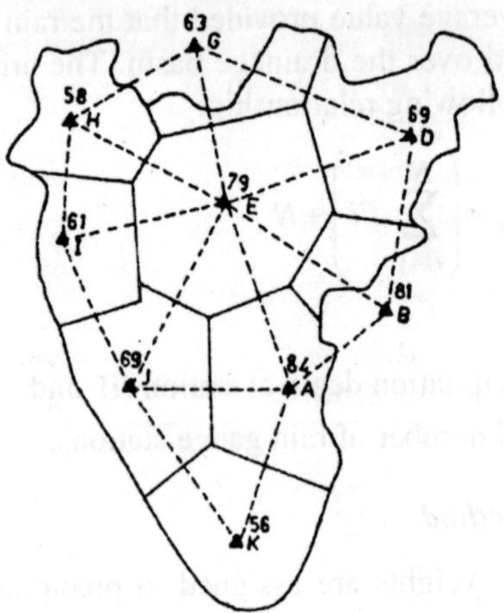

Fig. 2.29: Thiessen polygon method of assigning weights

Isohyetal method

This method is found be superior to Thiessen Polygon method as the effect of topography and other characteristics are taken into consideration. It is the most preferable method for determining mean precipitation over an area. Lines of equal precipitation known as isohyetes are drawn on the map; direction of storm movement, observed runoff and other relevant information is also marked on the map. The mean precipitation is determined by the following relationship:

$$P_{av} = \frac{\sum_{i=1}^{n} (P_i + P_{i+1})(A_i/2)}{\sum_{i=1}^{n} A_i} \quad \text{....(2.13)}$$

where,

P_i = the precipitation at contour *i*,

n = the total number of different contours, and

A_i = the area included between precipitation contours *i* and *i* + 1.

Linear interpolation along the lines joining adjacent stations is not possible as in case of Thiessen polygon method, otherwise the topographic effects cannot be accounted for. If linear interpolation procedure is adopted then the results obtained by Thiessen and isohyetal method would be the same.

Example 2.8: Determine the mean annual precipitation for the river basin shown in Fig. 2.29 by (i) Arithmetic Mean method; (ii) Thiessen Polygon method; and (iii) Isohyet method. The location of the various raingauge stations and the point precipitation values are given in Table 2.10.

Table 2.10. Location of raingauge stations and rainfall

Station	**Coordinates**		**Rainfall P_i**	**Station**	**Coordinates**		**Rainfall**
	X	***Y***	**(cm)**		***X***	***Y***	**P_i (cm)**
A	0	0	84	H	-40	38	58
B	15	13	81	I	-40	20	61
D	17	40	69	J	-24	-2	69
F.	-14	27	79	K	-7	-25	56
G	-20	52	63				

Solution:

Step I: Arithmetic Mean of precipitation is determined on using equation (2.12), as,

$$P_{av} = \frac{84+81+69+79+58+61+69+56}{8}$$

$$= \frac{557}{8} = 69.625 \text{ cm.}$$

Step II: Thissen Polyson Method

The polygons are drawn as shown in Fig. 2.29, Areas 'A' enclosed by each polygon are also shown in this Fig.

ΣA_i = 7.88 + 2.80 + 8.67 +12.57 +3.02 + 5.79 + 5.22 + 8.45 + 8.05 + =62.45

$\Sigma P_i A_i$ = [84 x 7.88 + 81 x 2.8 + 69 x 8.67 + 79 x 12.57 + 63 x 3.02 + 58 x 5.79 + 61 x 5.22 + 69 x 8.45 + 56 x 8.05)

= 661.92 + 226.80 + 598.23 + 993.03 + 190.26 + 335.82 + 318.42 + 583.05 + 450.80

= 4358.33

The mean precipitation is determined on using equation (2.13) as,

$$P_{av} = \frac{\Sigma P_i A_i}{\Sigma A_i} = \frac{4358.33}{62.45} = 69.789 \text{ cm.}$$

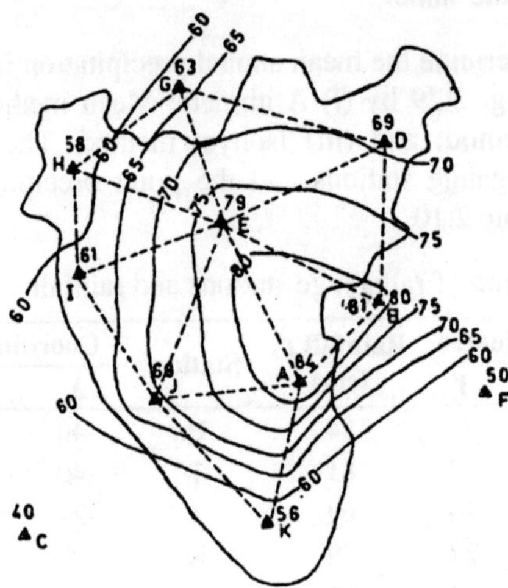

Fig. 2.30: Isohyetal method of determining mean precipitation

Step III: Isohyetal Method

Isohyetal are drawn as shown in Fig. 2.30. Net areas between the isohyets are shown in this figure. The procedure of determination is shown in Table 2.11.

Table 2.11: Mean precipitation-isohyetal method

Between Isohyets (cm)	**Net area within basin boundary (km^2)**	**Average precipi-tation (cm)**	**Precipitation volume Col. 2xCol. 3 ($10^3 x m^3$)**	**Cumulative volume ($10^3 x m^3$)**	**Equivalent uniform depth (cm)**
(1)	(2)	(3)	(4)	(5)	(6)
60	4.26+6.56 =10.82	60.0	649.20	649.20	60
60-65	8.37	62.5	523.15	1172.35	140.06
65-70	13.79	67.5	930.82	2103.17	152.51
70-75	12.77	72.5	925.82	3028.99	237.19
75-80	10.37	77.5	803.68	3832.67	369.59
80	6.05	80.0	484.00	4316.67	713.50

$\Sigma A = 62.17 \quad \Sigma AP = 4316.67$

Mean depth = 4316.67/62.17

= 69.43 cm.

2.9.7 Depth-area curves

It is the graph plotted between rainfall depth versus the drainage area or it is curves show the relationship of rainfall depth and area of drainage basin. For plotting such curves, point precipitation data is not enough. Information on average precipitation over an area is also needed; average precipitation determined by isohyetal method provides the necessary data for plotting depth-area curve. Typical depth-area curves are shown in Fig. 2.31.

For a given duration the depth-area curves expressed by the following expression:

$$\overline{R} = R_o e^{-KA^n} \qquad \text{....(2.14)}$$

where,

$\overline{R}$ = The average depth of rainfall (cm) over an area A,

A = Area (km^2),

R_0= The maximum amount of rainfall (cm) at the centre of the storm, and

K and n = constants for the region.

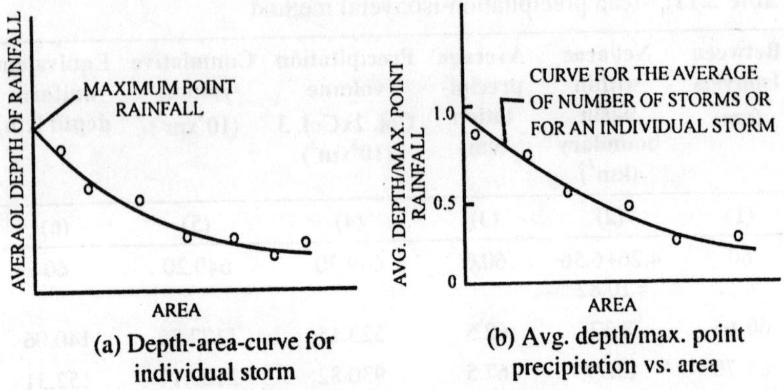

(a) Depth-area-curve for individual storm

(b) Avg. depth/max. point precipitation vs. area

Fig. 2.31: Typical depth-area curves

2.9.8 Depth-area-duration curves

These curves indicate accumulated average precipitation versus area at selected times during a storm period.

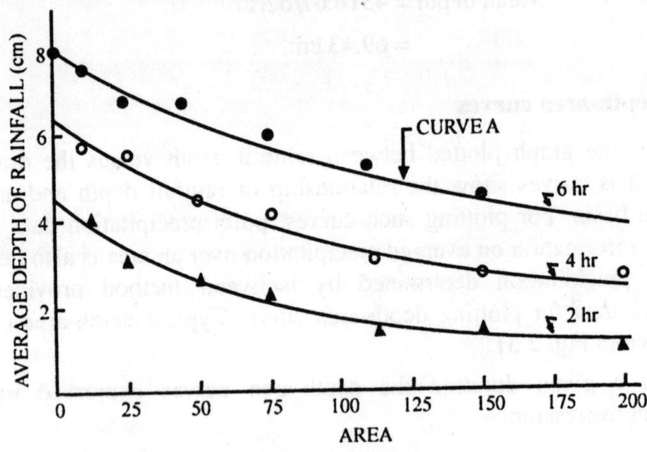

Fig. 2.32: Typical depth-area-duration curves

It is seen from these that the average depth of rainfall for the given storm decreases with the increase in area; and for a given area, the average depth of rainfall increases with the increase in duration.

In the same manner, *maximum depth-area-duration* curves may be prepared. Such curves are needed for determination of severest floods. The depth-area-duration (DAD) curves is prepared as in the following steps:

Step I: Decide the duration of storm for which DAD Curves are to the plotted.

Step II: Select the storms representative of a region. In case maximum-depth area duration curves are to be plotted, then select the severest storms that have occurred in the region.

Step III: Prepare the mass curves for the recording stations. In this hourly amounts of rainfall are accumulated at each hour and plotted against time to get mass curves.

Step IV: Time distribution of rainfall at non-recording stations is developed with the help of a base recording station associated with it. If the times of beginning and ending of rainfall at the non-recording gauge station are known then the duration of rainfall is readily known. When only time of beginning of the storm is given, duration of rainfall is estimated to be the same as that at the base recording station, thus fixing the time when the storm ended.

Step V: Prepare the isohyets with the knowledge of point precipitation (total) at the various recording and non-recording gauges situated in the drainage basin.

Step VI: Determine the area enclosed by the isohyets. In case the same isohyets encloses separate areas, total up the two areas.

Step VII: Determine the difference of areas of adjoining isohyets.

Step VIII: Determine the average rainfall depth, which is mean of the isohyets values of two adjoining isohyets.

Step IX: Determine the product of values in step VII and VIII. This will give volume of rain between one set of isohyets. Then work out the accumulated volumes.

Step X: Divide each accumulated value obtained in step IX by the corresponding area (obtained in step VI) to obtain average depth of rainfall.

Step XI: Plot average depth of rainfall (values obtained instep X) versus area enclosed by each set of isohyets (values obtained in step VI). This will yield a depth-area curve for selected duration of rainfall (step I).

The procedure of determination of depth-area-duration curve is illustrated by taking a hypothetical case where isohyets representing 8 cm, 7 cm,....... 3cm, 2 cm and 1 cm rainfall have been plotted on a drainage basin and area enclosed. The rest of the table can be determined as shown in Table 2.12 and depth-area- duration curve is drawn.

Table 2.12: Calculation for depth-area-curve for a hypothetical storm

Isohyets (cm)	Area enclosed by isohyets (km^2)	Area between isohyets (km^2)	Average depth of rainfall (cm)	Volume of rainfall between isohyets (km^2 cm)	Accumu-lated volume (km^2 cm)	Average depth (cm)
(1)	(2)	(3)	(4)	(5)	(6)	(7)
Duration of rainfall 6h ending at 10 PM Aug. 1990						
8	0	-	-	-	-	8.00
	-	10	7.5	75.0	-	-
7	10	-	-	-	75.0	7.50
	-	15	6.5	97.50	-	-
6	25	-	-	-	172.5	6.90
	-	20	5.5	137.5	-	-
5	45	-	-	-	310.0	6.89
	-	30	4.5	135.0	-	-
4	75	-	-	-	445.0	5.93
	-	35	3.5	122.5	-	-
3	110	-	-	-	567.5	5.16
	-	40	2.5	100.0	-	-
2	150	-	-	-	667.5	4.45
	-	50	1.5	75.0	-	-
1	200	-	-	-	-	-

2.9.9 Probable maximum precipitation

For the design of small structures, flood frequency analysis is pre requisite to determine design flood. The transposition of rainfall may be defined as the transfer of rainfall from the place where it occurred to a water shed/basin for which the design of a hydrological event is proposed. By doing so, the meteorological experience of the basin under consideration is increased. The physical maximum value of rainfall which may occur over a drainage basin is called probable maximum precipitation (PMP). Determination of such a value is particularly useful in flood deign studies *e.g.*, design of spillway for large dams. No probability can realistically be attached to the worst possible event because as the probability of occurrence approaches zero, the flood magnitude would become infinitely large. But flood magnitude have to have a finite maximum value because flood results directly or indirectly at precipitation which shall have a finite value. The decision, therefore, has to be a policy decision: that maximum protection has to be provided.

References

A.T. Hjelimfelt, Jr. and J.J. Cassidy. 1975. 'Hydrology for Engineers and Planners', Iowa State Univ. Press, Ames Iova.

Alexander Binnie, "The Variation of Rainfall" Proc. Inst. Civil Engrs. (Longdon), Vol. 109.

Chow, V.T. 1964. Ed. 'Handbook of Applied Hydrology', McGraw Hill, New York.

Linsely, R.K. Jr., M.A. Kohler and J.L.H. Paulhus. 1958. "Hydrology for Engineers", McGraw Hill Book Co., Inc. New York.

Linsely, R.K., M.A. Kohler and J.L.H. Paulhus. 1949. "Applied Hydrology", McGraw Hill Book Co., Inc. New York.

Mutreja, K.N. "Applied Hydrology" Tata McGraw Hill Co. Ltd., New Delhi.

Nemec J. "Engineering Hydrology", Tata McGraw Hill Co. Ltd., New Delhi.

Parthasarthy, B. and O.N. Dhar. 1975. "Hydrological Science", Bulletin XX 26/1975.

Raghunath, H.M. 1985. "Hydrology-Principles, Analysis Design", Wiley-Eastern Ltd., New Delhi.

Raw, K.N., C.J. George and K.S. Ramassastri. 1971. 'Potential Evapotranspiration (PE) over India', Proceedings of symposium on "Water Resources", I.I.Sc. Bangalore, India, pp. A 2-1-14.

Schulz, E.F. 1974. "Problems in Applied Hydrology", Water Resources Publications, Fort Collins, Colorado, USA.

Subramanya, K. "Engineering Hydrology", Tata McGraw Hill Co. Ltd., New Delhi.

Vander Leeden, F. 1975. 'Water Resources of the World', Water Information Centre, Port Washington, N.Y., U.S.A.

Varshney, R.S. 1979. "Engineering Hydrology", Nemi Chand and Bros., Roorkee, India, 2nd Ed.

Wisle, C.O. and E.F. Brater. 1959. "Hydrology", John Wiley and Sons Inc., New York.

EXERCISES

2.1 Differentiate the following-

(i) Precipitation and rainfall

(ii) Recording and non-recording rain gauge

(iii) Isohyets and contours

(iv) Wet cycle and dry cycle

(v) Mass curve and double mass curve

(vi) Intensity-time graph and isohyets graph

(vii) Extreme precipitation and probable maximum precipitation

(viii) Overland how and inter flow

(ix) Orgraphic and conventional precipitation

2.2 Explain the following-

(i) Hydrologic cycle

(ii) Intensity of rainfall

(iii) Rain gauge

(iv) Precipitation gauge network

(v) Consistency of rainfall records

(vi) Moving mean average

(vii) Distribution graph of rainfall

(viii) Mass curve

(ix) Season

(x) Probable maximum precipitation.

2.3 Explain the following in brief-

(a) Rainfall measurement at a place

(b) Factors affecting rainfall at a place

(c) Hydrologic cycle in India

2.4 Describe various types of rain gauges with the help of neat and clean sketches.

2.5 Explain the relative advantages and disadvantages of recording and non-recording gauges.

2.6 (a) What considerations should be kept in mind while selecting a site for a rain gauge?

(b) What are the different methods of measurement of rainfall?

2.7 Discuss briefly the following-

(i) Radar measurement of precipitation

(ii) Satellite measurement of precipitation

2.8 (a) Discuss the various approaches for the design of precipitation gauge network.

(b) Give the ISI recommendations for minimum densities of precipitation network.

2.9 In a drainage basin, there exist 5 rain gauges, and the average rainfall at these stations is 95, 85, 70, 55, 61 cm. Determine, whether the number of rain gauge stations is enough to yield a representative and reliable picture of intensity and duration of rainfall on the basin with 10% error.

2.10 Describe the following in brief-

(a) Method for estimating missing rainfall data at a station

(b) Method for testing the consistency or homogeneity of precipitation data

2.11 Write a short note on the following-

(i) Intensity-duration-frequency curves.

(ii) Theoretical probability distribution curves of rainfall.

(iii) Depth-area-duration curves.

(iv) Depth-area curve

2.12 Explain the Station-year method for determining the recurrence interval of larger rains.

2.13 (a) Why it is the essential to determine mean rainfall over a drainage basin?

(b) What are the different methods for determination of mean rainfall over a basin? Discuss the relative merits and demerits of each.

2.14 Discuss briefly the important characteristics of precipitation in India.

2.15 Define probable maximum precipitation and discuss how is it evaluated. Indicate the areas of watershed management in which it finds application.

2.16 The normal annual precipitation at four rain gauge stations *A, B, C* and *D* is 122, 100, 80 and 115 cm, respectively. During a particular storm, the rain gauge at station *D* was found to he non-functional, however, the rainfall depths recorded at stations *A, B* and *C* are 15.5, 10 and 8 cm, respectively. Determine the rainfall depth at station *D* during the storm.

2.17 The following table gives the ordinates of a rainfall mass curve of a particular storm. Plot the hyetograph for the storm using a uniform time interval of 10 minutes. Determine the maximum intensities of rainfall for 10, 20, 30, 40, 50 and 60 minutes duration and plot the intensity-duration curve. The rainfall dates were recorded as follows:

Time from the beginning of storm (minutes)	**Cumulative rainfall (mm)**	**Time from the beginning of storm (minutes)**	**Cumulative rainfall (mm)**
10	0.0	80	75
20	10	90	100
30	15	100	115
40	25	110	118
50	35	120	120
60	45	130	120
70	60	140	120

2.18 Assume that the intensity-duration curve obtained in Prob. 2.18 can be fitted by an equation of the form $i = b/(D + a)^n$, where i = the maximum intensity of rainfall in mm/h and D = the duration of rainfall in minutes, and *a, b* and *n* are constants.

Determine the constants *a, b* and *n*.

2.19 Explain the procedure for determining depth-area-duration curves. Discuss in brief the practical uses of such curves in hydro-meteorology.

2.20 The annual maximum for 24 hour rainfall data for station *A* is given below for the last 27 years. Determine the magnitude of rainfall at the station corresponding of a recurrence interval of (a) 25 years and (b) 50 years.

Year (cm)	Annual rainfall (cm)	Year (cm)	Annual rainfall (cm)
1960	90.3	1974	66.5
1961	75.5	1975	72.1
1962	61.2	1976	80.0
1963	70.0	1977	59.1
1964	74.5	1978	52.8
1965	56.8	1979	51.6
1966	62.2	1980	72.5
1967	68.0	1981	85.3
1968	96.0	1982	82.2
1969	56.3	1983	69.3
1970	64.3	1984	67.6
1971	75.0	1985	65.2
1972	68.0	1986	61.8
1973	55.1	-	-

2.21 Annual rainfall data at station *A* and the average of the annual rainfall for 30 years for the 10 nearby rain gauges are given below. Test the consistency of the rainfall data at station *A*.

Workout the year in which change occurred, commenting on the possible reasons for this change.

Year	Annual rainfall (mm)		Year	Annual rainfall (cm)	
	Stn. A	10-Station average		Stn. A	10-Station average
1950	465	416	1965	295	339
1951	384	370	1966	331	427
1952	388	394	1967	374	393
1953	405	373	1968	344	337
1954	400	393	1969	364	340
1955	387	334	1970	320	337
1956	448	420	1971	384	408
1957	481	386	1972	374	360

Year	Annual rainfall (mm)		Year	Annual rainfall (cm)	
	Stn. A	10-Station average		Stn. A	10-Station average
1958	400	367	1973	317	347
1959	368	350	1974	332	370
1960	594	428	1975	335	435
1961	373	374	1976	303	340
1962	347	350	1977	305	419
1963	328	340	1978	320	304
1964	372	408	1979	382	383

■■■■

Chapter 3

RAINFALL AND WATER BALANCE ANALYSIS

3.1 Introduction

The 'rainfall' is the main source of water in rainfed farming and decisions in water management are often based on rainfall analysis. However, it varies with time and space, therefore its analysis is important for efficient utilization of rainwater for increasing productivity of rainfed areas. From Historical data of rainfall atleast last 30 years, it is possible to get some insight into problems about the availability of water in a region. For that one must know the stochastic process underlying the sequence of rainfall data. This process is however very complicated as it usually exhibits seasonal variations and when working with small increments, one encounters the problem of serial correlations.

Generally, rainfall observations are made once in a day and time increment of a day is suitable for solving many problems in Hydrology. Daily rainfall is generally characterized by serial correlations and many zero rainfall values. While analyzing rainfall amounts one must be aware of non-homogeneity in the data due to changes in the measuring instruments, change of site etc. The rainfall series is said to be homogeneous if for each year rainfall on a particular calendar day or month is a realization of the same variable. A homogeneous series may not be stationary as it exhibits seasonal variations. In this view, data must be tested for homogeneity before subjecting it to statistical analysis. A number of non-parametric tests such as Run test, Mann-Kendall tests are available for it. Once the data are found to be homogeneous, then it is subjected to further statistical analysis. Theoretical Distributions are applied to the data to get insight into these rainfall processes and to understand as well as to predict future events. The accuracy of the model depends upon the correct choice of the statistical distribution that represents the data.

When rainfall amounts are taken on small time increments (<24 hrs) distribution of rainfall amounts is markedly skewed, because zero values and some very large values occur with relatively large probabilities.

While rainfall amounts over large period are less skewed because of the effect of central limit theorem. Therefore, annual and seasonal totals can be assumed as normally distributed. Monthly totals are generally skewed therefore some non Gaussian distribution in particular positively skewed distributions are applied to them.

3.2 Different Types of Statistical Distributions

There are following distributions as described under:

(1) Probability Distributions of Discrete Random Variable, and

(2) Continuous Probability Distribution

3.2.1 Discrete probability distribution

1. Binomial distribution

Let us consider that X denotes the no. of successes in 'n' trials. Let 'p' be the probability of success in each trial. Then distribution of X is Binomial with 'n' and 'p' parameters denoted by; X-B(n, p).

It's Probability Mass Function (pmf) is given as;

$$\text{Prob}(X=x) = nCx.\ P^x\ .(1-p)^{n-x} \quad x = 1,2,....\ n \text{ and } 0 \leq p \leq 1 \quad ....(3.1)$$

$$E(X)=n.p, \quad V(x)=n.p.(1-p) \quad ....(3.2)$$

2. Geometric distribution

In Binomial distribution number of trials are fixed and number of successes is a random variable. But in this case number of successes is fixed and number of trials required to get these successes is a random variable. Probability Mass Function (pmf) is given by,

$$\text{Prob}(X=x) = (1-p)^{(x-1)} \quad x = 1,2,....\ n \text{ and } 0 \leq p \leq 1 \quad ....(3.3)$$

$$E(X)=1/p; \quad V(X)=(1-p)p^2 \quad ....(3.4)$$

3. Poisson distribution

In Binomial Distribution if 'n' is large and 'p' is small such that product n.p is nearly constant say 'λ', then X follows Poisson distribution with parameter λ pmf of random variable X, is given by,

$$P(X=x) =e^{-\lambda}\lambda^x/\text{fact}(x) \quad x=0,1,2 \text{ and } \lambda.>0. \quad ....(3.5)$$

This distribution can be applied to study the rare events such as floods, droughts etc.

3.2.2 Continuous probability distributions

1. Normal distribution

Let X follows Normal Distribution with location parameter 'μ' and scale parameter 'σ' then its Probability Density Function (pdf) is given by,

$$f(x) = 1/(\sigma.\sqrt{2\pi}).\exp\{-(x-\mu)^2/\sigma^2\} \quad -\infty < \mu < \infty \quad -\infty < x < \infty, \sigma > 0 \quad ...(3.6)$$

It can be seen that 1. Normal curve is a bell shaped curve. 2. Curve is symmetric about its location parameter. 3. It is Unimodal.

The distribution is also known as Gaussian Distribution.

$$P(\mu - \sigma < x < \mu + \sigma) = 0.68$$

$$P(\mu - 2\sigma < x < \mu + 2\sigma) = 0.95$$

$$P(\mu - 3\sigma < x < \mu + 3\sigma) = 0.99$$

Moments of the distribution

$$E(x) = \mu$$

$$V(X) = \sigma^2$$

Skewness = 0; and

Kurtosis = 0

Methods of parameter estimation

Different methods of parameter estimation are available but many these, method of Moments and Method of Maximum Likelihood are commonly used. It can be seen that, for Normal distribution both the methods give same estimators. They are given by,

$$Est(\mu) = \bar{x} \text{ and } est(\sigma^2) = S^2 = \Sigma(xi-\bar{x})^2)/(n-1)$$

where, $\bar{x}$ and S^2 are sample mean and variance.

The normal distribution is the mostly used statistical distribution. The principal reasons are:

(i) Normality arises naturally in many physical, biological, and social measurement situations.

(ii) Normality is important in statistical inference.

Normal probabilities: Often we want to compute the probability that a given observation lies in a specified interval, i.e., Prob(a < X < b) = F(b) - F(a); where F(.) denotes the cumulative distribution function of 'X'. It is given by, $F(x) = {}_{-\infty}\int^{y} f(x).dx$

2. Log-normal distribution

Some times, the original data are not symmetric but, some transformations such logarithmic makes the data symmetric. In such case Normal distribution is fitted to logarithms of the data. i.e. X is said to be Log-Normal Distributed if Ln(X) is normally distributed.(Ln represents the natural log).

3. Gamma distribution

It is seen many times that zeros and some high values in the data make the underlying distribution asymmetric. They may be positively or negatively skewed depending upon the long tail of the distribution or concentration of the values is to the right or left accordingly. The Gamma distribution has been found to be good fit to the daily depression rain, monthly rainfall totals. It has been shown that simple square root, cube root, logarithmic transformations arc of limited utility in normalizing the rainfall values. As Gamma distribution covers a wide range of skewness and kurtosis, it has wide applicability in rainfall analysis. The pdf of the gamma random variable X is given by,

$$f(x) = x^{(a-1)}.e^{-b.x}.b^2/\Gamma(a) \quad x>0 \; a,b>0 \qquad ....(3.7)$$

Here 'a' and 'b' are shape and scale parameters. As 'a' $\rightarrow \infty$, Gamma distribution approaches to normal distribution.

First two moments of the distribution are given by;

$$E(X) = a/b; \; V(X)=a/b^2$$

Method of moments

$$Est\,(b) = \bar{x}/s^2$$

$$Est\,(a) = \bar{x}.est(b)$$

Where $\bar{x}$ and s^2 are sample mean and sample variance.

Method of maximum likelihood estimation

When deviation from normality increases. The efficiency of moment estimates decreases. In such cases ML estimates arc consistent and sufficient estimates of the parameters.

Let 'L' denote the likelihood of the sample x_1, x_2....x_n. Then $L=\pi\Gamma(xi)$ then by differentiating w.r.t. parameters and equating to zero we will get ML estimates. It can be seen that these equations are not explicit to solve so iterative procedures are to be used. The equations are given by,

$$\text{Est (b)} = \text{est(a)}/\bar{x} \text{ and}$$

$$\Psi(a) - \ln(a) - \Sigma - \ln(xi) / n + \ln(\bar{x}) = 0 \qquad ....(3.8)$$

where,

Ψ is as digamma function.

$\Psi = d(\ln(a))/da$

4. Truncated gamma distribution

Some times it is observed that, observations lie into a specific range therefore in such cases by fitting complete gamma distribution, unnecessarily some probability is assigned to the range where observations do not occur. In such cases distribution can be truncated at those points and that interval can be considered as the real range of the variable.

Suppose 'l' is the lower limit of the observations or $0 < l < xi < \infty$ for all 'i'

Then Pdf of X is given by;

$$f(x) = x^{(a-1)}.e - {}^{b.x}.b^2/\Gamma(a, l)\ 0<l<\infty \qquad ....(3.9)$$

where $\Gamma(a, l)$ denotes the Upper Incomplete Gamma Function is which is given by;

$$\Gamma(a,l) = \int^{\infty} t^{a-1}.e^{-t} \qquad ....(3.10)$$

Similarly Lower Incomplete Gamma Function is given by;

$$\Gamma(a,x) = \int^{x} t^{a-1}.e^{-t} \qquad ....(3.11)$$

$$\text{It can be seen that } \Gamma(a,x) + \Gamma(a,x) = \Gamma(a) \qquad ....(3.12)$$

5. Extreme value type I or Gumbel distribution

Theoretically it can he shown that if the underlying process follows distribution from Gamma family then extreme arising from them follows Extreme value type I distribution or also Known as Gumbel Distribution. Its pdf is given by,

$$f(x) = 1/\alpha.\exp\{-(x--u)/\alpha - \exp[-(x--u)/\alpha]\} \ldots \quad .(3.13)$$

where, u and α are location and scale parameters of the distribution. 'u' is the modal value of the distribution. It can be seen that mode<mean for this distribution.

For the transformation y = (x-u)/ α pdf is given by;

$$F(y) = \exp\{-y-\exp(-y)\}\ y>0 \qquad \text{....(3.14)}$$

which is known as standard Gumbel Distribution. For further transformation $z = e^{-y}$ it reduces to standard exponential distribution. Or $f(z) = e^{-z}$

Parameter estimation

Method of moments

Est(α) = √6 . σ/π and est(u) = $\bar{x}$ - γ. Est(α) where γ is Euler's constant = 0.5772

Method of maximum likelihood

Likelihood Function is given by,

$$L = \prod f(x_i) \qquad \text{....(3.15)}$$

Maximize L w.r.t. parameters u and α, we get equations;

$$\text{Est}(\alpha) = \bar{x} - (\Sigma x_i . e^{-xl/\alpha}) / (\Sigma e^{-xl/\alpha}) \text{ and}$$

$$\text{est}(u) = -\text{est}(\alpha). \ln\{(\Sigma e^{-xl/\alpha})\}$$

These equations are not in explicit form and are to be solved iteratively.

Once the parameters are estimated from the data then it is tested for the goodness of fit and then it is used further to study the properties of the rainfall distribution.

3.3 Return Period or Recurrence Interval

It is the average time interval between the occurrence or accidence of an event.

If 'p' is the probability of occurrence of an event. Then return period T = l/p

For Gumbel distribution,

$$F(x) = \exp\{-\exp(-(x-u)/\alpha)\} \qquad \text{....(3.16)}$$

$$\Rightarrow x = u + \alpha\{-\ln.\ln(1/F)\} \text{ as } F=(T-1)/T$$

$$x = u+\alpha.\{-\ln.\ln(T/(T-1))\}$$

3.4 Testing the Goodness of the Fit of the Statistical Distribution

The procedure is involved in the following steps:

Step 1: Estimate the parameters of the distribution by any one method mentioned above.

Step 2: Divide the observations into no of classes of equal width. Count the frequency of each class. Denote them by observed frequency(o_i).

Step 3: Calculate the expected frequency of each class (a,b) by e_i= N.Prob ($a<x\leq b$) where Probability is calculated using the flitted statistical distribution.

Sten 4: Use the Chi-square test of goodness of fit test where test statistic is given by

$\chi^2 = \Sigma(o_i-e_i)^2/ei$ and then compare it with table value for Chi-sq. with (n-1) d.f where n is no of classes.

3.5 A Case Study

Jeevarathnam and Jayakumar (1979) analysed, the annual maximum daily rainfall data of Ootacamund for 60 years (1918 to 1977) (Government Botanical Gardens, Ootacamund)

The return periods and corresponding daily maximum rainfall have been plotted on normal probability paper for all the three different methods namely, Gumbel Extreme Probability Distribution, Log Pearson III and Log-Normal Distribution and then compared with the observed data. The acceptable distribution amongst the three has been selected by testing statistically the goodness of fit to the observed values.

The annual daily total rainfall values are arranged in descending order of magnitude starting from the highest values and given in Table 3.1.

Table 3.1: Annual maximum daily total (24 hours) rainfall at Ootacamund, Tamil Nadu

Order No.	**Year**	**Date**	**Annual maximum daily rainfall X, mm**	**Return period** $T_r = \frac{n+1}{m}$ **Years**
1.	1939	15/11	226.6	61.00
2.	1924	16/7	210.8	30.50
3.	1957	22/5	207.0	20.33

Order No.	Year	Date	Annual maximum daily rainfall X, mm	Return period $T_r = \frac{n+1}{m}$ Years
4.	1955	19/5	189.0	15.25
5.	1930	8/5	186.7	12.20
6.	1958	21/11	163.8	10.16
7.	1944	25/11	155.7	8.714
8.	1941	3/12	153.9	7.625
9.	1954	15/7	152.9	6.777
10.	1964	8/8	142.0	6.100
11.	1966	10/11	114.8	5.545
12.	1943	4/1	107.4	5.083
13.	1934	8/11	105.9	4.692
14.	1946	3/11	105.7	4.357
15.	1925	11/11	104.1	4.066
16.	1969	18/8	102.9	3.812
17.	1937	1/10	101.8	3.588
18.	1932	27/10	101.3	3.388
19.	1919	3/11	96.5	3.210
20.	1928	14/10	95.5	3.050
21.	1959	29/10	95.3	2.904
22.	1971	20/18	91.9	2.772
23.	1952	2/12	90.7	2.652
24.	1947	13/4	89.7	2.541
25.	1923	28/10	87.6	2.440
26.	1962	17/5	81.3	2.346
27.	1956	16/10	74.9	2.259
28.	1953	21/6	73.2	2.178
29.	1948	20/11	73.1	2.103
30.	1940	5/5	71.6	2.033
31.	1929	30/5	69.1	1.967
32.	1972	14/5	67.8	1.906
33.	1926	28/1	67.6	1.848

Order No.	Year	Date	Annual maximum daily rainfall X, mm	Return period $T_r = \frac{n+1}{m}$ Years
34.	1977	28/5	66.0	1.794
35.	1973	8/7	63.5	1.742
36.	1951	28/9	61.7	1.694
37.	1922	1/12	61.5	1.649
38.	1970	2/6	60.5	1.605
39.	1936	28/6	59.2	1.504
40.	1933	19/8	58.4	1.525
41.	1942	25/4	57.4	1.487
42.	1935	6/11	56.9	1.452
43.	1961	3/7	56.1	1.419
44.	1976	18/10	55.9	1.386
45.	1974	11/5	55.6	1.355
46.	1963	22/10	55.4	1.326
47.	1965	25/9	54.6	1.298
48.	1950	21/8	53.6	1.271
49.	1920	2/1	53.3	1.245
50.	1921	1/11	51.3	1.220
51.	1975	16/9	50.8	1.196
52.	1938	4/9	48.0	1.173
53.	1927	17/6	47.2	1.151
54.	1945	10/7	46.5	1.130
55.	1960	11/6	45.7	1.109
56.	1949	20/5	45.0	1.089
57.	1918	15/11	44.7	1.070
58.	1931	4/6	42.2	1.052
59.	1967	30/10	41.4	1.034
60.	1968	20/5	38.1	1.017

The return period (T_r) values are obtained by using Weibull formula, $T_r = \frac{n+1}{m}$.

Gumbel extreme value distribution

The probability density of this type of distribution is:

$$P(X < x) = e^{-h}(x)$$

where h (x) e^{-y} and y = a (x - x_0) with a and x_0 as parameters. Substituting h (x) and y into the above equation, the Gumbel Extreme distribution is:

$$P(X < x) = e^{-e^{-a(x-x_o)}}$$

Using the method of moments. Gumbel determined the values for x_o and a. These values are:

$$x_o = \bar{x} - 0.57722\frac{S_x}{a}$$

$$\text{and, } a = \frac{1}{0.7797 S_x}$$

where $\bar{x}$ and S_x are the sample mean and sample unbiased standard deviation respectively.

The values for $\bar{x}$ and S_x obtained from Table 3.1 are as follows:

$$Mean, X = \bar{x} = \frac{5288.6}{60} = 88.14$$

$$Std.dev., S_x = \sqrt{\frac{\Sigma X^2 - \frac{(\Sigma X)^2}{n}}{n-1}}$$

$$= \sqrt{\frac{595736.1 - \frac{(5288.6)^2}{60}}{59}}$$

$$= 46.86$$

The probable rainfall (X_{Tr}) for different return periods are computed by using the equation:

X_{Tr}, = $\bar{x}$ - 0.45 S_x + 07797 S_x $\log_e$ Tr are furnished in Table 3.2.

Log Pearson type III distribution

The general and basic equation to define the probability density of a Pearson distribution is given as,

$$P(X)=e\int_{-\infty}^{x(a+x)/(b_o+b_{1x}+b_{2x^2})dx}$$

where, a, b_o, b_1 and b_2 are constants.

Table 3.2: Computed rainfall for different return periods by Gumble distribution

Return period Tr	Probability % P_{Tr}	$\log_e$ Tr	Rainfall X_{Tr} mm
1.01	99	0.01	67.41
1.50	66.6	041	82.03
2.33	42.9	0.25	98.11
5.00	20	1.61	125.88
10.00	10	2.30	151.09
25.00	4	3.22	184.71
50.00	2	3.91	209.92
100.00	1	4.61	235.50
200.00	0.5	5.30	260.71
500.00	0.2	6.21	293.96
1000.00	0.1	6.91	319.54

For Log Pearson Type III distribution, the probability density with the origin at the mode is:

$$P(X)=P_o\left(1+\frac{x}{a}\right)^o e^{-ex/a}$$

where, $c=\frac{4}{\beta_1}-1, a=\frac{c}{2}.\frac{\mu_3}{\mu_2}$

$$P_o=\frac{N}{a}.\frac{C^{e+1}}{e^o\sqrt{(e+1)}}$$

The Log Pearson Type III distribution is based on the use of log transformed data. Hence, the required statistical parameters for this distribution are log mean, log std. dev. and log skew coefficient. Values for these parameters obtained from Table 3.1 are as follows:

$$\text{log mean} = \log \bar{x}=\frac{\Sigma \log X}{n}=\frac{113.72}{60}=1.90$$

$$\text{log std. dev.} = S_{\log x} = \sqrt{\frac{\Sigma(\log X - \log \bar{x})^2}{n-1}}$$

$$= \sqrt{\frac{2.41}{59}} = 0.20$$

$$\text{log coef. skew} = g_{\log x} = \frac{n\Sigma(\log X - \log \bar{x})^2}{(n-1)(n-2)(S_{\log x})^3}$$

$$= \frac{60 \times 0.2768}{59 \times 58 \times (0.20)^3}$$

$$= 0.61$$

The values of X_{Tr} for assumed recurrence intervals are computed using the equation:

$$\log X_{Tr} = \log \bar{x} + K_{Tr}\,(S_{\log x})$$

where, K_{Tr} is a skew curve factor (K) for a particular recurrence interval (Tr) obtained from the table given by Schulz (1974) using the values of g_{logx}.

The probable rainfalls (X_{Tr}) for assumed recurrence intervals are furnished in Table 3.3.

Table 3.3: Computed rainfall for different return periods by Log Pearson Type III Distribution Return

Return period Tr	Probability % P_{Tr}	K_Tr	K_Tr (S log x)	log X_{Tr}	Rainfall mm X_{Tr}
2	50	-0.099	-0.0198	1.8802	75.90
5	20	0 800	0.1600	2.0600	114.80
10	10	1.328	0.2656	2.1656	146.40
25	4	1 939	0.3878	2.2878	194.00
50	2	2.359	0.4718	2.3118	235.40
100	1	2.155	0.5510	2 4510	282.50
200	0.5	3.132	0.6264	2.5264	336.00
1000	0.1	3.960	0.7920	2.6920	490.00

Log normal distribution

The probability density of this distribution is given as,

$$P(X)=\frac{1}{Sye^{y}\sqrt{2\pi}}e^{-(y-\mu_y)^{2/2S_y 2}}$$

where, y=In x, x is the variate, μ_y is the mean of y and S_y is the standard deviation of y.

Chow (1964) has derived the statistical parameters for X as:

$$\mu = e^{\mu y} + S_y^{2/2}$$

$$S = \mu\ (es_y^{2} - 1)^{1/2}$$

$$\alpha = (e^{3}s_y^{2} - 3es_y^{2} + 2)\ Cv^{3}$$

$$M = e^{\mu y};\ \frac{\mu}{M} = e^{s}{}_{y}^{2/2}$$

$$Cv = (es_y^{2} - 1)^{1/2}$$

$$Cs = 3Cv + Cv^{3}$$

The values of, the mean rainfall $\bar{x}$, the mean of the transformed variate y, the standard deviation Sy, the coefficient of variation Cv and the coefficient of skewness Cs, were computed from Table 3.1, are as follows:

$$\text{Mean rainfall} = \bar{x} = \frac{\Sigma X}{n} = \frac{5288:6}{60} = 88.14$$

$$\text{Mean y} = y = \frac{\Sigma y}{n} = \frac{261.86}{60} = 4.364$$

$$\text{Std. dev.} = y = Sy = \sqrt{\frac{\Sigma y^2 - \frac{(\Sigma y)^2}{n}}{n-1}}$$

$$= \sqrt{\frac{1155.52 - \frac{68570.66}{60}}{59}}$$

$$= 0.4635$$

$$Cv = (e^{0\ 2148} - 1)^{½} = 0.4899$$

$$Cs = 3\ (0.4899) + (0.4899)^{3} = 1.59$$

The value of frequency factor K_{Tr} were taken from the table given in Chow (1964). X_{Tr} is calculated by using the following equation:

$$X_{Tr} = \bar{x}\ (1+Cv\ K_{Tr})$$

The probable rainfall (X_{Tr}) for different intervals are computed and furnished in Table 3.4.

Table 3.4: Computed rainfall for different return Log Normal Distribution

Return period Tr	Probability % P_{Tr}	Frequency factor K_{Tr}	Rainfall X_{Tr} mm
1.01	99	- 1.414	27.08
1.05	95	- 1.183	3606
1.25	80	- 0.801	53.55
2	50	- 0.209	79.10
5	20	+ 0.671	117.14
20	5	+ 1.890	169.75
100	1	+ 3.355	233.01
1000	0.1	+ 5.645	331.88

2. Test for goodness of fit

Comparison of the three p.d fs to find out the goodness of fit to the observed values, has been done by using the Chi-Square test. It has been shown that the Chi-Square parameter:

$$x^2 = \sum_{i=1}^{k} \frac{(N_1 - N_{P1})^2}{N_{P1}}$$

will have Chi-Square distribution with the number of degrees of freedom equal to K-h-1. The Chi-Square test prescribes the critical value X_0^2 for a given α, so that, for $\chi^2 < \chi_0^2$, the null hypothesis of a good fit is accepted. In the comparison of the three distribution (Table 3.5), the Log Pearson Type III distribution gives close fit to the observed values ($\chi^2 = 6.46 < \chi_0^2 = 7.815$).

Table 3.5: Chi-square test

Return period year	Observed value N	Probable values Np			$(N-Np)^2/Np$		
		Gumbel	Log P. III	Log N	Gumbel	Log P. III	Log N
2	70.35	93.00	75.90	79.10	5.52	0.40	0.91
5	107.10	125.88	114.89	117.14	2.80	0.52	0.86
10	162.91	151.09	146.40	145.00	0.92	1.86	2.21
15	188.81	165.00	170.00	68.00	3.43	2.08	2.58
25	207.75	184.71	194.00	185.00	2.87	0.97	2.79
30	210.61	198.00	208.00	190.00	0.80	0.03	2.23
60	225.09	220.00	237.00	215.00	0.12	0.60	0.47
				$\chi^2 =$	16.46	6.46	12.11
		For $\alpha = 0.05$, the χ_0^2 value =			9.488	7.814	7.815

It is obvious that all hydrological events do not have a practical significance in the design of hydraulic structures. The design of these structures mainly depends upon a few extreme events and the prediction of these few events is extremely essential for safe design of structures. The selection of a particulär cumulative distribution function to predict the extreme event is very important and depends upon experience and verification to observed data (Chow, 1964). The statistical comparison by Chi-Square test for goodness of fit clearly shows that the Log Pearson Type III distribution is the best fit to the observed values.

3.6 Markov Chain Model

While applying the statistical techniques we, generally, assume the independence between the observations of the process but this may not be the case always. It has been observed that in case of natural processes like daily rainfall, may depend on each other or in other words 'persistence' may be there in these observations. In such cases distribution theory is not applicable as it assumes observations to be independent identically distributed. An immediate generalization of the scheme of independent trials is that of a Markov chain which was first systematically investigated by Renowned Russian Mathematician Prof. A.A. Markov. Theory of Markov chain permits this dependence only through preceding trials. When observations are independent, probability of occurrence of any outcome say 'as' is same for all the trials but when lwrsistoncc exists in the observations then probability of occurrence of each outcome does not remain same at every trial but to every pair

$\{a_i, a_j\}$ there corresponds a conditional probability P_{ij} given that event a_j has occurred in previous trial. In addition to that initial probability it should be known, which is given by,

$$\pi_I(0) = \text{Prob}(X_0 = i),\ \pi_i \geq 0 \text{ and } \Sigma\pi_i = 1 \qquad ....(3.21)$$

Generally initial state of the process is denoted by X_0.

Then Prob $(a_i\ a_j) = \pi_i.p_j$

Probabilities $\{\pi_i\}$ and $\{p_j\}$ completely describe the process.

Markov Property: Given the current state of the system the future evolution is independent of its history i.e.

If X_1, X_2 are successive trials then;

Prob $(X_{n+1}=a_{n+1}/X_a=a_n, X_{n-1}=a_{n-1}=a_{n-1}.... x_1=a_1)$ Prob $(X_{n+1} = a_{n+1}/X_n = a_n, X_{n-1}=a_{n-1})$ ....(3.23)

Then the sequence $\{X_n\}$ is said to possess Markov property.

3.7 Types of Markov Chains

1. Finite Markov Chain

Let $\{X_n)$ represent the observed sequence of a process, which has finite number of outcomes $\{a_1,a_2....,a_s\}$. If the probability that the process will enter the state a_k on the kth step depends on the sequence of states in steps (k-t) through (k-1) i.e. previous 't' steps then the process is said to be t-th order s-state process.

In this case the possible outcomes are finite in number, therefore this is a finite Markov chain.

2. Denumerable Markov Chain

If the number of possible outcomes are denumerable then the process $\{X_n\}$ is said to be Denumerable Markov Chain.

3. Infinite Markov Chain

If possible outcomes are infinitely many, then the corresponding Sequence of outcomes possessing Markov property is said to be Infinite Markov Chain.

One Step Transition Matrix

Let X_1,X_2 are successive trials. If

$\text{Prob}\,(X_{n+1}=a_{n+1}/X_a=a_n, X_{n-1}=a_{n-1}.....x_1=a_1)$
$= \text{Prob}\,(X_{n+1} = a_{n+1}/X_n= a_n)$

and if p_{ii} denotes the probability of going from state 'i' to state 'j' in one step.

Then the matrix of probabilities $P= \{p_{ii}\}$ is called one step transition matrix,

Where $p_{ii} \geq 0$ and $\Sigma_j\, p_{ii} = 1$. (row totals add up to one).

Such a matrix is called 'Stochastic Matrix'. Any stochastic matrix with above properties can be seen as transition probability matrix.

Higher Transition Probabilities

Denote by $p_{ii}\,(n) = \text{Prob}\,(X_{n+1} = j/X_n, X_{n-1},......,X_1 = i)$ ....(3.24)

It is the probability of a transition from 'i' to 'j' in exactly n steps. It is the conditional probability of entering j at the nth step given the previous state i. This is the sum of all possible combinations given starting state i and ending with j. or in other words we can write,

$$P_{ii}\,(n) = \Sigma_k p_{ik} \,.\, p_{kj}) \quad ....(3.25)$$

Similar to matrix P defined above we can construct a matrix of n step transition probabilities P^a of $\{p_{ii}(n)\}$. Note that here also $\Sigma_j\, p_{ii}\,(n) = 1$ for all i.(row totals of transition matrix are all one.)

This matrix is also a stochastic matrix. Note that for independent trials all the rows of the matrix are identical and $P^a = P$.

By induction we get general recursion formula,

$$\mathbf{P_{ij}(m+n) = p_{ik}(m).p_{kJ}(n)} \quad ....(3.26)$$

$$\text{or} \quad \mathbf{P^{m+n} = P^m \,.\, P^u} \quad ....(3.27)$$

These equations are called as Chapman-Kolmogorov equations.

Properties of Markov Chain

1. Homogeneous Markov Chain

If $p_{ij}(n\text{-}m) = \text{Prob}\,(X_a = j \,/\, X_m = i)$ (m<n) which depends only on the difference n-m. such transition probabilities are called stationary or lime homogeneous Markov chain

2. Irreducible Markov Chain

The Markov Chain is said to be irreducible if and only if every state can be reached from every other state.

3. Ergodic Markov Chain

If the system is in state j is practically independent of what state it was in the remote past. i.e. lim p_{ij} (n) = π_j for all i. as n→∞.

Classification of states

A state a_j is called persistent if the system starts with a aid will ever pass through aj or with probability one it infinitely often returns to a1 and a state is called transient if system enters to it only finite times.

A state a1 is said to be periodic with period 't'>0 if p (n)=0 for all n = kt (multiple of t) i.e. periodically it returns to state, a_j.

A state a_j is called absorbing if p_{jk} = 0 for all k(k # j)

Illustrative examples

1. Consider a process such that there are only 2 outcomes say, a_1 and a_2. Then one step transition probability matrix of the process will be given by,

$$P = \begin{bmatrix} 1-p & p \\ \alpha & 1-\alpha \end{bmatrix} \quad \text{....(3.29)}$$

where,

p = probability {X = a2/X_{n-1}=a1} and α = probability {Xa = a1/X_{n-1} = a2)

(i) Random walk with absorbing states

let $a_1.a_2$,...., a_p be the possible states of a process. Consider the one step transition matrix of the process, which is given by;

$$P = \begin{bmatrix} 1 & 0 & 0 & 0 & 0 \\ q & 0 & p & 0 & 0 \\ 0 & q & 0 & p & 0 \\ 0 & 0 & 0 & 0 & 1 \end{bmatrix} \quad \text{....(3.30)}$$

From each of the interior states transition is possible to left or right, but once the system reaches state a_1 and a_p it remains there forever. States a_1 and a_p are absorbing states.

3.8 Utility

Markov Model is most powerful tool for analyzing complex systems and used in estimation of dry and wet spell probability analysis. Analyses yields results for both time dependent evolution of the system and steady state of the system.

It also yields a variety of useful performance measures describing the operation of the system. These performance measures include the following:

System reliability, mean time to failure, mean time between two failures, probability of being in the given state at a given time, average no. of visits to a given state in given time.

Practical situations where Markov model can be applied

1. Dry and Wet Spells
2. Pentad rainfall
3. Days with thunder storm activities.

3.9 To Fit the Markov Model and to Test the Goodness of Fit

Past weather events exert an influence on the future weather. The notion of persistence in the sequence of weather events is very important Markovian models have been applied to climatological processes exhibiting sequential patterning. Most of the studies dealt exclusively with the occurrence of wet and dry days.

The sequence used to estimate the model is called training sequence and the sequence, which is used to evaluate, or to test the performance of the model is called test sequence.

Steps to be followed in the estimation of Markov model

Step 1: Estimation of one step Transition Probability Matrix

Let $\{a_1,a_2,...a_p\}$ be the outcomes of the process. One step transition probabilities are calculated by the formula,

$p_{ij} = n_{ij} / ni.$, where n is the number of times process enters from state i to state j and n_i is the number of epochs when process enters from state i to any other state. $\{p_{ij}\}$ forms the one step transition probability matrix.

Step 2: Estimation of n step Probability Matrix

'n' step transition probability matrix can be obtained by raising the stochastic matrix of one step transition probabilities 'P' by power 'n'.

Step 3: Estimation of steady state probability

Steady state probabilities of a process are obtained by raising the stochastic matrix to that power when difference between successive probabilities of respective states becomes negligible. The state for which probability is highest can be taken as characteristic feature of the system.

Step 4: Mean time Period for State 'j'

It is calculated by the formula;

$E_j = 1/1(1-p_{ii})$ where p_{ij} is the probability of persistence of state 'j'.

Step 5: Expected First Passage time and Recurrence time.

It is defined as the no. of transitions needed to go from state 'r' to 's' for the first time. If 'r'='s' then it is called expected recurrence time for state 'r', which is the reciprocal of the steady state probability.

3.10. Water Balance Analysis

Water is a limiting factor for crop growth and yield in the semi-arid tropics. Relationships have been developed between crop growth and soil moisture by many research workers. However, large scale applications of these relationships have not been possible, because of lack of adequate soil moisture data.

A number of meteorological methods, for estimating soil moisture from weather data are widely used. Similarly, water budgeting models based on meteorological data have been developed and are extensively being used, both for research and for applications in crop planning.

Concept of water balance

The general water balance equation can be expressed as;

$$(P+I) = ET + R + D + \Delta S \qquad ....(3.31)$$

where,

P = rainfall,

I = Irrigation,

ET = Evapotranspiration,

R = Surface runoff,

D = Deep drainage, and

ΔS = Change in soil moisture.

If moisture storage capacity of the soil is known, the water balance equation can be solved by comparing rainfall and irrigation water with evapotranspiration. The water balance can be computed on a daily, weekly or monthly basis. For long range water resources planning, the monthly computation may be adequate. However, for crop planning, weekly computations are more suitable and useful.

3.10.1 Thornthwaite's water balance method

Thornthwaite's water balance model is one of the early developed models. The climatic water balance was first introduced and used by Thornthwaite (1948) for the climatic classification in 1948 and subsequently improved in 1955 by Thornthwaite and Mather (1955). In the original bookkeeping procedure as assumed that the soil mantle has a capacity to hold 100 mm of water for purpose of evapotranspiration and whenever precipitation (P) falls short of the water need, i.e., potential evapotranspiration (PE), the shortage can be made good from the stored soil moisture so long as it is available, When P is in excess of PE, the excess will go first to recharge the soil to its field capacity, and the surplus will be available for runoff and deep percolation. Thornthwaite and Mather (1955), subsequently increased the storage capacity of soil to 300 mm in deep loam silt soils. However, we can determine the storage capacity for our soil under consideration.

Climatic water balance consists of precipitation, potential evapotranspiration, actual evapotranspiration, soil moisture storages, surplus and deficit. The model takes the difference between monthly precipitation and evapotranspiration and carries forward a balance of water surplus or deficiency. A first requirement is the water holding capacity of the soil relative to soil type and land use.

Computation of components of climatic water balance

1. **Precipitation (P):** Precipitation at given place is measured by rainguage.
2. **Potential evapotranspiration (PE):** In this method potential evapotranspiration is calculated according to Thornthwaite method. By this method PE is calculated as under:

$$E = 1.6(10T/I)^a \qquad \text{....(3.32)}$$

where,

E = Unadjusted potential evapotranspiration in cm (unadjusted value for 30 days each of 12 hrs.)

T = mean monthly air temperature, °C

a = Constant

$$= 0.000000675\, I^3 - 0.0000771\, I^2 + 0.01792\, I + 0.49239$$

I = annual or seasonal heat index, the summation of 12 values of monthly heat indices (I) when,

$$i = (T/5)^{1.514}$$

The potential evapotranspiration (PE) is obtained by multiplying E by a correction factor which depends on latitude of the station and the month of the year.

3. **Precipitation minus the potential evapotranspiration (P-PE):** To determine the period of moisture excess and deficiency it is necessary to obtain the difference between P and PE. The negative value of (P-PE) indicates the amount by which the precipitation fails to supply the potential water need of a vegetation covered area. A positive value of (P-PE) indicates the amount of excess water which is available during certain period of a year for soil moisture recharge and runoff.
4. **Storage (ST):** In this study, the storage (ST) is the actual computed value of moisture, whose upper limit is the storage capacity of the soil.
5. **Change in soil moisture storage (ΔST):** As an aid in later computation, it is desirable to obtain the difference in the amount of soil moisture storage from one month to the next.
6. **Actual evapotranspiration (AZ):** When the precipitation is greater than the PS, the soil remains full of water and actual evapotranspiration will equal the potential. When precipitation drops below PE, the soil begins to dry out and actual evapotranspiration becomes less than potentially possible. In these months, the actual evapotranspiration equals the precipitation plus the change in storage.

   Thus, If, $P > PE$, then $AE = PE$

   If, $P < PE$, then $AE = P + \Delta ST$

7. **Moisture deficit (D):** The amount by which the actual and potential evapotranspiration differs in any month is the moisture deficit for that month

   $$D = PE - AE$$

8. **Moisture surplus (S):** In water balance computation, surface runoff and deep percolation are lumped together as surplus. Surplus occurs whenever the rainfall exceeds the moisture storage capacity. Hence,

   $$S = [(P + ST) - PE] - \text{Storage capacity.}$$

In this equation, ST value of previous week is taken.

The estimation of surplus water is of considerable practical importance to farmers. It gives not only an indication of possible extent of soil erosion, but also of the intensity of leaching of nutrients in the soil. The difference between rainfall and surplus is a measure of the effective rainfall.

The sample calculations of the above components are shown in Table 3.6.

Table 3.6: Computation of water balance by Thornthwaite's method (Storage capacity = 100 mm)

Met. week (mm)	P (mm)	PE (mm)	P-PE (mm)	ST (mm)	ΔST (mm)	AE (mm)	D (mm)	S (mm)	MAI (AE/PE)
21	28.4	37.2	-8.8	0	0	28.4	8.8	0	0.8
22	25.7	34.8	-9.1	0	0	25.7	9.1	0	0.7
23	34.3	30.4	3.9	3.9	3.9	30.4	0	0	1.0
24	34.2	31.4	2.8	6.7	2.8	31.4	0	0	1.0
25	19.2	28.4	-9.2	0	6.7	25.9	2.5	0	0.9
26	28.9	31.4	-2.5	0	0	28.9	2.5	0	0.9
27	33.2	26.0	7.2	7.2	7.2	26.0	0	0	1.0
28	140.3	27.3	113.0	100	92.8	27.3	0	20.2	1.0
29	19.3	30.9	-11.6	88.4	11.6	30.9	0	0	1.0
30	111.2	33.8	77.4	100	11.8	33.8	0	65.8	1.0
31	13.5	32.8	-19.3	80.7	19.3	32.8	0	0	1.0
32	13.4	37.2	-23.8	56.9	23.8	37.2	0	0	1.0
33	20.3	35.8	-15.5	41.4	15.5	35.8	0	0	1.0
34	27.4	30.4	-3.0	38.4	3.0	30.4	0	0	1.0
35	52.7	27.9	24.8	63.2	24.8	27.9	0	0	1.0
36	56.6	27.4	29.2	92.4	29.2	27.4	0	0	1.0
37	45.5	30.5	15.0	77.4	15.0	30.5	0	0	1.0
38	7.4	29.4	-22.0	55.4	22.0	29.4	0	0	1.0
39	5.7	30.5	-24.8	30.6	24.8	30.5	0	0	1.0
40	6.5	31.0	-24.5	6.1	24.5	31.0	0	0	1.0
41	10.3	31.3	-21.0	0	6.1	16.4	14.9	0	0.5

Application of water balance method in Thornthwaite's climatic classification

On the basis of surplus, deficit arid water need, Thornthwaite (1948) determined the humidity index and aridity index as under.

Humidity Index lh = 100 S/n

Aridity Index Ia = 100 D/n

Where n is the water need (i.e. PE).

By giving more weightage to humidity index than aridity index, he determined the moisture index as follows.

Moisture Index Im = (100 S - 60 D)/n

On the basis of the computed values of I_m, then he gave the climatic classification as under.

Climatic type	Moisture index (I_m)
A – Perhumid	100 and above
B_4 – Humid	80 to 100
B_3 – Humid	60 to 80
B_2 – Humid	40 to 60
B_1 – Humid	20 to 40
C_2 – Moist Subhumid	0 to 20
C_1 – Dry Subhumid	- 20 to 0
D – Semi-arid	- 40 to – 20
E – Arid	- 60 to – 40

Thornthwaite Mather (1955) further improved the book-keeping procedure and gave equal weightages to humidity and aridity indices for determining the moisture index. On the basis of this moisture index, they gave the revised climatic classification. Based on Thornthwaite's method, Subramaniam (1956) classified India's climate into six regions, i.e. arid, semi-arid, dry subhumid, moist subhumid, humid and perhumid. Krishnan and Mukhtar Singh (1968) classified the arid and semi-arid zones in India.

Computation of Moisture Availability Index (MAI)

The ratio of computed values of AE and PE gives the Moisture Availability Index (MAI), which can be used for crop planning.

Thus, MAI = AE/PE

The MAI value is an indication of the extent to which the water need of the crop is met. Hence, the MAI values can be used to determine the moisture availability periods, which gives the length of the growing period.

Moisture availability periods on the basis of MAI:

Thornthwaite's method of computing water balance and MAI can be effectively used for crop planning purpose, by deciding the moisture availability periods for different types of soils with the help of normal values of rainfall and potential evapotranspiration. The Centre of Advanced Studies in Agricultural Meteorology has done the crop planning on the basis of MAI for Pune and Ahmednagar districts of Maharashtra, wherein total 27 stations were considered and data of 20 years (1971-90) was used (Raskar, 1994). Crop planning has also been done by CASAM for plain zone of Maharashtra by using MAI and moisture availability period based on assured rainfall at 50 and 70 per cent probability levels (Saudagar, 2000).

The crop growth period was considered as the period during which the MAI was more than 0.5 at the time of sowing and active vegetative growth period and more than 0.3 at the time of maturity. All periods during active vegetative growth for which MAI was less than 0.5 were considered as stress periods. Cessation of rainy season does not mean the end of crop season. Crop can thrive on stored soil moisture. It is, therefore, necessary to examine and find out the amount of moisture stored in the soil at the end of season.

The probable earliest date of sowing of *kharif* crops was assumed to be the day when rainfall received along with the stored soil moisture met half evaporative demand during the moist period following a sub-humid period and the probable date of sowing of *rabi* crops was assumed to be the day when the soil moisture storage was sufficient to meet the full evaporation demand of the climate. Based on these assumptions, the moisture availability periods (number of days in the moist period and the following sub-humid period) were worked out for planning the cropping pattern. For deciding the growing period at different stations following two assumptions were made that the active growth, the demand is at least 0.5 MAI (ii). At the physiological maturity stage, the crop demand may fall upto 0.3 MAI. The crop growth periods were worked out for different stations by considering the above assumptions.

Example: Moisture availability periods at Pune.

The Moisture Availability Periods based on weekly normal rainfall at Pune were computed and are presented in Table 3.7. Different cropping patterns were suggested on the basis of these moisture availability periods for different soil types.

Table 3.7: Moisture availability periods for different soil types at Pune

Soil	Season	Moisture availability periods		
		From (MW)	To (mw)	Duration (weeks)
Shallow	*Kharif*	23	40	18
Medium	*Kharif*	23	40	18
	Rabi	35	48	14
Deep	*Kharif*	21	40	18
	Rabi	35	52	18

From the above discussions, it is obvious that Thornthwaite's Water Balance Method is very useful in determining the moisture availability periods for crop planning for a given region.

3.10.2 FAO method

Soil moisture is the limiting factor under dryland ecosystem for successful crop productivity. The risk involved in agriculture therefore needs to be brought out to know the chances of success and failure of crop at a place under prevailing climatic conditions. Considering this, a systematic monitoring of rainfall during growing season in the drought prone areas in order to detect early date of anomalies in the quantity and distribution of rainfall which might affect the crops. For this a model was developed in FAO, the crop monitoring and forecasting based on a cumulative weekly or ten-daily (decade) crop water balance (Frere and Popov, 1979) which at a given moment of the crop growing cycle gives an index expressing the degree of satisfaction of the crop water requirement. This index is strongly correlated with the yield and gives a very good idea of the yield to be expected. This method is popularly known as FAO method to workout water balance.

The FAO model:

In this model, crop monitoring is performed by finding the following:

i. Potential evapotranspiration,

ii. Water requirement of crop, and

iii. Moisture availability index or water requirement satisfaction index,

Potential evapotranspiration is calculated by using the formula described by Penman (1948). Since crop yield is proportional to the moisture availability (and water used) it is important to study the water requirement of crop. Hence, water requirement is calculated by considering the crop coefficient of the crop.

Moisture availability index (MAI) is a measure of the adequacy of precipitation in supplying crop water needs (Hargreaves, 1971). Hargreaves defined MAI as the ratio of the assured rainfall expected with 75% probability to the estimated potential evapotranspiration for the concerned period. Whereas. Frere and Popov (1979) defined an index in percentage the extent to which the water requirements of the crop had been satisfied in a cumulative way at any stage of crop growth. This index is called Moisture Availability Index (MAI) or Water Requirement Satisfaction Index (WRSI). As WRSI has direct relationship with crop yield, it becomes imperative to study the relationship between WRSI and yield.

Procedure

It is based on a cumulative water balance established over the growing season of a crop and established for successive periods of 10 days (decades) or weeks. The water balance is the difference between precipitation received and the water lost by the crop and soil. The water retained by the soil is also taken into account in the calculation. The sample calculation of water balance for pearl millet is presented in Table 3.6. Here it was noted that the normal length of cropping season (sowing to physiological maturity) for pearl millet crop was 75-80 days. Hence, the calculations for water balances were carried out for twelve weeks only.

The steps involved for estimation of different components of water balance are as follows:

a. Normal precipitation (PN): It is calculated from long term series of climatological data for the station.

b. Actual precipitation (Pa): It is the total precipitation which fell in each decade or week.

c. Number of rainy days (da): It is a day on which 2.5 mm or more of rain was recorded.

d. Potential evapotranspiration (PET): It is calculated as described by Penman (1948).

e. Crop coefficient (Ke): It is the ratio of actual crop evapotranspiration (ETo) to the potential evapotranspiration (PET). Already determined Ke values for the crop under consideration for different growth stages may be used.

f. The water requirement (WR): It is arrived at by multiplying the potential evapotranspiration for time decade by the respective crop coefficient for the same decade.

g. Pa – WR: Difference between actual precipitation and water requirement (Pa - WR) is calculated.

h. Water reserve in the soil (RS): it expresses the water stored in the soil which could readily be used by the crop. In other words, it is the water reserve between the field capacity (FC) and the permanent wilting point (PWP) for a effective root zone depth.

i. Surplus and deficit of water (S/D): Surplus refers to any quantity of water above the selected retention level (i.e. storage capacity). Surplus is indicated by positive sign. Deficit refers to any water requirements under the zero level of the water storage capacity. Deficit is indicated by negative sign.

j. Water Requirement Satisfaction Index (WRSI): It indicates in percentage the extent to which the water requirement of the crop was satisfied in a cumulative way at any stage of crop monitoring and yield forecasting.

References

Chow, V.T. 1964. Hand Book of Applied Hydrology, McGraw Hill Book Company Inc: 8-2 to 8-35.

Dahale S. D. and Puranik P. V. 2000. Climatology and Predictability of the Spatial Coverage of 5-days Rainfall over Indian Subdivisions, *International J. Climatology*, 20, 443-453.

Edgar B. Forester, 1949. Rainfall and Runoff. The Macmillan Company, New York: 217.

Feller, W. An Introduction to Probability Theory and its Applications, Vol. I. John Willey.

Glenn O. Schwab, 1966. Soil and Water Conservation Engineering. The Ferguson Foundation Agricultural Engineering Series, New York: 39-45.

Gnedenko, B. V. 1963. The Theory of Probability. Chelsa Publi., New York.

Gumbel, E.J. 1958. Statistics of Extremes. Columbia University Press, 375.

Jeevarathnam, K. and Jay Kumar, M. 1979. Probability Analysis of Annual Maximum Daily Rainfall for Ooctacamund. *Indian J. Soil Conser.*: 7(1): 11-16.

John C. Rodda, 1967. A country-wide study of intense rainfall for the United Kingdom. J. Hydrol, Amsterdam 5: 58-59.

Krishnan, A. 1980. Agroclimatic classification methods and their application to India. In: Climatic Classification. Proceedings of the Consultants Meeting held at ICRISAT, Patancheru, Hyderabad, 14-16 April, 1980. pp 59-88.

Krishnan, A. and Mukhtar Singh. 1968. Soil climatic zones In relation to cropping patterns. In: Symposium on Cropping Patterns, ICAR, New Delhi. pp 172-185.

Medhi, J. 1992. Statistical Methods: An Introductory Text. Wiley Eastern Ltd. New Age International Ltd., publication.

Mooley, D. A. 1974: Research Report: Suitable tables for application of Gamma Probability Model to rainfall, IITM.

Raskar, S.M. 1994. Crop planning on the basis of moisture availability index for Pune and Ahmednagar districts of Maharashtra state. An M.Sc. (Agri.) Thesis.185 p.

Ronald A. Fisher and Frank Yates, 1957. Statistical Tables for Biological. Agriculture and Medical Research, Oliver and Boyd, London: 42-50.

Saudagar, S.T. 2000. Study on cropping pattern for plain zone of Maharashtra on the basis of soil and rainfall characteristics. An M.Sc. Thesis. Mahatma Phule Krishi Vidyapeeth, Rahuri, 147 p.

Schulz, B.F. 1974. Problems in Applied Hydrology. Water Resources Publication, Colorado, U.S.A: 413-435.

Subramanium, V.P. 1956. Climatic types of India according to rational classification of Thornthwaite. *Indian J. Meteorology and Geophysics.* 7(3): 253.

Thornthwaite, C.W. 1948. An approach towards a rational classification of climate. Geographical Review. 38: 55-94.

Thornthwaite, C.W. and J.R. Mather. 1955. The water balance. Publications in Climatology. Vol. 8(1). Drexel Institute of Technology, Laboratory of Climatology, Centertron, N.J. pp. 104.

Victor Mockus, 1957. Engineering Hand book of Hydrology. U.S. Department of Agriculture. 4: 279-307.

Vujica Yevjevich, 1972. Probability and Statistics in Hydrology. Water Resources Publications, Colorado: 224-227.

Wilks, D.S. 1995. Statistical Methods in the Atmospheric Sciences. Academic Press.

EXERCISES

3.1 Define a random variable, continuous variable and a discrete variable.

3.2 Distinguish between deterministic and stochastic solution of a hydrological problem.

3.3 Discuss the role of probability theory and statistics in the processing of data and hydrological design.

3.4 Define recurrence interval, probability density function, and probability distribution function.

3.5 Discuss the different methods of fitting a theoretical probability distribution function to the empirical frequency distribution?

3.6 Discharge of a river is measured at discrete times, then why is it still called a continuous variable?

3.7 Explain the visual/graphical method of checking the goodness of fit of theoretical probability distribution to empirical frequency distribution of a random variable. Indicate how mean and standard deviations are determined in case of normal distribution.

3.8 Why is it thought expedient to use a special kind of graph paper, known as probability paper in statistical analysis of data?

3.9 When the goodness of fit of any theoretical probability distribution to empirical frequency distribution can be judged visually (graphically), then why it is checked by means of numerical tests such as Chi-square test? Explain.

3.10 For large recurrence intervals fitting of historical data by theoretical probability curves is almost a must. Comment.

3.11 Work out the mean, standard deviation, skewness coefficient and Kurtosis for the following theoretical probability distribution (see data Table).

(i) normal (ii) log-normal (iii) log-Pearson type III
(iv) Extreme Value Type I (Gumbel)

Data table for Problem 3.11

S. No.	Year	Flood peak	S. No.	Year	Flood peak	S. No.	Year	Flood peak
1	1930	198.00	12	1941	518.53	23	1952	130.01
2	1931	100.00	13	1942	352.66	24	1953	512.55
3	1932	332.00	14	1943	321.28	25	1954	177.83
4	1933	472.00	15	1944	171.85	26	1955	218.17
5	1934	222.00	16	1945	150.93	27	1956	104.60
6	1935	421.00	17	1946	361.62	28	1957	496.11
7	1936	334.00	18	1947	439.33	29	1958	249.56
8	1937	47.00	19	1948	491.63	30	1959	279.44
9	1938	279.45	20	1949	325.76	31	1960	203.23
10	1939	388.52	21	1950	351.16	32	1961	52.30
11	1940	124.03	22	1951	449.79			

For the data given in Problem 3.5., work out the flood floss with recurrence intervals of 5, 10, 25, 50 and 100 years assuming the distribution as, normal, tog-normal, log-pearson type III and Gumbel's distribution.

On 15th November probability of rainfall is 0.20 what is the probability of no rainfall occurring during the next 15 days?

Calculate the probability of 30-year flood occurring exactly thrice in a 30-year period.

From the observation of past records of precipitation at a rain gauge station, it is guessed that probability of having a rainy day is 0.40. What is the probability that in the next year there would be at least 100 rainy days?

What is the probability of rainfall x, with a return period of 250 years to occur four times in 1000 years?

3.12 (a) Illustrate with the help of neat sketches;

(i) Trend, (ii) seasonal component and (iii) random component.

(b) Discuss briefly the moving mean method of finding dry and wet cycles of a long historical rainfall data.

3.13 Discuss the double-mass curve method of adjustment of precipitation data.

3.14 Write short notes on

(a) Limitations of probability methods

(b) Outlier points of data

(c) Fitting of a straight line to the plotted data by least square method.

3.15 In a river, 10-year and 25-year flood values are 73 and 84 mm. Calculate the 100-year rainfall assuming that rainfall at a 70% probability follows a normal distribution.

3.16 It has been observed that the annual runoff volume of a river can be represented by normal distribution with mean = 30 million cubic metres and standard deviation = 6 million cubic metres. Determine the probability of annual runoff exceeding 35 mm^3 in any year. What is probability that annual runoff volume in any year will be between 20 and 32 mm^3.

3.17 It is assumed that a rainfall depth series covering 23 years of record can be fitted by a Pearson Type III distribution. The rainfall series has mean = 14.90, standard deviation = 33.55 and third moment μ_3 = -27.32. Compute the rainfall for return periods of 5, 10, 25, 50 and 100 years using Chow's frequency factor method.

3.18 Annual maxima stream flows may he assumed to be log-normally distributed with untransformed values of mean μ= 300 m^3 / and standard deviation σ = 100 m^3/s. Compute the 100-yr. flood.

■■■■

Chapter 4

DROUGHT INDICES

4.1 Introduction

Drought has been a matter of serious concern to man since ancient times, and even today it is an outstanding example of man's helplessness before nature's large scale and formidable phenomena. It is a natural phenomenon that occurs fairly periodic all over the globe. Throughout the world drought has become a central problem of concern today to meet the ever increasing demand of increasing population. Further, the water demand is increasing in different sectoral such, domestic purpose, industrial expansion and power generation every day due to change in life style, industrialization. Drought is climatic anomaly characterized by deficient supply of moisture such deficiency may result from sub-normal rainfall, erratic rainfall distribution, excessive water need or a combination of these factors.

Drought ranks as one of the foremost among earth's natural hazard representing the adverse effect due to shortage of water mainly from rainfall. Droughts are threat to food production through agriculture and are still an unconquered ill. They are an outstanding example of man's limitation in the face of the broad scale formidable forces of nature. Therefore, knowledge of drought will help in crop planning and scheduling irrigation delivery to make efficient use of rainwater for crop production and other uses.

4.2 Definitions of Drought

According to WMO (1997), drought is the consequence of a natural reduction in the amount of precipitation over an extended period of time, usually a season or more in length, often associated with other climatic factors (*viz.* high temperatures, high winds and low relative humidity) that can aggravate the severity of the event. Drought implies dryness.

Drought is a situation when the seasonal rainfall is deficient by more than twice the mean deviation (Ramdas, 1960). In the USSR, drought was defined as period of 10 days with total rainfall not exceeding 5 mm. Gibbs and Maher (1967) made a study of drought in

Australia by using annual rainfall deciles as drought indicators. The U.S. Weather Bureau defined drought as lack of rainfall so great and long continued as to affect injuriously the plant and animal life of a place. According to Yevjevich (1967) drought is the depletion of water supplies both for domestic purpose and for the power plants, especially in those regions where rainfall is normally sufficient for such purposes.

There is no universally accepted definition of drought. Thus, there are number of definition of drought as studied by researchers. Early workers defined drought as a prolonged period without rainfall. According to India Meteorological Department (IMD), drought is defined as the year with annual rainfall of 75 percent or less than normal and if 50 percent or less as compared to normal than it is called as severe drought year.

Palmer (1965) defined drought is the prolonged period without rainfall and also normal moisture deficiency.

Failure of the monsoon is the most important cause for famines in India and Rajasthan while occasional failure of crops due to attack of insects and pests is also responsible for drought condition.

According to State Revenue department, a year is known as drought year when more than 50% crop losses occur during the year.

However, drought is evaluated in a more meaningful manner by water balance computation taking into account the rainfall pattern, evapotranspiration, moisture characteristics of a soil, rooting depth and stage of crop growth.

Sen (1970) analysed the data on droughts and famines in India from 1800 to 1966 and concluded that severe and widespread drought usually led to a sharp fall in food production in the country as a whole and gave rise to famine conditions. According to him famine is due to wide spread drought condition. Lack of sufficient water to meet the normal water requirements of a locality may be considered as a common and fairly satisfactory definition of drought.

Drought can result not only from precipitation deficiency, but also from excessive evapotranspiration. A dry spell is not particularly harmful to vegetation, although there may be serious hydrological consequences.

4.3 Classification of Droughts

Drought is broadly classified into three categories as given below:

(i) Meteorological drought,

(ii) Hydrological drought, and

(iii) Agricultural drought.

4.3.1 Meteorological drought

According to Meteorological Department (IMD), it is a situation when the seasonal rainfall received over the area is less than 75% of its long term average value. It is further classified as "moderate drought" if the rainfall deficit is between 26-50% and "severe drought" when the deficit exceeds 50% of the normal.

4.3.2 Hydrological drought

Linsley *et al.* (1975) defined hydrological drought as a period during which the stream flows are inadequate to supply established uses of water under a given water management system.

It can also be defined as year in which the aggregate run off is less than the long-term average runoff. Most of the criteria developed for hydrological drought are location specific or individual streams or river basins.

4.3.3 Agricultural drought

It occurs when available soil moisture is inadequate for healthy crop growth and causes extreme stress and wilting.

Classification of agricultural drought

Agricultural drought is classified into different categories based on different criteria:

(i) Based on time of occurrence in the season,

(ii) Thornthwaite (1947) criteria, and

(iii) Singh (1989) criteria.

4.3.3.1 Classification based on time of occurrence in the season

(a) Early season drought: It is the drought which occurs in association with the delay in commencement of sowing rains. Sometimes, early rains may occur tempting the farmers to sow the crops followed by a long dry spell leading to withering of seedlings and poor crop stand establishment.

(b) Mid season drought: Mid season droughts occur when there is break in southwest monsoon. Drought conditions during vegetative crop growth phase might result in stunted growth, low leaf area development and even reduced plant population.

(c) Late season drought: It occurs due to early cessation of rainy season. Sometimes crops may encounter moisture stress during its reproductive phase which leads to forced maturity of the crop due to rise in temperature. The grain yield of the field crops is highly correlated with the water availability during the reproductive phase of crop growth (Ramana Rao *et al.*, 1983, and 1984). Therefore, late season droughts greatly damage the prospects of a good harvest.

(d) Permanent drought: These droughts are associated with inadequacy of soil moisture/rainfall to meet the water requirements of the crop during most of the year. In the arid and even in some of the seasonally dry tropical regions, rainfall is somewhat assured for a shorter period of just 6 or 7 weeks or even less. In these regions, crops are grown for subsistence, although the rainfall is not adequate to grow even a short duration crop of 75-80 days. Therefore the crops will be often be subjected to moisture stress during most of the years.

(e) Apparent drought: It is the situation when the rainfall in the region may be adequate for one crop but may not be so for others. For example, paddy is grown in some of the sub-humid regions where it should not be grown.

4.3.3.2 Classification based on Thornthwaite (1947) criteria

According to Thornthwaite (1947) drought is classified into the following categories:

(a) Permanent drought: It is defined as characteristic of the driest climate where sparse vegetation growing is adapted to drought and agriculture is possible only by irrigation during the entire crop season.

(b) Seasonal drought: This type of drought occurs in climates which have well defined rainy seasons. The natural vegetation is made up of plants that produce seeds during the rainy season and then die, plants that remain alive but become dormant in the dry season. Most of the arid and semi-arid zones fall in this category. The sowing dates and the duration of crop and their varieties in these regions should be so adjusted that their growing season should fall within rainy season followed by the period having residual moisture storage.

(c) Contingent drought: It is dependent upon irregularity of rainfall and so it is not regular to occur in any definite season.

(d) Invisible drought: This type of drought occurs when rains do not supply enough water to counteract the water loss by evapotranspiration, the result is a border line water deficiency in soil resulting in less than optimum crop yields. It occurs usually in humid zones and in areas which are receiving frequent rainfall.

4.3.3.3 Classification based on Singh (1989) criteria

Singh (1988) proposed the following classification:

(a) Atmospheric drought: This type of drought is frequently caused by low humidity and hot, dry desiccating winds even under conditions of relatively high soil moisture. Wilting due to atmospheric drought is always reversible and it occurs because of either an under developed root system or their physical inability to conduct water fast enough to compensate the water losses from the leaves and tender portions of plants during the stress period. In this type of drought plants may recover from injury when they are supplied with water, though the dried up portions do not recover.

(b) Soil drought: It is defined as when soil moisture lags behind evapotranspiration at times of greatest need of water by plants, such as during the grand growth period.

(c) Physiological drought: It often occurs due to some physiological aberrations. Plant roots fail to absorb water and nutrients from soil due to increased concentration of soil solution. Presence of soluble salts in soil water increases the concentration of the solution and hence availability of water to plants is reduced, although sufficient water remains in the soil. Similarly presence of certain chemicals in the soil interferes with the water absorption by plants *e.g.* humic acid restricts the water absorption in bog soils and the term is used as physiological dryness of the soil; however, plenty of water remains in the soil.

4.4 Difference between Aridity and Drought

Generally, people confuse between aridity and drought. However, these are two distinct entities. Persistent of drought for years leads to aridity whereas drought means dryness. Like drought, aridity is also caused by lack of water. But the similarity ends there. Aridity is a permanent climatic situation of a region. It is usually restricted to area of low rainfall and generally high temperature though there are also cold arid zones like Ladakh. Human activities in the arid zone get adapted to the perpetual lack of moisture. On the other hand, drought is a temporary phenomenon and often subsides once rainfall occurs in sufficient magnitude. The arid zone of India comprises of West Rajasthan, Kutch and parts of Saurashtra in Gujarat, some pockets in Tamil Nadu also have arid climate.

4.5 Climate Change and Drought

The most important mechanisms of climate change occurred due to changes in the supply of solar energy, changes in the transmissivity of the atmosphere for both incoming and outgoing radiation and changes in land use that alter the radiation balance. According to Schneider and Lender (1984) there are following causes of climate change:

1. The release of heat (thermal pollution) that warms the atmosphere directly,
2. The upward transport of chlorofluorocarbons (CFCS) and nitrous oxide into the stratosphere, where photochemical reaction of their dissociation products destroys stratospheric ozone with a consequent increase in ultraviolet radiation of the surface and its inhabitants; and
3. The release of trace gases, like nitrogen oxides, carbon-monoxide, or methane, that increase ozone concentration in the troposphere (lower 15 km atmosphere, approximately) by photochemical reactions.

The main reason of climate change may be due to large scale nuclear war potential cause. Such an event could lead to very large scale injections of root and dust into the troposphere and stratosphere, causing transient cooling that might last for weeks to months.

Rapid industrialisation and human activities are found to be exacerbating drought problem. Bryson (1973) found that increasing atmospheric turbidity and green house warming (due to CO_2) both of which are attributed to man mists be implicated as the causes of the Sahelian droughts or due to global warming. Deforestation, largely contributed by the anthropogenic factors, also found to intensity drought conditions. Charney (1975) reported that an increase in land albedo (reflectivity for solar radiation) such as occur when land is denuded of vegetation leads to a decreases in net incoming radiation and an increase in the relative cooling of the air. This would cause the air to sink resulting in suppressed convective activities and its' associated rainfall. Low rainfall further reduces plant cover, raises albedo and amplifies the effect.

4.6 Reasons of Drought

There are following reasons for the occurrence of drought:

(i) Extraterrestrial forces

There is an 11 year cycle in the variation of annual mean sunspot number. But there is no positive evidence of such a cycle in the energy output of the sun. If such cycle exists, the atmospheric circulation would be in a state of constant readjustment in accordance with the greater or lesser amount of solar energy reaching the earth. As the readjustments evolve, drought would probably appear first in one region then in another.

(ii) Volcanic activity

The injection of large amount of ash and dust into the atmosphere, by violent volcanic activity may alter earth's radiation balance and thereby create compensating circulation adjustments which include fluctuations resulting in drought.

(iii) Atmospheric composition

Changes in the atmospheric composition e.g. water vapour CO_2 and O_3 would modify the heat balance of the earth and in turn would produce fluctuations in large scale circulation pattern resulting in drought. Rapid industrialisation and population aggravated this problem, which also polluted the environment and resulted into global warming.

(iv) Interaction in earth–ocean atmospheric complex

The biggest natural climate fluctuations occurring globally on time scale of a new years is the southern oscillation and associated El-Nino phenomenon, together they called El Neno South – Oscillation of ENSO the ENSO phenomenon is the most notable example of inter annual climate variability. Major ENSO episodes lead to massive displacements of rainfall regimes of the tropics, bringing droughts over vast areas and torrential rains to other dry regions. ENSO normally occurs once in every 2 to 7 years. ENSO shows strong links with droughts over India, Indonesia and Australia.

4.7 Drought Indices

Drought is assessed with the help of drought indices. A drought index enables the description of drought to be expressed quantitatively.

Drought indices are broadly classified into the three categories based upon type of drought:

(i) Meteorological drought indices,

(ii) Hydrological drought indices, and

(iii) Agricultural drought indices.

4.7.1 Meteorological drought indices

(i) Banerji and Chabbra (1963) classified meteorological drought condition based on rainfall deficit as given below.

Drought condition	Rainfall deficit
Slight drought	11-25%
Moderate drought	26-50%
Severe drought	>750%

(ii) Yevjevich (1971) proposed the following classification of meteorological drought as given below:

(a) Mild drought– when rainfall or inflow is below the mean.

(b) Moderate drought– when rainfall or flow is below 0.75 mean.

(c) Sever drought– when rainfall or inflow is below 0.55 mean monthly values.

(iii) IMD criteria: According to the definition proposed by the India Meteorological Department (IMD) meteorological drought is characterized as under:

(a) Moderate drought– It occurs when rainfall variance from the normal is 26-50 per cent.

(b) Severe drought– It occurs when the deficiency exceeds 50 per cent.

The recurrence of drought in various parts of India is presented in Table 4.1

Table 4.1: Recurrence of drought in India

Recurrence	Regions
Once in 4-5 years	Rajasthan, Punjab, Gujarat and Telenagana
Once in 6-8 years	Haryana, Jammu and Kashmir, Uttar Pradesh (east), Sub-Himalayan West Bengal, Vidarbha (interior), Karnataka (South) and Coastal Andhra Pradesh

Recurrence	Regions
Once in 10 years	Himachal Pradesh, Uttar Pradesh (West), M.P., Bihar (plains), Marathwada, Konkan, Royal Seema and Kerala
Once in 15 to 20 years	Madhya Maharashtra, South Assam, Tamil Nadu (interior), Karnataka (north) and Orissa
Very rare	Coastal Karnataka, Bihar platteau, Bengal

Source: Mavi, 1987

(iv) According to Krishnana, (1979) meteorological droughts are classified as follows:

Rainfall	Drought
Less than mean (x)	Mild
Less than 0.75 times mean (0.75 x)	Moderate
Less than 0.55 times mean monthly	Severe

(v) Sharma *et al.* (1979) proposed the following classification for meteorological drought:

Drought year - Any year receiving rainfall less than or equal to $x - \sigma$.

Abnormal year - Any year receiving rainfall less than or equal to $x + \sigma$.

Normal year - Any year receiving rainfall between the $x \pm \sigma$, where x is the mean yearly rainfall and σ is the standard deviation.

4.7.2 Hydrological drought indices

Yeujevich (1967) defined drought as a lack of rainfall so great and long continued to affect injuriously the plant and animal life of a place and to deplete water supplies both for domestic purpose and for the power plants especially in those regions where rainfall is normally sufficient for such purpose.

Hydrological drought is classified as under:

(i) Mild drought– It is defined as when inflow is below the mean is called mild drought.

(ii) Moderate drought– It occurs when inflow is below 0.75 times mean.

(iii) Severe drought– It is defined as when inflow is below 0.55 times mean is called as severe drought.

4.7.3 Agricultural drought indices

(i) The drought can be better defined in terms of moisture deficit or aridity index. Aridity is the ratio of moisture deficit to annual water need. Subramanyam (1964) used this parameter and defined drought intensities as follows on using standard deviation (σ) of aridity index and analysed occurrence of droughts in arid zone stations.

Departure of aridity index from the median	Drought intensity
< ½ σ	Moderate
½ to σ	Large
σ to 2 σ	Severe
> 2 σ	Disastrous

(ii) Krishnan and Thanvi (1971) proposed the criteria of aridity index for *kharif*-cropping season to characterize drought as under:

Aridity index in *kharif* season	Drought intensity
50 to 60	Slight
60 to 70	Moderate
70 to 80	Severe
Greater than 80	Disastrous

The study revealed that droughts were more common (60-90%) than good years in arid zone. Drought years constituted 6-20 % in semi-arid zone and 5% in sub-humid zone of Rajasthan.

(iii) Palmer drought index (PDI): Palmer (1959 & 1965) defined drought in various ways and designed a quantative technique which took into consideration the combined effect of rainfall and other parameters. Palmer drought index is developed on the basis of hydrological accounting. In this technique, moisture supply in the form of rainfall, soil moisture demand by way of evapotranspiration, current and antecedent weather and climate of the place are taken into account for obtaining a drought index numerically. He used rainfall anomaly as a crucial parameter.

(iv) Subramanyam and Shastri (1969) proposed following classification of agricultural drought based on aridity index as given below:

Departure of aridity index from the median	Drought intensity
< ½ σ	Moderate
½ to σ	Large
σ to 2 σ	Severe
> 2 σ	Disastrous

where, σ is the standard deviation of aridity index.

(v) Aridity anomaly index (AI)

$$AI = \frac{PE - AE}{PE} x100 \qquad ...(4.1)$$

where,

AI = weekly aridity index,

AE = Actual evapotranspiration, and

PE = Potential evapotranspiration.

Weekly Anomaly of AI	Associated Drought Intensity
1-25	Mild
26-50	Moderate
>50	Severe

For the sake of computing anomaly, long term weekly values of aridity index are computed. The difference between actual AI in any week of the crop season and the normal AI expressed as a percentage called aridity anomaly which represents water stress condition. Based on the AI the incidence, spread intensification and cessation of different drought intensities on weekly basis is monitored. Following criteria suggested by Choudhary *et al.* (1977) is used in defining various drought intensities.

(vi) Normalised difference vegetation index (NDVI)

National Remote Sensing Agency (NRSA), Hyderabad of the Department of Space have been using this index for monitoring agricultural drought over the country on a fortnightly time scale using daily NOAA satellite data. NDVI is determined as,

$$NDVI = \frac{NIR - VIS}{NIR + VIS} \qquad ...(4.2)$$

where,

NDVI = Normalized Difference Vegetation Index,

NIR = Reflected radiation in near Infra Red Channel, and

VIS = Reflected radiation in Visible Channel.

(vii) Modified scheme of drought characterisation

Shastri and Rama Krishna (1982) proposed modified scheme of drought charactersation for western Rajasthan given as below:

Departure of aridity index from the median	Drought intensity
$< ½\ \sigma$	Moderate
½ to σ	Large
σ to 1½ σ	Severe
$>$ 1½ σ	Disastrous

where, σ is the standard deviation of aridity index.

(viii) Assessment of duration and severity of drought spells

Folly (1957) developed cumulative deviation on technique to assess duration and severity of drought. In this technique departures of weekly water deficiency from the normal were taken and expressed as ratio of the normal water need in units of thousands. These units were cumulated from an arbitrary origin and plotted against successive weeks. The core period of drought is then estimated. The analysis of drought consists of three phases, namely definition of drought, choice of unit time and the choice of truncation level. To define the drought and visualize, three important core parameters are necessary namely, severity (S), duration (D) and magnitude (M). These three parameters are related by the following relationship.

$$M = \frac{S}{D} \qquad \text{...(4.3)}$$

Since any one of the parameters is completely determined by the other two in equation (4.2), only two of the three parameters are required to specify drought event. The steepest portion of rise in the mass curve and the gradient consideration gives the drought severity index (DSI) during the period.

Subrahmanyam and Subrahmanyam (1967) classified the drought spells based on DSI into four which categories are given below:

Drought severity index (DSI)	Category of drought
0-10	Moderate
10-20	Large

20-40	Severe
> 40	Disastrous

4.8 Drought Indices in India

In India, 99 districts in 13 states covering a geographical area of a little more than one-third of the country have been identified as drought prone (CWC, 1982). More than 150 definitions of drought exist in literature. National Institute of Hydrology has conducted a review of available drought indices (NIH 1986-87) on the basis of data available, merits and appropriateness. Central Water Commission selected 10 indices for identification of drought in the 99 drought prone districts. A brief description of these indices is described below:

(i) Inadequacy of rainfall (IR)

Irrigation Commission (1972) observed that an area, having annual normal rainfall less than 750 mm, would not be able to sustain a staple crop. Hence, any taluka/district satisfying this criterion is considered as drought prone.

(ii) Criterion of India Meteorological Department (IMD)

IMD defined drought is a situation occurring in any area where annual rainfall is less than 75% of normal in 20% or more of the years examined.

(iii) Variability of rainfall (VRF)

Based on the coefficient of variation (CV) or rainfall, the normal practice adopted by IMD is as:

(a) When annual or seasonal CV is 30 percent or more, the rainfall is said to be erratic and the area is classified as drought prone.

(b) When monthly CV is 50 percent or more, the rainfall is said to be erratic and the area can be classified as drought prone even if CV in any one of the months of rainy season is 50 percent of more.

(iv) Frequency of occurrence of different ranges of rainfall (FDR)

This criterion is synonymous with IMD criteria of drought identification. The probability distribution curves of various ranges of rainfall of 100 mm are drawn and if the probability of 75% of the normal rainfall or more was found to be so in 80% or less of the years considered, the area is classified as drought prone.

(v) Erratic distribution of rainfall (ERF)

According to National Commission on Agriculture (1976), drought is defined as an occasion when rainfall in a week is half of the normal or less, when the normal weekly rainfall is 5 mm or more. The agricultural drought is a period of four such consecutive weeks in monsoon season (generally from middle of May to middle of October) or six consecutive weeks in the rest of the year. Based on this, the areas where such drought conditions persist in 20% or more of the years examined, the area may be considered as drought prone.

(vi) Deficiency of Runoff (DRO)

The CWC introduced drought index based on runoff as the crop production is closely tied up with runoff. An area is identified as hydrologically drought affected if annual runoff is less than 75% of normal annual runoff in more than 25% of the years considered.

(vii) Soil moisture stress (SMS)

The Central Arid Zone Research Institute, Jodhpur proposed the following criteria on the basis of soil moisture stress during the crop period.

Drought free crop period: when AE < (PE/2)

Moderate drought crop period: When (PE/2) < AE < (PE/4)

Severe drought crop period: When AE < (PE/4)

where, AE and PE are actual and potential evapotranspiration for the period.

The analysis is to be carried out for the period of 11th June to 20th October and for a period of 10 years. If the analysis indicates that drought occurred in more than 20 percent of the years considered, the area is classified as drought prone.

(viii) Aridity index (AI)

The monthly and annual water surplus and deficit in the soil are worked out by using Thornthwaite and Mather (1957) method. The Aridity Index (AI) is an expression of the water deficit as percentage of annual water need. A criteria based on departure of Aridity Index (AI) from normal, for drought identification is given as under:

Departure of AI from normal	Drought classification
Less than 0	No drought
0 to ½ d	Moderate drought
½ d to d	Large drought
D to 2d	Severe drought
More than 2d	Disastrous drought

where, d is the standard deviation of Aridity Index. Based on this, if drought occurred in 20 percent or more of the years considered, the area is classified as drought prone area.

(ix) Irrigation facilities (IF)

The Irrigation Commission (1972) accepted the criteria of drought of IMD and further mentioned that if 30% of the cropped area is irrigated in drought prone area (taluka/tehsil), it can reasonably sustain of tehsil/district having irrigation facilities may be examined.

(x) Crop loss (CL)

Crop loss has also been used as an indicator of drought in various states. This is also known as Annawari or Gardawari criterion. If there is a fall in the annual production of crops by more than 25 per cent from the normal, the area is considered as suffering from drought in that year. When this happens in 20 percent or more of the years considered, the area is classified as drought prone.

4.9 Suitability of Drought Indices for Different Regions

The Central water Commission applied the above mentioned ten indices in section 4.8 to study 99 districts to identify as drought prone in India. Chaube *et al.* (1992) have made studies on inter-consistency of the above ten indices in identifying 99 drought prone districts as preferable for each of the four identified agro-climatic zones of India. Based on this regional analysis some indices were recommended for identification of drought which are presented in Table 4.2. The recommendation is based on a very high degree of inter-indices consistency in drought identification.

Table 4.2: Recommended indices for drought identification for different agroclimatic zones of India. (Broad Classification)

Agroclimatic zone	Region considered	Preferred indices
Most	Eastern part of India sub-humid zone	VRF, IF, DRO, ERF & Al
Dry Sub-humid zone	J&K, Southern part of Ganga basin	i) DRO & Al (ii) ERF & VRF
Semi-arid zone	Peninsular region and parts of Central and Central – West region	ERF & Al
Arid zone	Western region of India bordering boundary with Pakistan	IRF, IMD, VRF, ERF DRO, SMS & Al

Recommended indices

Chaube (1999) proposed indices based on the consistency analysis, suitable drought indices which have a very high degree of inter consistency to identify drought are recommended zone wise (Table 4.3).

Table 4.3: Recommended indices

Zones	Region	Recommended Indices
A.B. and C (Arid zones Very low)	Gujarat, Rajasthan and Haryana	IRF,IMD,VRF, DRO, ERF. SMS. Al AND CL
Zone-D (low crop potential) MAI r 0.3<10 weeks r0.7<1 week	Karnataka, Andhra Pradesh	IRF, DRO, ERF,SMS AL AND VRF
Zone – E MAI 0.3>10 weeks r0.7>1 week	Haryana, from Ahmednagar distt. In Maharashtra upto Cuddapah distt. In A.P. through Satara, Pune, Solapur and Sangli districts (Maharashtra) and Bijapur, Bellary and Tumkur (Karnataka)	
Zone-F MAI r0.3>11 weeks r0.7>4 weeks	Parts of Gujarat, Rajasthan, Uttar Pradesh Haryana & Punjab. In South from Nasik to Kanyakumari and large portions of Karnataka and Andhra Pradesh	DRO, ERF and Al

Zones	Region	Recommended Indices
Zone –G MAI r0.3>20 weeks r0.7>9 weeks	Parts of Uttar Pradesh Madhya Pradesh, Gujarat, Tamil Nadu, Maharashtra, A.P., Karnataka and small portion of Himachal Pradesh	ERF & AI
Zone – H MAI r0.3>18 weeks r0.7>9 weeks	Parts of Karnataka, Tamil Nadu, Andhra Pradesh, Orissa and West Bengal	VRF, IF ERF, SMS AND AI

Crop loss index

Crop production loss during drought situation not only varies with deficit rainfall but also due to various factors such as crop species, climate, crop diseases, soil fertility etc. Even if the rainfall during a year is above normal, the production may fall by 10 to 15% of the normal production due to any of the above factors.

Chaube *et al.* (1992) found that crop loss indicator is poorly consistent and required modification.

4.10 Drought Monitoring Indices

IMD proposed the following indices are combined together for drought monitoring for operational purpose:

(i) Standardized precipitation index (SPI)

McKee *et al.* (1993) proposed standardized precipitation index (SPI) for drought monitoring and is expressed as,

$$SPI = \frac{X_i - \overline{X}}{\sigma} \quad \ldots(4.4)$$

where,

X_i = rainfall for the period under consideration for i year,

$\overline{X}$ = mean rainfall for the period i year, and

σ = standard deviation of the corresponding period

SPI is determined based on the probability distribution of precipitation. A long-term precipitation record is fitted to a probability distribution, which is then transformed into normal distribution so that the mean SPI is zero. SPI may be computed for different time steps. This index could be useful for identification emerging droughts sooner than other drought indices.

(ii) Dryness index (Id)

This index uses normal temperature and precipitation in classifying drought. It is expressed as,

$$Id = 56 \times \log_{10} (T20 \times 10) / p \qquad \ldots(4.5)$$

where,

T = annual mean temperature, (^{0}c), and

p = annual mean precipitation, (cm)

(iii) Water requirement satisfaction index (WRSI)

It is expressed as,

$$WRCI = [\Pr ev. week\, WRSI] + [SMS / ACCAET] \times 100 \qquad \ldots(4.6)$$

where,

SMS = soil moisture storage

Monitoring the departure of rainfall from the long period average indicates deficiency or scanty rainfall situation in a region which may indicate starting of drought episode. Rating of various drought indices is presented in Table 4.4.

Table 4.4: Rating of drought indices

Code	Category	SPI	% departure of rainfall	Dryness index	WRSI
A	Normal	7 to 0.5	7 to 15	< 25	75 to 100
B	Moderate	-0.5 to – 0.07	-15 to -25	25 to 50	51 to 75
C	Extreme	-0.71 to -1.2	-25 to -50	51 to 75	25 to 50
D	Severe	< -1.2	< -50	> 75	< 25

4.11 Declaration of Drought by State Government

Situation of scarcity or drought exists if the following criteria are satisfied in an area:

- Mandal should be taken as unit for declaration of drought.
- Significant deficiency in rainfall, gross uneven spread of rain and very much prolonged dry spells.

- Steep reduction in area sown and also heavy damage to standing crops.
- Fall in the estimated yields of crop based on crop cutting experiments in the area with the average yields of preceding 5 years. In case yield values for the principal crops for the last five years are not available for a particular mandal for examining normal yield i.e. average for preceding five years, then yield values of similar crop adjacent mandal for the remaining years should be obtained and normal worked out for examining the drought conditions;
- Considerable fall in extent of grain and fodder non-agricultural wages as compared with normal times;
- Extent of unemployment position with reference to agricultural operations and on going works of Government, local bodies an big employers; and
- Unusual movement of labour in search of employment.

Assessing rainfall deficiency

- 15 per cent deficiency in case of mandals having average annual normal rainfall of less than 750 mm (i.e. up to 749.9 mm).
- 20 per cent deficiency in case of mandals having average annual normal rainfall of 750-999.9 mm.
- 25 per cent deficiency in case of mandals having average annual normal rain fall of 1000 mm and above.

Rainy day

Rainy day is considered as one if the rainfall received per day exceeds more than 5.0 mm in case of mandals with average rainfall more than 750 mm or if exceed, 2.5 mm of mandals with average rainfall less than 750 mm.

With predominant red soils 5 mm of or more rainfall per day be treated as a rainy day and for black soils area 2.5 mm or more rainfall per day will be treated as rainy day.

Dryspell and drought

According to Central Research Institute for Dry Land Agriculture, a dry spell can be considered if the rainfall received is less than 50 per cent of its normal for a consecutive period of more than 20 days during a season and that if the deficiency occurs for 21 to 28 days mild drought, 29 to 42 days as moderate drought and more than 43 days severe drought. Dryspell condition of different stages of crop in different soil is given as under:

Stage of crop growth	Minimum rainfall required	Minimum no. of rainy days	Dry spells that can be tolerated
Sowing			
a) Red soil	-	20 mm in two days	-
b) Black soil	-	40 mm in ten days	-
Seedling stage (3 weeks)	25 mm to 30 mm	2	15 days in red soil 20 days in black soil
Vegetative phase (4 to 5 weeks)	120 mm	2 to 3	20 days
Reproductive phase (5 weeks)	150 mm	3 to 4	20 days
Maturity stage (4 to 5 weeks)	120 mm	3	20 days

4.12 Prediction of Drought

There are three main components of the drought plant (Svoboda, 2000) are prediction, monitoring and early warning, risk and impact management, mitigation and response. There is need to investigate the prediction and monitoring systems, the EWS allows for the early detection of drought conditions. It enables us to respond in a proactive rather than reactive manner (Wilhite and Svoboda, 2000). However, the most important and difficult aspect is the prediction of drought with reasonable accuracy sufficiently well in advance. For this, it is necessary to identify with a reasonable degree of success the years of poor rainfall which lead to drought. This problem finally emerges out to be forecast of deficient monsoon, seasonal rainfall and its distribution. A number of scientists attempted to forecast drought during the monsoon season over India from antecedent meteorological parameters and synoptic features. But the long term applicability of these studies to predict monsoon rainfall is almost negligible. Moreover, major studies attempt to forecast total seasonal rainfall or its departures from normal which are of little use in the fields like agriculture. The parameters considered are many but some of them notably are the pre-monsoon rainfall, surface pressure anomalies, surface temperatures and winds, sea surface temperatures and oceanic features; stratospheric and tropospheric mean circulation features and associated dynamic patterns; solar activity and solar radiation etc. (Banerjee *et al.* 1976, Sen *et al.*. 1978, De *et al.*. 1984. Thapliyal, 1984). World Meteorological Organization (WMO) Geneva and its member countries have been actively engaged in searching methods for giving reasonably satisfactory drought prediction (Sarker, 1995).

4.13 Drought Management Strategies

The drought management strategies include following interdisciplinary approaches to mitigate drought effects which is described as under (Singh, 1988):

(i) Engineering approaches,

(ii) Physiological approaches,

(iii) Genetic approaches, and

(iv) Agronomic approaches

(i) Engineering approaches

These approaches are aimed at soil and moisture conservation through regulated run-off, collection of surplus rainwater, checking evaporation and seepage losses of water and efficiently recycling of stored water for providing life saving irrigation etc. These could be considered as long-term measures for managing drought effectively. For achieving these objectives, following practices are recommended:

(a) Contouring across the slope,

(b) Scooping of land,

(c) Opening ridges and furrows,

(d) Compartmental bunding,

(e) Summer deep ploughing,

(f) Deep summer ploughing followed by surface tilling during rest of the year, and

(g) Water harvesting.

(ii) Physiological approaches

These approaches aimed to alleviate the drought effects by the reducing transpiration, encouraging root growth, protecting cytoplasmic proteins with the use of suitable chemical compounds which are according to their roles may be classified as under:

(a) Anti-transpirants

Any substance, that on application to the surface of the plants reduces transpiration is termed as anti-transpirant *viz.*, phenylmircuric acelate, film forming substances and silicones, and also P.M.A. mobi leaf.

(b) Growth retardants

There are certain chemicals which is used to modify the plant structure or dwarfing the plant reduce the total water requirement of the plant. The most important chemical of this group is cycocel or CCC-2 (2-Chloroethyl trichloromethyl ammonium chloride).

(c) Radiation reflectants

Use of certain chemicals such as hydrated lime, chalk and kaolin over the foliage results in partial reflection of solar radiation which helps in keeping a lower leaf temperature and thereby reduce transpirational loss of water.

(d) Plant hardening compounds

There are certain compounds such as calcium chloride, agrosan etc., when used for seed treatment to bring about drought hardiness in plants right from seedling stage. Plants treated with hardening compounds exhibit increased protoplasmic viscosity, higher photosynthesis and lower respiration rates, higher root/top ratio and less yield reduction when exposed to drought.

(iii) Genetic approaches

Crop should be selected in such a way as to satisfy the following criteria:

(a) Shorter crop growth duration and early vigour,

(b) Deeper root system with maximum ramification of roots at deeper zone,

(c) Dwarf plant types with lesser number of erect leaves and lesser number of stomata,

(d) Thick stems,

(e) Moderate tillering with synchronous growth habit. Profuse tillering causes competition; therefore, it is not good from the view point of drought tolerance.

(f) Resistance to diseases,

(g) Lesser period between flowering and maturity so that the grain filling is least affected by moisture stress condition, and

(v) Agronomic approaches

It includes approaches like re-orientation of cropping pattern, use of suitable crops and cultivars, moderate fertilizer use, adoption of intercropping, protective or life saving irrigation, contingent crop planning etc. can be considered as very effective short-term measures in alleviating the adverse effects of drought on crops.

Contingent crop planning should be made to cope up with the following situations arising out of vagaries of monsoon:

(i) Delayed onset of monsoon,

(ii) 'Breaks' in monsoon during cropping season,

(iii) Uneven distribution of monsoon rains in space and time, and

(iv) Early cessation of monsoon rains.

References

Air Ministry 1936. British Rainfall. London Meteorological Office, London.

Allred, E.R. and Chen, R. 1953. Evaluating irrigation needs in humid areas. *Agric. Engg*. 34:611-615.

Association of British Insurers (ABI). 2005. Financial risks of climate change, Summary Report, 40 pp., www.abi.org.uk.

Baier, W. 1967. Recent advancements in the use of standard climatic data for estimating soil moisture. Ann. Zone. 6:1-21.

Baier, W. and Robertson, G.W. 1966. A new versatile soil moisture budget. Can. J PL Sci. 46:299-315.

Baier, W. and Robertson, G.W. 1968. Estimating irrigation requirements from climatological data Can. Agric. 9: 46-50.

Banarji, S. and Chhabra, B.M. 1963. Drought conditions in the Telengana Division (AP) during south west monsoon season. Indian Journal of meterology and Geophysics, 14(4):403-415.

Barger, G.L. and Thorn, H.C. 1949. Evaluation of drought hazard. Agronomy Journal. 41(11):519-526.

Bohra. A.K., Basu, S., Rajagopal, E.N., Iyengar, G.R., Das Gupta, M., Ashrit, R. and Athiyaman, B. 2006. Heavy rainfall episode over Mumbai on 26 July 2005. Assessment of NWP guidance. *Current Science*. 90. No. 9, 10 May 2006.

Brown, L.H. and Cocheme, J. 1969. A study of the agro-climatology of the highlands of eastern Africa. Tech Note 125, Wld Met. Org. Geneva.

Budyko, M.T. 1956. Heat balance by earth's surface. Translated by NA Stepanova, US Weather Bureau, Washington, 1958.

Central Water Commission. 1982. Reports on identification of Drought prone Areas in the 99 Districts", Government of India, New Delhi.

Chaube, U.C. 1999. Drought indices and appropriateness for agroclimatic regions of India. *Journal of Indian water resources*, Vol. 19(5), No. 2:10-16.

Chaube, U.C., Sinha, C.P., Saxena, R.P. 1992. Regional Analysis of Drought Indices" World Congress on Natural Hazard Reduction, Institution of Engineers, India, New Delhi (India), 10-14 January, 1992.

Chowdhury, A., Ramasastri, K.S. and Rentala, G.S. 1977. A study of agricultural drought of 1965 and 1966 *Kharif* season. IMD Pre-published Scientific Report 77/4.

Cocheme, J. and Franquin, P. 1967. An agro-climatological survey of a semi-arid area in Africa, South of Sahara. Tech Note 86. Wld Met. Org. Geneva.

De Martonne. 1926. Une nouvelle function climatologique l' indica a' artidite la Meteorologia 68:449-458.

Dilley, M. 2002. The use of climate information and seasonal prediction to prevent disasters, WMO Bull, 51, 1, 4248.

Emanuel, K. 2005. Increasing destructiveness of tropical cyclones over the past 30 years Nature, 436, 686-688.

Emberger, L. 1955. Une classification biologeographique des climates et zoologie de traveus des laboratories de botanique geologi et zoologie de la faculte's des sciences del' universtite de Montpellier.

Foley, J.C. 1957. Drought in Australia, Book No. 45, Melbourne, Bureau of Meteorology.

Gibbs, W.J. and Maher, J.V. 1967. Rainfall decile as drought indicators. Bull. 48, Bur. Met., Melbourne.

Gray, W.M. 2006. Global warming and hurricanes, Paper 4C.1. 27th Conference on Hurricanes and Tropical Meteorology, 23-28 April 2006, Amer, Meteor. Soc., Available at http://w.w.w. ametsoc. Org.

Hargreaves, G.H. 1971. 'Precipitation dependability and potential for agricultural production in North-east Brazil' EMBRAPA and Utah State University Publication No. 74-D159, pp. 123.

Hargreaves, G.H. 1974. "Precipitation Dependability and Potential for Agricultural Production in North East Brazil", Publ No. 74-SD-159, EMBRAPA and Utah State University (USA), 123 pp.

Hargreaves, G.H. 1975. "Water requirement manual for irrigated crops and rainfed agriculture' EMBRAPA and Utah State University, Publication No. 75-D158, pp. 40.

IPCC, 2001. Climate Change 2001: The Scientific Basis, Contribution of Working Group 1 to the third Assessment Report of the Intergovernm-ental Panel on Climate Change (IPCC). J.T. Houghton *et al.*, Cambri-dge University Press, 944 pp. [Available from Cambridge University Press, The Edinburgh Building. Shaftesbury Road, Cambridge CB2 2 RU England. Also available online at http://w.w.w. ipcc.ch/]

Irrigation Commission. 1972. Report Volume 1 Chapter VIII, Ministry of Irrigation and Power, Government of India, New Delhi.

Ivanov, N.N., 1948. 'Landscape-climatic zones of the earth surface'. Proceeding of All-Soviet Geogr. Cong. Vol. 1, Academic Sciences of USSR, Leningrad.

Jat, M.L. Drought predication under different agro-climatic zone. Daya Publishing House, New Delhi, pp. 343.

Katakam, A.L., Bavadam, and Bunsha, D. 2005. High water and hell, Frontline, 22, 17, Aug. 13-26.

Kennedy, D. 2006. "Acts of God", Editorial Science, 311, 5759, p 303.

Knutson, T.R. and Tuleya, R.E. 2004. Impact of CO2-Induced Warming on Simulated Hurricance Intensity and Precipitation: Sensitivity to the Choice of Climate Model and Convective Parameterization. J. Climate, 17. 18, 3477-3495.

Koppen, W. 1936. Das Geographische system der Klimate. In Handbuche der Klimatologie. Vol. I Part C (ED.) W. Koppen and R. Gieger, Glereuder Bomtiager, Berlin.

Krishanan 1979. Key paper on definition of droughts and factors relevant to specification of agricultural and hydrologic droughts. International symposium on hydrological aspects of droughts, pp. 67-102, III, Delhi.

Krishnan, A. and Thanvi, K.P. 1971. Occurrence of droughts in Rajasthan during 1941-1960. Proceedings of All-Indian Seminar on Dryland Farming, New Delhi.

Lal, B., Jayanthi, N., Prasad, T., Rajeeven, M., Devi, S., Srivastava, A.K. and Pai, D.S. 2006. Monsoon 2005: a Report, IMD Met. Monograph No: Synoptic Meteorology–3/2006, National Climate Centre, India Meteorological Deptt., Pune.

Lanberg, H.E. 1974. Drought-summary of lectures and discussion, Geneva, WMO, Bulletin, XXIII(4):pp-220.

Land Berg, H.E. 1974. Drought summary of lectures and discussion Ganeva, WMO Bulletin XXIII (45 pp. 226).

Lang, R. 1920. Verwitterung Und Bodenbuilding als Einfuhrung in die Bodenkund Stuttgart, pp. 107-123.

Laska, S. 2005. Presentation at Seminar on New Orlens. Hurricanes and Climate Change: A Question of Resilinency, American Meteorological Society (AMS), Environmental Science Seminar Series, 20 June 2005, Available at http://www. Ametsoc.org/.

Linsely, Jr R.K., Kohler, M.A. and Panthus, J.C.H. 1959. Applied Hydrology, New York. McGraw Hill, P 581.

Mckee, T.B., Doesken, N.J. and Kleist, J. 1993. The relationship of drought frequency and duration to time series. Preprints 8th conference on applied climatelogy, Anaheim CA., Amer, Met. Soc., 179-184.

Michaels, P.J., Knappenberger, P.C., and Landsea, C. 2005. Comments on Impacts of CO_2– Induced Warming on Simulated Hurricane Intensity and Precipitation Sensitivity to the Choice of Climate Model and Convective Scheme, J. Climate, 18, 23, 5179-5182.

National Institute of Hydrology. 1986-87. Comprehensive Review of Drought Indices, Report Number RN 40, Roorkee (UP). India.

Palmer, W.C. 1965. Meteorological drought. Res. Pop. No. 45. US Department of Commerce, Weather Bureau pp 58.

Papadakis, J. 1957. Experiences con especies exptìcas due cultives de paste de paste Y forraje en al. Altipiano due puno (Peru). ILO Geneva, Switzerland.

Papadakis, J. 1970. Agricultural potentialities of world climates. Buenos Aires.

Papadakis, J. 1975. Climates of the world and their potentialities. Buenos Aires.

Parthasarthy, B., Sonatakke, N.A., Monot, A.A. and Kothwala, D.R. 1987. Droughts and floods in the summer monsoon season over different meteorological sub-divisions of India for the period 1871-1984. *Journal of Climatology*, 7(11): 75-70.

Pielke, R.A. Jr, Rubiera, J., Landsea, C., Fernaindez, M.I. and Klein, R. 2003. Hurricane vulnerability in Latin America and The Caribbean: Normalized Damage and Loss Potentials, Natural Hazards Review. 101-114.

Pielke, R.A., Jr and Landsea, C.W. 1998. Normalized hurricane damages in the United States 1925-95. Wea Forecasting, 13, 621-631.

Pielke, R.A., Jr. Landesa, C.W., Mayfield, M., Laver, J. and Pasch, R. 2005. Hurricanes and global warming, Bull. Amer. Meteor. Soc., 86, 11, 1571-1575.

Popov, V.P. 1948. 'Moisture balance in the soil and the dryness indices of climate in the Ukrainian'. SSR Scientific Report of the State University of Kiev. Vol. 7.

Raghavan, S. 2003. Tropical cyclone impact in Andhra Pradesh, J. Appl. Hydrology Visakhapatnam, XVI, 4, 21-26.

Raghavan, S. 2006. Increasing Economic import of floods and Cyclones: cause and solution. J. of Applied Hydrology. XIX (3): 15-22.

Raghavan, S. and Rajesh, S. 2003. Trends in tropical cyclone impact: A study in Andhra Pradesh. India, Bull. Amer. Meteor. Soc., 84, 635-644.

Ram Mohan, H.S., Juliet, S.M. and Valsala, P. 1984. An agro-climatic approach to the assessment of tropical drought. Mausam, 35(3): pp. 291-294.

Ram Mohan, H.S., Valsala, P. and B.V. Appa Rao 1983. Drought spells over north-central India during the 1979 south-west monsoon, Mausam, 34(3): pp. 299-302.

Ramakrishna, Y.S., Sastri, A.S.R.A.S., Rao, G.G.S.N. 1985. Water balance and crop planning. A case study of Western Rajasthan. Annals of Arid Zone. 24(2):114-119.

Ramakrishna, Y.S., Sastri, A.S.R.A.S., Rao, G.S.S.N. and Ramana Rao, B.V. 1984. Prediction of drought in the Indian Geographical J of India. 26(1-2):44-47.

Raman, P.K. and Satakopan, V. 1934. Evaporation in India calculated from other meteorological factors, India Meteorological Department, Scientific Notes 61:1-51.

Ramana Rao, B.V., A.S.R.A.S. Sastri and Ramakrishna, Y.S. 1984. Variation in the moisture regime of the Indian arid zone. *The National Geographical Journal of India*, 30(2): pp. 73-80.

Ramana Rao, Sastri, A.S.R.A.S. and Ramakrishnana, Y.S. 1984. Variation in the moisture regime of the Indian Aril. Zone. *The Natural of Geographical J. of India*. 30(2):73-80.

Ramdas, L.A. 1960. Crops and Weather in India. Indian Council of Agricultural Research, New Delhi, pp. 127.

Rao, K.N., George, C.J., Moray, P.E. and Mehta, N.K. 1971. Special analysis of drought index (palmer) for India IMD pre-published Scientific Report No. 169.

Sarkar, R.P., Biswas, B.C., 1986. A New Approach to Agroclimatic classification to Find out Crop Potential. *Mausam Journal* (1986), 57, 1, 27-38. India.

Shastri, A.S.R.A.S. and Ramakrishna, Y.S. 1980. A modified scheme of drought classification applicable to the arid zone of Western Rajasthan. Annals of Arid Zone. 19(1&2):65-77.

Shastri, A.S.R.A.S. and Ramakrishna, Y.S. 1980. A modified scheme of drought classification applicable to the arid zone of western Rajasthan. Annals of Arid Zone. 19(1&2) 65-72.

Savoboda, M.D. 2000. Development of a climate monitoring to the drought monitor. In: Proceedings of the Central and Eastern European Workshop on Drought Mitigation. Budapest, pp. 205-212.

Sen, S.R. 1970. Drought in India certain dimensional conditions in growth and instability Indian Agriculture. Pp. 29-35.

Sharma, H.C., Chauhan, H.S. and Sewa Rama 1979. Probability analysis of rainfall for crop planning, *Journal of Agricultural Engineering*, ISAE, 16(3): pp. 87-94.

Singh, R.P. and Ramana Rao, B.V. 1988. Agricultural drought management in India (Principles & Practices), Central Research Institute for Dryland Agriculture, Hyderabad.

Sly, W.K. 1970. A climatic moisture index for land and soil classification in Canada. Ca. *J. Soil. Sci.* 50:2911-301.

Subrahmanyam, V.P. and Subarahmaniam, A.R. 1964. Application of water balance concepts for the climatic study of drought in south India, Annals of Arid Zone, 4(1): pp. 46-55.

Subrahmanyam, V.P. 1964. Climatic water balances of the Indian arid zone. Proceedings of Symposium on Problems of Indian Arid Zone, Jodhpur, pp. 405-411.

Subrahmanyam, V.P. 1967. Incidence and spread of continental drought, WMO, IHD reports on WMO/IHD projects. Rep. No. 2, pp. 51.

Subrahmanyam, V.P. 1969. Some aspects of drought climatology of the dry sub-humid zones of south India. *Journal of Meteorological Society (Japan)*; 47(4):239-244.

Swaminathan, M.S. 2005. Report of the Committee to review the Coastal Regulation Zone Notification, 1991, February 2005, 122 pages pdf file http:// mpcb.mah.in/mczma/swaminathan-1pdf.

Thapliyal, V. 1984. Prediction of Indian droughts with lower stratospheric spheric winds, Mausam, 35(3): pp. 367-374.

Thornthwaite, C.W. 1947. Climate and moisture conservation. Ann. Ass. Am. Geogr. 37:87-100.

Thornthwaite, C.W. 1948. An approach toward a rational classification of climate. Geogr. Rev. 38:85-94.

Thornthwaite, C.W. and Mather, J.R. 1955. The water balance climate, 8(1), Dreseal Inst. of Tech., New Jersey, U.S.A., p. 104.

Troll, C. 1965. Seasonal climates of the earth. In World maps of Climatology. (Eds) E Rodenwalt and H Jusatz Springer Verlag, Berlin, pp.28.

Webster, P.J., Holland, G.J., Curry, J.A. and Chang, H.R. 2005. Changes in Tropical Cyclone Number, Duration, and Intensity in a Warming Environment, Science, 309 1844-1846.

Wilhite, D.A. and Svoboda, M.D. 2000. Drought early warning systems in the context of drought preparedness and mitigation. In: Early warning systems for drought preparedness and drought management. W.M.O. Geneva, pp. 1-16.

Yevjevich, V. 1967. Objective approach to definitions and investigation of continental hydrological droughts, Hydrology Papers, Colorado State University, Fort Collins, Colorado, pp. 1-18.

EXERCISES

4.1. Define drought.

4.2. Classify drought.

4.3. Differentiate the following:
   - (i) Agricultural and meteorological drought
   - (ii) Apparent and permanent drought
   - (iii) Soil and contingent drought
   - (iv) Soil and atmospheric drought
   - (v) Drought and aridity

4.4. Classify the drought based on the following criteria:
   - (i) Thornthwaite (1947)
   - (ii) Singh (1989)

4.5 Discuss the causes of climate change.

4.6 Describe reasons of drought.

4.7 Describe the drought indices.

4.9 What is the Palmer drought indices?

4.10 Explain the normalized difference vegetation index.

4.11 What is the aridity index?

4.12 Explain the dryness index.

4.13 Describe the drought monitoring indices.

4.14 What is the criteria for declaration of drought by the State Government?

4.15 Explain rainy day and dryspells.

4.16 Discuss the prediction of drought.

4.17 Describe the strategies for drought management.

■■■■

Chapter 5

CLIMATE, WEATHER FORECASTING AND AGRO-ADVISORY SERVICES

5.1 Introduction

Crop growth and development are mainly governed by environmental conditions, *viz.*, climate and soil. Weather decides success or failure of agricultural crop production. Weather manifests its influence on agricultural operations and farm production through its effects on soil and plant growth. Agricultural operations can be advanced or delayed with the help of advanced weather forecast from three to ten days also with agronomic practices to mitigate the adverse effects of weather. Weather has significant influence on each agricultural operation from preparatory tillage to harvesting and storage. A sound knowledge of the climatic factors is prerequisite for successful agriculture. An agricultural relevant forecast is not only useful for efficient management of farm inputs but also leads to precise impact assessment (Gadgil, 1989).

Weather is a part of climate or condition of atmosphere at a given place and at a given time. It is daily variations in lower layers of the atmosphere. Weather pertains to smaller area like village, city or a district for smaller duration of time *viz.*, part of a day or complete day, and is expressed by numerical values of meteorological parameters. Some examples are hot day, rainy day, cloudy weather, dry weather etc. whereas, climate is a generalised weather or summation of weather conditions over a given region comparatively for a longer period. It is weather conditions related to larger areas like zone, state, country, part of continent or whole of continent for longer duration of period like month, season or year and best described by the normal and averages, *viz.*, cold season, sub tropical, tropical climate etc.

Climate is one of the key components influencing agricultural production in Indian small holder systems, which accounts for two-thirds of variation in production.

India is a vast country having different type of physiographic conditions, hence, climatic conditions differ from place to place. The central part of the country experiences continental type of climate while coastal part has equitable type of climate around the year. Among all the natural resources, climate is a key component for all around development of the any region. It plays a decisive role on the type of farming system and other allied agricultural practices. Weather influences agricultural production and accounts for two-third of the variation in the yield of a crop. In spite of improved varieties and cultivation practices, favourable weather is must for good harvest (Rao *et al.*, 1999).

Meteorology is the science that deals with laws and principles as they apply to atmospheric phenomenon. It is also known as science which deals with water atmosphere. It is referred as a science of atmosphere and its attendant activities. It is also defined as a branch of physics that deals with physical processes in the atmosphere that produce weather. Agricultural meteorology is application of meteorology which deals with the response of crops to the physical environment. In this physical process of the atmosphere that produces weather in relation to agricultural production, and also deals with agroclimatology, instrumentation and weather forecasting. Climatology is the science which deals with the factors which determine and control the distribution of climate over the earth's surface. Agroclimatology deals with the relationship between climatic regimes and agricultural production and includes instrumentation and weather forecasting.

5.2 Classification of Atmosphere

Atmosphere consists of colourless, tasteless and odourless mixture of gases that surround the earth. It extends up to a height of about 1,600 kilometres. It is having 99 per cent of total mass of the atmosphere within 40 kilometres from the earth. The gaseous envelop of invisible film of air surrounding the earth is known as atmosphere. There is no definite upper boundary layer but fades away gradually into interplanetary space. With increasing altitude it becomes lesser dense until there is no longer any detectable amount of air left. Most of living organisms are present at the bottom of the atmosphere. The mechanical composition of different gases in the atmosphere does not react with each other. These gases have weight and exert pressure. These gases are compressible.

On an average pure air consists of 75.5 per cent nitrogen, 2.31 per cent oxygen, 1.28 per cent argon, and 0.045 per cent carbon dioxide by weight (Table 5.1). Very minor amounts of neon, helium, methane, krypton, nitrous oxide, hydrogen, ozone, xenon etc. comprise the remaining fractional part. The carbon dioxide content in atmosphere is less. However, it is an important constituent with a considerable influence on the origin of life on earth as it is vital ingredient in photosynthesis.

While, the lower part of the atmosphere contains water vapour ranging from 0 to 4 per cent by volume. Most of the water vapour content of the atmosphere is derived from evaporation from water bodies. It also contains impurities like dust particles, salt crystals, bacteria, pollen etc.

Table 5.1: Composition of atmosphere

Gases	Volume (%)	Weight (%)
Nitrogen	78.088	75.527
Oxygen	20.948	23.143
Argon	0.930	1.282
Carbon dioxide	0.033	0.045
Other gases	Traces	Traces

5.2.1 Atmosphere structure

Atmosphere is classified based on vertical temperature differences, into four major layers or strata (Fig. 5.1). However, the boundaries between them are never sharp, more or less indistinct with gases mixing upwards and downwards cross boundaries.

(i) Troposphere

It is the lower layer of the atmosphere extending up to a height of 8 to 18 km from the earth depending on latitude. It is thicker at the equator than at poles. The layer is the densest layer of the atmosphere having 85 per cent of the atmosphere's mass. It also affects all weather phenomena like clouds, fog, dew, mist, rain etc. that occur in this layer. The most important characteristic of this layer is that the temperature decreases with increase in altitude. The temperature at the boundary of troposphere is roughly –60°C.

Tropospause

It is a thin layer of transition that separates the troposphere from the above lying stratosphere. Tropopause is a zone of transition from one set of physical phenomenon to the other. There is no change in temperature with height in this transition zone.

(ii) Stratosphere

It is a layer of atmosphere which lies above the tropopause. This layer lies above the height of 8 to 18 km extending up to 50 km depending on latitude and it is dust free cloudless and warmest layer. The most of the photochemical reactions in air takes place in this layer. Temperature increases with increase in altitude up to about 50 km in this layer. At this height the air, although very thin, is nearly as warm as the air at sea level. This layer has 15 per cent of the atmosphere's mass. Therefore, the mass of the air up to 40 km altitude is more than 99 per cent of the total mass of the atmosphere. The air in the stratosphere layer is thin and highly rarefied as compared to air at ground level.

Though, this layer having low density, the stratosphere absorbs heat from the sun in the form ultraviolet radiation. It does so since oxygen seeping up from the troposphere below absorbs energy from the sun. The solar radiation breaks the two atoms of oxygen in a molecule and leaving each one free to link up with another molecule of diatomic oxygen. It results into formation of two molecules of ozone. The term ozone layer or, ozonosphere is synonymous with stratosphere. Therefore in the process of formation of ozone layer, the harmful ultraviolet radiation is filtered which otherwise makes life impossible on the earth. This is due to that the stratosphere is warmer at higher altitude; hot air no longer rises there. Since the air above is hotter than the air below, convection cannot occur. This layer acts as a lid on the troposphere keeping convection and weather, below the tropopause.

(iii) Mesosphere

It lies above the stratosphere. The stratosphere and mesosphere are separated by a narrow transitional layer called stratopause. In this layer, there is a strong decrease in temperature with increase in altitude up to about 80 km, warms up again in thermosphere.

(iv) Theremosphere

It is separated by mesopause from the mesosphere. This layer is the outermost layer, extending from the top of the mesosphere (Fig 5.1) and is characterised by a steadily increasing temperature with altitude. The lower layer of thermosphere is called ionosphere. Long distance radio communication is made possible through this ionised layer.

Atmosphere has tremendous influence on climate. Generally, the changes that take place in the lower layers of the atmosphere are called as the climatic changes.

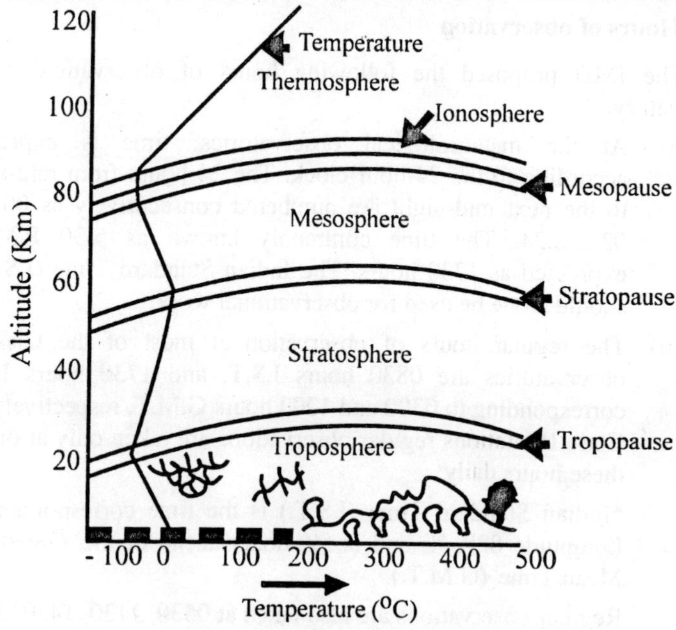

Fig. 5.1: Structure of atmosphere

5.3 Meteorological Elements and Hours of Observation

The meteorological elements are generally recorded in Class-III observatories and categories as per recommendation of the IMD:

(i) Barometric pressure: the pressure of the air that is measured with the barometer.

(ii) Dry bulb temperature: the temperature of the air as measured by the dry bulb thermometer in the Thermometer Screen at the time of observation.

(iii) Wet bulb temperature: which gives in conjunction with the dry bulb temperature, the humidity of the air inside the Thermometer Screen and its dew point temperature.

(iv) Maximum temperature: the highest temperature of the air indicated by the maximum thermometer in the thermometer screen since its last setting.

(v) Minimum temperature: the lowest temperature of the air indicated by the minimum thermometer in the Thermometer Screen since its last setting.

(vi) Amount of rainfall between successive observations.

5.3.1 Hours of observation

The IMD proposed the following hours of observation in the observatory:

(i) At the meteorological observatories, time is expressed according to the 24-hour clock. The 24 hours from mid-night to the next mid-night are numbered consecutively as 00, 01, 02,......24. The time commonly known as 5-30 P.M. is expressed as 1730 hours. The Indian Standard Time (I.S.T.*) should alone be used for observational work.

(ii) The regular hours of observation at most of the Class II observatories are 0830 hours I.S.T. and 1730 hours I.S.T. corresponding to 0300 and 1200 hours G.M.T. respectively. At Class III stations regular observations are taken only at one of these hours daily.

*Indian Standard Time (I.S.T.) is the time corresponding to Longitude 82½ ºE and is 5½ hours ahead of the 'Greenwich Mean Time' (G.M.T.).

Regular observations are also taken at 0530, 1130, 1430, 2030, 2330 and 0230 hours I.S.T. (0000, 0600, 0900, 1500, 1800 and 2100 hours G.M.T. respectively) at Class I observatories and some Class II observatories manned by Departmental Observers

The expressions 'morning observations' and 'afternoon observations' refer to the observations taken at 0830 hrs. I.S.T. (0300 G.M.T.) and 1730 hrs. I.S.T. (1200 hrs G.M.T.) respectively.

(iii) Extra observations may be requisitioned at any hour by various forecasting centres. The standard of time adopted for reporting these special observations is always the Greenwich Mean Time (G.M.T.) which is 5½ hours less than the corresponding Indian Standard Time (I.S.T.)

5.3.2 Weather elements

The state of weather at any particular time and at particular place comprises of the following elements viz., (1) solar radiation (2) air temperature (3) atmospheric pressure, (4) wind movement, (5) relative humidity, (6) preparation and (7) clouds.

5.3.2.1 Solar radiation

Solar radiation is the source of energy for all the physical processes taking place in the atmosphere which accounts for 99% at earth surface from sun. The small quantities of energy are available from different sources (Table 5.2).

Table 5.2: Thermal energy sources at earth surface relative to average solar energy of 263 kilo Langley per year

	Sources	Energy
(1)	One fourth solar constant	1
(2)	Heat flux from interior of the earth	18×10^{-5}
(3)	Radian from full moon	3×10^{-5}
(4)	Sun's radiation reflected from the moon	1×10^{-5}
(5)	Energy from lightning discharge	6×10^{-7}
(6)	Total radiation from stars	4×10^{-8}

(Source: Sellers, 1967)

Radiation mainly depends upon temperature. This energy also acts as driving the process of photosynthesis, evaporation, heating the soil air and hydrologic cycle. It is well known that all objects with temperature emit radiation. The sun and earth are having temperature of 6000°C and 273°K, respectively. Almost a constant amount of solar radiation (1.94 cal/cm^2/min) is emitted by the sun continuously. It is also called as solar constant, and is referred as the energy falling in one minute on a surface area of one square centimetre at the outer boundary of the atmosphere held normal to the sunlight, at the mean distance of the earth from the sun. It comes through the space without any change or loss. When it enters the atmosphere then changes take place due to absorption by atmosphere, reflection by clouds and scattering by dust particles. It is received in the form of electromagnetic waves and comprise of stream or flow of particles. These particles are called as quanta or photons if the energy is spread uniformly over the full surface of the earth, the amount received per unit area per unit time is determined as,

$$Qs = \frac{\pi r^2 S}{4\pi r^2} = \frac{S}{4} = 0.5\, ly\ \text{min}^{-1} \quad \text{....(5.1)}$$

$$= 263 \text{ kly year}^{-1}$$

Radiation from the sun is of short wave while from the earth is of long wave. This exchange is critical in hydrology.

Forms of solar radiation

Solar radiation is mainly classified into the following forms (i) Direct solar radiation (ii) Diffused solar radiation and (iii) Global solar radiation.

Direct solar radiation: It is the radiation received directly from the sun by a surface normal to the incident radiation.

Diffused radiation: It scattered by the suspended particles in the atmosphere depends on latitude, season and cloudiness. In high latitudes, diffused radiation is a very important source of solar radiation. It forms a substantial part of solar radiation during winter months. Cloudiness also increases the ratio of diffused to direct radiation. Before sunrise and sunset, all the energy received is diffuse or sky radiation. Diffuse radiation penetrates into plant canopies more effectively than does the direct solar radiation. The spectral distribution of radiation is shown in Table 5.3. Diffused radiation is having 65 per cent photosynthetically active radiation (PAR) as compared to direct radiation (45 per cent).

Table 5.3: Different forms of solar radiation

Type of radiation	Per cent of total radiation		
	Ultra violet	**Photosynthetically active radiation**	**Near infrared**
Direct radiation	0.02	0.42	0.56
Diffused radiation	0.10	0.65	0.25
Global radiation	0.03	0.50	0.47

Global solar radiation: It is the sum total of direct and diffused solar radiation received on a horizontal surface from the sun directly and from the sky as scattered radiation. Solar radiation that is reflected without any change in its quality is called reflected solar radiation or albedo. Clouds are very effective reflectors as little solar radiation reaches the earth's surface on a cloudy day. Snow is very effective reflector, particularly when it is fresh. Water surface and sea are poor reflectors and thus serve as a good sink for solar energy whereas rocks, sand, soil and vegetation reflect 10 to 30 per cent of the incident solar radiation. Reflected solar radiation is very important for prediction of drought and its management.

Thermal radiation: It is emitted by the earth and is called terrestrial radiation. Actually the earth absorbs solar radiation which is in short waves and emits back in the form of long waves. Terrestrial radiation heats the atmosphere. The radiation balance between global and reflected solar radiation is called net radiation.

Solar radiation is measured in units of watts per square metre. The CGS unit is most commonly used in meteorology. One watt energy is equivalent to one joule per second.

1 watt = 1 joule/s

697.93 W/m^2 = 1 cal/cm^2/min

10000 lux = 3.47 Mega joules

Calorie = 4.7868 joules

The ratio of amount of solar radiation falling on the earth's surface to the amount of radiation reflected by surface varies from 0.08 to 0.3 for soil, 0.05 to 0.18 for forest, 0.05 to 0.55 for water and is 0.78 for snow.

The efficiency of solar radiation utilization value in different crops is given in Table 5.4.

Table 5.4: Drymatter production and efficiency of radiation utilization for several crops

Crops	Yield (t/ha)	Efficiency (%)
Potato	9.6	0.5
Winter wheat	10.45	0.52
Sugarbeat	16.0	0.92
Carrot	6.86	0.39
Turnips	3.60	0.51
Corn	15.52	1.05
Sugarcane	129.48	1.43

(Source: Chang, 1968)

Pyranometer is used for measurement of total incoming radiation. There are different types of pyranometers manufactured by different firms. Bellani type pyranometer comprising of a long narrow glass tube with graduation. The top end of the glass tube is attached to a spherical glass which is coated with metal outside and contains alcohol inside. The spherical glass is coated with metal outside and contains alcohol inside. The spherical glass is enclosed in a glass dome which allows only short wave radiation. When solar radiation falls on a metal coated sphérical glass, it absorbs radiation and heats up. The alcohol present in the spherical glass evaporates and escapes into the narrow tube where it condenses. The amount of alcohol collected in the tube indicates the amount of solar radiation. Direct solar radiation is measured by Phrhelio meter.

Solar radiation: The solar radiation may be measured by the Bellami spherical pyranometer installed as shown in Fig. 5.2. It consists essentially of a glass sphere filled with alcohol, which when warmed by solar radiation falling on the sphere distils into a graduated glass tube below. The radiation received during a given period is directly read on the graudeated tube at 12.30 IST and half an hour after sunset and reset both the times.

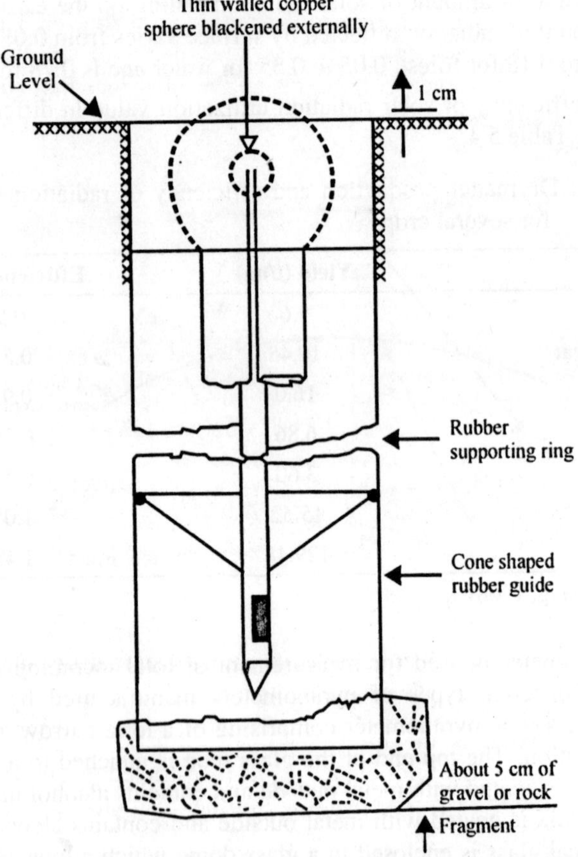

Fig. 5.2: Gunn-Bellani radiation integration

5.3.2.2 Air temperature

It is the temperature of air and is defined as the degree of hotness or coldness of a day. Solar radiation is the main source for air temperature. It is measured by thermometer. Solar radiation received at the outer limits of the earth's atmosphere is quite constant. However, the amount that reaches the earth's surface is highly variable depending on the amount of the clouds and is lost to space. In the absence of colour most of the solar radiation passes directly through the atmosphere and reaches the surface. Some solar radiation is scattered in the atmosphere by molecules and by minute particles of solid matter. Of this scattered radiation some is lost to space. When cloudiness is average, the absorption of solar radiation by earth's surface is about 43 per cent and by atmosphere about 57 per cent. About 35 per cent is reflected back (Fig 5.3) into space.

The reflected solar radiation is unchanged in character. However, solar radiation absorbed either by the atmosphere or by the earth, converted into thermal energy. It warms up the substance that absorbs it is reradiated as long waves. The air temperature is influenced by altitudes and cloudiness.

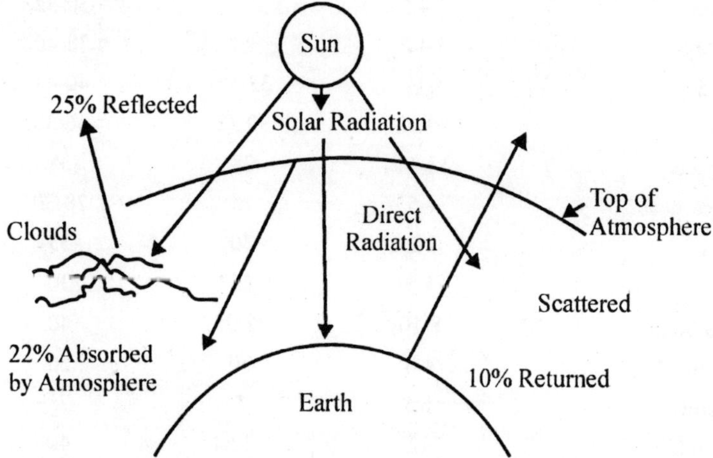

Fig. 5.3: Approximate distribution of incoming solar radiation during average cloudiness

Air temperature is the most important climatic variables which affect plant life. The growth of higher plants is restricted to a temperature between 0°C and 60°C and crop plants are further restricted to a narrower range of 10 to 40°C.

Cardinal temperature: Three temperatures of plants have been recognized and are often known as cardinal temperature.

(i) A minimum temperature below which no growth occurs. For cool season crop, it ranges from 0 to 5°C while for hot season crops between 15 and 18°C.

(ii) An optimum temperature at which plant growth occurs. For cool season crop, it ranges from 25 to 31°C while for hot season crops from 31 to 37°C.

(iii) A maximum temperature above which the plant growth stops. For cool season crop, it ranges from 31 to 37°C while for hot season crops from 44 to 50°C.

The cardinal temperature for some of the plants for germination are given in Table 5.5.

Table 5.5: Cardinal temperatures for the germination of some important crops

Crops	Cardinal temperature (°C)		
	Minimum	**Optimum**	**Maximum**
Wheat	3-4.5	25	30-32
Barley	3-4.5	20	38-40
Maize	8-10	32-35	40-44
Rice	10-12	30-32	36-38
Tobacco	13-14	28	35
Sugar beat	4-5	25	28-30
Peas	1-2	30	35
Oats	4-5	25	30
Sorghum	8-10	32-35	40
Lentils	4-5	30	36
Carrot	4-5	8	25
Pumpkin	12	32-34	40

(Source: Bierhuizen, 1973)

Maximum thermometer is used to measure the highest temperature attained during the day. It consists of a graduated glass tube tapered into a bulb. The glass tube has a narrow bore through the length of bulb which is filled with mercury. When temperature increases, mercury rises with increase in temperature. When temperature falls, mercury above the constriction cannot enter the bulb because the force is not sufficient to push mercury into the bulb. Thus, maximum temperature attained during the day can be recorded by maximum thermometer. Maximum thermometer is set in the morning when the temperature is low. The setting is done by whirling the thermometer so that mercury above the constriction can be forced into the bulb.

Minimum thermometer is used to measure the lowest temperature of a day. It consists of graduated glass tube with alcohol in the bulb. A dumb-bell shaped index is present inside the alcohol in the narrow bore. When the temperature rises, alcohol expands, flows over the dumb-bell shaped index. As the temperature falls, alcohol contracts and pulls back the index. When the temperature rises again, alcohol flows over the index without moving it. Minimum thermometer is set in the evening by inverting the thermometer which brings the index to the end of the alcohol in the bore. The air temperature is expressed in terms of °C, °F and K. Following relation is used for conversion of temperature in different units of temperature.

$$C = (F-32)\ 5/9$$

$$F = C\ (9/5) + 32$$

$$K = C + 273$$

Thermograph is an automatic device to record air temperature continuously. It comprising of a metallic element whose expansion varies with temperature. One end of the curved sensitive elements is connected to a long movable lever arm which contains an inked pan at its end. The arm rests on a cylindrical drum which rotates by means of an inner clock-work. A sheet of graph paper is wrapped around the drum and is divided into days and hours horizontally and in degrees vertically. Changes in temperature cause variations in curvature. The pen inscribes an inked line on the sheet surrounding the drum. As it does so, the drum is slowly rotating by clock-work mechanism and a line is traced which indicates the temperature at any time.

The relevant temperatures are measured in meteorological observatory and marked on maps for certain period. Then the points having the same temperatures are joined and known as *isotherms*.

The following thermometers are used for measuring temperature:

(i) Dry-bulb Thermometer- For obtaining temperature of the surrounding air.

(ii) Wet-bulb Thermometer- This helps to find out the relative humidity of the surrounding air.

(iii) Maximum Thermometer- To indicate the highest temperature reached, since the time of its last setting.

(iv) Minimum Thermometer- To indicate the lowest temperature reached, since the time of its last setting.

Grass minimum thermometer

The grass minimum or terrestrial radiation thermometer is used mainly to obtain information about 'ground frost' at night. It is a sheathen minimum thermometer in which the graduations of the stem are protected by an outer glass jacket. The bulb is link-shaped and provides a larger surface for exposure than a spherical bulb. This instrument is installed in the Thermometer Screen enclosure on a plot covered with short grass 2.5 to 5.0 cm. high. It is kept on two Y-shaped wooden supports fixed in the ground to a depth of 12 cm. with the bulb just touching the tips of the blades of grass.

5.3.2.3 Soil temperature

Soil temperature is an important environmental factor in plant growth and distribution. It is more pronounced as compared to air temperature due to the varying characteristics of composition of the soil. It is more important in plant life than air temperature. Extremely low temperature are equally determined low temperatures inspede the intake of nutrients and stops at 1°C while, excessively high temperatures are also harmful to the roots and also cause lesions on the stem.

The surface of the earth plays an important role in meteorology, by absorbing solar radiation and warming the air as well as the soil layers adjacent to it. About 40 per cent of the incoming solar radiation is absorbed by earth's surface. The heat absorbed from the sun, is however, accumulated in the upper few centimeters of the earth's surface owing to its very low conducting power. As a result of this, the diurnal variation of temperature is maximum at the surface of the earth and decreases rapidly downwards becoming almost negligible at a depth of 30 cm. For germination of seed different soil temperature is required for different crops. The shallow depth-soil thermometer is shown in Fig. 5.4.

When the earth surface absorbs solar radiation, it not only heats up the air above but also the lower layers of the soil. The heat transfer into the soil is mainly by conduction and to a lesser extent by convection. The heat flow into or out of the soil is called soil heat flux. The rate of heat flux into soil depends on temperature gradient and thermal conductivity. Thermal conductivity is defined as the quantity of heat flowing in unit time through a unit cross-section of soil in response to a specified temperature gradient. It has units of Watts/m depth of soil/°K temperature gradient.

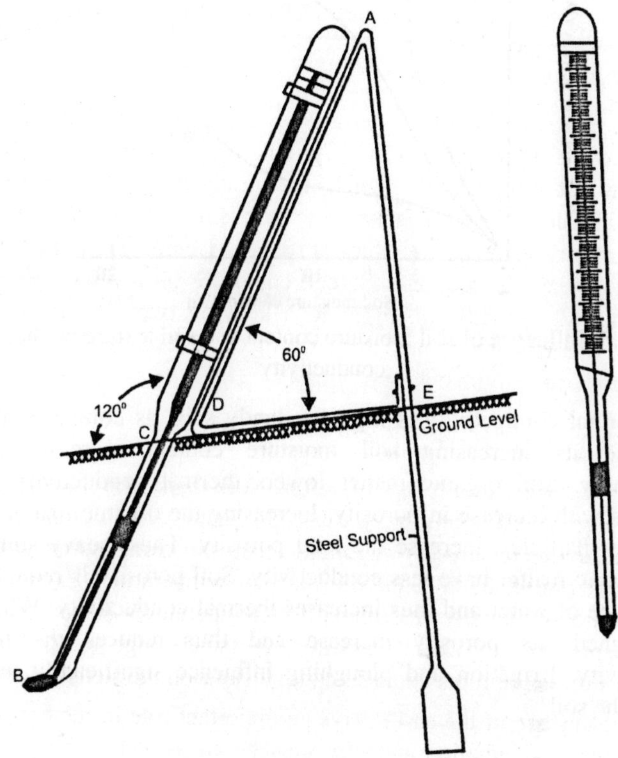

Fig. 5.4: 120° Bend soil thermometer

Soil temperature is measured by soil thermometer. It is having mercury, therefore, with a bend at the bulb end. The portion of the glass tube between bent ends of the glass tube to the beginning of graduation is called stem. The stem length varies depending on the depth at which temperature is to be measured. The lower end to the thermometer is bent so as to have maximum content with the soil at that particular depth. Soil thermometer is buried in the soil up to its stem inside the soil while graduated tube is outside. When the soil temperature increases, mercury expands and rises in the graduated tube and measured.

The thermal conductivity or movement of temperature into the soil depends on texture, moisture and organic matter content of the soil (Fig. 5.5).

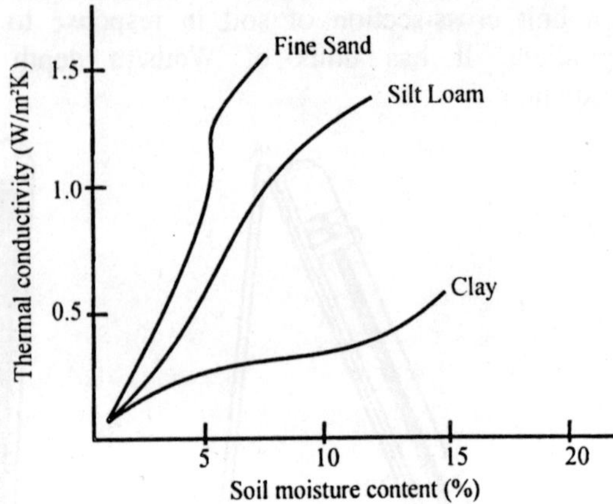

Fig. 5.5: Influence of soil moisture content and soil texture on thermal conductivity

Thermal conductivity is high for sandy soils as compared to clay soils and its increasing soil moisture content increase thermal conductivity. Soil organic matter lowers thermal conductivity and it decreases with increase in porosity. Increasing the organic matter which is in finer particles, increase the total porosity. Thus, heavy soils with high organic matter have less conductivity. Soil porosity is reduced due to presence of water and thus increases thermal conductivity. When soil is ploughed, its porosity increase and thus reduces the thermal conductivity. Irrigation and ploughing influence significantly the heat flow in the soil.

5.3.2.4 Atmospheric pressure

The pressure exerted by the atmosphere on the earth's surface is called atmospheric pressure. Atmosphere contains air molecules which have a certain weight. This results into atmosphere pressure on the earth's surface. Pressure is defined as the force per unit area. The atmospheric pressure is defined as pressure exerted by a column of air with a cross–sectional area of a given unit extending from the earth's surface to the upper-most boundary of the atmosphere.

Generally, increase in air temperature, decreases the atmospheric pressure of an area and low temperature results into high pressure. In the lower layers of atmosphere, the atmospheric pressure is more as compared to other upper layers rest on it. Increasing in elevation decreases the atmospheric pressure. Atmospheric pressure does not have direct influence on crop growth. It is, however an important parameter in weather forecasting. It is measured by the standard instrument known as aneroid barometer and barograph. Aneroid barometer consists of cylindrical chamber with corrugated tops and sides known as sylphon chamber. A strong spring within the chamber prevents it from collapsing under the air pressure but allows expansion of contraction vertically. As the outside pressure changes, the chamber either expands or contracts which is magnified by means of levers and pulleys and conveyed to the pointer. The pointer swings around the dial that is calibrated in atmospheric pressure are reflected on a vacuum chamber either by contraction or expansion. These changes are magnified by a system of levers and communicated to a long pen-arm which marks on a graph paper. The graph sheet is wrapped around a drum which rotates through clock-work mechanism. The graph sheet is calibrated vertically in pressure units and horizontally in time. The atmospheric pressure is expressed in millibars.

Pressure

The pressure of the atmosphere at any point is the weight of the air column which stands vertically above unit area with the point as its centre. For meteorological purposes, atmospheric pressure is usually measured by means of a mercury barometer where the height of the mercury column represents the atmospheric pressure. The C.G.S. unit of pressure is the dyne per square centimeter but this is inconveniently small. The unit employed for reporting pressure for meteorological purposes, is, therefore, the millibar which is defined as equal to 1000 dynes/sq. cm.

Its essential parts are (i) A glass tube about 90 cm. long closed at the top and open below (ii) A cup or cistern and (iii) a brass scale. The glass tube is filled with mercury and its open end is dipped in the mercury in the cistern, which prevents air from entering the tube. Above the mercury column in the tube is an empty space and great care as taken to remove all air from this space as its presence even in every minute quantities, will vitiate the reading of the instrument. The mercury column in the tube is supported by the pressure of the air on the surface of the mercury in the cistern. This forms the basic principle on which a barometer is constructed. As the mercury in the barometer tube rises or falls due to changes in the atmospheric pressure, the mercury level in the cistern changes in the opposite direction. This change of level in the cistern is taken into account in the graduation of the scale itself.

5.3.2.5 Wind

Wind is defined as air in motion and is expressed in terms of direction and speed. It serves as a medium for transfer of moisture heat to and from the surfaces with which it comes in contact and therefore, it influences evaporation and precipitation considerably wind is a rector having both speed and direction.

Wind direction is regarded as the direction from which it blows and speed as the rate of movement of air in its instantaneous direction.

Wind direction is determined with reference to true north and is expressed to the nearest 10 degrees or to 16 points of the compass (Fig. 5.6). Wind speed is measured in knots (nautical miles per hour) or in kilometres per hour.

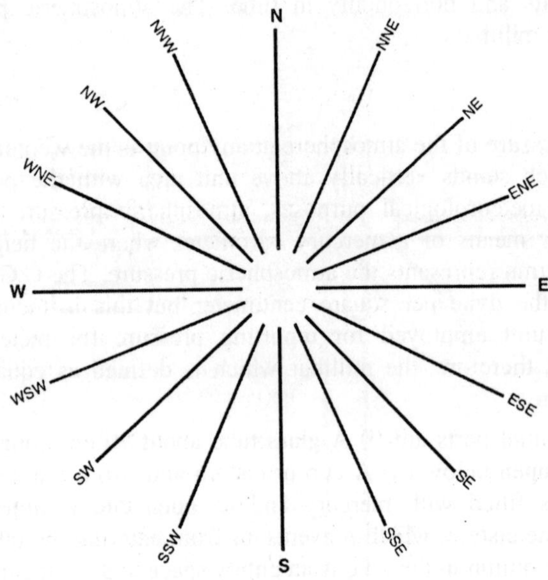

Fig. 5.6: Points of the compass

Wind direction

Air from high pressure areas rushes towards low pressure areas causing horizontal movements of wind. Wind direction is the direction from which it flows. For example, the wind flowing from south to north is called south-ward wind. Windward refers to the direction from which the wind comes and leeward refers to the direction towards which it blows. Wind direction is most commonly indicated by wind vane.

The wind direction is given by an instrument called the Windvane (Fig. 5.7). It is a balanced lever which turns freely about a vertical axis. In the most common type, one end of the lever exposes a broad surface to the wind, whilst the other end is narrow and points to the direction from which the wind blows. Under this moveable system, is fixed a rigid cross, the arms of which are set to the four cardinal directions – North, East, South and West. Some windvanes are provided with eight direction indicators, N, NE, E etc.

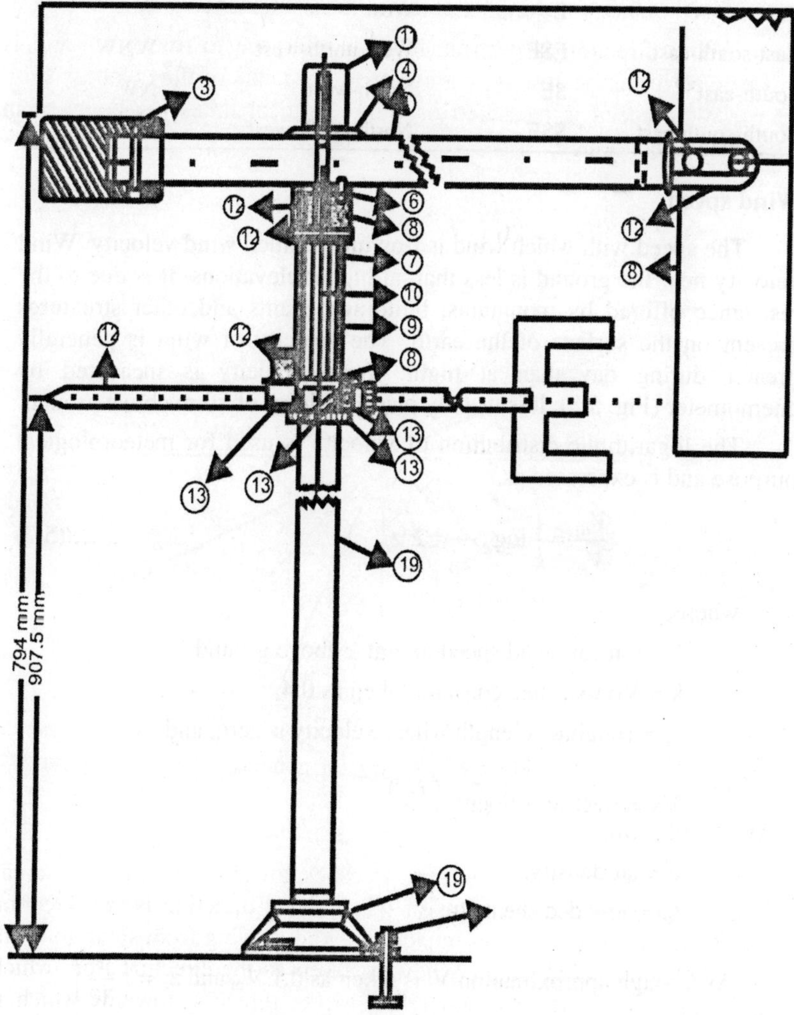

Fig. 5.7: Windvane MK II – Details

Table 5.6 Direction and code for wind direction

Direction	Code	Direction	Code
North	N	South	S
North-north-east	NNE	South-south-west	SSW
North-east	NE	South-west	SW
East-north-east	ENE	West-south-west	WSW
East	E	West	W
East-south-east	ESE	West-north-west	WNW
South-east	SE	North-west	NW
South-south-east	SSE	North-north-west	NNW

Wind speed

The speed with which wind is flowing is called wind velocity. Wind velocity near the ground is less than at higher elevations. It is due to the resistance offered by mountains, buildings, plants and other structures present on the surface of the earth. The velocity of wind is generally greater during day than at night. Wind velocity is measured by anemometer (Fig. 5.8). It is usually placed at level of 10 m above ground.

The logarithmic distribution of velocity is used for meteorological purpose and is expressed as,

$$\frac{V_m}{V_o} = \frac{1}{k} \log_e \frac{z}{z_o} \; z \geq z_o \qquad \text{....(5.2)}$$

where,

V_m = mean wind speed height ≥ above ground,

k = Von-karman constant taken as 0.4,

z_o = roughness length where velocity is zero, and

Vx = traction velocity $\left(\frac{t_o}{e}\right)^{1/2}$

e = air density,

t_o = bounded shear stress

As a rough approximation V_t is taken as 0.1 V_m and $z_o = \left(\frac{1}{30}\right)^z$

The Wind speed is measured by an instrument called the Anemometer (Fig. 5.8). The instrument consists of three large semi-conical cups with beaded edges fixed at the ends of three rods. The cups are mounted symmetrically about a vertical axis so that the diameter plans of each cup is vertical. As the force on the concave side of any cup, due to the wind is greater than that on a convent side in a similar position, the cup wheel rotates.

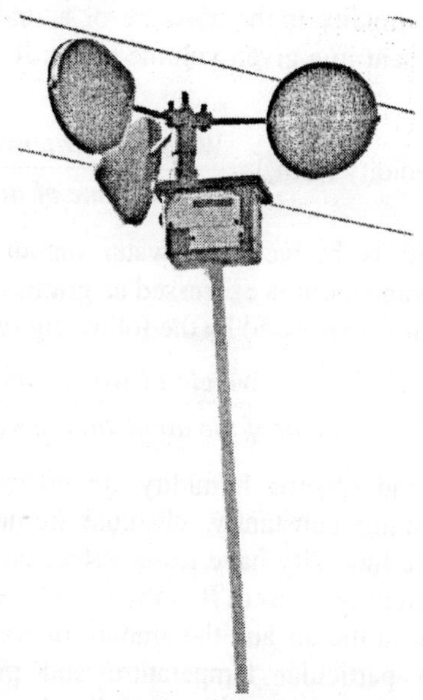

Fig. 5.8: Cup counter anemometer MK II

Wind speed in horizontal direction is the travel of wind from a high pressure area to a low pressure area. Due to difference in surface conditions of the earth and angle of incidence of sun's rays, different amounts of solar radiation are absorbed at different areas. This results in high and low temperature areas thereby creating high and low pressure areas.

5.3.2.6 Humidity

Humidity is the water vapour content of the atmosphere. Liquid water is converted into water vapour by evaporation for which necessary energy is provided by solar radiation in the form of temperature. Water vapour is colourless, odourless and tasteless and consists of minute droplets of water suspended in the air. The amount of water vapour in the atmosphere depends upon wind and temperature. Higher the temperature, more is the amount of water vapour that can be held by the atmosphere. The air is said to be saturated when it held maximum amount of water vapour at a particular temperature. It the temperature rises, the atmosphere becomes unsaturated as it can accommodate some more water vapour.

The quantity of water vapour held in the atmosphere at any particular time can be expressed as absolute, specific or relative humidity. Absolute humidity is the absolute or actual quantity of water vapour by weight present in a given volume of air. It is expressed by the following relation.

$$\text{Absolute humidity (g/m}^3\text{)} = \frac{\textit{Weight of water vapour}}{\textit{Volume of air}} \quad \text{....(5.3)}$$

Specific humidity is the weight of water vapour per unit weight of air (including water vapour). It is expressed as grams of water vapour per unit kilogram of air. It is expressed by the following relation:

$$\text{Specific humidity} = \frac{\textit{Weight of water vapour}}{\textit{Weight of the air including water vapour}} \quad \text{...(5.4)}$$

Both absolute and specific humidity are difficult to measure. As temperature is changing constantly, absolute humidity also changes. Absolute and specific humidity have lesser effect on crop growth. Thus the terms relative humidity is used. It is the ratio between the amount of water vapour present in the air and the amount of water vapour required for saturation at a particular temperature and pressure. It can be expressed as percentage or ratio and by the following relation:

$$\text{Relative humidity (RH)} = \frac{\textit{Water vapour present in the air}}{\textit{Water vapour required for saturation}} x100 \quad \text{....(5.5)}$$

Vapour pressure deficit (VPD) is the difference between the saturation vapour pressure and actual vapour pressure. It is expressed in pressure terms as bars. When the vapour pressure deficit is around or less than 1.5 kPa. Air is said to be humid and over and above 2.5 kPa it is dry air. It gives a rough estimate of drying power of the air.

It is measured by dry and wet bulb thermometer with tables. The instruments used for measuring RH are called spectrometers or hygrometers. Hair hygrometer is generally used to measure relative humidity inside the rooms and Hygrograph is used to record humidity continuously.

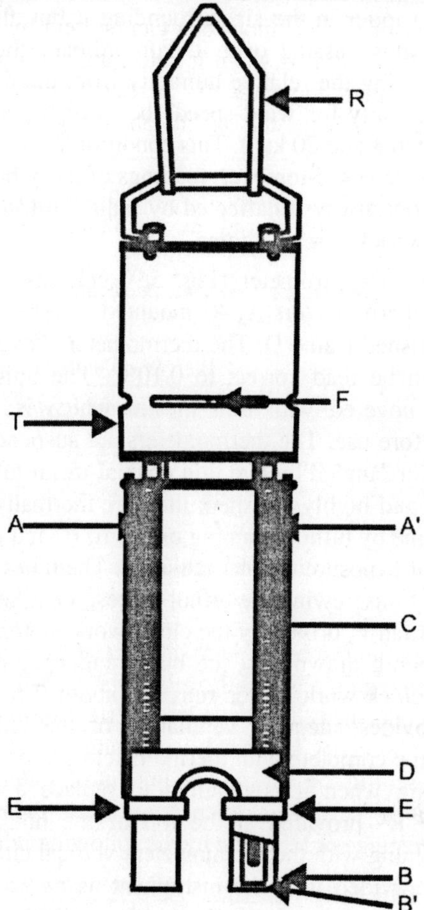

Fig. 5.9: Assmann Psychrometer

Assmann Psychrometer (Fig. 5.9) is a portable instrument with gave more accurate readings of the wet and dry bulb temperatures of the air. In this instrument air is drawn past the dry and wet bulbs by means of a clock-work exhaust motor. Each bulb is protected from external radiation by two highly polished coaxial tubes so that the instrument can be held even in strong sunshine without risk of solar radiation affecting the readings.

If the air is not saturated with moisture, the wet bulb thermometer will always read less than the dry bulb thermometer. This is due to evaporation of water from the muslin round the wet bulb and the consequent cooling. The wet bulb temperature depends not only on the amount of water vapour in the air surrounding it but also on the speed with which the air is passing over it. For ordinary thermometers, the formula for calculating the relative humidity from the dry and wet bulb readings is correct only for wind speeds between 4 and 10 meters per second (*i.e.* between 8 and 20 kts.). This condition is not always obtained in a Thermometer Screen. Moreover, readings of a dry bulb thermometer in the screen are not always unaffected by radiation true temperature of the air. These drawbacks are overcome.

The Assmann Psychrometer (Fig. 5.9) consists of two sensitive mercury-in-glass thermometers A, A^1 mounted side by side in a nickel-plated, highly polished frame D. The thermometers have a range of -10° to +60°C and can be read correct to 0.10°C. The bulb of one of the thermometers is covered with thin muslin which is moistened with distilled water before use. The thermometers are suspended in the frame with their bulbs surrounded by two thin coaxial metal tubes B, B^1 which are nickel plated and highly polished; they are thermally insulated from the rest of the frame by tufnol retaining pieces E, E^1 and protect the bulbs from the effects of exposure to solar radiation. The tubes can be removed from the frame by unscrewing the tufnol pieces. Ventilation of the bulbs is effected by the fan F, driven by the clock-work motor encashed in the housing H, air being drawn past the bulbs and up the hollow central column C. The clock-work motor runs for about 7-8 minutes on one winding and provides adequate ventilation of the bulbs for about 5 minutes. The frame complete with thermometers can be unscrewed from the motor housing when it is required to replace a thermometer. A suspension ring "R" provided at the top of the housing enables the instrument to be hung with the thermometers vertical either from a metal support fitting, provided with the instrument as an accessory, or from one's hand. The instrument is supplied with muslin for the wet bulb, a container for distilled water, the metal support fitting and an injector for moistening the wet bulb.

It is defined as the water vapour in air. Only a certain amount of water vapour can be contained by air at a given temperature and barometric pressure. In hydrology vapour pressure is defined as the pressure exerted by vapour molecules and is expressed in millibars.

Saturation vapour pressure (V_s): It is the maximum vapour pressure in a saturated space and it is dependent upon temperature alone.

The vapour pressure is indicative of the moisture content of the atmosphere. The values of saturation vapour pressure are generally read from psychometric tables. The following empirical formula can give values of the saturation vapour pressure over water within 1 percent approximation (in the range of -50 to 55°C can be determined with the following relationship:

$$V_s = 33.8639\,[(0.00738T+0.8072)^8-0.000019\,|1.87+48|+0.0013161] \quad \text{....(5.6)}$$

where,

Vs = Saturation vapour pressure (millibars), and

T = temperature (°C)

The water is converted into vapour by evaporation from water surfaces. There is a continuous exchange of water molecules to and from the atmosphere. However, the term evaporation as used in hydrology can be defined as the net rate of vapour transfer. Evaporation is that phase of hydrological cycle in which the precipitation reaching the earth's surface is returned to the atmosphere.

Condensation: It is the process by which vapour changes to the liquid or solid state. Whereas *sublimation* is the process by which a substance in solid state is directly converted in to vapour state and vice versa. Vaporisation and condensation go on taking place at the same time in space. If the space is in contact with saturated medium, e.g. water. Otherwise the rate of vaporisation will exceed the rate of condensation. Vaporisation and condensation are balance in a saturated space, provided the air and water temperatures are the same. In case of evaporation, molecules of water which have sufficient kinetic energy in excess of attractive forces that tend to hold them within the body of liquid water, leave the water surface. Vaporisation, therefore, removes heat from the liquid being evaporated and in the reverse process *i.e.* in condensation heat is added to the system.

Latent heat of vaporisation: It is the amount of heat absorbed by a unit mass of a substance while passing from liquid to vapour state, temperature remaining unchanged. The change from vapour to liquid state releases an equivalent amount of heat.

With the increase in temperature, kinetic energy of the molecules escaping water surface increases and surface tension decreases, and as a result the rate of evaporation increases. The heat of vaporisation (H_v) in calories per gram varies with temperature and can be determined upto 40°C on using the following relationship:

$$H_v = 597.3 - 0.564\,T \quad \text{....(5.7)}$$

Latent heat of fusion: It is the amount of heat required to convert one gram of ice to liquid water at the same temperature. In the reverse process of conversion of one gm of liquid water at 0°C into ice, the temperature remaining the same, there will be release of latent heat of fusion (79.7 cal/g).

The amount of heat required to transform one gram of ice into vapour without passing through the intermediate state of liquid, and temperature remaining unchanged is called the *latent heat of sublimation.* Obviously it is equal to the sum of heat of vaporisation and latent heat of fusion. Direct condensation of vapour into ice (without passing through the intermediate liquid state) at the same temperature will release an equivalent amount of heat.

The *specific humidity (SH):* It is defined as the ratio of mass of water vapour (grams) to the mass of moist air (kilogram) and determined as,

$$S_H = 622 \frac{V}{P_a - 0.378V} \cong 622 \frac{V}{P_a} \qquad \text{....(5.8)}$$

where,

V = vapour pressure (millibars), and

P_a = atmospheric pressure of air (millibars).

Specific humidity tends to remain almost constant in an air mass until a storm occurs which removes large quantities of the atmospheric water.

Saturation vapour pressure (V_s): It is vapour pressure of the molecules at upper limit of vapour amount of water vapour which a volume of air can contain. It has been found that V_s is a non-linear function of air temperature. The *relative humidity (R_H)* is considered to be a better way of specifying the quantity of water present in the atmosphere. The *relative humidity (R_H)* of air is defined as the ratio of the actual to the saturation vapour pressure and is usually expressed in percentage:

$$R_H = (V/V_s)100 \qquad \text{....(5.9)}$$

The difference between actual and maximum humidity is called the *saturation deficit (d), and* is sometimes called the *drying power* of the air. The hydrological balance equation is strongly influenced by saturation deficit of moist air.

Dew point: It is the temperature at which the saturation is reached if cooled at pressure and constant vapour content.

The *relative humidity* is inversely proportional to temperature and as such it increases with increasing latitude and is maximum in the early morning and minimum in the afternoon. Atmospheric moisture tends to decrease with increasing latitude, it also decreases with elevation; and is minimum in winter and maximum in summer.

The air moisture is measured by *psychrometer,* which consists of dry and wet bulb thermometers. The bulb of wet bulb thermometers is covered by a jacket of clean muslin cloth saturated with water. Since evaporation from the saturated surface will give rise to cooling effect, the wet bulb thermometer reads lower than the dry bulb thermometer. The air moisture can also be determined by measuring the temperature of the dew point. This is done by dew gauge or condensation hygrometer. A more convenient instrument but not so accurate for measuring air moisture is the *hair hygrometer.*

The vapour pressure (V) is determined by the following relationship:

$$V = Vs - 0.00066\,(t - t_w)\,(1 + 0.00115\,t_w) \qquad ....(5.10)$$

where,

t = dry bulb temperature (air temperature) (°C), and

tw = wet bulb temperature (°C)

The difference $t - t_w$ is known as the *wet bulb depression.*

5.3.2.7 Sunshine hour

The hours of bright sunshine are usually recorded by the Campbell Stokes pattern sunshine recorder which consists essentially of a glass sphere about 10 cm in diameter, (Fig. 5.10). The hours of bright sunshine are measured by the length of the burn of the cards due to the heat of the sun rays focused on the card by the glass sphere and the cards are changed each day.

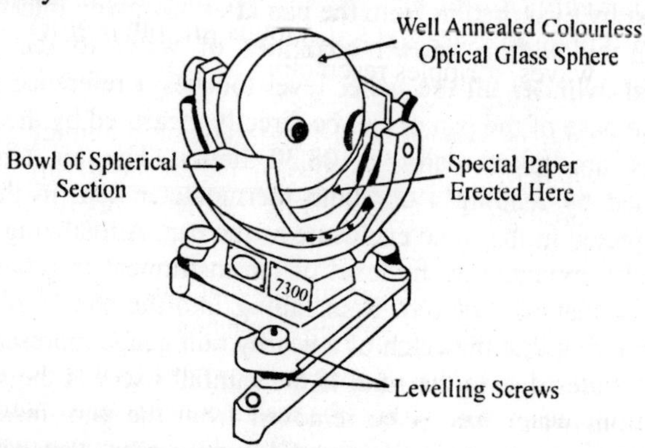

Fig. 5.10: Sunshine recorder (Campbell-Stokes pattern)

5.3.2.8 Evaporation

It is the loss of water from soil as well as from water bodies. Different types of evaporimeters are available for measuring evaporation *viz.*, USWB open pan evaporimeter, Sunken screen evaporimeter, Can evaporimeter and Portable evaporimeter.

USWB open pan evaporimeter is most widely used evaporimeter developed by the United States Weather Bureau. It comprising of a cylindrical vessel of 25 cm height and 122 cm diameter. It is filled with water and is covered with a wire mesh to prevent animals from drinking. The evaporation rate is measured by the change in level of water per unit time. The level of water is measured with a hook-gauge. It is installed on a wooden frame. Evaporation is expressed as mm/day.

Estimates of evaporation from free water surfaces is measured by USWB Class A pan.

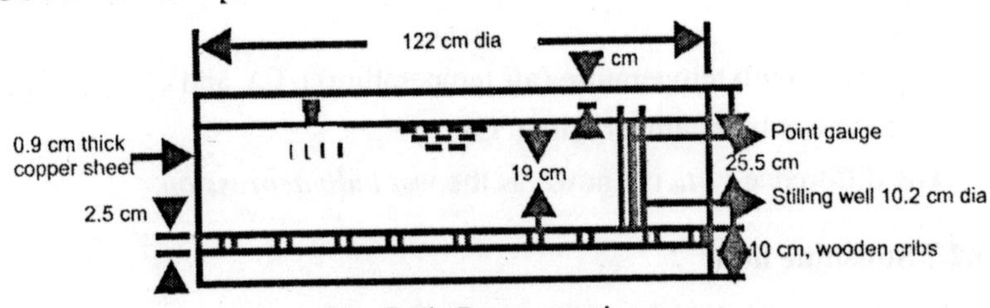

Fig. 5.11: Pan evaporimeter

The USWB class A pan is 122 cm diameter and 25.5 cm deep with stilling well, vernier point gauge, is shown in Fig. 5.11 and measuring cylinder. It is made of brass metal. The pan is placed on a wooden frame work of 10 cm thickness and covered with a wire screen. The amount of water lost by evaporation from the pan after any given interval of time is measured by adding known quantities of water to the pan from a graduated cylinder till the water level touches a reference point 19 cm above the base of the pan or can be directly measured by the point gauge. Readings are taken daily at 08.30 hours. The air temperature is determined by reading a dry bulb thermometer kept in the Stevenson screen erected in the same enclosure as the pan. A totalizing anemometer is normally mounted at the level of the instrument to provide the wind speed information required. Rain falling into the pan is allowed for on the assumption that the catch of a nearby rain gauge represents the added depth of water due to the rain. If the rainfall exceeds the water lost by evaporation, water has to be removed from the pan, instead of being added. The pan is protected from GF wire screen. Pan with a GI wire screen shows that, on the average, the unscreened pan reading is 1.144

times that of the screened one, since the screen affects the free air movement and also prevent the losses by birds and other animals.

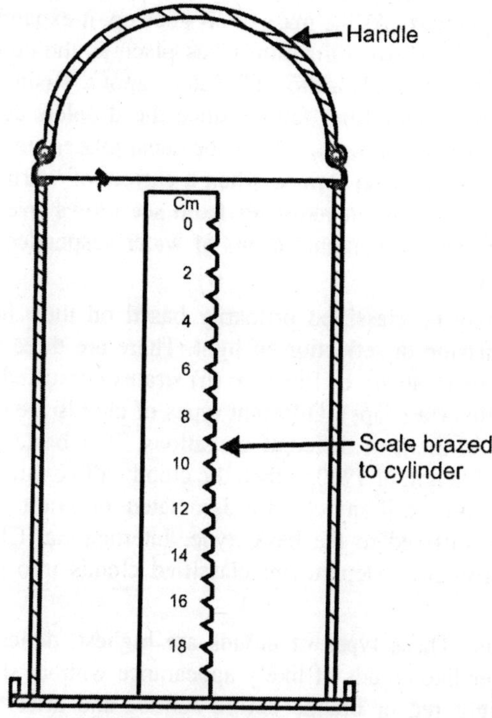

Fig. 5.12: Measuring Cylinder

5.3.2.9 Rainfall

Rainfall may be measured by a network of rain gauges. Following types of rain gauges are used for measurement of rainfall:

(a) Non recording type,

(b) Recording type,

(c) Automatic radio pestering rain gauge

(d) Radars

5.3.2.10 Precipitation

The precipitation is the mainly in the form of rainfall though there is appreciable snowfall at high altitudes in the Himalayan range (Chapter 4).

5.3.2.11 Clouds

Air mass with moisture (evaporation) rises upwards on account of increasing temperature. As the mass of air goes up, it expands due to low pressure and cools. Cloud formation takes place as the cooling reaches up to saturation and condensation of water vapour. Rising air currents tend to keep the clouds from falling since the droplets composing the clouds are very small and light. This is the usual process in which clouds are formed. These are also formed when a current of warm air strikes a parcel of cool air, or when moist air from sea blows over cold. Thus, cloud is an aggregate of minute drops of water suspended in the air at higher altitudes.

Clouds may be classified primarily based on their height, shape, colour, transmission or reflection of light. There are three basic type of clouds: (i) cirrus (feathery or fibrous), (ii) stratus (stratified or in layers) and (iii) cumulus (in heaps). Different types of clouds are either pure or combinations of them at different elevations. If a basic cloud occurs above a normal height of 1950 m then the cloud will be thin and the word 'alto' is prefixed to it. If any cloud is associated with rain, then 'nimbus' is prefixed or suffixed to the basic type. International Cloud Atlas of World Meteorological Department classified clouds into the following types:

1. Cirrus: These types of clouds are highest, delicate, detached, fibrous, feather like clouds of likely appearance without shading. These appear as bright red or orange before sunrise and after sunset. These clouds have thin crystals or needles of ice and most droplets of water.

2. Cirro-cumulus: These are appear in patches of small white flaky or sealy globular masses covering small or large portions of the sky and have no shading. Such types of clouds are often arranged in bands or fused into waves or ripples resembling those of and on sea shore.

3. Cirro-stratus: These are typically as a thin whitish veil of sheet, often covering or a good portion of the sky. These clouds may be very thin, giving the sky a slight milky white appearance. This type of Cirro-stratus clouds are responsible for halos and often occurs but do not blur the outlines of sun or moon.

4. Alto-stratus: These are uniform bluish or greyish-white clouds sheets cover all or large portions of the sky. Sometimes they ring may occur in uniform broad bands. The sun may be totally obscured or may shine weak or watery conditions. Alto-stratus clouds give good amount of rainfall the middle and high latitudes.

5. **Alto cumulus:** These elliptical globular units occurring individually in groups. Individual alto-cumulus clouds are frequently elongated ellipsis or lenticular units without vertical doming.

6. **Stratus:** It is a uniform grey cloud sheet or layer. Stratus clouds have particular form or structure and usually completely cover the sky. This type of cloud is normally thicker and darker than the may higher alto-stratus which be covering.

7. **Nimbo-stratus:** These are thick, dark grey, shapeless cloud sheets with red broken clouds beneath and surrounding them. Clouds are the normally associated of steady precipitation either rain or snow.

8. **Strato-cumulus:** These are large, heavy roils or elongated globular masses arrange long grey parallel bands that usually cover all or most of the sky. Cloud constitutes the flattening of cumulus clouds which may be arranged in bay or may develop as continuation of alto cumulus occurring at low altitude. In the later case, the strato-cumulus appears darker, lower and heavier the related also cumulus.

9. **Cumulus:** These are the majestic, billowing white clouds so promine in summer time. However, they may occur at any season. Such type of clouds are typically flat based with a pronounced vertical thickness which exited upwards as a dome or cauliflower shaped.

10. **Cummulo-nimbus:** These are developed from cumulus that have developed into tremend towering clouds with a vertical range from base to top of 3 to 8 kilometer. When grown to this height, such clouds form the well-known thunderstorm.

5.4 Classification of Surface Observatories

According to IMD, the surface observatories are classified into the following six class:

Class I (Special stations with autographic instruments)	Taking three or more observations per day at fixed times, having autographic instruments for recording at least three of the elements, pressure, temperature, wind and rainfall and communicating (by telegram, telephone, teleprinter or W/T) atleast two sets of observations daily.
Class II (Airway and synoptic stations managed by Departmental Observers)	(a) Stations managed by departmental staff (combined surface and pibal and aviation stations) taking and communicating (by telegram, telephone, teleprinter or W/T) atleast two sets of observations daily.

Class III (Synoptic stations managed by Part-time Observers)	(b) Part-time observer stations taking and communicating (by telegram, telephone, teleprinter or W/T) atleast two sets of observations daily.
	(c) Part-time observer stations taking atleast two sets of observations daily but communicating (by telegram, telephone, teleprinter or W/T) only one set of observations.
	(d) Part-time observer stations taking atleast two sets of observations daily and reporting them by monthly registers only.
Class III (Synoptic stations managed by Part-time Observers)	(a) Observatories taking and communicating (by telegram, telephone, teleprinter or W/T) only one set of observations daily.
	(b) Observatories taking only one set of observations daily and reporting the same by monthly registers only.
Class IV (Climatological stations managed by Part-time Observers)	(a) Observatories not equipped with barometers but taking two sets of all or most of the remaining observations daily and reporting them by monthly registers.
	(b) Observatories not equipped with barometers but taking all or most of the remaining observations daily and reporting the same by monthly registers only.
Class V (Rainfall stations managed by Part-time Observers)	(a) Observatories taking once daily instrumental observations of rainfall only and communicating (by telegram, telephone or W/T) the same alongwith non-instrumental observations in the brief code.
	(b) Observatories recording only rainfall once, daily and communicating the same (by telegram, telephone or W/T) during certain seasons and by monthly or weekly registers during the rest of the year.
	(c) Observatories recording only rainfall once daily and reporting the same by monthly or weekly registers.

Class VI (Miscellaneous non-instrumental stations and also those with a few instruments which do not come under any of the above classes)	(a) Non-instrumental stations recording clouds, wide directions and speed and visibility without any instrument and reporting the same by monthly registers. (b) Stations recording observations with a few instruments and which cannot be shown under class IV or V. (c) Stations recording observations with a few instruments, and communicating (by telegram, telephone or W/T) the same throughout the year or part of the year and reporting the same by monthly registers during rest of the year, and which cannot be shown under class IV or V.

5.5 Weather Abnormalities

Aberrant weather like cyclones, storms, cloud bursting, hailstorms, frost, hot and cold winds, droughts and floods lead to natural disasters affecting agricultural production and national food security adversely. Scenario of climate change due to doubling of CO_2 concentration in the atmosphere, indicates the mean global temperature has increased by 0.6°C (IPCC, 2001). Observations also indicate that all the warmest years' during the past century have occurred in the last two decades. Among these years, 1998 was the warmest one. Model indicates that the mean global temperature could rise by 1.4-5.8°C in next century by doubling of CO_2 concentration with significant regional variations and increased variability in rainfall.

Understanding climatic conditions and impact of weather on crops in advance will help farmers immensely to take the advantage of favourable weather and reducing the crop losses under aberrant weather conditions. Weather based Agro-advisory Bulletins issued by Agro-meteorological Field Units (AMFUs) of India Meteorological Department on real time basis contain 5 days advance weather information along with suggestions to farming community based on forecast in order to inform the farmers to conduct agricultural and allied operations timely and safely.

Weather of particular location is not always normal. Some abrupt or sudden changes take place in weather which are called weather abnormalities. These abnormalities reduced the crop yield *viz.*, floods, droughts, untimely rains, thunder storms, hail storms and dust storms, cold waves along with frost, heat waves, high winds, and cyclones. Out of the total annual crop losses, a substantial portion is because of aberrant weather. These losses could be minimized by making adjustment with coming weather through timely and accurate weather forecasting.

5.5.1 Floods

Rainfall of high intensity seals soil pores and results surface runoff if the intensity of rainfall is more than the infiltration capacity of soil. If the rainfall continues for longer periods then the runoff increases and causes floods. When the catchment areas of rivers receive very high rainfall then rivers cannot carry the entire runoff. It overflows and submerges the neighbouring fields. Submergence of field with muddy water, silt is deposited. Floods cause considerable damage to the crops, livestock and human life. These losses may be controlled by construction of multipurpose dams, afforestation in catchments areas and provision of adequate drainage.

5.5.2 Droughts

Drought is complex phenomena of nature under which crops fail to grow and mature due to deficit moisture. Occurrence of drought is mainly due to delay in the onset of monsoon, prolonged dry spells during the crop period, and/or early withdrawal of monsoon. The adverse effects of drought may be mitigated to an extent by water harvesting either *in-situ* or *ex-situ*, wind breaks and agronomic practices to overcome moisture stress.

5.5.3 Other abnormalities

Rainfall during crop period at pollination, fertilization and harvesting causes damage. These are frequent in the sub–mountainous tract of North India. Frequency of occurrence is once or twice in five years. When enough moisture is present in the atmosphere, thunder storms occur and in the absence of moisture, a dust storm may occur. Hail storms may cause losses to standing crops. Cold waves and frost occur in North-India, causes losses to crops. Due to high temperature and hot winds, most of the water bodies dry up quickly resulting in water shortage during the period from March to May.

High winds are associated with tornadoes, water–spouts and also with cyclones. It is the most destructive among the weather abnormalities and occurs in middle latitudes. They are gigantic, whirling funnel of air extending earthward from heavy black cumulonimbus clouds. It causes heavy down-pour of rain, hail and terrific winds. If tornado moves over water bodies, it is called water spouts.

5.6 Weather Forecasting

Weather is the main factor of natural phenomena determining the success or failure of agricultural crops/cropping systems. Therefore, farmers have no control over this natural phenomena. Weather manifests its influence on agricultural operations and farm production through its effects on soil, plant growth as well as on every phase of animal growth and development. Aberrant weather gave the maximum loss out of the total annual crop losses. Thus, weather forecasting of modern type can play a vital helpful role provided that such predictions are correct, reliable, adequately available well in time and the effects of future weather are understood.

Noffsinger *et al.* (1964) reported that annual losses in agricultural are about 13 billion dollars; out of this amount about 11 per cent are caused directly by weather hazards like floods, hailstones and storms. At the same time, losses due to conditions affecting harvesting, storage, parasites, crop and animal diseases are highly influenced by weather. In all, directly and indirectly, weather makes a contribution of approximately three fourth of the annual losses in the farm production.

Crop losses in India are more as compared to other countries. Every year, there is an extensive damage due to floods in one part of the country and a severe drought damage crops in another part. In one season, crops are damaged by a severe attack of pests and diseases, in the other season the damage may be more due to some other factor. Atwal (1983) estimated the total annual pre-harvest losses for the various crops in the country as a whole, are estimated to range between 10 and 100 per cent while the post-harvest losses are estimated to range between 5 and 15 per cent; a 7 per cent average is considered an appropriate figure for all the crops. At these levels, the total loss of food grains and some other important for the year 1982-83 have been estimated and are given in Table 5.7.

It is an established fact that a major part of the crop losses is primarily due to direct and indirect effects of weather. Assuming three fourth of the total losses due to weather, as one study conducted in US, it revealed that the annual crop losses due to weather in India for the six major crops during the year 1982-83 were estimated at about 26 million tonnes. This loss is worth about Rs. 49570 million.

In spite of substantial gains in food grain production in recent decades, 26% of India's population lives in poverty. Although poverty levels declined from 51% in 1970 to 26% at present, India continues to face a persistent challenge of feeding a growing population against a background of climatic uncertainties (Shukla *et al.*, 2002).

Table 5.7: Crop losses in India during 1982-83

Crop	Total production (mt)	Per cent losses		Total losses (mt)	Value of losses (mrupes)
		Before harvest	After harvest		
Wheat	42.5	10	7	7.22	10974
Rice	46.5	20	7	10.50	20097
Millets	27.8	25	7	8.94	10906
Pulses	11.6	10	7	2.02	8080
Groundnut	5.6	10	7	1.21	4840
Cotton	7.7	50-100	7	4.39	11194
Total	141.7	10-100	7	34.28	66093

The destructive meteorological phenomena are depending upon different weather elements (Table 5.8). The dangers which have a meteorological content (variety, insects, diseases, weeds, water, soil crop climate, nutrients etc.) have good correlation with crop production. Inter-annual variability of the South Asian monsoon affects more than 60% of the Earth's population (Webster *et al.* 1998), and has large-scale impacts on food production, power generation, drinking water supply and overall economy (Parthasarathy and Pant, 1985). Irrespective of production systems, climate-related events such as droughts, cyclones, floods, hailstorms, frost, high winds and extreme temperature contribute to farmer's vulnerability and impact on national food security. Improved climate risk management based on skillful seasonal forecasts can play an important role in reducing rural poverty.

Table 5.8: Possible effects of abnormal weather variables in temperate fruits

Parameters	Remarks
Precipitation during March to May (150-200 mm)	▪ Recharge and maintaining of soil moisture regime. ▪ Control of canopy temperatures. Maintaining optimum humidity.
During winter temperature should be below 20°C but it must be below 7°C for 1200-1600	▪ Very crucial for induction of dormancy, bud break and ensuring flowering in apple.

Parameters	Remarks
hours, and 150-200 mm well distribution precipitation	▪ Precipitation in the form of snow is most beneficial for chilling, nitrogen and soil aeration. ▪ Early rains/snow (>100 mm) (November-January) are more beneficial in fulfilling chilling requirement.
High Temperature in Spring (>25°C)	▪ Affecting chilling requirements. Increase the pace of development. Enhanced sprouting.
High temperature during April (>26°C)	▪ Dry and unsticky pollen grains. ▪ Poor pollination by insects. ▪ Poor Germination of pollen grains. ▪ Slow growth of pollen tube. ▪ Fruit setting reduced. ▪ Reduced yield up to 25 per cent.
Low humidity (<40%) at flowering and setting	▪ Desiccation of pollen/bud. ▪ Killing of pollen grains. ▪ Low fruit settings.

From historic time, farmers have adapted their farming systems to climate variability, changing economic conditions, technologies, and resource availability. The experience of centuries has taught the traditional farmer an 'empirical art of practicing agriculture' that is in tune with nature's vagaries (Randhawa *et al.* 1961).

5.7 Indigenous Techniques of Weather Forecasting

Indigenous knowledge and tradition still play a role in climate risk management. In Kangra area of Himachal, for instance, if there are good *Marigsana* (scorching heat during June) and *Sango* (rain in first 8 days of *Sawan* month), it is believed that the rainfall in the succeeding season will be good. If the breeze is towards the north-west during July, the summer monsoon will be good; if towards south-east the winter monsoon will be successful in the Kullu and Mandi areas. Although some of these beliefs are loosely based on climate phenomena that we are now beginning to understand, indigenous knowledge is slowly losing its importance and is being replaced by climate forecasts based on our physical understanding of the climate system. The process of incorporating new information into the existing knowledge base is an iterative approach (Pinners and Balasubramanian, 1991). Our extensive discussions with the stakeholders provided ample opportunity to document local knowledge on climate prediction. We observed limited, highly location-specific indigenous climate knowledge. Farmers were more inclined to incorporate climate information into their risk management strategies when the traditional wisdom and scientific climate forecasting coincided. The iterative risk management model involves a cycle of analysis, implementation, testing, modification, evaluation and redesign of risk management strategies and helps to assimilate scientific knowledge with local knowledge (Fig. 5.13).

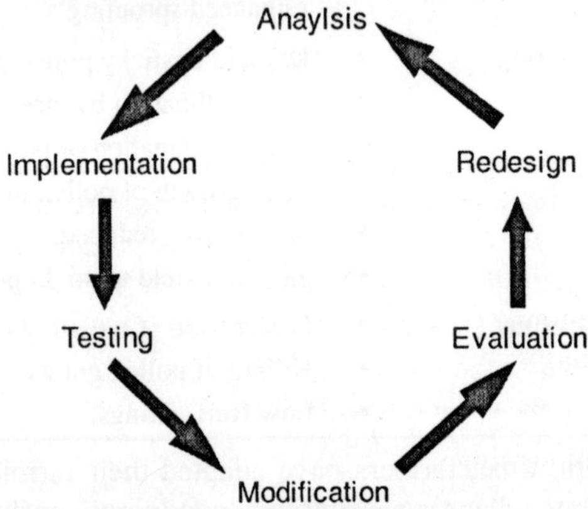

Fig. 5.13: Participative climatic risk management processes

Participatory methods served as a source of information about local practices, problems, impacts, recommendations, policies (Butler and Butler, 1987) and facilitate spontaneous responses and high level involvement of farmers. Participatory co-learning processes offer greater scope for identification of cropping system determinants and farmers' real and perceived need. Therefore, it is necessary to design climate educational programmes that take socio-economic factors such as farmer's perception, social behaviour, and attitude to risk and their knowledge level into account. In the present chapter it is attempted to collect and document the indigenous knowledge and traditions in weather prediction through participatory co-learning processes and incorporate it with medium range weather forecasting to improve its reliability.

Traditional weather prediction

Man's need for water has been increasing from his very appearance and existence on the earth. However, the unpredictable nature of the rainfall did not changed much. Adequate availability of water through rainfall has, therefore, always been the major worry of man. In primitive time tribal chiefs were responsible for bringing rain for their people. Like the politicians of today, the tribal chiefs delegated the responsibility of rain making to someone else, obviously as a hedge against failure. If there was no rain, it was the luckless assistant who was punished rather than the chief. These early rain makers, forerunners of the present day meteorologists, were perhaps among the most intelligent of the primitive man. Again, it was these very people who set the stage for scientific inquiry into the causes of weather and rainfall. Some of the approaches they have used for the prediction of ensuing weather are discussed below.

Magical approach

The initial pioneer rainmakers were magicians who used imitations, appeals, supplications, intimidation and various strange ceremonies as approaches to make rain. Even today, in many parts of the world rainmaking is practised through though there are many variations to this magic approach. Some of these approaches used by the early rain maker magicians are as:

- Invoking the spirit of dead.
- Mock ploughing.
- Sprinkling of water on soil hoping heavens would do the same.
- Drum beating to imitating thunder.
- Use of firebrands to stimulate lighting.
- Even hurling causes at the Gods.
- Blew mouthfuls of water into the air like rain.
- Blowing of water into air through special pipes.
- Sprinkling of blood on the soil to bring rain.

Hanging of frogs from trees to induce the heavens to rain over them.

Religious approach

It is difficult to mention the exact time when the rainmakers turned from magic to the Gods. Imitative magic gave way to supplications, a systematic inducing approach directed, not towards clouds and the heavens, but towards Gods representing these phenomena. This respect subsequently shifted to the dead. Thus supplications were first directed at the tomb of a departed ancestor. Men prayed for rain to the dead and then to the Gods. It was believed that prayed itself could bring rains. Praying for rain is firmly routed in Indian culture even today.

In the next stage of its evolution rainmaking was centered round the bestowing of gifts through scarified offerings. Initially these were in the form of live human sacrifices, and subsequently involved animals. Intimidation was also an important tool in religious practice. In several European countries statues of saints were uprooted and made to stand on their heads when prayers before them failed to bring rain. As late as 1893, Italians, affected by drought, banished statues of saints from the country. On one occasion the statues were chained and their wings were clipped off when the rain failed.

Indigenous knowledge system

Development and refinement of the art of weather prediction has been essential since time immemorial. In present time we have many improved technologies for making weather forecast as well as for their dissemination. Previously when there were no such technologies available farmers based their prediction on many natural phenomena, birds and animals, activities, and cultural and social beliefs. It is worth mentioning that the animals and birds activities are more sensitive and accurate in predicting the ensuing weather but, only problems with human being to understand, analyze and express them.

(a) From astronomical phenomena

Based on various conditions and situations of the atmosphere, sky, clouds amount and movement, wind speed and direction, pressure, relative humidity and temperature ranges etc. people predicted the future weather.

- People said that the spectrum around the sun with greater diameter accompanied with a spectrum around the moon with low diameter is an indication of forth coming rainfall and vice-versa.
- During rainy season more intensity of red colour in rainbow predicts forth coming weather will be fair as water vapours in the air are more.
- In the same rainbow if the intensity of green colour is high then rainfall is predicted in the forth coming days.
- Accumulation of clouds in the south-east direction accompanied by southern wind brings rainfall within a day or two.
- Redness of the sky during the time of sunset predicted rainfall on next day.

- Bright sunshine at evening after having rain throughout the day predicted rainfall on the next day also.

(b) From birds' activities

As mentioned above birds are sensitive to weather and change their activities with the change in the ensuing weather. By closely observing these activities farmers made prediction about the forthcoming weather in their region and took action accordingly in their farm and home. Few of the ITK collected during the survey related to these activities are discussed below:

- During any year if the ***tatihari*** (lapwing) bird lays eggs on the higher portion of the field then good monsoonal rainfall is predicted in the year. But if eggs are laid on the lower portion of the field then farmer predicted the year will be dry with very low rainfall.
- Emerging of brown colored centipedes from their holes and carrying their eggs predicted early rainfall.
- Cry of bird ***Papiha*** and ***Neulu*** for water in hot summer days brings rainfall as farmer know that the birds are very thirsty and heaven will crunch their thirst.
- Farmers predicted heavy rainfall if the ***Maitilari*** bird flies high in the sky.
- Bathing of *Mama* in water indicated early rainfall.
- Frolicking of a group of sparrow in the sand predicted rainfall but playing in water predicted dry weather is the next day and rainfall after two three days.
- When dogs start eating grass then strong winds with rain are predicted.
- When the peacocks start dancing in groups then good rainfall is predicted.

(c) From animals activities

- During grazing if the bull and cow start jumping with joy then farmers predicted rain and a pleasant forth coming weather.
- If there is swelling on the lower portion of the camel's leg then the farmers of predicted rainfall is good.
- Immobile leeches at the surface of pond water indicated dry weather but if they are moving rapidly upward and downward direction then rainfall is predicted.

- During the hot summer season if farmers see the snakes frequently then they predicted forthcoming rainfall.
- Peculiar type of barking by dogs indicated earthquake or any major natural hazard.
- If sheep start feeding milk to their young ones at relatively higher hill then early rainfall was predicted.

(d) From social beliefs

Many cultural, social, religious beliefs and activities and superstitions pertaining to the prediction of future weather prevail since generation to generation. Their forefathers had deep beliefs in them so people are adopting these and doing these regulority.

- If the 15 ***pervisthas*** of the month ***jeth*** are very hot then good rainfall was predicted in the forth coming monsoon season.
- People in Solan area of Himachal Pradesh firmly believe that rainfall will occur on the ***Janki*** of mother ***Sooluni*** during the ***Sooluni mela*** every year during the month of May-June.
- It is also believed that rainfall will also occur around the national festivals of 26 January and 15th August every year.
- People of Nauni areas predict rainfall when grey colored cloud descend below the hill of ***Dako*** fort.
- If the *khejari* tree bears good fruit in a particular year then farmers predicted good rainfall and vice-versa.
- When ***tambaku bags*** show more moisture in then then the farmers expected rainfall in near future.
- The leaking of ***karachhi*** (cooking spoon) indicated rainfall during marriage of an individual.

In Bengal and Punjab people threw filth on the homes of the neighbours who in turn abused them. This was considered auspicious for rainfall occurrence.

(e) From folklores

The folklores of the popular poet ***Gag*** and his wife ***Badari***, who lived during the 17th century, regarding weather forecasting are still very popular and valid in north India. Some of their folklores and some others locally collected from the farmers are given below:

- When strong eastern winds are blowing continuously over a region then one can understand that the rainy season has set on in the foremationed areas.

- When days are very hot and dew formation takes place at night then according poet to Gag under these atmospheric conditions the rainfall is far away *i.e.* very low chances of occurring rainfall.
- When the light spectrum (halo phenomena) around the sun formed with lager diameter then rainfall is expected in one or two days.
- When the rainbow is formed towards Bengal *i.e.* to east direction then rainfall is predicted in the evening or the next morning.
- When days are partly cloudy and nights are clear and light winds from eastern direction are blowing then according to Gag there are no chances of rainfall, rivers may dry out so washer men may be forced to use the water from wells to wash their clothes.
- This is the famous folklore prevailing in Mandi areas of Himachal Pradesh according to the people of Mandi if clouds are going towards Chamba then heavy rainfall is predicted but if the clouds are going toward Kuhn then weather will remain so dry that even river may dry out and no water is available in houses for domestic use.

Deep knowledge and traditions are exist among the elderly age farmers regarding estimation of forth coming weather and its management which wash out slowly and slowly. There is great need to collect and document it. Climate sensitive, agricultural risk management requires more rigorous and systematic approaches than normal agricultural extension. It appears that there are substantial opportunities for resource-limited smallholder farmers in developing countries to better manage production through use of climate forecasts. However, communicating the risk/opportunities of alternative management options is a major challenge. Transferring simple rules of thumb into a comprehensive modering framework may answer some of the 'what if' questions faced by the farmers. This also facilitates translating climate information into diverse outcomes, a fact particularly relevant and important for the farmer, planners and other users. Incorporating this technology know-ledge in weather prediction into the present day medium range weather forecasting can improve its regional specific reliability and validity.

5.8 Weather Forecasting Organisations

Several organisations are working on weather forecasting all over the world. Norms are developed for measuring, assigning values and codes for different countries. Indian Meteorological Department was established in 1975 at Pune. Agricultural meteorological division was started in 1932 for conducting research on crop weather relationships. A major step was taken in the early forties to set up specialized meteorological observatories in a crop environment to include weather consciousness among farmers and to farmers' agro advisory. There are many applications of weather forecasts to improve crop yield and to minimize the crop losses.

5.9 Weather Elements and Time Factor

The weather elements which influence the agricultural operations may be used for forecasting of crop production up to different spans of times. It increases as time span decreases the accuracy of the forecast. For weather forecasting following weather elements forecasted with different time span for agroclimate purposes:

i)	Hail	Less than 12 hours in advance
ii)	Tornados	Less than 12 hours in advance
iii)	Flash floods	Less than 12 hours in advance
iv)	Heavy rainfall	24 hours is advance
v)	Thunderstorms	24 hours is advance
vi)	Wind velocity	36 hours in advance
vii)	Rainfall amount	36 hours in advance
viii)	Occurrence of rainfall	5 days in advance
ix)	Temperature intensity	3 months in advance
x)	Temperature departure	3 months in advance

Decisions may be taken in advance for agricultural operations and planning is made accordingly to make the best use of favourable weather conditions and adjustment with the adverse weather (Smith and Strim, 1973).

5.10 Modern Weather Prediction

Weather prediction or forecasting of modern type can play a helpful role, provided that such predictions are reliable, adequately available and the effects of future weather are understood. There are various methods of weather forecasting like Synoptic Forecasting, Numerical Weather Prediction and Statistic Forecasting etc. are most commonly used. The basic notion on which the weather forecasting rests is extremely simple. At any movement of time, there is a discernible pattern of weather covering the earth. This pattern is observed at thousands of weather stations in all parts of the world. Ships at sea, aircraft aloft, balloons and in recent year's weather satellites are making weather observations. All of these data are collected by radio and displayed on weather maps (synoptic charts). And, in this way one judges from his scientific knowledge and experience are to how the pattern 'is likely to change.

The forecast that is disseminated by radio, television, and print media or by any other means, can be helpful to everyone, but special and much more technical forecasts are issued to special users. For example, no air craft will take off without receiving a weather forecast at ground level and at flight level along the flight path. This information is becoming important for making many operational both strategic and tactical decisions in agriculture, which are largely influenced by the past, present and expected weather. Thus, weather forecast must include the following weather elements which immediately affect them.

Sky cover	-	clouds and sunshine percentage.
Rainfall	-	amount, intensity and probability.
Temperature	-	Maximum, minimum, dry bulb and wet bulb temperature.
Wind	-	speed and direction.
Humidity	-	lowest and highest.
Dew	-	amount and duration.
Snow	-	amount and depth.
Drying conditions	-	droughts.
Hail	-	size, amount and intensity.
Cyclones	-	direction, speed, intensity, duration and destructiveness.

Weather forecast for agriculture can be divided into three categories *viz.,* short range forecast (up to 48 hours), medium range weather forecast (up to 10 days) and long range weather forecasting (>10 days). Each type of forecast has its own role and importance in farming operations and planning of agricultural activities. However, the reliability of a forecast decreases with increasing period of its validity and decreasing area.

Weather forecasts for agriculture may be grouped into three groups *viz.*, *short range forecast* (upto 48 hours), *extended forecast* (upto 5 days) or medium range and *long range forecast* (4 weeks to the season). Each forecast plays an important role for planning farm operations and agricultural activities (New man and Stearns, 1957; Noffsinger *et al.*, 1964; Jensen and Davis, 1974; Hudson, 1972).

Based on time or duration of forecasting period, the weather forecasting can be divided into short range, medium range and long range weather forecasts. Short range weather forecasts are for a day or two. These daily forecasts are useful to irrigation engineers. Mariners, aviation engineers and farmers. Medium range forecasts are for a period of 3 to 4 days to two weeks, while the long range forecasts are for periods of more than four weeks. The long range forecasts are issued thrice in a year. The first forecast is for the period June-September, second forecast is not accurate.

5.10.1 Application of short range forecasting in agriculture

It emphasis on the following weather elements *viz.*, High and low temperature, wind velocity and direction, sunshine duration, time and amount or precipitation, relative humidity. Its accuracy is 70-80 per cent.

The short range weather forecasting helps in the following operations:

i. Irrigation scheduling,
ii. Timing of field operations,
iii. Protection of plants from frost,
iv. Spray of chemicals,
v. Labour efficiency-workable hours,
vi. Insect-disease effects,
vii. Soil workability,
viii. Livestock protection from cold and heat,
ix. Animal production rate,
x. Drying rate of straw, and
xi. Drying rate of soil.

5.10.2 Application of extended forecast in agriculture

It emphasis on change of weather type, sequence of rainy days, normal weather hazards in farming such as strong winds, extended dry or wet spells. It is also known as medium range weather forecast (MRWF). Its accuracy is 60-70 per cent.

Both synoptic and statistical approaches are used to forecast medium range forecasts. These are based on recent climatological data, which give sufficient advance notice on the overall seasonal impact of a particular disease and pest. The purpose of medium range forecasts is to enable the planners to make a complete programme for taking control measures, and for taking such steps which may suppress the spread of the disease and pest etc.

This type of forecasting helps in the following farm operations:

i. To determine the depth at which seed be sown to achieve an optimal rate of seedling emergence,

ii. To determine whether or not to sow a crop in this period,

iii. To take account of expected rainfall to plan irrigation,

iv. To decide whether or not to harvest a crop in this period,

v. To ensure maximum efficiency of spray programme,

vi. To prepare in time for the protection of crops against frost,

vii. To the management of labour and equipment, and

viii. To animal feed requirements.

5.10.3 Long range forecasting (LRF)

The long range forecasting of weather has a profound socio-economic impact on management of natural resources. This forecast will help farmers to select crop and cropping systems, it will help to take preventive measures either for flood or drought to minimize adhesive effects on soil. Many countries in the world have stretched issuing LRF of weather on operational basis. India has the distinction of being the first country in the world for developing LRF techniques for monsoon rainfall over India. The first LRF of monsoon rainfall over whole of India and Burma was issued on June 4th, 1986. It was based on analysis of Himalayan snow cover. Subsequently, India introduced various techniques for LRF as correlation, multiple regression (MR), auto regressive integrated moving average (ARIMA), and dynamic stochastic transfer (DST). These techniques have also been used by different researchers for LRF or rainfall in various tropical countries.

Various LRF techniques used in India for the past hundred years have been reviewed (Thapliyas, 1984, 1986) mainly two approaches have been followed for increasing the accuracy of LRF. First approach In this is the predictors having physical linkages with the Indian monsoon were identified using the correlation approach and were used for LRF of monsoon. But due to temporal variation be increased beyond a certain limit. In second approach in this approach efforts have been made to increase the accuracy of LRF of monsoon rainfall developing better techniques. Various techniques used in LRF of rainfall over different regions of the tropics/subtropics with special emphasis on Indian rainfall are as follows:

(i) Multiple regression technique

The prediction equation derived from the multiple regression analysis has the following form

$$R = a_0 + a_1 x_1 + a_2 x_2 + \ldots + a_n x_n \qquad \text{....(5.11)}$$

Where, R is predict or and or the percentage departure of monsoon rainfall x is the predicator, n is the number of predictors employed in the determination by the method of least square. The types of possible predictors considered can be divided into two types of possible predictors considered can be divided into to two groups-station values and derived values. The former include directly observable atmospheric variable as temperature, sea level pressure and the geopotential heights of pressure surface at individual stations; whereas the latter includes 200-500 mb thickness at individual stations. Temperature differences, sea level pressure difference and geopotential height differences between two stations.

(ii) Parametric and power regression technique

The accuracy of multiple regression models is not always very satisfactory. In view of this, Gowarikar *et al.* (1989) suggested parametric and power regression models for predicting monsoon rainfall over India well in advance. The parametric model gives a qualitative forecast whereas the power regression model gives a quantitative forecast.

Gowarikar *et al.* (1989) analysed 15 regional and global meteorological and oceanic parameters from the data of the period 1951-1987 which were divided into four main groups-temperature, pressure, wind and snow cover (Table 5.9). These parameters were selected on the basis of their possible physical linkages with the monsoon rainfall over the country, out of which some were inter-related. Inferences on subsequent monsoon rainfall based on signals from 15 parameters are presented in Table 5.10.

Table 5.9: Analysis of 15 parameters for LRF of monsoon

		Temperature						**Wind**			**Pressure anomaly**				**Snow cover**		
Year	**Monsoon condition**	**EI Nino in current year**	**EI Nino in previous year**	**Northern India (March)**	**East coast of India (March)**	**Central India (May)**	**Northern Hemisphere (Jan & F...)**	**500 hp a ridge (April)**	**50 hpa ridge through extent (Jan ...)**	**10 hpa (30 km) westenly wind (Jan)**	**Tahiti – Darwin (Spring)**	**Darwin (Spring)**	**South America, Argentina (Apr.)**	**Indian Ocean Equational (Jan-...)**	**Himalayan (Jan-March)**	**Eurasian (Previous Dec.)**	**No. of parameters favourable/ Total No. of parameters**
		(-)	(+)	(+)	(+)	(+)	(+)	(+)	(-)	(+)	(+)	(-)	(-)	(-)	(-)	(-)	
1951	D	F	U	U	U	F	U	U			U	U	U	F	U		3/12
1952	N	F	U	U	U	F	F	U			F	U	U	U	F		5/12
1953	N	U	U	F	F	F	F	F			U	U	U	U	F		6/12
1954	N	F	F	F	F	F	U	F			F	F	U	F	U		8/12
1955	N	F	U	F	F	F	F	F			F	F	F	F	F		11/12
1956	N	F	U	F	U	F	U	F			F	F	F	F	F		9/12
1957	N	U	U	U	U	U	U	F			U	U	U	U	U		1/12
1958	N	F	F	F	F	F	F	F	U	U	U	U	F	U	F		9/14
1959	N	F	U	F	U	F	F	F	F	F	F	F	F	F	U		11/14
1960	N	F	U	U	U	U	F	F	U	U	F	F	F	F	F		8/14
1961	N	F	U	F	F	F	F	U	F	F	U	U	F	F	U		9/14
1962	N	F	U	U	U	U	F	U	F	U	F	U	U	F	F		6/14
1963	N	F	U	U	U	U	F	U	F	F	F	F	F	F	F		9/14
1964	N	F	U	F	F	U	U	F	F	F	F	F	U	F	F		10/14
1965	D	U	U	U	U	U	U	U	U	U	U	U	U	U	U		0/14
1966	D	F	F	U	F	U	F	U	F	F	U	U	F	U	F		8/14
1967	N	F	U	U	U	U	U	F	F	U	F	F	U	F	F		7/15
1968	D	F	U	U	U	U	U	U	U	U	F	F	U	F	U		3/15
1969	N	F	U	F	F	F	U	F	F	F	U	U	F	U	F		9/15
1970	N	F	U	F	F	F	F	F	F	U	U	U	F	U	F		10/15
1971	N	F	U	U	U	U	U	F	F	F	F	F	F	F	U		9/15

Year	Monsoon condition	Temperature						Wind			Pressure anomaly				Snow cover		
		El Nino in current year	El Nino in previous year	Northern India (March)	East coast of India (March)	Central India (May)	Northern Hemisphere (Jan &	500 hp a ridge (April)	50 hpa ridge through extent (Jan	10 hpa (30 km) westenly wind (Jan)	Tahiti – Darwin (Spring)	Darwin (Spring)	South America, Argentina (Apr.)	Indian Ocean Equational (Jan-	Himalayan (Jan-March)	Eurasian (Previous Dec.)	No. of parameters favourable/ Total No. of parameters
		(-)	(+)	(+)	(+)	(+)	(+)	(+)	(-)	(+)	(+)	(-)	(-)	(-)	(-)	(-)	
1972	D	U	U	U	U	F	U	U	U	U	U	U	F	F	U		3/15
1973	N	F	F	U	F	F	F	F	F	F	U	U	F	U	U		9/15
1974	D	F	U	F	F	U	U	U	U	U	F	F	U	F	U		6/15
1975	N	F	U	U	F	U	F	F	F	F	F	F	F	F	U		12/15
1976	N	U	U	F	F	U	F	F	F	F	F	U	U	F	F		9/15
1977	N	F	F	F	F	U	U	U	F	F	U	U	F	U	F		9/15
1978	N	F	U	U	U	F	F	U	U	F	F		U	U	F		6/15
1979	D	F	U	U	U	U	U	U	U	U	U	U	U	U	F		2/15
1980	N	F	U	F	F	F	F	U	F	F	U	U	F	U	U		9/15
1981	N	F	U	F	F	U	F	F	F	U	U	U	U	U	F		8/15
1982	D	U	U	U	F	U	U	U	U	F	U	U	U	U	F		3/15
1983	N	F	F	U	F	F	F	U	F	F	U	U	F	U	F		9/15
1984	N	F	U	U	U	F	F	U	F	U	U	U	U	U	F		6/15
1985	N	F	U	F	U	F	U	U	F	F	U	F	F	F	U		10/15
1986	D	F	U	F	F	U	F	U	U	U	U	U	F	F	U		6/15
1987	D	U	U	F	U	U	F	F	U	U	U	U	F	U	F		5/15
1988	N	F	F	F	F	F	F	U	F	F	F	F	U	F	F		13/15

Note: +ve and –ve signs indicate direct and inverse relationship of predictors with monsoon; N and D indicate normal and deficient monsoon rainfall; F and U indicate favorable and unfavorable signs with predictors for normal monsoon.

(**Source:** Gowarikar *et al.*, 1989.)

Table 5.10: Inferences based on signals from 15 parameters for the period: 1951-1987

Percentage of parameters favourable	No. of occasions	Subsequent monsoon condition	
		Normal	**Deficient**
90 (11 out of 12)	1	1	0
80	2	2	0
70	6	6	0
60	18	18	0
50	21	20	1
50	16	7	9
40	12	3	9
30	7	1	6
20	6	1	5
10	2	1	1
0	1	0	1

(Source: Gowarikar *et al.*, 1989)

This model is purely qualitative where equal weightage is given to all the 15 parameters. But the relationship between the monsoon rainfall with individual parameter is non-linear.

Gowarikar *et al.* (1989) developed a power regression model where, they determined the magnitude of influence of each parameter. They used, a curvilinear equation of the form,

$$R = C_0 + \sum_{i=1}^{15} c_i \, x_i^{pi} \qquad ...(5.12)$$

where, R is the monsoon rainfall over India, X_1, X_2 .... X_{15} are different parameters and C_s and p_s are the model constant. The model constant on using the 23 years data (1958-80) and the final equation had the form,

$$R = 621.0 + \sum_{i-1}^{15} C_i x_i^{pi} \qquad ...(5.13)$$

These parameters were arranged in order of their decreasing correlation coefficient with monsoon rainfall. As regard the performance of power regression models, it was observed that it best predict the droughts of 1982, 1986 and 1987 during the independent test period of 1981-88. The performance of linear regression model and power regression model during, the period (1975-88) is shown in Fig. 5.14 which suggested that the power regression model is superior multiple regression model not only on account of its better performance but also on its capacity to utilize some part of the nonlinear relationship for prediction.

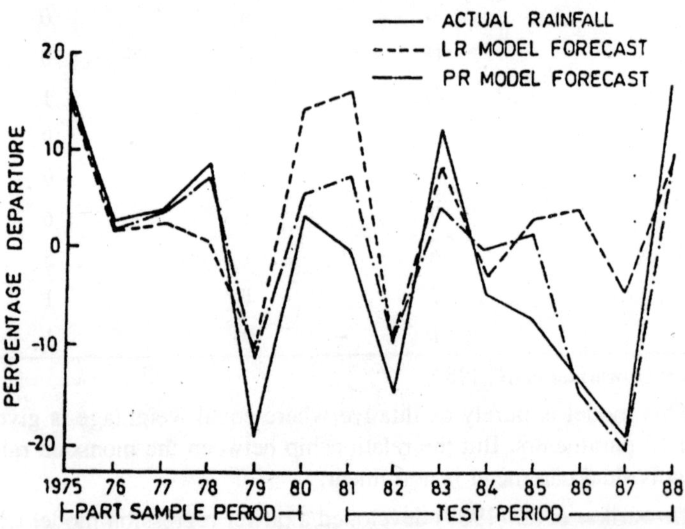

Fig. 5.14: Performance of linear regression (LR) and power regression (PR) model during part sample period (1975-1980) and test period (1981-1988). (Source: Gowarikar, *et al.*, 1989)

(iii) ARIMA technique

It is evident from the studies that ARIMA (Auto Regressive integrated Moving Average) technique may be used to forecast the weather in India and abroad (Delleur and Kavvas, 1978; Jackobson, 1979; Wenjeet *at el.*, 1980; Hongxing *et al.*, 1986; Thapliyal, 1981; 1986). The ARIMA technique fits the model to the past and projects forward a single time series like rainfall, temperature etc.

The general form of model is denoted by ARIMA (p,d,q), which can be expressed as suggested by Box and Jenkins (1976).

$$\phi(B)W_t = \theta(B)\,a_t \qquad \ldots(5.14)$$

where,

$$W_t = (1-B)^d z_t$$
$$\phi(B) = 1-\phi_1 B-\phi_1 B^2$$
$$\theta(B) = \theta_0-\theta_1 B-\theta_2 B_2 ...\theta_q B^q$$

Z_t, is the original non stationary time series, W_t is stationary time series derived from Z_t B is backward shift operator, p is the highest order of auto regressive term, q is the highest order of moving average term, d is the highest number of time for which original non stationary series is to be differenced, a is the random shock, and θ and ϕ_s are unknown constants.

(iv) Dynamic stochastic transfer technique

Thapliyal (1082, 1984) and Barnett and Somerville (1983) proposed the technique of combination of deterministic and statistical approaches. This technique might improve the accuracy of operational LRF. The general form of the Dynamic Stochastic Transfer (DST) model is expressed as suggested by Thapliyal (1981, 1982).

$$\gamma_t = \frac{w(B)}{\delta(B)} x_{t-b} + \frac{\phi(B)}{\theta(B)} a_t \;; b \geq 0 \qquad ....(5.15)$$

where,

$$W(B) = W_0 - W_1 B - W_2 B^2 ...... W_S B^S$$
$$\delta(B) = 1 - \delta_1 B - \delta_2 B^2 ....\delta_r B^r$$

$\phi(B)$ and $\theta(B)$ have already been defined in eqn. (5.14), Y_t and x_t are the output (say. Monsoon rainfall over India) and input (say, a meteorological predictor, a_t is random shock or white noise).

Fig. 5.15 Indicates that the forecast of rainfall for peninsular India together with the actual rainfall. The DST model has correctly predicted the droughts in 1979, 1982, 1984 and 1985. However, the model could only predict the sign of the deficiency in 1986 and 1987 but not the actual amount of deficiency.

The LRF issued by India Meteorological Department is given in Table 5.11.

Table 5.11: Long Range Forecast prepared and issued by India Meteorological Department.

Long Range	**Forecast for technique(s) used**	**Forecast issued in Ist week of**	**Forecast area**
Monsoon onset	MR	March	Kerala (South thern most state)
Rainfall during entire monsoon season (June to September)	DST and MR	May June	NW India & peninsula
Rainfall during Second half of monsoon season (August & Sept)	DST and MR	Aug.	-do-
Precipitation during winter season (January to March)	DST and MR	Jan	NW India
Rainfall during each calendar month	ARIMA	Each month	All 31 met. Sub div. of India.

(Source: Bulletin NTP 70 of NIH, 1989-90)

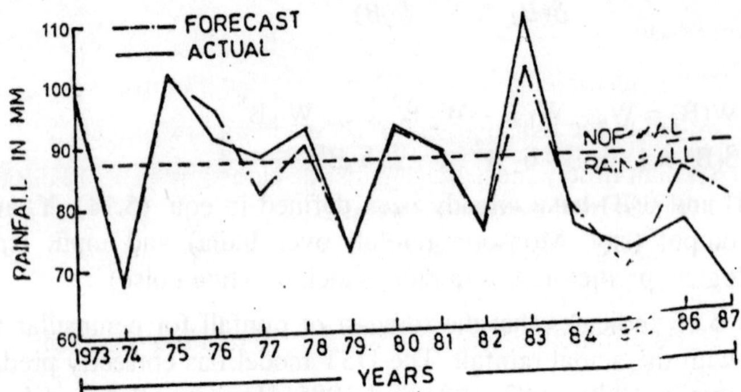

Fig. 5.15: Verification of DST model forecast for Peninsula.
(Source: Thapliyal, 1988)

5.10.3.1 Application of long range forecast

The forecasts of crop diseases and pests are based on laboratory research and long-term analysis of diseases and climatological data. These forecasts are concerned with probable development, spread and modification of a disease or pest, over a few months or a year. The purpose of long range forecasts is to give warnings of the probable nature, extent and time of the spread of disease or pests, and to prepare a control programme for them. A few examples of long-range forecasts of disease and pests are given as follows. It emphasis on abnormalities in temperature and precipitation. Its accuracy is 60 per cent.

This type of forecasting helps in the following farm operations:

(i) In soil moisture management,

(ii) In pasture management,

(iii) In determining irrigation frequency,

(iv) In harvesting crops for short term storage where adverse conditions are likely to interfere,

(v) To decide whether to put perishable products into short term storage to even out supplies to market, and

(vi) In avoiding chemical sprays when diseases or pests are unlikely to be troublesome.

Seasonal outlook

It emphasize on Abnormalities in temperature and precipitation. Its accuracy is 60 per cent.

Applications

(i) To decide whether or not to grow marginal crops,

(ii) To aid in the management of limited water resources,

(iii) To plan timely measures against diseases and pests likely to be favoured by the expected weather,

(iv) To choose varieties most likely to thrive in the expected pattern of the weather,

(v) To determine the acreage needed to give a required tonnage of crop, and

(vi) To determine crops yield.

5.11 Preparation of a Weather Outlook for the Farmers

Smith and Walter (1973) reported that the weather outlook for farmers is an interpretation of as how expected weather in the future and weather conditions accumulated upto the present, will effect crops, livestock and farm operations. A crop-livestock operations calendar is prepared and also used for agronomic operations, diseases and insects at proper time. This prepared farm summary, is known as synopsis of weather systems expected through the next two to five days period.

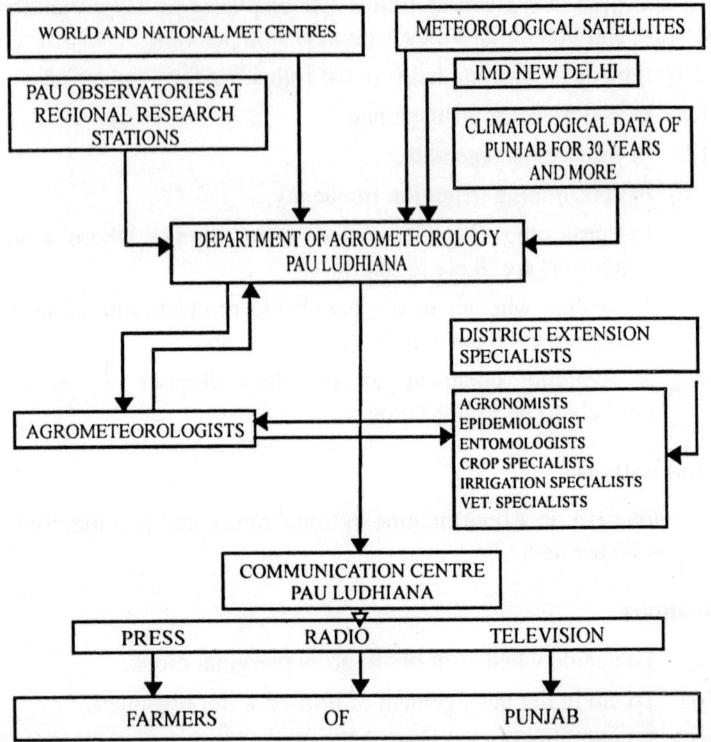

Fig. 5.16: Weather and crop outlook for the Punjab (Source: Mahvi, 1987)

Synopsis

It is a description of locations and movements of low pressure areas, high pressure areas, fronts and upper air troughs, and the associated weather with these systems. These informations would be derived form guidance charts of National Meteorological Centres. For this prognostic charts for surface and upper air conditions are used for preparing this synopsis.

Sunshine

This parameter gives an idea as to how much of possible sunshine will be available to dry soil, hay, evaporation of water and to complete farm operations. Based on the clod conditions sunshine hour is classified as below (Table 5.12).

Table 5.12: Classification of cloud based on sunshine hour

S. No.	Cloud classification	Sunshine hour
(i)	Minor cloud	90-100%
(ii)	Partly cloud or scattered cloudiness	50-90%
(iii)	Considerable cloudiness	20-50%
(iv)	Mostly cloudy	< 20%

The average cloudiness for the entire day-light period may be broken down into morning and afternoon periods. *Forecasting of cloud cover* is based on the moisture content of air between 1000-500 mb, stability or lapse rate, vorticity field and the type of average flow (cyclonic anticyclonic or vertical motion type of advection). Generally, it has positive vertical motion with mean RH 70 per cent or more and conditional instability, with cold air advection at 500 mb and warm surface advection. It will produce cloudiness and precipitation. Surface lifting of moist air is essential, either frontal or convectional. Clear sky will be seen when negative vertical motion, vorticity 10 or less and warm air advection aloft–prediction may be made by satellite charts, weather depiction charts, synoptic cloud observations and radar echos.

Precipitation

A forecast for the amount of precipitation is most important to farmers in planning their operation and estimating its effect on crop health and crop yield. For forecasting precipitation, rainfall analysis is necessary for a weather system. It is performed on using cloud pictures, upper air moisture, dew points, vertical motions, stability and temperature information (Anonymous, 1971). Rainfall is described in the forecast in the following manner:

(i) Isolated rainfall,

(ii) Scattered rainfall, and

(iii) Widespread rainfall.

Dew

Dew formation depends on cloud cover, wind, surface moisture and temperature-dew point relationships clear sky, winds under 8 kmph and moist soil surface will usually produce heavy dew formation. Dew formation results from rapid outgoing traditional cooling, very little surface air mixing and absence of positive vertical motion. Increasing wind and cloudiness decreases dew potential. Usually cloudy skies and speed of 15 km per hour or more, then no dew formation take place. High pressure areas and associated with dense air is more conductive for dew formation as compared to areas under low pressure systems. Dew forecast is classified into six categories *viz.*, very light, light, moderate, heavy, very heavy and wet.

Relative humidity

For forecasting of relative humidity at air temperature for minimums and maximums is determined for high and low RH values. A dew point analysis is made with an observed synoptic charts, then moved with pressure patterns and pressure systems. Maximum and minimum relative humidity is determined with the help of air temperature as dry-bulb and dew-point values. A diurnal curve may be drawn.

Drying rate

This forecast is performed by establishing a relationship between wind, temperature, radiation, % sunshine and humidity (or moisture content of air). It may be determined using an appropriate equation based on actual observed evaporation and projected conditions to cause changes in intensity.

Thresholds for spraying

The efficiency of spraying insecticides and fertilizer depends on the weather conditions during which the operations are conducted. For ideal condition of spray temperature should be varied from 5 to 27°C. At higher temperatures there is heavy loss by evaporation.

Table 5.13: Threshold wind speeds for spraying

Wind speed	Condition
Less than 8 kmph	Ideal
8-15 kmph	Good
16-22 kmph	Fair
23-28 kmph	Marginal
> 28 kmph	Unfavourable

For spraying six–hour sunshine duration is sufficient to spraying or application. However, any rainfall (even of one mm or less) is significant and should be taken into consideration as the chemicals used can get diluted or washed away. The threshold wind speeds for spraying is given in Table 5.13.

Spraying should be undertaken early in the morning or late in the evening, to ensure a minimum three hours of ideal weather under normal weather conditions.

5.12 Agroadvisory Services

The influence of weather is in fact operative even before the crop is sown, as the moisture availability and the thermal regime of the seed zone determine the date of sowing and the appropriate crop genotype to be sown. In spite of cultivation of high yielding varieties, improved cultural practices and plant protection measures, favorable weather is a must for good harvest. In India, NCMRWF's medium range weather forecast type (MRWF) based agromet advisory service in terms of farm operations by the farmers has become very successful in the country.

Agrimet advisory bulletins are an aid to Indian farmers for efficient utilization of resources for increasing the crop production and minimizing the adverse weather effects. The accuracy of the forecast for a long period is a great challenge but the medium range weather forecasts in recent years were improved to a great extent due to the modern technology. The emerging ability to provide timely, skillful weather forecasts offers the potential vulnerability to weather vagaries (Hansen, 2002). The National Centre of Medium Range Weather Forecasting provides the location specific weather forecast for four days in advance with the help of forecast models. The accuracy and reliability of weather forecast depends upon the performance of the model output. The verification of forecast is a difficult and controversial problem but it improves the confidence of farmers for its reliability and applicability in day to day field work.

Satellite takes pictures of the earth and atmosphere in the visible and infra-red spectra and transmits them towards the earth for the meteorological purposes. These pictures may be received at any station by a Meteorological Satellite Ground Receiving Station. Infrared images are used to determine the temperature of earth's surface and cloud temperature and its height. The pictures in the visible spectrum indicate cloud formations and enable to determine wind velocity from the movement of small clouds as tracers. The cloud pictures received at short intervals are used to provide accurate information on the development and movements of the clouds, their type, height, temperature as well as the wind velocity. The weather forecast is made with greater accuracy by super-imposing the synoptic charts and cloud pictures with respect to the occurrence of the following weather phenomena. The following informations may be obtained from satellite pictures:

- Occurrence of hail storms and squalls,
- Flash floods,
- Occurrence of rainfall, its amount and time,
- Intensity and persistence of cold and heat waves,
- Rate of evaporation and evapotranspiration,
- Range of wind speed,
- Duration of cloudy and sunny hours, and
- Frost intensity.

The forecast of above weather phenomena would enable agricultural scientists to give specific recommendations to farmers for the following operations:

(i) To either complete sowing operations or to withhold them,

(ii) To irrigate a crop or to withhold the irrigation,

(iii) To start and complete the harvesting operation or to with hold it,

(iv) To apply fertilizer or not,

(v) The appropriate time to spray pesticides and fungicides,

(vi) To take measures against frost damage,

(vii) Measures to protect young livestock from cold,

(viii) Management of farm labour,

(ix) Transportation of farm produce to the market, and

(x) Handling, transportation and storage of food grains.

5.12.1 Farm advisories services

Farmers appreciated weather based farm advisory service and adopted improved agricultural practices like irrigation scheduling, judicious and economic use of fertilizers, timely and effective management of pest and diseases and balance feed and timely vaccination for their animals.

Dissemination of farm advisories in right time to the farmers is most essential. Mass communication of Agromet Advisory (AAS) bulletins through All India Radio, Doordarshan, and local News TV channels and newspapers in local language has been found very useful. Similar findings have also been reported by Patel, *et al.* (1998) and Rana *et al.* (2005).

5.12.2 Utility of forecast

Validation of rainfall forecast revealed that the forecast was usable from 42 to 96.6 per cent in Bundelkhand region of Uttar Pradesh in different seasons of the year. Usability of rainfall forecast during winter season was observed to be 96.7 per cent followed by post monsoon and pre monsoon season. The rainfall forecast during SW monsoon period was usable in only 42 percent cases. Similar findings have also been reported by Singh *et al.*, (1999).

Season wise average ratio score of rainfall varied from 70.5 to 94.8 per cent. High percentage of ratio score was computed in post monsoon (96.7%) followed by pre monsoon, winter and SW monsoon. However, there were variations among different seasons. H.K score of rainfall also varied between 0.20 to 0.53 during various seasons. The RMSE values were also varied from 2.2 to 26.52. The lower values being indicative of minimum variability between observed and predicted rainfall forecast (Table 5.14).

Table 5.14: Analysis of skill scores of rainfall forecast for Bundelkhand zone of Uttar Pradesh (average of 7 years)

Season	Ration Score	H.K. Score	RMSE
Pre monsoon	94.8	0.21	2.36
SW monsoon	70.5	0.53	26.52
Post monsoon	96.7	0.20	2.20
Winter	88.7	0.38	7.90
Entire year	87.7	0.39	16.0

Forecast for maximum temperature was usable from 54 to 75 per cent in the region. The usable forecast was higher in post monsoon season (75% followed by SW monsoon season (73%) but the usability for maximum temperature during winter was comparatively low. The maximum percentage (90%) of usable forecast for minimum temperature was obtained in post-monsoon season followed by monsoon season. The accuracy of wind speed forecast was between 81 to 100 per cent whereas the usability of wind direction remained low (41%) with higher variability and needs substantial improvement.

The accuracy of forecast was very good for rainfall, maximum and minimum temperature, wind speed except wind direction. Rainfall forecast (other than no rain) especially in SW monsoon season still needs improvement in forecast accuracy. In all the forecasted variables except wind direction, the overall reliability of forecast being about 60 per cent. Weather based farm advisory to be quite profitable in farm management decisions provided its occurrence at acceptable level. Similar findings were given by Singh *et al.* (1999).

Table 5.15: Comparative statements of input cost, output and net profit (Rs/acre) and C:B ratio of AAS farmers over Non-AAS farmers for rice crop

S. No.	Year	Input name	AAS (Rs/acre)	Non-AAS (Rs/acre)	Difference (Rs/acre)	% saving due to AAS
1	2004-05	Rice yield (q/acre)	19.4	17.5	1.19	9.8
2		Total input cost	5125.9	5150.0	24.1	0.5
3		Total output	11366.7	10318.8	1047.9	10.2
4		Net profit earned	6240.8	5216.1	1024.6	19.6
1	2005-06	Rice yield (q/acre)	21.7	20.0	1.7	7.8
2		Total input cost	5146.6	5634.2	487.6	8.7
3		Total output	13194.7	12162.2	1032.4	8.5
4		Net profit earned	8048.1	6528.1	1520.0	23.3
1	2006-07	Rice yield (q/acre)	20.9	19.8	1.1	5.5
2		Total input cost	4989.2	5283.1	293.9	5.6
3		Total output	13073.4	12326.4	746.9	6.1
4		Net profit earned	8084.2	7043.3	1040.9	14.8
1	2007-08	Rice yield (q/acre)	21.0	19.9	1.1	5.5
2		Total input cost	5463.0	5998.8	535.8	9.8
3		Total output	13758.3	13020.7	737.6	5.7
4		Net profit earned	8295.4	7022.4	1273.0	18.1

5.12.3 Economic impact assessment of forest

The outcome of this study is based on analysis carried out on weekly feedback data collected from the farmers for rice and wheat crops. A comparative statement of input cost, output, net profit and cost benefit ratio of AAS farmers over non-AAS farmers is given in Table 5.15 for rice crop and in Table 5.16 for wheat crop.

Table 5.16: Comparative statements of input cost, output and net profit (Rs/acre) and C:B ratio of AAS farmers over Non-AAS farmers for wheat crop

S. No.	Year	Input name	AAS (Rs/acre)	Non-AAS (Rs/acre)	Difference (Rs/acre)	% saving due to AAS
1	2004-05	Wheat yield (q/acre)	18.6	17.8	0.8	4.3
2		Total input cost	3526.3	3521.4	4.9	0.1
3		Total output	12878.0	11865.0	1013.0	8.5
4		Net profit earned	9351.7	8343.6	1008.1	12.1
1	2005-06	Wheat yield (q/acre)	19.8	18.6	1.2	6.1
2		Total input cost	2828.9	3132.6	303.8	9.7
3		Total output	15996.7	14924.2	1072.5	7.2
4		Net profit earned	13167.8	11791.5	1376.3	11.7
1	2006-07	Wheat yield (q/acre)	19.6	18.4	1.2	6.1
2		Total input cost	3261.3	3810.8	549.5	14.4
3		Total output	19200.5	17913.1	1287.4	7.2
4		Net profit earned	15939.2	14102.3	1836.9	13.0
1	2007-08	Wheat yield (q/acre)	20.5	19.8	0.7	3.5
2		Total input cost	3602.0	4290.6	688.6	19.1
3		Total output	22002.0	21156.6	845.4	4.0
4		Net profit earned	18400.0	16866.9	1533.1	9.1

The results thus obtained indicated that AAS farmers harvested 19.4, 21.7, 20.9 and 21.0 q acre^{-1} as compared to 17.5, 20.0, 19.8 and 19.9 q acre^{-1} yield of rice crop which was 9.8, 7.8, 5.5 and 5.5 per cent more than Non-AAS farmers for three kharif seasons of 2004, 2005, 2006 and 2007, respectively (Table 5.15).

The AAS farmers were able to reduce input cost by 0.5, 8.7, 5.6 and 9.8 per cent as compared to Non-AAS farmers for all these respective years. Finally the AAS farmers gained 19.6, 23.3, 14.8 and 18.1 per cent more income from rice crop over Non-AAS farmers for the year 2004, 2005, 2006 and 2007, respectively. The AAS farmers harvested 18.6, 19.8, 19.6 and 20.5 q acre^{-1} as compared to 17.8, 18.6, 18.4 and 19.8 q acre^{-1} yield of wheat crop which was 4.3, 6.1, 6.1 and 3.5 per cent more than Non-AAS farmers for three Rabi seasons of 2004-05, 2005-06, 2006-07 and 2007-08, respectively (Table 5.16). The AAS farmers were able to reduce input cost by 0.1, 9.7, 14.4 and 19.1 per cent as compared to Non-AAS farmers for all these respective years. Finally the AAS farmers gained 12.1, 11.7, 13.0 and 9.1 per cent more income from wheat crop over Non-AAS farmers for the year 2004-05, 2005-06, 2006-07 and 2007-08, respectively.

Thus, it is evident from the assessment that AAS farmers were able to reduce the input cost because they were receiving the AAS bulletins in real time by the use of which they were able to manipulate the agricultural practices according to the expected weather condition likely to prevail in the Tarai and Bhabar agro-climatic zone of Uttarakhand. The National report of National Centre for Medium Range Weather Forecasting also indicated the economic benefit of the advisories for different agromet field units and it ranged between Rs. 330 and 3750/- in Maize and Rice (Anonymous, 2002).

5.13 Operational Forecasts

The short–range forecast is used for general farm activities or farm operations. The forecast period covers two days with an outlook for the third, fourth and fifth (Anonymous, 1971). These forecasts are used to guide the farmers for carrying out their farm operations with minimum risk and minimize crop losses and cost of inputs.

Soil preparation

The important parameters in the forecasts are soil moisture, soil temperature, precipitation and wind. Based on these parameters an idea may be got for the workable condition. It is found that the soil should be in a workable condition with moisture less than 80 per cent of field capacity; soil temperature should be above 0°C, the amount of rainfall less than 1 mm; and the wind speed should be less than 45 kilometres per hour.

Sowing crops

The parameters in the forecast are soil moisture, soil and air temperature, precipitation and wind. Based on this information sowing of crop is forecasted.

It is found that sowing operation should be performed when soil moisture varied from 40 to 80 per cent of field capacity, soil temperatures at depths of 50 to 100 mm should be above 4.5°C, though this varies with the type of crop – for cotton it should be above 15.5°C, and for wheat about 5°C; the rainfall should be less than 1 mm; and the wind speed less than 30 kilometres per hour.

While for transplanting crop the requisite condition is that the soil moisture should be 60 to 90 per cent of field capacity up to a depth ranging from 10 to 20 centimetres; the soil temperature should be above 10°C; the air temperature should be above –2°C and the wind speed below 25 kilometres per hour.

Fertilizer application

The parameters in the forecast are soil moisture, soil and air temperature, precipitation and wind. The requisite condition for fertilizer application in crop is that: (i) soil moisture -30 to 80 % of field capacity, (ii) The soil temperature should be below 10°C; for higher nitrogen levels. (iii) Precipitation should be less than 1 mm and, (iv) the wind speed below 40 kilometres per hour for granular fertilizers, 30 kilometres per hour for spraying fertilizers and 15 kilometres per hour for dusting fertilizers.

Irrigation

The weather parameters are soil moisture, precipitation, air temperature, wind, evapotranspiration and radiation. For application of irrigation upto a depth of 15 cm the moisture should be less than 50 per cent of field capacity in the root zone. There should be no precipitation and the wind speed should be less than 25 kilometres per hour.

Spray and dust

The important parameters are cloud ceiling, visibility, low level temperature inversion, air temperature, precipitation, wind and dew. The prerequisite condition for spray aerial application and dusting is that:

(i) The cloud ceiling should be greater than 150 metres and visibility greater than 1.5 kilometres;

(ii) A surface inversion is desirable;

(iii) Air temperature should be below 30°C; and

(iv) The wind speed should be less than 15 kilometres per hour.

While for ground application is that:

(i) The soil moisture should be below 90 per cent of field capacity.

(ii) No rainfall; and

(iii) The wind speed should be less than 25 kilometres per hour.

Harvesting

The weather parameters are soil moisture, air temperature, precipitation, wind, humidity, sunlight, dew and evapotranspiration.

a. Moisture-sensitive crops: Harvesting should be performed when the soil moisture is less than 90 per cent of field capacity; no rainfall; and the wind speed 8 to 30 kilometres per hour. The relative humidity should be below 75 per cent.

b. Temperature-sensitive crops: Harvesting is performed when the soil moisture is below 90 per cent of field capacity up to a depth of 15 centimetres, and the air temperature below 13°C. The rainfall is less than 1 mm.

c. All hardy crops: Harvesting is performed when the soil moisture is below 90 per cent of field capacity and the rainfall is less than 1 mm.

5.14 Methods of Weather Forecasting

The techniques used for weather forecasting may be divided into three broad categories: (i) Conventional or synoptic methods, (ii) Statistical methods and (iii) Numerical weather prediction techniques.

5.14.1 Synoptic method

This method involves detailed analysis of current weather reports from a large area. The current weather patterns are related with the past analogous situations and forecasts are prepared on the assumption that a current situation will behave on the lines of the past analogous situation. It is based on the experience and memory of the forecaster, but with the advent, of computers, picking analogues has become faster and more objective. It is useful for short range forecasts.

5.14.2 Statistical methods

Regression equations of other sophisticated relationships are developed between different elements and resulting climate. Normally selection of predictors or weather parameters is based on possible physical relationship with the predictant. This technique is useful for short as well as for long range forecasting. Multiple regression equation is developed for prediction of annual rainfall based on 16 parameters in India.

5.14.3 Numerical weather prediction method

The behaviour of atmosphere is represented by equations based on physical laws governing air movement, air pressure and other information. The developed equations are used for weather forecast. This technique is found suitable for medium range forecasts. Meteorological data coupled with modern weather prediction techniques and transmission of data through telemetric networks. With suitable backing agricultural support systems, it will imply a sea-change in ability to plan agriculture in optimal way for a given set of weather conditions.

5.15 Weather Modification

It is an uphill task. Some of the possible weather modifications are: artificial rain making, cloud dissipation, frost management maintenance of high soil temperature, reduction of evapotranspiration losses and wind velocity etc.

5.15.1 Artificial rain making

Principles of rain making

Clouds are classified into warm and cold clouds based on cloud temperature. If the cloud temperature is positive, these clouds are known as warm clouds and if negative then known as cold clouds. The nucleus is needed for precipitation, which differs with the type of clouds. Hygroscopic materials are essential as a nucleus for warm clouds. While, cold clouds, there are two approaches in cloud seeding *viz.* static and dynamic approach. For static approach, it is assumed that deficiency of ice crystals as the reason for non-precipitation. So dry ice or silver iodide is applied at one nuclei for a litre of cloud air. However, in dynamic cloud seeding, massive quantities of silver iodide or rice crystals of 100 to 1,000 nuclei for a litre of cloud are applied.

Cloud seeding

The basic objective of cloud seeding is to correct the deficiencies of nuclei in the cloud. The water droplet in a cloud grows a million times its original volume on each around a nucleus present in the atmosphere to become rain drop. Application of foreign material to clouds to induce precipitation is known as cloud seeding.

Cloud seeding depends on type of clouds. In the case of cold clouds, precipitation does not occur if the ice crystals do not grow in sufficient quantities. Water droplets in super cooled clouds do not necessarily freeze at temperature below 0°C. Application of silver iodide stimulates additional upper-cooled water droplets to from ice particles. Hygroscopic materials are used to initiate coalescence of water droplets for warm clouds. When water droplets become larger, they break up and fall down. Silver iodide and sodium chloride (common salt) is used for cloud seeding to cold and warm clouds respectively.

5.15.2 Reduction of wind velocity and its losses

The adverse effect of high wind velocity may be minimized by wind-breaks. Wind breaks or shelter-breaks such any structures as reduce wind speed and shelter–belts are long rows of plants across the winds. The plants useful for shelter-belts are casurina, subabul, bamboo, phalsa etc. Every crop like castor, maize, sorghum, pigeonpea etc. can be used as temperature shelter–belts to protect crops like turmeric, onion, groundnut, sugar, soybean etc. It reduces wind velocity considerably thus overcomes the adverse effects of high winds. Evapotranspiration losses are decreased which help in avoiding moisture stress for rainfed crops. Shelter-belts create microclimate. They reduce potential and actual evapotranspiration. Crop relations are improved with higher leaf water potential and lower strata resistance. Shelters improve opportunity for photosynthesis and yields increased substantially (20 per cent) especially in dry land ecosystem.

Though shelter-belts of 2 to 4 rows occupy valuable cropped land, it competes with crop for water and nutrients and shade the crop nearer to the shelter-belts. Despite this total crop yield was increased. If the value of wood, fruits etc., of trees of shelter-belt are considered the total economic returns are as high as compared to non- sheltered land.

References

Adeoge, K.B. 1984. Effect of shelter belts in the Sudan Savanna Zone of Nigeria on microclimate and yield of millet Agric Forest Meteorol. 33:99-107.

Agarwal, A. 1988. Stochastic model of monthly rainfall for Nainital tarai. *J. Agric Engg*. 25(3):20-31.

Allen, R.G., Pereira, I.S., Daes, D. and Smith, M. 1988. Crop evapotranspiration, Guide line for computing crop water requirements. FAO Irrigation and Drainage Paper 56, Rome Italy, p. 300.

Anonymous, 2002. Background paper for 11th Annual Review Meeting of National Centre for Medium Range weather Forecasting (NCMRWF) New Delhi. In: Status of economic impact. pp 1-3.

Aparna R. 2001. The 29 Commandments of Jambeshwar. German Research 1: 8-11.

Barnett, T.P. and Somerville, R.C.J. 1983. Rev. Geo physics and Space Phys. 21:1098-1102.

Barry, R.G. and Chorley, R.J. 1976. Atmosphere, Weather and climate. Edition, English Language Book Society and Methuen and Co. Suffolk.

Berry, F.A. and Bollary, E. 1945. Handbook of Meteorology McGraw Hill Company Inc., New York.

Bhakar, S.R. and Raj Vir Singh. 2005. Stochastic Modelling of Minimum and maximum Relative Humidity at Udaipur. *Journal of Water Management*. 13(1):27-35.

Bhakar, S.R., Bansal, A.K., Chhajed, N., and Singh, R.V. 2006. Rainfall analysis for drought estimation of Udaipur Region. *Journal of Agricultural Engineering*. 43(3):40-43.

Bhakar, S.R., Rajvir Singh and Hari Ram, 2006. Stochastic modeling of wind speed at Udaipur *Journal of Agricultural Engineering*, 43(1):1-7.

Bhakar, S.R., Rajvir Singh, Neeraj Chhajed and Anil Kumar Bansal. 2006. Stochastic modeling of monthly rainfall at Kota region. *ARPN Journal of Engineering and Applied Sciences*. www.arpnjournals.com 1(3):36-44.

Bhatia, P.C. 1999. Crops and weather resources. In: 50 years of Natural Resource Management Research, (Eds. Singh, G.B. and Sharma, B.R.), ICAR Publication, New Delhi, pp 49-62.

Box, F.E.P. and Jenkins, G.M. 1976. The time series analysis forecasting and control, Holden-Day.

Butler, L.M. and Butler, R.O. 1987. Needs assessment in international development: The context. In: Needs assessment Theory and Methods (Eds. D.E. Johnson *et al.*) Lowa State University Press, *Ames,* p. 213-243.

Chang, J. 1968. Climate and Agriculture. Aldine Publishing Company, Chicago

Crtenfield, H.J. 1975. General climatology. Third edition. Prentice – Hall of Private Limited, New Delhi.

DAS, L.A. 1964. Crop and Weather in India ICAR, New Delhi.

Datta, R.K. 1989. Weather Forecasting–Role of medium range techniques. In Sci. of India Agricultural 1989. The Hindu, Madras.

Delleur, J.M. and Kavvas, M.L. 1978. *J Appl. Moteo*, 17, 1528-1536.

Demnis, A.S. 1983. Augmentation of rainfall from summer cumulus clouds. Water Manage., 7: 3-14.

Donn, W.L. 1965. Meteorology, McGraw Hill, New York.

Durbin, J. 1960. The fitting of time series models Review of the International Statistical Institute. 28:233-244.

Fuller, W.A. 1976. Introduction to Statistical Time Series.

Goulden, C.H. 1959. Methods of Statistical Analysis.

Gowarikar, V., Thapliyal, V., Sarkar, R.P., Mandal, G.S. and Sikka, D.R. 1989. Parametric and Power regression models' New Approach to longrange forecasting of monsoon rainfall in India, *Mausam*, 40,2 pp. 115-122.

Harpal Singh, M. 1974. Agricultural Meteorology, Punjab Agricultural University Ludhiana.

Hippel, K.W. and Mcleed, A.I. 1994. Time series modelling of water resources and environmental system Elsevier Science, B.V. Amsterdam. P. 1013.

Hongxing, Cao, Wenjie, H. and Mianmian, W. 1986. WMO/TD 87 pp. 775.

ICRISAT, 1980. Proceedings of International Workshop on the Agro-climatology Research Needs of the Semi-arid Tropics, 22-24 November, 1978. Hyderabad India.

Indian Institute of Tropical Meteorology. 1976. Proceeding of the symposia on Tropical Monsoons held at Pune-5, India, 8-10 Steeper, 1976.

INH, 1990. Long range forecasting of most of drought conditions in Tropics and Subtropics Bulletin No. 77, p 70.

Jackobason, T. 1979. Arch. Met. Geophys. Biokl Ser. A 28 pp. 149-158.

Jaishwal, R.K. 1992. Stochastic modelling of pan evaporation. Unpublished ME Thesis submitted in Department of Soil and Water Engineering, CTAE, Udaipur pp 125.

Jat, M.L., Rajvir Singh, Bhakar, S.R., Gupta, A. and Sharma, S.K. 2006. Stochastic Modeling of water deficit under climatic condition of Jodhpur. *Indian Journal of Agric. Research and Development*. 21(1):31-39.

Kakde, J.R. 1985. Agricultural Climatology, Metropolitan Book Co., New Delhi.

Kaledhonkar, M.J., Kamra, S.K. and Yadav, R.K. 1996. Probability of historic droughts for South-Western Orissa in India. *Jr. of Indian Water Resources Society*, 2(4):19-23.

Kashyap, R.L. and Rao, A.R. 1967. Dynamic stochastic models from empirical data. Academic press, New York.

Khanna, S.S. 1989. The agro-climatic approach 'In Survey of Indian Agricultural 1989. The Hindu, Madras.

Khichar, M.L. and Bishnoi, O.P. 2003. Accuracy of weather forecast for western agroclimatic zone of Haryana during kharif season. *Haryana Agric. Univ. J. Res.*, 33: 97-101.

Kottegoda, N.T. 1980. Stochastic water resource technology. The McMillan Press Ltd. London.

Lock Wood, J.G. 1979. World Climatology – An environmental Approach English Language Book Society and Edward Arnold (Publishers) Ltd. Be Square.

Matalas, N.C. 1967. Mathematical assessment of synthetic hydrology. *Water Resources Research*, 3(4): 937-945.

Mather, J.R. 1974. Climatology – fundamentals and Applications. McGraw Book Co., New York.

Mavi, H.S. 1987. Introduction to Agrometerology. Oxford Publishing Hose. Pp. 237.

Mavi, R.H. 1965. Introduction to the Atmosphere, International Student Edition, McGraw Hill Book Company, New York.

McBride, J. and Ebert, E. E. 2000. Verification of quantitative precipitation forecast from operational numerical weather prediction models over Australia. *Weather and Forecasting*, 15: 103-121.

McIntosh, D.H. and Thom, A.S. 1969. Essentials of Meteorology. The Ly Publication (London) Ltd. London.

Munn, R.E. 1970. Biometeorological Methods Academic Press, New York.

Narayana Iyengar, R. 1982. Stochastic modelling of monthly rainfall. Journal of hydrology, 57 (3/4): 375-387.

Nelson, R.A, Holzworth, D.P., Hammer, G.L. and Hayman, P.T. 2002. 'Infusing the use of seasonal climate forecasting into crop management practice in North East Australia using 2 discussion support software'. *Agric. Systems*, 74: 393-414.

Pande, A. 1970. Modem Hygrometry. Somaiya Publications Private Ltd., B

Parthasarathy, B. and Pant, G.B. 1985. Seasonal relationship between Indian summer monsoon rainfall and the southern Oscillation. *Intern. J. Climatol.* 5: 369-378.

Partiia, A.T. 1954. An Introduction to Climate McGraw Hill Book Company. Inc. New York.

Patel, H.R., Saikh, A.M. and Venkatesh, H. 1998. Status of present day weather forecasting to the farmers: A case study of Middle Gujarat region. *Annals Agric. Res.*, 19: 285-289.

Patterson, S. 1956. Weather Analysis and Forecasting Sec. Edition McGra Book Company Inc., New York.

Pinners, E. and Balasubramanian, V. 1991. Use of the iterative diagnosis and design approach in the development of suitable agroforestry systems for a target area. *Agroforestry Systems,* 15: 183-201.

Prasad, M., Sharma, A. and Bhakar, S.R. 2006. Rainfall Modeling for Bhilwara Climatic Conditions: A case Study. Journal of Water Management. 14(1): 30-40.

Ramdas, L.A. 1996. Weather and crops. In: Hand Book of Agriculture, Indian Council of Agricultural Research, New Delhi, pp 1-31.

Rana, R. S., Prasad, R. and Kumar, S. 2005. Reliability of medium range weather forecast in mid hill region of Himachal Pradesh. *J. Agrometeorol.,* 7(2): 297-303.

Randhawa, M.S., Sivaraman, M.S., Naidu, U. and Vaidya, Suresh 1961. Farmers of India, Madras, Andhra Pradesh., Mysore and Kerala, Volume II. Indian Council of Agricultural Research, New Delhi, pp. 428.

Rao, G.G.S.N., Kesava Rao, A.V.R., Ramakrishna, Y.S. and Victor, U.S. 1999. Resource Characterization of Drylands: Climate. In a book, Fifty Years of Dryland Agricultural Research in India (Eds. H.P. Singh *et. al*), CRIDA, Hyderabad.

Rijks, D. 1978. Agrometeorological network over Africa. Proceedings Agrometeorological Research Needs of Semi-Arid, Tropics, ICRISAT Pantancheur, A.P., India.

Salas, J.D., Delleur, J.W., Yevjevich, V. and Lane, W.L. 1985. Applied modelling of hydrologic time series. Water resources publication, littletan, Colorado. P. 482.

Shukla, P.R., Sharma, S.K. and Venkata Ramana, P. 2002. Climate change and India- issues, concerns and opportunities, Tata McGraw-Hill Publishing Company Limited, New Delhi, 317 p.

Singh, S.V, Rathore, L.S. and Trivedi, H.N. 1999. Verification of Medium range weather Forecast (in) Guide for agrometeorological Advisory Services, pp 73-81 DST/ (NCMRWF), GOI, New Delhi.

Singh, S.V., Rathore, L.S. and Trivedi, H.K.N. 1999. Guide for Agrometeorologic al Advisory Service. *DST-NCMRWF,* New Delhi.

Subbarayappa, B.V. 2001. The roots of ancient medicine: A historical outline. *J. Biosci.,* 26: 135-143.

Thapliyal, V. 1981. 'Stochastic dynamic model for long range prediction of monsoon rainfall in peninsular India, *Mausam* 33, pp. 399-404.

Thapliyal, V. 1984b. 'State of Art Report on Long Range Prediction of monsoon rainfall; South Asian Association for Regional Cooperation, Meteorology, pp. 80 (Submitted).

Thapliyal, V. 1986. Prediction of India monsoon variability–evaluation and prospects including development of new model. Climate of China and global climate. Ed. Ye Duzhing Fu Congbin, Chao Jipin and M. Yoshino, China Ocean Press Beijing, 397-416.

Thapliyal, V. 1986. WMO.TD/87. 723-732.

Thornhwaite, C.W. 1948. An approach towards a rational classification of climate. Geographical Review, 38: 55-94.

Thornthwaite, C.W. and Mather, J.R. 1955. The water balance climate 8(1). Dreseal Institute of Technology, New Jersey, USA PP 104.

Thornthwaite, C.W. and Mather, J.R. 1957. Instructions and tables for computing potential evapotranspiration and the balance. Publication in climate lab. of cli., Elmar., New Jersey. 10(3):45-50.

Thotiam Reddy, J. and Bhaskaram K. 1982. Agricultural profile. In 'Agrocultural in Andhra Pradesh Vol. I. The Farmers' Welfare Trust and Society for Advancement of Agriculture, Hyderabad.

Webster, P.J., Magana, V.O., Palmer, T.N., Shukla, J., Tomas, R.A., Yanai, M. and Yasunari, T. 1998. Monsoons: processes, predictability, and the prospects for prediction. *J. Geophy. Res.*, 103: 24-45.

Wenjie, H.C., Hongxing, G. Lau and Jintian, X. 1980. ARIMA Seasonal Model of Time series and its application to Long-Range Weather Forecasting, Proc. WMO Symp. On Probability and Statistical Methods in Weather forecasting, Nicc. Sept. 281-285.

Wilks, D.S. 1995. Statistical methods in the Atmospheric sciences: An Introduction. *Academic Press, New York:* 467.

WMO. Technical Note: Climate Change, TN No. 79.

Yevjevich, V.M. 1972. Stochastic processes in hydrology. Water resources publication, Fort Collins, Colorado.

EXERCISES

5.1 Differentiate the following:

(i) Weather and climate

(ii) Troposphere and tropospause

(iii) Direct and diffused solar radiation

(iv) Air and soil temperature

(v) Absolute and relative humidity

5.2 Explain the classification of atmosphere.

5.3 How wind speed is measured?

5.4 Describe the method of measurement of evaporation.

5.5 How rainfall is recorded in the observatory?

5.6 Classify the surface observatories?

5.7 What are the weather abnormalities?

5.8 Explain weather forecasting.

5.9 What are the indigenous techniques of weather forecasting?

5.10 What are the weather elements?

5.11 Discuss the application of short range weather forecasting in agriculture.

5.12 Describe the techniques of long range forecasting.

5.13 What is the stochastic transfer technique?

5.14 Discuss the application of long range forecast.

5.15 What is the weather outlook for the farmers?

5.16 Explain the threshold limit for spraying.

5.17 Explain the agro-advisory service.

5.18 Discuss the impact of the forecast.

5.19 Explain the operational forecasts.

5.20 Write short note on the following:

(i) Synoptic method

(ii) A cloud seeding

5.21 Explain the principle of rain making.

■■■■

Chapter 6

IN-SITU RAINWATER MANAGEMENT

6.1 Introduction

Dryland contributes about 45 per cent of the country's food production. This contribution from drylands to India's total production is by no means insignificant. More than 70 per cent of the raw materials needed for industries, oil seeds and cotton are obtained from drylands. However, under rainfed condition productivity is very low and unstable. Moisture conservation was concentrated upon construction of various types of bunds and other techniques across the land slope to control erosion and conserve rainwater and soil otherwise rain drop falling on the earth surface goes as waste in the form of runoff and create the problem of soil erosion or nutrient losses. *In-situ* water harvesting play a key role to increase productivity and reduce runoff, soil loss and nutrient loss. The *in-situ* water harvesting is defined as conserving the every drop of rain where it falls on the earth surface for reducing runoff and soil loss thereby increasing crop production. Rainfed areas which constitute nearby 70 per cent of the total cultivated areas in India suffer from not only scarcity of moisture but also from poor fertility on account of heavy rainfall which results into the loss of the fine available particles and nutrient.

Any practice which helps to control the runoff will naturally make available more water and nutrient to crop. The principle behind the recommendation of different practices is to increase the infiltration by reducing the rate of runoff temporarily and impounding the water on the soil surface to increase the opportunity time for infiltration by modifying the land configuration for inter-water harvesting and also by tillage and residue etc. Other practices which include improving soil physical conditions so as to increase the rate of infiltration and water retention in the soil system. Adoption of suitable practices for the local specific help improve the soil moisture availability and thereby result into higher crop yields. The management of rainwater in the dryland areas has been discussed in this chapter with special reference to *in-situ* water harvesting for increasing moisture availability to the maximum possible extent through different techniques which takes into account the realities of relationship between land-water-plants and cultivation practices.

6.2 Factors Affecting Rainwater Management

In planning the measures for rainwater management are to be controlled runoff and soil erosion. As water and land management are inseparable and successful strategies on rainwater management are location specific and depend upon the rainfall, slopes, soil type and texture etc.

There are following factors which affect rainwater management:

1. Climatic factors- It includes precipitation, wind, humidity and solar radiation. The total amount and intensity of precipitation affect runoff as well as erosion.
2. Wind- It also affects erosion by water and wind characteristics influence raindrop velocities and angles of impact ultimately on runoff. Temperature and wind affect evaporation and transpiration and thus affect the vegetation of the area. Humidity and solar radiation influence temperature and thus have an indirect effect on rainwater management erosion.
3. Soil- The properties which influence erosion and runoff include soil structure, texture, organic matter content, moisture content, and density or compactness. The chemical and biological characteristics of the soil also have an influence on runoff or rainwater management.
4. Vegetation- Vegetation plays an important role in rainwater management and water conservation and acts in the following ways: (i) transpiration of moisture from the soil back into atmosphere thus decreasing soil moisture increasing storage capacity; (ii) binding effect of the root system in surface layer of soil and thus retarding erosion; (iii) roots of vegetation after decay provides numerous tubular cavities which promotes infiltration; (iv) improvement of soil structure by addition of organic matter; (v) increased surface friction which reduces volume and velocity of runoff; and (vi) increased surface friction to the soil which helps in soil bacterial activity providing conditions for better plant growth. Thus, vegetation is useful in several ways. The extent of benefit provided depends upon the type of vegetation on the land surface.
5. Topography- Topographical features that influence runoff and erosion are degree of slope, length of slope, size and shape of the watershed.

6.3 Some Definitions

Erosivity- It is defined as the potential ability of the rain to cause erosion. It is a function of the physical characteristics of the rainfall. For a given soil conditions one storm can be compared quantitatively with another and a numerical scale of the values of erosivity can he developed.

Erodibility- It is defined as the vulne*rabi*lity or susceptibility of the soil to erosion. Erodibility is the function of both the physical characteristics of the soil and the land management practices. For a given rainfall condition, one soil condition can be compared quantitatively with the other.

Erosivity of rainfall- It involves work or expenditure of energy of rainfall event which causes the soil erosion. Energy is required to break the soil aggregates splashing them and subsequently carrying them with run off. The raindrops while falling acquire kinetic energy and on impact, the kinetic energy is used up in detaching the soil particles. When surface runoff flows down the land slope gain kinetic energy which is responsible for the scouring action on the land surfaces.

Hundson (1971) gave a comparison of the kinetic energies gained by the falling raindrops and surface runoff (Table 6.1).

Table 6.1: Effects of tillage implements and depth on maize yield in Alfisols of Bangalore

	Rain	Runoff
Mass (M)	Assume mass of falling rain is M	Assuming 25 per cent runoff, of runoff is M/4
Velocity (V)	Assume Terminal velocity of 8m/Sec	Assume velocity of surface flow is 1m/sec.
Kinetic energy $\frac{1}{2} MV^2$	32 M	M/8

Soil Erodibility- It depends upon the physical properties of the soil, the land, and crop management practices adopted. The effect of the physical properties can be evaluated more precisely compared to the effects of the management practices.

Attempts have been made by soil scientists to relate the amount of soil erosion as measured in the field to some physical characteristic of the soil that can be measured in the laboratory. Bouyouces (1935)

suggested that erodibility of a soil depends upon its mechanical composition and is proportional to the ratio of

$$\frac{\%\,sand + \%\,silt}{\%\,clay}$$

The term dispersion ratio based on the changes in silt and clay content before and after dispersion in water was also used. To quantify the effects of land management and crop management practices on soil erodibility is difficult as the variables involved are too many. The study of runoff and soil loss using runoff plots both experimental and field size is the only way to arrive at some acceptable values.

6.4 Raindrop Characteristics

The rainfall characteristics which influence runoff are the intensity and duration of rainfall, whereas the characteristic of raindrops influence erosion. The characteristics of the raindrops that are important from soil erosion point of view are: (i) raindrop size, (ii) drop size distribution, (iii) terminal velocity, and (iv) kinetic energy.

6.5 Basic Principles for *In-situ* Rainwater Management

There are following basic principles for *in-situ* rainwater Management:

i. Increasing time of concentration or opportunity for arresting rainfall at the site of occurrence by increasing infiltration rate with the help of deep tillage, profile modification, vertical mulch and by keep the soil surface rough to facilitate more infiltration,
ii. Reducing the velocity of runoff by creating barrier,
iii. Intercepting a long slope into several short ones, so as to maintain less than a critical velocity for the runoff water,
iv. Reducing evaporation from the soil surface, and
v. Protection against damage owing excessive runoff.

The important components of the technology given by Bhan and Uttam (1998) are: (i) maximum retention of rainfall at the site of occurrence with minimum loss through runoff and deep percolation, (ii) preservations of the conserved moisture for longer time by controlling evaporation and weed growth, (iii) water harvesting and its recycling and (iv) efficient utilization of the conserved moisture by selecting suitable crops and their varieties, cropping systems and adoption of appropriate soil and crop management practices. Though individual practice may bring about 20-30 per cent increase in yield of crops but appropriate combination of several practices may increase the yield by 100 per cent or even more.

6.6 Building *In-situ* Soil Moisture

There are following techniques for building *in-situ* soil moisture in soil profile:

i. Tillage,
ii. Residue,
iii. Tillage with residue,
iv. Mulching,
v. Land treatments,
vi. Agronomic measures, and
vii. Mechanical measures.

6.6.1 Tillage

Tillage is the soil manipulation for creating suitable edaphic environment for raising crops. Different types of soil require different tillage practices for cultivation of different crops. Light soils require lesser tillage as compared to heavy soils (Venkatsewalu, 1981).

Tillage is an important consideration under rainfed conditions for ensuring fine soil tilth, adequate moisture availability and weed control, leading to good crop stand and productivity. Rough and cloddy tillage before the onset of monsoon rains enhances infiltration, and decreases runoff and soil toss. Post-establishment inter-crop cultivation, locally known as *haloding,* decreases weeds, conserves more moisture in the soil profile and increases maize yield significantly over control (Sur and Ghuman, 1994). Harvesting of *kharif* crops and ploughing of the fields thereafter affects storage of moisture in the soil profile for utilization by the succeeding crop. Delayed harvesting or ploughing results in loss of moisture through evaporation from the capillaries formed in the soil (Maskina *et al.,* 1998). The continuity of the capillaries is destroyed due to ploughing and thus more moisture is conserved in the soil. Early sown *rabi* crops utilize residual soil moisture more effectively for better initial crop stand and establishment, whereas the performance of the late-sown crops is dependent on the rains received after sowing.

Ploughless tillage systems have achieved considerable interest among researchers and farmers due to high costs, unprotected soil surface and degradation of soil properties. Conservation tillage is associated with conventional tillage. The use of herbicides to control weeds, high costs of fuel in tractor dominated agriculture and public awareness of the environmental consequences of soil erosion have stimulated farmers to adopt modified tillage methods for cultivation of crops in the agriculturally advanced countries. The concept of conservation tillage was developed during the 1960s based on zero tillage keeping a protective cover of plants and/or plant residues on the soil surface to reduce soil losses and conserve moisture by allowing more water to infiltrate (Blevins and Fyre, 1992). Conservation tillage systems save time and require less labour and energy. Sowing of wheat with zero-till seed drill is fast becoming popular in the irrigated rice wheat areas of north-western India, and presently >1 million ha area is covered under this system.

6.6.1.1 Response to tillage

Modification of the edaphic environment by tillage affects growth development and economic yield of crops. Tillage influences both root and shoot growth in pre and post establishment phases, tillage effects the runoff and soil loss and input use efficiency.

Crop yield

An important step in successful cropping is obtaining an adequate crop stand consistent with available supplies of other inputs determining potential yields. The maize yield in alfisols of Bangalore was found highest with 30 cm deep M.B. ploughing (Table 6.2).

Table 6.2: Effects of tillage implements and depth on maize yield in Alfisols of Bangalore

Implement	Tillage	Maize yield (q/ha.)
Wooden plough	10	2430
Iron plough	15	4600
Disc harrows	20	5580
M.B. plough	30	4860
Chiseling	30	4780

Similar, study was conducted at Dryland farming Station, Arjia, Bhilwara, Rajasthan state to find the tillage requirement for maize cultivation. The results presented in Table 6.3 revealed that disc plough followed by twice cultivator gave highest maize yield as compared to other tillage system and reduces runoff by 30%, and soil loss by 66.6% over twice cultivator at the time of sowing. The time of application of tillage practices also affect the maize yield.

Table 6.3: Effects of different tillage practices on maize yield and water use efficiency

Treatment	Crop yield (q/ha)		Stover yield (q/ha.)		Percent runoff		Soil loss (t/ha)		Water use efficiency (kg/ha/cm)	
	1996	1997	1996	1997	1996	1997	1996	1997	1996	1997
Summer tillage										
Disc plough followed by cultivator	15.37	10.74	29.77	22.52	10.35	9.96	0.204	0.181	0.22	0.26
Cultivator followed by cultivator	14.00	6.67	31.14	15.96	12.78	10.75	0.311	0.245	0.20	0.16
Deshi plough followed by cultivator	13.88	8.68	30.68	21.66	12.78	10.03	0.305	0.286	0.20	0.21
Deshi plough followed by deshi plough	14.80	8.48	30.74	20.06	12.88	10.02	0.334	0.302	0.22	0.20
After Pre-monsoon										
Disc plough followed by cultivator	19.14	8.41	36.86	22.00	12.70	12.85	0.365	0.328	0.28	0.22
Cultivator followed by cultivator	20.28	5.64	39.14	16.13	13.10	14.19	0.399	0.359	0.30	0.21
At sowing										
Cultivator followed by cultivator	21.42	5.37	41.43	15.39	14.85	14.49	0.613	0.539	0.32	0.14

(Source: Jat *et al.*, 2007)

In chronic drought prone areas, deep tillage (20-30 cm) was found specifically applicable to soils having textural profiles or hard pans (Table 6.4). Under uni-modal (<500 mm) rainfall situation in semi arid regions with shallow Alfisols, sowing across the slope and ridging later was useful. It assisted in greater rainwater absorption and control of pernicious weeds.

In black soils, deep tillage alone was of no avail. Tillage combined with compartmental bunding was found to the most effective soil management practice for Vertic Inceptisols. With rainy season cropped Vertisols (uni-modal rainfall regions), water surplussing is an integral part on in situ moisture conservation. In bi-modal medium (500-750 mm) rainfall representing semi- arid Alfisol, graded border strips were found advantageous. Extensive tillage of sandy soils (common to desert areas) made them more vulnerable to wind erosion. Superficial scraping to eliminate weeds was found adequate.

Table 6.4: Crop response to primary deep tillage in drought proofing approach under three types of severity

Crops (seasons)	Grain (Mg/ha)		Response (%)
	Shallow tillage (10-15 cm)	Deep tillage (20-30 cm)	
	Severe Drought		
Coarse cereals (5)	0.92	1.30	41
Legumes (6)	0.61	0.76	25
	Moderate Drought		
Coarse cereals (16)	2.14	2.37	11
Legumes (2)	1.21	1.38	14
	Less Prone to Drought		
Coarse cereals (4)	1.55	1.72	11
Legumes (1)	0.25	0.29	16

(Source: Vittal *et al.*, 2005)

Soil moisture

Summer tillage helped in reducing the runoff losses thereby increases the depth of penetration as given in Table 6.5 (Dixit *et al.*, 1981) shows the increase in soil moisture with increase in depth of tillage.

Table 6.5: Influence of deep tillage on the profile moisture status

Depth (cm)	Moisture (%)	After 81 mm rainfall
	Plough area	Unplough area
0-15	10.74	3.59
15-30	13.22	7.13
30-60	12.27	8.59
60-90	13.44	Dry

Runoff and soil loss

No tillage practice has been found to be very effective in controlling runoff and soil loss than the conventional tillage (Triplett *et al.*, 1973; Oschwald and Slemens, 1976). The data of Free and Bay 1976 illustrates reduction in runoff and soil loss by no tillage (Table 6.6)

Table 6.6: Runoff and soil loss as influenced by no-tillage

Treatments	Runoff (% of rain)	Soil loss (t/ha.)
No tillage	23	2.5
Conventional tillage	29	17.5

Although soil loss was more curtailed yet runoff reduction conversely moisture conservation was of no less importance. Oschwald and Siemens (1976) reported that no-tillage practices conserved 20% more moisture in the soil profile than that of conventional tillage. This was attributed to higher rainwater infiltration and better structure conditions.

6.6.1.2 Special tillage systems

In order to alleviate soil physical constraint to cropping and/or to meet some specific crop requirements (as in rice), special tillage practices have also been devised.

Puddling

Puddling is an important soil management technique in weltand rice cultivation (Sanchez, 1973). Puddling defined as destruction of soil aggregates into ultimate soil particles at a moisture level near saturation (Sharma and De Datta, 1986) and it is important practice in rice cultivation. It comprises pre-puddling (dry) tillage followed by soaking irrigation and stirring the topsoil layer with water ponded at the surface. The soil is worked repeatedly in a liquid, semi-liquid, or saturated state until larger clods are broken down into smaller aggregates and primary particles. A puddled soil thus is visualized as one whose structure has been destroyed, the aggregates lose their identity, and soil is converted into a structurally homogeneous mass of textural separates. The destruction of soil structure through puddling results in decreased macroporosity and increased water retention, decreased saturated hydraulic conductivity, and reduced percolation and nutrient leaching losses. Puddling also helps in land leveling and maintenance of uniform floodwater depth, in creating reduced soil conditions. It softens the soil for ease of seedling transplanting and reduces soil strength, which enhances root proliferation and control weeds.

Compaction

Sandy soils in arid and semi-arid regions pose specific management problems for rainwater management to augment crop production because these soils having excessive permeability, high potential for nutrient leaching, and low available soil moisture storage capacity. Due to low organic matter content and cation exchange capacity, these soils have a lower productivity index than medium-to fine-textured soils. Crops raised on these soils are more prone to yield-reducing water and nutrient stressors. Compaction of the surface layers in such soils has been found to benefit crop production (Gupta *et al.,* 1989; Ohu and Folorunzo, 1989; Onwualu and Anazodo, 1989; Kayombo and Lal, 1993; Agrawal *et al.*, 1987; Agrawal, 1991). Soil compaction increases micro porosity at the expense of non-capillary pores. It reduces infilte*rabi*lity and increases the retention of water and nutrients in the surface layers and also enhances the surface water flow during flood irrigation.

Profile modification

Profile modification consists of digging and mixing soil to depths greater than those of ordinary ploughing (0.25 m), with or without the addition of physical or chemical amendments (Unger, 1979). It is also defined as tillage operation that alters any part of the soil profile. The main objective of profile modification is to alleviate soil constraints to root growth, proliferation, and activity. Unger (1979) identified the following three soil problems that limit crop production and may be alleviated by profile modification –

(1) soil layers that restrict root growth and water movement; (2) undesirable substances (salts, toxic materials, radioactive fallout); and (3) coarse-textured materials at the surface or at greater depths.

Eck and Unger (1985) extensively reviewed soil profile modification for crop production under diverse soil, climate, and management situations. Soil profiles having high strength zones, which restrict root growth and water movement, may be modified by deep ploughing, chiseling, trenching, or slip ploughing, depending upon the depth and density of the restricting layer. Deep ploughing may help alleviate the effect of undesirable substances by mixing horizons and/or by placing them deeper in profile. Where coarse sand in underlain by fine textured material, deep ploughing helps improve the water relationship of surface soil by mixing it with fine textured materials. Exportation of surface sandy layers to expose find/medium-textured soil has been practised on large areas in the Punjab.

Off-season tillage

It is another simple practice which is usually employed by farmers for conservation of rainwater. Deep tillage is resorted to conserve the monsoon rainfall, whereas shallow tillage by chiseling is considered useful during *rabi* season. Deep tillage in *rabi* after 3-4 years of shallow tillage is also reported to be useful for moisture retention. It is, however, now argued that one of the major causes of excessive land degradation in areas of high intensity rainfall is conventional soil preparation by plough, which leaves the soil exposed to rain, wind and sun. Conventional tillage may cause long term negative effects as a result of increased oxidation of organic matter due to increased aeration and prolonged exposure of bare soil to solar radiation, increased, water and wind erosion etc. Recent research shows significant improvements in crop yield, reduced soil erosion and reduced labour requirements after the introduction of conservation tillage. More research is, however, needed to quantify the decline in organic matter vis-à-vis tillage operation and to establish usefulness of conservation tillage.

Deep ploughing/summer ploughing

In rainfed areas continuous shallow ploughing with country plough over the years result in the development of hard pan at about 15 cm depth which hinders root growth and percolation of rain water. Thus the valuable rain water is flushed out from the field as runoff. It is, therefore, essential to break the hard pan by deep ploughing or chiselling. The results of a field study conducted on sorghum and cotton at Kanpur convincingly prove the utility of deep ploughing in increasing the yields of both the crops (Table 6.7 and 6.8).

Table 6.7: Effect of deep ploughing on sorghum

Method of ploughing	Grain (q/ha)	Additional yield over country	
		q/ha	(%)
Two ploughing with country plough	26.4	-	-
Two ploughing with cultivator	26.5	-	-
Two ploughing with harrow	28.8	2.4	9.1
One ploughing with victory plough + one ploughing with cultivar	29.9	3.5	13.3

(Source: Tripathi and Solanki, 1989)

Table 6.8: Effect of summer ploughing on yield of seed cotton

Type of summer ploughing	Yield ($q\ ha^{-1}$)	Increase over control	
		$q\ ha^{-1}$	(%)
No. of ploughing	4.42	-	-
One ploughing with tractor drawn cultivator	6.48	2.06	46.6
Two ploughing with tractor drawn cultivator	7.17	2.75	62.2
One ploughing with tractor drawn disc harrow	7.52	3.10	70.1
One ploughing with tractor drawn disc plough	8.71	4.29	97.1
One ploughing with tractor drawn M.B. plough	8.23	3.81	86.2

(Source: Tripathi *et al.*, 1995)

6.6.2 Crop residues

Crop residues play an important role in controlling runoff and soil loss. They dissipate energy of falling rain drops, reduce surface sealing and help to maintain higher infiltration rates and reduce runoff velocity (Mannering and Meyer, 1963; Meyer *et al.*, 1970; Foster and Meyer, 1973). Effectiveness of crop residue for controlling runoff and soil loss depends upon kind, amount of residue and mode of application. Kind of crop residue (Coarse v/s find material) influences soil loss because of differential surface cover provided per unit weight of these materials. Wischmeier (1973) found that more than 95% surface coverage is obtained with 5.6 t/ha of wheat straw but for similar coverage 8.2 t/ha of chopped maize stalks were required. The amount of residue needed for effective control of soil loss is closely related with the extent of surface coverage provided. This is because of the fact that the raindrop interception is a direct function of percent soil surface covered. In general, runoff and soil loss are found to be inversely related to mulch rate (Mannering and Meyer, 1963; Lattanzi *et al.*, 1974; Lal, 1975; Sood, 1975; Hadda, 1983).

Surface residues are an essential component of no tillage system, therefore, effectiveness of these practices greatly depends upon the amount of residue present at the surface (Moldenhauery, 1977), this is also evident from the results (Ketchesen, 1977) shown in Table 6.9. The results emphasize that without surface residues no tillage practices are less effective in controlling runoff and soil loss.

Table 6.9: Effect of surface residues on runoff and soil loss

Treatments	Runoff, mm		Soil loss t/ha	
	Residue present	No residue	Residue present	No residue
No tillage	11	37	3.1	38.1
Conventional tillage	25	25	24.6	40.3

A study conducted by Mannering and Meyer (1963), typically illustrate the effect of mulching rate (Table 6.10). The results revealed that even mulch rates as low 0.56 t/ha could reduce soil loss by 70%. A mulch rate of 4.48 t/ha reduces soil loss to zero.

Table 6.10: Runoff and soil loss from various mulch rates on a silt loam soil with 5% slope

Malching rate (t/ha)	Surface covered	Runoff (cm)	Soil loss (t/ha)
0	0	7.2	27.6
0.56	40	6.4	7.2 (70.3)
1.12	60	4.0	3.2 (88.4)
2.24	87	0.8	0.7 (97.4)
4.48	98	0.2	-
8.96	100	0	-

(Figure in parenthesis indicate percent reduction in runoff and soil loss relative to no match treatment)

6.6.3 Crop residues with tillage

Sandal & Acharya (1977) conducted study at Palampur. Results revealed that deep tillage to either of the crops in maize-wheat cropping system was not beneficial in improving productivity (Table 6.11). The yields of maize and wheat were highest under minimum tillage when Lantana was applied at harvest of wheat or to standing crop of maize. Conservation tillage involving minimum tillage (in wheat) with Lantana mulch (in standing maize or at its harvest) conserved more moisture and resulted in higher grain yield of wheat than conventional tillage without mulch. Wheat yield due to mulching in standing maize, at maize harvest and after the sowing of wheat were, respectively, 1.5-2.4, 1.3-1.5 and 1.1-1.7 times higher than conventional tillage, and depended on the rainfall events in different years. Mulching increased grain yield of maize significantly after two cropping cycles due to improvement in soil physical and chemical properties.

Table 6.11: Effect of different tillage and mulching treatments on productivity of maize and wheat crops at Palampur (mean of 3 years)

Treatments		Grain yield (t/ha)	
Maize	**Wheat**	**Maize**	**Wheat**
Lantana mulch at harvest of wheat and sowing with minimum tillage	Lantana mulch at harvest of maize and sowing with minimum tillage	3.55	2.47
Conventional tillage + Lantana mulch in standing maize	sowing with minimum tillage	2.62	2.81
Conventional tillage + Lantana mulch in standing maize	Incorporation of Lantana residue at sowing	2.64	2.24
Conventional tillage	Deep tillage	2.61	1.50
Deep tillage	Conventional tillage	2.24	1.42

6.6.4 Mulching

Mulching is a traditional practice in dryland ecosystem areas for controlling soil loss, runoff, weed growth and conserving moisture in the soil profile *viz.* soil mulch, straw and plastic and also gravel mulch. The primary objective of mulching is to reduce the evaporation loss. Numerous soil physical properties and conditions are influenced either directly or indirectly. Soil mulch is practical both in *kharif* as well as *rabi* situations. In *kharif* it helps in reducing the weed population and increasing infiltration whereas in *rabi* its main function will be to reduce evaporation. The surface mulch was incorporated into the soil in the subsequent years and the addition of organic matter had an advantages on soil physical properties and crop yields.

Application of organic material as surface mulch or timely interculture operation to create adequate dust mulch would reduce evaporation losses and improve the moisture availability, thereby resulting is higher yields as is evident from Table 6.12.

The maximum grain yield was obtained by dry grass against the control (728 kg/ha). Under bellary conditions, application of surface mulch and dust mulch resulted in 1760 and 1833 kg/ha. Jowar grain yield, respectively against control 934 kg/ha.

Table 6.12: Effect of surface mulch on the yield of Jowar (kg/ha)

Treatment	**Grain**	**Straw**
Control	728	3430
Jowar stubbles	1035	4720
Wheat straw	1120	5290
Redgram stalk	1182	5620
Dry grass	1265	6240

*Average of three years

Inter-cultivation and mulching for moisture conservation

Surface evaporation and transpiration through weeds are the two most important unwanted sources of loss of soil moisture under rainfed condition. Inter-cultivation by *kassi* (hand hoe), harrow, cultivator etc. in between the crop rows removes weeds and helps check evaporation losses by forming a dust mulch on the soil surface. In various field experiments conducted at this University inter-cultivation has resulted in increased yield of crops to the extent of 1.03 q ha^{-1} (13.6%) in bajra, 1.3 q ha^{-1} (53.3%) in sesame, 5.40 q ha^{-1} (16.0%) in sorghum 3 80 q ha^{-1} (33 3%) in mustard, and 1.03 q ha^{-1} (8.47%) in barley over control (unintercultivated) plots.

Plastic mulching is one of the most promosing practices for reducing evaporation. However, complete to cover of the soil surface with such mulch may not be desirable as it prevents the entry of rainwater into root zone. Ma Shijun (1988) has described two methods of plastic mulch in North China.

i. Sowing before mulching and perforating the film at the early seedling stage.

ii. Mulching and then perforating the film and then sowing.

The first method is applicable to all kinds of sowing, especially to machine work. It is necessary that within row or in hill seeding, the rows or hills be 3 to 4 cm under mulch to prevent young seedling from touching the plastic film directly thus avoiding burn damage. The second method is good for soil and water conservation. However, this method requires more manual labour at seedling. The good performance was due to increased soil temperature, better utilization of light and suppression of evaporation losses.

Mulching may be defined as the process of covering the soil surface around the plants to create congenial conditions for the crop growth. This may include moisture and soil conservation, temperature moderation, salinity and weed control etc. It exerts a decisive effect on earliness, yield and quality of the crop. The different mulch materials have the effect of modifying heat exchange between the covered and atmospheric environment. The temperature difference is the resultant effect between the accumulation and loss of energy.

Accumulation of energy	-	Shortwave radiation (Incoming) generates heat: 380-780 nm-Most effective 500 nm. Some contribution by 280-380 UV and 780-2500 I.R. also.
Loss of energy	-	Long wave radiation (5000-35000 mm) Terrestrial Radiation maximum losses 9000-1000 nm

Area of application

(i) Mulching is mainly employed for

a. Moisture conservation in rainfed areas,

b. Reduction of irrigation frequency in irrigated areas,

c. Soil temperature moderation,

d. Weed control,

e. Soil solorisation for control of soil borne diseases,

f. Harvest of better quality produce,

g. Reducing the rain impact, prevent soil erosion and maintain soil structure, and

h. Increasing over all crop production.

Types of mulch materials

The materials those may be used for laying over the soil may be organic plant residues, inert material like pebble etc. or and synthetic materials like plastics. Every material has got its merits and demerits.

Types of plastic materials

The plastic materials may be either PVC or polyethylene. Owing to its greater permeability to long wave radiation which can increase the temperature around the plants during the night times, polyethylene is preferred.

Basic requirement of a mulch film

(i) Should be air proof so as not to permit any moisture vapour to escape.

(ii) Should be thermal proof for preservation of temperature and prevention of evaporation.

(iii) Should be durable atleast for one crop season.

Important parameters of the plastic film

a. Thickness- Normally the thickness of the film does not affect the mulching effect except when it is used for solarisation. But some of the recent references do indicate the impact of film thickness on crop yield. Since it is sold by weight it is advantageous to use as thin a film as possible but at the same time due consideration should be given for the longevity of the film.

b. Width- This depends upon the inter row spacing. Normally a one meter width film can be easily adapted to different conditions.

c. Perforations- The perforations may be advantageous under some situations and disadvantageous for some other situation. The capillary movement of water and fertilizer distribution will be better and more uniform under imperforated conditions. But for prevention of water stagnation around the plants perforation is better. But it has got the disadvantage of encouraging weed growth.

d. Mulch colour- The colour of the mulch affects

i. Soil temperature,

ii. Temperature of air around the plants, and

iii. Soil salinity.

- Due to lesser quantity of water used
- Due to reduction in evaporation and prevention of upward movement of water

Transparent film - Deposits more salt on soil surface

Black film - Restricts water movement and upward movement of salt is reduced.

iv. Weed flora

v. Insect control- Opaque white film acts as insect repellent while yellow/golden colour attracts insects.

Selection of mulch

The selection of mulches depends upon the ecological situations and primary and secondary aspects of mulching.

Rainy season	- Perforated mulch
Orchard and plantation	- Thicker mulch
Soil solarisation	- Thin transparent film
Weed control through solarisation	- Transparent film
Weed control in cropped land	- Black film
Saline water use	- Black film
Summer cropped land	- White film
Insect repellent	- Silver colour film
Yield	- Appropriate colour mulch
Early germination	- Thinner film

Mulch laying technique

i. Mulch should be laid on a non windy condition.

ii. The mulch material should be held tight without any crease and laid on the bed.

iii. The borders (10 cm) should be anchored inside the soil in about 7-10 cm deep small furrows at an angle of 45°.

iv. Pre planting mulch: The mulch material should be punctured at the required distances as per crop spacing and laid on the bed. The seed/transplants should be sown/transplanted in the holes.

Main disadvantages are-

a. Probability of 'burning' of the plant due to high temperature of black film.

b. Difficulty in application of top dressed fertilizer.

c. Can not be used for another crop with different spacing.

v. Laying of mulch after establishment

The mulch material should be cut about 5 cm more than the inter row spacing and the two adjacent sheets should be stapled.

Precautions for mulch laying

a. Do not stretch the film very tightly. It should be loose enough to overcome the expansion and shrinkage condition caused by temperature and the impacts of cultural operations.

b. The slackness for black film should more be as the expansion, shrinkage phenomenon is maximum in this colour.

c. The film should not be laid on the hottest time of the day, when the film will be in expanded condition.

Water saving

Polythene and petroleum mulches conserved more moisture to an extent of 45 and 15 per cent, respectively, than control was observed by Rajput and Singh (1970).

Abdel-Hafeez *et al.* (1977) observed that plastic sheet (transparent and black) conserving more soil moisture than the control plots and differences were 47 and 50 per cent for transparent and black polythene, respectively. A two year experiment conducted at Navsari (Gujarat) has shown (Table 6.13) that about 50 per cent of irrigation water can be saved by adopting alternate furrow method of irrigation alongwith the use of plastic mulch in cotton. (Raman *et al.,* 1990).

Table 6.13: Effect of plastic mulching on yield of cotton and water saving

Treatment	Seed cotton yield (q/ha)	No. of irrigations	Percent saving in irrigation water
Irrigation in flat bed	17.3	4	-
Irrigation in all furrow	17.5	4	-
Irrigation alternate furrow	17.7	4	-
Irrigation alternate furrow + plastic mulch	18.2	2	50

Singh and Bhan (1990) found lowest water use in mulched plots in which 28.36 percent water was saved as compared to unmulched plots. Maximum water use-efficiency was observed in Polyethene mulched plots (Table 6.14).

Table 6.14: Effect of mulches on water use, water use efficiency in spring cotton.

Treatment	Seed cotton yield (q/ha)	Water use (mm)	WUE (mg/ hamm	Percent saving of water over control
Control	10.39	482.6	2.2	-
Straw mulch	13.70	392.3	3.6	18.71
Polyethene mulch	14.63	345.7	4.4	28.36
Jalshakti	11.19	452.4	2.6	6.25
Polyethene + Jalshakti	14.87	338.8	4.5	29.80

Table 6.15: Average soil temperature between 11 AM and 3 PM during the spring trial (Average Temp. ºC)

Mulch	Soil depth (cm)	Non Mulch	7.6	Band width of 15.2	Mulch (inches) 30.4	60.9
Petroleum mulch	30.4	15.1	20.7	22.6	57.66	23.6
Clear polythene film	30.4	20.6	20.8	22.6	59.70	24.7
Black Polyethene film	30.4	20.6	20.3	20.8	53.85	22.0

Effect on soil temperature

Plastic mulch when applied on soil surface works as insulator and thus keeps control on the abrupt fluctuations in soil temperature. It is the most important factor in the seed germination, plant growth. Colours of plastic mulches generally effected soil temperature. White or reflective plastic mulch decreased temperatures (Pusztai, 1963) or had no effect (Schales, 1964 and Adams, 1966) because the soil directly absorbed most of energy from incoming solar radiation with black plastic, the soil received only a portion of the incoming energy absorbed by the film. Maximum temperature was found in plastic mulch which enhanced seedlings emergence, increased plant growth and promoted earliness in corn plot than control (Willis *et al.,* 1957 and Clarkson, 1960), Takatori *et al.* (1964) reported that petroleum mulch and clear plastic mulch increased soil temperature during the daylight hours to a depth of 15 cm and retained some soil heat during the night (Table 6.15). Bahram (1970) studied the effect of petroleum mulch on cotton and found minimum soil temperature by upto 5.4°C and maximum temperature by upto 3.5°C. A study on the effect of black rubber mulch on soil temperature on cotton revealed that the diurnal variation of soil temperature ranged from 20.7 to 44.2°C (Venkatachalam, *et al.*, 1980).

Effect on growth and yield of crops

The favourable effect of plastic mulch on erosion control, moisture conservation, weed control has been found which resulted in increased growth and yield of cotton crop (Lawes, 1961). Bahram (1970) reported that the petroleum mulch accelerated the germination, early growth, fruiting and yield by as much as 7.65 gm seed/plant. Black and Grab (1962) reported that the plastic mulch increased the crop yields in irrigated and non-irrigated areas as compared with bare soil. Chaudhary and Chatterjee (1967) observed that the mulching with Encap (Esso) increased the yield of wheat, barley, gram and linseed on eroded soils (Table 6.16). Rajput and Singh (1970) observed that the black Polyethene mulch increased the plant growth and yield of cotton. Black Polyethene mulch increased watermelon yields over the non-mulched plots (Schales *et al.*, 1990).

Table 6.16: Grain yield (q/ha) as influenced by mulches

Mulches	Wheat	Barley	Gram	Linseed
Straw	5.90	8.60	10,28	7.98
Encap	5.58	7.70	11.43	6.58
Control	4.65	6.68	7.88	5.88
L.S.D.5%	0.72	1.38	1.97	1.38

Singh and Bhan (1990) found that the Polyethene mulch and Jalshakti mulch increased the growth, number of boils per plant and yield of seed cotton in irrigated spring cotton. The seed cotton yield was increased upto 40.81 and 7.70 per cent, respectively, over the unmulched plots (control) Table 6.17.

Table 6.17: Seed cotton yield (q/ha) as affected by mulch

Treatments	Boils per plant	Seed cotton Yield	Increase yield per cent
Control		10.39	-
Straw mulch	9.50	13.70	31.86
Polyethene mulch	10.37	14.63	40.81
Jalshakti	6.20	11.91	7.70
Polyethene mulch+Jalshakti	10.65	14.87	43.12

Economics of plastic mulch

Polythene mulch is not popular in agriculture because of their prohibitive cost and difficulty in their application. Plastic mulch has been used experimentally and in the commercial production of many fruit crops. Plastic mulch is economical and advisable in the area where the irrigation facilities are not available and nature of crops is perennial. Bhan and Singh (1979) reported that the Polyethene mulch increased the yield of maize and mustard and an additional return of Rs. 2019/h was obtained Singh and Bhan (1990) reported that the black polyethlene mulch and Jalshakti increased the yield of seed cotton and an additional return of Rs. 1071/ha was obtained with Polyethene mulch (Table 6.18).

Table 6.18: Economics of different mulches in spring cotton

Treatments	Seed cotton yield (q ha^{-1})	Additional yield over control	Add. yield value Rs.	Add. Expenditure (Rs.)	Add. Return (Rs. ha^{-1})
Control	10.39	-	-	-	-
Straw mulch	13.70	3.31	1490	460	1030
Polyethene mulch	14.63	4.24	1908	837	1071
Jalshakti	11.19	0.80	360	543	-
Polyethene mulch+ Jalshakti	14.87	4.48	2016	1380	636

Cost of seed cotton @ Rs. 450/quintal Bajra stover mulch @ Rs. 6/quintal Black Polyethene @Rs. 24/kg Jalshakti @ 106/kg

Gravel mulching

Large pores between gravel permit rapid infiltration of water to the soil but retard evaporation. Water can return to the atmosphere only in vapour phase. Thus, gravel tends to act as a one, way water valve for the soil. Gravel sand mulching as practised in China is described by Ma Shijun (1988) as follows.

Before mulching, the field should be ploughed deeply, manured with large amounts of farm yard manure, harrowed and leveled in the summer fallow time as for preparing a seed bed. Gravels and sands of mixed mulch pebbles, gravels and sand in different sizes taken from river alluviums and glacier deposits are spread to a thickness of 10-16 cm to the frozen soil in winter. During the growing season, the soil or mulched fields are not tilled. There are special tools for sowing and manuring to prevent the mulch layer from being mixed with the soil.

The porosity of the mulch can be maintained for about 30 years. When the gravel and sand mulch is mixed with soil to an extent of about 25 per cent by weight, its functioning will diminish gradually. Then the old mulch should be removed.

Series of experiments and observations on the gravel-sand mulched fields indicated that water content (March-September) increased by 2-7 per cent in the surface 20 cm depth, evaporation reduced by 30 per cent and temperature below the mulch increased by 1.5 to 2.3°C.

In Gansu and Qinghai provinces, soil water content understand mulch was higher at any time compared to no mulch (Table 6.19).

The main difficulty of putting gravel-sand mulch farming into practice is that it requires much strenuous manual labour, that is *why* the practice is limited to a small area. The practice appears to be suitable for tree and fruit crops where small area of the basin can be manually maintained.

Table 6.19: Water content (g/kg) in sand covered land

Treatment	Mar.	Apr.	May	June	Jul.	Aug.	Sep.
Sand Covered	134	123	133	157	126	138	189
General area (not covered)	68	62	71	79	88	85	102

(Source: Li *et al.*, 1988)

6.6.5 Land treatments

Land treatments are practices that promote maximum conservation of rain water where it falls *i.e. in-situ* rainwater management. Water and land management are inseparable and successful strategies on water handling are location specific and dependent on rainfall intensity, slopes, soil type and texture. In general *in-situ* water harvesting can be achieved by increasing infiltration rate with the help of deep ploughing, profile modification, vertical mulching and by keeping soil surface rough. On lands having slope up to 1 to 2 per cent, water conservation could be ensured by field bunding, land levelling, contour ditching, and cultivation along contour. On lands having 2 to 6 percent slope, graded contour bunds can be constructed and on slopes ranging from 6 to 33 per cent bench terraces can be made.

Water harvesting or runoff recycling is defined as the process of collecting water resulting from surface runoff from an area for some beneficial purpose. Interplot water harvesting (Fig. 6.1) is used under very low rainfall conditions. The surface runoff from the upper area gets collected in the lower part and thus more water is available in the lower part. In case of enough rainfall is expected in the area, a crop like maize requiring good drainage may be grown in the contributing area and crops requiring more water can be put in the receiving area. Chemical sprays, compaction etc. are sometimes used in the contributing area to increase the runoff.

6.6.5.1 Land levelling and grading

Land levelling or grading is the process of preparing or modifying (*i.e.* reshaping) the land surface to a penned grade to provide a suitable surface, Land levelling usually requires cutting of high areas and raising of low spots, in order to remove the surface irregularities and unevenness to make a plane surface, It makes suitable field surface to control the flow of water, to check soil erosion and provide better surface drainage.

Factors governing land levelling

The type and amount of levelling required depends on the following factors:

Soil characteristics

The depth of soil governs the depth of cut. Excessive cut pose the strata that is not desirable for cultivation. The soil properties like infiltration and permeability are also considered while deciding the length of irrigation runs and its economic feasibility.

Topography

The amount and cost of levelling depends on the prevailing land slope and the land treatments.

General smoothening of lands

Prior to bunding, small gullies are formed all over the land due to spread and flow of water. At the time of bunding these need to be smoothened so that there is better water conservation over the land surface. The gullies which are not to be used for grass waterways may be also leveled or smoothened. This helps in conserving moisture *in-situ*. If this is not done, water quickly collects near the bunds and water stagnation occurs.

Land levelling in deep alluviums help to increase yields especially in sub-normal season (Table 6.20)

Table 6.20: Effect of land shaping on the yield of wheat and chickpea in sub-montane Ballowal-Saunkhri and pearlmillet in alluvial soils of Agra

Land shaping	Wheat and Chickpea mixture, 3 seasons (Mg/ha)	Pearmillet (Mg/ha)	
		Rainfall 320 mm	Rainfall 660 mm
Natural	1.05	1.34	2.22
With land levelling	1.65	1.63	2.31

Source: Vittal *et al.,* 2005

Levelling and bunding

Despite being well known and efficient soil and water conservation practices, higher cost involved prohibited their large scale adoption by the farmers. The efficiency of these practices are amply demonstrated through the information in Table 6.21, which are based on the experiments conducted at Rahamankheda, Lucknow for 5 years.

Table 6.21: Effect of levelling and bunding on yield of maize and sorghum

Treatments	Maize			Sorghum		
	Grain yield (q/ha)	Additional yield over control (q/ha)	Additional yield over control (%)	Grain yield (q/ha)	Additional yield over control (q/ha)	Additional yield over control (%)
Control	12.84	-	-	13.13	-	-
Bunding	17.85	5.01	39	17.2	3.99	29
Levelling	21.21	8.37	65	18.93	5.80	42
Bunding + levelling	26.09	13.25	1.0	23.19	10.06	74

(Source: Anonymous, 1981)

Vegetative bunding/vegetative barrier

Apart from higher cost another problem with earth bunds is that they are breached frequently and soil gets washed away with the runoff water. Such problem could be avoided with vegetative propagation on earth bunds for their stabilization or putting vegetative barriers across the slope (Table 6.22 and 7.23).

Table 6.22: Runoff and soil loss as influenced by various barriers

Vegetative barriers	Runoff (mm ha^{-1})	Soil loss (t ha^{-1})	Reduction over control (%)	
			Runoff	Soil loss
Control	215.6	7.5	-	-
Sessbania sesban	131.4	3.5	39.1	53.3
Leucaena leucocephala	133.2	3.7	38.2	50.7
Cajanus cajan	143.3	4.1	33.5	45.3
Vetiveria zizanoides	127.0	3.5	41.1	53.3
Cenchrus ciliaris	140.1	4.0	35.0	46.7

(Source: Chand, 1997)

Table 6.23: Effect of vegetative barriers on the yield of rainfed cotton and sorghum

Vegetative barriers	Yield (q ha^{-1})		Increase over control (q ha^{-1})	
	Cotton	Sorghum	Cotton	Sorghum
Control	6.60	15.62	-	-
Vetiveria zizanoides	8.22	18.21	1.62	2.59
Leucaena leucocephala	9.77	19.94	3.17	4.32
Cenchrus ciliaris	8.07	17.03	1.47	1.41
Sesbania	10.51	20.55	3.91	4.93
Cajanus cajan (Perennial)	10.04	19.29	3.44	3.67

(Source: Dubey and Bhan, 1992; Chand, 1997)

In-situ moisture conservation measures

In order to increase the moisture availability to the agricultural or tree crops, it is necessary to adopt *in-situ* moisture conservation techniques in addition to the above measures utilized for water conservation. Various measures utilized for *in-situ* moisture conservation are described below:

(i) Forming ridges and furrows

In this each ridge of furrow acts as barrier and the channel acts as storage space for the rainwater. This practice becomes very effective when furrows are laid out along the contours in the field.

Planting row crops on preformed ridges or preparing furrows in between the rows after crop establishment across the slope helps control soil erosion and runoff and increases crop yields (Table 6.24). Such a practice has been found useful even on flat lands in saving the crops both during drought as well as wet season. Furrows provide more opportunity time to runoff water to be soaker by the soil while the ridges provide adequate aeration to the crop during water inundation or flooding of short duration. The efficiency of conserved moisture in furrows may further be improved if ridges and furrows are superimposed with application of organic residue mulch (Table 6.25).

Table 6.24: Effect of ridging and furrowing on *kharif* crops at Kanpur

Crops	Yield (q ha^{-1})		Increase over control		Reference
	Control (flat)	Ridging & furrowing	q ha^{-1}	(%)	
Jowar 'Varsha'	13.60	20.10	6.50	47.8	Bhan *et al.*, 1995
Jowar 'Varsha'	26.60	35.10	8.50	32.0	Tripathi & Bhan, 1993
Cotton 'Vikas'	5.60	8.26	2.66	46.0	Tripathi *et al.*, 1995
Bajra 'WCC 75'	12.60	21.79	9.20	73.0	Bhan & Parihar, 1987
Maize 'Azad Uttam'	11.75	14.15	2.40	20.4	Gupta & Bhan, 1997

(Source: Bhan and Uttam, 1998)

Table 6.25: Grain yield of sorghum as influenced by organic residue mulch

Treatments	Grain yield (q ha^{-1})	Additional yield over flat and unmulched plot	
		q ha^{-1}	(%)
Flat and unmulched	26.6	-	-
Ridging and furrowing+mulching	41.9	15.30	57.52

(Source: Tripathi and Bhan, 1993)

(ii) Forming of ridges

This is a practice in which holes are dug to a depth of 10-15 cm having a diameter of 10-30 cm. These holes can also be square or rectangular in shape. The spacing between these holes is generally kept as 50-100 cm depending on the area of the field. Basin liter can be used to form the ridges.

(iii) Forming water spreads

The v·ater from a *nallah* or stream is diverted and spread over the land surface on either side during rainy season. This spread water will get infiltrated into the soil profile and will improve the crops for a large time.

(iv) Forming V-ditches

In this 'V' shaped ditches are formed on contours at every 4-6 m interval either mechanically or manually. A small earthen bund is formed just below the ditch which helps in retaining water. The water stored in these ditches percolates in the soil profile and adds moisture availability to the crops.

(v) Furrow diking

Furrow diking is an alternative to land levelling for retaining precipitation. It includes construction of small dikes or dams at intervals along furrows to retain potential runoff and allow it to infiltrate. Furrow diking is also known in the United States as furrow damming, basin tillage, basin listing, and microbasin tillage. In the semi-arid tropics of Africa, this practice is called tied-ridging. Furrow diking has been used to reduce runoff and soil loss and to increase crop yields in arid and semi-arid areas. In high rainfall years/periods, it may result in water logging.

(vi) Zingg-terracing

In contour bunds, runoff water is not well distributed; on the other hand, it concentrates near the bunds leading to damage of the crop if stagnation is allowed for longer period. To avoid this difficulty, A.W. Zingg, Agricultural Engineer, USA developed a practice of levelling the lower one-third portion of the land between contour bunds to help spreading of runoff water in the larger portion of the inter-bunded area and thus ensure better availability of water to crops.

(vii) Compartmental bunds

In deep black soils depending upon the land slopes, the entire field is laidout into small bunded compartments of varying size from 6m x 6m to 10m x 10m. Rains received during *Kharif* are collected in these bunded areas which slowly harrowed and land prepared into a good seed bed.

(viii) Tied ridging

After one or two harrowings, converting the land into ridges and furrows and tying them septa, so as to make the rainwater collect in these furrows is a successful method of moisture conservation. The water collected in these furrows will slowly percolate into the soil profile.

(ix) Scooping

Scooping out soil to form small basins with similar implements, helps in retaining the water on the land surface for a longer time. At Hagari in Bellary district, inter-cultivation hoes with ropes tied around the prongs have been used successfully for scooping purpose. Small basins can be prepared in the field without much expenses. Scooping helped in reducing the water loss from 13 to 5.3 cm thus saving more than 50 per cent of runoff and soil losses from 8.6 to 3.0 tonnes/acre. Yields of crops also increased by about 11-12 per cent at Bijapur. This method has been found to be very effective when done within compartments.

(x) Ridge and furrow system

Cultivation of crops under ridge and furrow systems across the land slope with a gradient of 0.2 to 0.4 per cent and in land having 2-4 per cent slope will greatly help in moisture conservation *in-situ*. This is suitable for widely spaced crops with 60 cm or more. A length of 60-90 m is optimum. Studies conducted on moisture conservation for cowpea-ragi double cropping system revealed that ridging up after flat on a grade sowing is more advantageous under Bangalore agro-climatic conditions (Table 6.26).

Table 6.26: Effect of ridging on moisture and yield of crop

Treatment	Yield (q/ha)	
	Cowpea	Ragi
Flat on a grade sowing	7.76	34.84
Flat on a grade sowing but later ridging up	7.54	36.66
Sowing on beds (135 cm for cowpea and 150 cm for ragi)	6.73	32.83

(xi) Bedding system

This is a system having a furrow at every few rows of crops across the slope on a grade of 0.2 to 0.4 per cent. The bed width could be 3 to 6 meters. This is suitable for narrow spaced row crops. Even if a few rows are lost due to the furrow, the yields are made up due to better *in-situ* moisture conservation. There is no water stagnation near the bunds. Hence, this system acts both as disposal and conservation method.

Studies on inter-terrace management for *in-situ* moisture conservation by Channappa (1978) revealed that bedding system of land management with a furrow opened at the time of sowing the crop either at 1.5 or 3 m intervals was found to increase/stabilize yield levels over years by 8-10 per cent, apart from better rainwater management at times of low as well as high intensity rains. Modified technique known as paired row redgram-ragi intercrop with a furrow in between the redgram rows and 8/10 rows on bed found to be the best intercrop as well as inter-terrace management practice for the red soil regions of Karnataka state. This practice is being adopted in all the watershed management programmes taken up in the state (Fig. 6.1).

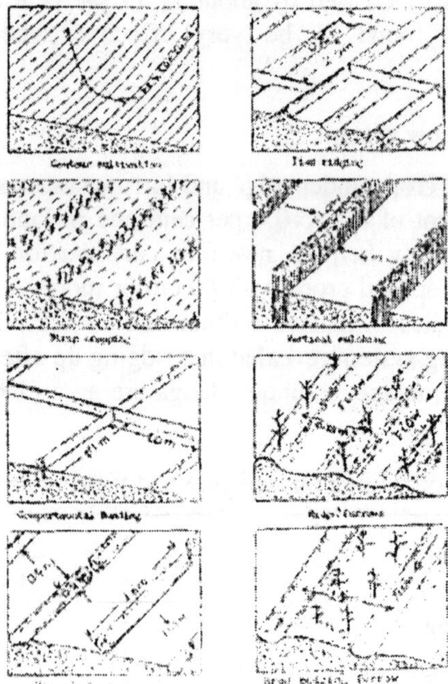

Fig. 6.1: *In-situ* moisture conservation practices for black soil

(xii) Broad-bed and furrow system

The broad-bed and furrow system refers to modifying the land surface into alternating broad (1.5 meter [rn]) beds and narrow (0.15 m) furrows. This system has been found useful with vertisols in low to medium rainfall regions (mean annual rainfall less than 1,000 mm) (El-Swaify *et al.*, 1985), wherein erosion and aeration stress are the main constraints crop production. The beds, laid on a minor gradient, act as minichecks against the free flow of rainwater. Also, as a large quantity of rainwater does not stagnate over the bed for long periods, it helps to maintain favorable aeration status on the bed.

Another system of rainwater management, termed a raised and sunken bed system, is suitable for vertisols in high rainfall regions (mean annual rainfall greater than 1,000 mm). This system consists of an array of alternating raised and sunken beds, The runoff from raised beds, where an upland summer crop is seeded, is arrested in the adjacent sunken beds supporting a crop with a relatively high water tolerance, *e.g.*, rice. The widths of raised and sunken beds vary with annual rainfall (Gupta and Sharma, 1990). In high rainfall regions (greater than 1,200 mm y^{-1}), 3-6 m wide raised beds with an elevation difference of 200 to 300 mm alternate with about 4 m wide sunken beds. However, in medium rainfall regions (less than 1,000 mm), 8 m wide raised beds alternate with 4 m wide sunken beds having an elevation difference of 150 to 200 mm. Both of the previous soil management systems were found suitable in assured unimodal rainfall pattern or double-cropped regions (Katyal and Hegde, 1998).

Tomar *et al.* (1996) conducted the long-term experiment on raised bed and sunken bed (RSB) land configuration in vertisol. Results revealed that the rainfall received in different years varied considerably Total rainfall during the rainy season ranged from 627 to 1626 mm while during the single month of August (the month of most intense rainfall), if varied from 29 to 715 mm. Grain yield of crops grown on raised and sunken beds during *kharif* are presented in tables 3.26 to 3.28. The highest average grain yields of soybean, blackgram, pigeonpea and sesame were observed on 6 or 9 m. wide raised beds in different seasons. The grain yield of all the crops on 6 to 9 m wide raised bed was significantly higher than the other is widths of raised beds or flat plots. In case of sesame pigeonpea and blackgram all the widths of raised beds gave significantly higher yields than the flat plots. The results further indicate that the seed yield of soybean (Table 6.27) on 12 or 15 m wide raised bed did not differ significantly with each other. In most of the seasons pigeonpea could not survive its flat plots due to continuous saturated/flooded conditions at the early growth stages.

Table 6.27: Long-term effect of land configuration on the grain yield (kg ha^{-1}) of the *kharif* crops in vertisols

Year	Soybean yield width of raised bed (m)				Flat	CD5%	Paddy yield sunken bed (m)
	6	9	12	15			
1985	2253	2605	2234	2178	1000	-	2500
1986	2131	2328	2096	1782	647	262	4220
1987	1840	1330	1230	1258	715	440	1716
1988	3260	2862	2817	2788	1731	180	1877
1989	2779	2499	2317	2197	1544	266	1520
1990	2425	2329	2292	2014	849	139	3185
1991	2523	2287	2147	1972	125	182	2200
1992	3107	2738	2497	2144	581	390	3001
1993	1733	1600	1444	1493	1050	88	4758
1994	1969	1904	1736	1628	1469	242	3922
1995	1755	1714	1691	1254	1113	206	4473
Average	2343	2197	2045	1882	984	-	3033

Better yield of all the crops on 6 or 9 m. wide raised bed was mainly due to better surface drainage and hence, better soil aeration conditions. This was corroborated by higher soil moisture suction and deeper water table depths in 6 or 9 m. wide raised beds (Table 6.28 and 7.29). Temporary water logging in 12 or 15 m wide raised beds or in control plots, particularly during continuous heavy rains, caused localized root development in soybean. Under such conditions blackgram, pigeonpea and sesame could not survive in some of the years. The positive response of soybean to deeper water table was also reported by Ghowail and Ghavery (1978) and Gupta *et al.* (1979). These results are further confirmed by the root data presented in Table 6.30 which show that highest root density was observed in 6m raised beds in all the crops.

Table 6.28: Effect of land configuration on the grain yield (kg ha^{-1}) of blackgram in vertisols

Year	Width of raised bed (m)				Flat	CD at 5%
	6	9	12	15		
1986	1156	881	834	714	384	287
1987	1098	723	508	494	394	315
1988	1280	1240	1132	1076	478	102
1989	993	773	778	748	708	91
1990	1023	991	904	714	500	130
1991	872	721	700	572	112	102
1992	936	826	937	604	-	197
Average	1046	893	813	703	429	-

Table 6.29: Effect of land configuration on the grain yield (kg ha^{-1}) of pigeonpea and sesame in vertisols

Raised bed	Pigeonpea						Sesame		
	90-91	91-92	92-93	93-94	94-95	95-96	1987	1988	1989
6	2012	2004	2532	1825	1548	1174	360	321	421
9	1861	1709	2327	1800	1481	1114	329	235	383
12	1687	1496	2003	1714	1283	1091	241	236	340
15	1003	945	1580	1746	1223	654	217	214	319
Flat/control	128	167	Nil	Nil	Nil	Nil	41	25	262
CD 5%	297	178	253	NS	304	243	154	24.3	29.6

As rains cease during the later growth period, the soil moisture stress is expected to affect the yield adversely because particularly the crop like soybean is more susceptible to moisture stress at flowering and seed development stages (Tomar *et al.*, 1985). But this was not the case with crop grown on raised beds, because favourable moisture regime was maintained through lateral movement of water from the adjoining sunken beds. Soil water balance studies in this experiment have clearly indicated that the lateral movement of water into raised beds from sunken beds was effective only up to 3 to 4 m distance from the sunken beds. The average yield of paddy planted in sunken beds was 3033 kg ha^{-1} (Table 6.27). The variation in paddy yields in different years is explainable on the basis of amount and distribution of rainfall received in different years. In most of the growing seasons there was no need of additional irrigation to paddy crop in sunken beds but thereafter due to rain free periods additional water was needed. Accordingly, two irrigations of 6 cm each were given to paddy crop in 1989 and 1991 seasons from the excess run-off water collected in a farm pond.

Table 6.30: Effect of width of raised beds on the rooting density (mg lit^{-1}) of soybean and blackgram in three seasons

Raised bed	Year	Soybean				Blackgram			
		1987	1988	1989	Av.	1987	1988	1989	Av.
6	0-0.2	6.05	8.57	6.96	7.19	2.64	3.81	2.78	3.08
	0.2-0.4	0.30	0.68	0.48	0.49	0.12	0.35	0.49	0.32
	0.4-0.6	0.13	0.11	-	0.12	0.07	0.10	-	0.08
12	0-0.2	4.43	6.83	5.48	5.58	2.01	2.82	2.29	2.38
	0.2-0.4	0.43	0.46	0.40	0.43	0.12	0.35	0.52	0.33
	0.4-0.6	0.26	0.30	-	0.15	0.80	0.07	-	0.08
Flat	0-0.2	1.60	3.43	4.61	3.22	0.78	1.31	2.03	1.37
	0.2-0.4	0.45	0.32	0.38	0.38	0.17	0.19	0.40	0.25
	0.4-0.6	0.05	0.02	-	0.04	0.15	0.08	-	0.12

Performance of winter *rabi* crops

After harvest of summer crops, winter crops *viz.*, wheat, chickpea, safflower, and linseed were grown on raised and sunken beds using zero/minimum tillage in different years. As shown in Table 6.31; the highest yield of all the *rabi* crops was recorded in raised beds as compared to sunken beds and flat plots. Due to favourable moisture conditions in raised and sunken beds the germination was better than in the flat plots. The lowest yields were recorded in flat plots. There was more than two fold increase in grain yield of all the crops in RSB system as compared to flat beds. In the years 1989 and 1991, due to low and early withdrawal of monsoon rains, a come up irrigation was given to the crops grown in raised beds and flat plots. Latter, sufficient winter rains were received. Water use data indicated that wheat, chickpea and safflower had higher water use as compared to linseed crop. Among the cropping systems, crops grown after soybean on raised beds and after paddy in sunken beds had more or less the same water use except safflower and linseed.

Table 6.31: Grain yield (kg ha^{-1}) and water use (cm) of rainfed *rabi* crops grown after soybean and rice under raised and sunken bed technology (average of 10 years)

System/factor	Crops			
	Wheat	**Chickpea**	**Safflower**	**Linseed**
Raised bed				
Yield	2304	1807	1382	1104
Water use	28.8	27.54	2848	23.4
Sunken bend				
Yield	1976	1599	1132	725
Water use	29.19	28.42	22.03	24.58
Flat				
Yield	989	825	639	566
Water use	23.5	22.78	22.0	20.63

(xiii) Ridge & furrow and bedding system

Crops like maize are either planted on a ridge and furrow system across the slope or are earthed up after 45 days of sowing to form ridges. In any case, the ridges thus formed have been found to reduce runoff and thereby improve moisture availability to crop growth. Observations with maize crop during 1980 at Bangalore have shown that the runoff with ridges and furrows on a gradient of 0.3 to 0.4 per cent was 17.77 per cent of the total rainfall as compared to 31.21 per cent in bedding system. This system is useful for widely spaced crops at 60 to 90 cm between rows. Opening furrows after two seed drill widths (3 m interval) on a gradient of 0.3 to 0.4 per cent was found to be advantageous for finger millet crop in red soils.

Studies on inter-terrace land treatments by Katama Reddy (1979) revealed that forming a dead furrow and forming compartments with bund former conserves rainwater and improves grain yields by 15-30 per cent (Table 6.32).

Alfisols and vertisols, the two predominant soil types of semi-arid tropical regions, need different land treatment plans. Important land configurations for drylands included the broad bed and furrow (BBF) system, flat on grade and ridging later, graded border strips, etc. The BBF system developed at ICRISAT is useful with deep vertisols, while flat sowing and ridging later is a simple scheme that works well on alfisols. Graded border strip, a method developed at Banglore, was rated highly efficient for controlling runoff and stabilizing crop yields in relatively deep alfisols.

Table 6.32: Bajra grain yield (kg ha^{-1}) as affected by different land treatments

Treatment	1976-77	1977-78	1978-79
Furrows at 2m interval across the slope (broad ridges)	380	1025	152
Treatment 1+ filling with *bajra* straw	390	1009	146
Compartmental bunding with compartments of 16m x 9m formed with bund former	390	997	58
Treatment 1 and 3	470	1230	197
Cultivation across the slope	350	896	177
Cultivation along the slope	290	885	147

The relative performance of different land configurations namely flat-bed system (FB), broad bed and furrow (BBF), narrow bed and furrow (NBF) and raised-sunken bed systems (BSB) was studied at Indore (MP) in terms of rainwater conservation *in-situ*. The results indicate that the highest seed yield of 20.95 q/ha and water use efficiency of 8.81 kg/ha/mm was realized due to BBF followed by RSB and FB systems in black soil region.

(xiv) Corrugations

Whenever sowings are done across the slope on a grade, it is useful to open up small furrows between the crop rows at the time of last inter-cultivation so that the rainwater can be conserved *in-situ*. This can be done by the use of a ridger.

Contour cultivation

Carrying out farming operations namely, ploughing, planting and cultivation along contour lines is one of the recommended practices for arable lands on mild slopes. By doing so every furrow acts as a miniature reservoir to hold the excess runoff, dissipates the energy of flow and gives increased time opportunity for runoff absorption. On 8% slope at Dehradun, the contour cultivation of maize reduced the runoff by 23.8% and soil loss by 31.6 percent (Tejwani *et al.* 1975) compared to up and down cultivation of maize, which recorded 54.1 per cent runoff and 28.5 t/ha/yr soil loss. On the same plots reduction in runoff and soil loss due to contour sowing of maize (Table 6.33) was 13.8 per cent and 18.3 per cent respectively, (Khybri, 1983).

Table 6.33: Effect of mulching on runoff, soil loss and yield of maize at 8% slope

Conservation measure	Runoff (%)	Soil loss (t ha^{-1})	% reduction		Crop yield kg ha^{-1}
			Runoff	Soil loss	
Maize (up & down)	42.8	21.8	-	-	2433
Maize (on contour)	36.9	17.8	13.8	18.3	2433
Mulching 2 t ha^{-1}	17.9	7.3	51.3	59.0	2317
Mulching 4 t ha^{-1}	14.7	5.4	60.2	69.7	2395

6.7 Agronomic Measures

The agronomic measures are of great significance to dissipate the erosive rain energy in combination of mechanical measures. The effectiveness depends upon the per cent coverage, height and thickness of the canopy. On arable lands canopy cover can be managed through selection of cropping system, crop rotations, plantation techniques, agroforestry and vegetative barriers. For non arable lands, cover management is linked with tree and grass management under various combinations which include forestry, silvipasture and agri-horticulture system etc. There are following measures:

Cropping systems

Inter planting of erosion resistant leguminous crops, such as cowpea, soyabean etc., which develop quick canopy cover has been found efficient in reducing soil erosion from corn. The cultivation of pure maize on 8 per cent slope at Dehradun lost 33.3 per cent of rainfall as runoff and produced 25.1 t/ha of soil loss. Interplanting of *til* (Sesame) marginally reduced runoff and soil loss while cowpea was quite effective (Khybri, 1988) and reduced the runoff and soil loss to 26 per cent and 20.6 t/ha, respectively (Table 6.34).

Strip cropping

Strip cropping of erosion resistant legumes with cereal is an established conservation. Beneficial effects of such mixed cropping have been reported at 4% slope by Singh *et al.* (1979) while soyabean and its strip cropping with maize (Table 6.35) reduced the runoff and soil loss considerably. Soyabean strips with maize in the ratio of 2 rows of maize and 8 rows of soyabean.

Table 6.34 Effect of legume inter-plantation on runoff and soil loss

Conservation measure	Runoff (%)	Soil loss (t/ha)	% reduction	
			Runoff	Soil loss
Maize	33.3	25.1	-	-
Maize+til	32.15	22.05	3.45	12.15
Maize+Cowpea	26.0	20.6	21.92	17.92

Table 6.35: Efficacy of legume intercropping on runoff and soil loss reduction

Measure	Runoff (%) (t/ha)	Soil loss	% reduction		Grain yield	
			Run off	Soil	Maize	Soyabe an
Maize	61.2	15.76	-	-	1999	-
4 rows of maize + 6 rows of Soyabean	43.1	9.53	21	398	1101	593
2 rows of maize + 8 rows of Soyabean	33.5	7.27	42	54	1136	678
Soyabean	27.2	5.46	55	65	-	1331

Table 6.36: Effect of spacing in maize on runoff and soil loss

Spacing	Runoff (%)	Soil loss (t/ha)	% reduction	
			Runoff	Soil loss
45 x 40 cm	47.8	22.6	-	-
60 x 30 cm	37.4	20.6	21.8	8.8
75 x 24 cm	32.8	18.5	31.4	18.1
90 x 20 cm	32.1	13.9	32.8	38.5

Crop geometry

Manipulation crop geometry in order to achieve closer spacing within the rows across the slope provides greater resistance to over land flow resulting in reduction of soil loss and runoff. At 4% slope, at Dehradun 50,000 plants/ha of maize with conventional spacing of 45 x 40 cms produced 47.8% runoff and 22.6 t/ha soil loss (Table 6.36). With the same plant population, spacing of 90 x 20 cm decreased the runoff by 32.8% and soil loss by 38.5% without any monetary input (Bhardwaj *et al.* 1984).

Date of sowing

Date of sowing effect the crop canopy, which is a great of significance to dissipate erosive rain energy and runoff. Yadav and Varshay (2005) conducted experiment during *kharif* (1998-99) at Pune. Results revealed that the early sowing of S_1 produced significantly higher grain yield as compared to delayed sowing S_2 and S_3 (Table 6.37) Similar results regarding grain yield, fodder yield and total biological yield were observed by Gautam and Kaushik (1984), Bhagchand and Gautam (2000). Muchow (1989) also found higher biomass in S_1 and S_2 at maturity compared to S_3 and S_4 due to more favourable thermal and radiative conditions with early sowing with maize and pearl millet. However, fodder and biological yield of pearl millet was not significantly influenced due to different sowing dates (Andhale *et al.*, 2003).

Table 6.37: Effect of sowing dates and methods on grain, fodder and biomass yield

Treatment	Grain yield (kg ha^{-1})	Fodder yield (kg ha^{-1})	Biomass yield (kg ha^{-1})
Sowing dates			
S_1 (5th July)	1955	5594	7549
S_2 (13th July)	1835	5393	7228
S_3 (20th July)	1821	5201	7022
S.E.	20.86	141.53	133.57
C.D. at 5%	62.53	-	-
Sowing methods			
Local (L_1)	1783	5134	6917
Border (L_1)	1906	5451	7374
Ridge and furrow (L_1)	1923	5603	7509
S.E.	20.86	141.53	133.57
C.D. at 5%	62.53	-	-

Agroforestry

Growing multipurpose trees or shrubs in agricultural fields not only satisfier fuel and fodder needs of the farmer, but also has a great potential for erosion control through the tree canopy, litter and root effects. Tree rows may also act as barrier to surface runoff. The plantation of *Leucaena* or *Eucalyputs* in maize field in paired rows at 4.5 x 1.5 m spacing reduced soil loss by more than 70% (Narain *et al.*, 1988). Eucalyptus rows were more effective than *Leucaena* when both are managed as trees (Table 6.38). Although trees had adverse effect on the crop yields but tree products compensated the loss making system economically viable.

Table 6.38: Runoff and soil loss under agroforestry systems

Land use	Runoff (%)	Soil loss (t/ha)	% reduction	
			Runoff	Soil loss
Maize	18.3	17.7	-	-
Maize+Subabul	8.9	5.0	51.9	71.8
Maize+Eucalyptus	3.6	0.91	80.3	94.9
Chrysopogon fulvus	1.6	0.33	91.3	98.1
Grass+Subabul	0.6	0.13	96.7	99.3
Subabul	0.4	0.04	97.8	99.8
Grass+Eucalyptus	0.1	0.02	99.5	99.9
Eucalyptus	0.1	0.01	99.5	99.9

Grasses

Grasses are perhaps the best friend of soil conservations. Low and evenly distributed canopy and fibrous root systems with high soil binding capacity make grasses highly effective in controlling soil erosion. Chrysopogon grass alone could reduce the runoff and soil loss by 91.3 and 98.1 per cent respectively (Narain *et al.*, 1988), but when grown with trees reduced the runoff and soil loss completely (Table 6.39).

At 9 to 11 per cent slopes at Dehradun, efficacy for erosion control of various grasses was more than 98% (Tejwani *et al.*, 1975) hence their selection shall be based on the production potential of these grasses under given edaphic conditions and local preferences. Pueraria hirsute and Dicanthium annulatum have shown the highest production at Dehradun (Table 6.39).

Table 6.39: Runoff and soil loss under different grass species

Land slope	Conservation measures	Runoff (%)	Soil loss t/ha	% reduction Runoff	Soil loss	Yield kg/ha
9%	Bare & ploughed	59.6	155.95	-	-	-
	Bare fallow	71.1	92.42	19.3	40.75	-
	Cynodon dectyion	27.1	2.10	54.5	98.70	4355
	Natural grass	21.2	1.02	64.4	99.3	-
11%	Cultivated fallow	16.2	18.45	-	-	-
	Pueraria hirsuta	1.80	0.11	88.89	99.40	149625
	Dicahthium annulatum	1.90	0.23	88.27	98.75	6805
	Chrysopogon fulvus	2.50	0.30	84.57	98.40	19170
	Eulaliopsis binata	5.20	0.29	67.1	98.43	16290

Table 6.40: Effect of vegetative borriers on runoff and, soil loss

Treatment	Runoff %	Soil loss (t/ha)	% reduction Runoff	Soil loss
Guenia grass	36.7	6.95	24.4	84.55
Bhabar grass	42.7	9.79	17.9	78.24
Khus grass	39.6	8.10	23.8	82.00
Cultivated fallow	52.0	45.0	-	-

Grasses as vegetative barrier

Vegetative barriers include contour rows of perennial grasses, hedges, wind brakes and shelter belts etc. As live bund of perennial grasses use of Guatamala grass *(Tripsicum lamms)* on lateritic soils of Ootacamund was an excellent example for formation of Puertoricon or California type terraces through vegetative barriers (Chinnamani and Rege, 1965), which should be considered as an alternative to contour bunds. Khus (*Vetveria zizaniodes)* has been recommended widely by the World Bank in India and South East Asian countries for growing as live bunds. In Doon Valley on 4% slope Bhardwaj (1900-91) reported that live bunds for Guenia grass, bhabar grass and khus grass reduced the runoff by more than 18% and soil loss by more than 78 per cent compared to cultivated fallow which produced 52 per cent of rainfall as runoff and 45 t/ha soil loss (Table 6.40). The vegetative barriers are cheap and equally effective as compared to mechanical measures on milder slopes.

Vegetative barrier of paired rows of Leucaena planted at 1 m V.1. were effective in controlling erosion at 4 per cent slope in Doon Valley (Narain *et al.*, 1992).

Horticultural plantation

A wide range of temperate and tropical fruits trees can be grown on uncultivable waste lands of this region. Fruit trees are generally widely spaced and have slow growth rates than forest species and may cause substantial erosion if orchards are maintained as clean weeded (Table 6.41). It is therefore recommended that only immediate viginity of trees may be weeded and either weeds or crops may be grown in vacant space in orchard atleast during monsoon period (Ghosh and Ram Babu, 1972).

Table 6.41: Effect of different fruit plants on runoff and soil loss at 11% slope

Conservation measure	Runoff %	Soil loss (t/ha)	% reduction	
			Runoff	Soil loss
Cultivated fallow	16.20	18.45	-	-
Pomegranate clean	33.50	16.38	106.80	11.22
Pomegranate with weed	7.24	1.34	5.3	92.74
Pineapple clean	10.50	8.49	35.19	54.25
Pineapple with weed	5.80	1.69	65.43	90.84

6.7.1 Mechanical measures

These measures essentially consist of construction of barriers across the direction of the flow to retard or to retain the runoff and thereby reduce the soil and water losses and increase productivity.

There are following mechanical measures:

1. Contour bunding
2. Graded bunding
3. Broad base terraces
4. Bench terracing
5. Contour trenching

6.7.1.1 Contour bunding

It consists of construction of small earthen embankments or bunds on or along the contours (across the slope and land).

Types of bunds

- Contour bunds- These are constructed along contours or with permissible deviations from contours.
- Side bunds- These are constructed at extreme ends of the contour bund, running along the slope.
- Lateral bunds- These are constructed along the slope in between two side bunds in order to prevent concentration of water along one side and to break the length of contour bund into convenient bits.
- Supplemental bunds- These are constructed between two contour bunds so as to limit a horizontal spacing to the maximum required are called supplemental bunds.
- Marginal bunds- These are constructed along margins of the water-shed, road margins, river or stream margins, gully margins.

Contour bunding (narrow base terracing)

Contour bunding is recommended for gentle slope ranging from 0.5 to 6% with the average annual rainfall not exceeding 750-1000 mm. It has been extensively recommended in Maharashtra, Andhra Pradesh and Uttar Pradesh. The practice can be adopted on all types of relatively permeable soils except the clayey or deep black cotton soils, where it may cause water stagnation and reduction of yields on upstream side. Such bunds intercept slopes length, maintain flow under critical velocity and increase infiltration opportunity. The contour bunding at 2, 4 and 8 per cent slopes reduced the runoff to 8.5, 20.1 and 32.4 mm and soil loss to 3.14, 4.85 and 8.75 t/ha compared to corn on contours (Sastry, 1994). Thus the practice reduced 80 per cent soil loss at all slopes and could retain the soil loss within tolerable limits.

On large areas for control of soil erosion, it is necessary that land treatment between the bunds *e.g.* partial levelling, contour cultivation offers better protection as well as moisture conditions favourable for higher production. It is also necessary in contour bunding system to remove excessive runoff resulting from high intensity storms, surplussing arrangements being provided wherever necessary.

Soil suitable for bunding

Contour bunding may be recommended on all types of relatively permeable soils (*e.g.* alluvial, red, laterite, brown soil, shallow and medium black soils) except the clayey or deep black cotton soils.

Contour bunding in black soils was associated with the following drawbacks:

Prolonged water stagnation and consequent loss of area for crop production, breaches in the bunds due to soil cracking and pressure of water, instability of masonry waste weirs, man made breaches consequent to water stagnation and extensive rill erosion and gulleying due to concentrated flow in the series of wasteweir located one below the other.

6.7.1.2 Graded bunding

It is also known as graded terraces or channel terraces are laid along a pre-determined longitudinal grade instead of along contour.

Function

Graded bunds consist of constructing series of wide and relatively shallow channels across the slope very near the contour ridges and at critical intervals. The primary function of these terraces as drainage channels for including and regulating the excess runoff water and draining the same with a mid and non-erosive velocity. The purpose of graded bunding is to make runoff water to trickle into the soil rather than to rush out.

Suitability for graded bunding

It is suited in areas where annual rainfall is more than 80 cm and particularly in clayey soils even with lesser rainfall. Graded bunds are recommended upto 10% slope for areas, where annual rainfall exceeds 800 mm particularly on clayey soils with poor drainage.

Factors to be considered

The first item to be considered in making preliminary field plans is to locate the most desirable terrace outlets. Secondly it provides a safety factor. Short terraces with several outlets in natural water courses are preferred as compared to a system having long terraces and one outlet. In graded bunding, strengthening of natural waterways has advantages as terrace outlets. Where vegetation alone is not able to provide protection to carry concentrated water down the slope, construction of structures may be done, especially in the following situations:

- sudden drops at the channel alignments,
- too great velocities to be borne by grass cover, and
- establishment of good grass cover is not possible.

Specification of graded bunding includes vertical interval, grade of channel and cross-section. The graded bunding reduced the runoff by 62, 67, 57 per cent and soil loss by 74, 71 and 67 per cent over contour maize at 2, 4 and 8 per cent slopes, respectively (Table 6.41). Although practice is recommended upto 10 per cent slopes but results indicated that beyond 4 per cent slope, efficacy of graded bunding tended to reduce gradually.

Table 6.41: Runoff (mm) and soil loss (t/ha) on different slopes treated with mechanical measures

Mechanical measure	2% slope		4% slope		8% slope	
	Run off	Soil loss	Run off	Soil loss	Run off	Soil loss
Contour maize	40.4	16.08	100.8	23.90	175.6	43.08
Graded bunding	15.3	4.24	33.6	6.87	75.6	14.33
Contour bunding	8.5	3.14	20.1	4.85	32.4	8.75
Bench terracing	5.7	1.65	9.6	2.05	18.2	3.11

6.7.1.3 Broad-base terraces

These are also called as ridge terrace or broad-base bunds. It consists of a channel with a ridge on its lower side, whose function is to drain surface runoff or to absorb runoff. These terrace are constructed on land upto 10% slope. There are following types of terraces –

(i) Drainage terrace

(ii) Absorption terrace

Suitability

Drainage terrace - It is constructed where,

- the slope of land is 3-10%,
- the permeability of soil is slow,
- the field is dissected by gullies or depressions; and

- contour cultivation is insufficient to control surface runoff and soil erosion.

Absorption terrace - It is constructed where,

- rainfall is low and rainfall intensities are not excessive;
- the slope is less than 6%; and
- infiltration and permeability are high.

The absorption terrace differs from drainage terrace in being level along its length, and in having a longer cross-section. The channel and ridge side slopes should be 1:6 to 1:8. Ridge width on the upper side should be not less than 4 m when machinery is used for construction. The ridge should be 0.45 m higher as compared to the channel.

6.7.1.4 Bench terracing

Bench terracing is one of the most popular conservation measures adopted by farmers of hilly areas. Though the practice has been recommended for slopes upto 33 per cent, farmers are having terraces even at 60-70 per cent slopes in NWH. The practice consists of construction of step like fields or benches along contour by cut and fill method, thus breaking the slope length and impounding the rainwater for rice cultivation or channelising it for safe disposal. Bench terracing at 2, 4 and 8 per cent slopes reduced the runoff to 5.7, 9.6 and 18.2 mm, respectively reducing soil loss by 89.7, 91.4 and 92.8 per cent, respectively under maize, when compared with maize on contours (Sastry, 1994). It is one of the best practice among all mechanical measures but expensive for poor farmers unless state subsidies are arranged.

It consists of construction of series of platforms along contours cut into hill slope in a step like fashion. These are constructed at regular intervals by vertical drops or by steep sides and protected by vegetation and sometimes by packed stone retaining walls. Bench terraces convert the long unintercepted slope into several small strips and make protected platforms available cultivation.

Classification of bench terraces

Bali (1978) classified the bench terraces on the basis of purpose they are used as,

1. Hill-type bench terraces- It is used for hilly areas with a reverse grade towards the hill.
2. Irrigated bench terraces or level benches- It is recommended under irrigated conditions.

3. Orchard bench terraces or narrow width terraces (about 1 m)- It used for individual trees.
4. Gradual bench terracing - It is the conversion of land into bench terraces over a period of time.

Rama Rao (1974) also classified the bench terraces based on the slope as shown in Fig. 6.2. (1) Bench terraces sloping outwards, (2) Bench terraces sloping inwards, and (3) Bench terraces with level top.

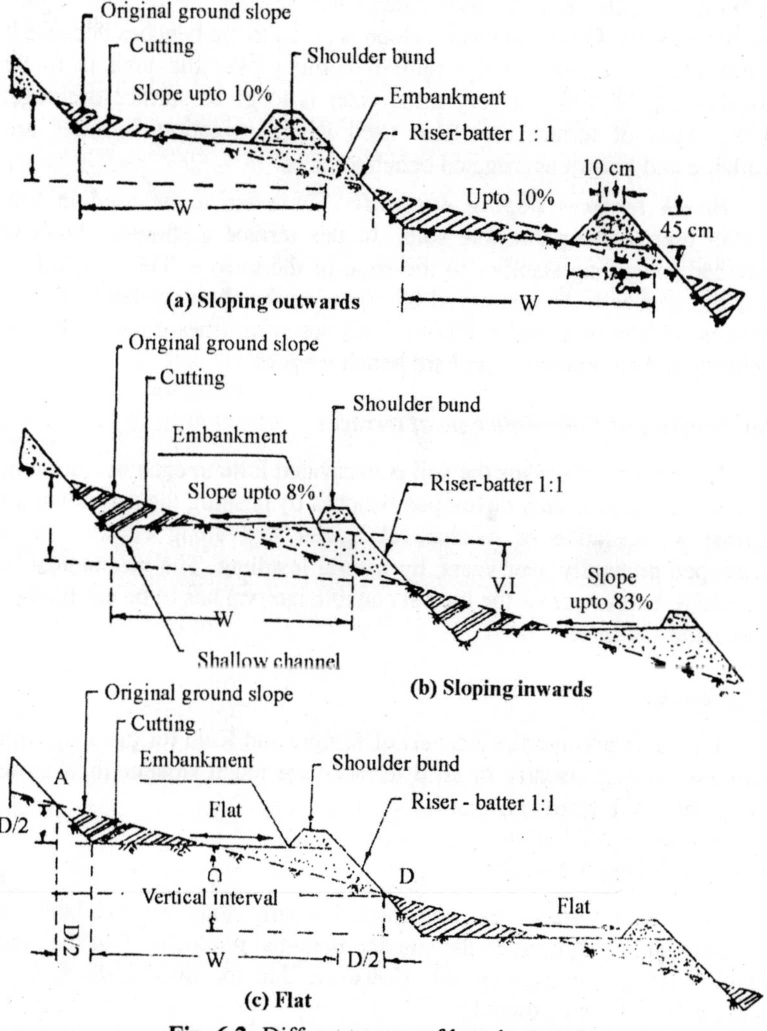

Fig. 6.2: Different types of bench terraces

Suitability of bench terraces

Bench terraces with slopes inside: These are suited in heavy rainfall areas where a major portion of the rainfall is to be drained as surface runoff. It required a suitable drain out at the inward end of each terrace provided to drain the runoff. These are also known as hill-type bench terraces.

Bench terraces with level tops- These are suitable for areas of medium rainfall, evenly distributed and having deep and highly permeable soils. Owing to this no slope is given to the benches because it is expected that most of the rainfall coming over the area is to be absorbed by the soil and very little water is to go as surface drainage. These types of terraces are also used where irrigation facilities are available and known as irrigated bench terraces.

Bench terraces sloping outwards- These are to be used in low rainfall areas with permeable soils. In this terrace a shoulder bund is provided for giving stability to the edge of the terrace. The rain falling over the area is to be conserved by retaining the shoulder bund. Bench terraces with narrow width (about 1 m) are sometimes constructed for orchards and are known as orchard bench terraces.

Purertorican or California type of terraces

In this type of terrace the soil is excavated little in each operation of ploughing and gradually developed benches by pushing the soil downhill against a vegetative or mechanical barrier laid along contour. It is developed gradually over years, by natural levelling. The mechanical or vegetative barrier across the land at suitable interval has to be established to provide stability.

Strip terrace

It is generally used in districts of Kangra and Kulu for growing fruit trees on contour. Widths of strip terraces are much smaller than those recommended for growing paddy.

Stone walls bench terraces

This type terraces are constructed where stone are available in adequate quantities close to the site and potential productivity of the land justifies the expenses incurred. However, due to stony surface area, cultivation may be restricted.

Bench terraces with stone walls may be recommended for annual crops and perennial tree plantations. The latter are likely to produce the highest return on the investment, except in cases of special high value annual crops. Furthermore, tree crops require less cultivation operation and cost.

Conservation bench terraces (CBT)

The terrace system was originally designed (Zingg and Hauzer, 1959) to stabilize yields of rainfed crops by inter plot water harvesting. At Dehradun, 3:1 was the optimum ratio of runoff contributing to receiving area. Lower ¼ area is terraced to collect runoff from upstream ¾ area. Terraced part is used for rice cultivation while the area maintaining natural slopes is used for maize or any crop which prefers good drainage. In a study at 2 per cent slope in Doon Valley, conservation bench terracing with 20 cm. depth of impoundment at the end, resulted only 13 per cent runoff and 0.18 t/ha of soil loss conserving 96 per cent of runoff and 93 per cent soil loss as compared to control plots (Sharda, *et al.,* 1992). The practice could be of extension value in Bihar, Chootanagpur plateau, where Haveli system of cultivation is prevalent, choosing appropriate crops.

6.7.1.5 Contour trenching

It implies excavating trenches along the contour or along a uniform level. A typical cross-section of the trench is shown in Fig. 6.3. These are used both on hill slopes as well as on degraded and bare waste lands for soil and moisture conservation and afforestation purposes.

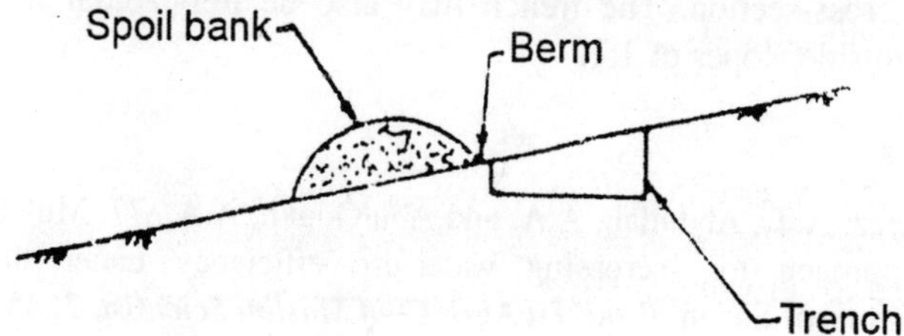

Fig. 6.3: Cross-section of a contour trench

Function of trench

1. Break the slope lengths,
2. Reduce the velocity of surface runoff, and

3. Consequently retard its scouring action and carrying capacity. The water retained in the trenches help in conserving the moisture and essential environment for sowing and planting.

Contour trenches have been used on all slopes. However, it is not recommended on slopes exceeding 20% either technically or economically. It has been profitably used in both high and low rainfall conditions, varying soil types and depths, for soil and water and conservation and afforestation.

Types of trenches

Contour trenches may be classified as continuous or interrupted as shown in Fig. 6.4. The interrupted trenches may be constructed in series or in a staggered manner. While, continuous trenches are essentially used for moisture conservation in low rainfall areas and require careful layout. Intermittent trenches are adopted in high rainfall areas.

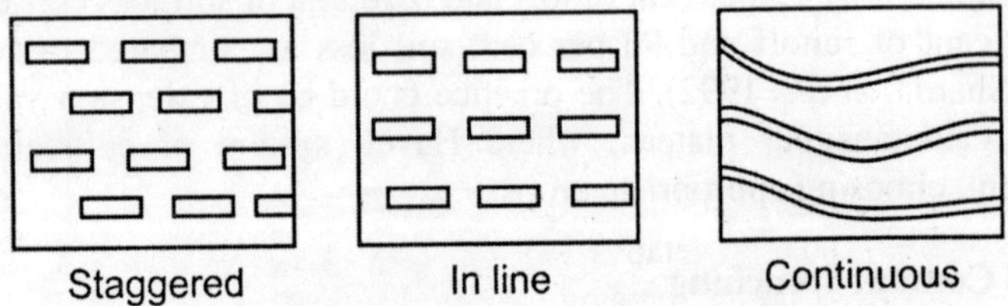

Fig.: 6.4: Layout of contour trench

The size of the trenches depends upon the soil depth but normally size varies from 1000 sq. cm to 2500 sq. cm in cross-section. The dimensions of the trench may be 30 cm base with 30 cm top width and square in cross-section. The trench may also be trepezoidal in cross-section with side slopes of 1:1.

References

Abdel-Hafeez, A.T., Abdullah, A,A. and Abu Goukh, A.A,. 77. Mulching as an approach for increasing water-use efficiency under the arid conditions of Sudan. *Proc. 1st Agri. Conf Muslim Scientist.* **2**: 453-65.

Adams, J.E. 1970. Effect of mulches and bed configuration on soil temperature, growth and yield responses of grain Sorghum. *Agron. J.* **62**(6): 785-90.

Agrawal, R.P., Jhorar, B.S., Dhankar, J.S. and Raj, M. 1987. Compaction of sandy soils for irrigation management, *Irrigation Science,* **8**: 227-232.

Allinaras, R.R. and Dowdy, R.H. 1985. Conservation tillage system and their adoption in the United States *Soil and Tillage Research,* **5**: 197-222.

Andhale, R.P., S.H. Sinde and Pawar, V.S. 2003. Effect of sowing dates on growth and yield of pearl millet during summer season. *J. Agrometerol.*, 5(2): 102-105.

Anonymous, 1995. Training manual on plastic culture, Deptt. of S.W.E., C.T.A.E., Udaipur.

Auerswald K. 1998. Vertisols and technologies for their management *Soil Sci.* **163**(6): 510-511.

Bahram, H.J. 1970. Effect of petroleum mulch on cotton growth. *Acta Agron. Hung.* **19**: 117-122.

Bali, J.S. 1978. Design of bench terrace in the hills, Paper presented at the 16th Annual Meeting of ISAE, Kharagpur.

Bhagchand and Gautam, R.C. 2000. Effect of organic manure, bio-fertilizer and inorganic fertilizers on growth, yield and quality of rainfed pearl millet. *Ann. Agric. Res.*, 21(4): 459-464.

Bhan, S. and Singh, S.R. (1979). Water harvesting and moisture conservation practices for dry farming of maize in U.P *Annals of Arid Zone* **18**(l& 2): 100-107.

Bhardwaj, S.P 1994. Vegetative barriers as an effective economic and eco friendly measure of erosion control of agricultural lands. Proc. *8th ISCO Conference on Soil and Water Conservation Challenges and opportunities,* New Delhi, India, Dec., 4-8: pp 204-206.

Bhardwaj, S.P and Sindhwal, N.S. 1998. Zero tillage and weed mulch for erosion control on sloping farm land in Doon valley. *Indian J. Soil Cons.* **26:** 81-85.

Bhardwaj, S.P., Khybri, M.L., Sewa Ram and Prasad, S.N. 1984. Effect of atrazine application in maize on soil and water losses, *Indian J. Soil Cons.,* **7**(1): 5-9.

Black, A.L. and Greb, S.W. 1962. Nitrate accumulation in ,;oils covered with plastic mulch. *Agron. J.* **54**: 366.

Channappa, T.C. 1978. Annual Report of AICRPDA, UAS, Bangalore Centre, Karnataka, India.

Chaudhary, P.C. and Chatterjee, S.N. 1967. Moisture conservation under mulches in the eroded terraced soils of Ranchi. *J. Soil and Water Cons. India* **15**: 61-71.

Chinnamani, S. 1989. Effectiveness of soil conservation measures. *J. Agric. Engg.*, ISAE **24**: 148-54.

Choudhary, M.R., Gajri, P.R., Parihar, S.S. and Khera, R. 1985. Effect of deep tillage on soil physical properties and maize yields on coarse textured soils. *Soil and Tillage Research,* **6***:* 31-34.

Clarkson, V.A. 1958. Watermelon yields increased with polyethene mulch. *Oregon Veg. Digest* (April, 1958).

Clarkson, V.A. 1960. Effect of black polyethene mulch on soil and microclimate temperature and nitrate level. *Agron.* J. **52**: 307-309.

Clarkson, V.A. and Frazier, N.A. 1957. Plastic mulches for horticultural crops. *Oregon Agric. Exp. Sta. Bull.,* 562.

Dixit, L.A., Hegde, B.R. aid Channappa, T.C. 1981. Soil and Moisture Conservation Practices in Red Soils. Panel Discussion on SWC in Red and Black soils, CSWCR & TI, R.C., Bellary and UAS, Bangalore.

El-Swaify, S.A., Singh, A. and Pathak, P. 1987. Physical and conservation constraints and management components for SAT Alfisols. In Proceedings of the consultants workshop on the state of art and management alternatives for optimizing the productivity of SAT Alfisols and related soils. ICRISAT, Patancheru, A.P., India, 33-48.

Foster, G.R. and Meyer, L.D. 1972. Erosion mechanics of mulches, ASAE Paper No. 70-754. Amer. Soc. Agr. Eng. St. Joseph, Michigan, p. 20.

Free, G.R. and Bay, C.E, 1969. Tillage and slope effects en runoff and erosion. Trans. ASAE **12**(2): 209-211.

Gautam, R.C. and Kaushik, S.K. 1984. Role of split application of nitrogen doses on the growth and yield of pearl millet hybrids. *Indian J. Agron.,* 29(3): 327-330.

Gupta, J.P., Agrawal, R.P. and Somani, L.L. 1989. Managing soil physical environment, In Proceedings of International Symposium on Managing Sandy Soil, Jodhpur, India: Central Arid Zone Research Institute, pp. 175-192.

Gupta, R.P., Kumar, *S.* and Singh, T. 1984. Soil management to increase crop production. Consolidated report 1967-1982 ICAR, New Delhi.

Hadda, M.S, 1983. Effect of soil management practices on nutrient, sediment and water yield from an agricultural sub-catchment in sub-mountaneous tract of Punjab. M.Sc. Thesis, P.A.U., Ludhiana.

Hudson N.W. 1971. Soil Conservation, B.T. Batsford Limited, London.

Jat, M.L., Sharma, S.K., Balyan, J.K., Jain, L.K. and Sharma, R.K. 2007. Economic evaluation of tillage practice for maize cultivation in dryland areas of Rajasthan. Current Agril., 31(1&2): 35-39.

Katama Reddy, B.C. 1979. Note on the soil and moisture conservation structures at Agricultural Research Station, Ananthapur (Unpubl.).

Katcheson, J.W. 1977. Conservation tillage eastern Canada. *J. Soil and Water cons.* **32**(1): 57-60.

Katyal, J.C., Singh, R.P., Sharma, S.K., Das, M.V. Padmanabhan and Mishra, P.K. 1995. Field Manual on Watershed Management, CRIDA, Hyderabad, p. 223.

Kayothbo, B. and R. Lal 1993. Tillage systems and soil compaction in Africa, *Soil and Tillage Research,* **27**: 35-72.

Khybri, M.L. 1988. To find out 'C' factor or some important crops on Doon Valley, Annual Report CSWCRTI, Dehradun, UP, pp. 57-58.

Lal, R. 1975. Role of mulching techniques in tropical soil and water management, IITA Tech. Bull. No. 1, p. 38.

Lal, R. 1985. No-tillage and surface-tillage system to alleviate soil-related constraints in the tropics. In M.A. Sprague and G.B. Triplen (Eds.). No-tillage and Surface-tillage Systems Related to Soils in the Tropics, Wiley, New York, 261-317.

Lal, R. 1989. Conservation tillage for sustainable agriculture: Tropics versus temperate environments, *Adv Agron.* **42***:* 85-180.

Lal, R. 1994. Soil and tillage research, 30: 169-185.

Lattanzi, A.R., Meyer, L.D. and Baumgardner, M.F. 1974. Influence of mulch rate and slope steepness on inter-drill erosion, Soil Sci. Soc. Am., Proc. **38**(6): 946-950.

Lawes, D.A. 1961. Rainfall conservation and yield of cotton in Northern Nigeria, *J. Exptl. Agric.* 116 307-318.

Linden, R. 1963. Soil protection with plastic mulches. *Ann. Gembl.* **69**: 601-608.

Ma Shijun 1988. Advances in mulch farming in China. In Proc. Int. Conf. on Dryland Farming; Amarillo, Bush land, pp. 510-513.

Mahapatra, R. 1998. Down to Earth 6 (18) 1998.

Manescu, Band Ciofu, A. 1970. The influence of mulching with plastics on the thermal and water conditions of soil. *Horticulture (Seria)* **13**: 63-72.

Mannering, J.V. and Meyer, L.D. 1963. The effect of various rates of surface mulch on infiltration and erosion. Soil Sci. Soc. Am. Proc. **27**(1): 84-86.

Mannering, J.V., D.L. Scherttz and B.A. Julian 1987, Overview of conservation tillage, In effects of conservation maize tillage on groundwater quality, eds. TJ. Logan, J.M. Davidtron, J.L. Baker, and M.R. Overcash, Chelsea, MI: Lewis Publishers, pp. 3-17.

Meyer, L.D., Wischmeier, W.H. and Foster, G.R. 1970. Mulch rates required for erosion on steep slopes. Soil Sci. Soc. Am. Proc. **34**(6): 928-931.

Mishra, P.J. and Sahu, D. 2001. Comparative performance of vegetative barriers in north eastern ghat zone of Orissa. Indian *J. Soil* Cons. **29**(1): 50-52.

Moldenhauer, W.C. 1979. Erosion control obtainable under conservation practices (In Universal Soil Lose Singh Equation: Past Present and Future). Soil Sci. Soc. Am. Special Pub. No. 8.

Muchow, R.C. 1989. *Field Crops Res.*, 20:191-205.

Murthy, V.V.N. 1982. Land & Water Management Engineering, Kalyani Publishers, Ludhiyana, pp. 590.

Ohu, J.O. Folorunzo, O.A. 1989. The effect of machinery traffic on physical properties of a sandy loam soil on Singh, the yield of sorghum in north-eastern Nigeria, *Soil and Tillage Research,* **13**: 399-405.

Onwualu, A.P. and U.G.M. Anazodo 1989. Soil compaction effects of tillage on maize production under various tillage methods in a derived savannah zone of Nigeria, *Soil and Tillage Research,* **14**: 99-114.

Oschwald, W.R. and Siemens, J.C. 1976. Conservation tillage. A perspective SM-30, Agronomy Facts, Univ. of Illinois, Urbana.

Parihar, S.S., Glildayal, B.P., Painuli, D.K. and Sur, H.S. 1985. Physical properties of mineral soils affecting rice based cropping systems. In: Soil Physics and rice, Proceedings of workshop on physical aspects of soil management in rice based cropping systems held at IARI, New Delhi, pp. 57-60.

Patnaik, U.S., Chittaranjan, S. and Rama Mohan Rao, M.S. 1987. Performance of conservation ditching at a glance. In Conservation Ditching for Soil and Water Conservation in Low Rainfall Vertisol Regions *Bulletin No.* T- 19/D- 15 CSWCRTI, Dehradun, 26-40.

Pusztai, A. 1963. The effect of plastic mulch on the soil and plant. *Agrokemt. Talalt* **12**: 351-360.

Rajput, R.K. and Singh, M. 1970. Efficacy of different mulches in conserving soil moisture in cotton. *Indian J. Agron.* **15**(1): 40-45.

Rama Rao, M.S.V. 1974. Soil Conservation in India, Indian Council of Agricultural Research, New Delhi.

Roy, K. and Singh, R.K. 1983. Effect of mulches and nitrogen on moisture use pattern and moisture use efficiency in rainfed wheat. *Indian J. Agric. Res.* (1794): 203-208.

Sadams, J.E. 1966. Influence of mulches on runoff, erosion and soil moisture depletion. *Soil Sd. Soc. Am. Proc.* **30**: 110-114.

Sandal, S. and Acharya, C.L. 1997. Effect of conservation tillage on moisture conservation, soils-physical conditions, seedling emergence and grain yield of rainfed maize (*Zeamays*) and wheat (*Triticum aestivum*) *Indian J. Agri. Science* **67**: 227-231.

Sastry, G., Dhruvanarayana, V.V. and Prasad, R. 1994. Watershed resources and conservation measures and degree of land slope. Paper communicated to Earth Surface Process and Landform, UK.

Schales, F.D. 1964. A study of some physical and biological effects of various mulching materials when used with several vegetable crops. *Diss. Abst.* **24**: 3487.

Schales, F.D. Liu. Cixian and Wong. D. 1990. Plastic mulch research in Gaungdong province, People Republic of China. *Proc. XI International Congress-Symp.* on the use of plastic in Agri. 26 Feb-2 March, New Delhi.

Shah, T. 1996. Concept paper on India water sector initiatives. UNDP India Water Sector Strategies, pp. 30.

Shaw, R.H. 1956. Water use from plastic covered and uncovered corn plots. *Agron. J.* **51**: 172-173.

Sheld rake, R. J. 1958. Black plastic, maize carpet for truck farmers, *Farm. J.* May 1 (1958).

Sood, M.C. *1975.* Evaluation of tillage and residue mulching for soil and water conservation on slopping land. M.Sc. Thesis, P.A.U., Ludhiana.

Sur, H.S. and Gutam, B.S.. 1994. Soil management and rainwater conservation and use. IV Alluvial soil under medium rainfall. In soil management for sustainable agriculture in Dryland Areas, pp. 56-65, Bulletin of the Indian Society of Soil Science, No. 16.

Takatori, F.H. Li pert, L.F and Whiting F.L. 1964. The effect of petroleum mulch and polyethylene films on soil temperature and plant growth. *Am. Soc. Hofl. Sci. Proc.* 85:532-540.

Tejwani, K.G., Gupta, S.K. and Mathur, H.N. 1975. Soil and Water Conservation Research, 1956-71, ICAR, New Delhi, 1-358 p.

Tripathi, S.C., Nagarrajan, S. and Chauhan, D.S. 1999. Evaluation of zero tillage in wheat *(Triticum aestiuum)* under different method of rice transplanting, *Indian J. of Agronomy,* **44**(2): 219-272.

Triplett, G.B. Jr., D.M. VanDoren Jr., and S.W. Bone 1973. An evaluation of Ohio soils in relation to no-tillage corn production. Ohio Agricultural Research and Development Centre, *Research Bulletin* No. 1068, Ohio Agr. Res. And Dev. Centre, p. 20.

Urger, P.W. 1979. Effect of deep tillage and profile modification on soil properties, root growth and crop yields in United States and Canada, Geoderma, **22**: 275-295.

Venkatachalam, C. Iruthayaraj, M. and Sriniwasan, T.A. 1980. Influence of mulches on the temperature environment of cotton crop. *Indian J. Agric. Res.* 14(1): 29-34.

Venkateswarlu, J. 1981. Maximization of crop production in drylands. *Indian J. Soil Conservation* **9**: 124-40.

Vittal, K.P.R., Rao, K.V, Sharma, K.L., Victor, U.S., Ravindra Chary, G. Marutishanker, G.R., Sharma, J.S. and Singh, J. 2005. Agricultural Drought Condition, cropping practices and amdioration paths for rainfed production systems. Experiences of three decades of dryland network research in India, Indian *J. Dryland Agri. Res. Dev.*, **20**(2): 155-202.

Wastelands 1998. Traditional water harvesting systems: Statement of shared concern in wastelands newsletter XIII(3): 18-24.

Willies, W.O. Larson W.E. and D. Kirkham 1957. Corn growth as affected by soil temperature and mulch. *Agron. J.* **49**: 323-8.

Wischmeier, W.H. 1973. Conservation tillage to control water erosion. In: Conservation Tillage: The proceedings of a national conference: 133-141. Soil Conservation Society of America, Ankkeny, 10.

Wischmeier, W.H., Johnson, C.B. and Cross, B.V. 971. A soil erodibility monograph for farmland and construction sites. *J. Soil Cons.*, **26**(5): 189-192.

Yadav, S.T. and Varshney, M.C. 2005. Influence of sowing dates and sowing methods on growth and yield of pearl millet.

Yoo, K.H., Touchton, J.T. and Walker, R.H. 1998. Runoff sediment and nutrient losses from various tillage. systems of cotton. *Soil Tillage Res.* **12**: 13-24.

Zingg, A.W. and V.L. Houzer *1959.* Terrace benching to save potential runoff for semi arid land. *Agron. J.* **51**: 289-292.

EXERCISES

6.1 Explain the in-situ rainwater management and describe the factor affecting it.

6.2 Define the following:

a. Soil erodibility

b. Puddling

c. Profile modification

d. Tillage

e. Crop residues
f. Mulching
g. Tied ridging
h. Contour trenching
i. Vegetative barrier

6.3 What are the basic principles for in-situ rainwater management?

6.4 Differentiate the following:

a. Erosivity and erodibility
b. Tillage and special tillage
c. Puddling and compaction
d. Plastic and gravel mulch
e. Land levelling and land grading
f. Furrow diking and v-ditches
g. Bedding system and ridge and furrow system
h. Contour cultivation and strip cropping
i. Narrow base and broad base terracing
j. Contour bunding and bench terracing

6.5. Explain the factors governing the land levelling.

6.6. Describe the various in-situ moisture conservation measures.

6.7. Explain the bedding system for in-situ rainwater management with neck sketches.

6.8. Discuss the different agronomical measure for in-situ rainwater management.

6.9. What is the contour bunding and type of contour bunds and its suitability?

6.10. Explain different types of bench terracing and their suitability.

6.11. What is the contour trenching and give its classification?

■■■■

Chapter 7

EX-SITU RAINWATER MANAGEMENT

7.1 Introduction

In arid and semi-arid areas, the low and erratic rainfall normally occurs with high intensity of short duration resulting in high run-off and poor soil moisture storage. As a result, about 50 to 60 percent of rainwater is lost through runoff. A runoff loss varies from 10 to 30 percent of the rainfall depending on the amount and intensity of rain, soil characteristics and vegetation cover. This surface runoff, if harvested over a large area can yield considerable amount of water for storage providing life saving irrigation to the crop during the dryspell in the monsoon season and also for sowing a second crop in winter season. The major constraints that still limit the adoption of this technology on a macro scale are, the high initial cost, and non-availability of cheap and effective sealants for permeable Alfisols. Additionally, long breaks in the monsoon and low intensity rains limit the runoff flow into the ponds during the dryspells when water is needed most. Despite these difficulties, small water storage ponds seem to be the most viable strategy to stabilize productivity of the ecologically disadvantaged dryland regions. The surface runoff from an area can also be increased by reducing the infiltration capacity of the soil through vegetation management, cleaning, chopping surface vegetation and reducing soil permeability by application of chemicals.

7.2 Need of Water Harvesting

Water harvesting is the need of time and it is summarized as under:

i. Dryspells or recurrent droughts is common features of climate,

ii. Low fertility status of soil,

iii. Low water retention capacity of soils,

iv. For increasing productivity under rainfed conditions,

v. For drinking purposes for livestock, and

vi. Groundwater recharge.

To maximize profitability from the limited quantity of water stored in small ponds, planning for its judicious use is most crucial. Research conducted at different locations in India establish that a supplemental irrigation of 5-10 cm at the critical stage of crop growth substantially increases the yields of cotton, wheat, sorghum, tobacco, pearl millet, etc. over no irrigation.

In order to fulfill such demands in the future we will need to think on the various measures capturing and storing harvested water, received from rain, is a wise option. Rainwater offers advantages in water quality for both irrigation and domestic use. Rainwater is naturally soft (unlike well water) contains almost no dissolved minerals or salts, is free of chemical treatments, and is a relatively reliable source of water for households. One of the beauties of rainwater harvesting system is its flexibility. A system can be as simple as a bucket kept under the edge of a roof to collect the rainwater or as complex as an engineer, multi-tank, pumping and pressurized construction to supply residential and irrigation needs. Indigenous water harvesting systems are presented in Table 7.1.

Table 7.1: Indigenous water harvesting systems in different agro-ecological regions of India

Agro-ecological regions	Water harvesting structures
Hill Areas	Water was diverted with the help of simple engineering structures into artificial channels to carry it to agricultural fields. Known as *Kuhls* or *Kuls* in Himachal Pradesh and Jammu region, they are typical examples of water harvesting structures.
Arid and Semi-arid areas	Rivers and seasonal streams were tapped and diversion channels were directed into storage structures, *viz.*, *zing* in Ladakh, *keres* in Karnataka and *ahar* in Bihar.
Cold and Arid Ladakh	In Ladakh which depends on snow-melt for water, it is collected during the evenings in small reservoirs for use the following morning.
Eastern Himalayas	In Nagaland and Arunachal Pradesh villagers divert a channel through the cattleshed in order to harvet nutrients for agricultural fields.
	Angami Nagas conserve forest cover above the streams to ensure constant supply of humus to terraced agricultural fields.
Northeastern Hills	Bamboo network is used as pipeline to direct water to a convenient spot for collection as drinking water.

Agro-ecological regions	Water harvesting structures
Meghalaya	In Meghalaya a bamboo drip system is in vogue for ages. Water is delivered to betel leaf plantation in rocky terrain by use of a bamboo network which provides quantified water (20-go drops) directly to roots of the plant.
Tamil Nadu	In Tamil Nadu a big stream is often diverted to feed a chain of 25 to 30 tanks in a sequence called system tanks. The stream collects water over a large catchment considered more desirable than small single tank with small catchment. The Palar Anicut System supplies water to 317 tanks irrigating an area of 32,000 ha. in North Arcot and Chengalpattu districts. Ramanathapuram has scores of tanks visible even in satellite imagery.
Western and Central India	In MP water is collected in a tank (bandhis) to moisten the soil. In this traditional '*haveli*' system where water is stored in the field itself, crop is taken on residual soil moisture.
Arid Kutch	In Kutch. the *Maldhari* nomads collect sweet water for drinking from well in tanks called *virda*. Although ground water is saline, sweet water remains on top owing to its lesser density.
Arid Rajasthan	In Jodhpur wells and step wells (*Baoris*) which are actually wells in the tank bottom, provide drinking water even in summers. In Phalodi, rooftops have been used as catchments. In Churn district '*kundis*' artificial wells which store runoff from an artificially prepared catchment surroun-ding them, are used. The logic is simple; if the region receives only 100 mm rain and if land area is no constraint, large area is tapped to allow runoff to collect in the 'kunds'. If rain is harvested over a hectare of land it would provide one million litres of water.

(Source: Katyal *et al.*, 1998)

7.3 State Sponsored People's Cooperatives

- ***Role of Indian rulers in medieval India*:** The rulers of different states in pre-colonial era encouraged their people to undertake building of collective water harvesting structure. For example in the *Thar* the *Jagirdar* who collected revenue, encouraged people to build *talabs*. In Udaipur, the Pichola Lake was built by the nomadic gypsies. In Vijayanagar empire the state took the initiative of building irrigation structures. The Temples also played an important role in encouraging the construction of irrigation structures. For example, there are hundreds of old and historic temples in almost each village of Karnataka and every temple invariably has a *Kalyani* (Temple tank).

- ***Colonial Period- Dismemberment of traditional irrigation system under British rule*:** The British in their colonial zeal to grab the riches strewn all over rural India, progressively dismembered the irrigation system and impoverished the rural folks who had so studiously built prospering village societies through centuries of resource management, developing agriculture, optimal land use and enduring irrigation systems. The irrigation tanks and wells fell to disuse for want of revenue for their proper maintenance. This led to community-held property to turn into no man's land and total despair. However, the British later realised their folly in undermining the strength of indigenous irrigation system and their belated efforts to revive them failed in Madras Presidency and in Bengal. As a result, large dams and canals commanding extensive areas were built. However, where the influence of the British rule was limited, the traditional irrigation systems flourished.

Gradually the water bureaucracies, tube wells and the interference by politicians in village affairs and the influential real estate lobbies in urban areas, have all corroded the self-sufficient water resources utilization systems throughout indiced only some remote pockets remain outside their clutches.

- ***Post Independence Era*:** The post-independence political thinking contributed to the building up of capital intensive large-scale multi-purpose projects in place of small indigenous cost-effective projects. Legal changes in 1975 (Wastelands, 1998) 'took away the jurisdiction of individuals and village communities over all the water sources and placed it under the government. In Jammu and Kashmir, many traditional Kuhals were taken over by the government, but as it failed to manage them, these traditional systems were totally obliterated causing difficulties to the rural communities. In Kerala too, an adverse impact on water rights of individuals was felt when energised pumpsets were introduced in the 1960s'. The commercialization of agriculture has further influenced the traditional wisdom of sustainable agriculture. Once the water resource' is created, water loving crops viz., sugarcane and lowland rice replaces millets and pulses which formed a major part of the vegetarian diet and animal fodder. The intensity of water guzzling crops can be gauged from the fact that sugarcane in Maharashtra occupies only 3 per cent area, but uses around 76 per cent of the entire irrigation water.

7.4 Principle of Ex-situ Rainwater Management

There are following important principles for ex-situ rainwater management:

i. Increasing the time of infiltration into the soil by construction of barrier across the slope *viz.* contour bunding, terracing and check dam etc. to recharge groundwater,
ii. Reduce the length of slope to reduce the velocity of flow,
iii. Prevent excess soil and water losses, and
iv. Storing the runoff and its efficient recycling.

7.5 Water Harvesting

Water harvesting is one of the most promising and technically feasible technology specially in rainfed areas. Runoff harvesting and recycling for crop production is an essential component of watershed management. Watershed approach is the basic approach for rainwater harvesting.

Community tanks

In hilly regions, the runoff collected in the natural depressions can be recycled for irrigation to crops during the scarcity periods (Anonymous, 1988). This type of water harvesting however has to be followed on a community basis. In this system, the donor catchment does not benefit from the runoff water. Most of water harvesting reservoirs in the watersheds of hilly areas are constructed according to site convenience and not for the convenience of utilizing the water for irrigation purpose. Lifting water from the big pond for irrigation is feasible but expensive and troublesome most of the time. Huge seepage also results in too little water in the reservoir when it is actually required for irrigation.

Intra-terrace runoff harvesting

Considering the limitations associated with the community ponds, it is important that water harvesting tanks are located in a way that farmers make use of gravity for providing irrigation. Small dug out and lined tanks for intra-terrace water harvesting, constructed by individual farmer as per his resources and requirements, serve more useful purpose than large tanks constructed on community basis. Water can be effectively stored in small poly lined tanks. The polyethylene sheet (200-250 micron) is protected from direct sunlight by providing a covering of loose soil or round river boulders or bricks or thin fabric sheet. It is also ensured that the polyethylene sheet does not get damaged from protruding roots or pointed stones underneath by providing, a cushion of soil/sand or discarded synthetic empty cement bags (Acharya and Kapur, 1997). Series of such small farm ponds have been constructed in Palampur and elsewhere in Himachal Pradesh. This system of water storage is superior over community tanks because (i) it does not require pumping sets for irrigation (ii) tanki do not develop cracks unlike pucca cemented structures (iii) tank can be constructed in any soil, elevation or climatic conditions (iv) individual fanner has the descretion of deciding the size, shape, use and maintenance of the tank independently (v) cost of construction is low as the digging of the tank and the repair of the damaged sheet by heat sealing can be done by the farmer himself.

Auxiliary tanks for *kuhl* water harvesting

Traditional *kuhl* is a natural gravity irrigation system followed in the hilly and mountainous regions. The water supply from the *kuhl* is inadequate and inequitable most of the time and is surplus in some periods when farmers do not require water for irrigation.

Due to poor water control and lack of water distribution system, there is a tendency of head reach farmers to divert *kuhl* to their fields frequently while the farmers located at the mid and tail reach do not get sufficient and timely water. The flow in the *kuhl* during the lean periods (when the water needs of the crops are minimal) can be diverted in the auxiliary tanks constructed alongside the *kuhl* at its different reaches and recycled during the scarcity period that follows. The storage in the tanks would also be supplemented through harvesting rain-water. Storage tanks are, however, to be suitably lined with LDPE sheet. Once the water is stored in the tanks, the farmer has the freedom for its use. The water can again be replenished in the tank from the main source as and when available.

Hill spring out flow harvesting

There are innumerable natural springs in the Himalayan region most of which are in a neglected state. Their outflow goes waste and can be diverted to the polylined tanks. This harvested water may be utilized to raise economic crops and rehabilitate the vegetation of the area. The increased vegetative cover around the spring further rejuvenates its outflow (Acharya *et al.*, 1989).

Roof-top water harvesting

In the mountainous regions the houses are constructed at a higher elevation of the hillock and cultivation is done on a terraced land down below. Such situations are available for roof-top water harvesting and the water can be diverted to the poly-lined tanks. This can help farmers in mitigating the problem of acute water scarcity. Farmer, otherwise, has to transport water from long distances to raise nursery and establish seedlings of vegetables/cash crops.

Water harvesting in cold deserts

Spiti region of the state is designated as desert despite receiving heavy snow during winters which melts in Spiti river in summer. Its outflow goes unutilized without irrigating the valley. The soils are coarse and boulder ridden with very low water retentivity. These soils require very light and frequent irrigations to sustain plant growth. Some snow melt stream flow is also available in these areas. Thus, water storage is impossible unless tanks are suitably lined. The cement concrete tanks develop cracks due to sub-zero temperatures during winters. Farmers in the Spiti valley have developed their own traditional methods of storage of water in ponds locally called "Zing". However, there are high percolation losses from these ponds. Under hostile climatic, the H.P. Agricultural University, Palampur has constructed a tank at Kaza in Spiti valley (a cold desert at about 4200 m above msl). The tank of 80 m^3 capacity was constructed by raising walls of loose debris on the porous soil surface and lined with Silpaulin sheet. The tank is filled by lifting water from Spiti River through hydraulic rams. The stored water is being used through gravity for raising plantation of Seabucthorn- a wonder plant of the cold deserts.

7.6 Increasing Productivity of Stored Water Nadi – A Case Study

Water harvesting and storage have been successfully tried in labour and cost effective small poly-lined tank located at a convenient site to make use of gravity for providing minimal irrigation at HPKV farm, Palampur. One such dug out tank lined with LDPE sheet (200 micron thickness) constructed at the end of terrace has the following dimensions length at the top, 13.1 m; width at the top, 5.0 m; depth, 1.4 in, length at the bottom, 11.3 m; width at the bottom, 2.85 m. The tank has 68.5 m^2 capacity and 350 m^2 catchment area for water harvesting. The LDPE film has been provided with a cover of fabric sheet for protection from UV rays. The water from the tank is used by gravity through flexible garden pipe to raise two maize based crop sequences *viz.* maize- wheat and maize- radish (seeds sown between rows in standing maize after withdrawal of monsoon) - onion (green)- french beans (seeds. dibbled in growing onion about two months before its harvest).

The results from maize-wheat cropping sequence showed that in the absence of timely rain even 5.4 mm of irrigation as pre-sown increased the grain and straw yield of wheat significantly (Table 7.2). The data (Table 7.3) of second rotation, showed that it was possible to meet the water scarcity during post monsoon period if water was stored and judiciously used through light but frequent irrigation. Diversification of cropping schedules as a result of assured water availability this way can increase the income of the farmers. In addition to growing vegetable crops, component of pisciculture was also introduced successfully. This can further supplemented the farmers' income.

Table 7.2: Grain and straw yield (t ha^{-1}) of wheat in maize-wheat cropping sequence under minimal water supply from poly-lined tank and 50% of recommended NPK at Palampur

Irrigation	Irrigation applied (cm)		Grain		Straw	
	1995-96	1996-97	1995-96	1996-97	1995-96	1996-97
T_1-Rainfed	0	0	2.60	3.01	1.40	2.73
T_2-Pre-emergence (PEI)	0.75	1.25	3.62	5.16	2.23	4.08
T_3-Pre-sowing (PSI)	0.41	0.54	3.00	4.50	2.21	4.25
T_4-PSI-Crown root initiation (CRI)	1.41	1.54	3.05	4.64	2.65	5.71
T_5-PSI+CRI+Flowing (FL)	3.41	3.54	3.70	5.69	2.73	5.77
T_6-Maize-Ginger-Wheat (PSI)	0.41	0.54	3.08	4.40	2.75	5.73
C.D. 5%	-	-	0.30	0.31	0.39	1.13

Wheat crop sown on November 17, 1995 and November 20, 1996. Rainfall during growing season 1995-96: 232 mm. Recommended fertilizers, N 120, P_2O_5 = 90, K_2O = 60 kg ha^{-1}
(Source: Acharya *et al.*, 1998)

Table 7.3: Vegetable yield In maize-radish-onion-french bean cropping sequence under limited water supply from poly-lined tank at Palampur

Crop*	Irrigation applied (cm)		Rainfall (cm)		Yield (t ha^{-1})		Gross return (Rs/200 m^2)	
	1995	1996	1995	1996	1995	1996	1995	1996
Reddish	8.2 (10)	15.6 (10)	6.0	3.5	14.17	21.46	992	1716
Onion (green)	13.0 (2.1)	10.3 (14)	21.1	24.8	20.60	33.96	1650	4754
French bean	8.3 (10)	7.8 (7)	40.7	31.7	10.63	12.29	1275	1229

Growing period: Radish-Sept. 20 to Dec. 5, 1995 and Sept. 12 to Nov. 11, 1996; Onion-Dec. 13, 1995 to May 6, 1996 and Dec. 2, 1996 to April 29, 1997; French bean-March 6 to July 8, 1996 and Feb 28 to June 28, 1997.
(Source: Acharya *et al.*, 1998)

Table 7.4: Impact of soil conservation measures on water resources development cropping intensity and productivity of crops in Rendhar watershed

Particulars	83-84	87-88	91-92
Ground water table (m)	10.0	7.50	6.34
Surface water storage (00m)	-	177.63	180.10
Irrigated area through surface resource (ha)	86.26	178.00	180.00
Dug wells (No.)	10.00	25.00	30.00
Shallow tube wells (No.)	-	16.00	47.00
Irrigated area (ha)	56.26	516.00	690.00
Gross cropped area (ha)	696.10	1081.60	1282.60
Cropping intensity (%)	100.20	155.60	184.40
Crop productivity (q/ha)			
Barley	6.50	33.30	42.80
Wheat	10.00	352Q.	48.70
Lentil	8.80	14.80	19.50
Gram	6.20	21.50	27.30
Linseed	2.60	8.80	19.00
Mustard	250	14.30	26.80

(Source: Bhan, 1996)

7.7 Water Harvesting and Recycling

In rainfed areas the excess rain during the monsoon may be collected and stored in dug or embackment type of storage structures which may be utilized for supplemental and life saving irrigation. Percolation from the storage structures enriches the ground water reservoir which may be tapped through shallow tube wells or large diameter wells. The excellent results obtained at Rendhar watershed in district Jalaun of Uttar Pradesh are testimony to this fact. Water harvested in a series of 21 embankments constructed across 5 *nalas* recharged ground rising water table to 6.3 m during 1991-92 from 10 m during 1983-84 and increased irrigated area from 56.26 ha during 1983-84 to 690.0 ha during 1991-92 enhancing cropping intensity by 84% and resulting in 3 to 5 fold increase in productivity of crops (Table 7.4).

Similarly, application of organic residual mulch in different crops increased the yield by 2.02 to 7.50 q/ha (Table 7.5).

Table 7.5: Effect of organic residue mulch on yield (q/ha) of *kharif* crops

Crop	Control	Mulching	Increase over control	Reference
Jowar 'Varsha'	26.60	34.10	7.50	Tripathi and Bhan (1993)
Maize 'Azad Uttam'	12.56	14.61	2.05	Gupta and Bhan (1997)
Cotton 'Vikas'	7.57	10.03	2.46	Kumar and Bhan (1993)
Cotton 'Vikas'	5.60	7.89	2.23	Tripathi *et al.* (1995)
Bajra 'WCC-75'	14.58	18.26	3.68	Bhan and Parihar (1987)

(Source: Bhan and Uttam, 1998)

One of the biggest constraints in proper utilization of the water harvested in farm ponds is the high seepage losses of stored water. Therefore, it would be appropriate to select natural depressions having less permeable soil for constructing water harvesting structures. As the harvested water is subjected to seepage and evaporation losses it is advisable to use it as quickly as possible after the cessation of monsoon preferably for pre-sowing irrigation of winter crops or irrigation of summer or winter crops at the most critical stages.

Water harvesting system

A structure constructed for collection and storage of run off water from the upstream areas at some suitable location is known as Water Harvesting Structure. Harvested water may be utilized for human consumption, animal consumption and to provide supplemental irrigation during crop stress periods. The selection of site and potential runoff available for the structure is the prime considerations in construction of water harvesting structures.

In *Dhinkli* watershed, two sites were selected for planning of the water harvesting system. In east of the village, one site called '*Baneda ka Talab*' was found suitable for an embankment type farm pond. In north of this site, about 3 km upward, another site was found suitable for construction of Anicut, locally to be called as '*Hakdar Anicut*'. The sites were topographically surveyed and contour maps were drawn. The area of each contour was determined by planimeter. Then, depth capacity curves for '*Hakdar Anicut*' and '*Baneda ka Talab*' were drawn as shown in Fig. (7.1) and (7.2), respectively.

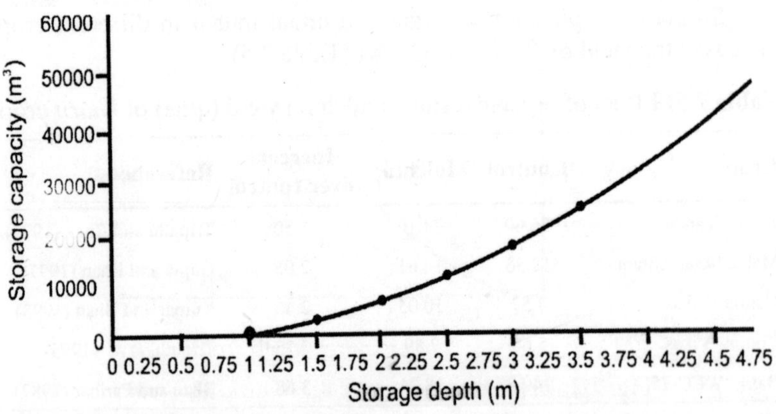

Fig. 7.1: Depth capacity curve for *Hakdar Anicut*

Anicut

Anicut is a drop type structure with rectangular weir to harvest the excess run off at suitable locations in downstream of catchment on lower reaches of drainage lines. It impounds the water against head wall and controls the damaging velocity of the run off water providing safe disposal. The word 'Anicut' is a local name of this water harvesting structure commonly known in Rajasthan.

Design considerations

i. The weir of Anicut is designed for sustaining the peak discharge resulting from the rainfalls in 25 years recurrence intervals,

ii. Structures should also be checked against the yield of the run off at 75 per cent probability level for its capacity of storage,

iii. The structures should be located at such a place where maximum crest length of the weir, and

iv. The slope of the nalla bed should be less than 2 per cent so that maximum water can be stored in the valley or nalla.

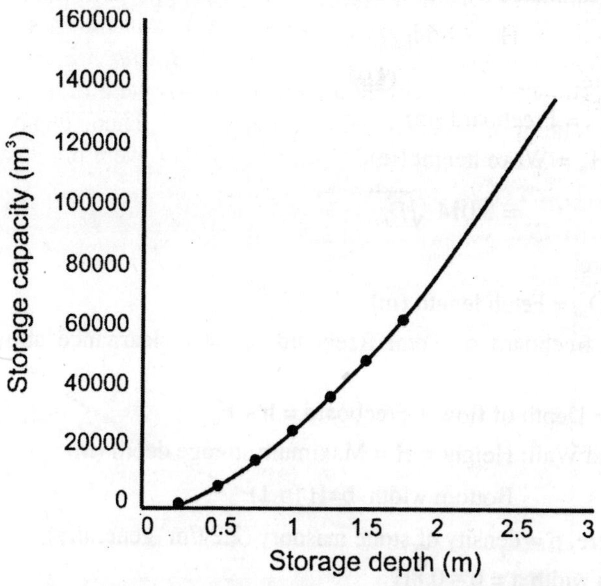

Fig. 7.2: Depth capacity curve for *Baneda ka Talab*

Design steps

Height: The height of Anicut (H) is decided on the basis of volume of run off expected from the catchment and volume of water required to be stored.

Estimation of peak runoff rate

Peak rate of runoff is estimated by SCS curve number method (equation 8.4).

Depth of flow over the crest

Depth of flow over the crest(h) is given by

$$h = \left(\frac{Q_p}{1.71L} \right)^{2/3} \quad .....(8.1)$$

Freeboard

It is estimated by the formula,

$$H_w = 1.5\ h$$

where,

F_b = Freeboard (m)

H_w = Wave height (m)

$$= 0.014\ \sqrt{D_m}$$

where,

D_m = Fetch length (m)

Net freeboard or Total freeboard or total clearnance above crest level,

H_1 = Depth of flow + Freeboard = h + F_b

Head Wall: Height = H = Maximum storage depth (m)

Bottom width, b=H (p-1)

where, p = density of stone masonry (2.2 t/m^3 generally)

Top width a = b – 0.8H

Head wall extension: Length H + h_f – 1 h_f

As per site condition, the length of head wall extension can be increased or decreased.

Height H + h_f, Top width = Bottom width = Top width of head wall

Side wall: length 1.7511 + 0.75 h_f + 0.45

Height = H + h_f top width = 0.6 m

Bottom width = 0.6 + 0.4 h_f

Wing wall: length = 2.25 h_f, Height = 1.5 h_f

Top width = Bottom width = 0.6 m

Approx: length = Crest length

Width 0.75 (H + h_f) + 0.45

Total length = crest length, Width = 0.45 m, Height = 0.3 m

Cut-off will: length = Crest length, width = 0.45 m, Depth = 1.0 m

Embankment type farm pond

An embankment type farm pond consists of an earthen embankment to retain the available run water. It is most common type of farm ponds and generally constructed across the stream or nalla. The pond is fed by surface runoff from the catchment area and the stored water is mostly taken out of the pond through gravity flow or pump for irrigation and other purposes.

Design considerations

a. The site should be such that the largest storage capacity is available with the least amount of earthwork.

b. A proper spillway site should be available for the safe disposal of excess water.

c. The effective height of embankment should be decided on the basis of values of run off expected from the catchment area and volume of water required to be stored.

Design steps

i. Peak runoff rate: It is estimated by SCS curve number method (equation 8.4)

ii. Freeboard and wave action:

Freeboard, F_b = 15% of depth of flow

Net freeboard, h_f = depth of flow + freeboard + settlement

$h + F_b + 5\%$ of $(h + F_b)$

Also wave height freeboard F = 1.5 h_w

where, h_w = wave height (m)

The higher of F_e and F can be adopted.

iii. Height of embankment: Height, H = storage depth + net freeboard

iv. Top Width of Embankment: Top width, T + H/5 + 1.5

v. Bottom width of embankment: Bottom width, B = Top width + u/s slope x H + d/s slope x H

vi. Design of rectangular waste weir:

a. Length of waste weir: L.= $Q_p/[1.71\ h^{(3/2)}]$

b. Stone pitching:

Top of the waste weir = 2

[Top width x 2.0]

Vertical trapezoidal sides

$= h_f \times (2 \times \text{Top width} + 5.5 \times h_f)$

Bottom of the waste weir

$= L \times (\text{Top width} + 5.5 \times h_f)$

Estimation and analysis of runoff

The weighed curve number of the watershed for AMC I, II and III conditions were worked out to be 67, 83 and 97, respectively. Then, by estimating the daily run off for thirty years (1971-2000), annual runoff and seasonal runoff were computed for each year. Out of these thirty; years, the run offs up to 10 per cent of rainfall were observed in 14 years, 10 to 20 per cent in 7 years and 20 to 30 per cent in 9 years. The annual run off more than 300 mm was observed in one year, between 200 mm and 300 mm in one year, between 100 mm and 200 mm in 8 years and less than 100 mm in 20 years. Annual run off and seasonal run off are same in 23 years out of the 30 years *i.e.* all the run off in these 23 years occurred in monsoon season only. The average annual and seasonal runoff were found to be 85.88 mm and 82.51 mm respectively. The average seasonal run off was computed to be 96.73 per cent of annual runoff. The expected available annual run off at 75 per cent probability level was found to be 28.47 mm. Hence, the annual run off generating capacity of the whole watershed is 185624 m^3 at 75 per cent probability level. The large difference between the average annual runoff value and its expected value at 75 per cent chance indicate that less run off occurred in most of the thirty years.

Design and analysis of water harvesting system

The design dimensions of the Anicut and Embankment type farm pond are presented in Table 7.6.

Table 7.6: Design dimensions of the anicut and the pond

A. Anicut

S. No.	Particulars	Value
1	Effective height	4.5 m
2	Peak runoff rate	18.02 m^2
3	Depth of flow over the crest	0.63 m

S. No.	Particulars	Value
4	Free board	1.0 m
5	Total height	5.5 m
6	Head wall	
	(i) Length	21.0 m
	(ii) Height	4.5 m
	(iii) Bottom width	4.2 m
	(iv) Top width	0.6 m
	(v) Depth of foundation	2.0 m
7	Heal wall extension	
	(i) Length	5.0 m
	(ii) Height	5.5 m
	(iii) Bottom width	0.6 m
	(iv) Top width	0.6 m
8	Side wall	
	(i) Length	9.1 m
	(ii) Height	5.5 m
	(iii) Bottom width	1.0 m
	(iv) Top width	0.6 m
9	Wing wall	
	(i) Length	2.25 m
	(ii) Height	1.5 m
	(iii) Bottom width	0.6 m
	(iv) Top width	0.6 m
10	Apron	
	(i) Length	21.0 m
	(ii) Height	4.6 m
	(iii) Depth	0.6 m
11	Toe wall	
	(i) Length	21.0 m
	(ii) Width	0.45 m
	(iii) Height	0.6 m
12	Cut-off wall	

S. No.	Particulars	Value
	(i) Length	21.0 m
	(ii) Height	0.45 m
	(iii) Depth	0.6 m

B. Embankment type farm pond

S. No.	Particulars	Value
1	Peak runoff rate	45.5 m^3/x
2	Storage depth	2.85 m
3	Free board	1.21 m
4	Height of embankment	4.1 m
5	Top width of embankment	2.32 m
6	Bottom width of embankment	24.87 m
7	Rectangular waste weir	
	(i) Length	26.6 m
	(ii) Stone pitching	65.43 m^2
8	Pipe spillway	
	(i) Pipe diameter	3 m
	(ii) Length	26 m

Anicut

The proposed Anicut has catchment area of 134 ha. The height of anicut is 4.5 m having the support of head wall extensions on both sides. The storage capacity of anicut is 3.60 ha m. After considering the different losses, the amount of water available for supplemental irrigation, drinking water supply etc., has been estimated to be 30299 m^3. The 75 per cent the submergence area (1.65 ha) may also be utilized for bed cultivation to raise the winter crops. The structure has been critically examined against various stability checks *viz.* overturning, rupture due to tension, sliding and crushing due to excessive compression. The structure was found to be safe with a fair degree of safety.

Embankment type farm pond

The proposed embankment type farm pond has a catchment area of 558.2 ha. The storage capacity of the pond is 14.37 ha m. A rectangular waste weir having crest length of 26.6 m with allowing depth of flow over the crest as 1.90 m is proposed for safe disposal of the peak discharge expected in 25 years recurrence interval's rainfall intensity. The stone pitching required for the waste weir is 65.423 m^3. A pipe spillway of 90 cm diameter and 26 m length with the regulator system has also been proposed to let out stored water in a regulated manner. After considering the different losses, the amount of water available for supplemental irrigation, drinking water supply etc., has been estimated to be 122467 m^3. About 70 per cent of the submergence (4.48 ha) may also be utilized for the cultivation to raise the winter Crops.

Economic analysis

The costs of the Anicut and the pond were estimated to be Rs. 6,53,000.00 and Rs.5.61,000.00, respectively. The cost per litre of stored water by the Anicut comes to Rs. 0.02, while that by the pond comes to Rs. 0.0039 per litre *i.e.* Rs. 3.90 per m^3; which are highly acceptable.

Conclusions

(i) The weighted curve number of the watershed for AMC II is 83; which shows that the watershed has good runoff potential.

(ii) The available annual runoff at 75 per cent probability level is 28.47 mm.

(iii) The storage capacity of the proposed anicut is 3.6 ha m and that of the farm pond is 14.37 ha m.

(iv) The amount of water available for supplemental irrigation, drinking water supply etc., is 30299 by the anicut and 122467 m3 by the farm pond.

(v) The cost per liter of stored water by the anicut comes to Rs. 0.02 while that by; the pond comes to Rs. 0.0039; which are highly acceptable.

7.8 Design of Water Harvesting System

It involves following steps-

Step-I

Estimation of runoff

The daily runoff is dependant upon rainfall of the day, antecedent moisture content and hydrologic soil cover. It was estimated by SCS.

Curve number method (Chow, 1964 and Vandersypen *et al.*, 1972)

According to this method runoff is determined by the following relationship:

$$Q = \frac{(P - 0.3S)^2}{(P + 0.7S)} \qquad \text{....(8.2)}$$

$$S = (25400/CN) - 254 \qquad \text{.... (8.3)}$$

where,

Q = Runoff depth, mm

P = Rainfall depth, mm

S = Maximum retention potential, mm

CN = Curve Number

The curve number is the characteristic of the catchment area and antecedent moisture content (AMC). If sum of rainfall of last 5 days is less than 35.0 mm then it is AMC I condition, if it is between 35.0 and 52.5 mm, it is AMC II condition, and if it is greater than 52.5 mm., it is AMC III condition (Singh, 2000).

The peak run off rate by SCS method is determined by the following formula,

$$Q_p = \frac{0.0208 \times A \times Q_d}{T_p} \qquad \text{.... (8.4)}$$

where,

Q_p = Peak runoff rate (m^3/S)

A = Catchment Area (ha)

Q_d = Runoff depth (cm

T_p = Time of peak (hr)

$= 0.6\ TC + \sqrt{T_c}$

T_c = Time of concentration (hr)

References

Acharya, C.L. and Kapur, O.C. 1991. Water storage in sloping lands. Low-External Input and Sustainable Agriculture (ILEIA Newsletter) Netherlands. 13(1) 30.

Acharya, C.L., Kapur, O.C. and Sharma, S.K. 1998. Management of rainwater and other natural water resources for sustainable hill agriculture. Proceeding of the National Seminar on "Water management for sustainable agriculture - Problems and prospective for the 21st century", IARI, New Delhi, pp. 249-253.

Acharya, C.L., Sharma, P.D., Kapur, O.C. and Kumar, R. 1989. Formation and how behaviour of selected Himalayan springs. *Indian J. Ecol.* 16: 119-123.

Anonymaus, 1988. Annual Progress Report 1987-58. Palampur Center of All India Coordinated Research Project on Water Management in High Rainfall Areas (ICAR).

Anonymous, 1995. Statistical Outline of Himachal Pradesh. Ed. Economics and Statistical Department, Himachal Pradesh, Shimla-1. 1995 58.

Chander, S. and P. Dana, 1991. Changing scenario of rainfed agriculture in India. Proc. International Conference on Extension Strategy for Minimising Risk in Rainfed Agriculture, Indian Soc. Extn. Education (ICAR,), April 6-9, 1991.

Chow, Y.T., 1964. "Handbook of Applied Hydrology", McGraw Hill, New York.

Katyal, J.C., Narayan Reddy M., Kaushalya Ramchandran and Mishra, P.K. 1998. Management of rainwater for crop production. Proceedings National Seminar on "Water management for sustainable agriculture – Problems and perspectives for the 21st century", IARI, New Delhi, pp. 221-234.

Lal, M. 2001. "Climatic change – Implications for India's water resources society", Vol. 21, No. 3: 101-119.

Schwab, G.O., Fangmeier, D.D., Eliot, W.T. and Frevert, R.K. 1993. "Soil and water conservation engineering", 4th edition, John Wiley and Sons, New York.

Singh, P.K. 2000. "Watershed Management (Design and Practices)", Ist Edition, E-media Publications, Udaipur, India.

Vandersypen, D.R., J.S. Bali and Y.P. Yadav, 1972. "Handbook of Hydrology", Soil Conservation Division, Ministry of Agriculture, Govt. of India, New Delhi.

EXERCISES

7.1 Explain what is the ex-situ rainwater management.

7.2 What is the need of water harvesting?

7.3. What was the role of Indian rulers in water harvesting in India?

7.4. Describe the different techniques of water harvesting.

7.5. Differentiate the following:

   (a) Community tank and farm pond

   (b) Embankment type pond and dug out farm pond

7.6. What are the principles of ex-situ rainwater management?

7.7. Discuss the water harvesting in cold deserts.

7.8. What is the auxiliary tanks?

7.9. Discuss the design consideration for anicut

■■■■

Chapter 8

WATER HARVESTING-CUM-ARTIFICIAL GROUNDWATER RECHARGE

8.1. Introduction

Water is a precious natural resource for sustaining life and environment. With incurring growth of population, rapid urbanization and industrialization, increased standard of living and life expectancy, improvement in sanitation etc. The demand for water is increasing day by day manifold. To overcome this problem it is essential to conserve every drop of rainwater that falls and use it for local needs throughout the year. There is urgent need to take steps to implement water harvesting techniques to ensure the availability of water in future for drinking, agricultural and industrial needs. Conservation of rainwater in rural areas is possible through different rainwater harvesting structures like ponds, check dams, percolation tanks, Injector wells, and the selection of these structures depends upon hydrogeological conditions of the area. Due to tremendous population and scarcity of available open space in urban areas, the rainwater goes waste by quickly being disposed into surface streams. One possible and simple technique is to divert the water falling on the rooftops to store underground spaces. Water harvesting forms an important component for the development and management of water resources for domestic, agricultural and municipal and industrial usages and there is a need for collecting and storing the rainwater during the monsoon period and to use the same for non-monsoon periods by means of water harvesting structures.

The dependability on groundwater has reached high in all time in recent decades due to reasons such as unreliable supplies from surface water due to vagaries of monsoon, increase in demand for domestic, agricultural and industrial purposes. This resulted into over exploitation all over the country and in certain places it has reached critical levels like drying up of aquifers, warranting immediate attention from engineers and planners for formulating plans on a war footing to augment the groundwater storage. Water harvesting and recharging of aquifer is one such technique which fulfill the above objective. As a large volume of surface runoff discharges into the sea, it can be judiciously harnessed to recharge aquifers.

8.2 Traditional Rainwater Harvesting Systems

India has a rich history of using traditional systems of rainwater harvesting in the semi-arid states (Table 8.1). Conservation of surface and groundwater has been an integral part of day-to-day living. Rainwater harvesting is a process of collection of runoff from treated or untreated land surfaces/catchments or rooftops and storing it in an open farm pond or closed water tanks/reservoirs or *in-situ* moisture storage in the soil itself. In fact, ponds and tanks represent an important community resource for drinking water in rural India even today. The storage of scanty rainfall, through simple as well as extensive types of traditional water harvesting of surface and groundwater has been the important source of water in arid regions of Rajasthan and Gujarat. The serious problems of water shortages in many parts of the country, particularly during drought years, which could at least partially cope with meeting the requirements of livestock and life saving irrigation, are being largely attributed to the discontinued use of traditional rainwater harvesting practices. There is a dire need for revival of these traditional rainwater harvesting systems. While the importance of providing safe drinking water is not questioned, the problem of water shortages and effects of droughts could be minimized by integration of drinking water schemes with cost effective rainwater harvesting technologies and practices.

Table 8.1: Summary of traditional rainwater harvesting systems in semi-arid region of India

System	Technology	Region
Kunta	Small size community pond for providing supplemental irrigation to paddy, chilies and tobacco during long breaks in rainfall	Prakasam, Guntur and Nizamabad (Andhra Pradesh)
Pacolation tank	Used for recharging the groundwater to bring stability to rainfed agriculture	Andhra Pradesh, Karnataka and Maharashtra
Haveli	Storing rainwater by raising field bunds, used in kharif fallow	Madhya Pradesh
Risers	Stone pitching of bench terrace risers to conserve rainwater	Hilly regions

System	Technology	Region
Stone pitched barriers	Stone pitching along with stubbles of pigeonpea / sorghum / cotton across the slope for rainwater conservation	Vidarbha region
	Field bunds with waste weirs, stone checks and stone / boulder boundaries to conserve rainwater	Andhra Pradesh, Gujarat and Karnataka

(Source: ICAR, 1999)

The *in-situ* moisture conservation practices are followed in medium to high rainfall areas. However, inspite of that surplus rainwater may likely to occur which can be harvested and recycled. The estimated potential volume of rainwater storage in small-scale water harvesting structures for different rainfall zones in the country is given in Table 8.2. On an average, 24 million ha-m out of the total 400 millions ha-m precipitation received annually in India is estimated to be available as harvestable runoff through rainwater harvesting (Katyal, 1997). The semi-arid region possess a harvestable runoff potential of 5.54 million ha-m.

Table 8.2: Estimated potential volume of rainwater storage for small-scale water harvesting structures

Rainfall zone (mm)	Area (million ha)	Rainfall for effective surface storage (mm)	Harvestable runoff (million ha m)
<500	52.07	5	0.78
500-750	40.26	6	1.51
750-1000	65.86	7	4.03
1000-2500	137.24	6	14.61

(Source: Katyal, 1997)

Although village level rainwater harvesting tanks have been traditionally in vogue in India, research during past 25 years has led to standardizing the variables with respect to runoff characteristics, optimizing the capacity of pond, lining materials, water lifting devices and suitable crops for utilizing the harvested rainwater efficiently (Singh, 1986). However. despite the advantages of rainwater harvesting and reuse through dugout ponds, the diffusion of this technology has been quite low due to high initial investments. Further research on controlling evaporative and seepage losses, water lifting devices, sealing materials and cost of dugout and need to be carried out to make the concept feasible and economically viable.

The DDP, DPAP and Integrated Watershed Development projects have an element of rainwater harvesting and involvement of local communities in the maintenance. An important aspect of the success of these traditional rainwater-harvesting systems has been the involvement of community at large. The community in many villages regulates rainwater harvesting, distribution and use. Before the onset of monsoon season, the catchments are cleaned of all waste and rubbish. Similarly, desilting of tanks, ponds, and cleaning of *baoris*, *nadis*, *kundies*, etc., and their repairs would be carried out by the community, in an annual or biannual operation.

Rooftop rainwater harvesting in towns and villages is innovative method of harvesting rainwater and is becoming increasingly popular in a number of towns and in cities, particularly following the thought of year 2000. The Central Ground Water Authority has begun intense awareness campaigns on popularizing rooftop rainwater harvesting across the country. The Ministry of Environment and Forests has brought out specific measures on rainwater harvesting including:

- All buildings constructed in urban areas should have a provision for rooftop rainwater harvesting.
- Institutional and commercial buildings should not thaw water from existing water supply schemes, which adversely affects water supply to villages and settlements.
- In rural areas, rainwater harvesting should be undertaken through such structures such as percolation tanks, storage tanks, etc.
- Rainwater collected through storm water drains should be utilized to clear water disposal drains.
- Groundwater aquifer recharge structures should be constructed wherever such structures do not lead to slope instabilities.

8.3 Classification of Water Harvesting Techniques

Water harvesting is the collection of surface runoff for its productive uses *viz.*, drinking water for human and livestock, crop production, industrial development and also artificially groundwater recharging. These techniques are classified as described below.

Water harvesting includes a wide range of techniques which can be classified (Dewan, 1987) based on the following criteria: source of water, required storage duration and the intended use as shown in Fig. 8.1.

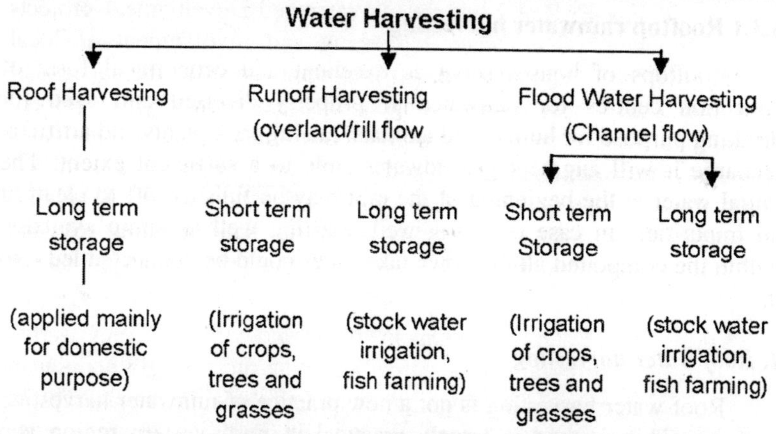

Fig. 8.1: Classification of water harvesting techniques

Other classification of water harvesting techniques

These techniques are also broadly classified as follows:

1. Rooftop rainwater harvesting for urban areas
2. Surface runoff harvesting for rural areas

These techniques are further classified as shown in Fig. 8.2.

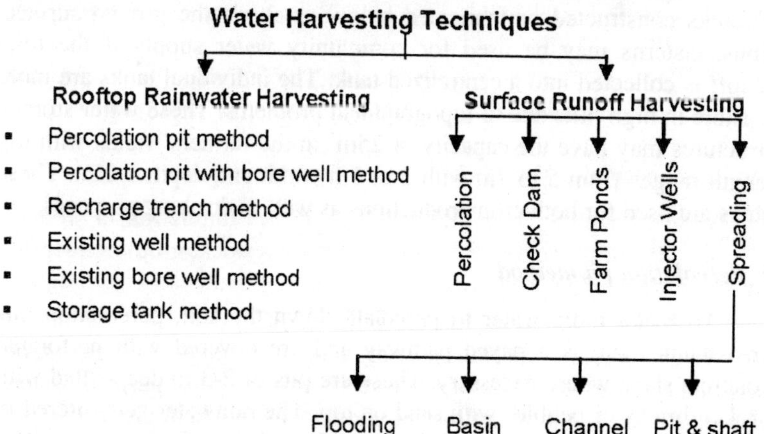

Fig. 8.2: Classification of water harvesting techniques

8.3.1 Rooftop rainwater harvesting

Rooftops of houses serve as excellent and economical form of collection centres for rainwater. If properly diverted and used for drinking purpose for human and animals, raising fruit plants and artificial recharge it will augment groundwater table to a sufficient extent. The initial water at the beginning of the rain may be finished off to clean of all impurities. In case of a dug well, existing well or sump available within the compound all the down take pipes could be connected led into it.

1. Roof water harvesting

Roof water harvesting is not a new practice of rainwater harvesting in India. This method is largely practical in north-eastern region and western Rajasthan in India. The rain water from the roofs of the buildings may be collected into underground cement or concrete made cisterns/tanks for household and kitchen garden purposes. Recently the importance of roof water harvesting has been realized greatly because of more demand of water for agriculture and household purposes for ever increasing population. The yield of rain water from roof surface depends upon the type of roof and material used in its construction. It has been observed that the runoff efficiency of roof rain water can be increased if the roofs are made of corrugated G-1-Sheet (85%) and stone slab and lime concrete (81%), while the traditional roofs made of clay soil and straw have as low as 40% runoff efficiency.

In high hills, it is common practice to collect rainwater into cisterns or tanks constructed or fabricated below or above the ground surface. Thus, cisterns may be used for community water supply if the total runoff is collected into a centralized tank. The individual tanks are more feasible in high hills due to topographical problems. These water storage structures may have the capacity of 25m^3 in the farmers' fields with top length ranges from 5 to 7m with 3 to 5m width and depth 1-2 m. These tanks are used for both crop productions as well as drinking purposes.

2. Percolation pit method

To make more water to percolate down the soil, percolation pits area when there is a payed pathway and are covered with perforated concrete slabs where necessary. These are pits of 2-3 m deep filled with broken bricks or pebbles with sand on top. The rainwater gets filtered in the process of flowing through layers and reaches underground aquifers and recharges it. In case of clayey soils percolation pit with bore well can be used.

3. Percolation pit with bore method

Whenever the depth of clay soil is more, recharge through percolation pit with bore is preferable. The bores are deep pits of about 20' to 25' deep depending on the soil condition. These bore can be at the center of the square pit and is filled with pebbles or brick jelly and the lop portion with river sand and covered with perforated concrete slab.

4. Recharge trench cum percolation pit (T.C.P.) method

The TCP is one of the most ideal recharge structures in limited open land, narrow streets and roads etc to trap the surface water and recharge the aquifer. The dimensions of the TCP depend on the site-specific condition and sub surface formation. One or two suitable shafts can also be provided inside the trench depending on the sub surface formation, and the availability of water for recharge purpose. The entire structure may be filled with pebbles or locally available boulders and sand and if necessary this structure can he covered with perforated slabs.

The TCP can be located on sides of roads, street corners, low lying areas inside the premises of individual houses. Multi-storey buildings, paved areas surrounding the houses, near the extraction structures like hand pumps, bore wells, dug wells etc.

5. Existing open well method

In case of a dug well available within the house/flat complex, all the down take pipes could be connected and led into it either through a filter or not, depending on the quality and quantity of water available in the said well.

6. Existing bore well method

In case of a bore well available within the house/flat complex, all the down take pipes could be connected and led into a filtering unit of size 21/2' x 21/2' x 21/2' and the filtered water can be allowed into the existing bore well.

7. Storage tank method

If a sump or storage structure is available within the site, then one or more down pipes could he directed into it through a sand filter containing a layer of sand sandwiched between two layers of pebbles/ blue metal for immediate use. Overflow water can be led into the well within the site, the down take pipes could be led into either a bore pit or a percolation well depending on the volume of roof top water.

8.3.2 Surface runoff water harvesting

Semi circular hoops consist of earth embankments constructed in the shape of a half-moon with the tips of the semi-circle on the contour. The water is collected within the hoop from the area just above it and impounded to a maximum depth defined by the height of the bund. Any excess water is discharged around the tips. In a field or hoops the next lower row is staggered so that the water overflowing from the upper row is readily intercepted. The height of the bunds ranges from 0.1 to 0.5 m and the radius may range from 5 m to 30 m. Semi-circular bunds are being used for irrigation of grass, fodder, shrubs, or trees.

Trapezoidal bunds

Trapezoidal bunds consist of earth embankments constructed in the shape of trapezoids. The tips of the bund wings are placed on the contour. The water is collected from the slopes above the bunded area and excess water overflows around the tips. The row of bunds is staggered to intercept over flow from above. The layout of trapezoidal bunds follows the same principles as those for semi-circular bunds, but they unusually enclose a large bunded area. The height of bunds ranges from 0.3 m to 0.6 m and their width across the tip ranges from 40 m to 160 m. Trapezoidal bunds were developed where rainfall intensities are high, causing surface water to flow at high discharges which would cause damage to contour bunds. Trapezoidal bunds are used extensively for irrigation of crops, grasses, shrubs or trees and enable intercropping within the large enclosed areas. They are also widespread on the west bank of the Indus in Pakistan.

Off-contour bunds or graded bunds

Off contour bunds consist of earthen or stone embankments built along 0.5-2.0 per cent slopes. They are varied from contour bunds and are used as an option where rainfall intensities and soils are such that large amounts of water, which require safe drainage through the field, are intercepted. Irrigation is improved and the excess water is led through the channel to the next field, this irrigates it as well. The height of the bunds ranges from 0.3 m to 0.5 m. The bund constructed below usually has a wing to help intercept water overflowing from the bund above. This makes it look like an open ended trapezoidal bund that is also known as Modified Trapezoidal Bunds. Off-contour bunding is used for irrigation of crops. It is widespread in many parts of Africa where runoff harvesting is being practise.

Contour bunding

A structure of bunds parallel to the countours of the slopes of the fields is one of the major activities of runoff control and prevention of soil erosion in a watershed development programme. These are more useful in the areas of highly and moderately dissected plateaus. The bunding works provide more soil moisture saturation and infiltration to ground water reservoir which helps in augmenting the recharge and reducing the soil erosion and siltation. This also helps in increasing the weathering and secondary porosity in the rocks.

Rock catchment

Exposed rock surfaces are used for the collection of water. The rainfall on the exposed rock is drained towards the lowest point, sometimes being channeled along low walls. Here it is collected in a storage tank or reservoir, of earthfill embankments, gravel stone-fill embankments, stone masonry dams or concrete dams. The catchment areas may range from a hundred sqm to a few thousand sqm with corresponding varying dimensions for the storage tank. The water from these reservoirs may be used for domestic, livestock or irrigation purposes. Many domestic water reservoirs are fed by rock catchments.

Ground catchments

Expanses of ground are used to collect water into storage tanks or reservoirs. The ground is cleared of vegetation and compacted. In Australia the ground is reshaped to create a series of channels leading to the reservoir. These are then compacted and sometimes covered with gravel. These are called Roaded Catchments. In Lesotho small ground catchments have been introduced for collection of water for the supplemental irrigation of vegetables. In Somalia large catchments of 5 km^2 are used to collect water into 30,000 m^2 reservoirs generally for livestock water. Ground catchments have been traditionally used for 4000 year in the Negev, where annual crops are grown and also some more drought tolerant spice like pistachio.

8.4 Water Harvesting Techniques in Arid and Semi-Arid Areas

Number of water harvesting techniques have been used for long times back in India and abroad specially for drought proofing. The old techniques have been modified scientifically to increase its efficiency and reduce the cost. Some of the important water harvesting techniques for drought proofing are given below.

1. Nadi/pond system

The nadi/pond is constructed at lowest point to store excess rain water from adjoining area or catchments. The capacity of a nadi may vary from 1200-1500 m^3 depending upon physiographic conditions and rainall pattern. In Rajasthan, it is considered as one of the oldest systems of water collection for its conjunctive use. At present, most of the villages in the region have one or more nadis. According to 1982 census, 1436, 592 and 1822 nadis were found in Nagaur, Barmer and Jaisalmer districts, respectively. Generally, the loss of water was quite high in traditional nadis through evaporation, seepage and heavy sedimentation. The research findings indicate that the surface to volume ratio should be <0.3 for reducing the evaporation loss. The silt traps may be used at inlet point to avoid siltation. The LDPE sheet or stone or concrete layer may be used for lining the bottom or side of the nadi to avoid seepage losses. The nadis/ponds may be square, rectangular or circular in shape but circular ponds have the highest storage capacity for a given surface area and side slopes but its curved area remains unused, hence, square or rectangular nadis/ponds should be preferred (Table 8.3).

Table 8.3: Surface area, perimeter and storage capacity of different shapes of ponds

Shape of pond	Side slope	Surface area (m^2)	Perimeter (m)	Storage capacity (m^3)
Circular	1.5: 1	900	100	1499
Square	1.5: 1	900	120	1464
Rectangular	1.5: 1	900	122	1458

2. Tanka

Tanka is an underground tank (cistern) constructed for collection and storage of runoff water from a natural catchment, artificially prepared catchment or from rooftops. The shape of *tanka* may be rectangular or circular, but circular shape is preferred from safety point of view. CAZRI has recently improved the designs of *tankas* by providing vertical surfaces with stone masonry or cement concrete structure; the base having 10-15 cm concrete layer. The top is covered with stone slabs and cement plaster, leaving opening of 60-45 cm for withdrawal of water. For design purposes suitable runoff availability estimates are required. Significant increase in runoff yield can be achieved by simple treatment of soil. Efficiency of surface treatments is given in Table 8.5.

Table 8.4: Expected water yield in farm pond

Region	Soil type	Annual rainfall (mm)	Land slope (%)	Expected water yield (%)
Bellary	Black	508	1-2	20
Ananthpur	Red	540	3-5	20
Hyderabad	Red	770	2-5	10
Banglore	Red	830	2-3	20
Karnal	Alkali	650	0-2	40
Chitradurga	Red	612	5-10	10

Table 8.5: Water harvesting efficiency of different surface treatments

Surface treatment	Water harvesting efficiency (%)
Compacted earth surface	30-60
Removing vegetation	7-21
Roaded catchments	24-41
Coating of bitumen on compacted earth	65-89
Sodium chloride	10-64
Sodium carbonate	35-71
Bentonite clay	19-56
Asphalt	60-90
Asphalt fibre glass sheet	85-95
Concrete membrances	56-80
Low density polyethylene sheets	60-85

In general, sodium carbonate has been found most reliable over years. University of Arizona has estimated indefinite life of the treatment. Sodium carbonate spray is also most economical. *Tanka* water is used for human consumption for drinking as well as for rearing of animals. In CAZRI *tanka* water was also used for raising fruit plants of *ber* and pomegranate.

3. Khadin system

In the natural catchment, the soil of lower area develops from the silt load carried in runoff from adjoining uplands and becomes fine textured. Generally, the lower part of the catchments has a rocky base that restricts deep percolation. Such areas are developed into productive agricultural lands by constructing earthen embankments at lower periphery of field with suitable spillway arrangement even at as low as 150-350 mm rainfall. This type of productive land or basin is known as *khadin*.

The *khadin* system is a very unique feature for soil and moisture conservation in arid areas of Jaisalmer in Rajasthan where average rainfall ranges from 100 to 225 mm. The history reports that Paliwal Brahmin of Jaisalmer were the first who originated the *khadin* system in Jaisalmer in 15th century. The scientific study reveals that the EC of *khadin* soil is 2-16 times less than the soil of upper part of the *khadin* because of leaching of salts through seepage water. The *khadin* soils store enough moisture and nutrients for successful crop cultivation.

This ancient practice of *khadin* cultivation and its development was neglected slowly during 19th century due to rivalry among the local inhabitants. This led poverty, scarcity of food grains and fodder and led to adverse socio economic impacts. Hence, the government realized its importance and renewed the old *khadins* and so far more than 500 *khadins* have been revived. Presently, CAZRI is working on scientific planning and development of *khadins* in arid areas of Rajasthan.

4. Anicut

In arid and semi-arid areas, the small streams carrying runoff water during excess rainfall are restricted with earthen fill structure with a proper spill way just to hold sufficient water upstream during rainfall. Such type of earthen structure is called Anicut. The main objective of anicuts is to conserve the soil and water and recharge groundwater table of the region. The studies indicated that the water levels in wells located down stream in Ujalion watershed areas increased from 1.8 to 2.2 m as compared to increase of only 0.5 m in wells located in adjoining areas. The moisture conservation upstream helps to establish silvipastoral systems that may fulfill the need of fuel and fodder of the region.

5 Percolation tank

It is the construction of an earthen dam on a suitable site in a natural catchment for collection of adequate quantity of surface runoff for replenishment of groundwater in arid and semi-arid areas as its utilization is not in proportion. The bedrock in the catchment area particularly in the submergence should be weathered rock and should allow the percolation of water through the beds by infiltration upto the aquifer storage it should be in hydraulic connectivity with the aquifer below in a shallow depth. These recharge techniques are useful in providing continuous recharge after the monsoon by augmenting the aquifers. The benefits of percolation tanks are observed in the wells in command areas. The Govt. of Maharashtra took a massive program for construction of percolation tank during the last two decades, major construction of work was done during scarcity period in the year 1970-1973. Nearly 485 percolation tanks were constructed in that period in various districts of the state. The capacity of the percolation tanks ranges between 0.05 and 1.00 mcm (million cubic meter) of storage of water. The studies of quantification of benefits of percolation tanks have been carried out by G.S.D.A. on the experimental basis, the evaluation was done by conventional water balance method.

The studies have indicated that the benefits of percolation tank for recharging ground water varies from 25 to 70% under different conditions of rainfall, storm pattern, runoff conditions and geology of the bed rock (Table 8.6). However, it has been established that the positive benefits in terms of large quantity of water infiltration into the aquifer achieved through percolation tank, the benefits have also been studied by G.S.D.A. in collaboration with B.A.R.C. by isotope tracer studies at Bhadji percolation tank in Aurangabad district and Bangarwadi percolation tank in Osmanabad district. These studies were useful in delineating the area of benefits and flow conditions of recharge water from the tank. The rise in the water level of the wells and reduction in the fluctuation of water levels between winter and summer have established the benefits of the projects.

Table 8.6: Water balance study results for the years 1976-77-78-79 for P.T.

Year	Name of the tank	Per-centage percolation	Per-centage seepage	Per-centage evapo-ration
1976-77	(a) Pachpirwadi Percolation Tank	38.79	16.10	45.10
	(b) Sanjuri Percolation Tank	29.97	39.97	30.06
1977-78	(a) Pachpirwadi Percolation Tank	60.74	13.33	25.93
	(b) Sanjuri Percolation Tank	50.28	21.35	28.37
1978-79	(a) Pachpirwadi Percolation Tank	45.91	20.26	33.83
	(b) Sanjuri Percolation Tank	64.90	9.60	25.50
	(c) Bangarwadi Percolation Tank	70.10	0.10	29.80

(Source: Uman Kar, 2002)

6. Check Dams/Nala Bunding

A small check dam constructed on a stream or a nala either with the earth material, stone and clay filled wall or by a rubble masonry dam wall structure near the agricultural fields. It would impound the runoff water in a small storage and such check dams could be constructed in series, placed at a suitable distance in the stream or Nala bed. These small impounded water bodies are useful in feeding the aquifer during post-monsoon period as influent seepage. It helps in longer retention of shallow water level conditions in the adjoining field area, providing better yields and recuperation of well for pumping out water for irrigation. The studies carried out by G.S.D.A on check dams constructed on a nala bed near village Ganori in Aurangabad district has indicated that about 77.26% of water stored due to check dam went as recharge to ground water. The well inventory in the command of check dam have also indicated the positive results of benefits (Table 8.7).

7. Sand filled Dam/Tank

Generally in arid and semi-arid areas, the sandy river beds have hard impervious subsurface and no surface flow except in rainy season. The sand filled dams are made across the streams as sub surface barriers to stop the runoff and store water in sand beds just for domestic water supply and recharge ground water in the vicinity. Initially the low head masonry dam is constructed across the stream and let the basin filled with sand and gravel and latter the dam height is increased.

Table 8.7: Benefits of nala bunding or check dams

District	Prior averag e water column	After average water column	Increase in water column in (mtrs)	Prior irrigated area (ha)	After irrigated area (ha)	Increase Irrigatio n (ha)
Nasik	3.90	4.90	1.00	56.2	194.5	38.3
Ahmadnagar	3.77	5.93	2.16	67.2	86.9	19.1
Sangli	0.45	8.72	2.37	35.2	88.5	53.3
Total	4.70	6.51	1.84	198.6	369.9	170.7

The basic principle of raising the darn height gradually to allow flood water to transport most of the fine sediments in suspension over the darn crest otherwise it gets deposited in the reservoir and leads to reduced storage capacity. The excess water of dam may be diverted for nearby *nadis* or ponds for irrigation and domestic uses. In one study made at Jodhpur reveals that evaporation losses of sand filled dams was only 8% and reduction in storage capacity was nearly 50% but quality of water was very high.

8. Water spreading

Water spreading involves deliberately diversion of flood water from stream source or catchments area to micro catchments with the help of dykes and ditches and held levelling to improve the soil moisture regime and also to facilitate the regeneration of ground water. It helps to regenerate the natural and artificial vegetation in the water spreading area. It has been found that the excess rainwater of 4-12 ha area can be spread in 1 ha area. Deep tillage also helps more infiltration of water. The water spreading practice is very common in Jalore district of' Rajasthan and locally is called 'rela farming'. The viability and economics are yet to be investigated for better acceptability of this system in arid and semi arid regions.

9. Water harvesting dams

The ephemeral streams carry heavy flood during rainy season. This flood water cause heavy soil erosion by making ravines and gullies. This excess water is collected by earthen or concrete dams across the streams and stored water is diverted through canals to larger areas to fulfill the irrigation and household water needs. This stored water also helps in regenerating the ground water table in adjoining areas.

10. Rock-fill dam

It is an embankment constructed across the waterway using variable sizes of stones. It is permeable structure of semi-permanent nature. Such structures slow down the velocity of the runoff and allow the settlement of sediment before it leaves downstream reach. They also help in vegetative growth and groundwater recharge.

11. Stop dams

These are masonry or concrete structures constructed across relatively flat *nalas* carrying high discharges of long duration. They create adequate head of water at upstream, producing low water falls at down stream.

They are usually made at narrow gorges of *nala* having beds underlain with murrum, rock out crops or hard strata. The level of water is raised for lifting of water for irrigation.

Tejpura watershed in Bundelkhand region is one of the success stories of stop dams across gullies. Due to construction of such structures, the watershed became double cropped area, whereas it had mostly fallow lands prior to taking up of watershed programme.

12. Sand/rock-filled tank

Sand/rock-filled tank is constructed in the lower valley of the rocky hill area or lower part of stony/gravelly or rock out cropped catchment area. The size of tank depends upon rainfall pattern, catchment characteristics and water requirements. The tank is first made through excavation; its surface is made smooth. The sides and bottom of tank are lined either with stone masonry and cement plaster or polyethylene sheets. Polyethylene lining if properly covered offers positive method of seepage control. A wire mesh reinforced coat of cement mortar or tiles or a 30 to 45 cm layer of soil could be used to protect polythene from getting punctured with rocks.

13. Sand-filled dams

In this system, a dam is constructed across sandy ephemeral streams for stopping the flow and storing water in sand bed for water supply and groundwater recharge in the near vicinity. The streambed gets filled with sand and gravel, deposited by floods. The rainwater is stored in voids between sand particles. The dam height is raised in stages to allow find sand particles to flush out and coarse particles to be deposited in the bed.

14. Sub-surface dams

These are artificial structures for intercepting or obstructing natural flow of groundwater. These barriers intercept the flow of an aquifer and reduce variation of level of groundwater at the up stream of dam. Such sub-surface structures are necessary across sandy bed streams to recharge the aquifer.

The sub surface barriers are constructed across the streams particularly to recharge the ground water aquifer. These sub surface barriers are useful in arid and semi-arid areas where recharging capacity of wells is very poor because of scanty rainfall. The studies indicate the rate of replenishment of groundwater table of wells situated in the nearby sub surface barriers increased and depletion rate decreased after construction of sub-surface barriers across the streams.

To recharge from 2.25 ha area and storage capacity of 15000 m^3 were observed nearly 1000 m^3 in artificial recharge study in alluvial soil during rainy season in Gujarat.

Table 8.8: Statement showing evaluation of underground check dam in terms of benefits

District	Location	Depth and length (m)	Observation wells monitored number	Benefits in terms of rise in water level
Pune	Karandj Tq.Shirur March 1985	D-5 m	12	Rise of S.W.L. from 0.10 m to 3.40
Sangli	Malewadi Tq.Miraj April 1984	D-1.90 m L-30	10	Rise 3.00 mtr water level in one well only in other, level remain constraint
Sangli	Ranjani Tq.K'mahankal January 1985	D-3.00 m L-29.00	16	No rise in water level (made 1 quarterains)
Sangli	Kunhi Tq. K'mahankal August, 1984	D- 3.00 L-29.00	10	No rise in water level (bade 1 quaterains)
Kolhapur	Halondigale Tq.Halkanan March, 1981	D-4.00 m L-18.00	10	Rise in water levels from 1 to 1.5m during summer. Also cropping

District	Location	Depth and length (m)	Observation wells monitored number	Benefits in terms of rise in water level
				pattern changed. Sugarcane and other cash crops taken.
Wardha	Jaspur Tq. Karanja	D-5.00 m L-11.00	IS	Rise in water levels in the wells from 0.5 m to 2.00 m.
Aurangabad	Shekta Tq. Aurangabad	D-4.5 m L-22.00	31	Rise in water level of 1 m & on an average per well hec-targe increased to 1.5 times

(Source: Umarikar, 2002)

15. Underground check dams

This is aimed at controlling the ground water discharge with the help of underground barrier to stop ground water outflow. It consist of a trench excavated across the *nala* bed upto the depth of aquifer bottom (4-5 mts) and filling of the trench with impervious material of clay and soil available locally in the adjoining fields. The clay till is compacted to form a barrier as a dam wall stopping the ground water outflow from the section of the aquifer. The check dams in the subsurface forms an artificial dyke separating the ground water body locally which gives rise to the water levels in the wells and better yields on a sustained basis for longer period. The structure can be constructed series at intervals similar to surface check dams for larger benefits. These structures help in controlling the depletion of ground water and retention of shallow water levels for longer period. These are useful in improving the drinking water supply source wells for their sustained water supply (Table 8.8).

8.5 Traditional Water Harvesting Systems

1. ***Harvesting of natural springs*:** There are very small perennial/ seasonal springs under hillocks with average flow rate of 0-5 liters per minute (LPM). They are used for drinking purposes and other household use. These springs may used for irrigation purpose by storing the water in small tanks and conveyed down to cultivated fields through pipelines. In case of high

gravity head availability, a small portable sprinkler or drip irrigation system can be used for irrigation. A small area up to 65 ha can be irrigated particularly for growing vegetables etc.

2. ***Harvesting of small sub surface water streams*:** The small water streams flow in the valleys in the upper portion of the watersheds round the year but the flow gradually reduces from September to June. The average flow rate of these streams in the contributing valley portion varies between 0.01 and 0.1 m^3. The small masonry check dam can be constructed across the stream to collect the sufficient water upstream site of the stream. If the stream is flowing through porous bed, the stream can be diverted to a storage structure by constructing a seepage trench and head wall with weir across the stream. The storage watcr can be either lifted for irrigation on the side bank cultivated area or conveyed through pipelines down stream fields.

3. ***Construction of hydram (Hydraulic Dam)*:** It is a water lifting device. It is self powered impulse pump using the energy of falling water for lifting water to higher altitudes. In hilly area, the hydrams may be constructed if the flow is 5 l/m and height of the waterfall is minimum one meter from the site of hydram. In this process, the water can be lifted tip to 150 m otherwise normal lift rarely exceeds 0.50 m.

4. ***Collection of roof water*:** In high hills, it is common practice to collect rainwater into cisterns or tanks constructed or fabricated below or above the ground surface. Thus cisterns may be used for community water supply if the total runoff is collected into a centralized tank. The individual tanks are more feasible in high hills due to topogrnphical problcms. These water storage structures may have the capacity of 25 m^3 in the farmers' fields with top length ranges from 5 to 7m with 3 to 5 m and depth 1-2 m. These tanks are used for both crop productions as well as drinking purposes.

5. ***Run-off water harvesting*:** In high rainfall areas, the excess runoff water is collected at central level into larger tanks of 150 m^3 capacity and water may be reused by two methods for irrigation purposes;

   a. **Intra-terrace water harvesting:** The runoff water is collected from a single terrace and recycled back to the same terrace,

b. **Micro-watershed water harvesting system:** The runoff water is collected from a micro-watershed having area <1000 m^3 and recycled to a community area in a down stream.

6. ***Construction of small canals/gules*****:** It is a very common practice of irrigation in hilly areas, The concrete and cement made small channels are constructed parallel to large streams. The small flow of water from large streams/rivers is diverted to these channels/canals/gules and used for irrigating both sides of pain area. The canals/ gules finally end into small streams/rivers.

7. ***Water spreading*****:** In this simple technique, the flood water is deliberately diverted from its natural courses (rivers or larger streams) over adjacent plains through ditches. dikes, small dams and bush fences. These wet flood plains/valleys are thereby used to grow crops successfully. The fertility area is enriched due to silting of line clay particles. Further, the water holding capacity of the area is also increased.

It is comprises to increase in surface area of infiltration to induce more infiltration. Spreading methods are suitable where large area for basin is available. Various spreading methods are as under.

i. **Flooding method:** Flooding method is suitable for relatively flat area where the topography is not spotted with large gullies and ridges. The water is spread as a thin sheet of water over the undisturbed native vegetation and soil cover to obtain higher rate of filtration. It is the cheapest method of spreading.

ii. **Basin method:** In this method water is improved in a series of basins formed by a network of small banks, check dams and sub-surface dykes. The basins are arranged so that the entire area is submerged during spreading operations. This is a most common method of artificial recharge, suitable for areas where the ground surface is irregular and has shallow gullies and ridges.

iii. **Channel method:** It consists of a system of closely spaced flat-bottomed channels to expose the maximum infiltrating area. This is suitable for irregular areas and where slopes are too steep for basin construction.

iv. **Pits and shafts:** In this method recharge is done through pits or shaft dug to puncture the impervious layer encountered at shallow depth between the surface and water table in the area, a condition that renders recharge by spreading methods unfeasible in the area. Pits are also practiced where sand and gravel are encountered.

8.6 Artificial Recharge Techniques

The recharge techniques are considered mainly on the basis of their regional extent and location specific benefits. These two categories are as follows:

1. Conventional methods
2. Unconventional methods

8.6.1 Conventional recharge techniques

The conventional methods includes the techniques of controlling runoff and water conservation techniques such as 1) Countour bunding 2) Nala bunding 3) Percolation tank 4) Check dams 5) Underground check dams (Dykes) 6) Gully plugging 7) Farm ponds etc.

8.6.2 Unconventional recharge techniques

The unconventional techniques are location specific with limited benefits such as 1) Boreblast technique 2) Jacket well method 3) Fracture seal cementation 4) Well injection technique 5) Hydrofracturing.

Unconventional projects of groundwater recharge are mainly location specific and provide localized benefits of improving the recharge and storage conditions of aquifer and improvement of hydraulic conditions of the area. The benefits are not on a regional scale but are useful in localized conditions. In Maharashtra 91% of area is occupied by hard rocks of which 82% is Deccan Traps. The shallow aquifers are moderate in storage conditions and aquifer parameters mainly due to secondary porosity. These Deccan trap rocks have the fracture porosity which is not very widely interconnected which gives the poor hydraulic conductivity reducing the yields of wells. In order to improve the fracture connectivity and increased storativity, the unconventional techniques are undertaken to improve the yields of dugwells aid borewells.

1. ***The bore blast techniques:*** Are designed by drilling of shallow borewells in a network of a smaller area and these bores are blasted with the help of specialized explosives creating fracture porosity in the areas. These projects were experimented by G.S.D.A. in the district of Pune, Ahemednagar and Nasik for solving the drinking Water

problem in smaller villages were found useful to improve the drinking water sources locally.

2. ***Jacket well technique:*** This technique is used for improvement of hydraulics of dugwell, shallow borewells upto depth of 10-15 m are drilled on a circular interval around the dugwell and these shallow borewells are blasted with special gelatin and detonators. This experiment produces fracture porosity in the hard rock aquifer and hydraulic connection of fractures so that more storage is created surrounding the welt which helps in better recuperation and yield of the well.

3. ***Fracture seal cementation and pressure injection grouting:*** These projects are implemented on the location of leakage of ground water in order to control the outflow of ground water from the aquifer. The locations of seepages and springs are closed with the cementing material either mechanically or with the help of cement slurry being injected under pressure with the help of mud pumps for sealing the fracture porosity and plugging of the fracture connections of the overflowing ground water locations. The pressure injection grouting is under taken similar to underground check dams in a nala bed by drilling shallow borewells upto ten m depth across the nala bed and each borewell is injected by the cement slurry under pressure in order to seal the porosity in aquifer face and a barrier is thus created forming the underground dam.

4. ***Hydrofracturing:*** The technique of hydrofracturing is used mainly for improving the yields of borewells. This technique is a modern technology utilizing the pressure techniques off water injection into the borewells or tube wells. The locations of hydrofracturing could be selected either from the borehole logs data or with the help of electrologging. A suitable section of aquifer is selected and the section is closed by the top and bottom by the inflating packers and the water is injected with high pressures ranging from 30-50 bars, under which pressure the rock strata selected undergoes fracturing which could be evidently observed by sudden reduction of pressure and water loss of the injected water in the well. The application of pressure depends on the selected section of the rock to be fractured. The technique of hydrofracturing has the following advantages.

   i. Widening of existing fracture,

ii. Removal of clogging in the fracture connectivity,

iii. Creation of interconnection of fractures,

iv. Extending the length of the old fractures, and

v. Creation of new fractures in the aquifer section under hydrofracture treatment.

The above conditions help in augmenting storage and improved yields of the wells/borewells. These techniques are deployed on a large scale in various districts of Maharashtra for improving the yields of drinking water borewells by the GSFDA. The status of bore wells before and after hydrofractunng are shown in Table 8.9.

Table 8.9: Classification of borewells on improvement in yield in Pune Region, Maharashtra

S. No.	Borewells Hydrofra-ctured Status	Classification of borewells after treatment								
		No.	**Da**	**Db**	**Py**	**My**	**Gya**	**Gyb**	**Hya**	**Hyb**
1	Da	31	1	5	21	4	-	-	-	-
2	Db	37	-	1	29	7	-	-	-	-
3	Py	68	-	1*	17**	18	19	6	4	3
4	My	1	-	-	-	-	-	1	-	-
5	Gya	11	-	-	-	-	3	1	5	2
6	Gyb	6	-	-	-	-	-	1	5	-
7	Gyb	1	-	-	-	-	-	-	-	-
	Total	155	1	7	67	29	22	9	14	6

Pa, Dh = DRY; Gya, Gya = Good yield; Hy = High yield; Py = Poor yield
(Source: Umarikar, 2002)

8.7 Increased Runoff

Increased run-off and reduced infiltration rate are more important principles in arid and semi arid areas for collecting runoff as much as possible during rainy season for life saving irrigation to crops during dry spells. Therefore, different materials and chemicals are used to collect more run off water. The sheet metal, butyl rubber, fiber glass, asphalt roofing bentonite paraffin, reinforced asphalt, plastic sheeting with gravel cover, Janta emulsion, sodium carbonate etc. may be used on catchments area to increase runoff volume for its further use. The Janta emulsion and sodium carbonate spray increased 68 and 66% runoff, respectively over control at CAZRI, Jodhpur.

Above water harvesting techniques and its principles have been found very useful not only to reduce soil erosion hut also to improve soil moisture status of the soil, that may facilitate better crop and animal production particularly in the arid and semi-arid areas.

8.8 Conservation Strategies

In planning water resource development projects, the needs of drought-prone areas should be given priority (National Water Policy, GOl, 1987). Drought prone areas can be made less vulnerable through soil water conservation measures, water harvesting practices, and transfer of surface water from surplus to deficit areas. A conservative strategy (Table 8.10) of runoff at 70% probability value can be considered for runoff water harvesting. Accordingly Marrigudem will have more water harvesting structures. As the long-term strategy even 30% probability can be considered for food for work programme during drought period. This can capture some intensive storms for additional storage and recharge of groundwater. With simple water and *in-situ* moisture conservation work in the farmers' field and water harvesting at 90% probability can easily be taken up by the farmers.

Table 8.10: Conservative strategies for runoff water harvesting

S. No.	Runoff probability (%)	Strategies
1	90% (runoff which can be managed at field level)	*In-situ* moisture conservation at field level, small farm ponds/dug outs and their net working (1% area to be devoted)
2	70% (runoff which can be stored at micro watershed level)	Permanent water harvesting measures like percolation ponds, check dams, CCTs, diversion drains, gully checks etc. De-silting of existing tanks
3	30% (runoff which can be stored at micro watershed level)	De-silting of existing tanks Creation of additional storage during drought years, MI tanks

References

Dewan, M.L. 1987. Water harvesting. *Indian J. Soil Cons.*, 15(1): 30-34.

Gould, J.E. and McPherson H.J. 1987. Bacteriological Quality of Rainwater in Roof and Groundwater Catchment Systems in Botswana, Water International, No.12, Page: 135-138.

Kanmani, S. and Karmegam, M. 1987. First Regional Seminar on Technology on Drinking Water and Related Water Management for Southern States and Union Territories, Department of Civil Engineering, Anna University, Madras.

Karanath, K.K 1987. Groundwater Assessment, Development and Management. Tata McGraw Hill, New Delhi, Page: 719.

Maggirwar C.N. 1984. Techo-economic feasibility of ground water development.

Murthy V.V.N. 1982. Land and Water Management Engineering Kalyani Publication, New Delhi.

Murthy, J.V.S. Watershed Management In India. Wiley Eastern Limited, New Delhi. Page: 55-77.

Ramasamy, S.M 2002. Designing of Artificial recharge schemes in Hard-rock aquifer systems, *Journal of Indian Waterworks Association*, No.2, Vol. XXXIV, Page: 14.

Schiller, E.J. and Litharm, B.G. 1987. A Comparison of Commonly Used Hydrologic Design Methods for Rainwater Collectors, Water Resources Development, No. 3.

Sharma, R.K and Sharma, T.K. A text Book of Hydrology and Water Resources Engineering. Dhanpat Rai Publications, New Delhi. Page: 913-928.

Umarikar, B.N. 2002. Artificial recharge and water conservation techniques in Deccan Trap Aquifers of Maharashtra, *J. of Applied Hydrology*, xv(1): 23-30.

EXERCISES

8.1 Define the water harvesting and discuss the traditional water harvesting techniques.

8.2 What is the potential volume of rainwater storage for small-scale water harvesting structures under different rainfall conditions?

8.3 Describe the classification of water harvesting techniques.

8.4 Explain the roof top rainwater harvesting.

8.5 Describe the different techniques of artificial groundwater recharge techniques.

8.6 Describe the various techniques used for water harvesting in arid and semi-arid areas.

8.7 Explain water harvesting efficiency and how it can be improved?

8.8 Differentiate the following:

(a) Anicut and percolation tank

(b) *Tanka* and *khadin*

(c) Conventional and unconventional method of artificial groundwater recharge

(d) Farm pond and nadi

8.9. Explain the conservation strategies for rainwater harvesting

■■■■

Chapter 9

DESIGN OF WATER HARVESTING STRUCTURES

9.1 Introduction

The unprediction of rainfall in terms of amount and its distribution may result in to scarcity of moisture in soil profile when it is critically needed by raised crops, despite of good seasonal rainfall received. Under such conditions, crop productivity declines drastically and in order to overcome such contingency rainwater must be harvested for its use when it is needed the most during crops season. To collect surplus water of any rainfall even numerous type of structures may be made. However, their design and capacity are largely governed by the various geographical factors including soil type, slope and terrain of the region.

Rainfall is the ultimate source of moisture and basic climatic factor and it is the single dominant weather element influencing location and itineraries of farming systems. It also determines the potential of any region in terms of crops to be produced, farming system adopted, the sequence of farming operations to be followed to achieve higher productivity (Singh and Dhillon, 1994). But variation in rainfall from year to year or even from one season to another season becomes critical for agricultural operation. Further, mid-season or terminal drought limits the productivity under dryland ecosystem. Therefore, water harvesting is the most viable option for increasing the productivity in such areas. To sustain the pressure of fast increasing population, the enhancement in the productivity level of arid and semi arid areas are to be the most viable alternative. This will also motivate the farmers of these areas to adopt modern scientific techniques as they will be helpful for better returns of their efforts and investments. Depending upon soil type, land slope, rainfall conditions and available resources etc. the selection of appropriate technique is to be made to harvest rainwater either in-situ or ex-situ in the field.

9.2 Rainwater Harvesting Structures

There are following engineering measures for rainwater harvesting and are described in subsequent sections-

9.2.1 Contour bunding

Contour bunding is the practice in which a low earthen embankment, about a meter in height with a specific side slope is constructed across the slope of land at certain intervals. It is a very effective practice to enrich the soil moisture on gently sloping lands (1 to 6%) particularly in light soils of low rainfall areas as it facilitates ponding of surplus rainwater and allows it to infiltrate into the soil profile providing more opportunity time. In this way the water is added to soil moisture which is utilized by plants for their growth.

Design of contour bunds

For designing a contour bunding system the following input data is generally prerequisite.

i. Land slope (S),

ii. Type of soil,

iii. Annual rainfall depth,

iv. Infiltration and vegetative cover conditions, and

v. Maximum 24 hour effective rainfall (RE) in the design recurrence interval.

Design parameters

The following parameters are required for designing a contour bunding system:

- Horizontal interval between two contour bunds (HI)
- Vertical interval between two contour bunds (VI)
- Net height of contour bund (d)
- Gross height of contour bund (D)
- Base width of contour bund (B)
- Side shape of contour bund (n)
- Top width of contour bund (T)
- Earthwork involved in construction (EW)
- Percent area loss due to contour bunding (%AL)
- Total area loss in contour bunding (AL)
- Cost of contour bunding (C)

Design procedure

The following relationships are used to determine the values of above design parameters -

Vertical interval (VI)	=	0.3 (X S + Y)	...(9.1)
Horizontal interval (HI)	=	(VI/S) x 100	...(9.2)
Net height of bund (d)	=	$[2VI(RE)]^{0.5}$	...(9.3)
Gross height of bund (D)	=	1.15 d	...(9.4)
Top width of contour bund (T)	=	D/2	...(9.5)
Base width of contour bund (B)	=	2nD + T	...(9.6)
Earth work volume (EW)	=	$\frac{T+B}{2} \times D \times \frac{A_L}{HI}$	...(9.7)
Percent area loss (% AL)	=	$\frac{B \times S}{VI}$	...(9.8)

The values of rainfall factor (X), infiltration and vegetative cover factor (Y) and recommended side slope as per the existing conditions can be selected from the Tables (9.1, 9.2 and 9.3).

Table 9.1: Values of rainfall factor (X)

Rainfall conditions	Value of X
Low rainfall (<500 mm)	0.8
Medium rainfall (500-1000mm)	1.0
High rainfall (>1000mm)	1.2

Table 9.2: Values of infiltration and vegetative cover factor (Y)

Infiltration rate of soil	Crop cover conditions	Value of Y
Below average	Low coverage	1.0
Below average	Good coverage	1.5
Average	Poor coverage	1.5
Above average	Good coverage	2.0

Table 9.3: Recommended side slope for different type of soils

Soil type	Side slope (V:H)	Value of 'n'
Light soil	1:2.0	2.0
Medium soil	1:1.5	1.5
Heavy soil	1:1.0	1.0

9.2.2 Bench terraces

Construction of bench terraces is the only option to put steep sloping lands (6-32%) under cultivation and conserve soil and water resources. Considering the existing conditions of land slope, soil type, soil depth, crop type, rainfall conditions etc. at a place, a proper selection criteria and design procedure should be followed as improper selection and design of these terraces will render them ineffective and uneconomical. India has about 93.06 mha mountainous region out of its total geographical area of 328 mha. Due to steep slopes, these lands can not be put under cultivation by following ordinary agricultural methods. Also, the high rate of soil erosion from mountainous lands is a very serious cause of concern among soil conservationists and environmentalists. In order to bring these lands under cultivation and also to reduce soil erosion and conserve water, the practice of bench terracing is very effective. Bench terracing basically involves the construction of step like fields across the general slope of the lands by cutting and filling of soil. A proper design procedure of bench terraces which should consider the local conditions like, land slope, soil type, depth of fertile soil, rainfall conditions, types of crop grown etc. at a place is required to make the system effective. Care is required to ensure while cutting and filling the soil, the infertile sub- soil does not get exposed at the top of the bench as it will render the top soil unproductive.

Classification of bench terraces

Bench terracing systems may be classified into the following categories based on the nature of the bench and its riser:

(i) Level bench terrace with vertical riser,

(ii) Level bench terrace with inclined riser,

(iii) Inward sloping bench terrace with vertical riser,

(iv) Inward sloping bench terrace with inclined riser,

(v) Outward sloping bench terrace with vertical riser, and

(vi) Outward sloping bench terrace with inclined riser.

Selection criteria for an appropriate type of bench terrace

A particular type of a bench terracing system is to be selected based on local conditions such as, land slope, rainfall amount, soil type, available soil depth, types of crops grown and type of agricultural operations adopted in the area. The guidelines may be followed to select an appropriate type of a bench terracing system at a place is summarized in Table 9.4.

Table 9.4: Selection criteria for bench terrace system

1	Level bench with vertical riser (a) Soil depth is medium (b) Land slope is medium (c) Rainfall is medium (d) Soil is stable in nature (e) Riser height is not more (f) Stone boulders are easily available at the site for stone pitching (g) Moisture conservation is required (h) Ponding of water is required for growing of crops (i) Soil has medium infiltration capacity	2	Level bench with inclined riser (a) Soil depth is medium (b) Land slope is medium (c) Rainfall is medium (d) Soil is stable in nature (e) Riser height is relatively more (f) Stone boulders are not easily available at the site for stone pitching (g) Moisture conservation is required (h) Ponding of water over the bench is needed (i) Soil has medium infiltration capacity
3	Inward sloping bench with vertical riser (a) Soil depth is good (b) Land slope is medium (c) Rainfall is high (d) Soil is stable in nature (e) Riser height is relatively less (f) Stone boulders are easily available at the site for stone pitching (g) Disposal of excess water is required	4	Inward sloping bench with inclined riser (a) Soil depth is good (b) Land slope is medium (c) Rainfall is high (d) Soil is unstable in nature (e) Riser height is relatively more (f) Stone boulders are not required as no stone pitching is needed (g) Disposal of excess water is required

	(h) Water susceptible crops are to be grown		(h) Water susceptible crops are to be grown
	(i) Soil has poor infiltration capacity		(i) Soil has poor infiltration capacity
5	Outward sloping bench with vertical riser	6	Outward sloping bench with in riser
	(a) Soil depth is shallow		(a) Soil depth is shallow
	(b) Land slope is high		(b) Land slope is high
	(c) Rainfall is less		(c) Rainfall is less
	(d) Soil is stable in nature		(d) Soil is unstable in nature
	(e) Riser height is relatively less		(e) Riser height is relatively more
	(f) Stone boulders are easily available at the site for stone pitching		(f) Stone boulders are not easily available at the site for stone pitching
	(g) Soil has good infiltration capacity		(g) Soil has good infiltration capacity

Design parameters of bench terraces

After the selection of an appropriate type of a bench terracing system at a particular place, the design parameters of bench terraces are determined based on the known variables such as, land slope (S), soil depth, soil type, rainfall conditions, type of bench terracing system and area to be bench terraced (A) at that site. The parameters of bench terraces which are required to be determined in the design of a terrace system include, width of bench, (W), vertical interval (VI), horizontal interval (HI), required depth of cut (D), inclination of riser (n), slope across the bench (g), slope along the bench(s), earthwork volume (EW) and percent area loss (%AL). The above design parameters are determined using following recommendations and mathematical relationships.

Determination of design parameters

The procedure to determine the values of various design parameters is described as under:

(i) Width of bench (W)

The minimum width of the bench is considered as 3.0m. If the benches are narrower than this width then it will become unsuitable for performing various agricultural operations. The width of bench decreases as land slope increases for a particular depth of fertile soil. The following are the recommendations for deciding width of bench based on the slope of the land.

Land slope, S (%)	Width of bench, W(m)
10-15	6.0 - 4.5
15-25	4.5
> 25	3.0

(ii) Vertical interval and horizontal interval (VI & HI)

The vertical and horizontal interval between two consecutive benches are related as follows:

$$(VI/HI) \times 100 = S\% \quad ...(9.9)$$

and,

$$HI = W + n(VI) \quad ...(9.10)$$

On using two equations (9.9 & 9.10) the expression for vertical interval in terms of bench width (W), land slope (S) and riser inclination (n) is expressed as,

$$V1 = (W \times S) / (100 - nS) \quad ...(9.11)$$

where, n is the riser inclination and taken as zero for vertical riser, 1 for inclined earth riser and 1/6 for stone boulder riser. The value of riser inclination (n) should be as small as possible, 'while ensuring the stability of riser, in order to minimize the loss of the cultivable land due to construction of the terraces.

(iii) Required depth of cut (D)

The required depth of' cut to construct bench terraces should not exceed the available depth of fertile soil at a particular place This is important to avoid the exposure of unfertile sub- soil at the top of the bench. The required depth of cut (D) is determined by using the following relationship.

$$D = (VI) / 2 \times g\ W/2 \quad ...(9.12)$$

Where, g is the slope across the bench (m/m). For level bench it is taken as zero and for inward sloping bench it ranges from 2 to 10% depending upon the drainage requirements. The bench is also provided a longitudinal slope of 0.1% along the length of the bench for quick disposal of excess runoff water in high rainfall areas.

(iv) Earthwork volume (EW)

The construction of bench terraces involves the filling of the soil in the lower half with the soil cut from the upper half keeping the cut-fill ratio preferable as one. The volume of earthwork (EW), therefore, depends on the depth of cut (D), width of the bench (W), vertical interval (VI), riser inclination (n) and area to be bench terraced (A) and is determined with the help of the following relationship:

$$EW = 1250\ (D \times W \times A) / (W + n\ (VI)) \qquad ...(9.13)$$

For vertical riser the value of n is zero and the above relationship becomes as,

$$EW = 1250\ D \times A \qquad ...(9.14)$$

where, EW is in cubic metres, D,W,VI are in meters and A is in hectares. For convenience, the earthwork is calculated for unit area and multiplied by the area to get total earth volume.

(v) Percent area loss (% AL)

The bench terraces should be so designed and constructed that the loss of the cultivable area due to their construction is minimum. The percent area loss is dependent on riser inclination (n) and is expressed as,

$$EW = \%AL = n \times S \qquad ...(9.15)$$

In case of vertical riser the value of n is zero and, therefore, the area loss is also zero. Thus, the riser should be designed as vertical as possible while satisfying the stability requirements.

A shoulder bund of appropriate size is to be constructed at the outer periphery of these bench terraces to check flow of water down the benches so that the riser is not damaged. Also, in case of inward sloping bench terraces, a channel is to be constructed along the inner side of the terraces to dispose off excess runoff safely. The channel is to be designed considering the expected maximum discharge during design period. If risers are of earth material these should be covered with some local grasses to make them stable. In case growing of grasses takes a longer time these should be covered with sods with well established grasses to provide immediate relief to the risers against their damage either due to erosion or other activities.

9.2.3 Temporary gully control structures

Temporary gully control structures are used in watershed management programmes for development of non-agricultural lands.

These structures control sediment storage, surface water disposal, water level control and improve groundwater recharge. These may be constructed by using the readily available local material like earth, sod, rocks or boulders, wood, brush, etc. They are helpful in supporting, protecting and enriching vegetative cover. Such temporary structures are recommended for gullies with small drainage areas. It is very essential to grow vegetative cover and barriers along with the construction of these structures. A series of temporary checkdams are more helpful and useful than a single large concrete checkdam. They should be constructed from top to bottom.

9.2.3.1 Checkdams

A series of small checkdams reduce the velocity of runoff leading to minimize runoff loan as sediment storage, moisture control, recharge groundwater and also more useful than a single large checkdam. Construction of such series of small checkdams from top towards bottom transforms the longitudinal gradient into a series of steps with low risers (Fig. 9.1).

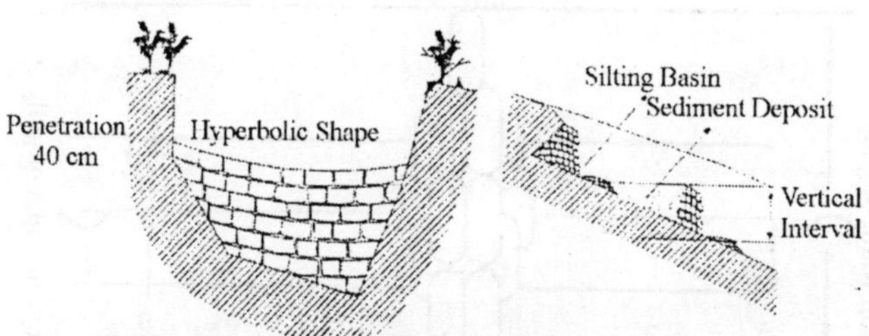

Fig. 9.1: Checkdam

These dams are required to be constructed in such a way that the crest level of one coincides with the base level of the next dam upstream. Generally, a grade of 0.1-0.5% may be provided. A series of earthen or boulder checks have shown effectiveness and sustainability over a span of years in gully control in Jhabua. Completion of checkdams on first line of drainage has been ensured before moving to the second, and third line of drainage. Earthen checkdams have proved to be the cheapest. They have been strengthened by vegetative cover especially of grasses, *jatropha* and *agave*. These checkdams are also constructed with loose rocks, boulders, brush, wire, etc. Regular checks of masonry or concrete structures may be essential to provide permanence. The temporary checkdam perform following basic functions: (a) collect sufficient soil and water to enable the proper growth of vegetative cover, and (b) check channel erosion until sufficient stabilizing vegetation can be ensured. The vertical interval between two checkdams should be equal to the height of a checkdam (Fig. 9.2).

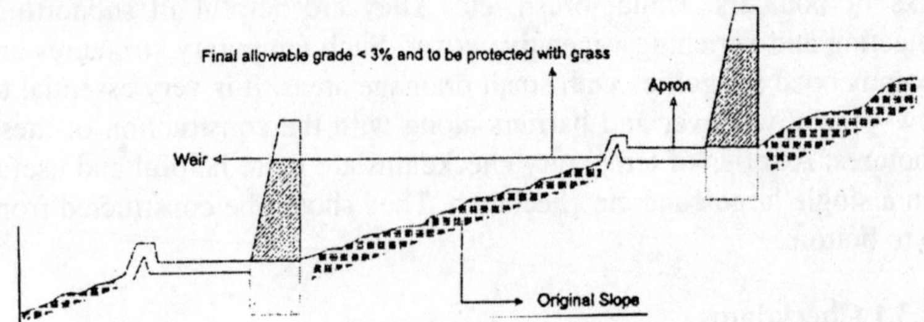

Fig. 9.2: Examples of checkdams. Gradient between head of lower weir and toe of next upper weir not to exceed 3% (Source: Rajora, 2002)

Earthen checkdams have been found effective in gullies with less than 1.8 m depth and 4 m width in Jhabua. These checkdams requires continuous repair and maintenance annually. They should also be inspected and repaired after heavy downpour (Fig. 9.3).

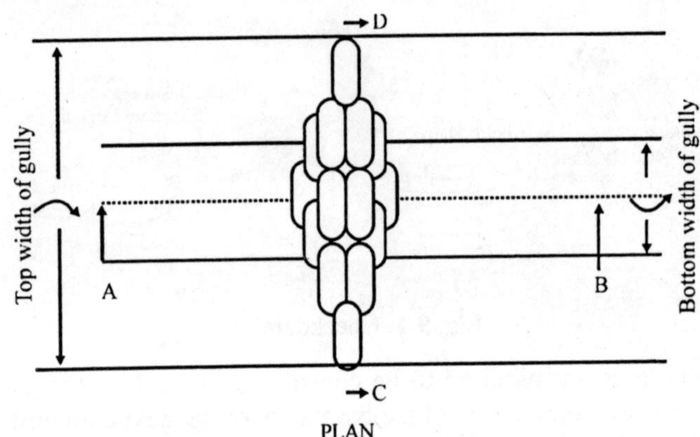

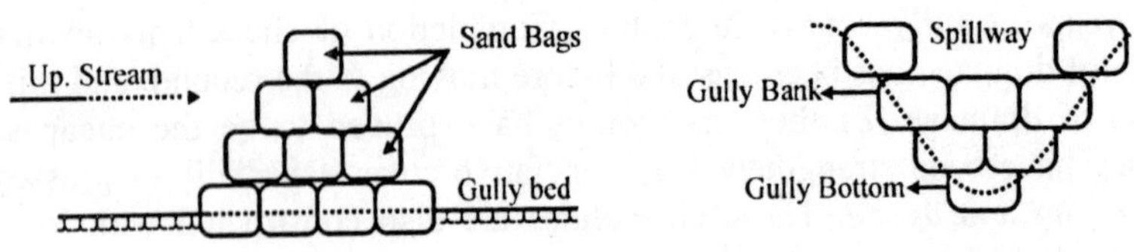

Fig. 9.3: Synthetic fibre bags filled with sand for checkdams (Source: Rajora, 2002)

9.2.3.2 Nalla bunding

Nalla bunding usually consists of constructing earthen bunds, loose rock dams and masonry check dams of suitable dimensions across the *Nallas* or gullies to hold the maximum runoff water. It creates flooding of up stream area which requires surplussing arrangements at suitable intervals to drain water. *Nalla* bunds should be avoided in isolation. Integrated *Nalla* bunding plan can be formulated and executed in Jhabua. It helps in water harvesting, moisture control and groundwater recharge. The number of such bunds primarily depends upon the slope of the gully or the *Nalla* and the quantity of runoff. It may not be advisable to construct *Nalla* bunds on *Nalla* with high gradient slope and where runoff of water flow is very high. The *Nallas* should be treated with drainage line treatment and gully control measures and *Nallas* with high runoff flow should be treated with gabionic structures and boulder checks with masonary work to curtail the runoff (Fig. 9.4 a, b).

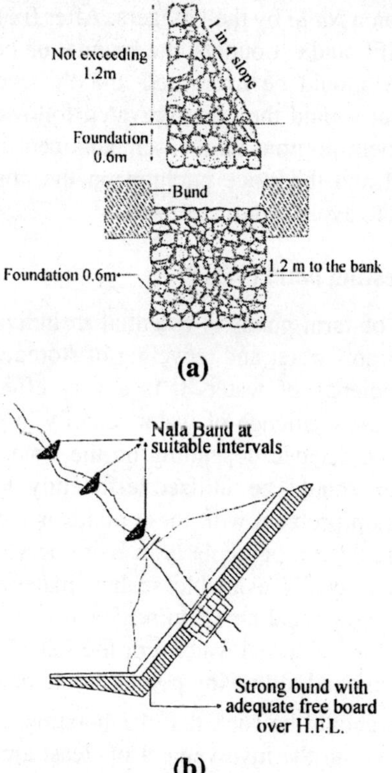

Fig. 9.4 (a, b): *Nalla* bunding (Source: Rajora, 2002)

The temporary storage of runoff against these *Nalla* bunds results in deposition of soil in upstream. The impounding of water facilitates the percolation of water to the deeper layers of soil. Excessive runoff may cause severe soil erosion and choking of natural drainage courses, lakes and reservoirs in absence of *Nalla* bunds. The *Nalla* bund reduces the runoff and before the water flow gains momentum, it encounters the next bund down below in the *Nalla*. The basin at the upstream side of the bund gets enriched by the progressive silt deposition. Construction of such *Nalla* bunds helps in reclamation of gullied lands and groundwater recharge.

In semi-arid regions with low rainfall, and relatively permeable soils, bunds may be constructed from the top of the catchment, from ridge to valley, at regular intervals. A detailed survey of the *Nalla* at suitable locations with a suitable grid of 20 m to a maximum of 90 m should be carried out. *Nalla* bunds should be located on the *Nalla* in the survey plan and a detailed estimate should be prepared. *Nalla* bunds should be located on a *Nalla* by the villagers. After fixing different points such as FSL and HFL and cut outlets, the foundation could be laid down. The puddle trench should be excavated and the core wall should be constructed. Cut-out should then be excavated followed by digging and filling of embankment portion. Earthen embankment is then constructed over the core wall and the stone pitching on the upstream side of the bund is carried out to avoid scouring of soil.

9.2.4 Water harvesting pond (WHP)

Construction of farm ponds is essential in individual farmers field for harvesting of runoff water and recycling of stored water for irrigation when there is deficiency of water. It is a very effective and efficient method for facing the challenge of water scarcity in dryland areas. The size of pond may be decided depending on the water availability in the area and its water should be utilized efficiently to draw maximum benefit. One common problem with these ponds is the loss of water due to seepage and percolation, particularly in more pervious soils. To check such losses cheaper locally available sealant materials *viz.*, clay lime, soil-cement (8:1) lining, coal-tar (Bitumen) and thick plastic sheets can be used to control loss of stored water. For the selection of a site for the-construction of farm pond, following points should be considered:

i. The site should be such that the maximum storage volume is available with the involvement of- least amount of earthwork. A narrow section of existing valley with steep slope should be preferred.

ii. There should not cause excessive seepage losses. Large areas with shallow water storage should be available to control evaporation losses and also to control the growth of water weeds.

iii. The selected site should be as near as possible to the place where the stored water is to be utilized. This will not only reduce the water conveyance cost but also reduce the loss of water in conveyance.

Following steps may be followed while designing a farm pond:

(i) Selection of an appropriate site (discussed earlier),

(ii) Design of the embankment,

(iii) Provision of a spillways, and

(iv) Provision for control measures.

Following points should also be considered while designing farm pond:

(a) Sediment: The runoff from the watershed contains sediment and they would deposit and accumulate in the reservoir over a period of time. Space must be provided to receive the sediment deposit otherwise the reservoir will have inadequate capacity before reaching the end of its design life. The volume required in the reservoir for sediment storage may be estimated on using the following formula:

$$V = \frac{ST}{1500} \quad \text{...(9.16)}$$

where,

V = Volume required for sediment storage, (acre ft),

S = Amount of soil loss from the watershed each year, (tons),

T = Expected design life of the pond, (years), and

1500 = Bulk density of the sediment, (tons/acre ft).

(b) Reserve pool: Certain volume has to be kept for reserve pool which will prevent the pond from drying out, cracking and water is always available for *nistar* purposes. Reserve pool depth depends upon the intended use of the farm pond. It represents minimum depth of the pool and should always exceed 90 cm to avoid weed problems. The volume of the reserve pool can be estimated from the volume versus elevation curve.

(c) **Evaporation:** The water loss from evaporation from the surface of the farm pond may exceed the amount replaced by the rainfall. The evaporation loss can be measurable from the evaporation from the open pan evaporation and multiplying it by 0.7.

(d) **Seepage:** Seepage rates of 2.5 cm per month for good construction and 5 cm per month for fair construction (or permeable soils) may be used. Evaporation losses depend upon the surface area of pond, which itself varies through the year. The area of the normal pool is the surface area of the pond when it is full. The volume of water loss is the product of the average area and the depth of loss.

(e) **Consumptive use:** Volume of water used for day-to-day consumption and activities.

(f) **Detention storage:** This usually is the volume contained between the principal spillway and emergency spillway elevations. The runoff water from very heavy rains can be stored temporarily in detention storage and is slowly drained through the principal spillway protecting emergency spillway from damage. Size of the drainage area is a crucial step for construction of farm ponds. It depends upon the amount of runoff especially when rainfall is limited. Land slope, soil infiltration, vegetative cover, surface storage. Amount, intensity and duration of rainfall determine the runoff with geographical and seasonal variations.

Capacity of farm pond

The capacity of the farm pond is determined from a contour survey of the site at which the pond is to be located. From the contour plan of the site, the capacity is calculated for different stages using the trapezoidal or Simpson's rule (Simpson's rule gives more accurate values than the trapezoidal formula).

For this purpose, the area enclosed between the contour is measured on using a planimeter. The volume V between two contours at an interval H and having areas A_1 and A_2 is estimated using trapezoidal formula.

$$V = \frac{H}{2}(A_1 + A_2) \quad \text{...(9.17)}$$

On using Simpson's rule the volume between any odd numbers of contours is estimated as,

$$V = \frac{H}{3}\left[\text{Twice the area of odd contours} + \text{4 times area of even contours} + \text{area of the first and last contours}\right] \quad \text{....(9.18)}$$

The formula is also known as the prismodial rule. For using this equation, the number of contours should be odd i.e. the number of intervals considered should be even.

Table 9.5: Calculate the capacity of a pond given the area enclosed by different contours at the site as follows

S. No.	Contour value (m)	Area enclosed (sq. m)
1	350	220
2	351	290
3	352	340
4	353	370
5	354	480
6	355	550
7	356	620

Contour interval = 1 m.

Solution:

Using trapezoidal formula

$$V = 1x100\left(\frac{2.2x6.2}{2} + 2.9 + 3.4 + 3.7 + 4.8 + 5.5\right) \quad \text{...(9.19)}$$

$$= 2450 \text{ cu. m}$$

Using prismoidal formula

$$V = \frac{1}{3}x100[2.2 + 4(2.9 + 3.7 + 5.5) + 2(3.4 + 4.8) + 6.2] \quad \text{...(9.20)}$$

$$= 2440 \text{ cu. m}$$

To plot the depth-capacity curve the following table can be prepared. Trapezoidal formula is used to calculate the volume increments.

Contour value	Area enclosed	Volume increment	Cumulative volume
350	220	-	-
351	290	255	255
352	340	315	570
353	370	355	925
354	480	425	1350
355	550	515	1865
356	620	585	2450

The storage capacity of an earthen embankment can be approximately calculated by the following formula:

$$\text{Storage (ha-m)} = 0.4 \text{ D} \times \text{A} \ (5.4) \qquad \text{...(9.21)}$$

where,

D = Maximum depth of water, (m), and

A = Area of water spread at the waste weir, (m^2).

Farm ponds are constructed to improve the availability of water for irrigation, livestock and drinking etc.

Table 9.6: Water requirements for farm users

Type of use	Average use in	
	(gallons per day)	(ha-m per year)
Household, all purposes, per person	50-100	0.01
Dry cow	9-18	0.002
Milk cow, 450 kg (1000 lb)	18-40	0.004
Horse or mule	8-12	0.0014
Chickens per 100 head	0-2	0.001
Swine per 45 kg (1001 b)	1-1.5	0.0002
Sheep per 45 kg (1001 b)	1-1.5	0.002
Orchard sparing, per year of tree age per application	1	
Irrigation (humid regions) per season		0.3-0.45
Irrigation (arid regions) per season		0.3-1.5

Procedure for the design of WHP

According to Sharma and Bhattacharya (2005) it involves following steps:

Step I: Rainfall analysis

For designing WHP two-day maximum rainfall of any week amounts to 70 per cent of that weeks total rainfall and runoff produced by this rainfall is sufficient to be considered for harvesting into the storage structures (Rao, 2001). The long-term daily rainfall data of 30 years were collected and analysed. These data were scrutinized for monsoon period only. These scrutinized data were then grouped into weeks (total 17 weeks) and finally two days maximum rainfall values were selected from each week.

Step II: Runoff estimation

The SCS (Soil Conservation Service) curve number method was employed to estimate runoff from rainfall. Two days maximum rainfall in a week was used for runoff estimation. The watersheds were considered without having any conservation practices and hydrologic soil group was decided based on soil information. The CN (Curve Number) values were selected from standard table. To take into account the effect of 48 hrs rainfall on runoff estimation, the selected CN values were reduced by 9 based on the work done by Rao (2001). The runoff estimated as above was considered as weekly runoff for its harvesting and storing in water harvesting ponds. This was repeated for all weeks (week no 24 to 40) and all years. A probability analysis of maximum weekly runoff of 30 years was done using Weibull formula. The estimated weekly runoff at probability levels ranging from 10 to 90 per cent were selected for designing the WHP.

Step III: Design of WHP

An inverted truncated pyramid shape pond is considered for the design, as it is a most practical shape to be constructed with ease. The side slope is taken based on type of soil (Table 9.7). A depth of 3 to 4 m is suitable in general for ponds (Samra *et al.* 1996). The bottom of pond is assumed to be square in shape. The estimated runoff is used as a storage capacity of the pond.

Table 9.7: Side slopes for preliminary design of earth dams

S. No.	Type of material	Up stream slope	Down stream slope
1	Homogeneous well graded material	2.5:1	2:1
2	Homogeneous core	3:1	2.5:1
3	Homogeneous silty clay or clay		
	ht. less than 15 m	2.5:1	2:1
	ht. more than 15 m	3:1	2.5:1
4	Sand or sand and gravel with clay core	3:1	2.5:1

(Source: Trerzaghi & Peck, 1948)

9.2.5 Design of different types of waste weirs and outlets

Surplussing arrangement is necessary for safely draining the runoff form watershed area, drainage area or farm pond. Usually masonry weirs are located in a contour, supplemental or marginal bund where such bund crosses a depression, with crest constructed at 0.3 m above contour. Once a masonry waste weir is provided in a contour bund, all the bunds situated below should be provided with outlets along the depression of water course. Successive masonry waste weirs are not to be constructed in a straight line but in staggered manner that they will not cause gulling of the field.

It is a good practice to have surplusing arrangements in all case except in areas having less than 350 mm of rainfall.

Clear overfall weirs

It is a permanent masonry structure and some what similar to drop spillway. It can be constructed by locally available material also.

The height the crest wall controls the depth of the water ponded up stream of the weir. The height of crest of weir notch above the contour is generally kept 30 cm. Length of weir is calculated by using weir formula.

$$Q = 1.71\ LH^{3/2} \text{ (for rectangular weir)} \qquad \ldots(9.22)$$

where,

Q = peak rate of runoff, (cumec),

L = length of weir, (m), and

H = head of flow, (m)

The runoff is estimated from the catchment area of weir. For catchment area beyond 40 ha, it is desirable to provide diversion instead of waste weir as the water flowing from field will be considerable. Ordinarily it is suitable for seasonal crops grown in scarcity area.

Pipe outlet

It consists of an upstream and downstream stilling wells and a conduit pipe. The stilling wells are constructed with brick or stone masonry or with cast concrete pipes. The pipe may be a rain forced concrete pipe, sewer pipe or corrugated metal pipe.

A 15 cm diameter pipe is usually recommended for a 3 ha catchment, a 22.5 cm diameter pipe for a catchment up to about 5 hectares and a 30 cm dia pipe for a catchment up to 8 ha. The crest level of the inlet is kept 30 cm above the ground level. The design and construction is similar to drop inlet spillway. The stilling well consists of 0.45 cm diameter pipe for outlet of 15 cm diameter and 0.60 m diameter pipe for outlet of 22.5 cm and 30 cm diameter. The well top is kept 0.30 m above contour level.

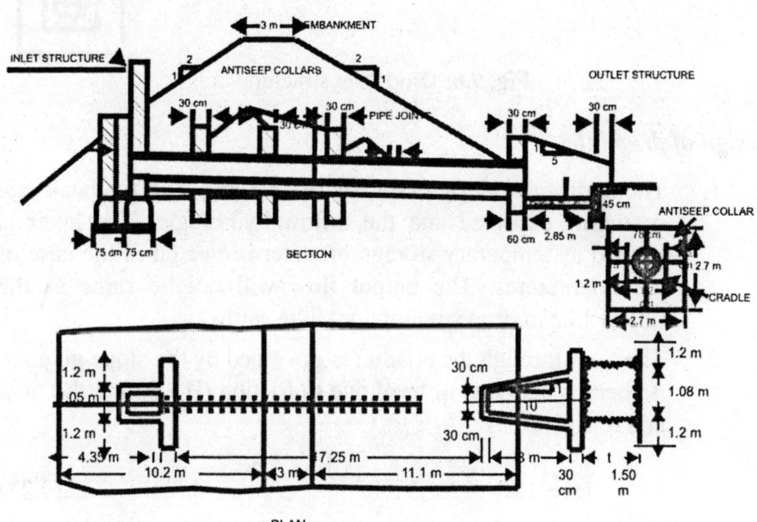

Fig. 9.5: Drop inlet spillway

Ramp cum waste weir

It is also called grass outlet. It consists of a simple opening in the bund protected with grass. This weir is simple in construction but cannot handle large volumes of water.

Drop inlet spillways

Drop inlet spillways are frequently used as gully control structures. This application is usually made where water may pond behind the inlet to provide temporary storage. A drop inlet structure consists of the inlet, conduit and the outlet (Fig 9.5 and 9.6). An earthen empanelment helps in storing the water and the drop inlet essentially lets out the excess water safely.

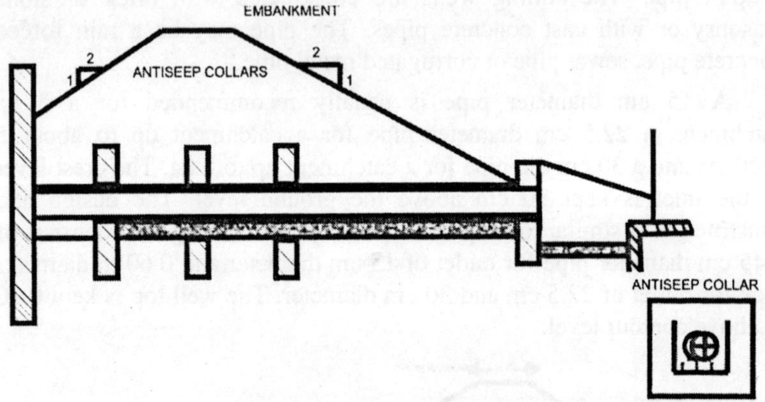

Fig. 9.6: Drop inlet structure

Design of drop inlet

1. The hydrologic design consists of knowing both the peak rate or runoff expected and the inflow hydrologic. The latter is needed as temporary storage of water is created in the case of these structures. The output flow will not be same as the inflow like in drop structure or chute spillways.
2. The flow through the conduit is governed by the slope given to it, there is also loss of head due to friction (H_f) during the flow and it is given by

$$Hf = LK_C \frac{V^2}{2g} \qquad \ldots(9.23)$$

where,

L = length of the pipe (m)

V = velocity of flow, (m/sec) and

K_C = head loss coefficient ≈ 0.03

The slope of the pipe is given by

$$S = \tan\theta = \frac{H}{\sqrt{L^2 - H^2}} \qquad \text{...(10.24)}$$

H is the difference in head of inlet and outlet neutral slope S_n is defined as,

$$Sn = \frac{K_C \frac{V^2}{2g}}{\sqrt{1-(K_C V^2/2g)^2}} \qquad \text{...(9.25)}$$

Actual slope should be less than neutral slope

3. The capacity of the conduit under full free outfall submerged pipe flow is give by

$$Q = \frac{a\sqrt{2gh}}{\sqrt{1+K_e+K_C L}} \qquad \text{...(9.26)}$$

where,

a = area of cross section of the conduit

K_C = coefficient of entrance losses = 0.3

If the conduit is at greater than neutral slope and the outlet is submerged, the flow will be controlled by the inlet section of the conduit, in such a case the discharge is calculated using the orifice formula as given by:

$$Q = ac\,\sqrt{2gh} \qquad \text{...(9.27)}$$

where,

c = coefficient of the orifice.

Using these formulae and the discharge characteristics of the drop inlet are to be prepared. From the site conditions, the information relating to the stage (water level above the level of the inlet) and the storage capacity can also be prepared. Using the inflow hydrograph, storage capacity curve and the discharge characteristics of the structure, the outflow hydrograph of the structure can be worked out. For the design storm hydrograph selected,

Chute spillways

It carries flow down slopes through a concrete lined channel rather than by dropping the water in a free overfall. Chutes may be used for the control of heads up to 6 meters. They usually require lens concrete than do drop inlet structures of the same capacity and drop. However, there is considerable dangers of undermining of the structure by rodents, and in poorly drained locations, seepage may threaten foundations. Where there is no opportunity to provide temporary storage above the structure, the flume with its inherent high capacity is preferred over the drop inlet pipe spillway. The capacity of a chute is not decreased by sedimentation at the outlet.

Design- The chute spillway consists of the three principal parts *viz.*, inlet, conduit and outlet. The different types of these parts are shown in Fig 9.7.

The hydrologic design of the structure consists in estimating the peak rate of runoff as in the case of drop structure.

In the hydraulic design the length of the crest is calculated using the weir formula as in the case of drop structure.

Velocity of flow at the toe (V_1) is given by

$$V_1 = \sqrt{2g(H - h_f)} \quad \text{...(10.28)}$$

where,

h_f = frictional head loss over the apron

H = total drop

for practical purpose $h_f = 0.1$ H

Once the velocity of the toe is known the depth of flow (d_1) is calculated from the relation.

$$Q = A_1 V_1 \quad \text{...(9.29)}$$

$$= b \times d_1 \times V_1 \text{ (for a rectangular conduit)}$$

A hydraulic jump will be formed on the horizontal part of the basin provided the Frond number (F) of the incoming flow and the corresponding downstream depth d_2 satisfy the equation

$$\frac{d_2}{d_1} = ½ \left(\sqrt{1+8F^2} - 1\right) \quad \text{...(9.30)}$$

where,

$$F = V1 \sqrt{gd_1} \quad \text{...(9.31)}$$

The above equation is also used for calculating the depth d_2 after hydraulic jump.

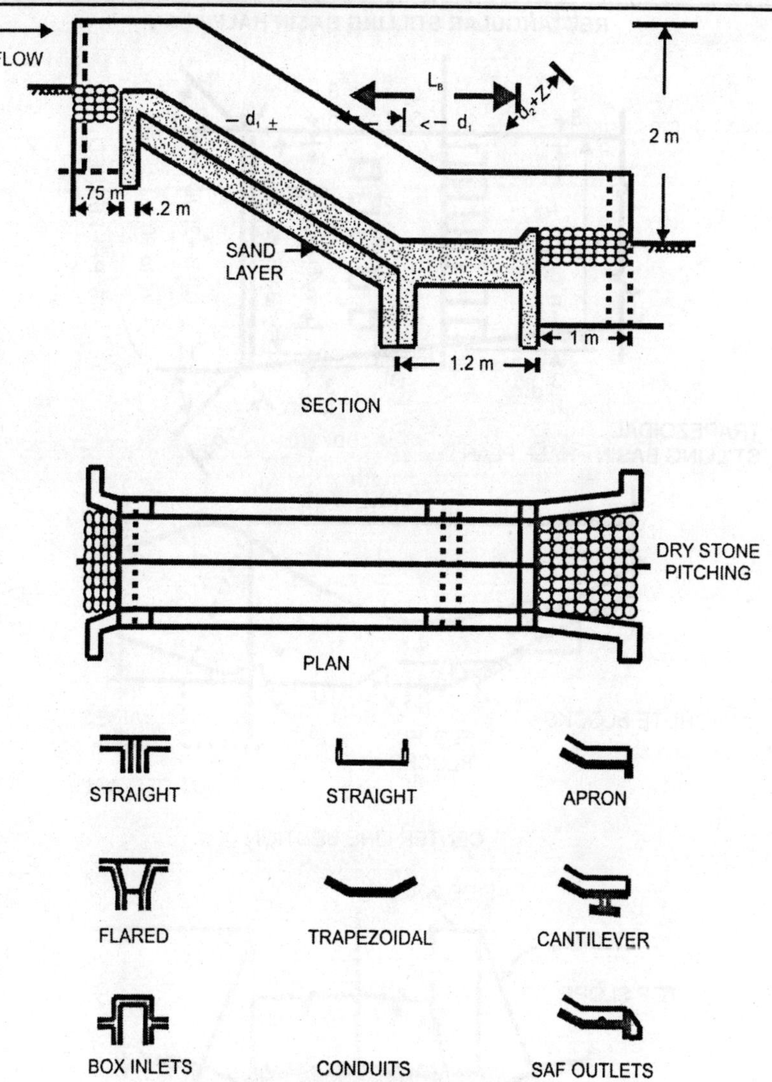

Fig.: 9.7: A typical chute spillway

The sidewalls of the basin may be parallel (rectangular basin) or may diverge (trapezoidal basin), the latter is preferred for higher flows. The side slope recommended is 1 in 10 depending upon the site conditions.

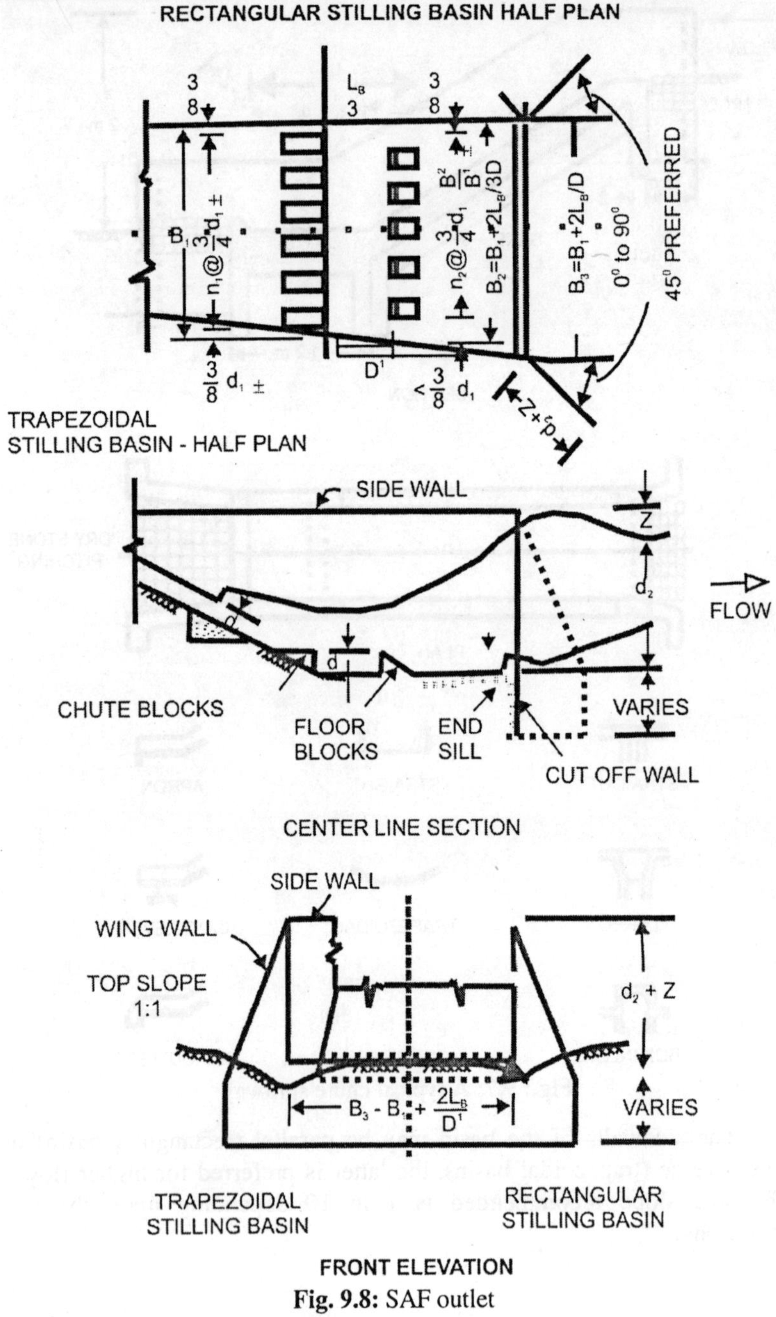

RECTANGULAR STILLING BASIN HALF PLAN

$\frac{3}{8}$ L_B $\frac{3}{8}$ $\frac{3}{8}$

$n_1 @ \frac{3}{4} d_1 \pm$ B_1 $n_2 @ \frac{3}{4} d_1 \frac{B_2}{B_1} \pm$ $B_2 = B_1 + 2L_B/3D$ $B_3 = B_1 + 2L_B/D$ 0^0 to 90^0 45^0 PREFERRED

$\frac{3}{8} d_1 \pm$ D^1 $< \frac{3}{8} d_1$ $d_2 + Z$

TRAPEZOIDAL
STILLING BASIN - HALF PLAN

SIDE WALL Z d_2 FLOW d_1 d

CHUTE BLOCKS FLOOR BLOCKS END SILL VARIES CUT OFF WALL

CENTER LINE SECTION

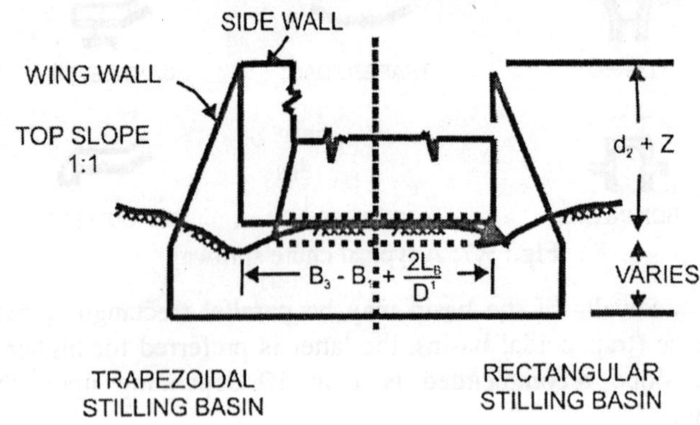

FRONT ELEVATION

Fig. 9.8: SAF outlet

The different recommended dimensions for the SAF stilling basin (Fig 9.8) are as follows.

Length of the stilling basin $L_B = \dfrac{4.5d_2}{F^{0.38}}$ ...(9.32)

$3 < F < 300$, $F = V^2/gd_1$ ...(9.33)

The height of chute blocks and floor blocks = d ...(9.34)

The width and spacing of the chute blocks and floor blocks

$= 3d_1/4$ ...(9.35)

Losses of water from farm pond

Losses of stored water from farm pond is a bottleneck problem is adoption of this structure for water harvesting. There are mainly two kinds of water losses from farm pond *viz.*, seepage and evaporation. The commonest cause of water loss is seepage and can sometimes be so serious that it may completely drain the reservoir or stored water. The seepage may happen from leaking basin or dam wall and may lead to water-logging and salination. In arid and semi-arid regions, the water loss by evaporation is high.

(i) Seepage losses

The seepage losses may be more in light texture soil and permeable sub-strata. These losses may be reduced by lining of the pond or reservoir. Cost, availability of material, ease of application, acceptability by the community. durability, soil and sub-strata conditions and annual maintenance and repair costs may decide the choice of the lining material. Different materials have been tried for lining to control seepage losses but almost all of them are quite costly. So, the use of these lining materials depends upon the cost of it. There are following techniques of reducing seepage from farm pond:

(a) ***Compaction:*** Pond areas having a high percentage of coarse-grained material can be made relatively impervious by compaction alone if the material is well graded from small gravel or coarse sand to fine sand, silt and clay. This technique for reducing seepage is the least expensive but its use is limited to the soil conditions described. The soil should have a depth of 20-25 cm and all rocks and tree roots should be removed. The loose soil should be composed under optimum moisture conditions into a dense and tight layer.

The thickness of the compacted layer should not be less than 8 inches for impoundments upto 10 ft in depth. Since seepage losses vary directly with the depth of water impounded, the thickness of the compacted layer should be increased by 1 inch for each foot of water over 10 ft. It will require compacting the soil in two or more layers, not exceeding 8 inches in thickness in case the water depth exceeds 10 ft.

(b) ***Clay blankets:*** Pond areas having high percentage of coarse-grained soils but lacking sufficient amount of clay to prevent high seepage losses by compaction alone can be sealed by blanketing. The blanket should cover the entire area over which water is to be impounded. It should consist of material having a wide range of particle sizes varying from small gravel or coarse sand to fine sand and clay in the desired proportions. Such material should contain approximately 20% clay particles by weight. The thickness of the blanket will depend on the depth of water to be impounded. The minimum thickness should be 12 inches for all depths up to 10 ft. It should be increased by one inch for each foot of water over 10 ft.

(c) ***Bentonite:*** The use of bentonite in well-graded coarse-grained soils reduces seepage in fine-textured colloidal clay by absorbing several times its own weight. At complete saturation, it will swell from 8 to 15 times its original volume. If bentonite is mixed with the coarse-grained material in correct proportion, then the mixture is thoroughly compacted and saturated, the particles will fill the pores in the material and make it nearly impervious. Bentonite is not adopted when a wide fluctuation in water level occurs, as it cracks when the reservoir dries. In case when the area is too wet, sealing operations should be postponed until moisture conditions are satisfactory. It is thor-oughly mixed with the soil to a depth of at least 15 cm with a rotary cultivator, disk, or similar equipment for compaction.

(d) ***Chemical additives:*** Excessive seepage losses often occur in soil having more than 30-40% fine fraction. Farm ponds situated in the foot hills and during the initial years after excavation. These soils have a relatively high permeability rate. The application of certain chemicals to these soils aggregates may result in collapse of the open structure and rearrangement of the clay particles. The chemicals used are called dispersing agents. For application of chemical treatment in the pond area it should contain more than 50% of fine grained material (silt and clay finer than 0.075 mm diameter) and at least 15% of clay fine than 0.002 mm diameter. The soil should have less than 0.5% soluble salt (based on dry soil weight). It is not effective in coarse-grained soils.

Kamra and Dhruvanarayana (1987) studied the chemical sealants for seepage control in farm pond at Central Soil Salinity Research Institute, Karnal. Alkali soil treatment was applied in the worst pond in which the seepage rates were as high as 18 cm/day (Table 9.8). However, over the years this treatment proved to be the most effective. It is seen from Table 10.8 that initially the reduction in the seepage rates included the effect of silting during monsoon, as a result of which seepage in control pond (T_1) also decreased. However, the seepage rates after 2 and 3 years are reported for the same month every year and they represent better comparative picture of the effectiveness of the lining treatments. Na_2CO_3 is more convenient to apply as compared to alkali soil which requires more labour for transportation and application. However, the seepage rates with Na_2CO_3 treatments had started increasing about 3 years after treatment as sodium was lost (Table 9.8). Therefore, a maintenance treatment @75% of the amount applied earlier reduced the seepage losses significantly. For sides, no maintenance treatment was applied and for the pond-beds the maintenance dose was added in ponded water. It could alternately be incorporated with the soil. So, on the basis of seepage rates it was observed that $Na_2\ Co_3$ is the best measuring for conservation of rainwater.

Table 9.8: The seepage rates of different lining treatments

	Seepage rates (cm/day)		
	Pond-1 (Control)	**Pond-2 (Sodium carbonate)**	**Pond-3 (Alkali soil)**
(a) Before treatment	10	14	18
(b) After treatment			
30 days	10	11	8
120 days	4.5	5.5	2.5
2 years	8.5	4.5	2.5
3 years	9.5	7.0	3.2
After maintenance treatment on bed (in T_2)	9.5	3.5	3.1

(Source: Kamra and Dhruvanarayana, 1987)

There are many other soluble salts that meet the requirement of a dispersing agent *viz.*, sodium polyphosphates, Tetrasodium pyrophosphate (TSPP) and sodium tripoly-phosphate (STPP) are most commonly used and effective. A laboratory analysis of pond soil is essential for selection of dispersing agents that will be most effective and its rate of application. The dispersing agent is mixed with the surface soil and compacted to form a blanket. For depths of water up to 8 ft, the blanket thickness should be at least 6 inches. For depths greater than 8 ft the thickness should be 12 inches and treated in two equal layers.

The soil moisture should be near optimum in the area to be treated for compaction to a depth of 12 inches. In case soil is too wet, treatment should be postponed and the soil is too dry water should be added by sprinkling. The dispersing agent should be applied uniformly over the pond area. Each chemically treated layer should be thoroughly compacted The treated area should be protected from puncture by livestock trampling.

(e) ***Calcareous soil lining:*** Calcareous soil is having low instillation rate and may be used for lining farm pond for reduction of seepage rate. In this technique, 10 cm thick layer of calcerous soil is spread in the farm pond at bottom and onsides. It is economic viable technique for reduction of seepage rate from the farm pond.

(f) ***Flexible membranes:*** For reduction of excessive seepage losses of flexible membranes of polyethylene, vinyl or butyl rubber may be used. Thin films of these materials are structurally weak but if kept intact, they are almost completely watertight. Agril films are less expensive and have better aging properties than watertight. Vinyl is more resistant to impact damage and is readily seamed and patched with solvent cement. Agri film can be joined or patched only by heat sealing. Butyl rubber can be joined or patched with special cement. The area to be lined should be drained and allowed to dry until the surface is firm and will support the men and equipment that must travel over it during installation of the lining. Certain plants roots penetrate vinyl and polyethylene end film. For avoiding, its surface should be sterlised with chemicals. Sterilisation is not required when butyl rubber membranes are used. The top edges of the lining should be anchored in a trench excavated completely around the lining. These trenches should have 8-10 inches deep and 12 inches wide. The films should anchor by burying 8-12 inches of the lining in the anchor trench and securing it with compacted backfill. The linings are usually laid in section or strips with a 6 inches overlap for seaming. Vinyl and butyl rubber lining should be laid smooth but in a loose state. Polyethylene should have upto 10% slack.

These thin films must be protected against mechanical damage if they are to be serviceable. All polyethylene and vinyl rubber membranes should be covered with soil of 6 inches thickness to protect against sunlight damage while butyl rubber membranes need to be covered for protection against livestock. In this case minimum cover of 9 inches should be used. The material used should be free from large clods, sharp rocks, sticks, and other objects that would puncture the film.

Even partial polythene lining on boulders layer for reduction in seepage losses from farm ponds was more effective in reducing seepage losses upto 60 to 48% at water levels in the range of 125-150 and 175-200 cm, respectively (Yadav *et al.*, 2006)

(g) ***Soil-cement/cement-concrete/asphaltic cement lining-*** Well graded sandy soils mixed with cement in the ratio of about 10:1 by volume and applied in the thickness of 7.5-15 cm reduce the seepage rates from reservoirs considerably. Proper watering and airing cement-concrete lining is the most effective but is very high capital intensive. Cement concrete of 1:3:6 mixture is applied to the thickness of 3-5 cm on the bed and sides of the pond after compaction. The maintenance cost of such lining is minimum and the life is long. Another lining mechanism consisting of use of mixture of sand and gravel bound together with asphaltic cement. The asphaltic cement lining also gives zero seepage results. But it is very costly, absorbs heat and encourages the growth of underlying vegetation, which may pierce the lining. Where stone slabs or boulders are easily available as in the case of western India, these slabs can be laid in cement mortar (1:3) directly on the bed and sides of the storage structures. Bricks may also be used but they are costly and need 1 cm plaster of rich cement mortar.

(ii) Evaporation losses

The losses from evaporation could be substantial in air and semi-arid regions due to high temperature, low rainfall, low relative humidity and high wind speed and water surface area. Several evaporation retardants such as fatty acids (cetyal alcohol), burnt oil, floating plastic membranes/foamed rubber sheets/polystyrene sheet, foamed wax blocks, asphalt-concrete blocks, plant residues like straw, saw dust, wood bark or soil residues, silicon, gum mixtures etc. for the control of evaporation from the water bodies are under consideration and evaluation. Maize stover (6 kg/sgm) is effective in reduction of evaporation from water surface bodies (DFRS, 2008). Tree plantation around water bodies break for reducing evaporation losses. Compartmentation techniques reduced the significant amount of evaporation losses. In this small resvoirs is constructed in big reservoirs. All these measures are very costly and are not very efficient. Still, the search for economical, efficient, practical and durable anti-evaporant is still going on. Evaporation losses can be reduced by modification in structure design and management. Reduction of surface area of the water body vis-à-vis depth of volume of water is most commonly measured for reduction of evaporation losses.

9.2.6 Percolation tank

Percolation tanks collect surface runoff in small storage structures to recharge groundwater. It is evident from studies that these structures release 50 per cent of the designed storage capacity to underlying groundwater to an influenced area of 1.5 sq. km. These are used for a impounding surface runoff to create small storage for inducing recharge the groundwater through percolation. They should be constructed at suitable places across *Nalla*s. The site should be relatively flatter *Nalla* reach, therefore, the slope of *Nalla* not exceeding 2 per cent. The catchment area required for construction of tank should be above 40 ha. To prevent breaching of the bund of the percolation tank, an emergency spillway may be provided. The emergency spillway should be constructed preferable in hard rock. However, the bund of the percolation tank require as to be adequately permeable.

For designing percolation tank, the top width of the bund should be above 1 m. A freeboard of one-third of impounding depth, but not less than 1 m. may be provided. The bund section should include a core wall and puddle trench on the lines discussed for earthen dams. The cut outlet type emergency spillway may be provided depending on the drainage of the area.

These tanks help in collection and storage of excess runoff water for recharging ground water in area. The recharged ground water can then be exploited for providing necessary life saving irrigations to the crops. Selection of site for a percolation tank should be based on the following considerations:

(i) Ponds should not be located in heavy soils or impervious soils. Sites with porous soils help quick percolation of stored water.

(ii) The selected site should not interfere with the existing other tanks and other water storage systems within a radius of about 1 km.

(iii) The selected site should have a number of irrigation wells m the zone of influence of the tank.

The following steps may be followed for designing a percolation tank:

(i) Appropriate selection of site.

(ii) Determination of catchment area from the topo-sheet or village man.

(iii) Determination of the required capacity of the percolation tank.

(iv) Based on the required capacity decide about the size of the tank to provide the required storage.

(v) Estimate the earth work involved and its cost of construction.

9.2.7 Sub-surface dykes

These structures are constructed below ground level to arrest the natural sub-surface groundwater flow. They conserve groundwater within aquifer itself. The design involves excavation of trench down to impervious layer and construction of a diaphragm wall. The crest of the dyke is kept few centimeters below ground level. Concrete plastered brick wall and clay are commonly used for construction of dykes.

9.3 Estimation of Groundwater Recharge

The groundwater budgeting for a particular basin is to be determined based on the guidelines recommended by the National Groundwater Estimation Committee. Following parameters are to be considered for determination of groundwater recharge in a particular basin:

i. Recharge through rainfall = Total area x water level fluctuation x sp. yield

ii. Recharge due to seepage from canals (unlined and lined)

iii. Return seepage from surface water irrigation

iv. Seepage from tanks /lakes/ ponds

v. Contribution from influent seepage normal monsoon rainfall

vi. Normalization of monsoon rainfall recharge =

$$\frac{\text{Normal monsoon rainfall}}{\text{Monsoon rainfall of the year}} \quad ...(9.36)$$

vii. Non -monsoon rainfall recharge

viii. Potential groundwater resources in flood prone areas

Total Annual Recharge (mcm) =

Recharge during monsoon

+

Non-monsoon rainfall recharge

+

Seepage from canals during non-monsoon

+

Return flow from irrigation during non-monsoon

+

Inflow from influent rivers

+

Recharge from tanks/lakes/ponds/percolation tanks ....(9.37)

Potential groundwater resources = Potential recharge in water logged and shallow water table areas + Potential GW resource in flood prone areas

Total groundwater resources = Total annual recharge + potential groundwater resource - (in mcm)

Annual utilizable groundwater = Total groundwater resources x 0.85

Resources for irrigation (mcm)

Gross annual groundwater draft (in mcm) through various structures

- Dugwells
- Dugwells with pumpsets
- Shallow tubewells
- Deep tubewells
- Others

Net annual groundwater draft (mcm) = 70% of gross annual groundwater draft/5

Groundwater balance = Annual utilisable groundwater resource for irrigation-net annual groundwater draft

Categorization of block/taluk based on the level of groundwater development:

Level of groundwater development as on date (%)

$$= \frac{\text{Annual net groundwater draft x 100}}{\text{Annual utilisable groundwater resource for irrigation}} \quad \ldots(9.38)$$

Yearly annual rate of groundwater development %=b ....(9.39)

$$\text{Net groundwater draft at year 5} = \frac{\text{Annual net }(1+5b)}{100} \quad \ldots(9.40)$$

Level of groundwater development at year 5 (%)

$$= \frac{\text{Net groundwater draft at year 5 x 100}}{\text{Utilisable groundwater resource for irrigation}} \quad \ldots(9.41)$$

Categorization of groundwater exploitation

Exploitation		Category
<65 %	=	White or safe
<65-85%	=	Grey or semi-critical
>85%	=	Dark or critical

References

Aggarwal, D. K. and Shukla, V. M. 1998. Development of graphical procedure for bench terrace design, A dissertation report submitted to G.B.P.U.A.T. Pantnagar (Unpublished).

Bhushan, L.S. 1979. Conservation bench terraces for rice in sub-humid climate. *Proc. Soil Science Society A*. 43(4), 754.

Das, D. C. 1976. Design of narrow base terracing (bunding Soil Conservation Digest 4(2). *Association of Soil and Water Conservationists*. Dehradun.

DFRS. 2008. Annual Report of Dryland Farming Research Station on DLA.

Kumar, Akhilesh and Das, Ghanshyam. 1988. Design of contour bunds under conditions of sediment deposition. *Journal of Agricultural Engineering*, I.S.A.E., Vol. XXX:1-7.

Rajora, R., 2002. Integrated watershed management, Field manual for equitable productive and sustainable development. Rawat Publications, Jaipur & New Delhi, p. 616.

Rao, B.K. 2001. Investigation on impact of rainfall distribution on runoff estimate using SCS curve number method. M.Sc. Thesis, Division of Agril. Engg., IARI, New Delhi.

Samara, J.S., Sharda, V.N. and Sikka, A.K. 1996. Water harvesting and recycling. *Indian Experiences*, ICAR Publication, New Delhi.

Sharma, K.K. and Bhattacharya, A.K. 2005. Development of models for designing water harvesting ponds in the rainfed areas. *Indian J. Soil Cons.*, 33(1): 27-30.

Singh, G. *et al.*, 1981. Manual of Soil and Water Conservation Practices in India. Central Soil and Water Conservation Research and Training Institute, Dehradun.

Singh, G.; Venkatararnan, Sastry, G.; and Josh, B.P. 1991. Manual of Soil and Water Conservation Practices Oxford & IBH publishing Co. Pvt.. Ltd. New Delhi.

Singh, J. and Dhillon, S.S. 1994. Physical determined of agricultural patterns. In: Agricultural Geography (2nd Ed.). Tata McGraw Hill Publication Co., New Delhi, pp. 60-70.

Yadav, R.P., Aggrawal, R.K., Arya, S.L., Tiwari, A.K., Singh, P., Prasad, R. and Bhattacharya, P. 2006. Partial polythene lining for reduced seepage losses from farm ponds, *Indian J. Soil Cons.*, 34(2): 118-122.

EXERCISES

9.1 What is the contour bunding?

9.2 Enlist the design parameter for contour bunding.

9.3 Describe the design procedure for contour bunding.

9.4 Determine the storage area required and the height of bund in medium soil having an average slope of 2%. The maximum expected rainfall during the 10 year recurrence interval is 5 cm and infiltration capacity is 40% of rainfall. The horizontal interval between bunds is 6 m and assume slope of seepage line to be 4%.

9.5 What is the bench terracing?

9.6 Discuss the selection criteria for an appropriate type of bench terraces.

9.7 Explain the design parameters of bench terrace and how these are determined.

9.8 Write short notes on the following:

(i) Temporary gully control structures

(ii) Check dam

(iii) Nalla bunding

9.9 What is the water harvesting pond?

9.10 Explain the steps involved in design of WHP.

9.11 How capacity of farm pond is determined.

9.12 Describe the procedure for design of farm pond.

9.13 What is the clear overfall weirs?

9.14 Explain the pipe outlet.

9.15 How drop inlet spillways is designed?

9.16 Write short notes on chute spillways.

9.17 Describe the measures to reduce seepage from farm pond.

9.18 What is the percolation tank?

9.19 How groundwater recharge is estimated?

9.20 Design a 150 m long bench terrace for land having an average slope of 20%. The soil is clay loam and terrace channel has the uniform grade of 0.5%. The maximum intensity of rainfall expected during the 10 year recurrence interval is 10 cm/hr.

9.21 Design a waste weir to discharge the excess runoff of 0-0 cum/sec. from a contour bund. Design the notch dimension of the drop structure required.

■■■■

Chapter 10

WATER LOSSES AND CONTROL

10.1 Introduction

In the design of any water harvesting structures for crop production under dryland ecosystem, two types of informations are essential: (i) water required for use and (ii) water that is likely to be lost through evaporation from free water surfaces, seepage from water harvesting structure or embankment, seepage losses from lined and unlined irrigation channels, evaporation from soil surface, transpiration losses from plant during the application of irrigation. The total water losses occurred from the water harvesting structures would be the sum of (i) and (ii). However, for interbasin transfer, watershed leakage should also be considered.

The evaporation losses may account upto 50% of total storage losses in open shallow reservoirs and 20% in deep reservoirs (Walton, 1969). High water losses including evaporation and seepage from the water harvesting structures is the bottleneck problem in recycling the scarce harvest runoff water during dryspell period to enhance productivity.

A hydrologist, engineer and agriculturist should have knowledge of monthly, seasonal, annual or long-term average yield that may be expected from any drainage basin to handle the water harvesting structure effectively. To enhance the water productivity a detailed knowledge of the water losses due to evaporation, seepage transpiration, interception etc. is pre-requisite.

For determination of water losses/abstractions for any portion of the drainage basin, the water which is artificially applied to the land should be added to the precipitation amount to determine the input to the system. Water losses *viz.*, interception, transpiration and evaporation for an irrigated area lumped together is usually called consumptive use of water. Consumptive use may be related to the growth of particular vegetation, it is then called consumptive use for that vegetation, when it is related to the whole of drainage basin then it is known as valley consumptive use. In case there is no watershed leakage then valley consumptive use can be taken as equal to total water loss.

Generally, it is not practicable to separate evaporation and transpiration and these two items are treated as a single item so far as the water balance studies of a drainage basin are concerned.

Evaporation from water harvesting structures *viz.*, lakes, reservoirs, etc., and from soil surface combined with transpiration from vegetation is called evapotranspiration or total evaporation. As this part of precipitation is not available as runoff, it may be termed as a loss.

10.2 Different Forms of Water Losses

There are following types of water losses which are countered:

(i) Interception loss: It occurs due to surface vegetation and rooftops, *i.e.* held by plant leaves.

(ii) Evaporation losses: It consists of evaporation from water surface, *i.e.* reservoirs, lakes, ponds, river channels, etc. and from soil surface, appreciably when the ground water table is very near the soil surface.

(iii) Transpiration: It occurs from plant leaves.

(iv) Evapotranspiration or consumptive use: From irrigated or cropped land. It is water used by crops.

(v) Infiltration: The entry of water into the soil at the ground surface.

(vi) Watershed leakage: It occurs when groundwater movement occurs from one basin to another or into the sea.

Different types of water losses are described as under:

10.3 Interception Loss

It is defined as fraction of precipitation intercepted by foliage (plant leaves, forests) and buildings and returned to atmosphere (by evaporation from plant leaves) without reaching the ground surface. Interception loss is high in the beginning of storms and decreases gradually. The loss is of the order of 0.5 to 2 mm per shower. It is evident from the Horton's mean curve of interception loss for different amount of showers (Fig. 10.1) is greater for light shower as compared to when rain is continuous.

$$\text{Effective rainfall} = \text{Rainfall} - \text{Interception loss} \quad \ldots(10.1)$$

However, if the rainfall is small then interception loss is large. While, it occurs with a constant value for large storms. The interception loss can be determined for any given storm by using the following relationship:

$$I_i = Vi + a.E.t. \quad \ldots(10.2)$$

where,

I_i = interception loss, (mm),

V_i = the interception storage whose value varies, depending upon the nature of vegetation, from 0.25 to 1.25 mm,

a = the ratio of vegetal surface area to its projected area,

E = the evaporation rate during precipitation, (mm/hr.), and

t = the duration of rainfall, (hours)

Its determination is essential for water-balance studies, while, for flood flow studies normally the contribution of interception losses is generally considered insignificant.

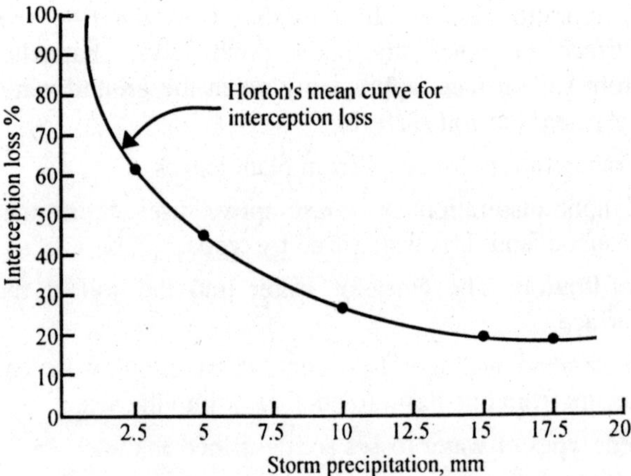

Fig. 10.1: Interception loss (Horton)

10.4 Evaporation

Evaporation is of great importance in hydro-meteorological studies and rainfed farming. It includes *viz.*, evaporation from water surface (farm pond) and from soils surface.

10.4.1 Evaporation from water surfaces

The rate of evaporation from the water surface is influenced by several factors *viz.*, radiation, air temperature, water temperature, wind velocity, atmospheric pressure, vapour pressure of the overlying air and surface area. The atmospheric pressure and salinity have a minor effect. The vapour pressure of sea water is about 2% less as compared to pure water at the same temperature or water bodies *viz.*, reservoir community tanks or lake, molecules of water vapour are convicted away from the water surface by molecular diffusion and wind into the atmosphere. This results in setting up of a gradient with water vapour content decreasing as the distance form the water body increases.

According to Dalton's law the rate of evaporation is proportional to the differences in vapour pressure at the water surface and in the atmosphere and is expressed by the following relationship:

$$E \propto (e_w - e_a) \quad ...(10.3)$$

This may be rewritten as,

$$E = K(e_w - e_a) \quad ...(10.4)$$

where,

E = daily evaporation,

e_w = saturated vapour pressure at the temperature of water,

e_a = water vapour pressure of the air (about 2m above), and

K = a constant.

A more general form of the Eq. (10.4) is written as,

$$E = K' (e_w - e_a) (a+bV) \quad ...(10.5)$$

where,

K', a, b = constants, and

V = wind velocity.

If higher the temperature and wind velocity then greater is the evaporation. While, greater the humidity and dissolved salts, smaller is the evaporation. The annual evaporation from irrigation tanks in South India is of the order of 160 to 180 cm. The highest evaporation occurs during the summer months of April and May due to higher air temperature, wind velocity and low humidity.

These losses are expressed with the basic water balance equation as stated below:

$$\Delta S = P + I - \text{losses } (E_o + ET_o + IC + R + D) \quad ...(10.6)$$

where,

ΔS = Change in soil moisture, (mm),

P = Precipitation, (mm),

I = Irrigation, (mm),

E_o = Evaporation form soil surface, (mm),

ET_o = Evapotranspiration, (mm),

IC = Interception loses, (mm),

R = Surface runoff, (mm), and

D = Deep percolation losses, (mm)

There are several methods available for determination of the evaporation from water surface. Each of the methods involves some assumptions and simplifications of the actual process. Therefore, estimated values by different methods may vary considerably. The most common methods of estimation of evaporation from reservoirs, farm pond, community tanks and lakes are as follows:

1. Pan evaporimeter,
2. Empirical formulae,
3. Water budget method,
4. Energy budget method, and
5. Combination method (combining energy budget and mass transfer/aerodynamic effect method-Penman's method).

10.4.1.1 Pan evaporimeters

Evaporation pans have been used for measuring evaporation from water bodies. A few commonly used pans are described as under.

US Class A Land Pan

This pan is having diameter of 1210 mm and depth of 225 mm (Fig. 10.2). It is the most widely used evaporimeter in developing countries. This pan is placed on a wooden platform above the ground to allow free circulation of air beneath the pan. The depth of water is maintained constant (18-20 cm).

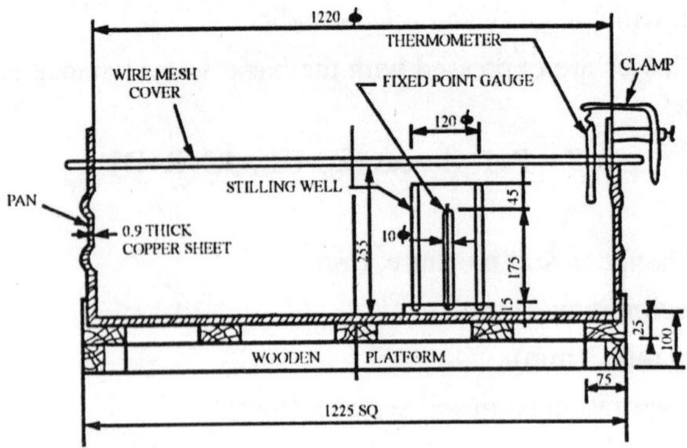

Fig. 10.2: US Class A Evaporation Pan

ISI (IMD) Standard Pan

This type of pan is having diameter of 1220 mm and depth of 225 mm. It is made of 0.9 mm thick copper sheet which is tinned inside and painted white on the outside (Fig. 10.3). The pan is placed on wooden grill 10 cm above the ground surface to allow free circulation of air beneath the pan. The pan is provided with a stilling well, fixed point gauge, and a thermometer. The fixed point gauge indicates that water level in the pan. It is maintained to a fixed mark in the pan. The observations are recorded once daily at 8.30 hours Indian Standard Time. A totalizing anemometer is usually mounted at the level of the instrument so that wind speed information if required may be obtained. The top of the pan is covered with a hexagonal wire mesh of galvanized iron to prevent falling in of the birds in the pan. The evaporation from the screened pan evaporimeters is 14% less as compared to that from unscreened pans. However, it makes the water temperature more uniform during day and night. The air temperature is found by reading a dry bulb thermometer kept in Stevenson screen mounted in the same enclosure of the pan.

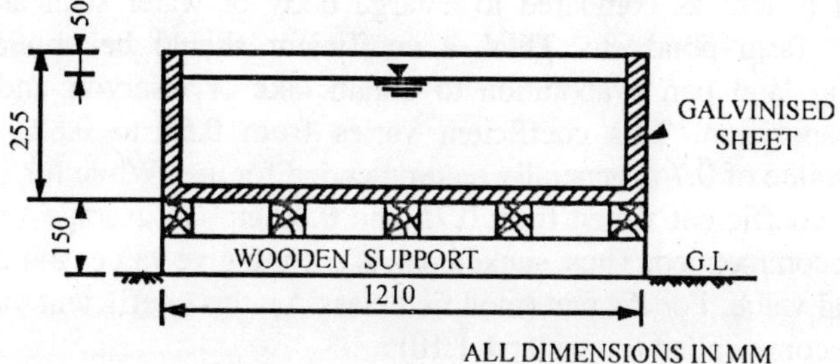

Fig. 10.3: IMD Evaporation Pan

Colorado Sunken Pan

It is installed near the water bodies to keep the same radiation and aerodynamic characteristics. These evaporimeters are known as sunken pan. This is having dimension of 920 mm in square shape and 420 to 960 mm depth. It is sunk into the ground such that 50 to 150 mm depth projects above the ground (Fig. 10.4). Water level is maintained almost at the ground level. This type of pan has the following drawbacks:

(a) Leakage of water cannot be detected.

(b) It is expensive to install, and

(c) It is difficult to keep the surrounding area free from dust and obstructions.

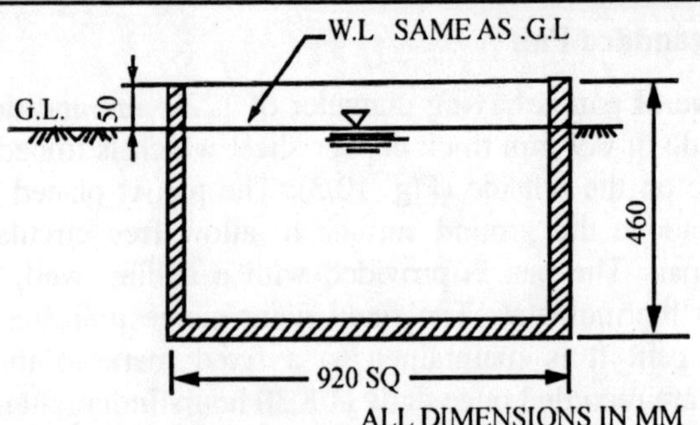

Fig. 10.4: Colorado Sunken Pan

Floating Pans

This type of pan is installed in the water bodies and is termed as floating pan. These evaporimeters are basically Class A pans and are mounted on floating barge in such a way that bottom and part of the tank are in contact with the water bodies surface.

It is evident from the results of the studies that evaporation from a land pan is less as compared to a large body of water such as lake, reservoir, farm pond etc. Thus, a coefficient should be applied for converting land pan evaporation to obtain lake or reservoir and farm pond evaporation. This coefficient varies from 0.60 to 0.82, on an average value of 0.7 is generally recommended for use. While for sunken pan, this coefficient varied from 0.75 and 0.86 and an average value of 0.78 is recommended. Thus, sunken pan is seen to give values more close to the real value. For ISI pan (modified class A), this coefficient value of 0.80 is recommended (range 0.64-1.10).

10.4.1.2 Evaporation formulae

A large number of evaporation formulae are available in literature. Most of these are based on Dalton's law and have the general form of the equation (10.4).

Out of the several evaporation formulae two are described below:

Rohwer (1931) developed the following relationship for determination of evaporation from water surface:

$$E_{day} = 0.771\,(1.465-0.000732B)\,(0.44+0.0733W) \times (e_w-e_a) \quad \ldots(10.7)$$

Mayer (1915) proposed the following equation for determination of evaporation from water surface.

$$E_{day} = C_m\,(e_w-e_a)\,(1 + W_9/16) \quad \ldots(10.8)$$

where,

E_{day}= the evaporation depth, (mm/day),

B = Mean barometric pressure at 0^0C (millibars),

W = mean velocity of the wind about 0.6 m above the ground (km/hr), and

W_9 = wind velocity at 9 m above the ground (km/hr)

C_m = Coeffiecient, and

e_w & e_a = As stated in equation (10.4)

While using Rohwer equation for determining evaporation from pan surface, coefficient 0.771 should be omitted. However, in case of Mayer equitation, the coefficient value (Cm) varied from 0.50 (small shallow tank, lakes etc.) to 0.37 (large deep lake). This is due to fact that large water bodies have greater capacity to store heat as compared to the shallow ones. General equations may be written and expressed as,

$$E = K (V_{sa} - V_a) (a + bW) \quad ...(10.9)$$

where,

k, a and b are empirical coefficients.

Evaporation losses from a reservoir varied from 150 to 200 cm of stored water annually in India. The maximum value may be 300 cm in arid zones of Rajasthan and Surashtra, the minimum value may be 50 cm in Jammu and Kashmir areas.

WMO recommends a network of evaporation stations as given in Table 10.1.

Table 10.1: Network of evaporation stations

Region	**Area served by one station (km^2)**
Arid	30,000
Humid temperature region	50,000
Cold region	100,000

10.4.1.3 Water budget method

It is not possible to have direct measurements of evaporation under field conditions. However, indirect determination of it on daily average basis, can be achieved with the help of storage equation as under:

$$E_{day} = I + P + \Delta S - O_g \quad ...(10.10)$$

where,

E_{day} = daily evaporation,

I = daily surface inflow,

P = daily precipitation (surface outflow),

Og = daily subsurface seepage, and

ΔS = the change in storage per day during a time interval (difference of storage S_1 at the beginning of time interval and storage S_2 at the end of time interval)

The units of all the parameters in Eq. (10.10) are either expressed in m^3 or mm depth over a reference area. It is assumed that all the parameters except E can be directly measured. This equation rarely gives reliable results because all the errors in measuring change in storage, gives precipitation, inflow and outflow are reflected directly in the computed value of the evaporation. It is extremely difficult to make accurate direct determination of ground water seepage which is estimated indirectly from measurement of ground levels, coefficient of permeability, etc.

10.4.1.4 Energy budget method

The energy budget for a reservoir or lake may be written according to law of conservation of heat energy within a water body as:

$$Q_n - Q_h - Q_e = Q_s - Q_v - Q_g \quad ...(10.11)$$

where,

Q_n = net radiation absorbed by the water body (or net heat energy received by the water surface),

Q_h = sensible heat transfer due to conduction to the atmosphere,

Q_e = energy used for evaporation
= $\rho L E_i$,

ρ = the mass density of water,

L = latent heat of vaporisation,

E_i = the evaporation (mm),

Q_s = increase in energy stored in water body,

Q_v = energy adverted into the water body, and

Q_g = the heat flux into the ground.

All the above mentioned quantities are expressed in calories/mm^2/day. It is evident that method based on a continuity equation (of energy)

similar to the water budget and determine evaporation as a residual to maintain a balance. Thus, an approximate budget is required since the energy values of inflow, outflow and storage with their respective temperature must be considered.

Conductive heat transfer to atmosphere (Q_h) cannot readily be determined. Bowen ratio (R), was conceived as a means of eliminating Q_h, from the energy budget equation. Therefore, Bowen ratio (R) is used to eliminate Q_h from the energy equation. This ratio is determined by the following relationship.

$$R = \frac{Qh}{Qe} \quad \text{...(10.12)}$$

or

$$R = 6.1 \times 10^4 \frac{T_w - T_a}{V_{sa} - V_a} \times P_a \quad \text{...(10.13)}$$

where,

P_a = atmospheric pressure, (mm of mercury),

V_{sa} = saturated vapour pressure, (mm of mercury),

T_{sa} = temperature of water surface, (°C),

T_a = tempeture of air (°C), and

Va = the actual vapour pressure, (mm of mercury).

Since it is not possible to determine the advective energy of a particular water body and temperature with the Eq. (10.11). This method gives satisfactory results when applied to periods less than a weak.

Solving Eq. 10.11, 10.12 and 10.13, we obtain as;

$$E_l = \frac{Q_n + Q_g + Q_s + Q_v}{\rho L(1+R)} \quad \text{...(10.14)}$$

10.4.1.5 Combination method

This method couples both the aerodynamic (wind and humidity) effects and energy (radiation) budget equations for determination of the evaporation from free water surfaces. This method requires climacteric data for determination of evapo-transpiration from water surface bodies. Several methods have been used for determination of evapotranspiration as discussed in Chapter 18. However, *Penman's* formula which is based on theoretical concepts is the one which is widely used in India, UK and in some parts of USA. The original Penman's formulae (1948) has undergone few changes since its first publication; the modified form of it is given by the following equation.

$$E_p = W.Q_{nl} + (1 - w).f(v)\,(V_{sa} - V_a) \quad \text{...(10.15)}$$

= radiation term + Aerodynamic term

where,

E_p = the potential evapotranspiration in mm/day,

w = the weighting factor for the effect of radiation $= \dfrac{\Delta}{\Delta + \gamma}$

Δ the rate of change of saturation vapour pressure with temperature and γ the psychometric constant. Value of w (for different temperature and altitudes) is given in Appendix A.

Q_{nl} = the net radiation in equivalent in mm/day, and is equal to difference between all incoming and outgoing radiations; *f*(v) = the wind related function:

$$fv = 0.27\left(1 + \frac{½}{100}\right) \quad \text{...(10.16)}$$

where,

V_2 = the mean wind speed at 2 m height, (km/day),

V_{sa} = the saturation vapour pressure at mean air temperature (mb) if vapour pressures are given in mm of mercury, multiply by 1.33 to find mb),

V_a = the mean actual vapour pressure, (mb),

1-w = the weighting factor for the effect of wind and humidity on E_p,

γ = 0.49 mm of mercury per °C.

Evaporation from free water surfaces of extended portions (such as lakes and reservoir) is determined by considering negligible heat storage capacity and is equivalent to potential evapo-transpiration. Potential evapotranspiration (PET) is defined as the water loss which will occur without deficiency of water in the soil for the use of vegetation. This definition is according to Thornthwaite. While, actual evapotranspiration is the real evapo-transpiration taking place in a particular situation.

Equation (10.15) consists of two terms radiation and aerodynamic, the relative importance of each term depends upon the climatic conditions. Weighting function has been assigned for relative importance. Aerodynamic term is usually less important than energy term when the weather conditions are calm. However, this term has great influence under windy conditions, and in arid regions. Mean daily climatic data should be used in Eq. (10.16). For calculating the wind velocity appendix may be used to determine the wind velocity at 2 m height.

The following relation may be used for wind velocity, upto a height of about 500 m above the ground level and is expressed as:

$$W_h = Ch^{1/7} \qquad \text{...(10.17)}$$

where,

C = constant,

W_h = the wind velocity at height h above the ground.

The net radiation Q_{nl} can be determined from solar radiation or sunshine hours or degree of cloud cover, temperature and humidity data with the following relation:

Q_{nl} = (Net short wave radiation)-(Net long wave radiation)

$$= Q_s \times (1 - r) - Q_{nl} \qquad \text{...(10.18)}$$

$$= Q_{nl} \times (Q_a\,(a+bn/N)\,(1-r) - \sigma^{T4}{}_{ak}\,(0.34-0.044\,(V_a)^{1/2}) \times (0.10 + 0.9\,n/N) \qquad \text{....(10.19)}$$

where,

Q = the amount of radiation received at the top of atmosphere. It is a function of latitude and time of year only and its values can be determined as given in Appendix A.

r = albedol (reflection) is a function of type of reflecting surface (and is determined on using Appendix A). It is equal to 0.25 for most crops. a = a constant which depends upon the latitude and can be determined from the relation:

a = 0.29 Cosθ. (a value of a = 0.25 is found acceptable for practical purposes). θ is the latitude.

b = 0.50 (average value),

n = the actual duration of bright sunshine in hours. This is measured by Campbell Stoke sunshine recorder and should be locally available.

N = the maximum possible hours of bright sunshine and hence depends upon the latitude of place, the value of N for different months and latitude is given in Table A-2.

σ = Stefan-Boltzman constant 2.01×10^{-9} mm/day,

T_{ak} = mean air temperature in degree Kelvin = 273 + °C,

V_a = actual mean vapour pressure in the air (mm of mercury),

Fig. 10.5 shows that how rQ_s represents part of Q_s which is reflected back by soil and crop and is lost to atmosphere. The value of *r* is a function of reflecting surface. Additional loss occurs when earth radiates back part of absorbed energy to the atmosphere as long wave radiation. This is greater than the down-coming long wave atmospheric radiation.

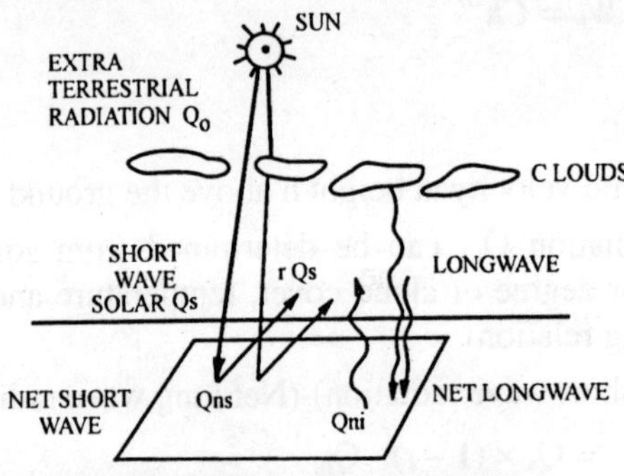

Fig. 10.5: Definition sketch

Example 10.1: (a) Determine the net solar radiation in mm/day for the following data and (b) potential evapotranspiration for a close crop. The following data were recorded:

Place: Situated in northern hemisphere

Latitude: 20°N; altitude 95 m; Month: July

T_{max}	:	33°C	Tmin	:	24°C
$R_{H\ max}$	:	80%	$R_{H\ min}$	:	30%

Actual duration of bright sunshine, n = 11 hrs.

Land is covered with close crops and r may be assumed = 0.25

wind velocity at 4 m height = 80 km/day (average over 24 hrs.)

(c) Effect of day night weather condition should be accounted for assuming adjustment coefficient = 1.02

Solution:

(a) Determination of Solar Radiation

Net solar radiation, Q_{nl} is given by the following relationship as given in Eq. 10.17.

$$Q_{n1} - Q_{ns} - Q_{nl} = Q_s (1 - r) - Q_{nl} \qquad \text{...(10.20)}$$

$= Q_a\,(a + b\,n/N)\,(1 - r) - \sigma\,(T^4_{ak})\,(0.34 - 0.044\,V^{0.5}{}_a)$
$\times (0.1 + 0.9\,n/_N)$

From Table A-1 Qa = 16.3 mm/day

(Lat: 20° N Month: July) a = 0.25*

b = 0.50

From Table 4.2 N = 13.2

$\therefore n/N = 0.833$

r = 0.25

$\therefore Q_{ns} = Q_a\,(a + b\,n/N)\,(1 - r) = 16.3\,(0.25 + 0.50 \times 0.833) \times (1 - 0.25)$
$= 8.15$ mm/day

Step II Determination of PET

$\sigma = 2.01 \times 10^{-9}$ mm/day

$$T_{mean} = \frac{T_{max} + T_{min})}{2} = \frac{(33+24)}{2} = 28.5^0\,C$$

$T_{ak} = 273 + 28.5 = 301.5^0K$

$\sigma T^4{}_{ak} = 16.61$

$$RH_{mean} = \frac{RH_{max} + RH_{min})}{2} = \frac{(80+30)}{2} = 55\%$$

From Appendix-A, V_{sa} @ 28.5^0C = 38.9 mb

$$V_a = V_{sa}\frac{RH_{mean}}{100} = \frac{(38.9 \times 55)}{100} = 21.4\,mb$$

$\therefore (0.34 - 0.44V^{0.5}) = (0.34 - 0.044 \times 21.4^{0.5}) = 0.14$ and

$(0.1 + 0.9\,n/N) = (0.1 + 0.9 \times 0.833) = 0.85$

$\therefore Q_{nl} = (16.61) \times (0.14) \times (0.85) = 1.98$ mm/day

$\therefore Q_{nl} = 8.15 - 1.98$ mm/day

* If one uses relation for computing 'a' as a = 0.29 cos θ where θ = latitude then a = 0.29 × cos 20° = 0.27

(b) Value of W read from Appendix-A for mean temperature of 20.5^0C=0.77

Correction for wind (From Appendix-A) = 0.88

$$\therefore f(v) = 0.27\left(1 + \frac{V_2}{100}\right) = 0.27\left(1 + \frac{0.88V_4}{100}\right)$$

$$=0.27\left(1+\frac{(0.88\times80)}{100}\right) = 0.46$$

$$(V_{sa} - V_a) = 38.9 - 21.4 = 17.5 \text{ mb}$$

Substituting in Eq. (11.15) we get,

$$E_p = (w.Q_{nl} + (1 - w) f(v) (V_{sa} - V_a)$$
$$= (0.77 \times 6.17 + (1 - 0.77)(0.46)(17.5)$$
$$= 6.63 \text{ mm/day}$$

(c) Adjusted value of $E_p = 1.02 \times 6.63 = 6.76$ mm/day

Example 10.2: Determine the net solar radiation in mm/day for the same data as given in Example 10.1 except that instead of relative humidity values (a) wet bulb and dry bulb values are given as under:

$T_{\text{dry bulb}} = 24°C$, $T_{\text{wet bulb}} = 8°C$

(b) dew point value is given $T_{\text{dew point}} = 30°C$

Solution:

$Q_{ns} = 8.15$ mm/day (as in example 10.1)

$\sigma.T^4_k = 16.61$ (as in example 10.1)

Wet bulb depression = 8^0C

(a) $T_{\text{dry bulb}} = 24^0C$

Against these values, the value of V_a is directly read from Appendix-A.

$$V_a = 12.8 \text{ mb}$$

$$(0.34 - 0.044\, V^{0.5}{}_a) = (0.34 - 0.044 \times 12.8^{0.5}) = 0.18$$

$$(0.1+0.9\times\frac{n}{N})=0.85 \text{ (Same as in Ex. 11.1)}$$

$$Q_{nl} = 16.61 \times 0.18 \times 0.85 = 2.54 \text{ mm/day}$$

$$Q_{n1} = Q_{ns} - Q_{n1} = 8.15 - 2.54 = 5.61 \text{ mm/day}$$

(b) $T_{\text{wet bulb}} = 30°C$

Appendix-A directly gives the values of V_a for T = 30°C

$$V_a = 42.4$$

$$(0.34 - 0.044\, V^{0.5}{}_a) = (0.34 - 0.044 \times 42.4^{0.5}) = 0.053$$

$$\therefore Q_{nl} = (16.61 \times 0.053 \times 0.85) = 0.75$$

$$\therefore Q_{n1} = Q_{ns} - Q_{n1} = 8.15 - 0.75 = 7.40 \text{ mm/day}$$

Example 10.3: Determine the daily potential evaporation from a reservoir situated in an area having data as given in example 10.1.

Solution:

The solution of this problem is similar to the method adopted for example 10.1 except that the value of albedo shall be taken as 0.05 (from Appendix-A); let this be designated by r_l.

$$Q_{ns} = (8.15)\frac{(1-r_1)}{(1-r)}$$

$$= 8.15 \times \frac{(1-0.05)}{(1-0.25)} = 10.32$$

$Q_{nl} = 1.98$ (as before)

$Q_{n1} = Q_{ns} - Q_{n1} = 10.32 - 1.98 = 8.34$

$W = 0.77$, $f(v) = 0.46$, and $V_s - V_a = 17.5$ (as before)

$\therefore$ potential evaporation, E_p

$= w.Q_{nl} + (1 - w).f(v).(V_{sa} - V_a)$

$= 0.77 \times 8.34 + (1 - 0.77)(0.46)(17.5)$

$= 8.27$ mm/day.

10.4.1.6 Measures for reduction of storage losses in farm pond

Seepage and evaporation constitute the main storage losses from tanks, ponds and reservoirs. To minimize these losses, storage volume should be maximized in relation to the exposed surface area.

10.4.1.6.1 Measures for reduction of evaporation from water surface bodies

Evaporation losses from large bodies of water such as water harvesting structures *viz.*, farm pond, community tank, lakes and reservoirs are a great economic loss. These losses assume alarming proportions in regions of low rainfall (arid zones) or dryland areas and it results into considerable loss of money. Therefore, this is bottleneck problem for large scale adoption of water harvesting structure and storage of harvested water by the individual farmers.

The following measures are recommended to control reservoir evaporation per unit of storage which help in increasing usable water supply.

1. Adopt the design of dam/reservoir site such that minimum surface area of reservoir is created per unit of storage. Thus, decreasing exposed surface area of water which results into less surface area by increasing depth.
2. Controlling the growth of water-loving weeds and plants like phreatophytes from the periphery of reservoir boundary.
3. Construction of wind breakers.
4. Establishment of wind breaks of tall trees *viz.*, Causerina on the windward side of the reservoirs. This will reduce the wind velocity and result into the reduction of evaporation.
5. Spraying certain chemicals or fatty acids and formation of films. By spreading a monomolecular layer of cetyl alcohol (hexadecanol) $C_{16}H_{33}OH$ over the reservoir surface (from boats)- a film is formed on the surface which is only 0.015 micron (approx.) in thickness. It is a polar compound and it has great affinity for water on one side (hydrophilic) and repels water on the other side (hydrophobic). The film will only allow precipitation from the top into it but will not allow water molecules to escape from it. This method is really effective when the wind velocities are less. If the wind velocity is more, it will destroy the thin film from the water surface and deposit it on the bank. However the film is pervious to O_2 and CO_2. About 2.2 kg (22 N) of cetyl alcohol is required to cover an area of 1 ha of reservoir surface. It is best suited for small and medium size reservoirs.
6. Compartmentation technique which allows flow of water. It results into reduction of temperature and evaporation place storage structures in such a way as the warmer surface water can be released by providing outlet.
7. Straightening the stream-channels, which reduces exposed area of the water surface (along the length) and hence evaporation is reduced.
8. Providing mechanical coverings like thin polythene sheets and shedding with an agrowaste to small agricultural ponds and lakes, it will reduce the water surface temperature.
9. Construction underground reservoirs because the evaporation from a groundwater table is very much less as compared to the evaporation from a water surface.

10. If the reservoir is surrounded by huge trees and forest, the evaporation loss will be less due to cooler environment.
11. Application of thin film of burnt oil (2 m thickness) on water surface.
12. It can also be reduced by floating box pieces of wax, wax rubber and light weight of concrete etc.

10.4.1.6.2 Measures for reduction of seepage losses

Application of appropriate sealant materials becomes necessary to reduce the seepage losses which otherwise have been found to increase with head, while minimizing the exposed area. Studies have shown that seepage losses are enormous (10 cm/day) in freshly constructed embankment cum dugout type of ponds (Sastry *et al.* 1982) and tend to stabilize after about 8 to 10 years. Grewal *et al.* (1982) reported that storage losses of 16 to 42 per cent of total stored monsoon rainwater have been reported with average seepage rates varying from 0.4 to 20 mm/day depending upon head of water, surface and sub soil strata in Shiwalik foot hills. Higher seepage rates are experienced in light textured soils with low bulk density. Plugging of pores with finer particles over a period of time has been found to reduce the seepage losses by 60 per cent in a span of about 10 years. Many sealant materials such as butyl rubber, polyethylene, asphalt, bitumen sheet, concrete, soil-cement, lime mortar, latex emulsion, polymers, sodium salts and cowdung have been tried to contain seepage losses. However, efforts are still needed to evolve the most economical and efficient sealant material under a given set of conditions.

Sastry and Mittal (1987) found that the lining with brick mortar has been found effective in reducing seepage losses by 20 per cent in Doon Valley. The seepage rates in newly dugout ponds increased from 0.28 cm/day to 10.4 cm/day as the head varied from 0.5 m to 2.0 m (Sharda *et al.*, 1986). Though LDPE lining was found effective, it gets punctured unless properly laid out and overlain with soil. Though Na_2CO_3 has been found to be the cheapest sealant materials, it requires frequent applications after 40-50 days periods.

Table 10.1 shows that calcareous soil plays an important role in reduction of seepage rate by 61.4-62% in 100% $CaCO_3$ and 52.6-55.8% by 100% $CaCO_3$ below 30 cm depth over control (without calcareous soil). It is seen that initially the reduction in the seepages included the effect of silting during monsoon, as a result of which seepage rate is reduced including control also. The small ponds are dugout in-situ where calcareous soil was present. If it is not present, then it may be transported from other site and laid in dugout pond in 5 cm. thick layer at bottom and also on sides. It does not require any maintenance cost as compared to other sealant materials.

Table 10.1: Seepage rate and seepage rate reduction under different formations of calcareous soil at different heads

Treatments	Seepage rate at different head ($cm/m^2/day$)			Seepage rate reduction (SSR) (%)		
	Head(m)			Head(m)		
	0-0.5	0.5-1.0	1.0-1.5	0-0.5	0.5-1.0	1.0-1.5
100% $CaCO_3$	1.05	1.06	1.08	61.4	61.90	62.10
<30 cm depth 100% $CaCO_3$	1.20	1.25	1.35	55.88	55.1	52.63
>30 cm depth 100% $CaCO_3$	1.60	1.65	1.75	41.18	40.70	38.60
Without calcareous soil (control)	2.72	2.78	2.85	-	-	-

[Source: Anonymous (2008-09)]

10.4.2 Soil evaporation

It is the loss of moisture through direct evaporation from soil surface and is also known as land evaporation. Soil evaporation is a function of same parameters which affect evaporation from free water surfaces coupled with a factor or known as evaporation opportunity. The evaporation opportunity is defined as the percentage that the actual evaporation from the land surface is of the rate of evaporation from the free water bodies, and its value ranges from zero (for thoroughly dry soils) to 100 per cent (for a bare soil immediately after a rain). It is expressed as,

$$\text{Evaporation opportunity} = \frac{\text{Actual evaporation from the land (soil) at a given time}}{\text{Evaporation from an equivalent water surface}} \times 100 \qquad (10.21)$$

Soil evaporation will continue at a high rate for sometime after the cessation of rainfall, then decreases as the ground surface starts drying, until a constant rate is reached which is dependent on the depth of the water table and nature of the soil in addition to meteorological conditions.

Nearly 25 to 50 per cent evapotranspiration of a crop includes evaporation from the soil surface. Evaporation is an important mechanism of water loss from the soil surface in the absence of plants or when they are still young and not covering the ground fully and from the area left between widely spaced row crops and trees. Vapour loss from the soil surface is controlled both by evaporative demand or 'sink strength' of the atmosphere and moisture transmissibility or hydraulic conductivity of the soil profile.

Soil evaporation is measured with lysimeters (tanks) filled with earth and with the surface almost flush with the ground (Fig. 10.6). The tank is usually having the dimension of 1.3 to 2.0 m^2 in square shape and 1+2 m depth. To measure the evaporation from a soil whose surface is within the capillary fringe, tanks are equipped with arrangement to maintain the water table at any desired elevation. The bottom is equipped with arrangements to drain the excess water from the soil mars in the tank. The difference between the rainfall and drainage is the soil evaporation. The soil evaporation is determined by weighing the tanks at fixed intervals of time and knowing the amount of water that was added in the soil or removed from it.

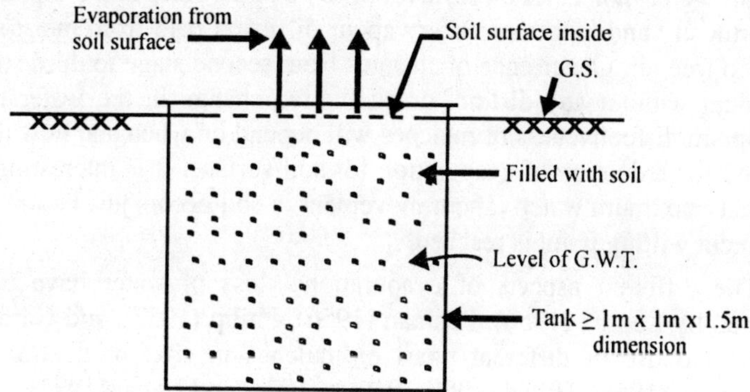

Fig. 10.6: Lysimeter for measurement of soil evaporation

10.4.2.1 Process of evaporation from soil

Process of evaporation from wet or irrigated soil takes place in three different successive stages during drying:

First stage: In this stage soil surface is moist and capillary conductivity or water transmissibility of the soil below the surface is not limiting, evaporation rate remains more or less constant and is governed by evaporability of the atmosphere. Soil evaporation is approximately equal to potential evaporation or open pan evaporation under similar conditions and it continues as long as the moisture content of the soil in the surface is within 90 to 95 per cent of field capacity.

Second stage: During this stage evaporational loss of water declines rapidly due to drying of soil surface and reduction of hydraulic conductivity. Intrinsic factors are more effective than atmospheric factors because they influence moisture flow to surface for longer period. The mechanism of moisture flow becomes more complex, especially under temperature gradients. As the soil progressively dries, instead of liquid flow vapour flow and distillation-condensation processes in the top layer become important. Vaporization occurs mainly a few cm below soil surface and the rate of water loss from the soil is influenced by vapour diffusion through the dry layer. Turbulent air flow continues to be of importance in the beginning of the second phase, especially in loose and cloddy structured soils. The rate of diffusion of water vapour through the soil is proportional to the square of the effective porosity, regardless of pore size.

Third stage: In this stage, soil evaporation occurs at extremely slow and nearly constant rate of water loss, influenced by strong absorptive force for water molecules on surfaces of dry soil particles in the top layer and tortuosity and length of water vapour diffusion path from moist soil layers to free air. Occurrence of changes from second stage to third stage is gradual without any distinct or qualitative change in the water loss mechanism. Effectiveness of mulches will depend on when and how they are used for reduction of evaporation for soil surface. It is interesting to note that maximum water vapour movement in soil occurs just before the permanent wilting point is reached.

The different aspects of evaporational loss of water have been reviewed by Lemon (1956), Penman (1956), Philip (1957), and Gardner (1960), and use of different types of mulches in crop production by Bhatia *et al.* (1966), Hilled (1968), Wendt (1971) and Unger (1975).

Unsaturated flow

When wet soil dries then larger pores are emptied of water the contribution of the hydraulic head or the gravitational component to total potential becomes progressively less important and the contribution of metric potential (Ψm) becomes more important. The effect of pressure is negligible because of the continuous nature of the air space. The osmotic potential is negligible and does not affect the potential gradient. Thus, soil moisture moves under the influence of the gravitational force and the force due to the difference in capillary potential (head) in the unsaturated zone. In horizontal movement, only metric potential (Ψm) influences. For downward movement, capillary and gravitational potential acts together, while, in upward movement metric gravitational potential opposes each other. The capillary flow is governed by Darcy's law as,

$$Q = k(\theta) ia \quad \text{...(10.22)}$$

$$i = \frac{\Psi_m}{l} + \frac{\Psi_g}{l} \quad \text{...(10.23)}$$

where,

Q = capillary flow,

K(θ) = coefficient of unsaturated permeability,

i = sum of the gradients of capillary potential (Ψ_m)and gravitational force (Ψ_g),

A = area of cross section of capillary flow,

ΔZ = difference in elevation of the two points A and B (if the flow is from A to B),

Δh = difference in capillary potential of the two points A and B, and

l = length of travel.

If the capillary potential becomes greater than the gravitational potential there is upward movement of soil moisture resulting in loss of water due to soil evaporation. When the flow is upward, the gravitational force must be subtracted from the capillary potential component. For horizontal flow the gravitational force is zero. For downward flow the two motivating forces act in the same direction. The coefficient of unsaturated permeability increases as the soil moisture increases. Movement of unsaturated flow ceases in the sand at lower tension as compared to finer textured soil and water film loose continuity sooner between the larger particles. In the 'moist range' the range of unsaturated flows in different textured soil is the following order:

Sand < loam < clay

10.4.2.2 Measures for reduction of evaporation from soil surface

The following measures may be recommended for reduction of moisture loss due to evaporation and transpiration found to be useful.

(a) Growing early maturing adaptable crop varieties with a deep and ramified root system and with a reduced number, size and horizontally orientation of leaves; such crop varieties should have wider adaptability with respect to the ever-dynamic thermic and hydrologic conditions of the atmosphere and soil, the chemical, physical and biological capacities of the soil and the higher potential of yield with moderate stability,

(b) Maintaining optimum plant population per unit area,

(c) Sowing crops either in dry soil anticipating rainfall suitable for early crop establishment with the first shower, and subsequent growth and development with subsequent rainfall, or in optimum soil moisture but with a minimum expenditure of seasonal moisture for land preparation and sowing,

(d) Keeping the field free from weeds,

(e) Adopting mixed or intercropping to utilize the slow growth phase of wide spaced crops, to restore soil fertility and to check soil and water loss,

(f) Sowing time and correct seed rate and also direction of sowing,

(g) Using mulches,

(h) Using agri-chemicals.

(i) Anti-evaporates for reduction of evaporation from the soil surface.

10.4.2.3 Use of mulches

A mulch is any material like straw, plastic, pebbles, dust and weeds & plants etc. which are spread on soil surface that to control evaporation and enhance water availability to crop plants and also control erosion the practice of applying mulches to soil is possibly as old as agriculture. Mulches are suitable where hand labour was readily available but impractical when mechanized agriculture is in practical. The current trend is to utilize crop residues as mulches on the areas where grown or to use transported and manufactured materials as mulches for some high value crops. Mid season drought during growing period is one of the main cause of low yields of rainfed crops. Mid season correction through various mulches and anti transpirants increase crop yields (Let *et al.*, 1974).

Reasons for using mulches

Mulches used for water conservation and erosion control are the most important for dryland agriculture. Other reasons for mulching include: soil temperature moderation, soil nutrient effects, soil salinity control, soil structure improvement, crop quality control and weed control.

Effect of mulches on soil properties and conditions

Several soil properties and conditions are influenced by mulching directly or indirectly. Among those influenced are:

Soil water	-	Through controlling by runoff, increasing infiltration and thereby decreasing evaporation and weed control.
Soil temperature	-	Through trapping radiation shielding, heat conduction and thereby produce cooling environment.
Soil nutrient	-	Through addition of organic matter, differential nitrification and mineral solubility.
Soil structure and biological regime	-	Through additions of organic matter additions, microbial and soil fauna populations and plant root distributions.
Soil erodibility and salinity	-	Through leaching and evaporation control.

Dust mulch created by frequent cultivations have the little influence in reducing evaporative loss of water on light textured soils. This is because by the time the field is ready for cultivation sufficient amount of moisture from the surface soil had already been evaporated. Cultivation followed by rolling or compacting the soil has been found to be more effective in reducing evaporation loss of water (Heinonen, 1971). However, cultivation decreases soil moisture loss by evaporation on deep clayey soils in which large amount of moisture evaporates from the cracks, by reducing and filling the cracks that have formed with the loose dry soils from the surface.

It is evident from the results of the study that mulching influenced the grain of crops and water storage in the soil profile under varying rainfall situations at two locations in Punjab. Mulching is found helpful to increase grain yield of maize and significantly also increase water availability to plants as compared to control in standing maize (Table 10.2). However in a high rainfall condition, mulching decreased the maize yield due to partial anaerobiosis of soil which restricted plant growth, reduced availability and uptake of nutrients and water. The mulched plots had invariably higher profile stored moisture in wheat sowing, more so in the low rainfall situation.

Table 10.2: Effect of mulching on maize yield, profile water storage at wheat sowing and wheat yield under variable rainfall conditions in the Shiwalik region

Rainfall situation	Rainfall (mm)			Maize grain yield (t ha^{-1})			Profile water storage (mm in 180 cm)			Wheat grain yield (t ha^{-1})		
	Kharif	*Rabi*	*Total*	*Mulched*	*Un-mulched*	*CD (0.05)*	*Mulched*	*Un-mulched*	*CD (0.05)*	Mulched	*Un-mulched*	*CD (0.05)*
Low rainfall	350	89	439	3.36	2.86	0.31	271	170	40	2.56	1.62	0.46
High rainfall	797	119	916	2.86	3.85	0.47	225	2.9	NS	2.47	2.33	NS

[Source: Sur *et al.* (1995)]

Example 10.4: Determine the PET and irrigation requirement for a crop which grows during the months: November, December, January and February in a locality situated in 20^0 North latitude. The mean monthly temperatures for the four months are 16^0, 12^0, 10^0 and 14^0C, respectively. The effective rainfall for the months is 2.5, 3.0, 3.2 and 2.2 cm, respectively. Assume water application efficiency as 60% and the value of *k* for the crop as 0.70.

Solution:

The solution is carried out in the Table 10.3.

$$U = k.F = 0.70 \times 42.17 = 29.52 \text{ cm}$$

$$\therefore \text{Irrigation requirement} = \frac{U - \sum P_e}{\eta} = \frac{29.52 - 10.9}{0.6} = 31.03 \text{ cm.}$$

Table 10.3: For Example 10.4

Month	Mean monthly temp. (0C)	Monthly percent of annual day. time hours. p	Effective rainfall (cm)	Monthly consumptive use factor
1	2	3	4	5
Nov.	16	7.58	2.5	11.74
Dec.	12	7.66	3.0	10.45
Jan.	10	7.74	3.2	9.42
Feb.	14	7.25	2.2	10.56
Total			10.9	42.17

This method uses only the pan-evaporation data and was developed by Hargreaves and Christiansan. Evapotranspiration or consumptive use is related to pan-evaporation *Ep* by the following relation:

$$E_T = kE_P \quad \text{....(10.24)}$$

Where the coefficient *k* is found to vary with the stage of crop growth. The values of k for various stages of crop is given in Table.

10.5 Transpiration

All moisture is discharged principally in the form of water vapour. Unlike evaporation which continues throughout day and night, transpiration is confined to daylight hours. Transpiration varies throughout 24 hours of the day. Transpiration is mainly affected by temperature, sunlight, available moisture wind, and stage of plant development. In the study of water balance of a drainage basin, both evaporation and transpiration are taken us a single item, because it is not practical to separate the two.

Different methods are used for the measurement of transpiration and one of the widely used methods is by closed phytometer. It comprises of a closed water tight tank with sufficient soil for plant growth with only the plant exposed; water is applied artificially till the plant growth is complete. The equipment is weighed in the beginning (W_1) and at the end of the experiment (W_2). Water applied during the growth (w) is measured and the water consumed by transpiration (Wt) is determined by the following relationship:

$$W_t = (W_1 + w) - W_2 \quad \text{...(10.25)}$$

The experimental values (from the protected growth of the plant in the laboratory) have to be multiplied by a coefficient to obtain the possible field results.

Transpiration ratio: It is the ratio of the weight of water absorbed (through the root system), conveyed through and transpired from a plant during the growing season to the weight of the dry matter produced exclusive of roots and expressed as,

$$\text{Transpiration ratio} = \frac{\text{weight of water transpired}}{\text{weight of dry matter produced}} \quad \text{...(10.26)}$$

The values of transpiration ratio for different crops vary from 300 to 800 mm and for rice it varies from 600 to 800 mm and on an average value is 700 mm.

Evaporation losses are high in arid regions where water is impounded, while transpiration is the major water loss in humid regions.

10.6 Evapotranspiration

The loss of water by evapotranspiration is an important factor in water resources. It is defined as amount of cropped (or irrigated) land due to evaporation from the soil and transpiration by the plants or used by the plants in building up of plant tissue and it is also known as consumptive use. It is the combined loss of water due to evaporation and transpiration. Potential evapotranspiration (E_{pt}) is the evapotranspiration from the short green vegetation when the roots are supplied with unlimited water covering the soil. It is usually expressed as a depth (cm, mm) over the area.

Evapotranspiration actually taking place in a given situation is known as actual evapotranspiration (AET). Potential evapo-transpiration can be considered equivalent to the evaporation from thin water surface of extended proportion. Different empirical formulaes are used to determinent PET and are discussed in Chapter 15.

10.6.1 Measures for reduction of transpiration

There are following measures recommended for evaporation and evaprotranpiration control:

1. Use of anti-transpirants or transpiration suppressants such as hydroxylarnine hydrochloride, phenyl mercuric acetate (PMA), silicone, cetyl alcohol, satiric acid, chiormequat chloride and alachior.
2. Application of plant modifiers or growth retardants such as abscisic acid, a-NAA, phosphon, daminozide, chiormequat chloride, mepiquat chloride, ethephon, diphenyl, alkenyl succinic acid, 2, 4-dinitrophenol, 8-hydroxyquinoline, chior-fiurenol, methyl ester, DMAS, TIBA, MH and CMH.
3. Application of desiccants or defoliants such as 2,4-D, paraquat, F12S04, Na-chlorate, diquat, dinoseb, metaxuron, ethephon. glyphosate, arsenic acid, Mg-chlorate, ametryn, endothal, Nachioroacetate, Na-pentachlorophenate, TCA, CCC, chi or (u reaol, N-dimethyl morpholine chloride and 2,3-dichloroisothia-zole.5-carboxylic acid.
4. Crop ripeners such as picloram, ammonium isobultylate, glyphosine, bromacil, CCC, tetrahydrobenzoic acid, cacodylic

acid, cycloheximide, mefluidide, cetyltrimethyl ammonium bromide, vanillin, carbamyl urea, isoaureomycin, penicillamine, disugran, isobultanol, paraquat, asulum, anisoniycin, enclothal. 2, 3-6 TBA, fluoridamide, bacitracin, mineral oil, ripenthal and desugran.

5. Anti-evaporants like synthetic or chemical mulch, for instatnce Encap Esso, plastic emulsion, Kaolin water.

10.6.2 Use of antitranspirant

An antitranspirant is any material applied to transpiring plant surface to reduce the water loss form the plant. Thus, any substance that will reduce the vapour pressure gradient or will increase stomal resistance to water vapour diffusion will act as an antitranspirant. Approximately 99% of water taken up by the plant roots is transpired to the atmosphere through stomal pores in the leaves. The stomata-bearing leaf surface is the most strategic sites for antitranspirant application. Antitranspirant such as phenylmercuic acetate and certain alkenlsucceinic acid act by inhibiting stomatal opening; film forming antitranspirant produce and external physical burner to retard the escape of water vapour from plants; silicones penetrate the leaf and may act directly on the wet cell walls; and white reflecting materials lowers leaf temperature and reduce vapour pressure gradient from leaf to atmosphere. Although reduction in transpiration is the primary goal of applying antitranspirant, that goal may be sought for different reasons:

(i) Conservation of water supplies when water is scarce resource and costly, and

(ii) Improved plant performance via an increase in plant water potential.

10.6.2.1 Types of antitranspirants

There are following types of antitranspirants and their mechanism as described below:

1. Inhibiting stomal opening: These anti-transpirant reduce stomal opening like phenylmercuric acetate, some alkenyl succinic acids and abscisic acid reduce transpiration by inhibiting stomatal opening when applied in millimolar and micromolar concentrations. In this inhibition of stomatal opening inhibits both the processes of transpiration and photosynthesis because diffusion of water vapour and CO_2 takes place through the same stomatal pore. But the extent to which photosynthesis is reduced in comparison to transpiration, will depend upon whether the chemical affects the site of photosynthesis in the mesophyll and whether mesophyll resistance is higher or less than epidermal and boundary resistances in the Co_2 pathway.

2. Increasing resistance to water vapour: These are film forming substances *viz.*, fatty alcohols and waxy or plastic emulsions which produce an external physical barrier that increases the resistance to water vapour diffusion. However, their effects on photosynthesis relative to transpiration will depend upon the native resistance of film to water vapour and CO_2 diffusion. Spray with film forming substances such as natural rubber, polyethylene, cellulose acetate, polyvinyl chloride and cellophanic were found to reduce photosynthesis more than transpiration because resistance to CO_2 diffusion was much higher than water vapour (Waggoner, 1966).

3. Reducing incident of solar radiation: These are substances such as silicones which on spraying penetrate through stomata and may reduce water loss from wet cell walls. But white substances like hydrated lime, chalk and kaolin when sprayed on drying reflect a part of incident solar energy, thereby, reduce the leaf temperature and vapour pressure gradient from stomatal cavity to the atmosphere, this results in reduced transpiration (Ray, 1968). Film forming and stomata closing type of antitranspirants on the other hand increase the leaf temperature (Patil, 1974) by reducing the evaporative cooling.

Table 10.4: Effect of antitranspirants on water use efficiency of rapeseed (*Brassica compestris* var. *Brown Sarson*)

S. No	Treatment	Water use efficiency 1972-73		(kg/ha – cm water used) 1973-74	
		Grain	Total dry matter	Grain	Total dry matter
1.	Control (water spray)	6.6	22.0	16.8	55.8
2.	P.M.A. one spray at 50 days @10x10^{-4} M.	7.4	23.8	18.1	60.6
3.	P.M.A. one spray at 70 days @1x10^{-4} M.	7.8	24.6	20.4	67.0
4.	Kaolin @6% suspension in water*	7.8	25.3	18.9	64.7
5.	Mobileaf @ 1:10 (v:v) commercial wax emulsion	8.5	25.8	20.5	72.3

*0.2 per cent emulsion of teapot was used in kaolin suspension as spreader sticker

Patil (1974) reported that when rapeseed (*Brassica compestris* var. *Brown Sarson*) leaves were sprayed with antitranspirants like phenylomurcuric acetate (PMA), kaolin and mobileaf (a wax emulsion of Mobil Oil Co. of U.S.A.), the seed yield as well as water use efficiency (in terms of seed yield and total dry matter) increased (Table 10.4). The possibility of using anti-transpirants under different situations and on different types of plants and crops was reviewed by Davenport and Hagan (1975).

The imperfect coverage, usually met with sprays of antitranspirants (Fisher and Lyon, 1972; Albrigo, 1972), may cause higher transpiration than unsprayed or untreated leaves (Brouwer, 1961), because partially covered leaves remain turgid and the stomata remain open for a longer time.

10.6.3 Use of growth retardants

Application of certain chemicals like (2-chloroethyl) trymethyl ammonium chloride (ccc or cycocel) and 2, 4-dichlorobenzyl-tributyl phosphonium (phosphon) etc. when applied plants to reduce their height and vegetative growth. Such types of chemicals maintain for a long time the turgidity of leaves during stress and reduce lodging in cereals. It was observed that when bean plants treated with CCC and phosphon once through the soil remained turgid and survived longer when subjected to moisture stress by withholding irrigation and produced more fresh and dry weight (Halevy and Kessler, 1963). But yield differences were not there when adequate soil moisture supply was maintained near field capacity through irrigation. Livne and Vaadia (1972) observed that the moisture stress tolerance of plants treated with growth retardants was due to the interference of these chemicals in biosynthesis and a reduced level of giberrellic acid that reduced stomatal opening and lowered transpiration. However no reduction was observed in transpiration rate per unit treated leaf area with growth retardants (Plautetal, 1964).

The yield was increased due to CCC treatments under conditions of even no or negligible lodging (Pinthus and Rudich, 1967, Humphries, 1968 a b and Papadakis,1972). Similarly, in case of wheat, Lowe and Carter (1970) and Philpitts (1972) attributed higher yields obtained with CCC due to the larger number of grains per Year head. Srivastava *et al.* (1968), noted that foliar applications of CCC increased significantly the grain yield of wheat. Gill *et al.* (1973) observed highest grain yield of oats when sprayed with CCC at the rate of 2 kg active ingradient/ha. In cotton (*Gossypium arboreum* L.) highest raw cotton yield was obtained by spraying cycocel at the rate of 0.2 kg a.i./ha (Singh and Singh, 1970).

Application of CCC as growth retardants increases root growth at all depths (Hanus, 1967; Singh and Singh, 1970) at the expense of vegetative growth. Deep and prolific root system of plants enables them to draw more water during the moisture stress period from the soil and maintain turgidity of leaves for a longer time consequently with more production and higher water use efficiency. There is little evidence that their beneficial effect is due to a reduction in transpiration rate of treated plants (Arnon, 1975).

10.7 Depression Storage

Before initiation of water flow as surface runoff, all the depressions on the ground must be filled. The water required to fill the depression before initiation of runoff is known as the depressions storage. This water could also be grouped under the category water loss as this water is also not available as runoff. Depression storage loss and interception loss coupled together are sometime known as *initial loss*. There is no general expression for determination of depression storage because of the great number of factors which influence it. Some of the more important factors are antecedent precipitation, type of soil, slope of the catchment. Antecedent precipitation is seen to have a pronounced effect on the loss of water due to depression storage. As a rough guide, depression storage loss for sand, loam, and clay may be taken as 0.50 cm, 0.40 cm and 0.25, respectively in case of intense storms.

For hydrological studies relating to flood, separate consideration of interception and depression storage is combined in terms of depths of precipitation and is deducted from the gross precipitation falling over the drainage basins.

10.8 Watershed Leakage

Every drainage basin is surrounded by a topographic divide that fixes the area from which the surface run off is derived. Similar to this, there is an underground or phreatic divide which demarcates the boundary of the area from which the groundwater is fed to each stream system (Fig. 10.7). Watershed leakage occurs when the topographic divide and phreatic divide do not coincide. It is equal to the ground water flow from the area between them. The watershed leakage always moves across the topographic divide.

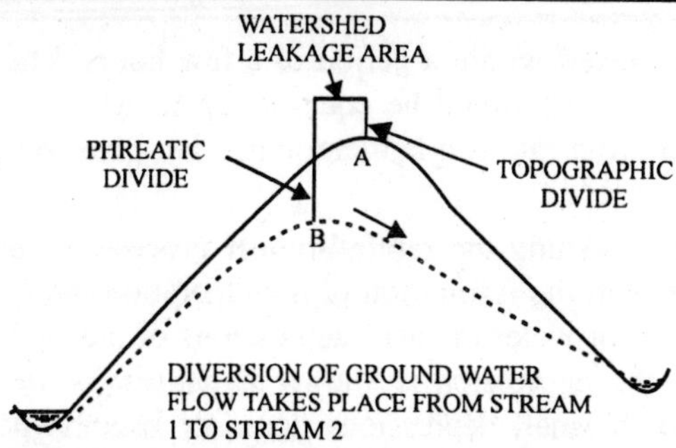

Fig. 10.7: Sketch illustrating watershed leakage

Generally, watershed leakage is considered relatively unimportant as compared to other losses, because the watershed losses from one drainage basin, are balanced by accretions from another. Thorough knowledge of the geological formations under the drainage basin in question usually is the best guide to the possibility of watershed leakage that may occur.

10.9 Infiltration

It defined as the movement of water from soil surface into the soil. The actual rate at which water is entering the soil at any given time is known as infiltration velocity. The infiltration rate decreases with passage of time. The rate of decrease is rapid initially and it trends to approach a constant value and nearly constant value after elapsed from start of irrigation is known as the basic *infiltration rate*.

This is very important element of hydrological cycle as greater portion of precipitation over the earth moves into the soil. The portion of the precipitation will first replenish the soil moisture deficiency and may bring it to field capacity. If the field capacity has been satisfied, then the excess water moves in downward direction and ultimately becomes a part of groundwater. It is one of the important soil characteristics influencing the irrigation. The rate at which the water may move into the soil is a function of many variables such as soil moisture, vegetative cover, temperature, compaction due to rain, etc. The maximum rate at which a soil in a particular condition can absorb water is called its *infiltration capacity* and is commonly denoted by f. *Accumulated infiltration* is the total quantity of water enters that the soil in a given time is also known as *cumulative infiltration*. Many factors influences the infiltration capacity, which varies with respect to time during the period of rainfall, and achieves an almost constant value of infiltration rate (f_c).

Generally, it achieved within a period of a few hours. The total amount of unfiltered water (*F*) would be equal to $\Sigma f.\Delta t$, where, Δt denotes the time step. Infiltration rate can he determined by measuring rainfall and run off.

It is useful to study the rainfall-runoff process. If the intensity of rainfall is more than the infiltration plus evaporation rates. It results into start of collection of water in small depressions of the region. The water required to fill the depression is known as *depression of storage* (V_d). The water held in small depressions does not become part of surface runoff but may contribute towards both infiltration and evaporation for a time period when water is held in depression. Some amount of rainfall is also intercepted by plants, rooftops, etc., before it reaches the ground. The water stored in this *way* is called *interception storage* (V_i). However, almost all of it is eventually evaporated back to atmosphere. The sum of the interception storage and depression storage is known as initial abstraction or loss (I_a) and is expressed as,

$$Ia = V_i + V_d \qquad \text{....(10.27)}$$

After the depression storage is filled, and if the rainfall continues and its intensity is greater than the infiltration capacity, then the water shall start accumulating on the surface as detention storage (V_{sd}). Rainfall leading to such a situation is called *rainfall excess* (P_e). It creates a sufficient head which result into overland flow to stream channel.

The various elements of rainfall-runoff cycle are shown in Fig. 10.8. This figure does not show the detention storage and channel storage. Thus these are part of rainfall accumulated prior to the initiation of direct surface runoff. However, these two are usually ignored in the study of infiltration process in natural drainage basin. No great amount of error is likely to be introduced by this assumption. The water which reaches by channel is this mentioned process is known as surface runoff. During rain water has also been infiltrating into the soil. The rate of rainfall in natural storms varies during a storm; and infiltration capacity also varies with respect to time. Thus, steady state conditions are seldom achieved in the overland flow process. The detention storage shall go on building up till the sum of rate of infiltration and rate of out flow by overland flow becomes equal to supply rate by excess rainfall A steady state condition is achieved when infiltration would be equal to rainfall excess minus surface runoff (Fig. 10.8).

Therefore, surface runoff can be determined using the infiltration approach. Under certain idealized conditions infiltration capacity curve for small and large watersheds can be obtained. This approach has practical difficulties in making use of this infiltration as given below:

1. It is a well known fact that time-intensity pattern of a storm is rarely uniform across a watershed. The infiltration capacity curve is a function of time and may also vary at various locations within the watershed depending upon antecedent moisture, cover, and soil. In that case rainfall intensity curve may fluctuate above and below the infiltration curve. Infiltration approach only helps to determine surface runoff and groundwater flow has to be ascertained by some other methods.
2. Another factor which hampers the application of the infiltration approach (except for small drainage basin). The ariel variation of rainfall rates hampers as runoff does not occur over the whole basin. The variation of rainfall rates over various parts of drainage basin results in periods when no runoff occurs with consequent decrease of infiltration capacity.
3. For a large watershed has surface storage element such as depression storage, detention storage and channel storage. With the increase in watershed size relative importance of channel storage increases *vis-a-vis* the other elements of storage which are aiding infiltration and as such the necessity for precise determination of infiltration decreases.

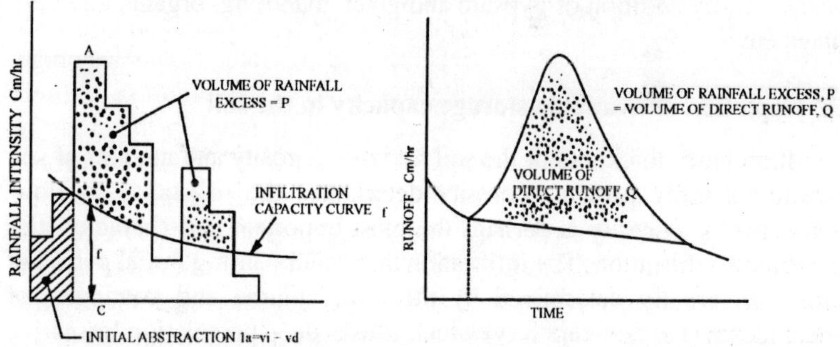

Fig. 10.8: Definition sketch showing relationship between rainfall, infiltration and initial abstractions

10.9.1 Factors affecting infiltration capacity

Infiltration capacity is influenced by the following factors:

(i) Factors which control surface entry of water,

(ii) Transmission of water through the soil,

(iii) Depletion of available storage capacity in the soil, and

(iv) Other factors.

(i) Factors which control surface entry of water

It includes soil texture, vegetation, crop residue tillage, organic manure and compaction. In case of fine textured soils (clay) mechanical compaction caused by raindrops reduces infiltration capacity; while, it is not significantly affected in a sandy soil. Vegetation is one of the most significant factors affecting the surface entry of water. Dense vegetation protects the soil surface from impact. Crop residue, addition of organic manure and forest litter, etc. not only protect the soil from rain impact but also help in the creation of high biotic activity in and beneath such layers, resulting in the opening up of soil and causing consequent increase in infiltration capacity. Moreover, transpiration by vegetation removes soil moisture and as a result infiltration rates are high during the initial period of rain. Compaction of soil surface due to vehicles traffic, such as farm tractors, and grazing of livestock or high pedestrian traffic would decrease the surface entry of water.

(ii) Transmission of water through the soil

The transmission can be improved by chemical or mechanical means *viz.*, by addition of gypsum and green manuring, organic and deep tillage etc.

(iii) Depletion of available storage capacity in the soil

It includes thickness of the soil horizon, porosity and amount of soil moisture already present. Porosity determines the resistance to flow. Non-capillary porosity is perhaps the most important soil characteristics that effects infiltration. The infiltration that occurs during initial period of rainfall is greatly determined by the size, volume and continuity of macrospores (*i.e.* non-capillary) which causes deep percolation losses.

Soil moisture present in the soil may increase or decrease the infiltration. If the soil is dry before the onset of the rain, the wetting of the top layer shall create a strong capillary potential just under the surface of the soil and this will enhance the gravitational force in causing infiltration. Colloids present in the soil would swell and reduce the infiltration rate during the initial period of rainfall due to wetting. As soil moisture is usually high in winter months and spring and low in other times. Thus, soil moisture is also responsible for seasonal variations in infiltration capacity. The current moisture content is dependent upon the time and amount of antecedent infiltration, evaporation rate and the transmission of underlying horizons. This factor may directly affect the infiltration rates during the storm.

(iv) Other factors

Factors like temperature, entrapped air, freezing of soil moisture, depth of surface detention and thickness of saturated layer may affect the infiltration capacity of soil. Flow through the soil is always laminar and changes in temperature would change the viscosity of water and hence the rate of infiltration will be affected.

10.9.2 Measurement of infiltration capacity

Broadly there are two methods of determining infiltration capacity:

(i) Infiltrometer method (discussed in Chapter 14), and

(ii) Hydrograph analysis method.

Infiltration capacities over a drainage basin is determined by hydrograph analysis. It is very useful for determination of runoff from rainfall on the same basin under similar conditions. For this purpose accurate data on the varying intensities of rainfall during any given storm together with continuous record of resulting runoff are needed.

10.9.3 Equation for infiltration process

It is evident that the infiltration capacity is maximum at the beginning of the rain due to initial soil moisture conditions and heavy demands of interception and depression storage. It will decrease rapidly during the early stages of the storm and finally assume a low but almost constant value (fc). Let us consider that interception and depression storage are included with infiltration. Thus, average infiltration value (f_a) is taken into consideration. Further the needs of the user and the availability of the basic data normally determine the degree of detailed consideration that can be given to the study of infiltration. For example, for the flood flow studies, the final nearly constant value, (f_c), may be the only value in which one may be interested but for irrigation studies information on initial infiltration, (f_o), or the knowledge of variation of f during the initial stages of storm is required. Knowledge of variation of f over the total storm period is helpful in determining the estimation of rainfall excess (surface runoff); in such cases, the depression storage is assumed at some constant value. Horton proposed the following relationship for determination of infiltration (Fig. 10.9):

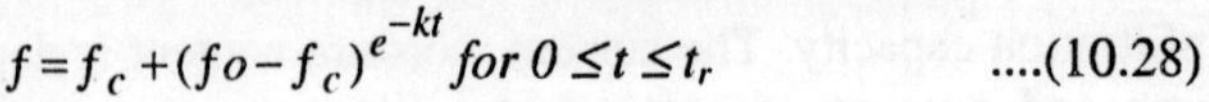

$$f = f_c + (fo - f_c)^{e^{-kt}} \quad for\ 0 \le t \le t_r \qquad ....(10.28)$$

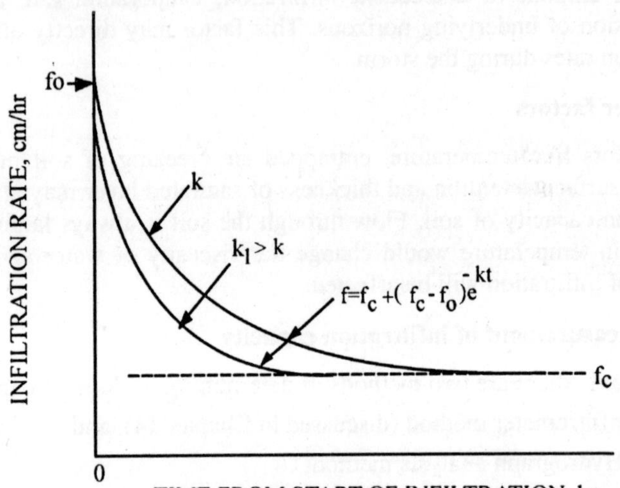

Fig. 10.9: Infiltration capacity curve

where,

k = empirical constant whose value depends upon soil characteristics, antecedent moisture conditions and types of vegetation cover,

f = infiltration capacity of the soil in cm/h. at any time (t) h,

f_0 = initial infiltration capacity at $t = 0$,

f_c= final value of infiltration capacity; and

t_r = duration of rainfall.

This equation is valid only when the supply rate (i_s) (rainfall intensity minus rate of retention) equals or exceeds (f), throughout the storm. Let us consider if $i_s \le f$ (actual infiltration rate is assumed equal to supply rate). Thus, the limiting value of f *i.e.* f is related to the permeability of the saturated soil. Eq. 11.29 can be rewritten as,

$$f - f_c (fo - f_c)^{e^{-kt}} \qquad ....(10.29)$$

Taking logarithms of both sides and solving for t, we obtain

$$t = \frac{1}{k \log_e} \log_e (f_o - f_c) - \frac{1}{k \log_e} \log_e (f - f_c) \qquad ....(10.30)$$

The above equation fits a straight line between t and log $(f\text{-}f_c)$ variables. The slope of this line is equal to - $1/k$ $\log_e$. It is plotted on one semilogarithmic paper and yield a straight line, and the slope of the line when equated to $-1/k$ $\log_e$. The value of k will be obtained. Consider that value of t is fo at time (t=0). This value can be obtained from the graph. Then the value of (t_c) is determined as all the constants are known.

Example 10.5: An infiltration experiment was carried out on a small plot having sandy soil. A small rainfall simulator was used to simulate rainfall over the plot, and rainfall was simulated at the rate of 10 cm/hour. The rainfall and the resulting run off data are given in Table 10.5. Temperature of water was 27°C. Derive and plot the infiltration curve.

Table 10.5: Data for Example 10.5

Elapsed time (min)	Simulated rainfall (cm)	Runoff (O)
0	0	0
1	0.20	0.06
3	0.40	0.16
5	1.00	0.47
7	1.41	0.75
10	2.00	1.10
15	3.00	1.71
20	4.0	2.27
25	5.0	2.90
50	10.0	5.46
75	15.0	9.90
100	20.0	14.09
130	26.0	19.10

Solution:

The computation has been carried out in the tabular form in Table 10.6. The various steps involved in the solution are as follows:

Step I: From Col. 1 and 7 of Table 10.6, plot the points (shown as blank circles on Fig. 10.10. There is a large scatter of points and passing of mean line through these points shall he very subjective. To obviate this suggestiveness adopt the following procedure.

Step II: Calculate the total infiltration F = simulated rainfall minus runoff for each elapsed time (Col.5 of Table). Then compute

ΣΔF (Col. 8) and plot it on a graph and draw a mean curve through the points (Fig. 10.11).

Step III: Measure the slope of tangent to the mean curve obtained in step II, at t = 5, 30, 70, 110 (shown as points A, B, C, D on mean curve and compute $f = \frac{\Delta F}{\Delta t}$.

Step IV: Plot these four points on Fig. 10.10 and pass a mean curve through these four points to obtain infiltration capacity curve.

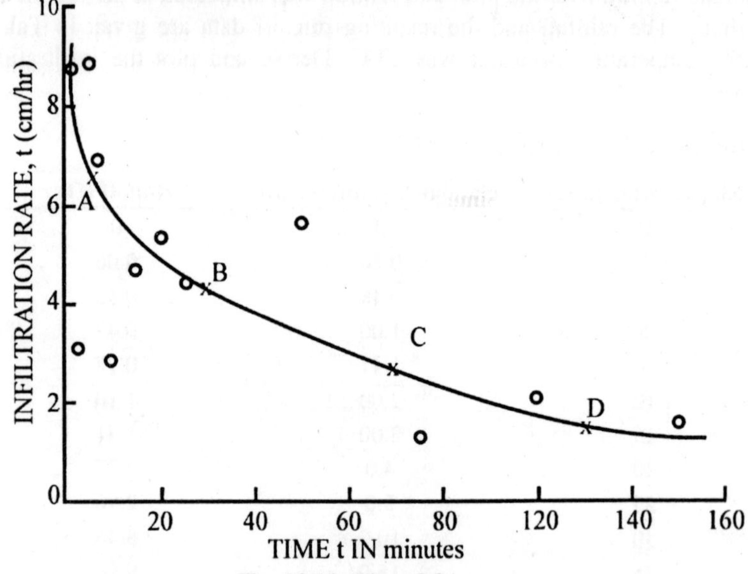

Fig. 10.10: Plot of *f* versus *t*

This is adopted to arrive at infiltration values (or infiltration curve) is as follows. It is noticed from the Table that, the interval between successive observations is very small and infiltration capacities worked out as $f = \frac{\Delta F}{\Delta t}$ would exhibit large scatter as is obvious would exhibit large scatter as is obvious from plot Fig. 10.10 (which is plot of Col. (7) and Col. (1). By first drawing a plot of accumulated infiltration versus time *i.e.* Col. (8) versus Col. (1), the large scatter can be visually evened out: the irregularities are better averaged on integral curse (we see that the mean curve passing through the plotted points is very smooth and scatter is almost negligible).

Table 10.6: Computations of F and f

Elapsed time (min)	*T* (hrs)	Simulated rainfall (cm)	Runoff (cm)	Total infiltration (cm)	*F* (cm)	$f = \frac{\Delta F}{\Delta t}$	$\Sigma\Delta F$
1	2	3	4	5	6	7	8
0	-	0.00	0.00	-	-	-	0.00
1	.016	0.20	0.06	0.14	0.14	8.75	0.14
3	.033	0.40	0.16	0.24	0.10	3.03	0.24
5	.033	1.00	0.47	0.53	0.29	8.78	0.53
7	.033	1.41	0.75	0.76	0.23	6.96	0.76
10	.050	2.00	1.10	0.90	0.14	2.80	0.90
15	0.83	3.00	1.71	1.29	0.39	4.69	1.29
20	.083	4.00	2.27	1.73	0.44	5.30	1.73
25	.083	5.00	2.90	2.10	0.37	4.46	2.10
50	.417	10.00	5.46	4.54	2.34	5.61	4.44
75	.417	15.00	9.95	5.05	0.51	1.22	4.95
100	.417	20.00	14.09	5.91	0.86	2.06	5.71
130	.500	26.00	19.10	6.90	0.89	1.78	6.60

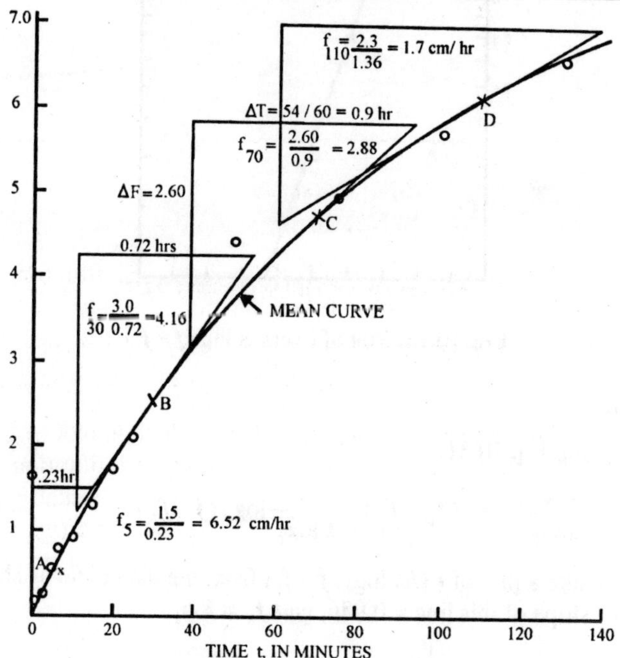

Fig. 10.11: Plot of *f* versus *t*

Example 10.6: The infiltration capacity of an area was studied and the following data was obtained.

Time (h)	0.25	0.50	0.75	1.0	1.25	1.50	1.75	2.0	2.25
f (cm/h.)	6.50	3.4	2.3	1.8	1.4	1.28	1.0	1.0	1.0

Derive an equation for infiltration capacity in the exponential form:

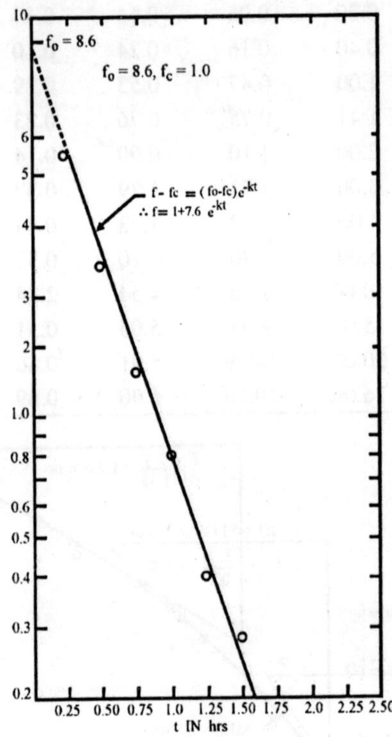

Fig. 10.12: Plot of t versus log $(f - f_c)$

Solution:

On using Eq. 10.31.

$$t = \frac{1}{k \log_e} \log_e (f_o - f_c) - \frac{1}{k \log_e} \log_e (f - f_c) \qquad \text{....(10.31)}$$

and make a plot of t v/s. $\log_e (f - f_c)$ from the data table as shown in Fig. 10.12 slope of this line = 0.986; and f_o = 8.6.

$$\therefore \quad \frac{-1}{(k \log_e)} = 0.986$$

$k = 2.33$

$\therefore \quad f = 1 + 7.6_e^{-2.33t}$

10.9.4 Infiltration indices

The infiltration indices would provide an average representative value of infiltration rather than the infiltration function/rate. There are following infiltration indices:

(i) φ-index: It is the difference between mass rainfall and mass runoff and shown in Fig. 10.13. It is evident from this figure that φ-Index, can be seen as average rainfall rate above which the volume of rainfall is equal to the volume of runoff, and is expressed as,

$$\phi\text{-}index = \frac{\text{Basin Recharge}}{\text{Duration of rainfall}} = \frac{P-Q}{t_r} \quad \text{....(10.32)}$$

where,

t_r = the duration of the storm,

P = the total precipitation corresponding to t_r,

Q = the surface runoff

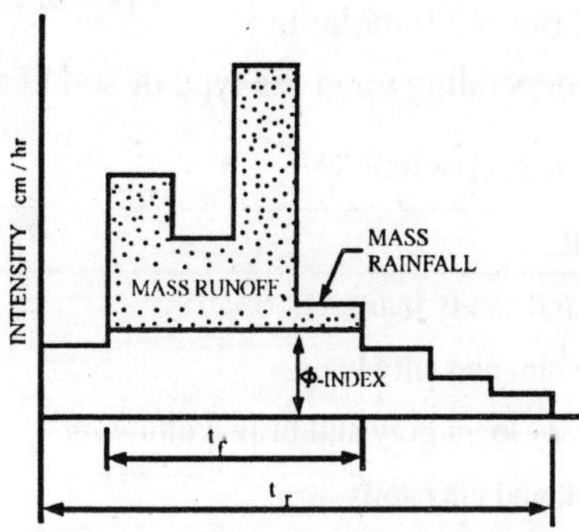

Fig. 10.13: Conceptual model of φ-index

(ii) w-index: It is defined as the ratio of basin recharge ($P - Q$) minus initial abstractions *i.e.* interception storage plus depression storage to duration of infiltration (t_r). It is expressed as,

$$w\text{-}index = \frac{P-Q-I_a}{t_f} \quad \text{....(10.33)}$$

$$= \frac{P-Q-V_i-V_d}{t_f} \quad \text{....(10.34)}$$

Thus, w-index is essentially equal to ϕ-index minus average rate of abstraction. However, if the rainfall continues for very long time (greater than 5-6 h) then infiltration capacity assumes a minimum value (f_c). Consequently the interception and depression storage is very low and ϕ and w-indices are almost equal.

The primary function of these indices is to establish minimum loss rates (*i.e.* rainfall-direct surface runoff) for large storms for use in the development of design floods from rainfall.

Central Water Commission, proposed the following relations for determination of ϕ-index for flood producing storms under prevailing soil conditions in India.

$$R = a\, I^{0.2} \qquad \text{....(10.35)}$$

and

$$\phi\text{-index} = \frac{I - R}{24} \qquad \text{....(10.36)}$$

where,

R = run off in cm from 24h rainfall,

I = intensity of rainfall (cm/day),

a = coefficient depending upon the type of soil (Table 10.7)

Table 10.7: Values of coefficient 'a'

S. No.	Type of soil	a
1	Sandy soil and sandy loam	0.17-0.25
2	Costal alluvium and silty loam	0.25-0.34
3	Red soils, clay loam gray and brown alluvium	0.42
4	Black cotton and clay soils	0.42-0.46
5	Hilly soils	0.46-0.50

In the absence of any data, the value of 'a' may be considered as 0.10 cm/h for determination ϕ-index for maximum design flood.

Example 10.7: The surface runoff from rainfall storm on a drainage basin was recorded as 4.5 cm. The area of the basin is 240 km^2. The equivalent uniform depth of rainfall on the drainage basin was 15 cm, and the time distribution of the rainfall is given as under. Determine the ϕ-index for this storm with assumption of data.

Hour (1)	9-10 (2)	10-11 (3)	11-12 (4)	12-13 (5)	13-14 (6)	14-15 (7)	Total
ppt (mm)	13	19	47	20	31	20	150

Solution:

The computation of φ-index involves the following steps:

Step I: Plot the bar diagram as shown in Fig 10.14 and let us assume that all the rainfall given in data table are effective rainfalls *i.e.* each one is contributing toward runoff.

Step II: Compute Basin recharge = P – Q = 15 – 4.5 = 10.5 cm

Step III: Determine the duration of rainfall (t_r)

Let t_r = 15 – 9 = 6 h.

Step IV: φ-index = (P – Q)/t_r (trial)

= 10.5/6 = 1.75 cm/h = 17.5 mm/h

Step V: Comparing this value with values in data table we see that rainfall in Col. 1 of data table is only 13 mm and is less than 17.5 mm/h and as such is not effective rainfall and hence would not contribute towards run off. Hence the φ-index value needs to be revised after rejecting the rainfall of 13 mm. and reducing t_r from 6 to 5h.

Step VI: Revised φ-index = [(15-1.3)-4.5]/5 = 1.54 cm/h.

Step VII: Checking again we see that now all the five remaining rainfalls are effective rainfalls.

Step VIII: Draw line AB on Fig. 10.14, which is the correct value of φ-index.

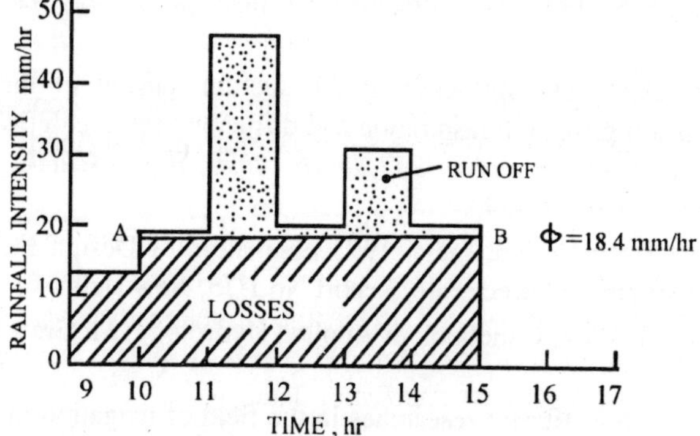

Fig. 10.14: Computation of φ-index

References

Aibrigo, L.G. 1972. Appearance and persistence of pinolene antitranspirants sprayed on 'Vallencia' orange leaves. *Hort. Sci.*, 7:247-248.

Arnon, I. 1975. Physiological principles of dryland crop production. In 'Physiological Aspects of Dryland Farming'. (Ed. Gupta, U.S.), Oxford & IBH Publ. Co., New Delhi, Bombay, Calcutta, pp. 3-145.

Asana, R.D. 1966. Physiological analysis of yield of wheat in relation to water stress and temperature. *J. Post Grad. Sch., Indian Agric. Res. Inst.*, New Delhi, 4:17-31.

Asana, R.D. 1975. Physiological Approaches to Breeding of Drought-Resistant Crops. *Indian Council Agric. lies., Tech. Dull.* (Agric.) No. 52, 48 pp.

Babu, R. C., Nguyen, B. D., Chamarerk, V., Shanmugasundaram, P., Chezhian P. Jeyaprakash., P., Ganesh ,S. K., Palchamy, A., Sadasivam, S., Sarkarung, S., Wade, L. J. and Nguyer, H.T. 2003. Genetic analysis of drought resistance in rice by molecular markers: association between secondary traits and field performance. *Crop Sci.*, 43 (4): 1457-1469.

Bhagchand and Gautam, R.C. 2000. Effect of organic manure, bio-fertilizer and inorganic fertilizers on growth, yield and quality of rainfed pearlmillet. *Ann. Agric. Res.*, 21(4): 459-464.

Bhatia, P.C., Singh, M. and Dastane, N.G. 1966. Use of synthetic products in the field of soil and water conservation. *Indian J. Agron.*, 11: 149-166.

Boerjan, W, Ralph, J. and Baucher, M. 2003. Lignin biosynthesis. Ann. Rev. Pl. Biol. 54:519-546.

Boonstra, A.E.H.R. 1929. Difference in vitality of leaves of four varieties of oats as connected with yield. *Meded Landb. Hoogech.*, Wageningcn, 33:1.

Boonstra, A.E.H.R. 1931. Plant breeding and plant physiology. *Zuchter,* 3:3 15.

Bronwer, R. 1961. The influence of the suction tension of the nutrient solutions on growth, transpiration and diffusion pressure deficit of bean leaves *(Phaseolus vulgaris.). Jaarb. Inst. Biol. Scheik. Onderz.*, Wageningen, pp. 11-24.

Central Water Commission, India, 1973 Estimation of Design Flood Peak, Flood Estimation Directorate, Report No.1/73, New Delhi.

Chow V.T. (Ed), 1964. Handbook of Applied Hydrology, McGraw Hill New York.

Dastane, N.G. 1968. Recent researches in the field of irrigation in India. *7th NESA Regional Irrig. Practs Semin.*, Lahore, Pakistan, pp. 221-237.

Davenport, D. C. and Hudson, J.P. 1967b. Meteorological observations and Penman estimates along a 17-km transect in the Sudan Gezira. *Agric. Meteoral.*, 4:405-414.

Davenport, D.C. and Hagan, R.M. 1975. Role of antitranspirants in arid agriculture. In 'Physiological Aspects of Dryland Farming' (Ed. Gupta, U.S.). Oxford & IBH Pub. Co New Delhi, Bombay, Calcutta, pp. 3 15-326.

Davenport, D.C. and Hudson, J.P. 1967a. Changes in evaporation rates along a 17-km transect in the Sudan Gezira. *Agric. Meteorol.*, 4:339-352.

David C. Davenport and Robert M. Hagan 1977. Role of anti transpirant in arid agriculture in ed. Gupta U.S. Physiological aspects of dryland farming.

Dungal, K.N. and Joshi, J.P. 1996. Elements of Water Resources Engineering. New Age International Publications, pp. 581.

Dungan, G.H., Sisodia, U.S. and Singh, G.D. 1955. The benefit of sowing maize for fodder in North South lines. *Allahabad Farms*, 29.8-15.

Fifty years of dryland agricultural research in India. Ed. Singh, H.P. *et al.*, Central Research Institute of Dryland Agriculture Santoshnagar, Hyderabad.

Fisher, M.A. and Lyon, T.L. 1972. Antitranpsirant film detection by scanning microceopy of cathode luminescence, *Horti. Sci.*, 7:24-27.

Gajri, P.R., Verma, H.N. and Parihar, S.S. 1981. Rainwater harvesting and its recycling for maximization of crop production. *Indian J. Soil Cons.* 9(2&3): 69-76.

Gardner, W.R. 1960. Soil water relations in arid and semi-arid conditions. *Arid Zone Res.*, UNESCO, XV: Plant-Water Relationships in Arid and Semi-Arid Conditions., pp. 37-61.

Gautam, R.C. and Kaushik, S.K. 1984. Role of split application of nitrogen doses on the growth and yield of pearlmillet hybrids. *Indian J. Agron.*, 29(3): 327-330.

Gill, A.S., Mainkar, N.D., Singh, Mukhtar and Abhichandani, C.T. 1973. Effect of cycocel (CCC) on plant height and yield of oats. *Indian J. Agron.*, 18:217-218.

Grewal, S.S., Juneja, M.L. and Singh, K. 1982. Soil water expense and water use efficiency of wheat in a system of supplemental irrigation through rainwater harvesting in foot hills of Northern India. *Indian J. Soil Cons.* 10(1): 43-45.

Grewal, S.S., Mittal, S.P., Gurmel Singh, Pratap Singh and Sud, A.D. 1990. Cropping practices for soil and water conservation in Shiwalik foot hills, Bull. CSWCRTI, Res. Centre, Chandigarh, pp. 55.

Gupta J.P. and Gupta G.N. 1981. A note on wind erosion from cultivated field in western Rajasthan. *Journal of Indian Society of Soil Science*, 29:278-279.

Gupta, J.P. and Gupta, G.N. 1983. Effect of grass mulching on growth and yield of legumes. *Agric. Water man.* 6:375-383.

Halevy, A.H. and Kessler, B. 1963. Increased tolerance of bean plants to soil drought by means of growth retarding substances. *Nature* (Lond.) 197:310.

Hameed Mohd. Shakeel 1978. Physico-chemical changes with different systems of organic recycling in semi arid red soil region. 217-227. In sustainable development of dryland agriculture in India (Singh, R.P. ed. 1995). Scientific Publishers Jodhpur.

Hanus, H. 1967. Die Beeinflussung dis Wurzelstems von Weizem durch CCC. *Z. Acker-Pflanzenbou,* 125:40-46.

Heinonen, K. 1971. Soil Management and Crop Water Supply. *Lantbrkshogskolans kompen dien a mud.*, 112 pp.

Hillel, D. 1968. Soil Water Evaporation and Means of Minimising It, *Rep. U.S.D.A , Hebrew Univ.* Jerusalem, Israel.

Howard, A. 1924. Crop Production in India. Oxford Univ. Press, Bombay, 100 pp.

Hsiao, T.C. 1973. Plant responses to water stress. *Ann. Rev. Plant Physiol*, 24:519-570.

Hsiao, T.C. and Acevedo, E. 1974. Plant responses to water deficits, water use efficiency and drought resistance. *Agric. Meteorol.*, 14:59-84.

Humphries, E.C. 1968b. CCC and cereals. *Field Crop Abstr.*, 21:91-99.

Humphries, E.C. 1968a The beneficial effect of CCC on wheat yields in dry conditions. *Euphytica* (Suppl. No. 1): 275-279.

Juyal, G.P. and Gupta, R.K. 1985. Construction of LDPE lined 'TANKAS' in hills - A case study. *Indian J. Soil Cons.* 13(1): 10-13.

Lal, R. 1974. Soil temperature, soil moisture and maize yield from mulched and unmulched tropical soils. *Plant and Soil* 40:129-143.

Leather, J.W. 1910-11. Water requirement of crops in India. *Mem. Dept:. Agr. India Client. Ser.*, 1:33-84, 133-134, 205-281.

Lemon, E.R. 1956. The potentialities for decreasing soil moisture evaporation loss. *Soil Sci. Soc. Amer. Proc.*, 20:120-125.

Linsley, R.K. Jr. M.A. Kohler and J. L. Paulhus, 1975. Hydrology for Engineers, McGraw-Hill Kogakusha, Limited Second Edition.

Lowe, L.B. and Carter, O.G. 1970. The influence of (2-chtorocthyl) trimethyl ammonium chloride (CCC) and giberrelic acid on wheat yields. Aust. *J. Exp. Agr. Anim. Husb.,* 10:354-359.

Mishra L.N. *et al.* 2002. Effect of various mulches on the physio-chemical properties of the sodic soil under Aonla-Guava cropping system. *Ann. Agric res.* New series 23(3) 377-380.

Mittal, S.P. and Grewal, S.S. 1983. To evaluate the effect of land treatment measures and runoff and productivity and to evaluate norms of erosion for various crops. Annual Report, CSWCRTI, Dehradun, U.P., pp. 62-64.

Mittal, S.P., Pratap Singh, Agnihotri, R.C. and Sud, A.D. 1988. Effect of strip cropping of maize and legumes on runoff, soil loss and productivity, *Indian J. Soil Cons.* 16(2): 12-16.

Moinuddin, Fisher, R.A., Sayre, K.D. and Reynolds, M.P. 2005. Osmotic adjustment in wheat in relation to grain yield under water deficit environments. *Agron. J.* 97:1062-1071.

Muchow, R.C. 1989. *Field Crops Res.*, 20:191-205.

Nguyen HT, Babu RC, Blum A. 1997. Breeding for drought resistance in rice: Physiological and molecular genetics considerations. *Crop Science* 37, 1426-1434.

Oh, S.J., Song, I.K., Kim, Y.C., Jang, H.J., Kim, S.Y., Kim, M., Kim, Y.K.. Nahm, B.K.H. and Kim, J. K. 2005. Arabidopsis CBF3/ DREBIA and ABF3 in transgenic rice increased tolerance to Abiotic stress without stunting growth. *Plant Physiol.* 138:341-351.

Papadakis, J. 1972. Auxins, Biochemical Plant interaction, Growth Retardants and Dense High Yielding Crops. Buenos Aires, 20. pp.

Pareek, S.S, Reddy, M.K. and Sopory, S.K. 2001. Transgenic approach towards developing Abiotic stress tolerance in plants. Proceedings of Indian National Science Academy 67: 265- 284.

Patil, S.B. 1974. Effect of nitrogen fertilization, row spacing and antitranspirants on growth, yield and water use by rapeseed *(Brassica campestris Var. Brown sarson)* under dry land conditions. Ph.D. Thesis., *Div. Agron., Indian Agric. lies. Inst.,* New Delhi.

Paul W. Unger 1977. Role of mulches in dryland agriculture in ed. Gupta U.S. Physiological aspect of dryland farming.

Pendleton, J.W. and Dungan, G.H. 1958. Effect of row direction on spring oats yield. *Agron. J.,* 50:341-343.

Penman, H. L. 1956. Evaporation; An introductory survey. *Netherl. J. Agri. Sci.*, 4:9-29.

Philip, J. R. 1957. Evaporation, moisture and heat fields in the soil *J. Meteorol.*, 14:354-366.

Philpotts, H. 1972. The effect of (2-chiororthyl) trimethyl ammonium chloride (CCC) and pre-sowing drought hardening on growth and yield of wheat. *Aust .J Exp. Agr. Anim. Husb.*, 12:70-74.

Pintlius, M.J. and Rudich, J. 1967. Increase in grain yield of CCC treated wheat *(Triticum* aestivum) in the absence of lodging. *Agrochhimica.*, 11:565-50.

Plaut, Z., Halevey, A.H. and Shmueli E. 1964. The effect of growth retarding chemicals on growth and transpiration of bean plants grown under various irrigation regimes. *Israel J. Agric. Pea.*, 14:153-158.

Pruitt, W.O. and McMillan, W.D. 1962. Some observations of downwind variation of evaporative heat flux on the surface. Investigation of energy and mass transfers near the ground including influences of the soil-plant-atmosphere system *Univ. Calif,* Davis, California, *Res. Rep.* pp 77-97.

Rabbani, M.S., Maruyama. K., Abe, H. Kha, M.A., Kastura, K, Ito, Y., Yoshiwara, K, Seki, M, Shinozaki, K. and Yamaguchi, K. 2003. Monitoring expression profiles of rice genes under cold, drought and high salinity stresses and ABA application using DNA microarray and RNA gel blot analysis. *Plant Physiol.* 133: 1755-1767.

Raghunath, H.M. 1985. Hydrology Principles, Analysis and design", Wiley Eastern Limited, New Delhi.

Rao, K.N., C.J. George and K.S. Ramasastri, 1971. Potential Evapotranspiration (PE) over India, Symposium on Water Resources, I.I.Sc. Banglore, India, pp. A 2-1-14,

Ray, S.B. 1968. Use of reflectant materials as antitranspirants. *Post Doct. Res. Rep., Univ. Calif* Davis, California.

Reddy, P.V. and Shah, C.B. 1973. Studies on growth response and transpiration behaviour of wheat S-227 to foliar application of antitranspirants, phenyl mercuric acetate and kaolin. *4th Ann. Workshop Dry Land Agr.* Indore., India, Feb., March, 1973 (Miiniogr).

Rider, N.E., Philip, J.R. and Bradley. E.F. 1963. The horizontal transport of heat and moisture- a micioclimatological study. *Quart. J. Roy. Meteorol. Soc.*, 89:507-531.

Rosenberg, N.J. 1966. Microclimate, air mixing and physiological regulation of transpiration as influenced by wind shelter in an irrigated bean field. *Agric. Meteorol* , 3:191224.

Saraf, C.S. and Dastane, N.G. 1970. Role of advective energy in water use by crops -A review. *India,, J. Agron., 15:89-95.*

Sastry, G. and Mittal, S.P. 1987. Water harvesting and recycling engineering aspects. *Indian J. Soil Cons.* 15(3): 17-22.

Sharda, V.N., Sastry, G. and Joshi, D.P. 1986. Studies on control of seepage losses. Annual Report, CSWCRTI, Dehradun, U.P., pp. 34-36.

Sharma, T.V.R.S., Attri, B.L., Suryanarayana, M.A. and Singh, D.B. 2000. Improving the coconut productivity by genetic manipulation. *India Coconut J.* 36(3): 10-13.

Singh, H.G., and Singh, B. 1970. Preliminary studies on the effect of cycocel on cotton (G. *arborium). Indian J. Agric. Sci.,* 40:562-565.

Singh, Mahendra, Pal, Mahendra and Kausik, S.R. 1975. Effect of dates of sowing on grain yield of sorghum. *Indian J. Agron.,* 20:103-105.

Singh, R.P. 1995. Sustainable Development of Dryland Agriculture in India ed. Singh, R.P. Singh, Scientific Publishers, Jodhpur.

Skidmore, E.L. and Hagen, L.J. 1970. Evaporation in sheltered areas as influenced by wind break porosity. *Agric. Meteorol.,* 7:363-374.

Skidmore, E.L., Jacobs, U.S. and Hagen, L.J. 1972. Microclimate modification by slat-fence wind breaks. *Agron. J.,* 64:160-162.

Srivastava, M.S., Sharma, S.M. and Singh, S.P. 1968. A note on the effect of cycocel on growth and yield of wheat. *Indian J. Agron.,* 13: 192-194.

Srivastava, S.P. and Bansal, O.N. 1975. Effect of delay in first irrigation, methods of N application and cycocel on wheat in mixed black soils. *Indian J. Agron.,* 20: 281-282.

Sturrock, J.W. 1975. Wind effects and their amelioration in crop production. In 'Physiological Aspects of Dry Land Farming'. (Ed. Gupta, U.S.). Oxford & NH Publ. Co., New Delhi, Bombay, Calcutta, pp. 283-313. 334

Turner, Neil C.1997. Further progress in crop water relations. *Adv. Agron.* 58: 292-324.

Turner. Neil C. 1986. Crop water deficits: A decade of progress. Adv. *Agron.* 39: 1-44.

Unger, P.W. 1975. Role of mulches in dryland agriculture. In 'Physiological Aspects of Dryland Farming'. (Ed. Gupta, U.S.) Oxford & 1131-1 Pub). Co., New Delhi, Bombay, Calcutta, pp. 237-258.

Van der Linde, J. 1962, Trees outside the forest. In 'Forest Influences'. *FAO Forestry, Forest Prod. Studies.,* 15: 141-208.

Van, Eimern, J., Karscon, R., Razumova, L.A. and Robertson, G.W. 1964. Wind breaks and Shelterbelt, *W.M.O.*. Genewa, Switzerland. *Tech. Bull.*, 59: 188 pp.

Venkateswarlu, J. 1981. Maximization of production in drylands, *Indian J. of Soil Cons.*, 9:124-140.

Verma, H.N. 1988. Rainwater management for crops in rained Kandi belt of Punjab. *Indian J. Dryland Agril. Res. and Dev.* 3(2): 228-235.

Vijayakumar Jacob Mathew, K., Amarnath, C.H., Nambiar, P.T.N., Jose, C.T. and Balakrishnan, T.K. 1987. Influence of weather on coconut yield, Agro meteorology of Plantation Crops. Proc. National Seminar, Kerela Agricultural University, Regional Agricultural Research Station, Pilicode– pp. 124-132.

Vincent, D., Lapierre, C., Pollet, B., Cornic, G., Negroni, L. and Zivy, M. 2005. Water deficits affect caffeate o-methyl transferees, lignifications and related enzymes in maize leaves a proteomic investigation.2005.*Plant Physiol.* 137:949-960.

Waggoner, P.E. 1966. Decreasing transpiration and the effect upon growth. In 'Plant Environment and Efficient Water Use', (Eds. Pierre, W.H. Kirkhan, D. Pesek, J. and Shaw, R). *Amer. Soc. Agron., Soil Sd. Soc. Amer.*, Madison, Wisconsin, pp. 49-68.

Walton, K. 1969. The arid zones, Hutchinson University Library Publication.

Weinser, C.J. 1970. Hydrometeorology, Chapman and Hall Limited London, U.K.

Wendt, C.W. 1971. A study of the mechanisms and suppression of evaporation of water from soils. Texas *A & M Univ. Water Res. Inst., Tech. Rep.* No. 33.

Wisler, C.O. and E.F. Brater, 1959. Hydrology, John Wiely and Sons Inc. New York Second Edition.

Yadav, S.T. and Varshney, M.C. 2005. Influence of sowing dates and sowing methods on growth yield of pearlmillet. *Journal of Agro-meteorology* 7(2):319-321.

EXERCISES

10.1 Differentiate the following:

(i) Consumptive use of a crop and valley consumptive use.

(ii) Evaporation and evapotranspiration.

(iii) Evaporation and potential evaporation.

(iv) ϕ-index and w-index.

(vi) Infiltration capacity and infiltration.

10.2 Write short notes on the following:

(i) Dalton's law of evaporation.

(ii) Infiltration indices.

(iii) Pan evaporimeters.

(iv) Lysimeters.

(v) Phytometers.

(vi) Watershed leakage.

(vii) Energy budget method of determining lake evaporation.

(viii) Methods for determination of interception losses.

10.3 Explain in briefly the factors affecting infiltration

10.4 Explain the following:

(a) Evaporation process.

(b) Suitable measures for reduction of water loss from water harvesting structures.

10.5 (a) Draw a typical infiltration curve and indicate on it the various components.

(b) State the Horton's equation (exponential type) for determination of infiltration capacity and describe the procedure for determining the empirical constants of this equation.

10.6 (a) Explain the method for determination of infiltration capacity by artificial rainfall simulators.

(b) The following table gives the infiltration capacity rates at various time intervals after the beginning of a storm on a drainage basin. Fit Horton's equation. $f = f_c + (f_o - f_c)^{e^{-kt}}$ to the observed data. Explain the utility of fitting empirical equations such as Horton's to be observed data.

(c) Determine the total rain and runoff.

Data Table for exercise 10.7

Time (min)	Rainfall intensity (cm/h)	Infiltration capacity (cm/h)	Time (min)	Rainfall intensity (cm/h)	Infiltration capacity (cm/h)
1	4	3.3	9	6	1.42
2	4	2.9	11	6	1.31
3	4	2.5	13	2	1.24
4	6	2.1	15	2	1.23
5	6	1.9	17	2	1.20
6	6	1.6	19	2	1.18
7	6	1.5	21	2	1.18

10.7 It is desired to determine run of from a catchment of 2.5 km^2 area from a storm which occurred for 24h resulting in total rainfall of 150 mm. Infiltration studies made some time hack yielded the following equation for determining the infiltration capacity:

$$f = 1.3 + 3.2\, e^{-9t}$$

in which e = base of Napierian Logarithm, and t = the time from the beginning of the storm. Water abstractions during the period of 24 hours amount to 0.5 cm depth of water.

10.8 Explain with the help of neat sketches the hydrograph analysis method for determination of the infiltration capacity of a drainage basin. Explain its advantages over other methods of determining infiltration capacity rate.

10.9 Determine the consumptive use for a crop in the month of October if the growing season for the crop is from September to December. Assume Pan evaporation for the month of October as 7.5 cm. Consumptive use coefficient at 40% stage growth may be taken as 0.50.

10.10 (a) Explain with neat sketches the following methods for determining the net storm rainfall:

(ii) Infiltration capacity method

(ii) w-index method

(b) A three-hour storm occurred over 62 square km^2 area. From the following data, estimate the net storm rain for the whole area and its hourly distribution.

Sub-area (km^2)	ϕ-index (cm/h)	Hourly rain		
		Ist	IInd	IIIrd
14	2.0	1.5	5.0	0.8
20	3.5	1.5	5.0	0.8
28	1.0	1.5	5.5	1.5

10.11 What is infiltration? How would you use infiltration capacity curve to determine run off from a small catchment?

Explain the following terms used in hydrologic cycle;

(i) Infiltration index

(ii) Time of concentration

Hourly rainfall of 3.6, and 25 cm occurred over an area of 25 ha. This area has three types of soil conditions`and antecedent soil moisture conditions leading to different value of ϕ for these three h are as under:

Sub-area (km^2)	ϕ-index (cm/h)
5	6
8	4
12	2

Derive the net rain in the successive hours.

■■■■

Chapter 11

EFFICIENT CROP PLANNING

11.1 Introduction

Precipitation diverse is the main source of water for all purposes on earth with the content reduction in availability of water. The need of its judicious use is reemphasized, particularly in the farm sector. A sustainable farming system would require management strategies with respect to crops and its cropping systems, tillage, land use and other drought management options in order to reduce the input cost and provide a sustained level of production and profit, from farming. Crop planning is an integral part of agricultural production system, which requires information on weather and climate under dryland agro-ecosystem. Water is vital and also a major resource for agricultural productivity and sustainability, especially in a stressed environment. It is necessary to develop farming systems those may give stable and sustainable production under changing climate and weather showing reasonable resistance to changing climate.

For an efficient farming system, precise temporal and spatial weather and soil information is needed so that the agriculturists can optimally plan their strategies to achieve the sought objectives under rainfed ecosystem due to limitation of predictability. Unfortunately the problem becomes more formidable with the extension in the temporal range and decrease in space. Considering these limitations, a method is presented for optimal use of information on climate assessment and weather prediction as available for economically and environmentally viable, agricultural planning and management.

Climate is defined as the long-term average conditions of weather at a particular place or region along with its statistical parameters, whereas weather relates to instantaneous atmospheric conditions at a place or region. The weather of a place is generally defined in terms of the various parameters i.e., temperature, humidity, wind speed and direction, pressure, rainfall, sunshine, and Cloud cover. However, rainfall is normally the single most important weather parameter for agriculture as it determines the moisture supply to crops yearly. Wide fluctuations in rainfall that occur from place to place result into a stressed environment in the form of drought and floods. The abnormal values of temperature, sunshine, or wind also affect the crops under rainfed agro-ecosystem.

11.2 Crop Planning

Crop planning is based on following parameters

1. Rainfall amount and its distribution,
2. Length of growing season and water availability period,
3. Water demand of crops/cropping system,
4. Matching the length of growing season and crop duration,
5. Probability of extremes of climate including dryspells and droughts, frost, high velocity winds etc.

11.2.1 Rainfall and its distribution

Rainfall characteristics like onset and withdrawal of monsoon, its distribution, length of growing season, critical dryspells are essential for selection of crops/cultivars and are determined as discussed in Chapter 2. The quantum and distribution of rainfall thus influences the choice of crops and cropping patterns in a region. The basic factor that decides the choice of crops in a given region is the length of the crop growing season. Therefore, attempts have been made to assess the length of the crop growing season on the basis of the period of moisture availability for crop production when temperature is not a limiting factor. The crop growing season thus depends not only on the rainfall distribution but also on the type of soil, soil depth, water holding capacity and moisture release characteristics of the soil, as the crops extract stored moisture from the soil profile during the rainless period. The length of the growing season therefore, is expressed as the sum of the length of the rainy season and post seasonal rainfall can meet .the water need of arable crops (Ramana Rao *et al.*, 1994).

11.2.2 Length of growing season

It is the sum of the length of rainy season and the period for which the stored soil moisture meets the crop water need after cessation of the rainy season. Thus, it depends on following parameter:

- Amount and distribution of rainfall,
- Water holding capacity of the soil, and
- The evaporative demand.

A simple graphical method has been suggested by Cocheme and Franquin (1967) to estimate the length of the crop growing season using only two parameters viz. mean monthly rainfall (P) and potential evapotranspiration (PET). For estimating the length of the growing season curves of monthly values of normal rainfall (P) and half (PET/2) and one fourth (PET/4) values of normal PET are plotted for the main rainy season. In this method the crop growing period starts when P is greater than 0.5 PET and ends when the rainfall curve crosses the 0.25 PET curve. This method, however, does not consider the soil water holding capacity and thus under estimates the length of the growing season. Therefore, Cocheme and Franquin method can provide reliable estimates of the length of the growing season in regions having shallow soils (Ramana Rao *et al.*, 1994). Hargreaves (1971) considered the ratio of dependable rainfall at 75 per cent probability to the PET and defined it as Moisture Availability Index (MAI). According to this concept the length of the growing season is the period when MAI is more than 0.33. This method also suffers from the deficiency of not considering the soil characteristics. Improvements of this concept (Higgins and Kessam, 1981) adopted by FAO consider the growing period to start as in the above method when P>0.5 PET and to end with utilization of assumed quantum of stored soil moisture (100 mm) after P falls below PET (subjected to minimum threshold temperatures of 5°C). A typical estimation of the crop growing season based on this method is shown in Fig. 11.1.

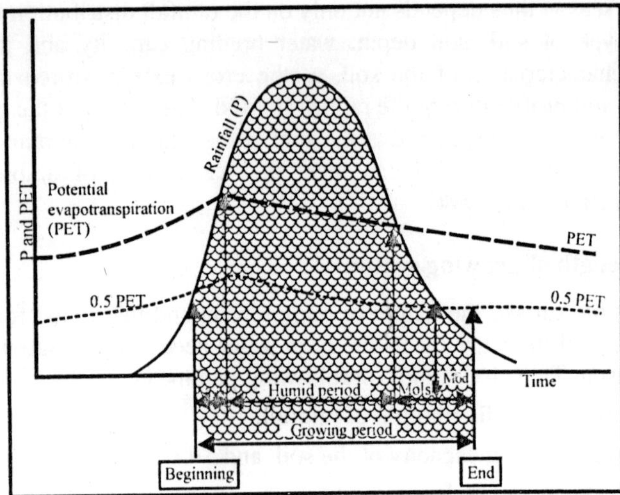

Fig. 11.1: FAO model for determining the length of growing period (LGP)

A useful method to assess the length of crop growing season is based on the water balance approach. This approach tries to assess the water availability (moisture received through rainfall) and water need (potential evapotranspiration demand) and the water holding capacity and moisture release characteristics of the soil, to assess the actual evapotranspiration (Actual crop water use). For this purpose Thornthwaite's (1955) method may be used to estimate the actual evapotranspiration (AET) on a monthly or weekly basis. Application of this method on weekly basis enables estimation of the ratio of AET/PET often referred as Index of Moisture Adequacy (IMA). Based on the IMA values the period of water availability is classified (Subrahmanyam *et al.*, 1963 and Ramana Rao *et al.*, 1994) as,

Value of IMA	Type of water availability
0.00 to 0.24	Dry
0.25 to 0.49	Semi-dry
0.50 to 0.74	Sub-moist
0.75 to 0.99	Moist
> 1.00	Humid

In this classification, the length of the growing season is assumed to commence from the period when IMA is 0.5 or more and extends till IMA reaches 0.25 and continues below 0.25 thereafter. The length of growing season estimated based on the above concept for soils of different water holding capacities (Ramana Rao *et al.*, 1994) are shown in Table 11.1.

Table 11.1: Characteristics of the length of the growing season

Station	Mean annual rainfall (mm)	Available water holding capacity (mm)	Start of growing season (Standard weeks)			Length of the growing season (weeks)			Probability (%) for growing season to exceed			
			Mean	Extremes		Mean	Extremes		10 week	16 week	22 week	30 week
				Early	Late		Low	High				
Jodhpur	369	100	28	24	40	10	0	20	63	10	-	-
Hyderabad	767	50	25	19	33	20	12	29	97	79	49	4
		150				24	15	36	97	86	62	23
		250				30	16	48	100	98	89	52
Indore	963	300	25	20	29	32	16	43	100	100	99	73
Bhubneshwar	1525	200	23	17	25	36	29	51	100	100	100	94

(Source: Ramana Rao *et al.*, 1994)

There is large variability in the length of growing season and also in its commencement (Table 11.1). It is also evident that soil depth and its water holding capacity determine length of growing season in a region for the similar rainfall situations. It enables to grow a long duration crop or to adopt intercropping or sequence cropping.

11.2.3 Water demands of crop/cropping system

Basic requirement of plants for water is meant to meet evapotranspirational needs that starts from the emergence of the plants. The leaves intercepting solar radiation have to transfer the water for maintaining favourable leaf temperature and facilitating gas exchange. Further, the water is also lost from soil surface in form of evaporation. The extent of water loss from plant and soil largely depends upon the evaporative demand of the atmosphere and on plant and soil characteristics to a minor extent.

Table 11.2: Length of humid period in relation to mean annual rainfall at different stations in India

S. No.	Stations	Mean annual rainfall (mm)	Length of the humid period (weeks)
1	Jodhpur	382	1
2	Hisar	400	1
3	Anantapur	590	4
4	Amiritsar	647	10
5	Jaipur	669	11
6	Rajkot	704	11
7	Hyderabad	733	13
8	Anand	909	14
9	Indore	1001	15

With the gradual multiplication and expansion of leaves, the crop canopy intercepts more solar radiation. Hence the transpiration from the canopy increases and gradually the soil evaporation decreases as the coiler of crop increases. When the crop canopy fully shades the ground, the evaporation from soil surface will be minimum and transpiration will be maximum. After crop completes its reproductive cycle, the ageing (senescence) of the leaves starts and the leaf area intercepting the solar radiation decreases. Measurements on the evaporation losses from dryland crop indicate that the water which crop needs will be about 0.3 to 0.4 times the evaporative demand in the beginning and it needs water almost at the potential rate and slightly more than potential rate in case of some crops after the crop canopy shades the ground. The crop continues to transpire water almost at the potential rate during its reproductive period and then onwards the water requirement declines to 0.5 times the evaporative demand when it attains maturity. The mean annual rainfall in relation to humid period is given in Table 11.2.

Water availability in relation to crop water needs

The approximate crop water requirements in relation to the mean annual rainfall for some of the rainfed crops grown in different agroclimatic regions are given in Table 11.3. In general, it can be seen that the actual water need of the crop is very much less than the mean annual rainfall. In spite of this, the dryland crops suffer from moisture stress intermittently during their growth cycle because of the erratic distribution of the rainfall during the cropping season. Therefore, rainwater management is the key element for successful crop production in the drylands. In these areas where the rainfall is slightly more than the crop water need, *in-situ* moisture conservation practised and crop cultivation in micro catchments may be advantageous, whereas in the moderate to high rainfall zones, rain water harvesting and recycling techniques have to be recommended in addition to the in situ moisture conservation practices.

Table 11.3: Mean annual rainfall in relation to crop water requirements at some dry-farming regions across India

Station	Mean annual rainfall (mm)	Crop water requirement		
		Crops	Duration of growth (days)	Water need (mm)
Jodhpur	382	Cowpea	70	280
		Pearl millet	80	320
Anantapur	590	Groundnut	110	352
Bangalore	827	Finger millet	120	384
Hyderabad	733	Sorghum	110	352
Indore	1001	Soybean	100	320
Raipur	1336	Rice	110	770

11.2.4 Matching the length of growing season and crop duration

Crops and varieties selected should match the length of growing season during which they are not subjected to soil moisture stress.

The choice of the crops and cropping systems will be guided by the following parameters:

- Distribution of rainfall and water availability in relation to the evaporative demand,
- Nature and type of the soil,
- Thermal regime during the period of water availability, and
- Socio-economic considerations of the farmers

Thus, selection of crops and their varieties for a given region depends on the pattern of the growing season and it becomes essential to match the crop peak water needs with the most probable periods of good moisture availability to achieve stable yields. Such an analysis not only helps in identifying the risks involved in production of a specific crop/varieties in a given region and enables identification of not only the choice of crops and the optimum dates of sowing (Ramakrishna *et al.*, 1988) but also in optimizing the cropping patterns for a given region (Ramakrishna *et al.*, 1985 and Ramana Rao *et al.*, 1994) for sustainable crop production.

Therefore, climatological analysis helps in identifying cultivars suitable for the region. Feasibility for monocropping intercropping, sequence cropping or double cropping can also be known from such analysis to enhance the productivity per unit of rainfall. Regions having cropping season less than 20 weeks, single crop during rainy or winter is recommended (Table 11.4). While, regions having more than 30 weeks and above the double cropping is feasible with relative ease. In regions having 20-30 weeks cropping season, double cropping may be risky, hence such areas are more suitable for intercropping.

Table 11.4: Length of effective cropping season in different areas

Category (weeks)	Effective cropping season (weeks) in different areas	Potential cropping system
<20	Bellary (8), Jodhpur (11), Anantpur (13), Hissar (17), Rajkot (17), Bijapur (17)	Sole cropping
20-30	Jhansi (21), Koralpatti (21), Hyderabad (22), Udaipur (22), Sholapur (23), Agra (24) Anand (25), Akola (27)	Intercropping
>30	Bhubneshwar, Hoshipur (35), Varanasi (32), Banglore, Indore, Rewa (36), Dehradun, Samba (44) Ranchi (45)	Double cropping

Water balance and water availability period for different agroclimatic regions has been worked out (Table 11.5). It shows that the regions with 350-600 mm rainfall having 20 weeks effective growing season are suitable for single cropping in rainy season (red and shallow black soils) or winter (black soil). Intercropping is possible in regions receiving 600-750 mm rainfall and having 20-30 weeks of effective growing season. Areas with more than 750 mm of rainfall or with more than 30 weeks are suitable for double cropping.

Table 11.5: Suitable cropping systems based on rainfall and length of water availability period

Rainfall (mm)	Soils	Water availability period (weeks)	Potential cropping system
350-600	Alfisols & shallow vertisols	20	Single *Kharif* cropping
350-600	Andisols & Entisols	20	Single cropping either *Kharif* or *Rabi*
350-600	Deep vertisols	20	Single *Rabi* cropping
600-750	Alfisols, vertisols & Entisols	20-30	Intercropping
750-900	Entisols, Deep vertisols, Alfisols & Inceptisols	>30	Double cropping with monitoring
>900	Entisols, Deep vertisols, Alfisols & Inceptisols	>30	Assured double cropping

Cropping systems

The cropping systems that can be recommended in a given agroecological region depends upon the length of the water availability period. The success of the succeeding crop in double cropping systems depends upon the soil water storage and winter rainfall as the crop is prone to face moisture stress during the later stages of growth.

Selection of crops

The selection of crops should be based on the length of the humid period i.e., the period for which the rainfall is greater than the evaporative demand. In the coastal regions as well as in the eastern parts of India wherever the average weekly rainfall is greater than twice the evaporative demand for a period of 12 weeks or more rice is grown under rainfed conditions (Table 11.6). However, in the regions having 1,000 to 1,200 mm rainfall and it is normally greater than twice the evaporative demand for about 10 to 12 weeks only. There, rice cultivation is restricted to lowlands.

Table 11.6: Approximate values of daily rates of potential evapotranspiration depending upon the climate and season in India

Climate	Daily rate of potential evapotranspiration (mm/day)during			
	Summer (March-May)	South-west monsoon (June-Sept.)	Pest rainy season (Oct.-Nov.)	Winter season (Dec.-Feb.)
Arid	8-12	4-6	4-5	2-4
Semi-arid	6-10	4-5	4	2-4
Sub-humid	5-8	3-4	4	2-4
Humid	4-6	3-4	4	2-4

Table 11.7: The length of the water availability period at some of the dry-farming locations in India

Station	Mean annual rainfall (mm)	Length of rainy season (days)	Available water holding capacity of the soil (mm)	Post rainy season rainfall (mm)	Water availability period (days)
Jodhpur	382	63	100	39	98
Hisar	400	63	100	61	103
Anantapur	590	77	50	32	98
Jaipur	669	77	100	571	126
Hyderabad	733	121	50	65	150
			250	65	200
Solapur	755	133	100	11	151
			250	11	198
Akola	840	112	100	36	144
			250	36	181
Indore	1001	112	300	34	194
Jabalpur	1373	112	100	38	151

However, the regions having annual rainfall greater than 1,200 mm, rice is grown in uplands also, as the length of the water availability period for about 12 to 14 weeks is recommended for rice cultivation. While, in the regions with annual rainfall ranging from 380 to 600 mm, the length of the humid period is about 1 to 4 weeks (Table 11.7). Thus, short-duration, drought-resistant pulse crops like mothbean, cowpea and mungbean and cereals of 10 to 12 weeks duration like pearl millet and minor millets are generally recommended.

The length of the humid period is about 10 to 15 weeks in the regions with annual rainfall of about 600 to 1,000 mm. However, sorghum, castor, sunflower and groundnut are grown in the regions with an annual rainfall upto 800 mm. For the regions having more than 800 mm rainfall crops like finger millet, maize, soybean, pigeon pea etc. are grown. In general, crops/varieties should be selected based upon their duration from vegetative to physiological maturity period to coincide with the length of the humid period so as to obtain good yields.

Selection of intercrops

The selection of the base crop and companion crop in an intercropping system is based upon the rainfall pattern. In areas having uncertain rains in the early part of the season, early planting of deep-rooted, long duration, drought tolerant crop like pigeon pea would be useful followed by the planting of medium-duration companion crop with the start of the active rainy season. While, in areas having uncertain rains in the later part of the season, the base crop should be of shorter duration with the companion crop of longer duration. In the areas having uniformly distributed rainfall, the selection of the crops should be depended upon the length of the growing season. The intercropping systems should be planned in such a way that this plan can make efficient utilization of rainwater for longer part of the growing season (Table 11.8).

Table 11.8: Length of the water availability period and suggested cropping system

S. No.	Water availability period	Cropping systems
1	Less than 75 days	Grasses, shrubs trees, short duration crops
2	75 to 130 days	Pulse crops
3	130 to 180 days	Monocropping
4	More than 180 days	Intercropping double cropping

Wherever, fruit crops are grown under rainfed conditions, the fruiting period should coincide as far as possible with the rainy season in shallow or lighter soils and with the post- rainy season in deeper or heavy soils to get higher productivity because of severe limitations of water availability It is important to note that the farmers are able to just grow a single crop in the regions with mean annual rainfall more than 800 mm in lighter soils though the actual rain water received is more than sufficient to grow two crops as appropriate technologies for rain water harvesting and management are not adopted by the individual farmers.

Climate and soil characteristics in different agroecological regions of the dryland tract strongly influence the water availability conditions and the choice of crops and cropping patterns. A simple methodology for approximate assessment of water availability conditions and crop growing periods is conveniently used for crop planning at micro level. It is nearly important to complete the sowing of the crops as early as possible with the commencement of the rainy season to take advantage of the available growing season and for maximum utilization of rainwater.

11.3 Drought Management Strategies

11.3.1 Crop planning under normal season

In case if the rainfall in the season is normal, deep tillage would be better and the operation may be correlation with tractor mould board plough/disc plough/cultivator/offset disk. This improves the soil depth and water holding capacity that improves the crop productivity. It will also be economically feasible. This is more so in pulses and oilseeds. With interculturing, dyking can be done. Seeding can be done by using improved multiple row seed (fertilizer) drills for covering the larger area under sowing with faster rate.

(i) Land preparation

The principal objective of land preparation is to improve moisture availability in the soil profile as and when a rainfall event takes place. This is accomplished by maximizing the infiltration opportunity time and minimizing the runoff. These twin objectives for making best use of limited rainwater are achieved by adopting practices that encourage maximum rainwater absorption into soil.

Guidelines for land preparation

The following practices for land preparation are recommended-

- Plug any breaches in the existing soil and water conservation structures or rodent burrows,
- Avoid costly land treatments and land preparation,
- Initiate off-season tillage as soon as possible,
- Shallow tillage with high field capacity implements as many times as possible in all soils,
- Graded line cultivation to create ridges and furrows,
- Dead furrows in shallow Alfisols and other light soils,
- Vertical mulching for low rainfall areas (Vertisols),
- Interculturing and dyking in furrows in all soils, and
- Compartmental bunding for moisture conservation on large scale mostly in Vertisols.

(ii) Soil fertility management

Rainfed tropical soils generally have low fertility due to high temperature induced exhaustion of organic matter. Therefore, management of soil fertility is essential to improve productivity. Fertilizers impart early seedling vigour with that initial root and above ground growth is promoted. Weeds are suppressed and healthy seedlings are able to withstand pest attack. For the predicted poor southwest monsoon rainfall, the costly fertilizer input has to be managed with greater efficiency. Following strategies are recommended for improving soil fertility-

- Oilseeds and cereals should be applied with 20-30 kg N/ha and 10 kg P/ha at the time of seeding.
- Leguminous crops should receive a basal application of 20 and 30 kg N-P/ha, respectively.
- Fertilizers and seed are to be applied together in the same row at required depths.
- Additional fertilizers have to be applied (top dress) depending upon the rainwater availability and type of crop.
- Organic manure is used. These supply N and several other micro-nutrients and improve water retention capacity of the soils as well.

- For maximum benefit organic manures are to be applied (FYM, compost etc.) in seed rows.
- INM practices should be evolved and to be used for aiming at reduction in cost and improving the soil fertility status.

(iii) Seeding

The following practices are recommended for seeding-

- Dry seeding is to be initiated following firm forecast of rainfall within the next 24 to 48 hours. Sowing is to be completed in shortest possible time.
- In the absence of firm forecast, seeding is to be done with the arrival of rains.
- The seed is to be treated appropriately to prevent seed and soil borne diseases.
- Wider row ratios are to be followed for accommodating more population of pulse crops in cereal, legume intercropping system and to facilitate mechanical interculture.
- Focus is to be given on short duration cultivars.

11.4 Crop Planning under Aberrant Weather Situations

Long term temporal and spatial analyses of rainfall indicate that the rainfall is seasonal, erratic and highly variable. There are numerous aberrations of weather those are to be kept in mind while planning for crops. Some of these aberrations have been discussed here.

1. Commencement of rains may be early or considerably delayed.
2. There may be prolonged dry spell during south-west monsoon. Rains may terminate earlier than normal cessation or may continue beyond normal rainy season.

Management options to mitigate the adverse effect of aberrant weather can minimize the risk of total crop failure.

After sowing of recommended cropping systems/crops, moisture stress for the crops can be anticipated. Depending upon time and intensity of moisture stress, management strategies have to be adopted. Moisture stress periods are usually classified as (i) early moisture stress, (ii) mid-season stress, and (iii) terminal stress.

11.4.1 Early season stress

Season starts with good rainfall followed by long dry spells and in this situation following practices are recommended-

(i) From the rainfall pattern prevailing till to-date off season tillage is missing. If rain occurs in May, off-season tillage needs to be practiced to reduce weed proliferation In view of this, blading of land is necessary. The primary tillage can be started with the onset of monsoon.

(ii) As initial lung dry spells are probable, deep tilled open soil loses more moisture leading to delay in sowing, poor germination and low yields. Therefore, instead of deep tillage, a shallow tillage to cover large area and for utilizing the available soil soaking rains for planting can be done. This also needs low energy than deep tillage. For faster coverage, use of tractor or improved implements with higher field capacity will be beneficial. With shallow tillage, the depth of soil will not be a problem, as is true with deep tillage.

(iii) If the onset or monsoon is with very heavy showers of 100 mm or more and this kind of showers occur over 2-3 weeks, probability of such events being generally low, deep tillage needs to be followed wherever practically feasible. This will assist in conservation of moisture and restricting weed emergence.

(iv) Graded line cultivation is beneficial; ridges and furrows are formed as a result of tillage operations. Ridges and furrows help in enhancing the rainwater infiltration opportunity time. With that more moisture is conserved and runoff of water and soil loss is reduced.

(v) Along with the practice of ridges and furrows dyking at 4-8 m interval of dead furrow is further beneficial in conserving runoff with light textured soils.

(vi) Compartmental bunding (3x3 to 5x5 m) on large scale is effective in low rainfall areas.

The early season moisture stress occurs due to the failure of rains after sowing of crops/cropping systems or due to delayed start of rainy season. If a prolonged dry spell occurs immediately after sowing of crops/cropping systems, the seedling may wither and adequate plant stand can not be maintained. Under this situation, if gaps are less than 25 per cent compared to the optimum, it is better to do gap filling of the same crop. If more than 50 per cent of the plants are withered, it is better to reshow the same crop of shorter duration variety or any other short duration crop as per recommendations for a given region.

11.4.2 Mid-season stress

Mid-season stress may occur due to the break in monsoon conditions after the establishment of the crops. Under this situation the following practices are recommended:

(i) Interculturing is essential for conservation of moisture by creating dust mulch and also weed free condition. For this graded line cultivation will be helpful. Under these conditions going in for land treatments like beds and furrows may be possible due to short duration of season and lack of appropriate farm implements.

(ii) Vertical mulching, dyking, compartmental bunding, dead furrows can be practiced during this time before planting.

(iii) Formation of dead furrows in between crop rows will help for better moisture conservation with subsequent rains.

(iv) Application of supplemental irrigation to the stressed crop wherever the facility of water harvesting pond exists.

(v) Apply top dressing after relief of the moisture stress during pre-flowering periods.

(vi) Removal of lower leaves of the plants,

(vii) Reduction of plant population by thinning,

(viii) Use of anti-transpirants etc.

In drought years, complete crop failure can be avoided by adopting mid-term corrections i.e.:

(a) Removing the susceptible component and allowing the resistant component in intercropping system.

(b) Removing main component and sowing of *Rabi* crop if rain received late.

(c) Rationing in case of sorghum and bajra can be done successfully.

Weed management

Weeds compete with crop for resources such as nutrients, water and such competition for soil water from weeds can be reduced by timely weed control.

Rationing/thinning

The rate of soil moisture depletion increases with increasing leaf area. If drought occurs at 40-50 days after sowing, reduction in leaf area either by rationing or thinning the plant population can mitigate the adverse effects of drought. Sorghum and pearl millet respond to rationing. Two per cent urea spray after drought period is useful for intermediate crops like castor, pigeon pea and groundnut.

11.4.3 Terminal stress

It occurs due to early termination of rainy season resulting in severe moisture stress during the productive and physiological maturity stages of a crop. In order to minimize the crop losses supplemental irrigation has to be provided to the standing crop wherever the farm ponds exist.

Mulching

Surface mulching with organic materials extend the period of water storage in soil profile by regulating the heat balance of soil. Use of mulches during rainy season will reduce the heat load, thus will cause reduction in evaporation loss of stored water. Repeated inter-cultivations in black soils results in formation of soil mulch on the surface. It acts as a barrier in capillary movement and check evaporation loss from the soil profile. Deep cracks formed due to drought increase the loss of stored water by evaporation, Soil mulch minimize deep cracking leading to reduced evaporation losses (Table 11.9).

Table 11.9: Effect of surface mulch on the yield of *Jowar* (kg/ha)

Treatment	Grain	Straw
Control	728	3430
Jowar stubbles	1035	4720
Wheat straw	1120	5290
Redgram stalk	1182	5620
Dry grass	1265	6240

Average of three years (Solapur, 1970-1975)

Water harvesting and protective irrigation

In-situ water harvesting and protective irrigation are the potent measures of crop life saving during periods of soil moisture stress (Table 11.10).

Table 11.10: Effect of life saving irrigation on grain and stover yield under different cropping systems

Treatment	*Kharif*		*Rabi*	
	Grain (q/ha)	**Stover (q/ha)**	**Safflower grain (q/ha)**	**Eq. yield stover (q/ha)**
Cropping system				
(i) Fallow-safflower			16.16	42.45
(ii) Sorghum (F)-Safflower	265.46	-	14.30	37.14
(iii) Maize + pigeon pea	19.45	46.08	20.76 (11.87)	67.53 (45.02)
SEm ±			4.59	12.70
CD at 5%				
Application of harvested water				
(i) 5 ha cm as presowing/early supplementation			14.20	43.15
(ii) 10 ha cm as presowing/early supplementation			15.96	48.13
(iii) 5 ha cm as presowing/early supplementation +5 ha cm at 45 DAS/late supplementation			21.26	55.34
SEm ±			0.80	2.19
CD at 5%			2.48	6.75

(Source: DFRS, Arjia, Bhilwara, 1994)

11.4.4 Delayed or late onset of monsoon

Delayed onset of monsoon brings in added dry spell over off season. One or two shower of rains but may occur and these may not be useful for planting. However, these are to be capitalized for moisture storage by wide blade harrowing. Primary tillage under delayed monsoon condition should have the objective of wider coverage for early planting to use available moisture. This can be accomplished by following shallow tillage practices.

Crops and varieties recommended for normal onset of monsoon may not be remunerative for delayed sowing and different crops and varieties may be selected. Two management options under these circumstances are transplanting and alternate crops/varieties.

Transplanting

Community nurseries under irrigation is the sure way of compensating delay in monsoon. Transplanting can be taken up with the arrival of rains. Finger millet, pearlmillet, sorghum and setaria (foxtail millet) are well adapted to transplantation.

Alternate crops/varieties

Certain crops and varieties can perform better even if sown late in the season. Depending on the arrival of rains, such crops and varieties can be selected. Castor is more remunerative in red soils of Telangana (AP than pigeon pea under late sown conditions, Sunflower, because of its shorter duration, performs better than the groundnut when sowing are delayed beyond mid of September in relatively deeper red soils. In western Rajasthan, short duration arid legumes such as moth bean and cluster bean are more remunerative than the pearl millet when sown late. Crops and varieties for different dryland regions have been identified for late of monsoon.

Dry spells during crop period is common in dryland agriculture. The adverse effect of moisture stress should be mitigated to avoid total crop failure.

11.4.5 Early withdrawal of monsoon

Early withdrawal of monsoon before winter seeding leads to problems of crop stand establishment and terminal drought. Therefore, winter season seeding should made be as far as possible in early September to avoid the above two problems. All the suggested measures to mitigate the adverse effect of prolonged dry spells during crop period are necessary to overcome the soil moisture stress due to early withdrawal of monsoon. Management options available for different drought situations are summarized in Table 11.11.

12.4.6 Extended monsoon

Extended monsoon is seldom a problem in dryland agriculture. It benefits long duration *Kharif* crops, taken as a component crops in intercropping system and also advantageous for *Rabi* crops in Vertisols.

Table 11.11: Management options available for different drought situations

	Rainfall Pattern	Management options
(i)	Delayed onset of monsoon	Transplanting of alternative crops and varieties
(ii)	Prolonged dry spells	
	(a) Immediately after sowing	Resowing of crops
		Reduce plant population
	(b) Mid season 'breaks'	Remove the sensitive crop component in mixed stand
		Mulching
		Weed control
		Land configurations (in-situ moisture conservation)
		Intercropping
		Recycling of harvested water
(iii)	Early cessation of rains	Reduce plant population
		Protective irrigation
		Mulching
		Defoliation of lower leaves

11.5 Crop Planning Based on Onset of Monsoon

Suitable crops and their varieties

Crops which do well under normal years may not do so under abnormal weather situations. For example, under south Rajasthan condition sorghum perform better than maize (traditional crop) during varied climate in summer and safflower, barley found better than Wheat (traditional crop) in normal years while taramira, gram and safflower during drought years in winter. Growing large areas under improved varieties of cereals, pulses and oilseeds in summer and winter. The following varieties have been found most suitable for dryland areas:

Maize	D 765 (low rainfall), Navjot (normal rainfall), PEHM 1, PEHM 2 Aravalli Makka 1, Surya. KH-510
Bajra	RCB 2, CM 16 CSH 6, CSH 9, SPV 245, SPV 346
Sorghum	JL 24, AK-12-24

Greengram	K 851, S 8, RMG-62
Blackgram	T 9, PU 30
Castor	Aruna, GAUCH 4
Sesame	RT 46, RT 125
Soybean	Monato, Gaurav, JS 71-105, Punjab 1
Clusterbean	Durgapura safed, Durga Jai RSG 936
Moth	RMO 40, Jwala, IPCMO 880, Maru Vardan
Wheat	Narbada 4, C 306, Mukta, Sujata
Barley	RD 297, RD 294, RD 31
Gram	Dahod yellow, BG 209, C 235, ICCV 10
Mustard	Varuna, Durgamani, Bio 902
Linseed	Chambal, RR 45
Safflower	Bhima, JSF 5, JSF 2
Pigeonpea	Gwalior 3 (long duration), T-21, ICPL 87, Prabhat
Taramira	T 27

Cropping systems

In dryland conditions intercropping systems showed superiority over sole cropping systems. Intercropping systems increased production and income also stability and assured successive crop production in addition to other indirect benefits in the soil health. Some important intercropping systems are recommended as follows:

(i) Paired planting of maize at 30/120 cm and growing of 2 rows of blackgram in the inter space is found better.

(ii) Alternate planting of maize and pigeon pea at 37.5 cm found remunerative.

(iii) Planting of maize and castor 1:1 row ratio gave higher returns.

(iv) Paired planting of sorghum 30/75 cm and growing of blackgram in the inter space was a better system.

(v) Paired planting of bajra at 30/75 cm and raising of greengram or cluster bean in the inter space was found most suitable intercropping system.

(vi) Growing of 2 or 3 rows of blackgram or greengram in the inter row space of cotton grown at 1 m. spacing was found remunerative,

(vii) Alternative planting of pigeon pea + groundnut at 37.5 was better under erratic rainfall conditions.

(viii) Growing of two rows of greengram in the inter row space of castor grown at 1 m row spacing was better than growing sole crops.

(ix) Groundnut + sesame in 6:2 row ratio at 30 cm spacing was better than growing of sole crops.

(x) Gram + safflower 6I row ratio at 30 cm spacing was better than growing of sole crops.

(xi) Gram + mustard in 4:1 row ratio at 30 cm spacing were found remunerative.

(xii) Intercropping combination like grass+mung and sunflower+ mung found better for western Rajasthan.

Alternate land use

In this process, marginal land not suitable for agricultural production has also been put to cultivation so much so that often at times the farmers can hardly get back his seed from the field to be cultivated. The discriminate use of land lead to severe soil erosion problems, silting-up of reservoirs and consequently the occurrence of flash floods. To achieve greater efficiency in utilizing the resources in dryland and also to combat the "Energy crisis" and deficiency of animal feed, major approaches in this direction should be –

(i) Putting the land (class iv & above) under suitable grasses, legumes and fodder crops and integrating with animal productive systems.

(ii) Using the same land or a portion of it simultaneously or sequentially for food, fodder and fuel, due attention being given to interaction among various uses aiming to obtaining greater sustained production and securing both immediate benefits and long term environmental concerns.

The various alternate land use systems which may be recommended are as follows:

a. Agro-forestry system,

b. Silvi-pastrol system,

c. Farm forestry system,

d. Agro-horticulture system,

e. Silvi-horticulture system, and

f. Alley cropping system.

For these systems the probable suitable crops, grasses, legumes and trees are given below for semi arid areas having 500-1150 mm annual rainfall.

Crops	**Grasses and legumes**	**Trees**
Sorghum	*Cenchrus ciliaris*	*Acacia Arabica*
Pearl millet	*Cenchrus setigerus*	*Acacia nilotica*
Maize	*Sehima nervosun*	*Aibizia lebbeck*
Crops	Grasses and legumes	Trees
Pigeonpea	*Panicum antidotale*	*Annona squmose*
Greengram	*Lasiurus sindicus*	*Azadirachta indica*
Blackgram	*Heteropogon controtus*	*Dalbergia sissoo*
Clusterbean	*Dichanthium annulatum*	*Prosopis cineraria*
Moth (Kidney bean)	*Atylosia scarabaeoides*	*Leucaena leucocephala*
Kulthi (Horsegram)	*Stylosanthes humilies*	*Ziziphus jujube*
Groundnut	*Clitoria ternatea*	
Mustard		
Safflower		
Barley		
Gram		

Other improved agronomical practices

It includes following -

a. Timely sowing at proper spacing,

b. Use of balanced fertilizers,

c. Proper and timely plant protection measures,

d. Timely hoeing and weeding operations,

e. Sowing improved and healthy seed with proper seed treatment, and

f. Seed soaking before sowing etc.

References

Anonymous. 1970-1975. Annual report AICRP on Dryland Agriculture, Solapur

Azpiroz, H.S.P., Vincourt and Serleys, H. 1988. Utilization of 'in vitro' test as an early screening technique for drought stress evaluation in sunflower. In: Proc. 12th Int. Sunflower Conf. Novi sad, Yugoslavia.

Blum, A. 1988. Plant breeding for stress environment. CRC Press, Boca Raton, Florida, 208 pp.

Bridge, R.R. and Chism, J.F. 1978. Registration of DES 24 cotton. *Crop Sci.* 18:523.

Cocheme J, and Franquin, O. 1967 . A study of the Agroclimatology of the Semi-arid Area of South Sahara in West Africa, FAO/UNESCO/WMO Inter Agency Project p.117.

DFRS. 1994. Annual report of Dryland Farming Research Station, Arjia, Bhilwara on Dryland Agriculture.

Dhar, O.N., Rakhecha, P and Mandal, B. N. 1981. Influence of Tropical disturbances on monthly monsoon rainfall of India. Monthly Weather Review 109, 188-190.

Flower, D.J. and M.M. Luldlow 1987. Variation among accessions of pigeonpea (*Cajanus cajan*) in osmotic adjustment and dehydration tolerance of leaves. *Field crops Res.* 17:229-243.

Garrity, O.P. and O' toole, J.C. 1995. *Agron. J.*, 87: 773-779.

Hall, A.E., Thiaw, S., Ismail, A.M. and Ehlers, J.D. 1997. Water use efficiency and drought adaptation of Cowpea. In: *Advances in Cowpea Research.* Edited by Singh B.B. Mohan Raj, K.E., Dashiell and L.E.N. Jackal, Co. Publications of IITA and JIRCAS, IITA, Ibadan, Nigeria. pp. 87- 98.

Hangreaves, G. H. 1971. Precipitation dependability and potential for Agricultural Production in Northeast Brazil. EMBRAPA and Utah State University publication 74 D 159, pp. 123.

Hanson, A.D. and Everson, E.H. 1978. *Agron. Abstr.* 75.

Higgins G.M. and Kasam A. H. 1981. The FAQ agro-ecological zone approach to determination of land potential, Pedologic XXXI, 2: 147-168.

Johnson, R.R. and Herro, M.P. 1981. Corn pollination under moisture and temperature stress. In: Proceedings of the 36th Annual Corn and Sorghum Research Conference. Chicago, Illinois, USA. pp. 66-77.

Lahiri, A.N. and Kumar, V. 1966. Studies on plant water relations III. Further studies on the drought mediated relations in the performance of bulrush millet. Proceedings of national Institute of Science, India, Part B. 32: 116-129.

Ludlow, M.M. and Muchow, R.C. 1988. Critical evaluation of the possibilities for modifying crops for high production per unit of precipitation. In: Bidinger; F.R. and Johansen, C. (eds.). Drought Research Priorities far the Dryland Tropics. Pattancheru, A.P. India. ICRISAT pp 179-211.

Mali, P.C. and Mehta, S.L. 1977. Phytochemistry, 16: 1355-57.

Purushothaman, M.G. 1998. Development of drought tolerant cell lines in Groundnut genotypes in vitro, *Indian J. Pl. Physiol.*, 3 (4): 280 - 289.

Quinsebarry, J.E. Jordan, W.R., Roark, B.A., and Fryear, D.W. 1981. Exotic cotton as genetic sources for drought resistance. *Crop Sci.* 21: 889 - 895.

Rainarkishna, Y.S. 1988. Climatic changes in relation to desertification in the Indian arid zone. In: Desertification and its control Ed. A.H. Tiwari. Scientific Publishers, Jodhpur. 99-113.

Ramage C.S. 1968. Problem of a monsoon ocean. Weather 23 (1): 28-36.

Ramakrishna Y.S., Ramana Rao, B.V. and Rao G.G.S.N. 1985. Water balance and crop planning. A case study of western Rajasthan. Ann. Arid Zone 24 (2): 114-119.

Ramakrishna Y.S., Ramana Rao, B.V., Rao, G.G.S.N and Joshi, N.L. 1988. Moisture regime variation and agricultural Productivity in Rajasthan. In Desert Ecology (Ed. I. Prakash) Scientific Publishers, Jodhpur: 65-88.

Ramana Rao, B.V. and Ramakrishna, Y.S. 1994. Water availability and it's requirement for crops in drylands. *Indian Farming Occ.* pp. 24-29.

Ramana Rao, B.V., Ramakrishna, Y.S. and Victor, U.S. 1994. Resource characterization for improved soil management Climate. In: Soil management: for sustainable Agriculture in dryland areas. Bull No.16 *Ind. Soc. Soil Science*, New Delhi: 26-32.

Rao, N.G.P. 1982. Transforming traditional sorghum in India. In: Sorghum in the eighties. Proceedings of the International Symposium on Sorghum at ICRIS 2-7 Nov. 1981, pp. 34- 59.

Reddy, S.R. 2004. Principles of Agronomy. Kalyani Publishers, New Delhi.

Richard, A. Morris 1988. The Mungbean crop: Abiotic stress and production practices. Multi-location testing on mungbean germplasm. International Training Course ARS-AVRDC, Kaset Dart University, Kemphaeng gaen, Nakhon, Pathom, Thailand. pp. 200-201.

Scientific Liaison Office 1974. Department of the Prime Minister. Rhodesia. Department of the University Library, Rhodesia University. Compliers, Rhodesia. Research India. 1973. Salisbury. Rhodesia: Government Printers 106 pp.

Seethararna, N. and Bidinger, F.R. 1977. Sorghum drought resistance crop presented at the International Sorghum workshop. ICRIS at 6-13 March 1977. Hyderabad, India.

Sheldrake A.R. and Narayanan, A. 1979. Growth, development and nutrient uptatkke in pigeonpea (*Cajanus cajan*). *Journal of Agricultural Science*, Cambridge 92, 513-526.

Singh, D. and Sahay, R.K. 1990. *Indian J. Agric. Sci.* 60: 739 - 41.

Singh, R.P. 1995. Sustainable development of dryland agriculture in India. Scientific Publishers, Jodhpur.

Singh, T.N., Aspinall, D. and Paleg, L.G. 1974. *Indian J. Genet. Pl. Breed.*, 34 A: 1074- 83.

Skonic, D. 1992. Sunflower and Breeding. Uljarsttvo 25(1): 1-90.

Subharamayya, I. Bhanu Kumar, O.S.R.V. and Babu, S.V. 1981. The Summa monsoon reins over South and East Asia and West Pacific. WMO Technical conference on climatic - Asia and Western Pacific pp 316-333.

Subrahmanyam, V.P., Subba Rao, B. and Suhrahmaniam, A.R. 1963. Moisture adequacy in relation to distribution of some crops in India. pp 462-467 In. Proc. 13th Assembly Int. Assoc. of Science and Hydrology (I.U.G.G) Berkelay, U.S.A.

Sullivan, C.Y., Norcio, V.V. and Eastin, J.D. 1977. Plant responses to high temperatures. Pages 301- 317 In: Genetic diversity in plants. (eds. A. Mohamed R. Aksel and R.C. Von Borstel). New York, USA: Plenum Publishing Crop.

Thornthwaite, C.W and Mather, J.R. 1955. The water balance. Publ. in Climatology. Drexel inst. Techn N.J. 8, 1-104.

Troedson, R.J. 1987. Effect of plant density on pigeonpea (*Cajanus cajan* L. *Mill Sp.*) in sub humid environments. In proceedings of the fourth Australian Agronomy conference. Australian Society of Agronomy, Melbourne, p. 22.

Troyer, A.F. 1983. Breeding corn for heat and drought tolerance. In proceedings of the 38th Annual crop and Sorghum Research Conference, Chicago, Illinois, USA, pp. 128-143.

Williams, T.V., Snell, R.S. and Cress, C.E. 1969. *Crop Sci.*, 9: 19-23.

EXERCISES

11. What is the crop planning?

11.2 Explain how crop planning is performed for a given region.

11.3 Define the following:
   - (a) Index of moisture adequacy
   - (b) Length of growing season
   - (c) Mid-season stress
   - (d) Delayed onset of monsoon
   - (e) Terminal stress
   - (f) Extended monsoon

11.4. Describe the FAO criteria for determination of length of growing period.

11.5. State condition in which intercropping and sequence cropping will be taken up.

11.6. Describe the factors on which crops and cropping system will be selected.

11.7. How intercrop is selected?

11.8. How early stress is managed?

11.9. Explain the mid-season stress and measured for its management.

11.10. Describe how delayed onset of monsoon is managed.

11.11. Explain the crop planning for alternate land use.

11.12. Discuss the crop planning for length of growing season of 130-180 days.

■■■■

Chapter 12

WATER LIFTING DEVICES

12.1 Introduction

Water lifting devices are used to lift the harvested rainwater from on-farm reservoirs for irrigation and drinking purposes. When the on-farm reservoirs is at a lower level than the field to be irrigated (e.g. reservoirs) and irrigation by gravity flow is impossible. This is known as flow irrigation. It also known as lift irrigation. Pumping is necessary to lift the water from water harvesting structures *viz.*, farmer's pond, nadi, community tank and anicut and also from streams or rivers etc. In case where free gravity flow is not possible water lifts are used to dispose off surface water.

Water lifting devices may be classified into following four groups based on the kind of power used to operate the water lifting devices,

(1) Human power operated,

(2) Animal power operated,

(3) Mechanical power operated, and

(4) Wind and solar power operated.

12.2 Human Power Operated

The human power operated water lifting devices are described as under:

(i) ***Scoop***: It is one of the oldest devices (Fig. 12.1) and operated by a single person. To give some mechanical advantage the scoop is sometimes suspended to a tripod.

(ii) ***Swing basket***: It is operated by two persons facing each other (Fig. 12.2). Each person holds two of the ropes and swing the basket, alternatively filling it from the source and emptying it into the delivery channel. This type of device may be used upto a lift of one meter.

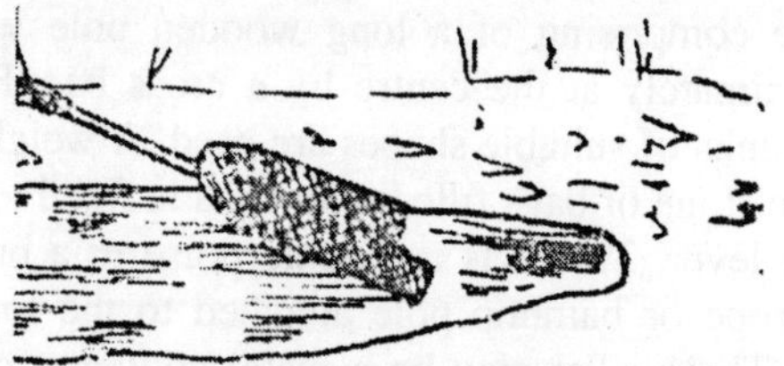

Fig. 12.1: Scoop

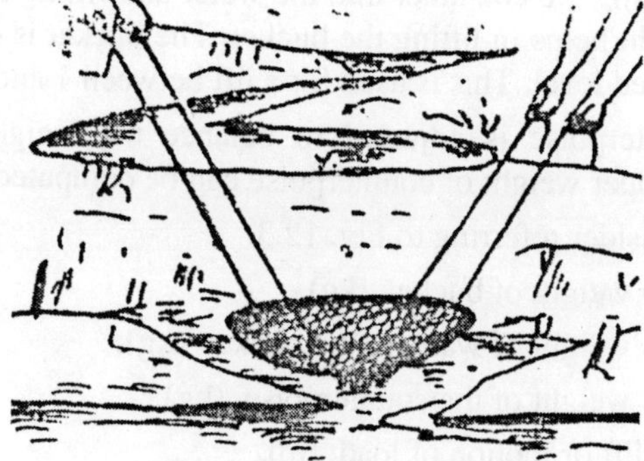

Fig. 12.2: Swing basket

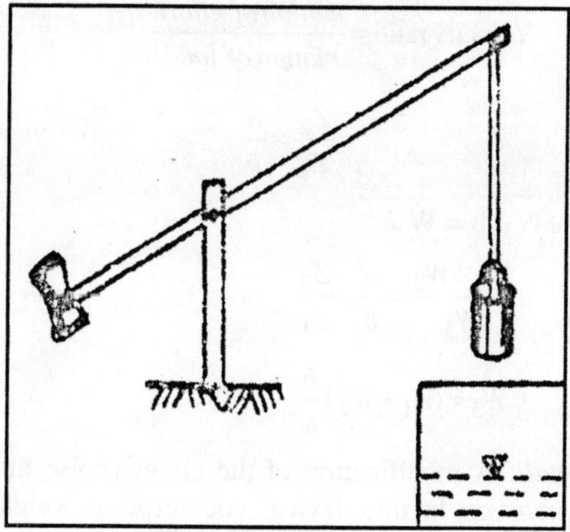

Fig. 12.3: Counter poise lift

(iii) ***Counterpoise***: It is human power operated water lifting device (Fig. 12.3). It operates based on the lever principle. Its output is more than the water lifting devices described earlier. This device comprising of a long wooden pole which is pivoted approximately at the centre by a cross bar. For this purpose tree trunks of suitable shapes are used. A weight in the form of a large stone or bags filled with soils is fixed to the shorter end of the lever. This acts as counterpoise to a bucket suspended by a rope or bamboo pole attached to the longer arm of the lever. The bucket may be a metal container or a leather bag. This device is operated by pulling down the rope or rod and bringing the container into the water and filling it. The counter weight helps in lifting the bucket. The bucket is emptied at the desired level. This is used for a lift between 1 and 3 m.

The counterpoise is adjusted to balance the weight of the full bucket. The proper weight of counterpoise can be computed as follows:

Let us consider referring to Fig. 12.3.

W_1 = weight of bucket, (kg),

W_2 = weight of water in the bucket, (kg),

W_3 = weight of the counterpoise, (kg),

Z_1 = lift or motion of load, (m),

Z_2 = motion of effort, (m),

$$\text{Velocity ratio} = \frac{\textit{motion of effort}}{\textit{motion of load}}$$

$$= \frac{Z_2}{Z_1} = \frac{a}{b}$$

Now, $(W_1+W_2)\,b = W_3a$

or
$$\frac{W_1+W_2}{W_3} = \frac{a}{b} = \frac{Z_2}{Z_1}$$

$$\therefore W_3 = (W_1+W_2)\frac{b}{a} \qquad \text{....(12.1)}$$

(iv) ***Picottah***: A modification of the counterpoise lift is known as the *picottah*. In this device, counterpoise weight, one or the two men walk back and forth along the top of the pivoted pole. The weight of the men shifts as they walk, thus causing the bucket to be raised or lowered. A third person fills and empties the bucket. This device is larger as compared to the counterpoise lift and having a lift of 5 to 8 m.

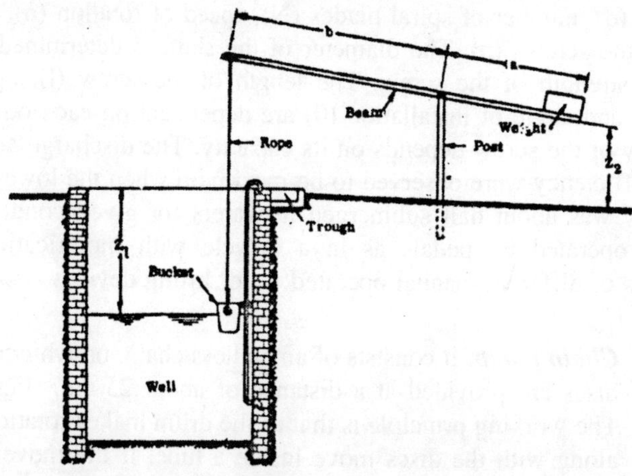

Fig. 12.4: Schematic stektch illustrating the principle of working of a counterpoise-bucket lift

(v) ***Archemedian screw***: It consists or a wooden cylinder with an auger like spiral (Fig. 12.5). The spiral is made of short wooden boards or battens fitted on a control shaft, one end of which serves as a handle. The spiral on the control shaft is enclosed in a cylinder of wooden planks. The battens form a double helix and help in the movement of water from the inlet end of the cylinder to the outlet end. This device is portable and is operated by one or two men facing each other in a sitting position. In this operators will have their feet submerged in water during the period of working. This causes water-borne parasite and consequent diseases.

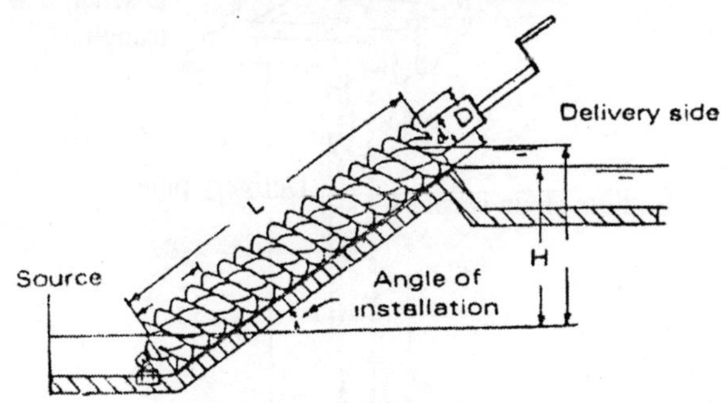

Fig. 12.5: Archemedian screw

Operation of the Archemedian screw depends upon the variables *viz.*, head (H), angle of installation (θ), screw diameter (D), shaft diameter (d), number of spiral blades (N), speed of rotation (n), and the pitch of the screws (p). The diameter of the shaft is determined by the required strength of the screw. The length of the screw (l), operating head (H) and angle of installation (θ) are dependent on each other. The efficiency of the screw depends on its capacity. The discharge and mechanical efficiency were observed to be maximum when the lower end of the screw was about half submerged in waters for given conditions. It may be operated by pedals as in a bicycle with modification. The capacities of different manual operated water lifting devices is shown in Table 12.1.

(vi) ***Chain pump***: It consists of an endless chain, on which circular discs are provided at a distance of about 25 cm. (Fig. 12.6). The working principle is that as the drum makes rotation chain along with the discs move inside a tube. If the movement of the chain is upwards, each disc brings up a volume of water and the water is discharged into a trough which in turn discharges into the outlet channel. The discs are made of metal or leather washers strengthened by metal washers. The later ones provide comparatively smoother operation. This pump operates satisfactorily upto a lift of 6 m lifts. The tube through in which the discs pass is of 10 cm dia. and larger the diameter greater effort will be required to operate the device. It may also be operated with animal power by adding a driving mechanism.

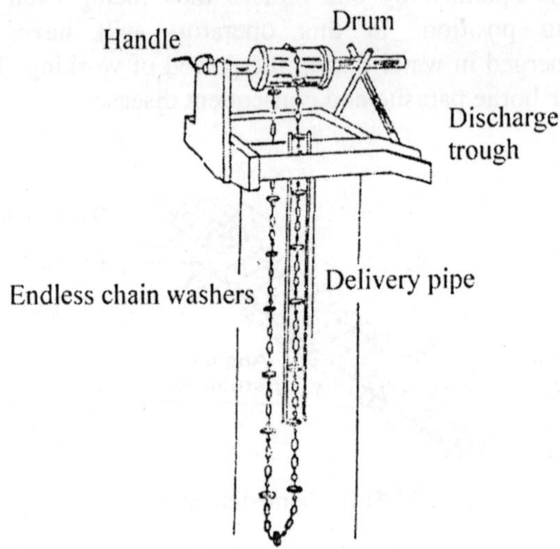

Fig. 12.6: Chain pump

(vii) ***Hand operated reciprocating pump***: It is also known as the Allahabad pump and operated by reciprocating pump (Fig. 12.7). It consists of a cylinder in which a piston is operated using a lever mechanism. The suction side is connected to a pipe which draws water from the source. This pump is used for a static lift of 4 m and gave a discharge of 3-6 l/sec.

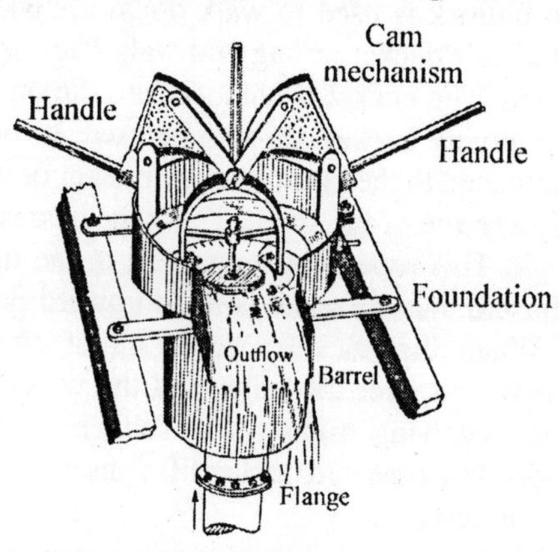

Fig. 12.7: Hand operated reciprocating pump

Table 12.1: Capacities of different manual operated water lifting devices

Name	Operating conditions	Capacity (Cu. m/hr)
Scoop	1 metre 1 man	0.8
Swing basket	4 men working in 2 shifts of two each Height 0.6 m	5.0
Counterpoise lift	Height lifted 2.5 m 1 man operating	3.0
Archemedian Screw	Height 0.45 m 1 man 185 cm length and 40 cm dia	4.0

(Source: Molenaar, 1956)

12.3 Animal Power Operated

Bullock may be used as a single or in pair and camel also for water lifting purposes. The animal power operated water lifting devices may be broadly grouped into two groups based on working principles as,

(i) Principle pulley, and

(ii) Roped bucket with rotating wheel

Rope and bucket: It consists of a bucket or bag of leather or canvas, attached to one end of a long rope. The rope passes over a pulley set in a suitable framework of wood installed over the water harvesting structure. A pair or a single bullock is used to walk down a sloped earthen ramp while pulling up the full bucket or bag and walk back up the ramp after the bucket is emptied. The bucket is emptied into the outlet channel by a person, or a self emptying bucket can also be used. In the self emptying bucket a spout is attached to the bottom of the bucket or bag. A light rope is fastened to the lower end of the spout. This rope passes over a roller on the receiving trough. This rope is also attached to the bullock yoke and its length is so adjusted that the spout is in an upward position while the bucket rises up. When the bucket reaches outlet channel, the spout follows the rope tied to it over the roller and the water in the bucket is discharged into the receiving trough. The self emptying bucket saves labour of one person. The rope and bucket lift is used to water upto a lifts of 10 m from on-farm reservoir.

This device has the drawback in its operation that animals have to walk backwards up an inclined earthen ramp while the bucket is lowered to the water source.

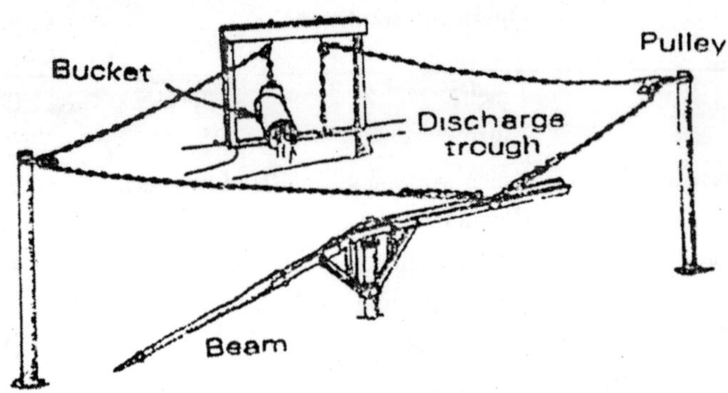

Fig. 12.8: Two bucket lift

Two bucket: In this device, the animal move in a circular motion which is relatively more convenient to the animals as compared to forward and backward movement. It comprising of two buckets which move up and down on guide rails. The raising and lowering of the bucket is achieved by a chain linkage (Fig. 12.8). The buckets move alternatively i.e. as one bucket moves up after filling the other one moves down after getting emptied. The emptying of the bucket is automatic. A flap valve at the bottom of the buckets allows the buckets to submerge and fill up as they are lowered in water. It is used to lift water upto a height of 5 m.

Persian wheel: It is one of the most popular animal operated water lifting device. It consists of an endless chain of buckets passing over a circular drum made from angle- irons (Fig. 12.9 a). The drum is rotated by a horizontal shaft which in turn is rotated by vertical and horizontal cogwheels. The large wheel has a pole extending from it horizontally. This pole is drawn by the animals moving in a circular manner. The metal containers have a very small opening at their bottom to drain out the water in them and thus not to provide large starting load. The main advantages are the rugged construction and reasonably good amount of discharges. The parts are durable and generally, it needs very little maintenance. The animals operating this device are sometimes blindfolded. This helps the animals to keep a steady pace, prevent them from getting giddy and also continue operation even when the operator is away temporarily.

Persian wheel is also known as the water wheel (Fig. 12.9 b). In this device, instead of endless chain of buckets, blades are fixed in the wheel itself. This device is, therefore, suited for small lifts, limited to the radius of the wheel.

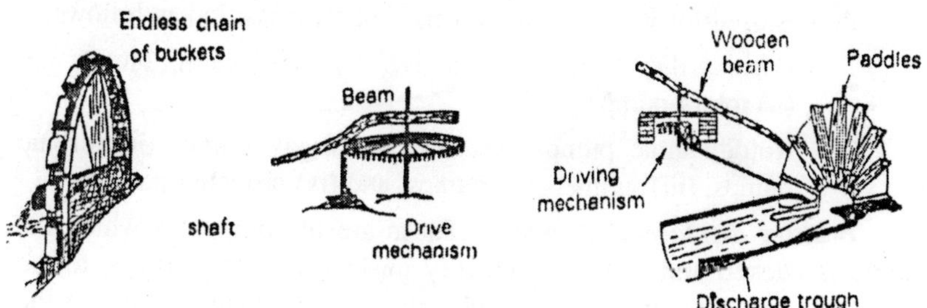

Fig. 12.9(a): Persian wheel; (b): Water wheel

Reciprocating pump: It is operated by animal. This pump consisting of two units of the Allahabad hand pumps as described earlier. The suction side of both the units is connected to a pipe through a T-joint. The pipe ends in the source of water. The units are driven by a gear mechanism similar to the Persian wheel. The animals move in a circle to drive the gear mechanism. The main advantage of this pump is that the suction pipe can be placed in a shallow bore for water lifting. It is used to lift of 6 m head with a discharge of 7 l/sec. The capacities of different animal drawn water lifting devices are summarized in Table 12.2.

Table 12.2: Capacities of animal water lifting devices

Name	Operating conditions	Capacity (Cum/hr)
Rope and bucket lift	9 metres lift Two pairs of bullocks	16 to 17
Two bucket lift	3 metres lift	22
Persian wheel	9 m lift	8-10
	6 m	10-12
	3 m	15-17

(Source: Molenaar, 1956)

12.4 Mechanical Power Operated

These devices provided much higher rates of discharge as compared to human and animal operated devices. Electricity, mechanical power, solar and wind power are used for some of the sources of energy which operate the pumps.

A pump is a machine which and is placed in a conduit, transfers energy from the external source to the liquid flowing through the conduit.

Pumps used for water lifting may be broadly classified as-follows:

1. Positive displacement pumps *viz.*, (i) reciprocating pumps and (ii) rotary pumps.
2. Rotodynamic pumps *viz.*, (i) centrifugal pumps, (ii) turbine pumps, (iii) submersible pumps, and (iv) propeller pumps.

Positive displacement pumps: These are the pumps in which the liquid is sucked and then it is actually pushed or displaced due to the thrust exerted on it by a moving chamber, which results in lifting the liquid to the required height.

Rotodynamic pumps: These pumps have a rotating element, called impeller, through which as the liquid passes its angular momentum changes, due to which the pressure energy of the liquid is increased. These pumps do not push the liquid as in the case of a positive displacement pump. The centrifuge pump is the most of common of example of rotodynamic pump.

Pumps may be classified, on the basis of the following: (i) mechanical principle of operation, (ii) type of power required, and (iii) type of service called for.

(a) Classification based on mechanical principle of operation:

Based on the principle of operation, pumps can be broadly classified into the following four types:

(i) Displacement pumps,

(ii) Centrifugal pumps,

(iii) Air lift pumps,

(iv) Miscellaneous pumps.

(b) Classification based on type of power required:

(i) Steam engine pumps,

(ii) Diesel engine pumps,

(iii) Electricity driven pumps.

(c) Classification based on depth of well:

(i) Low lift pumps,

(ii) High lift pumps,

(iii) Deep well pumps,

(iv) Booster pumps.

The selection of a particular type of pump depends upon the following factors:

(i) Capacity of the pump,

(ii) Number of pump units required,

(iii) Suction conditions,

(iv) Lift (total head),

(v) Discharge conditions, and variations in the load,

(vi) Floor space requirement,

(vii) Flexibility of operation,

(viii) Starting and priming characteristics,

(ix) Type of drive required,

(x) Initial cost and running costs.

12.4.1 Positive displacement pumps

Pumping is accomplished by alternatively filling and emptying one or more chambers with water. The rate of discharge in these pumps depends mainly on the speed of the device. These pumps discharge almost same volume of water.

Reciprocating pumps: Reciprocating pumps are more suitable for lifting relatively clean water, against high and fluctuating head. Centrifugal pumps are well suited to pump water under steady low head, though the water may be even turbid. Therefore, these pumps are mostly used for lifting raw water from the river to the treatment plants, and then lifting treated water to the distribution reservoirs. These pumps is consist of a piston or plunger operating in a cylinder. The other components of the pump are shown in Fig. 12.10. The crank is rotated either by the driving engine or the motor. Due this, plunger moves to and fro in the cylinder (also known as pump barrel). The outward movement of the plunger creates partial vacuum in the cylinder. Water is sucked into the cylinder due to the atmospheric pressure acting on the surface of water in the source. The suction valve remains open during the outward stroke and closes during the inward stroke of the plunger. During inward stroke of the plunger the delivery valve and water is forced up the delivery pipe.

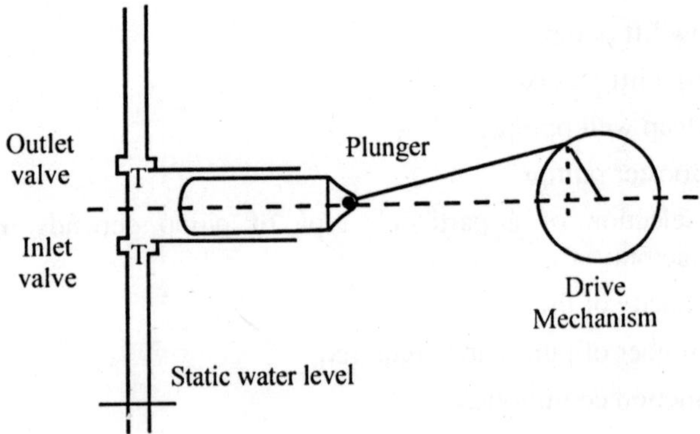

Fig. 12.10: Single cylinder reciprocating pump

However, irrespective of the head against which operates the input power varies in direct proportion to the head. Such pumps are used extensively in groundwater monitoring wells, hand pumps equipped wells and wind power wells. They are rarely used for large capacity water wells. This is most common type of positive displacement pump in the piston pump.

Discharge is determined as,

Let us consider:

D = diameter of the plunger

R = crank radius, and

N = speed of crank-shaft (rpm)

Volume swept per stroke, or

$$\text{displacement} = \frac{\pi}{4} \times D^2 \times \frac{2\pi RN}{60}$$

The average rate of discharge under ideal situations

$$Q = \frac{\pi}{4} \times D^2 \times \frac{2\pi RN}{60} \quad \text{....(12.2)}$$

Due to occurrence of leakage at the valves and plunger and due to lag in the closure of the valves, the actual discharge (Qa) is less than Q.

Types of reciprocating pumps

The reciprocating pumps may be classified according to the liquid water being in contact with one side or both the sides of the piston or plunger or according to the number of cylinders provided.

Classification according to contact of liquid or water with one side or both the sides of the piston or plunger:

(i) Single acting pump,

(ii) Double acting pump

(i) *Single acting pump*: In these pumps the liquid or water is in contact with one side of the piston or plunger only.

(ii) *Double acting pump*: These are the reciprocating pumps in which liquid or water is in contact with both sides of the piston or plunger.

Reciprocating pumps may be classified

According to the number of cylinders provided as,

(i) Single cylinder pump,

(ii) Double cylinder pump,

(iii) Triple cylinder pump,

(iv) Duplex cylinder pump,

(v) Quintuplex cylinder pump,

Work done by reciprocating pumps

The expressions for the work done by different reciprocating types of pumps are described as under:

(a) Single acting pump

The theoretical work done by the pump is expressed with the following relationship:

$$\text{Work done} = W\,Q_{th}\,(H_s + H_d) \qquad \text{....(12.3)}$$

where,

W = specific weight of liquid,

Q_{th} = theoretical discharge,

H_s = suction head, and

H_d = delivery head.

The theoretical discharge (Q_{th}) is obtained by the following relation:

$$Q_{th} = \frac{A\,L\,N}{60}$$

where,

A = cross-section area of the piston,

L = length of stroke, and

N = revolution per minute

since L = 2 r (r is the radius of the crank)

Thus, the theoretical discharge is then obtained as,

$$Q_{th} = \frac{A\,2\,r\,N}{60} \qquad \text{....(12.4)}$$

Theoretical power required to operate the pump is obtained with the following relationship:

$$P = \frac{WQ_{th}\,(H_s + H_d)}{75}$$

$$= \frac{W\,A\,L\,N\,(H_s + H_d)}{4500} \qquad \text{....(12.5)}$$

Maximum speed of reciprocating pump

The speed of a reciprocating pump is usually restricted by the pressure corresponding to the point considered. This is because higher the speed, the greater will be the acceleration head and lower will be the pressure in the cylinder at the beginning of the suction stroke. For water the value of this limiting pressure is about 2.6 m of water absolute or if atmospheric pressure is 10.3 m of water then (H_s + H_d) at any point during the suction stroke must not be greater than (10.3 – 2.6) = 7.7 metre of water.

The maximum permissible speed of a reciprocating pump may be obtained with the following expressions:

$$H_{sep} = H_{st}\,H_{as}$$

$$\text{Hsep} = H_s + \frac{ls}{g}\frac{A}{as}w^2 r$$

$$\text{Hsep} = H_s + \frac{ls}{g}\frac{A}{as}\left(\frac{2\pi N}{60}\right)^2 r$$

where,

Hsep = The pressure head below atmospheric pressure head at which the separation and cavifaction may occur.

Air vessels: The flow rate in the delivery pipe at any instant varies considerably in suction (zero to maximum in delivery stroke) for a single acting pump. In other words, the supply of liquid or water by a single acting pump will be quite intermittent. This difficulty may be overcome if the pump is double acting or it is a multi-cylinder pump so that a continuous supply may be obtained. But in case of most of the pumps the velocity of flow in both the suction and delivery pipes varies with the crank position, which resulted in a non-uniform rate of flow of liquid in these pipes. Air vessels are used in order to obtain a continuous supply of liquid at a uniform rate from a single acting pump and also to obtain a uniform rate of flow in both the suction and delivery pipes of a reciprocating pumps. A large vessel is fitted to the suction as well as the delivery pipe at a point close to the cylinder. It is a closed chamber having an opening at its base through which the liquid may flow into the vessel or it may flow out from the vessel.

Example 12.1: A single acting reciprocating pump has a plunger diameter of 25 cum and stroke of 35 cum. If the speed of the pump is 70 r.p.m. and it delivers 16.5 l/sec of water, find the theoretical discharge, coefficient of discharge, the slip and the percentage slip of the pump.

Solution:

Theoretical discharge of the pump is obtained as,

$$Q_{th} = \frac{A\,L\,N}{60} = \frac{\pi}{4}\left(\frac{25}{100}\right)^2 \frac{35}{100} \times \frac{70}{60}$$

$= 0.02 \text{ m}^3\text{/sec}$

$= 20.03 \text{ l/sec}$

Coefficient of discharge (cd) is determined as,

$$Cd = \frac{Q_a}{Q_{th}} = \frac{16.5}{20.03} = 0.82$$

Slip of the pump = $(Q_{th} - Q_a)$

$= 20.03 - 16.5$

$= 3.53 \text{ l/sec}$

Percentage slip is determined as,

$$\text{percentage slip} = \frac{Q_{th} - Q_a}{Q_{th}} \times 100$$

$$= \frac{20.03 - 16.5}{20.03} \times 100 = 17.6\%$$

Example 12.2: A double acting reciprocating pump have piston area of 0.1 m^2 and stroke of 0.3 m length. The pump is discharging 2.5 m^3 of water per minute at 50 r.p.m. through a height of 10 metre. Determine the slip of the pump and the power required to operate the pump.

Solution:

The theoretical discharge of the double acting reciprocating pump is determined with neglecting the area of the piston rod as,

$$Q_{th} = \frac{2ALN}{60} = \frac{2 \times 0.1 \times 0.3 \times 50}{60}$$

$= 0.05 \text{ m}^3\text{/sec}$

or $= 3 \text{ m}^3\text{/minute}$

Thus, slip of the pump $= Q_{th} - Q_a$

$= 3 - 2.5$

$= 0.5 \text{ m}^3\text{/minute}$

Power required to operate the pump is determined as,

$$= \frac{WQ_{th}(H_{st}\ H_d)}{75}$$

$$= \frac{1000 \times 0.05 \times 10}{75} = 6.67 \text{ h.p.}$$

Volumetric efficiency: It is ratio between the actual discharge and the discharge that could occur under ideal conditions and it is also known as coefficient of discharge. It is expressed with following relation:

$$\eta\, vol = \frac{Qa}{Q} \quad \text{....(12.6)}$$

Slip: It is the difference between the theoretical and actual discharge. Slip is often expressed in percentage slip and it is given by the following expression:

$$\text{Percentage slip} = \frac{Q - Qa}{Q} \times 100 \quad \text{....(12.7)}$$

or

$$Qa = \frac{\pi}{4} \times D^2 \times \frac{2\pi RN}{60} \left(\frac{100 - Slip}{100} \right) \quad \text{....(12.8)}$$

Once knowing the percentage slip, the actual discharge of the reciprocating pump may be determined.

Negative slip: The slip of the reciprocating pump is negative when the actual discharge (Qa) is less than the theoretical discharge (Qth) and cylinder of the pump is more than one.

Reciprocating pumps may be of multi cylinder to obtain continues discharge as compared to single cylinder. It gave more uniform discharge and to distribute the load on the power source more uniformly if two and three cylinder reciprocating pumps are used.

Variable displacement pumps: These pumps are used for large capacity wells. For these pumps, there is an inverse relationship between the discharge and the working head. Maximum input power is required when the pump has to operate at low heads delivering large volumes of water.

12.5 The Hydraulic Ram

It does not require an external source of energy and is also known as impulse pump. This device uses the dynamic pressure of water flowing under a low head to lift a portion of this water to a higher head (Fig. 12.11). It consists of a supply pipe and supply reservoir. The valve box has two automatic valves in which upper end of supply is connected to supply reservoir and lower end to a valve (O) and a waste valve (B) opening downwards and a delivery valve. Above the delivery valve an air vessel (F) to the foot of which the delivery pipe (D) is connected.

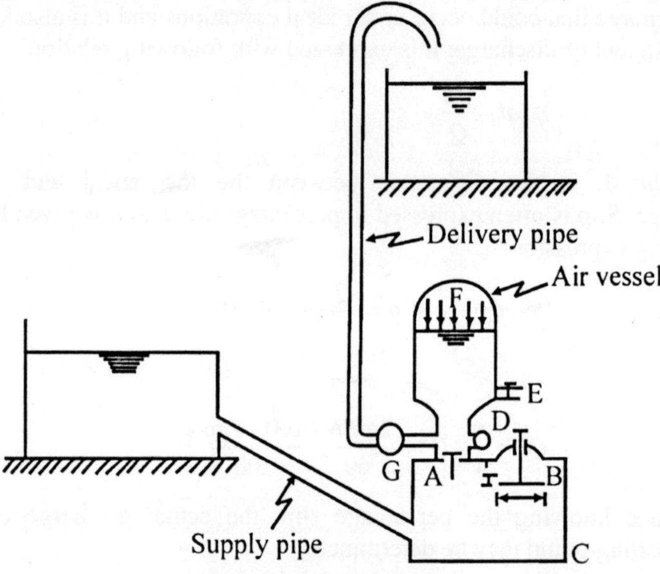

Fig. 12.11: Hydraulic ram

Principle of operation: The hydraulic ram is an impulse pump. The impulse is developed at the expense possessed by a moving water column of water of the dynamic inertia. When the waste valve is opened then the water from the supply pipe begins to move and starts overflowing through the waste valve. Increasing the rate of discharge through the valve the, pressure head in the valve box begins to rise because of the acceleration of the water column in the supply pipe. The total force acting on the lower face of the waste valve is greater than the downward force due to the weight of the valve then the waste valve gets closed. As soon as the waste valve gets closed, the pressure rises continuously, the delivery valve is forced to open and the water flows straight through from the supply tank into the air vessel and the delivery pipe. The flow remains continue until the original kinetic energy of water column in the supply pipe is exhausted. During this time, the pressure in the valve box momentarily falls below atmosphere. Then the waste valve opens automatically and the cycle of operation is repeated. A small quantity of air is drawn into the valve box through a small orifice in order to keep the air vessel charged during this negative pressure. The air vessel helps in maintaining a continuous discharge through the delivery pipe. Because water flows intermittently through the delivery valve, by temporally storing the excess water above the mean discharge.

Efficiency of hydraulic ram: The efficiency of the ram may be determined as, let us consider q is the volume of water delivered by the ram, and Q that escaping through the waste valve, H being the effective supply head, and h_d the effective delivery head measured from the level of the waste valve and including the friction in the delivery pipe line, the total input of energy to the ram is (Q+q)H, and the total output is qh_d, the ratio of output to total input energy is as expressed below:

$$\frac{qh_d}{(Q+q)H} \qquad \text{....(12.9)}$$

It is also known as D' Aubuission's efficiency ratio and gives the efficiency of the ram as a machine.

The main advantage of the hydraulic ram is that it did not require any energy for running after installation and it needs hardly any running cost. The machine can work continuously for 24 hrs and thus gives regular water supply.

Suitability of ram: Guhls and tanks have limited potential since they can irrigate lands below the water source. Most of the surface water sources in the hilly regions lie below the fields. The conventional methods of lifting water by engine and electric motor pumps etc. have their limitations because of uneven topography, inadequate support and maintenance facilities and higher costs. The hydraulic ram (hydram) is a self powered impulse pump which used the energy of the falling water to a level much higher than the original source as a result of water (expected life 30 years) and requires little attention except for change of rubber packing. The hydram pump generally lifts 1/7 to 1/10th of water delivered to it and works at the best efficiency (66-70%) if the drive head to delivery head ratio is 1:3 to 1:4. The hydram pump can lift water as high as 151 m but a more common lift is 45m. However, a minimum flow of 5 L per minute in the drive pipe and a minimum fall (drive pipe) of one m from the source to the ram is required.

Singh (1977) has summarized water lifting capacities for various sizes of hydram pump against lift magnification factors ranging from 2 to 30 L/sec.

If a larger quantity of water needs to be pumped, a battery of hydram pumps is used in parallel to discharge the flow in a common delivery pipe, affecting saving in the installation costs. It has been estimated that a battery of two hydrams requires only 30% extra cost whereas covers double the area. Shoulda battery of small capacity hydrams be installed; some of them can be shut off to handle reduced flow, whereas capacity of a larger capacity hydram would remain under utilized under such conditions. However, the size of the hydram units to be installed in a battery will depend upon the minimum flow rate available during the year.

Another effective means of judiciously applying this water for irrigation is through use of sprinklers with Hydrams (gravity fed) in hill. The water is raised and stored in the tanks at higher elevations through network of pipes. The fields located on lower elevations are irrigated using sprinklers.

Reciprocating pumps for tubewells: These pumps are used for lifting water from tubewells *viz.*, manually pump, deep well plunger pumps (Fig. 12.12).

The manually operated piston pump is also known as pitcher pump and it is widely used for domestic water supply facilities from shallow tubewells. The deep well plunger is operated by an internal combuston engine or an electric motor.

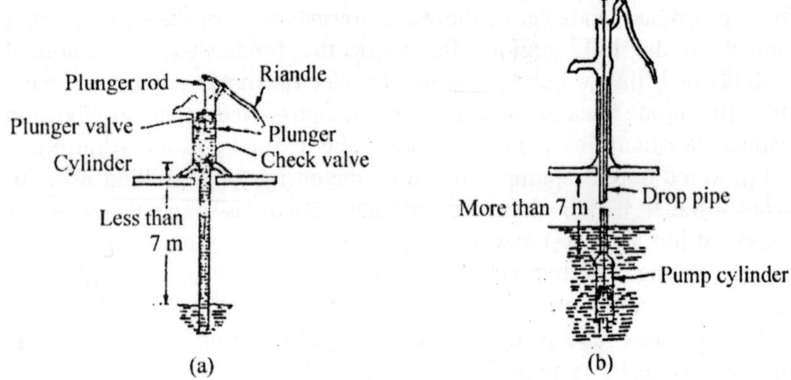

Fig. 12.12: Hand operated reciprocating pump (Hand boring pump)

12.5.1 Rotodynamic pumps

These pumps operate on the principle of centrifugal action. Pumping is accomplished by increasing the energy of water through a rotating impeller fixed with vanes. The impeller is enclosed in casing and water enters the casing at the centre. The rapid rotation of the impeller causes the water to flow to the outside of the rim and because of the casing a pressure head is built up. This pressure head causes the water to move up through the discharge. Because of the vacuum created at the centre of the impeller additional water is drawn in through the suction and the process continues.

The rotodynamic pumps may be classified based on construction into the following groups like, (1) centrifugal pumps, (2) vertical turbine pumps, (3) submersible pumps, and (4) propeller pumps.

(1) Centrifugal pumps

Centrifugal pumps are the most popular and are used for irrigation purpose. They are capable of delivering large volumes of water against high as well as low head with good efficiency. Beside, these pumps are relatively simple and compact. A typical centrifugal pump consists of (1) impeller, (2) casing, (3) fittings and (4) delivery side fittings. These are simple in construction, easy to operate and maintain as compared to other types of pumps and also are smooth even flow and adapted to high speed of operation (Fig. 12.13 (a and b)).

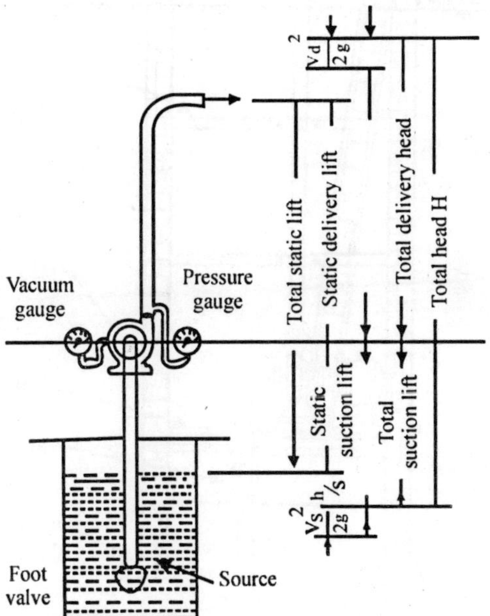

Fig. 12.13 (a): Typical centrifugal pump installation

The impeller rotated, by means of a prime-mover and it imparts kinetic energy to water. The kinetic energy is converted to pressure energy as water flows out through the casing. The impeller bearing provides the support for the rotating impeller. The stuffing box seals off the aperture around the rotating shaft where it passes through the delivery casing of the pump. The suction side consists of the pipe fittings and a foot valve at the bottom of suction line. The foot valve is a non-return or one way valve type which opens only in the upward direction. As such the liquid will pass through the foot valve only upwards and it will not allow the liquid to more downwards back to the sump. It prevents the trash from entering the suction side and at the same time retains water in the suction line to help the priming of the pump. The delivery side also has the pipe fittings. When there is a delivery line a gate valve is fitted on the delivery side to prevent the water rushing back into the pump when pumping is stopped. In some pumps, their construction is such that special arrangement is made to supply a liquid in the suction pipe, which facilitates automatic priming of the pump. Such pumps are known as "set priming pumps".

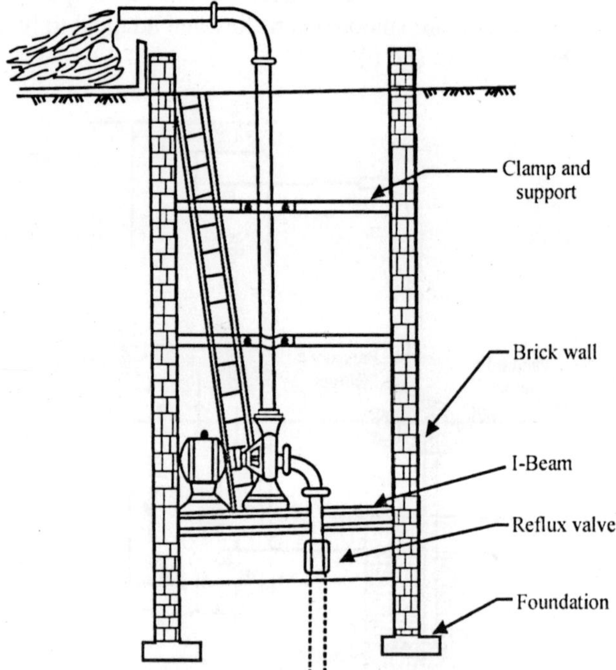

Fig. 12.13 (b): Centrifugal pump installed in well

Suction: Lift pumps create negative pressure at pump intake. The atmospheric pressure at the free surface of water in the well forces the well water into and up the intake pipe. The maximum suction lift depends on the atmospheric pressure (10.4 m of water head), vapour pressure of water, head loss due to friction and head requirements of the pump itself. Under field conditions, the average suction-lift capacity of a suction-lift + centrifugal pump is about 7.5 m (Driscoll, 1986).

Centrifugal pumps are classified based on the following criteria:

1. Based on axis of rotation
   (i) horizontal
   (ii) vertical
2. Based on type of energy conversion
   (i) volute
   (ii) diffuser
3. Based on impeller types
   (i) single or double suction
   (ii) open, semi-open or closed
4. Based on number of stages
   (i) single stage
   (ii) multi stage
5. Based on method of drive
   (i) direct
   (ii) gear
   (iii) belt drive.

A typical section of the horizontal centrifugal pump is shown in Fig. 12.14. The horizontal centrifugal pump is most commonly used for irrigation purpose. It is relatively less costly as compared to other type of pumps. These can easily be installed, accessible and required lesser maintenance. It required open pits for operating within suction limit when this type of pump is used for pumping deep groundwater. While, vertical type of centrifugal pump is driven by a vertical shaft. The motor is at the ground level and the pump is lowered to a level so that the suction lift is within the operating limits. Maintenance of these types of pumps is relatively more difficult as compared to the horizontal ones.

The casing around the impeller serves to convert the kinetic energy of water into pressure energy. Casings are classified as volute or turbine (Fig. 12.15). In the volute type pump, the casing is in the form of a spiral with a cross- sectional area increasing towards the discharge opening. Turbine-type pumps (also known as diffuser type) have stationery guide vanes surrounding the impeller. As the water leaves the impeller, it passes through gradually enlarging vanes, resulting in a reduction in velocity with the kinetic energy converted to pressure. The pressure energy makes the water to go through the discharge column.

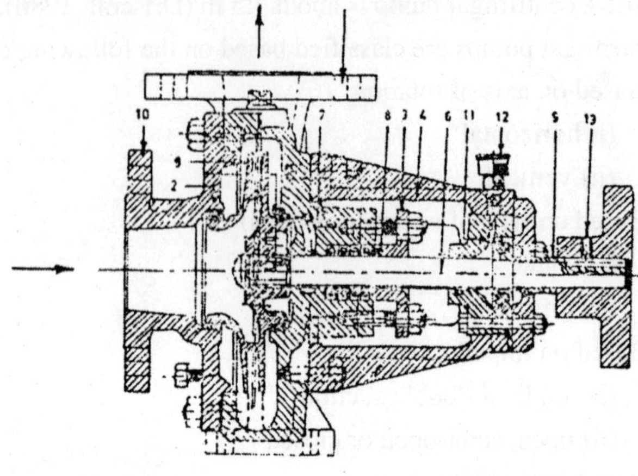

Fig. 12.14: Horizontal centrifugal pump

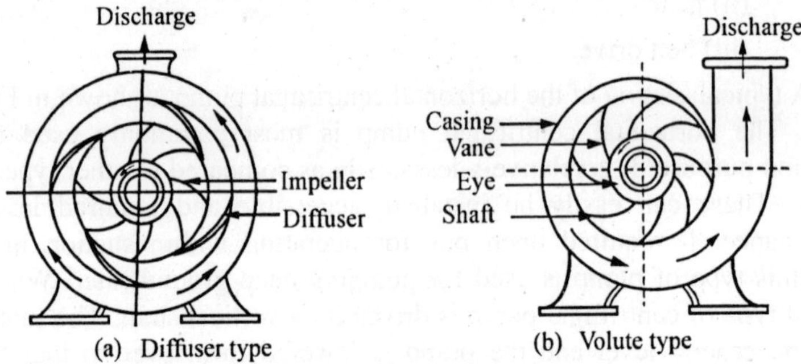

Fig. 12.15: Types of centrifugal pump casings

Centrifugal pumps may be classified based on impellers as open, semi-enclosed and enclosed (Fig. 12.16). The open type impeller has exposed blades which are open on all sides except where attached to the driving shaft. The semi-enclosed impeller has a cover on one side and the enclosed impeller has covers on both sides enclosing the blades completely. The open and semi enclosed impellers are suitable for pumping suspended material or trash in water. Enclosed impellers have higher capacity as compared to other types. However, these are not suitable where suspended sediment is carried in water as this material greatly increases the wear on the impeller.

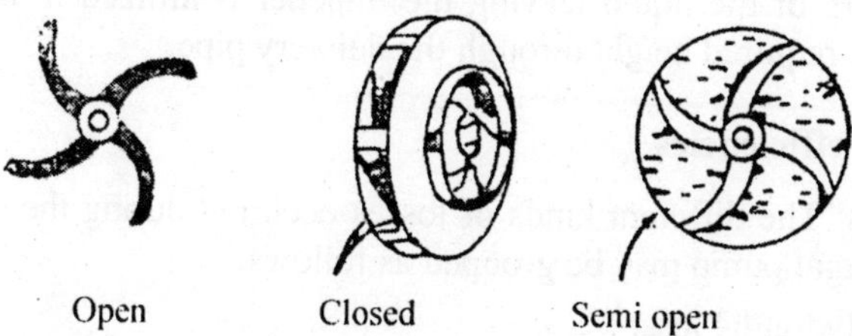

Open Closed Semi open

Fig. 12.16: Types of impellers

Single stage pump: It is a single impeller centrifugal pump whereas a multi stage pump comprising of more than one impeller mounted on a common shaft and operating generally in a single casing multi-stage pump. These pumps are used where high delivery heads are needed. A delivery from one impeller forms the suction of the next impeller which adds to the pressure head developed.

Working of centrifugal pump

The first step in the operation of a centrifugal pump is the priming. It is the operation in which the suction pipe, casing of the pump and their portion of the delivery pipe upto the delivery valve are completely pilled with the liquid which is to be pumped. So that all the air (gas or vapour) from this portion of the pump is driven out and no air pocket is left. If there is any small air pockets left in any portion of the pump may result in no delivery from the pump.

After the pump is primed, the delivery valve is still kept closed and the electric motor is started to rotate the impeller. The rotation of the impeller in the casing full of the liquid or water produces as forced vertex which imparts a centrifugal head to the liquid and thus results in an increase of the pressure throughout the liquid mass. The increase in the pressure in proportion and to the square of the angular velocity and distance of the point from the axis of rotation.

Now as long as the delivery valve is close and the impeller is rotating, it will just be churning the liquid in the casing. When the delivery valve is opened the liquid or water will be made to flow in an outward radial direction thereby leaving the vanes of the impeller of the out periphery with high velocity and pressure. A partial vaccum is created due to centrifugal action at the eye of the impeller. This resulted in movement of liquid or water from the sump (at atmospheric pressure) to the suction pipe to the eye of the impeller thereby replacing the liquid which is being discharged from the entire periphery of the impeller. The high pressure of the liquid leaving the impeller is utilized in lifting the liquid to the required height through the delivery pipe.

Losses and efficiencies

Losses: The different kinds of losses occurred during the operation of a centrifugal pump may be grouped as follows:

(i) Hydraulic loss,

(ii) Mechanical losses, and

(iii) Leakage losses

(i) ***Hydraulic loss***: These losses occur in a centrifugal pump installation and may be grouped as:

(a) Hydraulic losses in the pump $\left(\frac{WQH_{mono}}{75}\right)$,

(b) Other hydraulic losses $\left(\frac{WQHS}{75}\right)$

The hydraulic losses that may occur within the pump consist of the following:

- Shock or eddy losses at the entrance and the exit from the impeller,
- Friction losses in the impeller, and
- Friction and eddy losses in the guide vans and casing.

(ii) ***Mechanical losses***: These losses occur in a centrifugal pump on account of the following:

(a) Disc friction between the impeller and the liquid,

(b) Mechanical friction of the main bearing and glands.

(iii) ***Leakage losses***: These are losses due to leakage of liquid and known as leakage losses.

Efficiencies

The efficiency of a centrifugal pump is expressed in the following form:

(i) Manometric or hydraulic efficiency,

(ii) Volumetric efficiency,

(iii) Mechanical efficiency, and

(iv) Overall efficiency

(i) ***Manometric or hydraulic efficiency*** (η_{mano}): It is defined as the ratio of the manometric head developed by the pump to the head imparted by the impeller to the liquid and is expressed as,

$$\eta_{mano} = \frac{H_m}{H_m + \text{losses in the pump}}$$

(ii) ***Volumetric efficiency***: It is the ratio of the quantity liquid discharge per second from the pump to quantity passing per second through the impeller and is expressed as,

$$\eta_v = \frac{Q}{Q + DQ}$$

where,

Q = Discharge/second from the pump

(iii) ***Mechanical efficiency***: It is defined as the ratio of the power actually delivered by the impeller to the power supplied to the shaft by the prime mover or motor and is expressed as,

$$\eta_{mech} = \frac{W(Q + DQ)(V_{w1}\, u_1 / g)}{\text{Power given to shaft}}$$

(iv) ***Overall efficiency***: It is defined as the ratio of the power output from the pump to the power input from the prime mover operating the pump and is expressed as,

$$\eta_o = \frac{WQH_m}{\text{Power given to shaft}}$$

Minimum starting speed

When the pump is started then there will be no flow of water until the pressure difference in the impeller is not sufficient to overcome the gross or manometric head. If the impller is rotating but no flow occurs then the water is rotating in a forced vortex. Thus, a centrifugal head or pressure head caused by the centrifugal force on rotating water will be ($u_1^2 - u^2/2g$). The flow from the pump will commence only if,

$$(u_1^2 - u^2)/2g \geq H_m \qquad \text{....(12.10)}$$

$$\left[\frac{(\pi D_1 N / 60)^2}{2g} - \frac{(\pi D N / 60)^2}{2g}\right] \geq H_m \quad ....(12.11)$$

$$\left(\frac{(\pi N)^2}{60}\right)(D_1^2 - D^2) \geq 2g H_m \quad ....(12.12)$$

Thus, the minimum speed required for the pump to commence the flow will be determined on using the following relationship:

$$\left[\frac{(\pi N)^2}{60} - (D_1^2 - D^2)\right] = 2gH_m \quad ....(12.13)$$

Multistage pump

The head produced by a centrifugal pump depends on the rim speed of the impeller. To increase the rim speed, either the rotative speed or the diameter of the impeller or both must be increased. Increasing either of these has the effect of increasing the stress in the impeller material. For this reason it is usually not possible to produce very high head with one impeller. Generally, a pump with a single impeller can be used to deliver the required discharge against a maximum head of about 100 m. But if the liquid is required to be delivered against a still larger head then it can be done by using two or more pumps in series. These pumps consist of two or more identical impellers mounted on the same shaft and enclosed in the same casing. All the impellers are connected in series so that liquid is discharged with increased pressure from the impellers.

Pumps in parallel

The multistage pumps or the pumps in series are used for delivering a relatively small quantity of liquid against very heads. However, if a large quantity of liquid is required to be pumped against a relatively small, quantity then it may not be possible for a single pump to deliver the required discharge. In situations, two or more pumps are used which are so arranged as each of these pumps works separately to lift the liquid from a common sump and deliver it to a common collecting pipe through it is carried to the required height. In such cases arrangement of pumps are known as pumps in parallel. If Q_1, Q_2, Q_3.....Q_n are the discharging capacities of n pumps arranged in parallel then the total discharge delivered by these pump will be obtained as,

$$Q_t = Q_1 + Q_2 + Q_3 + ... + Q_4$$

Or if discharge capcacity of all the pumps are same then,

$$Q_t = nQ$$

Total head in a centrifugal pump

The total head against which the pump has to operate is represented as manometric head or dynamic head and is expressed by the following relation:

$$H_m = h_s + h_{fs} + h_d + h_{fd} + \frac{V^2}{2g} \quad ....(12.14)$$

where,

Hm = Total head, (m),

hs = static suction head, (m),

hfs = frictional losses in suction side, (m),

hd = static delivery head, (m),

hfd = frictional losses in delivery side, (m), and

$\frac{V^2}{2g}$ = velocity head, (m).

Equation (12.13) is correct if the suction and delivery pipes are of the same size. If these pipes are of different diameters then velocity head $\frac{V^2}{2g}$ is determined as,

$$\frac{Vd^2}{2g} - \frac{V_s^2}{2g} \quad ....(12.15)$$

where, V_d and V_s are the velocities in the delivery and suction pipes, respectively.

For the selection of the centrifugal pumps, total head (H_m) value is needed. Values of h_s and h_d can be measured for a particular site conditions. The maximum suction lift of centrifugal pumps is dependent upon the atmospheric pressure. Atmospheric pressure varies with elevation and at sea level, its value is 1.03 kg/cm^2 or 10.34m of water. Theoretically, pump should be able to operate with this suction lift at sea level. But because of air leaks past the impeller and through other openings the effective suction lift is of the order of 6.34m at sea level and about 4.5m for most inland conditions. The suction lift should be as small as possible for obtaining higher pump efficiencies.

Velocity head $\frac{V^2}{2g}$ is determined on using the following relationship:

$$V = Q/A$$

where,

Q = discharge, and

A = cross-sectional area of the pipe.

The head due to frictional losses is determined by knowing the lengths of the suction and delivery pipes and the pipe fittings. The frictional losses are determined in different pipe fittings as suggested by Anon (1976) and Chauhan *et al.* (1979). The values of velocity heads, head losses in galvanized iron pipes and equivalent lengths of pipe for different pump fittings are shown in Table 12.3 and 12.4.

Table 12.3: Head loss (due to friction) in galvanized iron pipes meters per 100 m of pipe length

Discharge L/sec.	**Pipe diameter (cm)**										
	4.0	**5.0**	**6.0**	**7.0**	**8.0**	**10.0**	**12.5**	**15.0**	**20.0**	**25.0**	**30.0**
1	2	3	4	5	6	7	8	9	10	11	12
1.0	3.7	1.1	0.43	-	-	-	-	-	-	-	-
1.2	5.0	1.6	0.58	0.27	-	-	-	-	-	-	-
1.4	7.3	2.2	0.83	0.37	-	-	-	-	-	-	-
1.6	9.2	2.8	1.10	0.50	0.23	-	-	-	-	-	-
1.8	11.8	3.7	1.40	0.62	0.29	-	-	-	-	-	-
2.0	15.5	4.5	1.70	0.73	0.37	-	-	-	-	-	-
2.2	16.2	5.2	2.15	0.90	0.44	-	-	-	-	-	-
2.4	20.5	6.4	2.50	1.07	0.52	0.16	-	-	-	-	-
2.6	23.5	7.5	2.90	1.27	0.62	0.18	-	-	-	-	-
2.8	27.5	8.7	3.30	1.47	0.70	0.22	-	-	-	-	-
3.0	32.0	10.0	3.80	1.68	0.83	0.25	-	-	-	-	-
3.5	42.5	13.5	5.30	2.30	1.10	0.33	-	-	-	-	-
4.0	56.0	17.5	7.30	3.00	1.50	0.45	-	-	-	-	-
4.5	71.5	22.5	8.80	3.80	1.85	0.55	0.17	-	-	-	-
5.0	87.0	28.0	10.80	4.70	2.30	0.68	0.22	-	-	-	-
5.5	-	33.0	12.40	5.70	2.70	0.83	0.26	0.095	-	-	-
6.0	-	40.0	15.50	6.80	3.20	0.96	0.32	0.118	-	-	-

Discharge L/sec.	Pipe diameter (cm)										
	4.0	**5.0**	**6.0**	**7.0**	**8.0**	**10.0**	**12.5**	**15.0**	**20.0**	**25.0**	**30.0**
1	2	3	4	5	6	7	8	9	10	11	12
6.5	-	47.0	18.30	8.00	3.80	1.15	0.36	0.140	-	-	-
7.0	-	54.0	21.50	9.30	4.50	1.30	0.42	0.17	-	-	-
7.5	-	62.0	24.00	10.60	5.20	1.50	0.47	0.18	-	-	-
8.0	-	70.0	28.00	11.60	6.00	1.80	0.55	0.21	-	-	-
8.5	-	80.0	31.00	13.30	6.80	2.00	0.62	0.23	-	-	-
9.0	-	90.0	36.00	15.00	7.50	2.20	0.68	0.27	-	-	-
9.5	-	100.0	38.00	17.00	8.30	2.50	0.76	0.29	-	-	-
10	-	-	43.00	19.00	9.40	2.80	0.85	0.32	0.065	-	-
12	-	-	63.00	27.00	13.00	4.00	1.23	0.47	0.10	-	-
14	-	-	78.00	37.00	18.00	5.50	1.65	0.63	0.13	-	-
16	-	-	-	47.00	23.00	7.20	2.20	0.83	0.17	0.055	-
18	-	-	-	60.00	30.00	9.00	2.80	10.50	0.22	0.068	-
20	-	-	-	72.00	37.00	11.00	3.31	1.30	0.27	0.080	-
22	-	-	-	86.00	45.00	13.50	4.10	1.60	0.33	1.100	0.038
24	-	-	-	-	52.00	16.00	4.80	1.90	0.40	0.120	0.045
26	-	-	-	-	62.00	18.00	5.60	2.20	0.47	0.145	0.055
28	-	-	-	-	72.00	22.00	6.60	2.50	0.55	0.165	0.060
30	-	-	-	-	80.00	25.00	7.50	2.80	0.63	0.19	0.070
35	-	-	-	-	-	33.00	10.30	4.00	0.85	0.27	0.095
40	-	-	-	-	-	-	13.50	5.20	1.10	1.10	0.128
45	-	-	-	-	-	56.00	17.00	6.50	1.40	0.43	0.16
50	-	-	-	-	-	70.00	21.00	8.30	1.70	0.55	0.20
55	-	-	-	-	-	85.00	26.50	9.70	2.20	0.65	0.24
60	-	-	-	-	-	100.00	32.00	11.50	2.50	0.75	0.28
65	-	-	-	-	-	-	37.00	14.00	2.90	0.90	0.34
70	-	-	-	-	-	-	42.00	16.30	3.50	1.08	0.40
75	-	-	-	-	-	-	48.00	18.50	3.90	1.20	0.46
80	-	-	-	-	-	-	55.00	22.00	4.30	1.40	0.52
85	-	-	-	-	-	-	60.00	23.00	5.00	1.50	0.58
90	-	-	-	-	-	-	68.00	27.00	5.60	1.70	0.65

Discharge L/sec.	Pipe diameter (cm)										
	4.0	5.0	6.0	7.0	8.0	10.0	12.5	15.0	20.0	25.0	30.0
1	2	3	4	5	6	7	8	9	10	11	12
95	-	-	-	-	-	-	75.00	30.00	6.30	1.90	0.72
100	-	-	-	-	-	-	85.00	33.00	7.00	2.20	0.80

(Adopted from Kirloskar Brothers 'Engineering Data')

Table 12.4: Length of straight pipe in metres giving equivalent resistance of flow in valves and fittings

Size of pipe in mm	Standard elbow	Medium elbow	Long radius elbow	45^0 elbow	Tee	Return bend	Sluice valve	Globe value open	Angle valve open	Foot valve or reflux valve
25	0.82	0.70	0.52	0.40	1.77	1.86	1.18	8.24	4.57	2.04
40	1.31	1.10	0.85	0.61	2.74	3.05	0.29	13.40	6.71	3.05
50	1.67	1.40	1.07	0.76	3.35	3.96	0.37	17.40	8.54	3.96
65	1.98	1.65	1.28	0.92	4.26	4.57	0.42	20.10	10.00	5.18
75	2.47	2.00	1.55	1.15	5.18	5.49	0.52	25.90	12.80	6.10
100	3.35	2.77	2.13	1.53	6.71	7.31	0.70	33.50	17.70	8.23
125	4.26	3.66	2.78	1.86	8.24	9.45	0.88	42.60	21.30	10.00
150	4.87	4.26	3.35	2.35	10.80	11.50	1.07	47.70	25.30	12.20
200	6.40	5.48	4.26	3.05	13.10	14.90	1.37	67.10	33.50	16.20

(Adopted from 'Jyoti Ltd. News' Issue 64, November, 1977)
(Source : Michael, 1992)

Selection of centrifugal pumps

Pump operating head and discharge vary according to installation. Therefore, it is essential to select such a pump that operates at maximum efficiency with the given head and giving the required discharge. This is obtained by characteristic curves, both for the well and the pump. The characteristic curves show the behaviour of the pump and on-farm reservoir of the well under various operating conditions.

Discharge capacity of pump

It is prerequisite be for any pump to be purchased that the pump discharge should meet the peak demand of water for the selected cropping pattern. The pump discharge depends on the area under different crops, this water requirement of the crops, rotation period (interval between two successive irrigations of a crop) and the duration the pump operation each day. It may be determined by the following relationship:

$$q = \frac{Ay}{RT} \times \frac{1000}{36}$$

$$= 27.78 \frac{Ay}{RT} \qquad \text{....(12.16)}$$

where,

q = rate of discharge of pump (litres/sec),
A = area of land under the crop (ha),
y = depth of irrigation (cm),
R = rotation period (days), and
T = duration of pumping (hours/day).

Example 12.3: A farmer wishes to purchase a new pump set for the following cropping pattern to be followed in his holding of three ha and his brother's holding of 3 ha. Determine the right size of the centrifugal pump to be purchased.

Season	Crop	Area to be irrigated (ha)	Intensity of irrigation (cm)	Rotation period (days)	Period of work (hrs/day)
Rabi	1. Wheat	2.5	7.5	12	10
	2. Cotton	0.5	7.5	20	10
	3. Vegetables	0.5	7.5	10	10
	4. Mustard	2.5	5.0	4.0	10

Solution:

$$\text{Discharge rate of pump} = 27.78 \frac{Ay}{RT}$$

$$= 27.78 \left(\frac{2.5 \times 7.5}{12 \times 10} + \frac{0.5 \times 7.5}{20 \times 10} + \frac{0.4 \times 7.5}{10 \times 10} + \frac{2.5 \times 7.5}{40 \times 10} \right)$$

= 6.55 L/sec

Loss of head in a pipe: The loss of head due to friction in a pipe, between two points at a distance apart, is obtained by the following formula:

$$h_f = \frac{4 fl\, v^2}{2gd} \quad \text{....(12.17)}$$

where,

hf = head loss (m),

f = coefficient of friction for pipe (m),

l = length of pipe (m), and

d = diameter of pipe (m).

The friction coefficient f is not simply a coefficient of friction between the water and the surface of the pipe, but includes a coefficient of resistance due to eddying motions in the water itself. It further depends upon the smoothness or roughness of the pipe surface. The value of f is less if the pipe is new and smooth. The approximate value of f are obtained by the following empirical relationships:

(1) For old pipes:

$$f = 0.01\left(1 + \frac{1}{4.725\, d}\right) \quad \text{....(12.18)}$$

(1) For new pipes:

$$f = 0.05\left(1 + \frac{1}{4.725\, d}\right) \quad \text{....(12.19)}$$

Since the factor $1 + \frac{1}{4.725\, d}$ is usually very small, it is often neglected and the value of f is assumed to be 0.01 and 0.005 for old pipes and new pipes, respectively.

Example 12.4: Compute the head loss due to friction in a pipe 7.5 cm in diameter and 120 m long when the water is flowing at a velocity of 1.8 m/sec. The value of f may be assumed to be 0.005.

Solution:

d = 0.075 m, 1 = 120 m

v = 1.8 m/sec, g = 9.81 m/sec^2

Head loss due to friction, $hf = \frac{4\, f\, l\, v^2}{2\, gd}$

$$= \frac{4 \times 0.005 \times 120 \times 1.8 \times 1.8}{2 \times 9.81 \times 0.075} = 5.28\,m$$

Table 12.3 presents the head loss due to friction in galvanized iron pipes for the range of discharge values and pipe sizes commonly encountered in irrigation pumping.

Head capacity curve

In this curve discharge against head is plotted as shown in Fig. 12.17. It shows the behaviour of the pump at one particular speed. The length AO gives the shut off head or the head developed when the discharge valve is closed (Fig. 12.17). There is considerable delivery head at pump installation. Thus, the pump is not operating against the total head at the time of starting. Since the head is low, the discharge tends to be high and the engine motor gets overloaded. Thus, at the time of starting, the discharge valve should be kept closed and gradually opened afterwards.

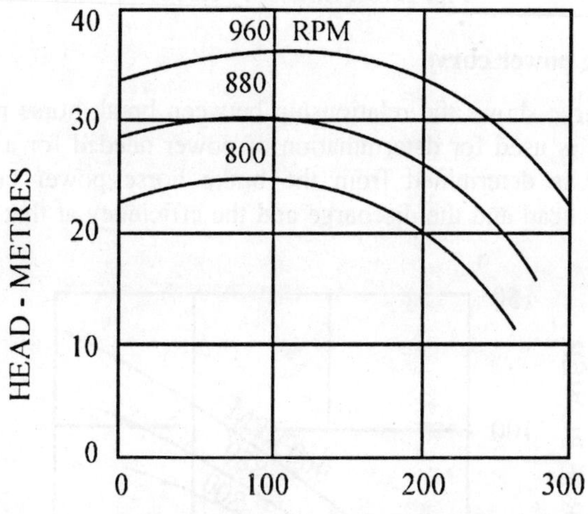

Fig. 12.17: Head capacity curves

Overall efficiency curve

It gave the relationship between the efficiency of the pump and the discharge at a particular speed and is shown by the efficiency curve. Efficiency curves may also be obtained for different speeds of the pump (Fig. 12.18).

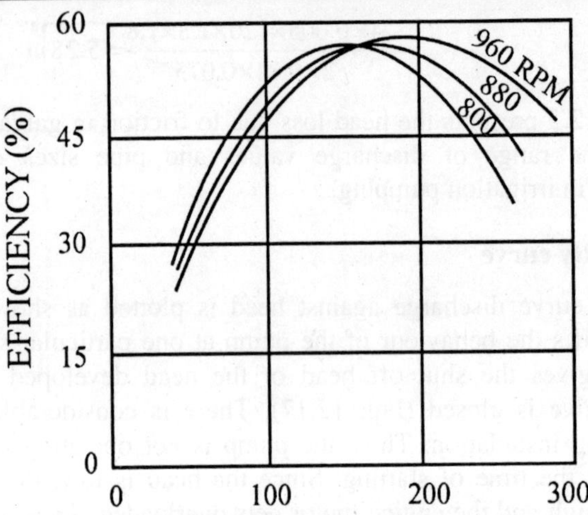

Fig. 12.18: Efficiency curves

Break-horse power curve

This curve shows the relationship between break horse power and discharge. It is used for determination of power needed for a particular pump. BHP is determined from the brake horse power curve once knowing the head and the discharge and the efficiency at that discharge (Fig. 12.19).

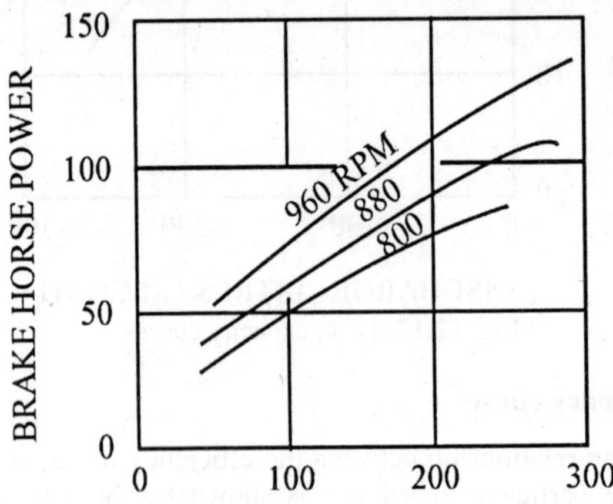

Fig. 12.19: BHP curves

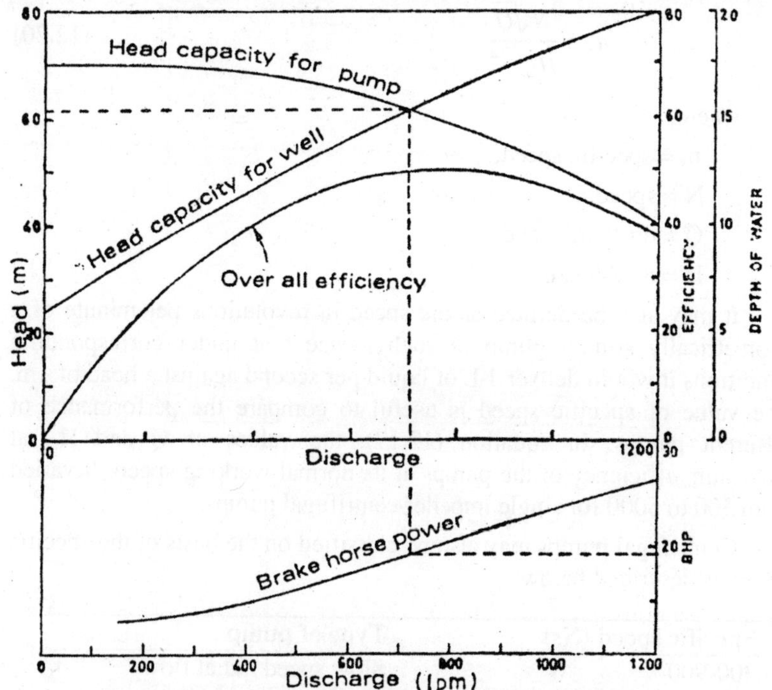

Fig. 12.20: Matching well and pump characteristics

For selection of centrifugal pump for pumping from an on-farm reservoir the total head has to be determined for selecting the suitable pump. In case of wells the head-capacity curve of the well is made use of in the selection of the pump. The head capacity curve for a given well is drawn from the data obtained after testing of wells. This curve shows the amount of water which the well will yield with different drawdowns. The same may also be prepared for on-farm reservoir. Pump is selected by matching the pump and well characteristics (Fig. 12.20). Let us consider that the head-capacity curves of the pump and the well intersect at point P. The efficiency can be read from the efficiency curve. The efficiency obtained should be maximum or near about it.

Specific speed of centrifugal pumps

An index to operating characteristics of pumps. It is and used to compare the performance of different pumps. The specific speed is determined by the following relationship:

$$n_s = \frac{N\sqrt{Q}}{H_m^{3/4}} \qquad ....(12.20)$$

where,

n_s = specific speed,

N = speed,

Q = discharge, and

H = total head.

It may also be defined as the speed in revolutions per minute of a geometrically similar pump of such a size that under corresponding conditions it would deliver 1 L of liquid per second against a head of 1m. The value of specific speed is useful to compare the performance of different pumps. In equation (12.17), the values of Q and H_m at maximum efficiency of the pumps at its normal working speed. It varied from 300 to 5000 for single impeller centrifugal pumps.

Centrifugal pumps may also be classified on the basis of the specific speed as described below:

Specific speed (Ns)	**Type of pump**
300-900	Slow speed radial flow
900-1500	Medium speed radial flow
1500-2400	High speed radial flow
2400-5000	Mixed flow or screw type
3400-15000	Axial flow or propeller pump

The characteristics of the impeller-type pumps may be expressed with the relationship between discharge (Q), speed (N), diameter (D), width of impeller (W), head (H), and power (P) with subscript zero referring to original conditions:

$$\frac{Q}{Q_0} = \frac{N}{N_0} = \frac{D}{D} = \frac{W}{W_0} \qquad ....(12.21)$$

$$\frac{H}{P_0} = \left(\frac{N}{N_0}\right)^2 = \left(\frac{D}{D_0}\right)^2 = \left(\frac{W}{W_0}\right)^2 \qquad ....(12.22)$$

$$\frac{P}{P_0} = \left(\frac{N}{N_0}\right)^2 = \left(\frac{D}{D_0}\right)^2 = \left(\frac{W}{W_0}\right)^2 \qquad ....(12.23)$$

Eq. 12.21 indicates that on increasing 50 per cent impeller speed, diameter or width will increase discharge by 50 per cent. Eq. 12.22 indicates that on increasing 50 per cent impeller speed, diameter or width will increase the head developed by (1.5) 2 or 2.25 times. While, Eq. 12.23 indicates that when speed diameter or width increased by 50 per cent then the power requirement increased by $(1.5)^3$ or 3.37 times.

Installation of centrifugal pumps

Installation of the pumps comparision of (1) location of the pump, (2) proper foundation, and (3) alignment of the coupling.

Location of the pump: The pump should be located as near as the water supply possible for minimizing suction lift. The pump should be accessible for frequent inspection. The location of the pump should be such that it does not get submerged in water. In case of open wells, water level rises during monsoon season, there should be a provision for pump to installed above the water level as water level. Thus, the pump should have two locations, one for the low water level conditions and one for the high water level conditions.

Proper foundation: The pump and the prime mover may be fixed on a trolley or on permanent foundation. Permanent foundation compr-ising of the base plate fixed to foundation bolts embedded in concrete.

Alignment of coupling: The pump shaft and the drive shaft should be aligned straight (Fig. 12.21). Wedges placed below the base plate are useful in alignment of the pump with the prime mover as the wedges may be used to raise or lower the pump unit.

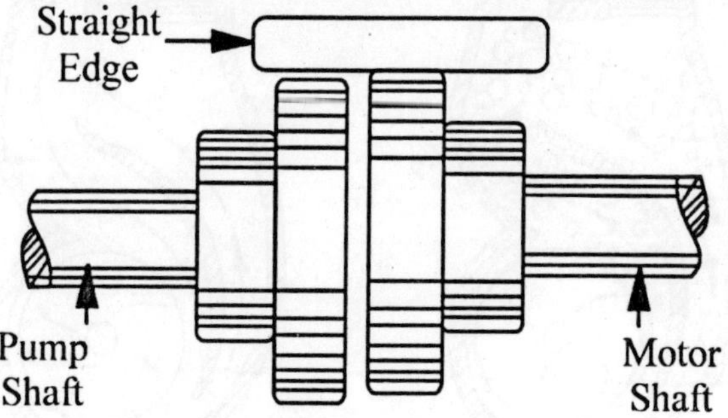

Fig. 12.21: Alignment of pump shaft

Operation of centrifugal pumps

For starting horizontal type centrifugal pump, it is prerequisite to fill the suction pipe and pump case with water to expel the air. This operation is known as 'priming the pump'. Priming is essentially achieved due to the non-positive action of the impeller. Priming is performed by adding water to the suction line from an outside source or by sucking out the air with various devices, thereby lowering the pressure in the suction line causing the water to rise from the source to the pump by atmospheric pressure. Hand primers and dry vacuum pumps are common devices used for priming.

Centrifugal pumps may be made self priming with a special design for the casing. In this pump, chamber has to be first filled with water. When the motor is started to rotate which causes water within the impeller to discharge upwards into the chamber (Fig. 12.22a). This action creates a vacuum at the centre of the impeller and as a consequence air from the suction line and water within the pump rushes into this void. These are mixed at the impeller periphery and discharged upward into the chamber where the air escapes from the water. The heavier air-free water moves down to the impeller due to the force of gravity. More air is entrained and the cycle repeated until the pump is primed. If pump is primed then pumping channels 1 and 2 become one common discharge channel and the water no longer circulate within the pump while pumping (Fig. 12.22b). The pump installation usually consists of a check valve (also known as reflux valve) on the suction pipe and it is also known as reflux valve. Thus, the pump is always full of water and priming is automatic after the pump is once filled by hand.

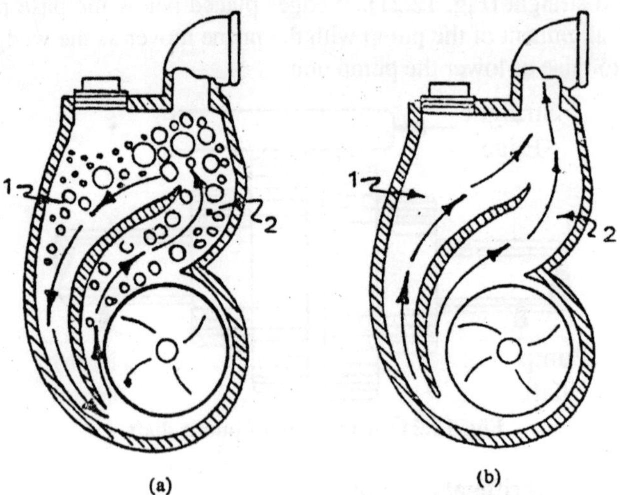

Fig. 12.22: Self priming pump

Common troubles reasons and remedies for operation of centrifugal pump

There are following common troubles that occur with centrifugal pump installations and their remedies:

1. Pump fails to deliver water

(i) There may be a possibility of an air leakage in the suction line. Threaded connections are a common source for it. These should be coated with white lead or pipe cement and tightened to make it air tight.

(ii) Defective gaskets at the inlet and outlet of the pump which may admit air. These should be tightened.

(iii) The foot valve or the reflux value may be defective, so that water is not retained in the suction side. Often the flap in the foot valve does not operate properly. This should be checked and replaced if required.

(iv) Suction lift too high to check with vaccum gauge the suction lift.

(v) Speed may be too low to check speed with tachometer.

(vi) Total specific head is much higher than for which the pump is designed.

(vii) Improper premised. Reprime the pump.

2. Pump fails to develop sufficient pressure or capacity

(i) The pump speed should be checked to see whether it is as per prescribed speed or not.

(ii) Suction line and foot valve should be checked for any clogging with debris or any other foreign materials. If it is clogged than should be cleaned.

(iii) The suction lift should be within the prescribed limits.

(iv) The impeller should be checked for its wear as the capacity of the pump is reduced due to this.

(v) Discharge head may be higher than anticipated.

(vi) Rotation may be in wrong direction. Arrow on the pump shows proper direction of rotation.

3. Pump takes too much power

(i) The speed of the pump may be higher than the rated speed.

(ii) The head may be lower than the rated head for the pump, thereby pumping too much water.

(iii) Liquid may have for higher specific gravity.

(iv) Mechanical defects such as bent shaft, tight stuffing box, misalignment of the pump and the driving unit should be checked.

(v) Pump may be operating in wrong direction.

4. Pump leaks excessively at the stuffing box

(i) The packing material used may be worn-out, incorrectly inserted or may not be of the right kind. These should be replaced.

(ii) The shaft may be worn-out.

5. Pump is noisy

(i) The suction lift may be too high.

(ii) Mechanical defects like bent shaft, improper alignment between the pumps and the driving unit, broken or worn-out bearing may cause noise during the operation of the pump.

(2) Turbine pumps

It consists of impellers placed below the water level and are driven by a vertical shaft. The turbine is having one or more impellers housed in a single or multi-stage unit called a bowl assembly. Each stage gives a certain amount of lift and sufficient number of stages are assembled to meet the total head requirement of the system (Fig. 12.23). The impellers are enclosed with stationery guide vanes. The guide vanes are gradually enlarging and convert the kinetic energy of water leaving the impellers to pressure energy. The water is discharged through the discharge column. The impeller and the guide vanes assembly is known as the pump bowl. It is surrounded by a screen to keep coarse sand and gravel away from entering the pump.

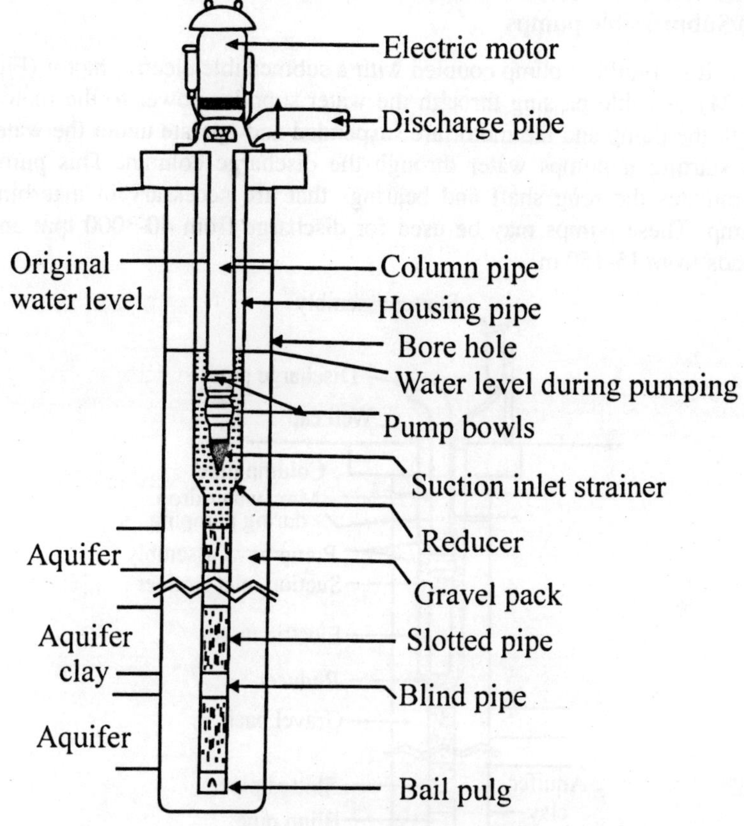

Fig. 12.23: Turbine pump

The turbine pumps are classified based on the type of flow produced by the impeller. The centrifugal type impeller pump discharges water at right angles to the axis of rotation. While, in axial flow type pump, water is given an upwards thrust by the impeller similar to a boat propeller. A combination of the centrifugal and axial-flow type are known as the mixed-flow type.

The turbine pump is selected as centrifugal pumps (discussed earlier). For this accurate data is required in respect of the well. Besides the head-capacity curve or the well, the seasonal fluctuations of water table should also be known so that the pump bowls are installed such that they are always under water. The vertical turbine pumps are well suited in high capacity wells reliable over long period of time. These pumps, however, cannot be used in wells which are out the alignment. The overall efficiency of turbine pumps range from 50 to 80 per cent.

(3) Submersible pumps

It is a turbine pump coupled with a submersible electric motor (Fig. 12.24). A cable passing through the water supplies power to the motor. Both the pump and the motor are suspended and operate under the water on starting it pumps water through the discharge column. This pump eliminates the long shaft and bearings that are necessary in a turbine pump. These pumps may be used for discharge from 40-3000 lpm and heads from 15-150 m.

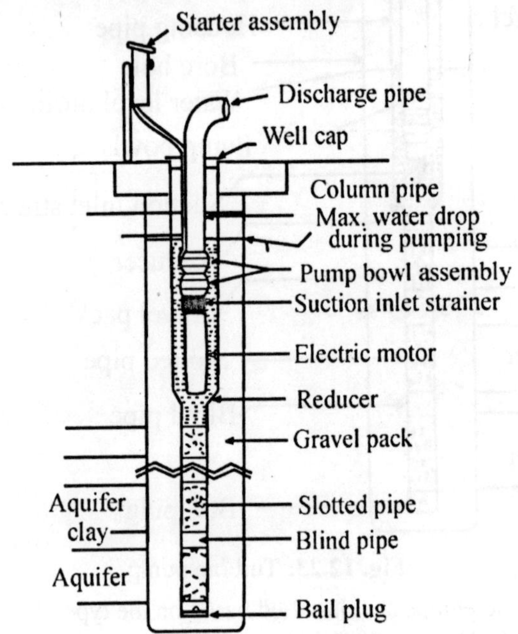

Fig. 12.24: Submersible pump

The submersible pumps are cheaper as compared to the vertical turbine pumps. These pumps are suitable for deep setting and also for crooked wells which not perfectly vertical. Its installation is easy with low initial cost of installation. The repairing of the submersible pumps is not easy; they go out of order, is not easy and require technical skill.

(4) Propeller pumps

Its principal part and method of operation is similar to the turbine pumps (Fig. 12.25). The main difference is in design of the impellers. The main feature of the design of the impellers is that the propeller pumps give high discharges at low heads.

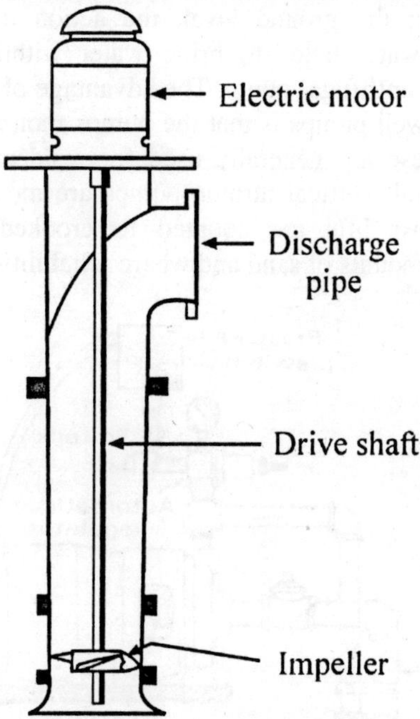

Fig. 12.25: Propeller pump

Two types of impellers *viz.*, axial flow type and the mixed flow type are propeller pumps. Single stage pumps require only one impeller whereas multistage pumps require more than one impeller.

It is selected based on head-capacity, efficiency, and horsepower for a given pump size, type of impeller, and discharge.

(5) Jet pumps

It is used when the suction lift of the centrifugal pump exceeds the permissible limits (Fig. 12.26). These pumps consists of a pump and a set of jet and gave small discharge of 40-90 lpm under low heads (15-45 m) when the water level is beyond 7-6 m from the ground surface. Their capacity reduces as the lift increases.

A portion of the water from the centrifugal pump is passed through the drop pipe to the nozzle of the jet assembly. This water is forced through the throat opening of the diffuser which creates a vacuum and causes water to be drawn from the well. The water mixed with the boost water is carried up through the diffuser where the high velocity energy is converted into useful pressure energy. It forces the water up through the delivery pipe to the centrifugal pump. Thus, even though the centrifugal pump is located at the ground level, the action of the jet element submerged under water helps to bring water within the permissible suction limits of a centrifugal pump. The advantage of the jet pump over other pumps deep well pumps is that the pumps motor may be set away from the well. These are generally used for residential buildings and hotels. The deep well vertical turbine pumps are most widely used for large tubewells. Air lifts are adopted to crooked wells, to wells discharging large amounts of sand and where reliability is more essential.

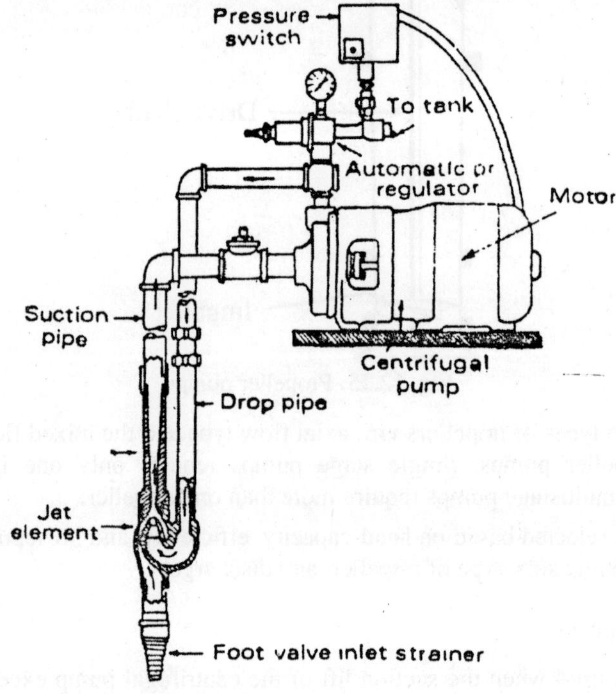

Fig. 12.26: Schematic diagram of a jet pump

(5) Air lift pumps

These pump use compressed air for lifting water (Fig. 12.27). Air is introduced through one or more nozzles, into water at the foot of the rising main fixed in the well. Air bubbles get mixed with column of water in the rising main. Consequently, the pressure due to the column of length L is less as compared to the static pressure (H). This result to flow begins through the outlet. Water and air are discharged through the outlet so long as the supply of air is maintained.

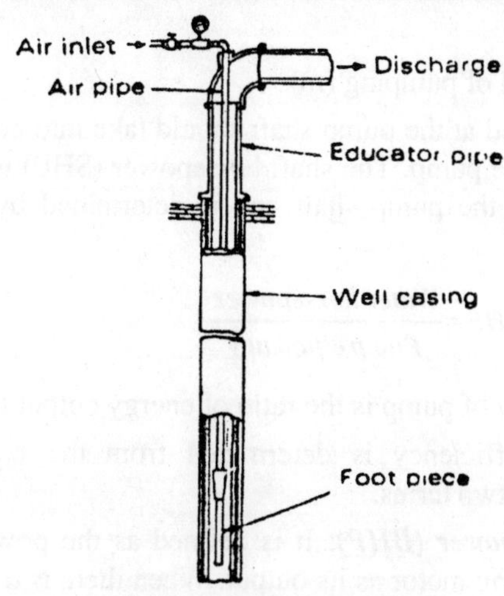

Fig. 12.27: Air lift pump

It is recorded that best results arc obtained if the useful lift (L-H) is less than the submergence. The efficiency greatly depends upon the percentage submergence and is reasonable when the percentage of submersible is 50 to 60%. The advantages of the air lift pump are that it has no moving parts below water level and it can raise more water through a bore hole of given diameter than any other type of pump. However, the efficiencies of the air lift pumps are low (20-40 per cent).

Power requirements for pumping

Power is defined as the rate of doing work. The unit of power commonly used is the horsepower (HP), 76 metre kg. per sec. Work is defined as the product of force and distance.

The horsepower of the electric motor or an internal combustion engine needed for a pumping may be determined from the following relations.

Water horsepower (WHP): It is the power required by a pump or other device for lifting a given volume of water in a unit time assuming no losses of power in the pump. This is expressed as,

$$WHP = \frac{Q(lit/\sec) \times H(m)}{76} \quad \text{....(12.24)}$$

$$or = \frac{Q(m^2/h) \times H(m)}{273} \quad \text{....(12.25)}$$

Where,

H = total head of pumping (m)

Power required at the pump shaft should take into consideration the loss of power in the pump. The shaft horsepower (SHP) is defined as the power required at the pump shaft and is determined by the following relation.

$$SHP = \frac{Water\ horsepower}{Pump\ efficiency} \qquad ....(12.26)$$

The efficiency of pump is the ratio of energy output to energy input.

The pump efficiency is determined from the equation (12.26) knowing the other two terms.

Brake horsepower (BHP): It is defined as the power supplied by the engine or electric motor as its output. When there is direct drive from the engine electric motor to the pump and the drive efficiency is 100 per cent, brake horse power is equal to shaft horsepower.

For belt or other indirect drives, brake horsepower is determined as,

$$BHP = \frac{WHP}{Pump\ eff. \times drive\ eff.} \qquad ....(12.27)$$

Similarly, for electric motors, the input horsepower (IHP) is determined as,

$$IHP = \frac{BHP}{Motor\ efficiency} \qquad ....(12.28)$$

$$or = \frac{Water\ horse\ power}{Pump\ eff. \times drive\ eff. \times motor\ eff.} \qquad ....(12.29)$$

The energy consumption in kilowatt hours per hour is determined as,

= input H.P. x 0.746

The product of the pump efficiency, drive efficiency and motor efficiency is known as the pumping plant efficiency. It is the overall efficiency of the entire pumping unit. It takes into account all losses in the pump and the drive unit. The pump efficiency is generally ranges from 60 to 70 per cent and the drive efficiency of motor is about 80 per cent. The overall efficiency of the system may be approximately 50 to 55 per cent.

Maximizing working efficiency of the tubewell and the pumps, proper selection of the pumps and the prime mover to suit the well conditions as well as care and maintenance of the pump and the tubewell components are essential. This is achieved by considering the following points.

1. The suction lift should be periodically checked and it should be within the permissible limits.
2. The gland packing in the pump should be checked and replaced whenever necessary. The gland packing should be inserted in small pieces of length equal to the circumference of the pump shaft. The ends of two adjacent rings should not coincide. The water should drip through the packing at a rate of 15 to 30 drops per minute.
3. Periodical inspection of the impeller of the pump is essential for wear. In case of cavity wells, small gravel is likely to get lodged in the impeller, reducing the capacity of the pump.
4. The r.p.m. of the prime mover should be at the rated value. Low voltage, improper pulley diameter and belt sag are some of the reasons for reduced r.p.m.
5. The alignment of the pump and motor shaft should be checked from time to time.

Wind power operated pumps

These pumps are used for lifting water where wind velocities are sufficient (6.5 km/hr on average and above). These pumps are known as wind milles and are used to lift water from on-farm reservoir for domestic purpose and kitchen garden due to low discharge. It consists of a large diameter vane wheel or rotor fixed on the top of a high steel or wooden tower. Usually a reciprocating pump is operated by the wind mill. The main disadvantage is that the functioning of the wind mill may not exactly coincide with the need for water. A storage tank for water is needed to store the lifted water.

Example 12.5: Determine the size of the electric motor needed for direct coupling with a centrifugal pump installation with the following conditions.

Static suction head = 5 m

Efficiency of pump = 70%

Static delivery head = 10 m

Efficiency of motor = 90%

Amount of water needed in 12 hr = 90 hec. cm

Solution:

$$\text{Rate of flow needed} = \frac{\textit{Volume of water needed}}{\textit{Working} \times \textit{eff.of pump} \times \textit{eff.of motor}}$$

$$= \frac{900 \times 10{,}000}{12 \times 70 \times 90 \times 3600}$$

= 33 litre/sec.

Total static head = 10 + 5 = 15 m

Assuming diameter of suction and

Delivery pipes = 100 mm

Frictional head loss

Frictional head loss in pipe line = (12.75/100 m pipe length from Table 12.4)

$$= \frac{1275}{1000} \times 100$$

= 1.91 m

For foot valve, equivalent length of pipe (Table 12.4)

= 8.23 m

Considering two bends, one on suction side and one on delivery side, equivalent length of pipe from Table 12.4

$= 2 \times 2.13$

= 4.26 m

Friction head loss in foot valve and bends

$$= \frac{(8.23 + 4.26) \times 12.75}{100} = 1.59$$

Total friction head = 1.91 + 1.59 = 3.5 m

Velocity head for 330 L/sec, discharge and 100 mm pipe size from (Table 12.3)

= 0.37 m

Total head required = 15 + 3.5 + 0.37

= 18.87 m

$$\text{Water horsepower from} = \frac{330 \times 18.87}{76} = 8.2 \qquad \text{....(12.30)}$$

$$\text{B.H.P.} = \frac{8.2}{0.60} = 13.7$$

$$\text{Input horsepower of motor} = \frac{13.7}{0.90} = 15.2$$

A motor of 15 HP will be required

(Note: Standard electric motors are made with horse power ratings of 1, 1¼, 1½, 1¾, 2, 2½, 3, 4 , 5, 7½, 10, 12½, 15, 20, 30....).

Example 12.6: For testing a centrifugal pump (10 cm × 7.5 cm) at discharge of 20 L/sec. The vacuum gauge on the suction side gave a reading of 20 cm of mercury. The pressure gauge reading on the delivery side was 1 kg/cm^2. The gauge on the delivery side was 0.76 m from the centre of the suction gauge. The pump was driven by an electric motor having 90 per cent efficiency. The watt meter on the line showed the input power as 4600 watts. Determine the efficiency of the pump.

Solution:

Suction head (since 760 mm of mercury ≈ 10.33 m head of water)

$$= \frac{20 \times 10.33}{76} = 2.72\,m$$

Delivery head (since 1 kg/cm ≈ 10 m of head of water) = 1 × 10 = 10 m

Distance between suction gauge and pressure gauge = 0.76 m

$$\text{Velocity head correction as suction} = \left(\frac{V_d^{\,2} - V_s^{\,2}}{2g}\right) = 0.57\,m$$

and delivery have different diameters

where, V_d = Velocity in delivery pipe

$$= \frac{20 \times 1000}{\frac{\pi}{4} \times 7.5 \times 7.5 \times 100}$$

$$= \frac{200000 \times 4}{3.14 \times 7.5 \times 7.5 \times 100} = 24.5\,m/s$$

V_s = Velocity in suction pipe

= 4.5 m/s.

Total head = 2.72 + 10 + 0.76 + 0.57 = 14.05 m

$$WHP = \frac{20 \times 14.05}{76} = 3.70$$

Output = 3.70 × 746 = 2758 watts

$$\text{Overall efficiency} = \frac{2758}{4600} \times 100 = 59.96\%$$

$$\text{Pump efficiency} = \frac{54.96}{0.90} = 66.6\%$$

References

Brooks, J.B. 1953. *Electric water systems for farms.* Univ. of Kentucky, College of Agr. and Home Eco., Cir. 452. p. 39.

Chauhan, H.S., Sewa Ram, Trivedi, S.K. and Sharma, H.C. 1979. Selection, installation and operation of horizontal centrifugal pumps for Irrigation, *Agricultural Engineering Tech. Bulletin 11*, G.B. Pant University of Agriculture and Technology, Pantnagar.

Church, A.H. and Jagdish, Lal 1973. *Centrifugal pumps and blowers.* Pub. Metropolitan Book Co. Pvt. Ltd., Delhi.

Dakshinamurti, C., Michael, A.M. and S. Mohan, 1973. *Water resources of India and their utilization in agriculture.* Pub. Water Technology Centre, IARI, New Delhi. pp. 269-285.

Das, G. and Michael, A.M. 1967. Evaluation of performance characteristics of circular two-bucket lift. Jl. of Indian Soc. of Agri. Engrs., Vol. 4, No, 2, pp. 1-8.

Driscoll, F.G. 1986. Groundwater and wells, Johnson Division, Minnesota, USA.

Fogel, M.M. and Myles, G.A. 1962. *Pumping from irrigation wells.* Univ. of Nevada Bull. No. 110. p. 23.

Gibson, A.H. 1924. *Hydraulics and its applications.* Pub. Constable and Co. Ltd., London.

Indian Standards Institution. 1970. *Indian standard specification for sewage and drainage pumps.* IS: 5600-1970. ISI New Delhi. p. 8.

Indian Standards Institution. 1970. Indian standard specification for vertical turbine pumps for clear, cold, fresh water. IS: 1971-72, ISI, New Delhi. p. 14.

Indian Standards Institution. 1972. Indian standard specification for horizontal centrifugal pumps for clear, cold, fresh water for agril. purposes. IS: 6595-1972, ISI New Delhi. p. 8.

Israelson, O.W. and Hansen, V.E. 1950. *Irrigation principles and practices.* 3rd ed. John Wiley and Sons, Inc., New York. pp. 52-74.

Johnson, Inc. 1950. *Irrigation pumps, their selection and use.* Uni. of California, Agr. Exp. Stn., Cir. 415. p. 54.

Johnson, Inc. 1966. *Ground water and wells.* Pub. Edward E. Johnson, Inc., Minnesota. pp. 375-394.

Khepar, S.D., Kaushal, M.P. and Katyal, A. 1975. Development of an Animal Drawn Lift Irrigation Pump, 12, 3 & 4. *Journ. of Agril. Engg.* (Indian Soc. of Agril. Engrs.).

Khepar, S.D., Kaushal, M.P. and Katyal, A.K. 1975. Development of an animal drawn lift irrigation pump. JI, of Agricultural Engineering. Vol. XII, No. 3-4.

Kill, D. 1973. Pumping water by the air lift method has practical applications. The Johnson Drillers JL, Nov.-Dec.

Kirloskar Brothers. 1962. *Pump bulletins,* Nos. 1 to 15. Pub. Kirloskar Brothers Ltd., Kirloskarvadi.

Kirloskar Brothers. 1962. *Pump bulletins.* Nos. 16 to 24, Kirloskar Brothers, Kirloskarvadi.

Kirloskar Brothers. 1967. *Engineering data.* Pub. Kirloskar Brothers Ltd., Kirloskarvadi.

LaL, J. 1969. *Hydraulic machines.* Pub. Metropolitan Book Co. (Pvt.) Ltd., Delhi. pp. 631.

Low, D.A. 1949. *Applied mechanics.* Pub. Longmans, Green and Co., London, pp.499-519.

Michael, A.M. 1972. Selection of pump sets. Madras Agril. Ji. pp. 147-151.

Michael, A.M. and Ojha, T.P. 1966. *Principles of agricultural engineering,* Vol. 11. Pub. Jam Brothers, Jodhpur. pp. 113-174.

Molenaar, A. 1956. *Water Lifting Devices for Irrigation,* Food and Agriculture Organization of the United Nations, Rome.

Patel, S.M. 1970. An easy and practical method for selection of lift irrigation equipment by the farmers. Personal correspondence.

Patel, S.M. 1970. Conditions for designing ideal lift irrigation plants for farmers. Agriculture and Agro-industries Jl., December 1970.

Rastogi, R. A. 1966. Performance characteristics of a washer rahat. Jl. of Indian Soc. of AgrI. Engrs., Vol. 3, No. 2,.pp. 23-27.

Schwab, G.O., Frevert, R.K., Edminster, T.W. and Barnes, K.K. 1966. *Soil and water conservation Engineering.* 2nd ed. Pub. John Wiley and. Sons, Inc., New York. pp. 183-505.

Sharma, S.C., Agarwal, R.L. and Lal, R. 1968. *Analysis of power consumption in Persian wheel. Jl. of Irrigation and Power*, Vol. 24, No. 4, pp. 375-382.

Stepanoff, A.J. 1957. *Centrifugal and axial flow pumps.* 2nd Ed. John Wiley & Sons, Inc., New York. p. 462.

United States Dept. of Agri. 1959. *Irrigation pumping plants.* SCS National Engg. Handbook, Section 15, Irrigation. Soil Cops. Service, U.S. Dept. of Agri., Washington, D.C. pp. 8-1 to 8-70.

White, R.G. and Kidder, E.H. 1962. *Tractor power for power-rake-off driven pumps.* Extn. Ser., Michigan State Univ, East Lansing, Cir. E-23. p. 23.

Wolf, J.W. 1950. *A pumping manual for irrigation and drainage.* Oregon State College, Agrl. Expt Stn. Corvallis, Bul. 481. p. 39.

Wood, I.D. 1950. *Pumping for irrigation.* Soil Conservation Service, U.S. Dept of Agr., Washington D.C. p. 40.

EXERCISES

12.1. Differentiate between rotodynamic and positive displacement pump.

12.2. Explain procedure of pump selection.

12.3. Explain the working of reciprocating pump.

12.4. Describe working principle of centrifugal pump.

12.5. Describe the procedure for selection of centrifugal pump.

12.6. Discuss the common problems in centrifugal pump operation and their remedies.

12.7. Describe propeller and airlift pumps.

12.8. A deep well single acting piston pump has a cylinder diameter of 10 cm and stroke 22.5 cm. The pump makes 300 strokes per minute. Calculate the discharge of the pump in litres per hour. If the total head is 36 m and the pump efficiency is 60 per cent, determine the horse power required to drive the pump. **(15900 litres/hr, 3.5 HP)**

12.9. A centrifugal pump is running at a speed of 1800 rpm at its best efficiency point and delivers 50 litres of water per second at a total head of 30 metres. Compute the specific speed of the pump. **(31.4 rpm)**

12.10. Determine the desired discharge rate of a centrifugal pump to be purchased by a farmer for the following irrigation schedule.

Season	Crop	Area to be irrigated (ha)	Intensity of irrigation (cm)	Rotation period (days)	Period of work (hrs/day)
Kharif	Paddy	1.6	10.0	15	10
	Cotton	2.0	10.0	20	12
Rabi	Wheat	3.2	7.5	12	12
	Cotton	2.0	7.5	20	12
	Vegetables	1.6	7.5	10	12
Hot weather I	*Bajra*	0.8	7.5	8	10

(9.14 L/sec)

12.11 Determine the annual consumption of electrical energy by a motor-driven centrifugal pump installed in a shallow tube well. The pump discharge is 16 litres per second against a total head of 7 m. The pump efficiency is 70 per cent and the motor efficiency 84 per cent. The drive efficiency may be assumed to be 100 per cent. The pump is operated for 3500 hours per year. **(6546 kilowatt-hours)**

12.12. Why is a reciprocating pump called positive displacement pump?

12.13. Explain the terms 'slip' and 'cavitation' with reference to reciprocating pumps.

12.14. Compare discharge curves for a single and double acting reciprocating pumps and indicate under what conditions are either type used.

12.15. Explain the working principles of reciprocating pump with sketches.

12.16. Explain with a neat sketch of its main parts, and a diagram of velocity v/s stroke, the working of a three throw pump. It an air vessel necessary for this type of pump? Explain your answer.

12.17. The work saved by fitting an air vessel on a single acting reciprocating pump is much more than in a double acting reciprocating pump. Why?

12.18. Explain the functions of air vessels in a reciprocating pump.

12.19. Show that the maximum inertia head in a reciprocating pump without air vessel is given by with usual notation.

$$H_a = \frac{l}{g} \times \frac{A}{a} \omega^2 r$$

12.20. A single acting pump has a piston diameter of 15 cm and a crank radius of 30 cm. The delivery pipe is 10 cm in diameter and 30 in long. The water is lifted 20 in above the centre of the pump. Find the maximum speed at which the pump can be run so that no cavitation takes place during the delivery stroke. Take H_{atm}=10.3 m of water and separation occurs at 2.6 m of water absolute.

(49.5 r.p.m)

12.21. A single acting reciprocating pump has a plunger of diameter 30 cm and a stroke of 40 cm. If the speed of the pump is 60 r.p.m. and coefficient of discharge is 0.97, determine the percentage slip and actual discharge of the pump. **(3%; 27.41 L/see)**

■■■■

Chapter 13

MEASUREMENT OF IRRIGATION WATER & CONVEYANCE STRUCTURES

13.1 Introduction

Harvested rainwater is the most valuable asset under dryland ecosystem. Accurate measurement of irrigation water permits more scientific use of this valuable natural resource. Such a measurement reduces excessive waste and allows the water to be distributed among users according to their needs. Accuracy in water measurement is, therefore, of prime importance in the operation of any water distribution system for application of irrigation and also for individual farmer's field to enhance water productivity.

13.2 Units of Measurement

Water is measured under two situations like (i) at rest and (ii) in motion. Water is at rest in reservoirs, farm ponds, nadi, tanks, community tanks etc. and is measured in units of volume *viz.*, litre, cubic metre, hectare-centimetre and hectare-metre etc. While, flowing in open channels, pipelines, field channels, and channel structures that is in a motion then it is measured in units of rate of flow *viz.*, litres per second, litres per hour, cubic metres per second, hectare-centimetres per hour, and hectare-metres per day etc.

Litre per second. It is defined as continuous flow amounting to 1 litre passing through a point each second (generally used to denote the discharge of a pump, small stream or pipeline).

Cubic metre per second. It is defined as flow of water equivalent to a stream 1 metre wide and 1 metre deep flowing at a velocity of 1 metre per second.

Unit conversion factors

$$\frac{1}{1000}\, cubic\, metre = 1\, litre$$

1 cubic metre = 1000 litres

1 hectare cm = 100000 litres

$$1\ hectare-metre = 1000\ cubic\ metres$$
$$= 10\ million\ litres$$

Example 13.1: A Persian wheel with an average discharge of 250 litres per minute irrigates I hectare maize crop in 45 hours. Determine the average depth of irrigation ?

Solution:

Discharge in 50 hours $= 250 \times 60 \times 45$

$= 6{,}75{,}000$ litres $= 675\ m^3$

$$\text{Depth of irrigation, cm} = \frac{Volume\ of\ water(cu\ m)}{Area\ of\ land\ (sq\ m)} \times 100$$

$$= \frac{675}{10{,}000} \times 100 = 6.7\ \text{cm.}$$

Example 13.2: Maize crop requires 40 cm of irrigation water during 120 days irrigating period. How much land can be irrigated with a flow of 25 litres per second for 20 hours a day.

Solution:

Total discharge during the irrigation period

$$= \frac{25 \times 60 \times 60 \times 20 \times 120}{1000} = 2{,}16{,}000\ m^3$$

Irrigation requirement per hectare

$$= \frac{40}{100} \times 10{,}000 = 4{,}000\ \text{cum}$$

$$\text{Area irrigated} = \frac{Volume\ of\ water\ available\ (cu\ m)}{Volume\ of\ water\ requ.\ per\ ha\ (cu\ m)}$$

$$= \frac{216000}{4000} = 54.0\ \text{ha.}$$

13.3 Techniques of Water Measurement

Several techniques are used for measurement of irrigation water on the farm. These techniques can be grouped into four categories as follows:

(i) volumetric measurements, (ii) velocity-area methods, (iii) measuring structures (orifices, weirs and flumes), and (iv) tracer methods (dilution).

13.3.1 Volume metric measurement of water

It is a simple method for measurement of small irrigation streams. In this technique, water in a container is collected (known volume) for a measured period of time (Fig. 13.1). An ordinary bucket or barrel is used as the container whose volume is known. The time required to fill the container is measured with a stop watch or a wrist watch. The rate of flow is measured by the following relationship:

$$\text{Discharge rate (L/sec)} = \frac{\textit{Volume of container}\,(L)}{\textit{Time required to fill}\,(\text{sec})}$$

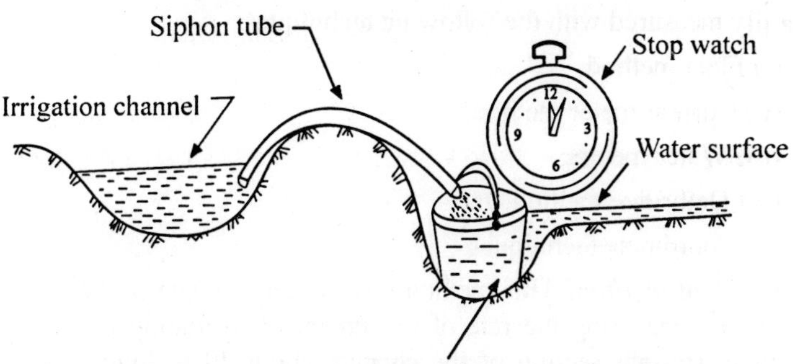

Fig. 13.1: Volumetric method of water measurement

***Example 13.3*:** 24-litre capacity bucket is filled in 10 seconds by the discharge from a 1 HP motor. Determine discharge

Solution:

$$= \frac{24}{10} = 2.4 \text{ lit/sec or } 144 \text{ lit/min.}$$

This technique is applicable to determine the discharge rate of pumps, Persian wheel and leather bucket lift (*charas*) etc. A barrel of 150 to 200 litres capacity is used to collect the flow. This technique is also suitable when the stream is discharging into a reservoir of sufficient capacity. The volume can be determined by direct measurements with a tape. The flow is determined by noting the time for the reservoir to fill to a certain depth, or for the water surface to rise from one level to another.

13.3.2 Velocity-area technique

The rate of flow passing a point in a pipe or open channel is determined with this technique. Then the discharge is determined by the following relationship:

Discharge = Area x Velocity

$$Q = a \times v \qquad \text{....(13.1)}$$

where,

Q = discharge rate, (m^3/sec),

a = area of cross-section of channel or pipe at right angle to the direction of flow, (m^2), and

v = velocity of flow, (m/sec).

The cross-sectional area is measured directly. The velocity is generally measured with the following techniques:

(i) Float method,

(ii) Current meter method,

(iii) Water meters,

(iv) Dethridge method, and

(v) Coordinate techniques.

(i) Float method: This method give rough estimate of the flow in a channel by measuring the rate of movement of a floating body. In this technique, straight section of the channel about 30 m long with fairly uniform cross-section is selected. The average cross-sectional area of this section is determined. A string is stretched across each end of the section at right angles to the direction of flow. The float (wooden block) is placed in the channel, a short distance upstream from the selected section. The time the float needs to pass from the upper to the lower section is recorded. Several trials are made to get the average time of travel.

The velocity of the float on the surface of the water will be greater than the average velocity of the stream. Therefore, it is essential to correct the measurement by multiplying by a constant factor of 0.85. The average velocity is determined as,

$$Q_v = 0.85 \times A_v \times V_v \qquad \text{....(13.2)}$$

where,

Q = Average discharge, (m^2/sec),

A_v = Average cross section area of channel, (m^2), and

V_v = Average velocity of flow (m/sec).

(ii) Current meter method: The velocity of water flowing in a stream or river may be measured directly with a current meter and the discharge is determined as under.

Discharge = Mean velocity × area of cross section ....(13.3)

This is a small instrument containing a revolving wheel or vane that is turned by the movement of water (Fig. 13.2). It is suspended by a cable in deep streams or attached to a rod in shallow streams for measurements of velocity of flow. The number of revolutions of the wheel in a given time interval will be recorded and the corresponding velocity is obtained from a calibration table or graph of the instrument.

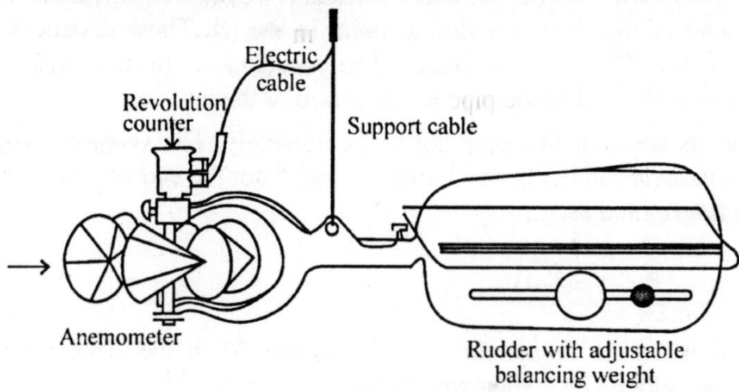

Fig. 13.2: Cup type current meter

Both the float and the current meter techniques have very limited application in field irrigation practice as these give only rough estimate of discharge.

(iii) Water meters: It consists of a multi-blade propeller and is made of metal, plastic or rubber, rotating in a vertical plane. Water meters are geared to a totalizer in such a way that a numerical counter can totalize the flow in any desired volumetric units. These are available in different sizes suiting the pipe sizes commonly used on the farm.

For accurate operation of the water meter, following requirements are prerequisite:

(a) The pipe must flow full at all times, and

(b) The rate of flow must exceed the minimum for the rated range.

(iv) Dethridge method: It consists of an under-shot water wheel turned by the discharge water passing through its implacement and made of a short concrete outlet. This is made in such a manner that it provides only the minimum practicable clearance of the lower half of the wheel at its sides and around its circumference. The meter gives a positive measurement of discharge. It is recorded directly in volumetric units by a revolution counter driven by the wheel axle. It works with 5 per cent accuracy under normal working conditions.

The main advantages of this technique is that it operates with small head loss. However, it is simple, robust in construction and it will pass ordinary floating debris without damage to or stoppage of the wheel.

(v) Coordinate techniques: This technique is used to determine discharge from pipe like pump discharge from well.

For measuring from discharge horizontal pipe, it is necessary to measure both a horizontal (*X*) and a vertical (*Y*) distance from some point on the end of the pipe to a similar point in the jet. These distances are known as the *X* and *Y* coordinates. The coordinates are measured from the centre of the end of the pipe to the centre of the jet.

Let us consider that pipe is filled completely. The vertical distance *Y* that a particle falls in time (*t*) after issuing from the end of a horizontal pipe is determined as:

$$Y = \frac{gt^2}{2} \quad \text{....(13.4)}$$

and it travels horizontally to a distance (*X*) in the same time (*t*) within the velocity (*V*). Then we obtained

$$X = Vt \quad \text{....(13.5)}$$

Combining these equations (13.4 & 13.5) for eliminating *t* and solving for *V*,

$$V = \frac{X\sqrt{g}}{\sqrt{2Y}} \quad \text{....(13.6)}$$

Then discharge from the pipe will be determined by the following expression as,

$$\mathrm{Q = av} \quad \text{....(13.7)}$$

We obtain discharge with equation (13.6) & (13.7)

$$Q = \frac{CaX\sqrt{g}}{\sqrt{2Y}} \quad \text{....(13.8)}$$

where,

Q = discharge, (m^3/sec),

C = Coefficient of contraction, (dimensionless)

a = Cross-section area of pipe, (m^2),

X = X-coordinate, (m), and

g = acceleration due to gravity, (m/sec^2)

The value of C commonly used in calculation is 1.0.

Fig. 13.3: A Dethridge wheel installed at the inlet of a branch water course

The discharge from partially filled pipes can be determined by multiplying the flow obtained from equation 16.8 by the percentage of the area of the pipe that is filled at the end of the pipe. The coordinates should be measured from the approximate centre of the jet at the end of the pipe instead of from the centre of the end of the pipe, as is the case with a full flowing pipe.

There techniques have low accuracy due to the difficulty in making accurate measurements of the coordinates of the jet. While, estimates of discharge from vertical pipes have very limited practical application on the farm.

It is used for measurement of well discharge from the horizontal pipes.

13.3.3 Measuring structures

The most commonly used devices for measuring water in farm irrigation practice are as under:

(i) Weirs,

(ii) Parshall flumes,

(iii) Orifices, and

(iv) Metergates

These structures or devices are used to measure rate of flow directly by making a reading on a scale. The discharge rate is determined by different formulas in different devices. The choice of structures to be used depends on the expected flow rates and site conditions.

(i) Weirs:

A weir is a notch of regular form through which the irrigation stream is made to flow. These are used to measure the flow in an irrigation channel, or the discharge of a well or canal outlet at the source.

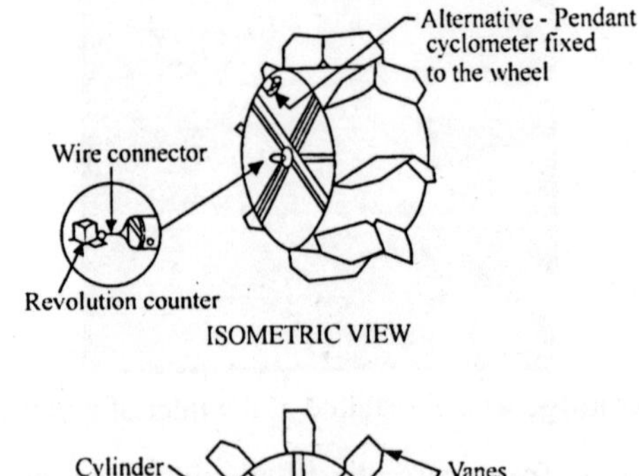

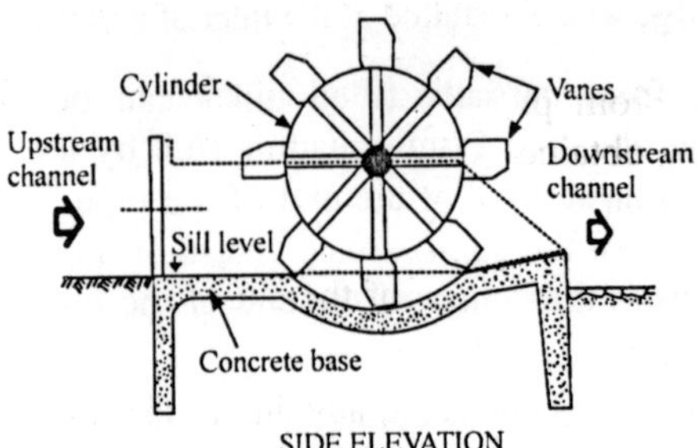

Fig. 13.4. Views of Dethridge meter showing parts details and installation

A weir consists of a weir wall which is made of concrete, timber or metal with a sheet metal weir plate fixed to it. These structures may be stationary or portable. The shape of notch may be rectangular, trapezoidal or triangular. Rectangular weirs and *90°-V* notch weirs are commonly used. It should be installed where there is a drop in the elevation of the channel bed. It will ensure that the flow through the weir fall freely to the downstream channel.

Important terminologies pertaining to measuring structures:

Weir pond: It is the portion of the channel immediately upstream from the weir.

Weir crest: It is defined as the bottom of the weir notch.

Head: It is defined as the depth of water flowing over the weir crest measured at sonic point in the weir pond.

Sharp crested weir: It is defined as thin-edged crest of war such that the overflowing sheet of water has the minimum surface contact with the crest.

End contraction: It is the horizontal distance from the ends of the weir crest to the sides of the weir pond.

Bottom contraction: It is defined as the vertical distance from the weir crest to the bottom of the weir pond.

Weir scale or gauge: It is the scale fastened on the side of weir or on a stake in the weir pond to measure the head on the weir crest.

Nappe: It is defined as the sheet of water which overflows a weir.

Weirs are mainly of two types as under:

(i) Sharp crested

These are commonly used for water measurement on the farm. Sharp crested weirs may be with end contractions and those without end contractions. End contractions are made when the crest length, or width of the notch is less as compared to the channel width.

(ii) Broad crested

In these weirs, the crest extends across the full width of the channel and no contractions are made. When the water surface downstream from the weir wall is far enough below the crest so that air has access around the nappe than the flow is referred as freefall, otherwise it is submerged.

The discharge from a weir notch is proportional to the head on the crest. This is influenced by several factors *viz.*, condition of the crest, contraction, velocity of the approaching stream and the elevation of' the water surface downstream from the weir. Fig. 16.5 shows the different parts of weirs.

The discharge is determined by the following formulae from a weir as expressed,

$$Q = CLH^m$$

where,

Q = discharge, (L/sec),

C = a coefficient, dependent on the nature of the crest and approach conditions, (dimensionless),

L = length of crest, (m),

H = head on the crest, (m) and

m = an exponent, depends upon the weir opening.

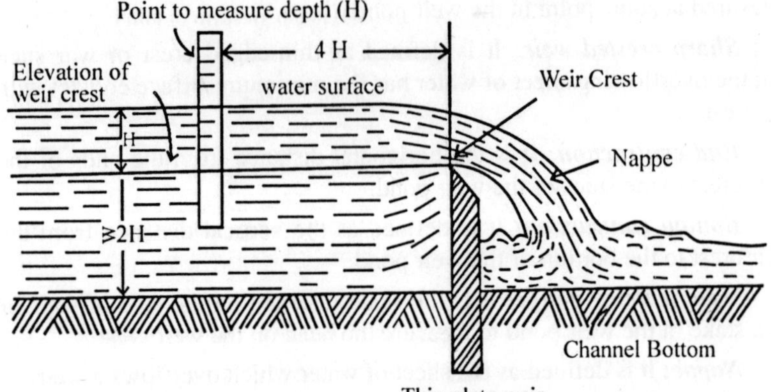

Fig. 13.5: Weir parts

Rectangular weirs

In this weir, the shape of the notch is rectangular. These are used for measurement of large discharges. It has a horizontal crest and vertical sides. This type of weir may be either contracted rectangular weirs or suppressed rectangular weirs.

The discharge through rectangular weirs may be determined on using the Francis' formula as under:

(i) Suppressed rectangular weir

$$Q = 0.0184\ LH^{3/Z} \qquad \text{....(13.3)}$$

where,

Q = discharge, (L/sec),

L = length of crest, (cm) and

H = head over the weir, (cm)

(ii) Contracted rectangular weir (with end contractions at both ends)

$$Q = 0.0184\ (L - 0.2H)\ H^{3/2} \qquad \text{....(13.4)}$$

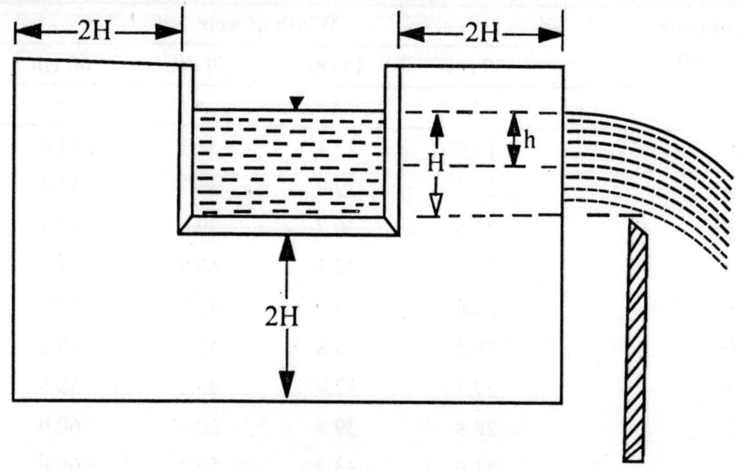

Fig. 13.6: Rectangular weir

The values of discharge through contracted rectangular weirs (assuming end contractions at both ends of the weir) under different widths and operating heads is summarized in Table 13.1.

Table 16.1: Discharge through contracted rectangular weirs (L/sec)

Head over weir (cm)	Width of weir			
	30 cm	40 cm	50 cm	60 cm
1	2	3	4	5
5.0	6.0	8.0	10.1	12.2
5.5	6.9	9.3	11.6	14.0
6.0	7.8	10.5	13.1	15.9
6.5	8.4	11.8	14.9	17.9
7.0	9.7	13.2	16.6	20.0
7.5	10.7	14.5	18.3	22.1
8.0	11.8	16.0	20.1	24.3
8.5	12.9	17.6	22.1	26.7
9.0	14.0	19.0	24.0	28.9
9.5	15.2	20.7	26.0	31.2
10.0	16.3	22.2	28.0	33.8
10.5	17.5	23.7	30.0	36.2
11.0	18.7	25.3	32.0	37.7

Head over weir (cm)	Width of weir			
	30 cm	40 cm	50 cm	60 cm
1	2	3	4	5
11.5	19.9	27.1	34.3	41.4
12.0	21.3	29.0	36.7	44.4
12.5	22.5	30.7	39.0	47.1
13.0	23.7	32.3	40.9	49.5
13.5	24.8	34.0	43.0	52.2
14.0	26.2	35.8	45.4	55.2
14.5	27.7	37.9	48.2	58.5
15.0	28.8	39.5	50.3	60.9
16.0	31.6	43.3	55.2	67.0
17.0	34.3	47.2	60.1	73.0
18.0	37.0	51.0	65.3	79.0
19.0	39.8	55.0	70.2	85.3
20.0	42.8	59.3	75.8	88.8
21.0	45.7	63.3	81.0	99.0
22.0	48.7	67.5	86.7	105.7
23.0	41.0	71.7	92.2	112.3
24.0	54.7	76.5	94.8	120.0
25.0	57.0	79.8	102.7	125.8
26.0	60.3	84.6	109.2	133.3
27.0	63.5	89.2	115.0	140.8
28.0	66.5	93.7	122.2	148.3
29.0	69.5	98.3	127.0	155.7
30.0	72.5	102.7	133.0	163.3

(Source: Michael, 1992)

Example 13.4: Determine the discharge of a rectangular weir of 40 cm length with a head of 15 cm under the following conditions:

(i) With no end contraction

(ii) With one end contraction

(iii) With two end contractions

Solution:

(i) With no end contraction

$$Q = 0.0184 \text{ L H}^{3/2}$$
$$= 0.0184 \times 40 \times 15^{3/2}$$
$$= 42.8 \text{ L/sec}$$

(ii) With one end contraction

$$Q = 0.0184 \text{ (L} - 0.1 \text{ H) H}^{3/2}$$
$$= 0.0184 (40 - 0.1 \times 15) 15^{3/2}$$
$$= 41.1 \text{ L/sec}$$

(iii) With two end contraction

$$Q = 0.0184 \text{ (L} - 0.2 \text{ H) H}^{3/2}$$
$$= 0.0184 (40 - 0.2 \times 15) 15^{3/2}$$
$$= 39.5 \text{ L/sec}$$

Cipoletti weir

It is a contracted trapezoidal weir in which each side of the notch has a slope of 1 horizontal to 4 vertical. The side slopes of 1 to 4 are just sufficient to provide correct for the end contractions of nappe. It does not require corrections for end contractions. The rate of flow is proportional to the length of the weir crest. The weir has sharp crest and share sides, beveled from the downstream side only. Generally, it is used to measure medium discharges. It is shown in Fig. 13.7

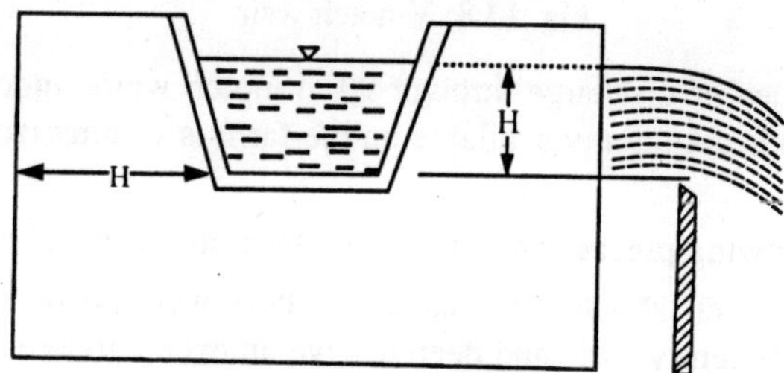

Fig. 13.7: Trapezoidal weir

The discharge from Cipoletti weir is determined by the following relationship:

$$Q = 0.0186 \text{ L H}^{3/2} \quad \text{....(16.5)}$$

where,

Q = discharge, (L/sec),

L = length of crest, (cm), and

H = head over the weir, (cm)

V-notch weirs

The 90° *V*-notch weir is generally used to measure small and medium size streams discharge. The advantage of the *V*-notch weir is to measure small flows accurately and is shown in Fig. 13.8. The discharge from a 90^0 *V*-notch weir may be determined by the following relationship:

$$Q = 0.0138\, H^{5/2} \qquad ....(13.6)$$

where,

Q = discharge, (L/sec), and

H = head, (cm)

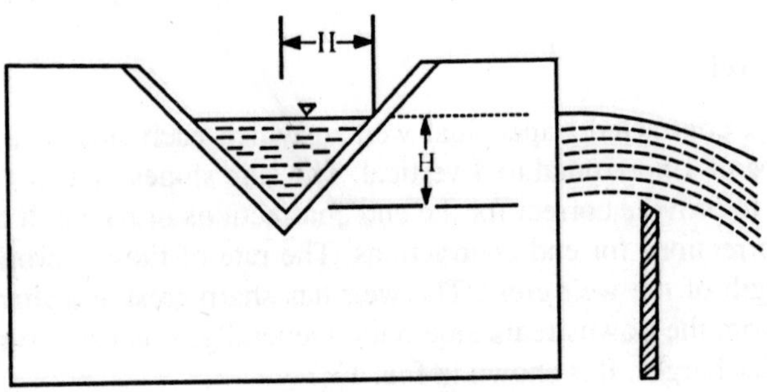

Fig. 13.8: V-notch weir

The values of discharge through 90° *V*-notch weirs under different operating heads commonly available on the farm is summarized in Table 13.2.

The following precautions should be essential for installing weirs:

1. The weir should be installed at the lower end of a long pool sufficiently wide and deep to give an even, smooth flow of the stream. The velocity of the approaching flow should not exceed 15 cm per second.
2. The weir wall or bulkhead must be vertical and it must extend far enough to provide security of bank.
3. The centre line of the weir should be parallel to the direction of flow.
4. The crest of the weir should be level so that water passing over it will have the same depth at all points along the crest.

5. The notch should be of a regular shape and its edges must be rigid and straight.
6. The weir crest should be above the bottom of the approach channel (weir pond) by at least about twice the depth of water flowing over the weir. A minimum distance of about twice the depth of flow should be maintained between the end of the weir crest and the bank of the upstream channel.
7. The crest of the weir is high enough so that water will fall freely below the weir.
8. The depth of water flowing over the rectangular weir should not be less than about 5 cm and not more than about two-thirds the crest width.
9. The scale or gauge used for measuring the head should be located at a distance of about four times the approximate head.

Table 13.2 Discharge through a 90° *V*-notch (L/sec)

Head of water over *V*-notch	**Discharge**	**Height of water *V*-notch**	**Discharge**	**Height of water *V*-notch**	**Discharge**
(cm)	**(L/sec)**	**(cm)**	**(L/sec)**	**(L/sec)**	**(L/sec)**
4.0	0.45	13.0	8.6	22.0	31.0
4.5	0.60	13.5	9.5	22.5	34.0
5.0	0.80	14.0	10.5	23.0	35.7
5.5	1.0	14.5	11.3	23.5	38.2
6.0	1.2	15.0	12.3	24.0	40.0
6.5	1.5	15.5	13.3	24.5	42.7
7.0	1.8	16.0	14.5	25.0	44.5
7.5	2.2	16.5	15.6	25.5	46.7
8.0	2.5	17.0	16.7	26.0	48.8
8.5	2.8	17.5	18.3	26.5	51.0
9.0	3.4	18.0	19.4	27.0	53.8
9.5	3.9	18.5	21.7	27.5	56.3
10.0	4.5	19.0	22.3	2.o	58.7
10.5	5.1	19.5	23.5	28.5	61.5
11.0	5.7	20.0	25.5	29.0	64.5
11.5	6.3	20.5	27.0	29.5	66.8
120	7.1	21.0	28.3	30.0	69.4
12.5	7.8	21.5	30.3		

(Source: Michael, 1992)

Demerit of weirs: Although weirs are easy to construct and convenient to use. However, they are not suitable at every place. These will not give accurate measurement unless proper conditions for their measurements are melt. These require a considerable loss of head which is not available in channels and easily not combined with turnout structures. Weirs are unsuitable if water carrying silt, which deposits in the approach channel and destroys the proper conditions for weir measurement.

Parshall Flume

It is an open channel type measuring device that operates with a small drop in head. The loss of head for free flow limit is only about 25 per cent of weir. It is a self-cleaning device. Thus, sand or silt flowing in the water does not affect its operation or accuracy. This device allows reasonably accurate measurement even under partially submerged condition. The velocity of the approaching stream has very little influence on its operation.

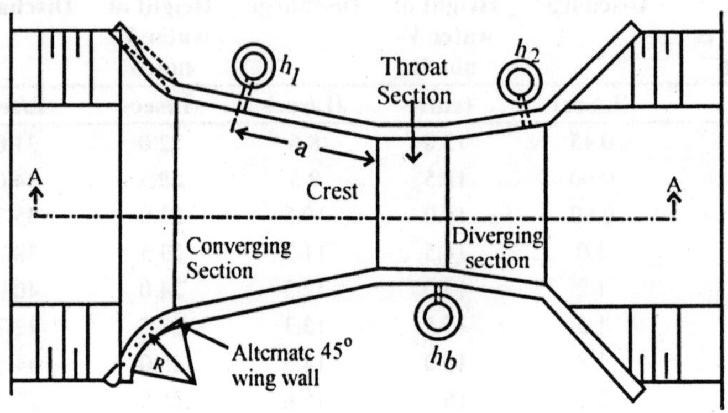

13.9: Parshall flume

The Parshall flume is made of metal sheet or concrete. The size of flume is expressed by the width of its throat (7.5 cm to several meters). The floor of the upstream section of the flume is level and the walls converge toward the throat section (Fig. 13.9). The walls of the throat section are parallel and the floor is inclined downward. The walls of the downstream section diverge toward the outlet and the floor is inclined upward. Large size flumes have an approach section and wing walls at the upstream end.

Stilling wells may be provided to measure the water surface elevations in the flume if turbulence is present. In case flow is smooth, a plastic scale is fixed to the inside of the flume will be sufficient to measure the head. The zero of the scales should be set at same level as the floor level of the converging section. There is no need to have a pond above the flume as the velocity of approach has little effect on the accuracy of management.

Standard dimensions of Parshall flumes and discharge values are summarized in Tables 13.3 and 13.4, respectively.

Table 13.3: Dimensions and capabilities of Parshall flumes of various sizes

Throat width (cm)	A (cm)	B (cm)	C (cm)	D (cm)	E (cm)	F (cm)	G (cm)	K (cm)
7.5	31	45.5	17.8	26	45.5	15	30.5	2.5
15	41.4	61	39.4	39.7	61	30.5	61	7.6
23	58.8	86.4	38.1	57.5	76	30.5	45.5	7.6
30	91.5	134	61	84.5	91.5	61	91.5	7.6

Throat width	N	X	Y	Free-flow capacity	
(cm)	(cm)	(cm)	(cm)	Minimum, L/sec	Maximum, L/sec
7.5	5.7	2.5	3.8	0.85	28.4
15	11.5	5.1	7.6	1.4	110.8
23	11.5	5.1	7.6	2.5	253
30	23	5.1	7.6	3.13	454.6

(Source : Michael, 1993)

Table 13.4: Free-flow discharge values for Parshall flumes

Head (cm)	Discharge (L/sec)			
	Throat width			
	7.5 cm	15 cm	23 cm	30 cm
3	0.8	1.4	2.6	3.1
4	1.2	2.3	4.0	4.5
5	1.7	3.3	5.5	7.0
6	2.3	4.4	7.2	9.6
7	2.7	5.4	8.5	11.4

Head (cm)	Discharge (L/sec)			
	Throat width			
	7.5 cm	15 cm	23 cm	30 cm
8	3.5	7.2	11.1	14.4
9	4.3	8.5	13.5	17.7
10	5.0	10.2	15.9	21.1
11	5.8	11.6	18.1	23.8
12	6.7	13.5	21.1	27.5
13	7.5	15.0	23.3	31.0
14	8.5	17.3	26.7	35.0
15	9.4	19.2	29.5	38.7
16	10.4	21.2	32.5	42.7
17	11.4	23.2	35.6	46.7
18	12.4	25.3	39.0	51.2
19	13.6	27.8	42.5	55.0
20	14.3	30.0	45.8	59.7
21	15.8	32.7	49.3	64.7
22	17.1	35.2	53.3	69.8
23	18.2	37.7	56.8	74.0
24	19.4	40.1	60.5	79.0
25	20.7	42.7	64.5	84.1
26	22.0	45.7	69.3	89.0
27	23.3	48.1	72.4	94.3
28	24.8	51.5	76.7	100.0
29	26.0	54.0	80.7	105.1
30	27.5	57.3	85.2	111.0

(Source : Michael, 1992)

Discharge from the flume can occur under either free flow or submerged flow conditions. If the elevation of the water surface downstream from the flume is high enough to retard the rate of discharge, the flow is considered submerged. For determination of discharge, two scales, H_a and H_b are provided at the upstream and downstream sections of the flume. Only H_a needs to be measured under free flow conditions. Free flow conditions are re-prepared by the ratio of H_b/H_a (degree or percentage of submergence) as under:

Width of throat	*Free flow limit* H_b/H_a
2.5 to 7.5 cm	0.5
1 5 to 22.5 cm	0.6
30 to 240 cm	0.7

One of the important characteristics of the Parshall flume is its ability to withstand a relatively high degree of submergence over a wide range of backwater conditions downstream from the structure and still it maintains free flow conditions.

Cutthroat flume

It is an improvement of the Parshall flume, mainly by simplifying the construction details (Fig. 13.10). The flume has a flat bottom, vertical walls, and a zero length throat section. Since it does not have throat section (zero throat length) and hence known as 'cutthroat flume'. The main advantage a cutthroat flume is economy and easy in construction. The flume length and threat width varied from 45 cm to 3 m and 2.5 cm to 1.8 m, respectively.

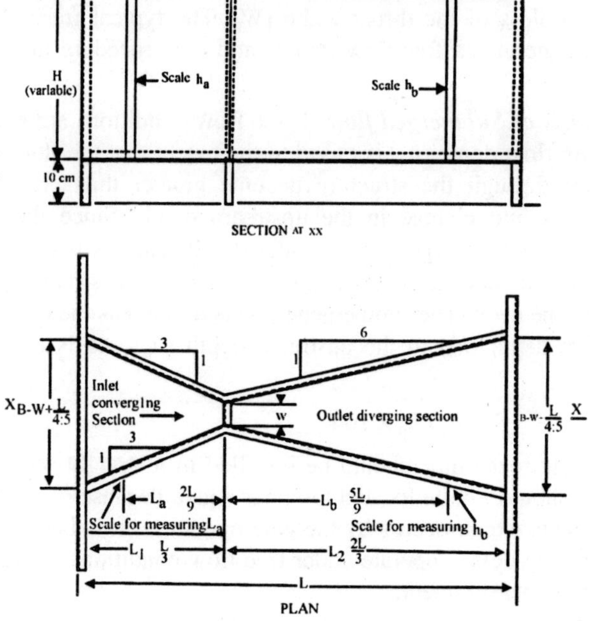

Fig. 13.10: Cutthroat flume

This flume can operate either as a free or a submerged flow structure. Under free flow conditions, critical depth occurs in the vicinity of minimum width (*w*). It is called as flume throat. The flow rate is determined by knowing only an upstream depth (ha) under critical flow condition (critical depth).

This is attributed to upstream depth (ha) which is not affected by changes in the down stream depth (*hb*) under critical flow condition. The ratio of inlet flow depth (h_a) to flume length should preferably be less than 0.4 under free flow condition. The discharge is determined by the following relationship:

$$Q = C_1 ha^{n_1} \quad \text{....(13.7)}$$

where,

Q = flow rate, (L/sec),

C_l = free flow coefficient, which is the value of Q when h_a is 1.0 foot, which is the slope of the free flow rating curve when plotted on logarithmic paper, and

n_l = Exponent, it value depends only on the flume length.

The value of n_l is a constant for all cutthroat flumes of the same length, regardless of the throat width (*W*). The typical free flow rating curve and generalized free flow ratings and the procedure in Appendix-C.

Analysis of Submerged flow: If the flow conditions are such as the downstream flow depth (h_a) is raised to the extent that the flow depths at every point through the structure become greater than critical depth, which results into change in the upstream depth. Since the flume is operating under submerged flow conditions. It requires two flow depths to be measured, one upstream depth (h_a) and down stream depth (h_b) from the flume neck. The submergence (*s*) is defined as the percentage of downstream depth (h_b) to the upstream depth (h_a) and is expressed as under:

$$S = h_b / h_a \times 100 \quad \text{....(13.8)}$$

The cutthroat flume should be installed in a straight section of the channel. It should not be located too near a gate because of unstable and surging effects which occurs by the gate operation. It is better to have a flow measuring device operate under free flow conditions because it will give accurate measurement.

Orifices

The shape of orifices is circular or rectangular. These are openings in a vertical bulkhead through which water flows in open channels. The edge of the openings are sharp and made of metal. The cross-sectional area of the orifice is small in comparison to stream cross-section. These conditions provide complete contraction of the stream flow and the velocity of approach becomes negligible.

Orifices may operate under free flow or submerged flow conditions. Under free flow conditions downstream water level is below the opening and the flow from the orifice discharges entirely into air (Fig. 13.11). While, in submerged flow condition, the downstream water level is above the top of the opening and the flow discharges into water. For measuring discharge, the measurements of head on free flow and submerged flow orifices are made. Partial submergence condition occurs when the downstream water surface is between the elevations of the top and bottom of the orifice. This condition is avoided to ensure accurate measurement of discharge with orifices.

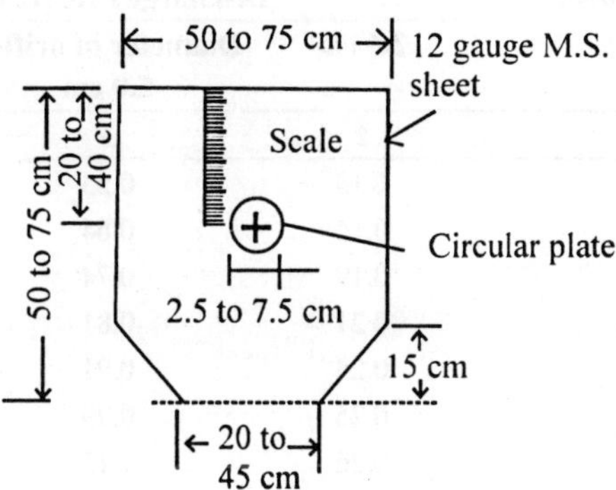

Fig. 13.11: Orifice plate

Free flow orifices:

These can be used to measure comparatively small streams *viz.*, flow into border, strips, furrows or check basins. It can also be used in restively level channels and no stilling and is required. It is made of sheet iron, steel or aluminum plates having circular openings or orifices and usually its size varied from 2.5 cm to 7.5 cm. The edge of the orifice need not be sharpened in case of thin walled orifice plates. A plastic scale may be used to measure head on the upstream face of the orifice plate with its zero reading coinciding with the centre of the orifice.

The discharge through an orifice is determined by the following relationship:

$$Q = 0.61 \times 10^{-3}\, a\sqrt{2gH} \qquad \text{....(13.9)}$$

where,

Q = discharge through orifice, (L/sec),

a = area of cross-section of the orifice, (cm^2),

g = acceleration due to gravity, (cm/sec^2) ($981 cm/sec^2$),

H = Depth of water over the centre of the orifice (on the upstream side) in case of free flow orifice, or the difference in elevation between the water surface at, the upstream and downstream faces of the orifice plate in case of submerged orifices, (cm).

The values of discharge through different size of circular orifices is given in Table 13.5.

Table 13.5: Discharge through circular orifices

Height of water over centre of orifice (cm)	**Discharge rate (L/sec)**		
	2.5 cm	**Diameter of orifice 5.0 cm**	**7.5 cm**
1	**2**	**3**	**4**
1.0	0.13	0.53	1.2
1.5	0.16	0.64	1.4
2.0	0.19	0.74	1.7
2.5	0.21	0.81	1.8
3.0	0.23	0.91	2.1
3.5	0.25	0.99	2.2
4.0	0.26	1.15	2.4
4.5	0.28	1.20	2.5
5.0	0.30	1.21	2.7
5.5	0.31	1.23	2.8
6.0	0.32	1.30	2.9
6.5	0.33	1.34	3.0
7.0	0.35	1.39	3.1
7.5	0.36	1.45	3.3
8.0	0.38	1.50	3.4
8.5	0.39	1.53	3.5
9.0	0.40	1.60	3.6
9.5	0.41	1.62	3.7
10.0	0.42	1.70	3.8

Height of water over centre of orifice (cm)	Discharge rate (L/sec)		
	2.5 cm	Diameter of orifice 5.0 cm	7.5 cm
1	2	3	4
10.5	0.43	1.72	3.9
11.0	0.44	1.75	3.9
11.5	0.45	1.80	4.0
12.0	0.46	1.83	4.1
12.5	0.47	1.87	4.2
13.0	0.48	1.90	4.3
13.5	0.49	1.93	4.4
14.0	0.50	1.96	4.5
14.5	0.50	2.10	4.5
15.0	0.51	2.50	4.6
15.5	0.52	2.80	4.7
16.0	0.53	2.10	4.80
16.5	0.54	2.15	4.83
17.0	0.54	2.20	4.92
17.5	0.55	2.22	5.00
18.0	0.56	2.25	5.13
18.5	0.57	2.30	5.15
19.0	0.58	2.31	5.20
19.5	0.58	2.33	5.30
20.0	0.59	2.37	5.32
20.5	0.60	2.40	5.00
21.0	0.60	2.42	5.47
21.5	0.61	2.45	5.53
22.0	0.62	2.50	5.60
22.5	0.63	2.51	5.65
23.0	0.63	2.53	5.70
23.5	0.64	2.57	5.77
24.0	0.65	2.60	5.83
25.0	0.66	2.63	5.95
25.5	0.66	2.65	6.00
26.0	0.67	2.67	6.10
26.5	0.68	2.70	6.12
27.0	0.69	2.75	6.18
27.5	0.69	2.80	6.23
28.0	0.70	2.81	6.30
28.5	0.71	2.82	6.37

Height of water over centre of orifice (cm)	Discharge rate (L/sec)		
	Diameter of orifice		
	2.5 cm	5.0 cm	7.5 cm
1	2	3	4
29.0	0.71	2.83	6.40
29.5	0.72	2.87	6.47
30.0	0.72	2.90	6.53

(Source : Michael, 1992)

Submerged orifices: These orifices may be divided into two types:

(i) those having orifices of fixed dimensions, and

(ii) those in which the height of opening may be changed.

However, standard submerged orifice has fixed dimensions. The opening is sharp edged and usually rectangular in shape and its width being 2 to 6 times the height. The discharge through a standard submerged orifice may be determined on using equation (16.9). The adjustable submerged orifice is one in which the height of opening and head may be changed to suit the requirements. The ordinary form is a simple head gate used to control flow from one channel to another. Adjustable submerged orifices should be calibrated to obtain the discharge under changing head conditions.

Meter gate: It is basically a modified submerged orifice. In these orifices, openings of orifice are adjustable. Meter gates are used to control the water flowing from one channel to another and manufactured commercially. These are also used to measure the rate of flow, if the head and opening area of orifice can be determined. Generally, these are used in canal irrigation at canal outlets where irrigation water is charged on volumetric basis. The head is considered same as under submerge orifice. However, its use has not become popular in India.

13.3.4 Tracer techniques

This technique of water measurement is independent of stream-cross-section and does not require any fixed structure. In this technique, a substance (tracer) is introduced in concentrated form into the flowing water and allowed to mix thoroughly. The concentration of the tracer is measured at a downstream section. Since only the quantity of water is essential for the dilution. There is no need to measure velocity, depth, head, cross-sectional area or any other hydraulic factors usually considered in discharge measurement techniques.

The tracer techniques is mainly of two types as described under:

(i) Dilution, and (ii) Radioisotope

(i) Dilution technique: In this technique, a large quantity of chemical or dye known as tracer is dissolved in a small quantity of water. Then it is filled in a bottle. For measuring discharge, the tracer solution can be discharged at a known rate into the water flowing in a channel or pipe.

For determination of discharge in open channel or pipe following procedure is adopted. Consider that the concentration of tracer $\frac{\text{Weight of tracer}}{\text{Weight of water}}$ in the bottle is a ratio C_1. The rate of injection is g_1 open channel or pipe. To account for any tracer amount which might already be in the upstream flow, the original tracer concentration may be considered as C_0. At the downstream of measuring station it is assumed that thorough transverse mixing of the tracer with the flow has taken place and assuming no loss seepage or evaporation in the section of the channel. The discharge Q at the upstream plus q_l which is the quantity of tracer added. The concentration of tracer at the downstream section is C_2. Then the discharge is determined as under:

Total discharge at upstream = Total discharge of downstream

$$QC_0 + q_1C_1 = (Q+q_1)C_2$$
$$QC_0 + q_1C_1 = QC_2 + q_1C_2$$

On rearranging we obtained,

$$QC_0 - QC_2 = q_1C_2 - q_1C_1$$

On simplification we obtained,

$$Q(C_0 - C_2) = q_1(C_2 - C_1)$$

$$\text{or } Q = q_1 \frac{(C_2 - C_1)}{(C_0 - C_2)} \quad \text{....(13.10)}$$

Salt concentration is difficult to measure. However, the electrical conductivity bears a linear relationship with the salt concentration and it is easily measured.

Many different substances *viz.*, salt (Nacl) and sodium dichromate is used as tracer. Dyes are used to colour the water and calorimeters and comparators are used to determine the dilution by comparing the samples with standard concentrated or dilute solutions.

(ii) Radioisotope technique: In this technique, radio-isotopes may be used in place of chemical or dye as in tracer technique and the degree of dilution is determined by counting the gamma ray emissions from the diluted isotope solution (the downstream flow) using Geiger counters or scintillation counters. A known amount of radioisotope is introduced into the flow in a relatively short time for measurement of discharge. For measurement of discharge concentration of the radioisotope tracer is determined by the gamma ray emissions at upstream and downstream of measuring section. The following mathematical relationship may be used to determine the discharge.

$$Q=\frac{F\,A}{N} \qquad ....(13.11)$$

where,

Q = volume of water flowing in the channel, (L/sec),

F = counts per unit of radioactivity per unit volume of water per unit of time,

A = total units of radioactivity to be introduced for each discharge measurement,

N = total counts.

However, this technique has drawback that radioisotopes influences public health. Thus, radioactive material is handled with great cares.

13.4 Farm Irrigation Distribution System

A properly designed water distribution system will make irrigation easy and efficient. Different types of structures are used to convey, divert and control irrigation water on the farm these structures are an essential part of water distribution system. Efficient structures will not only save water but also save labour and land.

Irrigation distribution systems at farm for surface irrigation methods can be classified into two groups *viz.*, surface channels and underground pipelines. Different types of structures are required for conveyance and control of irrigation water in different types of surface irrigation methods.

13.4.1 Open channels

It is defined as any conduit in which water flows with a free-water surface *viz.*, rivers, canals, and uncovered flumes. Pipes, drains, sewers, etc. may also act as open channels when flowing partially full. There are following types of flow in open channels.

Steady flow: It is the condition in which the flow characteristics at any point do not change with time like flows in lined irrigation channels tend to be steady under normal flow conditions. Mathematically it can be expressed as,

$$\frac{\partial v}{\partial t}0, \frac{\partial y}{\partial t}=0$$

where, Velocity of flow, *y* the flow depth and, *t* time of flow.

Unsteady flow: It occurs when the flow characteristics change with time at a channel cross-section, the flow is unsteady like intermittent runoff flow into terrace outlets, diversion channels etc. Mathematically it can be expressed as,

$$\frac{\partial v}{\partial t}\neq 0, \frac{\partial y}{\partial t}\neq 0$$

Uniform flow: When the channel flow is steady and the mean velocity is same at each succeeding cross-section, the flow is uniform like irrigation channels must have a constant cross-section and slope.

$$\frac{\partial v}{\partial l}=0$$

where, *l* denotes the length of the channel section under consideration.

For designing field channels the flow is considered to be uniform at the maximum discharge.

Non-uniform flow: It occurs when the mean velocity changes from cross-section to cross-section, the flow is non-uniform like flow in the upper portions of steep channels near the entrance is non-uniform and channel.

$$\left(\frac{\partial v}{\partial l}\neq 0\right)$$

Design of open channels: The parameters of an open channel are shown in Fig. 13.12.

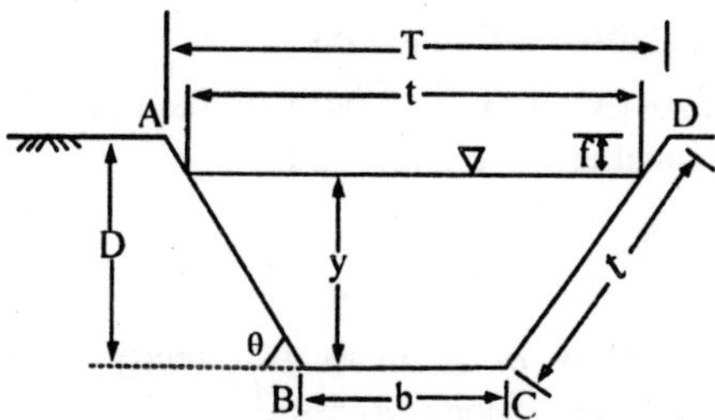

Fig. 13.12: Parameters of an open channel

T = top width of the channel

t = width of channel when the water is at depth (d), (m),

D = depth of channel after free board is added, (m),

d = depth of flow in channel, (m),

b = bottom width of channel, (m),

c = wetted sides of channel, (m), and

0 = angle between the sloping side and the horizontal, (degree).

Important terminology pertaining to channel design

Wetted perimeter (P): It is the sum of the lengths of that part of the channel sides and bottom which are in contact with water and is expressed as,

$$p = b + c + c$$

Frictional resistance offered to the flow of water-in channels is proportional to the area of the surface over which water passed (wetted perimeter).

(a) Area of cross-section: It is defined as the area of the wetted section of the channel and is expressed as,

$$a = \frac{(b+t)d}{2}$$

Hydraulic radius (R): It is the ratio between the cross- sectional area of the stream and its wetted perimeter and is expressed as,

$$R = \frac{a}{p}$$

The hydraulic radius is an important parameter in determining the velocity of flowing water in channels. Velocity of flow in channels is directly proportional to the square root of the hydraulic radius (*i.e.*, $v \propto \sqrt{R}$). The hydraulic radius decreases with the increase in wetted perimeter for a same cross-sectional area.

Hydraulic slope (S): It is the ratio of its vertical drop (h) for a length (l) of the channel and is expressed as,

$$S = \frac{h}{l}$$

The velocity of flow *in* channels is directly proportional to the square root of the hydraulic slope (*i.e.*, $v \propto \sqrt{S}$).

Free-board: It is the vertical distance between the highest water level anticipated in the design and the top of the retaining banks. Free board should be provided to overcome the problem of overtopping of structures due to wave action or damage due to trampling or other unforeseen conditions.

Angle of repose: It is the maximum angle of the sloping sides of the cone measured from the horizontal, when if a mass of loose dry soil is allowed to fall freely to the ground, the soil heap formed by it will attain a conical shape. Channel sides will be stable if its slope does not exceed the angle of repose of the soil in which it is constructed. Clay soils have steeper angles of repose as compared to sandy soils.

13.4.2 Velocity of flow in open channels

The design of channels is based on the mean velocity of flow of water in channel. This is obtained by different standard formulae as described under.

(i) Chezy's formula: It is expressed as,

$$v \propto \sqrt{R}.\sqrt{S}\,)$$

$$\text{or} \quad v = C\sqrt{RS}\,) \qquad \text{....(13.12)}$$

where,

R = Hydraulic radius,

S = Hydraulic slope, and

C = constant depending on the size, character of slope, depth of flow of stream and nature of surface of the channel.

Equation (13.12) is known as Chezy's formula. Determination of the value of C in equation (13.12) is comparatively difficult.

(ii) Manning's formula: It is expressed as,

$$v = \frac{R^{\frac{2}{3}} S^{\frac{1}{2}}}{n} \qquad \text{....(13.13)}$$

where,

v = mean velocity of flow in the channel, (m per sec),

R = hydraulic radius, (m),

S = hydraulic slope, (dimensionless), and

n = roughness coefficient of the channel.

(iii) Darcy-Weisbach formula: It is expressed as,

$$v = \sqrt{\frac{8\,g\,RS}{f}} \quad ....(13.14)$$

where,

g = acceleration -due to gravity, (m/sec^2),

f = Darcy-Weisbach roughness coefficient or friction factor (dimensionless).

Other factors are same as in equation (13.13).

The resistance coefficients C, n and f in equations (13.12), (13.13) and (16.14), respectively, can be related to each other as under:

$$\frac{C}{\sqrt{g}} = \sqrt{\frac{8}{f}} = \frac{R^{1/6}}{n\sqrt{g}} \quad ....(13.15)$$

The Manning's formula is frequently used for designing of water courses and field channels. The values of the roughness coefficient (n) for Manning's formula have been determined experimentally for various conditions of the channel sides and bottoms. The values of n under different conditions commonly met with in irrigation channels and structures (Table 13.6).

Table 13.6: Values of Manning's 'n' for irrigation channel

Type and description of channel	Manning's n
Farth channels	
Straight and uniform	0.023
Winding, sluggish	0.025
Stony bed, weeds on bank	0.035
Small drainage ditches	0.04
Lined channels	
Concrete	0.015
Masonary, rubble	0.017 to 0030
Metal, smooth	0.011 to 0015
Wooden	0.011 to 0014
Vegetated waterways	0.020 to 0.040
Pipes	
Cast iron	0.012 to 0013

Type and description of channel	Manning's *n*
Clay or concrete drain tile	0.011
Steel	0.015 to 0.017
Vitrified sewer pipe	0.013 to 0.015

(Source : Michael, 1992)

Discharge capacity of channels

The discharge capacity of channels is determined by the following relationship:

$$Q = a\,v \qquad \text{....(13.16)}$$

where,

Q = discharge capacity, (m^3/sec),

a = cross-sectional area of the wetted section of the channel, (m^2), and

v = mean velocity of flow in channel, (metres/second).

Unlined earth channels are frequently used in water conveyance on the farm. The advantages of earth channels are that these can be easily built and maintained by unskilled persons and require no special equipment or materials. These channels require low initial cost which is the major advantage of earth channels.

Earth channels should be constructed with stable side slopes and with banks strong enough to carry the desired flow of water safely. These channels should have ample capacity to carry the design discharge at non-erosive velocities. Side slopes should be flat enough to overcome the problem of cavins in and slide when these are saturated with water. Permanent irrigation channels should be provided side slopes less than 1½ horizontal to 1 vertical.

While, stiff clay should provided steep slopes upto to ½ to 1 and loose sand should be provided flat slopes of about 2 to 1. The top width of the main field channel should be about 60 cm.

Example 13.5: Assuming that an earth channel with grade of 0.11 per cent, depth of water 40 cm, bottom width 30 cm and side slopes 1½ to 1 is constructed. Determine the velocity of flow and carrying capacity of the channel when the Manning's coefficient (n) is 0.025.

Solution:

Side slopes 1½ to 1, depth of flow 40 cm

Length of wetted sides, (*c*),

$$= \sqrt{60^2 + 40^2} = 72 \text{ cm} = 0.72 \text{ m}$$

Wetted perimeter, (*p*) = 72 + 35 + 72 = 179cm = 1.79 m

Top width, (*T*) = 60 + 35 + 60 = 155 cm = 1.55 m

$$\text{Area of cross-section, } (a) = \frac{(55+160)40}{2} = 215 \times 20 \text{ cm}^2 = 0.39\text{m}^2$$

$$\text{Hydraulic radius, } (R) = \frac{a}{p} = \frac{0.39}{1.79} = 3900 \text{ cm} = 0.218 \text{ m}$$

$$\text{Hydraulic radius, } (S) = \frac{0.11}{100} = 0.0011$$

Manning's Coefficient, (*n*) = 0.025

$$\text{Mean velocity of flow, } (v) = \frac{R^{2/3}.S^{1/2}}{n} = \frac{(0.218)^{2/3} \times (0.0011)^{1/2}}{0.025}$$

$$= 0.458 \text{ m/sec}$$

$$\text{Carrying capacity, } (Q) = a\,v = 0.39 \times 0.458$$

$$= 0.183 \text{ m}^3\text{/sec}$$

$$= 183 \text{ L/sec}$$

Permissible slope of earth channels: The natural slope of the land is the deciding factor for determination of the channel bed slope. When channel is steeper then velocity and discharge will be more for the same cross-section. But excessive gradients produce very high velocities which cause erosion. Thus, a channel should have slope of about 0.1 per cent. On other hand, silting may take place in the channel if the slope is less than 0.05 per cent. For determination of the bed slopes of channels the velocities should be checked and the maximum grade designed so that the velocities should not exceed the non-erosive velocity (Table 13.7). When earth channels are to be built on steep slopes, it is essential to control the gradients and thus the velocity is controlled by drop structures or by building up the channel bed.

A change in the direction of the channel should be brought about by smooth curves and not abruptly. Similarly, when the cross-section is changed then the change should be gradual.

Table 13.7: Permissible velocities for various soil textures

Type of soil	Maximum permissible velocity (cm/second)
Bare channels	
Sand and silt	45
Loam, sandy loam, silt loam	60
Clay loam	65
Clay	70
Vegetated Channels	
Poor vegetation	90
Fair vegetation	120
Good vegetation	150

(Source : Michael, 1992)

13.4.3 Channel cross-sections

The cross-section of a channel which carries large flows should be such that it will give the maximum possible hydraulic radius under the prevailing natural surface conditions. Thus, the most efficient cross-section for an open channel is in semi-circle shape, as its wetted perimeter is minimum and its hydraulic radius is maximum. While for rectangular cross- section, half a square is the best discharging rectangular channel, *viz.*, depth = ½ x breadth *(i.e. d=b/2)*. The semi-circular and rectangular cross- sections are suitable only in small size *lined channels.*

Mostly lined channels are designed to carry large flows and have a trapezoidal section. The best hydraulic cross-section of a trapezoidal channel under favourable structural condition is given by the following relation:

$$b = 2d \tan \theta/2 \qquad ....(13.17)$$

where,

b = bed width, (m),

d = depth of flow of water, (m), and

θ = the angle between the side and the horizontal, (degree).

However, due to topographic and soil conditions and also practical problems of construction and maintenance make considerable departure from the above relationship. For lining of water courses of comparatively smaller sections, a rectangular cross-section having a depth of about half its width has been found economical. Because a trapezoidal section of the same capacity is occupied more land area as compared to the rectangular section and former section also requires wider embankments and its cost of construction is higher.

Lined channels: Earthen channels are lined to prevent excessive seepage and growth of weeds in channels. A huge portion of the water harnessed at a high cost is lost by seepage from unlined conveyance systems comprising of the main canal, branches, distributaries, water courses and field channels. The seepage losses may be as high as 20 to 40 per cent of the water delivered to the channel in permeable soils.

Water losses may occur by seepage, by breaches along the channel through rat holes, by ponding of water in depressions and irregularities in the channel sections and by evaporation. The seepage loss depends on the length of the water course, its wetted perimeter and the intrinsic permeability of the strata through which the channel passes. The main problems encounter by earth channels is the control of weeds. Weeds in a channel obstruct the rate of flow of water. Earth channels require continuous maintenance to control moss and weed growth, and to repair damage by livestock, rodents and erosion.

13.4.4 Lining materials

There are different types of materials used for lining of irrigation channels to reduce water losses through seepage. However, concrete and brick or stone masonary are commonly used for lining of irrigation channels. Well mixed cement concrete lining, single layer bricks or stones laid in cement, brick with agrilfilm or lime mortar reduces the water losses to the minimum. The bituminous mixtures, soil cement, chemical sealants, polyethylene film and impervious earth materials are used for lining for limited cases. However, most of these are susceptible to damage by trampling by livestock, insects, weed growth and erosion by high velocity flows. Thus, their life is short and often uneconomical unless special protection is provided.

There are following types of lining materials which may be used for reduction of seepage losses from irrigation channels :

(i) Cement concrete,

(ii) Brick masonry (cement),

(iii) Stone masonry (cement),

(iv) Brick masonry in lime mortar,

(v) Stone masonry in lime mortar,

(vi) Bitu minions mixture,

(vii) Soil cement,

(viii) Chemical sealant,

(ix) Agrifilms,

(x) Marble slurry,

(xi) Calcareous soil, and

(xii) Pre cast channels.

These lining materials may be used as single or in combination for reduction of more seepage.

Example 13.6: Determine the mean velocity of flow and carrying capacity of a lined canal water course. It is having rectangular section with a bottom width of 60 cm and depth of flow 25 cm (inside). Single layer bricks are laid in cement mortar with a cement plaster 6 mm thick. The slope of the channel bed is 2 m per Km and manning roughness coefficient is 0.015.

Solution:

Area of cross-section of the channel, (a)

$$= \frac{60}{100} \times \frac{25}{100} = 0.150 \text{ m}^2$$

Wetted perimeter, $(p) = 0.6 + 0.25 + 0.25 = 1.1$ m

$$\text{Hydraulic radius, } (R) = \frac{a}{p} = \frac{0.150}{1.1} = 0.145 \text{ m}$$

$$\text{Hydraulic radius, } (S) = \frac{2}{1000} = 0.002$$

Mean velocity of flow, (v) is determined on using Manning's formula as,

$$= \frac{R^{2/3} . S^{1/2}}{n} = \frac{(0.145)^{2/3} \times (0.002)^{1/2}}{0.015}$$

$$= 0.946 \text{ m/sec}$$

Discharge capacity, Q $= a \times v = 0.150 \times 0.946$

$= 0.01419 \text{ m}^3\text{/sec}$

$= 142$ L/sec

13.5 Erosion Controlling Structures in Channel

It is essential to construct irrigation channels on land slopes such that the water will net attain erosive velocities. Severe erosion will take place in high slopy unless slope controlling structures are not provided. Drop structures and chute spillway are commonly used to prevent erosion in field channels.

13.5.1 Drop structures

These structures are used to discharge water in a channel from a higher elevation to a lower one. Drop structures may be open type drops or pipe drops.

Open drop structures: These structures can be made of timber, concrete, or brick or stone masonary. Timber is usually not preferred due to its short life span. Drop structures built by using discarded drums or barrels of the coal tar etc. are of low cost.

A drop structure is consisting of three main components: (i) inlet, (ii) vertical overall section, and (iii) outlet. The check gate is provided at the inlet of the drop structure to control the water surface heights on the upstream stretch of the channel.

The inlet of a drop structure should have a minimum width equal to the bottom width of the irrigation channel. Water enters the structure through the inlet which is in the form of a weir or notch in a wall. Vertical walls are also known as cut-off walls which extend down into the soil under the inlet in order to prevent water seepage under the structure. Similarly, walls extending from the inlet to prevent seepage around the ends of the structure are known as headwall extensions. Water falls into a stilling basin it is an essential part of an erosion control structure by which control the erosive force of the water. The length of the stilling basin should be nearly twice the height of the drop. For constructing drop structures, following points should be considered:

1. When earth channels are to be constructed on steep slopes or on bench terraced lands it is essential to construct a series of drop structures to reduce the channel grade. So that the channel section between two drop structures is nearly flat. The crest of one drop structure and the level of the apron floor of the next drop structure should be nearly the same on sloppy land.
2. The depth of over pour does not exceed 30 cm then the water depth in this basin should be about 45 cm for drops upto 60 cm while 60 cm for drops from 60 to 90 cm.

3. A small cross wall known as the end still of about 10 to 12 cm height. It is placed at the end of the basin which increases the efficiency of energy dissipation in stilling basin.
4. Eddy currents often step up by drop structures in the irrigation stream. These currents tend to cause erosion of the channel section immediately downstream from the structure. Thus, to prevent channel erosion stones or brick-bats should be placed over a length of one to two metres from the structure.

13.5.2 Pipe drop structures

In some conditions construction of an open drop structure is not possible without disturbing an existing bund or dam. In these conditions water can be safely discharged from a higher level to a lower one by providing a pipe drop structure. This type of structure allows discharge of water through a pipeline, without disturbing of the bund or dam. Different type of pipes can be used as conduit *viz.*, vitrified sewer pipes or concrete pipes made with bell joints or corrugated metal pipes, PVC etc. A water-tight lid at the inlet will be provided to act as a check gate. The corner in the pipeline is made of a large radius bend. A stilling basin is provided at the outlet of the pipe conduit to dissipate the energy of the incoming stream. An apron made of masonry or concrete is provided around the inlet end of the pipe to prevent seepage around it.

The velocity of flow of water in pipe drop spillways may be determined by applying the Bernoullis' theorem as under:

Available head	=	Frictional loss in the pipe line	+	Velocity head	+	Head loss at the entrance of pipe	+	Head loss at the bend of pipe

This may be expressed as,

$$H = 4\frac{flv^2}{2\,gd} + \frac{v^2}{2g} + K_1 - \frac{v^2}{2g} + K_2 - \frac{v^2}{2g} \qquad \text{....(13.18)}$$

where,

H = difference in elevation between the water level at the upstream and downstream ends of the structure, (m),

v = velocity of flow in the pipe, (m/sec),

f = coefficient of friction for the pipe (usually assumed to be about 0.01),

l = length of pipe, (m),

g = acceleration due to gravity, (m/sec^2),

d = diameter of pipe, (m), and

K_1 and K_2 = constants (for pipe drops $K_1 = 0.5$ and $K_2 = 0.25$)

The discharge capacity: It is determined for the pipe drop structure by the following relationship:

$$q = av \quad \text{.... (13.19)}$$

where,

q = discharge, (cum/sec),

a = area of cross-section of the pipe, (sq m), and

v = velocity of flow, (m/sec)

Example 13.7: Determine the discharge capacity of the pipe drop spillway. Assuming H = 1m, d = 10 cm, f = 0.012, l = 5m

Solution:

On using equation (13.18)

$$H = \frac{4 + lv^2}{2gd} + \frac{v^2}{2g} + \frac{0.5v^2}{2g} + \frac{0.25v^2}{2g}$$

Substituting the values of the variables as given in the equation (13.18) we obtained:

$$H = \frac{4 \times 0.012 \times 5 \times v^2}{2 \times 9.81 \times \frac{10}{100}} + \frac{v^2}{2 \times 9.81} + \frac{0.5v^2}{2g} + \frac{0.25v^2}{2g}$$

$$v = \sqrt{6.26 = 2.5\, m/\text{sec}}$$

$$a = \frac{\pi}{4}d^2 = \frac{\pi}{4} \times \frac{10}{100} \times \frac{10}{100} = 0.0078\, sq.m$$

Q = av

= 0.0078 × 2.5 = 0.0195 cu m/sec = 19.5 L/sec.

13.5.3 Chute spillways

It is used to the flow down on steep slopes through a lined channel rather than by dropping the water in a free overfall. Chutes are more economical as compared to a series of drop structures on steep slope to take the flow down the slope. It consists of an inlet, channel section and outlet and made of concrete, or stone or bricks laid in cement mortar. A low cost chute spillway can be made of precast concrete channel sections with a stilling basin at the outlet.

A check gate is installed at the inlet in order to control the water level in the upstream channel. The following point should be considered for construction of chute spillways:

1. For minimizing the problems of settling and undermining, chutes are constructed on solid ground or on fill that has been carefully compacted,
2. Chute channels are usually rectangular in cross-section,
3. The length of the stilling basin varies from 1.0 to 1.5 m under the normal ranges of flow obtained in the farm irrigation system,
4. The depth of the stilling basin is about 10 cm below the bed level of the down-stream channel,
5. When the high velocity water is slowed down to a low velocity in a stilling basin, there is a sudden rise in the depth of flow which is known as a hydraulic jump,
6. The height of the walls in the downstream channel should be increased as compared to the channel section of the chute spillway in order to prevent overflow, and
7. Usually it is sufficient to keep the same depths for the upstream and downstream sections of the channel and to deepen the stilling basin floor level.

The discharge capacity of the channel section of chute spillways may be determined by using the Manning's formula (Equation 13.13).

13.6 Water Control and Diversion Structures

Water control and diversion structures should be provided to give easy and effective control of irrigation water on the field. It will reduce the labour required to irrigate field and check erosion and water loss. The following type of structures are used for controlling water and its diversion:

(i) check gates, (ii) portable check dams, (iii) diversion boxes, (iv) turnout boxes, (v) siphons, and (vi) pipe turnouts.

Check gates

These are placed in an irrigation channel to form an adjustable dam to control the elevation of the upstream water surface and do not allow water flow down stream. It is essential to raise the water level in the channel to divert the water flow from channels into the field. The water level in the channel should be at least about 8 to 12 cm above the ground surface so that siphon tubes or pipe turnouts may be used efficiently.

These structures are made of a masonry or metal wall built across the channel with a suitable check gate. The masonary wall is built in place while the check gate is usually precast and fixed to the wall. The crest of the check gate is at the same level as the bottom of the upstream channel. There are following types of check gates:

(i) Precast check gates, and

(ii) Steel or wooden gates

Portable check dams

These are made of canvas, plastic and sheet metal dams and are suitable to check water and raise of control water surface elevations.

Diversions

For carrying water to different farms, or to different parts of the same farm diversion are used. Two-way or three-way and four-way diversions are made of check gates at the channel junction will permit the diversion of all or any predetermined portion of the water flow.

Turnouts are used for diversion of water from a lateral channel into a field distribution channel or from a channel into a field. Turnouts may be portable or built-in and equipped with gates to control the flow of water. There are following types of turnouts (i) box turnouts, (ii) spilles and (iii) siphon tubes.

Box turnouts: These are portable wooden boxes provided with a gate at the inlet and a baffle board at the outlet. The baffle board divides the stream and reduces its velocity. The box should be long enough to extend completely through the channel bank and gate slides in saw-cut grooves in the box. It can he adjusted to divert only desired flow into a field or a secondary Channel. Its size will depend on the size of the stream to be diverted.

Spilles: These are made of bamboo, concrete or baked clay pipes. The size of the spiles depends upon on the flow of' water desired and the depth of water at the inlet and outlet of the pipe and it varies from 2.5 cm to 10 cm or more. The spille should be placed low in the bank to reduce erosion at the outlet.

Siphon tubes: These are used to carry water over the channel bank into a field or furrow. These tubes are made of plastic rubber or sheet metal (aluminum or mild steel) and are commercially available in different sizes *viz.*, usually 2.5 cm to 7.5 cm. The length of siphon tube may be about one metre or more depending on the height of the channel bank above the field. For using a siphon tube, it is dipped in water and filled completely with water. One end of the tube is closed tightly with the palm of hand and keeping the other end dipped in water. For operating it, the closed end is taken out and released at the ground surface.

The discharge from siphon tubes depends on the size of the tube and the elevation difference between the water surfaces at the upstream and downstream. It may be determined by the following relationship:

$$Q = 0.65 \times 10^{-3}\, a\sqrt{2gH} \quad \text{....(13.20)}$$

where,

Q = discharge from siphon tube, (L/sec),

a = Cross-section area of tube (inside of tube), (sq. cm),

g = acceleration due to gravity, (cm/sec^2), and

H = effective head causing flow, (cm)

In case when the outlet is not submerged then the effective head is the vertical distance from the water level over the inlet end to the centre of the discharge end.

13.7 Channel Crossing Structures

These structures are used for crossing irrigation channels across roads, hillsides and natural depressions or drainage ways. Siphon may be used to carry irrigation channels across natural farm obstructions. Generally, siphons are made of concrete or vitrified clay pipes with suitable masonry structures at the inlet and outlet. While, flumes are made of elevated pipes or open channels supported by pillars. The discharge capacity of siphons depends on following factors:

(i) size of the pipe, and (ii) difference in elevation between the water surfaces at the upstream and downstream ends of the structure.

The capacities of siphons may be determined similar to pipe drop spillways. For straight siphons the Bernoulli's theorem may be expressed as under:

$$H = \frac{4flv^2}{2gd} + \frac{v^2}{2g} + K_3 \frac{v^2}{2g} \quad \text{....(13.21)}$$

where,

K_3 = constant

The value of K_3 is 1.0 for a projecting pipe entrance. The discharge capacities of siphons at channel crossings, using different sizes of concrete pipes and at heads commonly available on the farm is summarized in Table 13.3.

Flumes

These are used to carry irrigation water across streams, canals, gullies, ravines or other natural depressions (Figs. 13.39 and 13.40). Flumes may be made of open channels or pipes which are often supported by pillars or may be affixed to bridges. Steel, concrete or vitrified clay pipes are used for its construction. The supporting structure may be made of timber, steel or concrete. In case timber is used then it should be treated with a preservative like creosote oil to extend its life span. Steel structures should be painted to increase life span. The capacities of open channels used on flumes may be determined on equation (13.21) while, discharge capacity of pipe flumes (flowing full) may be determined on using equation (13.21).

Table 13.8 : Discharge capacities of siphons at channel crossings using different sizes of concrete pipes Length = 5 m *(Discharge, L/sec)*

Diameter of pipe (cm)	Friction factor (f)	Head, (cm)								
		5.0	7.5	10.0	12.0	15.0	17.0	20.0	25.0	30.0
10.0	0.0095	3.9	4.8	5.6	6.2	6.1	7.4	7.9	8.8	9.6
15.0	0.0091	9.8	12.0	13.8	15.4	1.9	18.3	19.5	21.8	23.9
22.5	0.0071	24.3	29.7	34.3	33.4	42.0	45.0	43.5	54.3	59.4
30.0	0.0060	45.2	55.3	63.9	71.4	78.2	84.5	90.3	101.0	110.6
37.5	0.0056	72.1	88.3	102.0	114.0	124.9	134.9	144.2	161.3	176.6
45.0	0.0053	165.0	233.3	233.9	260.9	285.7	308.6	330.0	368.9	404.1

(Source : Michael, 1992)

Culverts

Generally, culverts are used at channel crossings when the road fill is sufficiently high and the channel bed lies below on the field surface on both sides. This structure consist of masonry headwalls at the inlet and outlet ends and these are connected with a buried pipeline. The depth of earth covering should not be less than 30 cm in any case. The stilling basin should be provided to control erosion at outlet of culverts.

For low cost it should be made of concrete pipes of adequate capacity and used for ditch crossing.

Inverted siphons

It is more economical used when a channel has to cross a wide depression or where the road surface lies close to the field surface. The structure consist of an inlet and an outlet tank and these tanks are connected at their bottom with a pipe. The common size of tank used is 60 cm x 60 cm.

A check gate should be provided at the inlet end to control the level of water surface in the upstream channel. The tanks of an inverted siphon act as stilling basins which slow down the velocity of the flowing water. Thus, silt carried by the water from upstream is likely to be deposited in the siphon. For collecting silt the bottom of the tank should be kept about 15 cm below the bottom of the pipe. The pipe should be protected against traffic damage, so it is placed below 30 cm or more from ground surface.

13.8 Underground Pipeline System

An underground pipeline is used for irrigation water distribution at the farm. It consists of buried pipes for conveyance of water to different parts of the farm with allied structures for the efficient functioning of the underground pipeline system.

Underground pipe offers the following distinguished advantages over open channel:

(i) Cultivation can be done above the pipe. Thus, wastages of land for cultivation is minimized and it saves 2-4% of land for cultivation as compared to the open channel.

(ii) Underground pipeline do not interfere with farming operations.

(iii) Its life span is more and required less maintenance cost.

(iv) It reduces the considerable water losses including seepage and evaporation as compared to open channels.

(v) Underground pipeline did not require any crossing structures as in case of open channels *viz.*, culverts.

(vi) Irrigation water can be carried out up and down as it is operated under pressure.

(vii) Underground pipeline is not clogged by vegetation and wind blown material.

(viii) It is mostly suited for undulating topography where open channel is not feasible.

(ix) In underground pipeline, well should be to located at high elevation but it should be located where more water is available.

(x) The underground pipeline is easily operated where water is lifted by pumps.

13.8.1 Types of pipe

The following types of pipes used for underground pipeline:

(i) Non-reinforced concrete pipe: These are suitable when operating head does not exceed 0.6 m (0.6 kg/sqm).

(ii) Reinforced concrete pipe: It is used when operating head exceed 0.6 m (0.6 kg/sqm).

(iii) Asbestos cement pipe (non pressure and pressure pipe).

(iv) Plastic pipes (PVC and LDPE etc.): These are available to with stand different operating head.

(v) Metal pipe: GI pipe and cast iron pipe etc.

Table 13.9: Specifications for standard concrete irrigation pipes

Internal diameter (*cm*)	Wall thickness (*cm*)	Minimum internal hydrostatic pressure (*kg/sq cm*)	Minimum three-edge bearing load (*kg/linear meter*)
15	2.2	3.5	1935
20	2.5	3.5	2010
25	2.9	3.5	2090
30	3.2	3.2	2235
35	3.5	3.2	2380

Internal diameter (*cm*)	Wall thickness (*cm*)	Minimum internal hydrostatic pressure (*kg/sq cm*)	Minimum three-edge bearing load (*kg/linear meter*)
37.5	3.8	3.2	2456
40	3.8	3.2	2530
45	4.4	3.2	2680
50	5	2.8	2750
52.5	5.4	2.8	2825
60	5.7	2.8	2980

(Based on American Society of Testing Materials Standard specifications, ASTM C118-66T)

(Source : Michael, 1992)

Testing irrigation pipes

There are two standard tests for irrigation pipes *viz.*, water absorption and strength test. The water absorption should not exceed eight per cent of the original dry weight of the pipe while, testing the irrigation pipes. For testing the water absorption, samples of the pipe are boiled for about five hours and the gain in weight by moisture absorption is recorded.

The hydrostatic pressure requirement prescribes the minimum hydrostatic pressure that the pipe may sustain without excessive leakage through the shell of the pipe. The three-edge bearing test is used to determine the strength of concrete pipes. The tests are performed on a standard testing machine. The load is applied parallel to and along a vertical plane passing through the axis of the pipe.

13.8.2 Discharge capacity of pipelines

It is determined on using the Darcy's equation as under:

$$V = \sqrt{\frac{Hdg}{2fl}} \qquad \text{....(13.19)}$$

where,

V = velocity of flow of water through the pipe, (cm/sec),

H = available head causing flow (difference between elevation of water level in the pump stand and the outlet point), (cm),

d = diameter of pipe, (cm),

g = acceleration due to gravity, (cm/sec^2),

l = length of pipe (cm), and

f = Darcy's roughness coefficient, (fraction).

The values of the Darcy's roughness coefficient f for different kinds of pipes at various values of velocity of flow is summarized in Table and the discharge capacities of concrete pipes at different operating heads and conveyance lengths is given in Table 13.10.

Example 13.9: Determine the discharge capacity of an underground concrete pipeline. Assuming diameter of pipe 22.5 cm, length of pipe line 150 m, difference in elevation between water levels at pump stand and discharge point 2 m and Darcy's roughness coefficient f is 0.006.

Solution:

On using Darcy's equation (13.19)

$$V = \sqrt{\frac{Hdg}{2fl}}$$

Value of putting in the variables

$$= \sqrt{\frac{200 \times 22.5 \times 981}{2 \times 0.006 \times 15000}}$$

$$= 10 \text{ cm/sec} = 1.1 \text{ m/sec}$$

Area of cross-section of pipe, (a) $= \dfrac{\pi d^2}{4}$

$$= \frac{\pi}{4} \times (0.225)^2 = 0.0266 \text{ sq m}$$

Discharge capacity, (Q) = $a\ v$

On putting the values of a & v, we obtain:

= 1.1 x 0.0266

0.0268

= 26.8 L/sec

13.8.3 Spacing of pipelines

Pipelines are spaced in such a way as to suit the field layout and soil type. Generally, one main line is usually laid along the upper boundary of the field. In case when the land slopes in nearly opposite directions, then the pipeline may be laid out in the middle so that one line can command fields on either side. In case, when the prevailing land slope is only in one direction then the spacing between pipelines varies from 100 to 200 metres. Light soil (sandy) required closer spacing of pipeline while, heavy soils (clay) required larger spacing of pipelines. Inspite of laying pipe along the field boundaries, it should be laed diagonally across the fields to reduce the total length and the cost.

13.8.4 Installation of concrete pipelines

Proper installation of the underground pipeline is prerequisite to make it trouble free and also to reduce maintenance cost. The following steps are involved for laying the pipeline.

(i) Selection of depth and grade of laying,

(ii) Digging to proper depth and grade,

(iii) Lowering the pipe and squeezing,

(iv) Sealing the joints, and

(v) Backfilling the trench.

The concrete pipes should be laid with their upper surface at a depth of at least *45* cm from the ground surface to avoid damage of pipelines from loads acting on it. The width of the trench should be about 70 cm so that it provides sufficient working space. The grade of the pipeline is kept uniform and is usually the same as the average land slope along the line the pipes are laid.

Table 13.10: Values of coefficient 'f' in Darcy's equation for pipe flow

Diameter of pipe (cm)	Velocity (cm/second)														
	30	60	90	120	150	30	60	90	120	150	30	60	90	120	150
	Galvanised iron pipe					*Cast iron pipe*					*Concrete pipe*				
15	0.00 56	0.00 47	0.00 43	0.00 42	0.00 38	0.00 61	0.00 59	0.00 58	0.00 57	0.00 55	0.01 02	0.00 94	0.00 9	0.00 86	0.00 84
22.5	0.00 51	0.00 44	0.00 41	0.00 38	0.00 36	0.00 57	0.00 55	0.00 54	0.00 53	0.00 52	0.00 8	0.00 74	0.00 7	0.00 67	0.00 64
30	0.00 47	0.00 42	0.00 38	0.00 35	0.00 34	0.00 54	0.00 52	0.00 51	0.00 5	0.00 48	0.00 67	0.00 62	0.00 59	0.00 56	0.00 54
45	0.00 44	0.00 37	0.00 35	0.00 33	0.00 32	0.00 52	0.00 51	0.00 5	0.00 48	0.00 47	0.00 56	0.00 51	0.00 48	0.00 46	0.00 44
60	0.00	0.00	0.00	0.00	0.00	0.00	0.00	0.00	0.00	0.00	0.00	0.00	0.00	0.00	0.00

Dia-meter of pipe (cm)	Velocity (cm/second)														
	30	60	90	120	150	30	60	90	120	150	30	60	90	120	150
	Galvanised iron pipe					*Cast iron pipe*					*Concrete pipe*				
	42	35	33	32	3	5	48	45	43	41	5	46	44	42	4
90	0.00	0.00	0.00	0.00	0.00	0.00	0.00	0.00	0.00	0.00	0.00	0.00	0.00	0.00	0.00
	37	33	31	29	29	42	4	37	36	34	44	4	38	36	34
120	0.00	0.00	0.00	0.00	0.00	0.00	0.00	0.00	0.00	0.00	0.00	0.00	0.00	4003	0.00
	35	32	29	28	27	36	34	32	32	31	39	37	35	3	3

(Source : Michael, 1992)

Table 13.11: Discharge capacities of concrete pipes of different diameters at different head

Diameter of pipe (cm)	Friction factor (*f*)	Operating head (m)	Conveyance length (m)						
			50	100	150	200	00	400	500
15	0.00900	2.0	31.8	22.6	18.4	15.9	12.9	11.2	10.0
		2.5	35.5	25.0	20.5	17.7	14.5	12.5	11.2
		3.0	38.9	27.5	22.5	19.4	15.9	13.7	12.3
		4.0	45.0	31.8	26.0	22.4	18.3	15.9	14.2
		5.0	50.0	35.5	29.0	25.1	20.5	17.7	15.9
22.5	0.00710	2.0	99.1	70.1	57.2	49.5	40.5	35.0	31.0
		2.5	110.8	78.3	64.0	55.4	45.2	39.2	35.0
		3.0	121.4	85.8	70.1	60.7	49.5	42.9	38.4
		4.0	140.1	99.1	81.0	70.1	57.2	49.5	44.3
		5.0	156.7	110.8	90.5	78.3	64.0	55.4	49.5
30	0.00595	2.0	222.2	157.1	128.3	111.1	90.7	78.6	70.3
		2.5	248.4	175.7	143.4	124.2	101.4	87.8	78.6
		3.0	272.2	12.4	157.1	136.1	111.1	96.2	86.1
		4.0	314.3	222.2	181.4	157.1	128.3	111.1	99.4
		5.0	315.3	248.4	202.8	175.7	143.4	124.2	111.1

(Source : Michael, 1992)

The bottom of the trench should be smooth by removing with large stones, clods, and other irregularities removed. Each pipe should be laid on firm, undisturbed soil. The fill should be properly compacted to produce a firm uniform foundation. The pipes are placed in a straight line at the centre of the trench. The different types of accessories are required for jointing the pipes.

(i) Bell jointed pipes

Pipes with bell ends are laid with their sockets (bell ends) facing upstream. The ends of the pipe are cleaned and wetted with a brush. Jute or hemp rope dipped in a cement paste is wrapped round the plane end of each pipe to avoid leakage from the joints.

(ii) Pipes with tongue and groove joints

Pipes with tongue and groove joints are laid with the groove ends facing upstream. The tongue or spigot end of the first section of pipe is cleaned and wetted with a brush.

(iii) Reinforced concrete pipes with collar joints

These concrete pipes are commonly available in lengths of 2.5 to 3 m. The pipes are lowered into the trench with the help of ropes by two men standing on the trench bank on either side of the pipe. Before an adjacent pipe is laid, a collar is slipped to its end. The pipes are laid with their ends placed about 1 cm apart. The gap is left for fixing a rope dipped in bitumen in the recession of the pipe ends.

In order to maintain a uniform clearance between the pipe joint and the collar, wooden battens of uniform thickness are plugged in on both sides of the collar. The gap between the collar and pipe is then filled with a dry mixture of cement and sand mixed in the ratio of 1: 1. Once the gap is sealed satisfactorily, the collar is lined with 1 to 1.5 cm thick plaster (1: 2 ratio) and beveled off at an angle of 45° with the outside edge of the collar.

Openings in all concrete pipelines should be covered to prevent air circulation except when work is actually in progress. Such openings should be kept closed until the pipelinc is filled with water. There should be an initial backfill of soil around the pipe, covering the pipe to a depth of at least 15 cm for the full width of the trench and not more than seven sections behind the laying. In backfilling, the space on either side of the pipe is filled first to avoid injury to the freshly applied mortar bands and not to dislocate the pipes. After the joints are cured, the pipeline is tested at low head (about 1 m). Any leakage has to be carefully observed and it marked for repairing .

13.8.5 Structures for underground pipelines

The underground pipeline requires special structures to control the water and protect the pipeline from damage. These structures include such as (i) inlet, (ii) water control, (iii) diversion structure, (iv) air release vents, and (v) end plugs.

Inlet structures

Water may enter a pipeline through inlet structures by gravity from an open channel or it may be pumped from a well or any other source. An inlet structure is required due the following reasons:

1. To develop the full flow capacity of the pipeline.
2. To maintain water surface elevations sufficient to distribute water to the different parts of the farm.
3. To ensure that excessive pressures do not cause damage.
4. Keep trash free pipeline.

(i) Pump stands: It is a vertical pipe extending above the ground and connected to the underground pipeline system is called as pump stand. Pump stands are located at the inlet of an underground pipeline system. It is larger (about 60 cm) than the size of the pipeline. The larger capacity of the stand permits dissipation of the high velocity stream and release of entrapped air before the water enters the pipeline and have an opportunity to escape at the pump stand. Entry of air into the pipeline can cause air pockets which restrict the flow of water and results into surging flow conditions and develop excessive pressures.

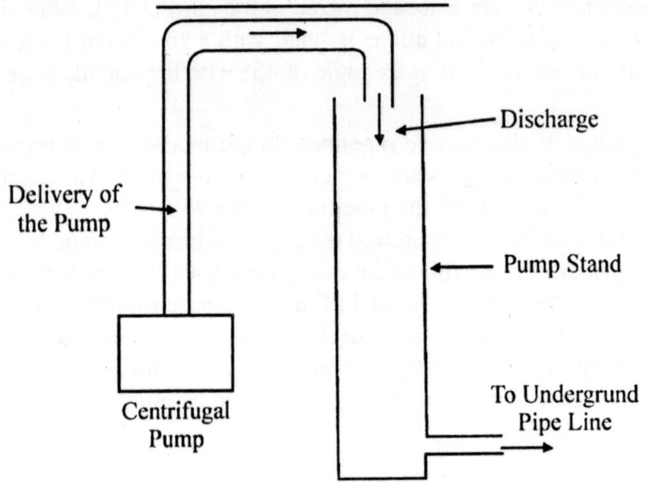

Fig. 13.13: Pump stand

(ii) Gravity inlets: This type of inlet is used when water enters the pipeline from an open channel. A screen is fixed to the inlet to keep trash out of the pipeline. The top of the structure is usually provided with a cover to prevent accidents and to keep trash from blowing in.

When irrigation water contains appreciable quantities of sand, a trap can be built into the inlet structure to remove most of the suspended material. If the stand functions as a sand trap also. Then extra-large diameter is used to ensure low velocity of water. A debris screen is provided at the inlet to clean the water before it enters the pipeline. If it contains considerable quantities of debris and weed seed.

(iii) Gate stands: These are used to control the flow into laterals when pipelines branch off in different direction. This type of structures prevent high pressures and act as surge chambers. These structures are made in the same way as pump stands, but are provided with slide gates or valves to control flow into the laterals.

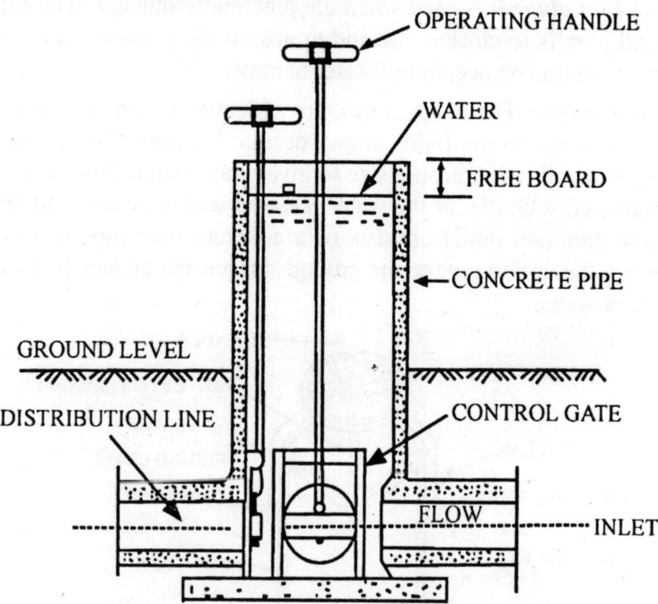

Fig. 13.14: Inlet of an underground pipeline water distribution

(iv) Air vents: These are vertical pipe structures and are used to release air entrapped in the pipeline and to prevent vacuum. Entrapped air must be removed to permit an even flow and avoid the danger of water hammer. The air vents is required at the following locations:

1. Near the pump stand,
2. At all high points in the line,

3. At sharp turns,
4. At points where there is a downward deflection of more than 10°,
5. Downstream from any structure that may entrap air, and at the end of the pipeline,
6. Immediately upstream from gates where closure of the gates would make such points the downstream ends of a lateral or line,
7. Vents are generally installed at about 150 m apart on straight pipelines with uniform slope.
8. Air relief values should not be used in any location where it may be necessary to relieve momentary high pressure surges.

(v) End plug: It is used when the pipeline terminates. The function of an end plug is to close a line and to absorb the pressure developed at the end of the line on account of water hammer.

(vi) Outlets: These structures are essential to provide controlled delivery of water to the fields at any desired location. These should be easy to open and close, proper size to given the desired flow and should be constructed such that as the water released will not cause soil erosion. The most common outlet consists of a concrete riser pipe to bring the water from the buried pipe to the ground surface and an attached valve to the flow of water.

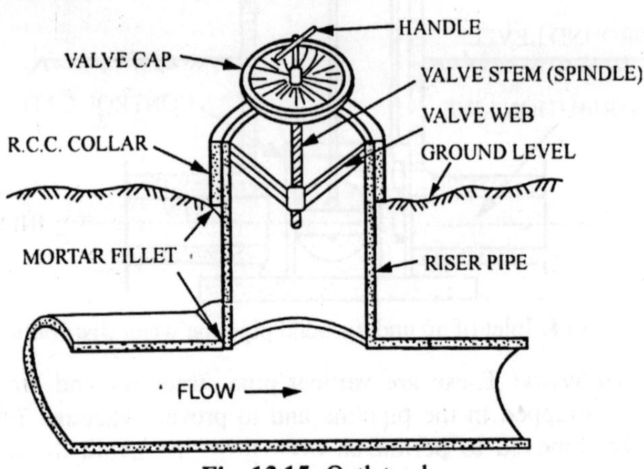

Fig. 13.15: Outlet valve

(vii) Riser valves: It is a screw valve installed on top of the riser pipe. A handle and cap plate are attached to a screw type shaft that moves up or down as the handle is turned.

Riser valves are of two types; the valve for releasing water directly to fields or open channels and hydrant valves which are adapted for fixing a portable hydrant on it. The hydrant valves are provided with a cage to seat the hydrant. The entire riser body may be built in a single unit and may be fixed to the top of the riser pipe with cement mortar. Hydrants are devices over riser valve outlets as a means of connecting portable gated pipe to the pipeline. These may have a single outlet which permits the flow of water in only one direction or may have outlets on two sides.

Hydrant and gated pipes

These are portable lines with uniformly spaced outlets and used for releasing irrigation water to furrows, border strips or check basins. When gates pipes are connected to buried pipelines through hydrants they permit conveyance of the water in an enclosed system from the source to the head end of the irrigation strip. These reduce seepage losses to a minimum. Gated pipe helps for regulating the flow of water into the furrows or border strips. Hydrates are placed ver valve outlets.

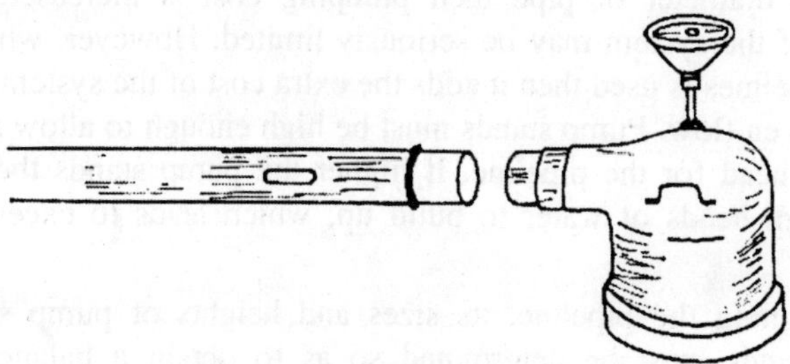

Fig. 13.16: Portable hydrant

Table 13.12: Discharge from standard gated pipes at heads varying from 30 cm to 150 cm

Head (cm)	Discharge (L/sec)					
	Gate opening					
	Full	**3/4**	**1/2**	**1/4**	**1/8**	**1/16**
With zero velocity in pipe						
30	3.1	2.2	1.4	0.66	0.30	0.15
60	4.2	3.2	2.0	0.96	0.50	0.20
90	5.1	3.8	2.5	1.20	0.55	0.23
120	5.5	4.4	2.9	1.30	0.60	0.26

Head (cm)	Discharge (L/sec)					
	Gate opening					
	Full	3/4	1/2	1/4	1/8	1/16
150	5.9	4.9	3.2	1.50	0.70	0.30
With a velocity of 30 cm/sec in pipe						
30	2.86	2.07	1.35	0.64	0.31	0.14
60	4.03	2.98	1.95	0.94	0.44	0.20
90	4.84	3.66	2.38	1.13	0.53	0.23
120	5.31	4.21	2.77	1.34	0.61	0.27
150	5.73	4.68	3.10	1.47	0.68	0.30

(Source : Michael, 1992)

Design of underground irrigation pipeline system

An underground pipeline irrigation system should be properly designed to handle efficiently the desired flow throughout the system. If smaller is diameter of pipe then pumping cost is increased and the capacity of the system may be seriously limited. However, when larger size of pipelines is used then it adds the extra cost of the system and may cause uneven flow. Pump stands must be high enough to allow sufficient operating head for the pipeline. If higher the pump stands then it may permit high heads of water to build up, which leads to excessive line pressures.

Designing the pipeline, its sizes and heights of pump stand and control stands must be determined so as to obtain a balanced water distribution and provide trouble-free and also ensure economical operation on the field. The working pressures in the pipeline should be kept within one-fourth the internal bursting pressures of the pipe. If it is necessary to design pipelines with higher heads then pressure pipes should be used. The size of the outlets should suit the flow required at diversion points in order to meet the require stream size.

For designing an underground pipeline system the head loss due to friction in the pipeline and other factors must be taken into consideration. Loss of head (h_f) between two points in a pipe line at l distance apart is determined by the following relationship:

$$h_f = \frac{fl\,v^2}{2\,g\,R} \qquad \text{....(13.19)}$$

where,

f = coefficient of friction for pipe, (dimensionless),

l = length of pipe, (m),

v = velocity of water in pipe, (m/sec),

g = acceleration due to gravity, [m/sec^2 (9.81 m/sec^2)], and

R = hydraulic radius, (m)

$$R=\frac{a}{p}$$

where,

a = area of cross-section of pipe, (sq m),

and p = wetted perimeter, (m),

$$a=\frac{\pi d^2}{4}$$

and p = πd,

where,

d = diameter of pipe, (m),

Also,

$$R=\frac{d}{4}$$

Therefore, we obtain

$$h_f=\left(\frac{2 f l v^2}{g d}\right)$$

The values of f for the common types of irrigation pipes is given in Table 13.13, while the values of friction loss in concrete pipes at various discharge rates is given in Table 13.14.

Entrance losses in pipelines may be determined from the following relationship:

(i) Pipe end flush with the wall of pump stand

$$h_e=0.5\frac{v^2}{2g} \quad \text{....(13.20)}$$

(ii) Pipe end protruding from wall

$$h_p = 1.0\frac{v^2}{2g} \quad ....(13.21)$$

(iii) Pipe with bell entrance

$$h_b = 0.1\frac{v^2}{2g} \quad ....(13.22)$$

where,

h_e = head loss at pipe entrance, (m) and v and g are the same as defined earlier.

The other losses which may be significant are the velocity head in the pipe and the head loss through riser valves. The velocity head due to flow in pipes is determined by the following relationship:

$$h_v = \frac{v^2}{2g} \quad ....(13.23)$$

where,

h_v = velocity head, (metres) and v and g are the same as defined earlier. Head loss through riser valves is given in Table 16.13.

Thus, total ahead required for operating pipe is expressed as under:

Total head (H) = $h_t + h_e + h_p + h_b + h_v$

It may be rewritten as,

$$H = \frac{2flv^2}{gd} + \frac{0.5v^2}{2g} + \frac{v^2}{2g} + \frac{v^2}{2g} + \frac{0.1v^2}{2g} + \frac{v^2}{2g}$$

$$= \frac{2flv^2}{gd} + \frac{2.6v^2}{2g}$$

$$= \frac{2v^2}{g}(\frac{fl}{d} + 0.65)$$

Table 13.13: Friction loss in concrete pipe in metres per 100 m

Flow rate (lt/sec)	Friction loss					
	Diameter of pipe					
	20 cm	25 cm	30 cm	35 cm	40 cm	45 cm
2.8	0.01					
5.7	0.03					

Flow rate (lt/sec)	Friction loss					
	Diameter of pipe					
	20 cm	25 cm	30 cm	35 cm	40 cm	45 cm
8.5	0.07					
11.3	0.12	0.04				
14.2	0.19	0.06				
17.2	0.26	0.09	0.04			
22.7	0.45	0.15	0.06			
28.3	0.68	0.23	0.09	0.04		
34.0	0.95	0.32	0.13	0.06		
39.6	1.23	0.43	0.18	0.08	0.04	
45.3	1.62	0.55	0.22	0.11	0.06	
51.0	2.02	0.68	0.28	0.13	0.07	
57.0	2.45	0.83	0.34	0.16	0.03	0.05
68.0	3.43	1.16	0.48	0.22	0.12	0.07
79.0	4.57	1.54	0.63	0.30	0.16	0.09
91.0	5.85	1.97	0.81	0.38	0.2	0.11
102.0	7.27	2.46	1.01	0.48	0.25	0.14
113.0	8.84	2.98	1.03	0.58	0.3	0.17
127.0		3.71	1.53	0.72	0.38	0.21
142.0		4.51	1.86	0.88	0.46	0.26
156.0		5.38	2.21	1.05	0.55	0.31

(Source : Michael, 1992)

The water surface elevations in pump stand and control stands should be sufficient to over-come the losses due to friction and other causes and distribute water through all riser in the field. The hydraulic gradient indicates the height to which water will rise in open standpipes or vents at any place along the line for a given flow. The hydraulic gradient line at any point in the pipeline may be determined. It is sum of the total losses in the pipeline. It includes changes in slope, sharp bends, and other factors will cause a shift in the hydraulic gradient should be considered.

Table 13.14: Head loss through riser valves

Flow rate (lt/sec)	Pressure loss in centimeters of water					
	Diameter of opening					
	15 cm	**20 cm**	**25 cm**	**30 cm**	**35 cm**	**40 cm**
10	2	1				
20	10	3	1			
30	26	7	3	1		
40	50	13	5	2	1	
50	83	22	8	4	2	1
60	126	34	12	6	3	2
80	250	64	24	11	5	3
100		105	40	18	9	s
125		165	65	29	15	7
150		250	97	44	22	12
175			136	65	30	17
200			185	83	41	22
225			245	100	55	30
250				137	69	37
275				170	83	45
300				210	100	56
350				300	145	77
400					195	103
450					250	135
500						170
550						210

(Source : Michael, 1992)

Common troubles of underground pipelines

Underground pipelines should be operated trouble free when it is properly designed and installed. Poor design and installation of pipeline may give rise to the following troubles in some situations:

1. Development of longitudinal cracks in the pipe, usually at the top or both at top and bottom,
2. Telescoping of sections,

3. Pushing of the pipe into the stands,
4. Development of circumferential cracks, and
5. Surging or intermittent flow of water.

The above first four troubles are caused mainly by the expansion and contraction of concrete pipes by wetting and drying. This is due to the temperature, it expands when temperature is high and contracts when temperature is low. Circumferential cracks occur due to drop in water or soil temperature or by drying of the pipe. Pushing of pipes into stands is not a serious trouble as the joint can be remade with cement mortar.

Surging: It is a common trouble in pipelines when adequate air vents are not provided. Air gets entrapped in water as it over- pours to the pump stand in absence of air vests. The entrapped air is carried in the pipeline and due to turbulence of the flow, the tendency for the air to separate out is minimised. After a short interval, the upstream portion of this reach of water becomes lighter than the downstream portion. This may cause a reversal of the hydraulic gradient until the water with entrapped air flows back to the stand and the air is dissipated Provision of adequate vents will eliminate the problem of surging.

13.8.6 Benefit cost analysis of underground pipelines

The cost of installing pipelines varies considerably from place to place. It may vary with the type of pipe, labour charges, depth of laying, type of soil and type of joint used. These may differ from place to place due to transportation cost and other factors.

References

American Society of Testing Materials 1965. Standard specifications for concrete sewer, storm drain and culvert pipes. ASTM: C 14-65.

American Society of Testing Materials 1966. Tentative specifications for concrete pipe for irrigation or drainage. ASTM: C 118-66 T.

Anonymous. 1962. *Measurement of irrigation water.* Chapter 9 in SCS National Engineering Handbook, U.S. Dept. of Agriculture, Washington, D.C. p. 72.

Booher, L.J. 1967. Surface irrigation. Food and Agriculture Organization of the United Nations, Rome. p. 161.

Christiansen, J.E. 1935. *Measuring irrigation water.* University of California, Agri. Expt. Stn. Berkeley. p. 91.

Code, W., 1961. *Farm irrigation structures.* Colorado State University. Bul, 496-S, p. 59.

Gupta, S.N. 1964. *Measurement of discharge at state tubewells by steel V-notch,* U.P. Irrigation Research Institute, Roorkee. p. 12.

Koluvek, P.K. 1970. Draft report on irrigation pipelines in India., U.S.A.I.D., New Delhi.

Michael, A. M., Arora, , D. R., Bhattacharya, A. K., Mandal, A. and Gupta, A. K. 1970. *Handbook of farm irrigation structures.* Indian Agricultural Research institute, New Delhi. p. 82.

Michael, A.M. and Ojha, T. P. 1966. *Principles of Agricultural Engineering, Vol. II.* Jam Brothers, Jodhpur. p. 534.

Michael, A.M.. Arora, D.R., Bhattacharya, A.K., Mandal, A. and Gupta, A.K. 1970. *Handbook of farm irrigation structures.* Indian Agricultural Research Institute, New Delhi. p. 82.

Normal Belonarme Ve Ilkel Gerilmeli Betonarme Kanallar-Istinat Egerlein-Ayaklar-Temeller 1964. *Prefabricke sulaona kanali tipleri CILT II.* Ankara DSI Mathaasi.

Parshall, R.L. 1926. *The improved Venturi flume.* Transactions, American Society of Civil Engineers, Vol. 89. pp. 841-880.

Peterka, A. J. 1967. *Water measurement procedures.* U.S. Bureau of Reclamation, Denver, Colorado, Report No. Hyd- 577. p. 153.

Portland Cement Association 1957. *Lining irrigation canals.* Portland Cement Association U.S.A., p. 33.

Portland Cement Association 1969. *Concrete pipe for irrigation.* American Concrete Pipe Association, U.S.A., p. 24.

Robinson, A.R., Lauritzen, C.W. and Muckel, D.C. 1963. *Distribution, control and measurement of irrigation water on the farm.* U.S.D.A., Misc. Pub. No. 926. p. 27.

Robinson, E.P. 1963. *Watertight seals for flow control structures.* Australian Civil Engineering and Construction.

Scott, V.H. and Houston, C.E. 1959. *Measuring irrigation water.* California Agri. Expt. Stn. Circular 473. p. 52.

Skogerboe, G. V., Bennet, R. S. and Walker, W. R. 1973. *Selection and installation* Of *cutthroat flumes for measuring irrigation and drainage water.* Pub. Colorado State University, Expt. Stn. p. 74.

Skogerboe, G.V. and Hyatt, M. L. 1967. *Analysis of .submergence in flow measuring flumes.* Journal of the Irrigation and Drainage Division, ASCE. Vol. 93, No. IR4, Dec., pp. 1-13.

Skogerboe, G.V. and Hyatt, M. L. 1967. Rectangular cutthroat flow measuring flumes. Journal of the Irrigation and Drainage, ASCE, Vol. 98, No. I R 4. pp. 569-583.

Skogerboe, G.V., Bennett, R. S. and Walker, W. R. 1972. *Generalized discharge relations for Cutthroat flumes. Journal of the Irrigation and Drainage Division*, ASCE, Vol. 98, No. IR 4, Dec., pp. 569-583.

Skogerboe, G.V., Hyatt, M. L. Anderson, R. K. and Elgleston, K. O. 1967. *Design and calibration of submerged open channel flow measurement structures: Part 3, Cutthroat flumes.* Report W G31-4, Utah Water Research Laboratory, College of Engineering. Utah State University, Logan, Utah.

Stock, E. M. 1955. *Measurement of irrigation water.* Bul. No. 5, Utah State Engineering Expt. Stn. p. 80.

The Concrete Association of India 1969. *Concrete pipes for irrigation.* New Delhi. p. 8.

U.S. Department of Agriculture 1966. Handbook of channel design for soil and water conservation. Still water Hydraulics Laboratory, Stillwater, Oklahama, U.S.A., SCS-TP-61. p. 22 .

Woffinden, D.S. and Clyde, C.G. 1971. *A Portable direct reading open channel flow measurement device.* Paper No. 2-6-190 presented at the Symposium on Flow-Its Measurement and Control in Science and Industry. Pittsburg, Pennsylvania, May 10-14, 1971. p. 15.

EXERCISES

13.1. A centrifugal pump has a discharge of 38 L/sec per second. One hectare of potato crop is irrigated in 6 h. What is the average depth of irrigation ? **Ans. 8.208 cm.**

13.2. Berseem requires 92 cm of water in 122 days of its irrigating period. How much land will be irrigated with a flow of 30 L/sec for 12 hr daily? **Ans. 17.185 ha.**

13.3. A farmer having 10 hectares of land grows wheat in 5 ha potatoes in 2 hectares, peas in 1 hectare and mustard in 2 ha during rabi. The irrigating season is 4 months. The water requirements of wheat, potatoes, peas and mustard are 45 cm, 50 cm, 15 cm and 10 cm, respectively. What size of irrigation stream will be required from a source which can supply for 8 hours each day during the season ? **Ans. 10.416 lit/sec.**

13.4. A Cipoletti weir has a breadth of 60 cm at its crest. The head of water flowing over the crest is 30 cm. Determine its discharge. **Ans. 183 litres/sec.**

13.5. Explain the various techniques of water measurement.

13.6. Describe the following in brief:

(a) Velocity area technique

(b) Parshell flumes

(c) Orifices

(d) Radioisotope technique

(e) Darcy-Weisbach formula

(f) Lining materials

13.9. Differentiate the following:

(a) Steady and unsteady flow

(b) Uniform and non-uniform flow

(c) Parshell flumes and cutthroat flumes

(d) Hydraulic radius and hydraulic slope

(e) Airvent and slope gate stand

(f) Drop inlet and chute spillways

(g) Spiles and siphon tube

(h) Flumes and culverts

13.7. Define the following:

(a) Weir crest

(b) Hydraulic radius

(c) Hydraulic slope

(d) Wetted perimeter

(e) Water control structures

(f) Diversion structures

13.8. Discuss different crossing structures and their suitability.

13.9. Describe in brief the underground pipe line system for irrigation.

13.10. What are the distinguished advantages offered by underground pipe line system.

13.11. Explain the procedure for designing underground pipe line system.

13.12. Determine the discharge of rectangular weir of 40 cm long with a head of 15 cm under all the following conditions.

(Ans. 34.4 L/sec, 33.5 L/sec. and 32.6 L/sec.)

13.13. An earth channel is having a grade of 0.11 per cent, depth of water 40 cm, bottom width 30 cm and side slopes 1½ to 1. Determine the velocity of flow and carrying capacity of the channel assuming the Mannings coefficient (η) is 0.25.

(Ans. 0.458 m/sec and 1.83 L/sec.)

13.14. Determine the discharge capacity of an under concrete pipe line. Assuming diameter of pipe 22.5 cm, length of pipe line 150 m, differences in elevation water levels at pump stand and discharge point 2 m and f is 0.006. **(Ans. 17.6 L/sec.)**

13.15. Determine the optimum bottom width, depth of flow and mean velocity of flow in a channel with 1.5:1 side slopes for a flow of 1.416 cu m/sec. Assume S0.2 m/km, n=0.04.

(Ans. v=0.289 m/sec, b=2.3 m, d=1.25 m)

13.16. Estimate the discharge from 10 siphon tubes, each of which is of 10 cm diameter, working simultaneously. The elevation of the water surface in the channel is 98.87 m and that of the water surface at the discharge point is/98.32 m. **(Ans. 13.4 L/sec.)**

13.17. A field channel having a discharge of 25 iitres/second has to cross a natural drainage-way. It is proposed to construct a semi-circular metallic flume at the crossing. Determine the size of the flume. The width of the drainage-way over which the structure is to be made is 20 m. There is a fall of 10 cm between the upstream and downstream ends of the structure. **(Ans. 25.1 cm diameter)**

13.18. A concrete pipeline is to be installed to carry a discharge of 100 L/sec. The friction loss in the pipeline should be equal to the gain in head due to the downfield slope of the land. The land slope along the line the pipes are to be laid is 1 per cent. What size pipe should be selected? **(Ans. 30 cm)**

■■■■

Chapter 14

SOIL PLANT WATER

14.1 Introduction

Water is crucial input for irrigation to plants to take varying quantities of water at different stages of plant growth. Thus, adequate and timely supply of water is to be ensured, otherwise physiological activities taking place within the plant are bound to be adversely affected. This resulted into reduction in crop yield. Soil-plant-water relationships help to understand the properties of soil and plants that influence the movement, retention and use of water. Soil provides the room for water to be stored and used by plants through the roots present in the same medium. Water also acts as a carrier for large amount of nutrients which required for the successful plant growth. Occurrence of moisture stress periods or dryspells due to inadequate and/or uneven distribution of rainfall during the crop growing period, it becomes essential to apply additional water to the soil for plant use in the form of irrigation to enhance crop yield or to save the total crop failure. The rate of entry of water into the soil and its retention, movement and availability to plant roots are all physical phenomena. These Influence the water requirement, irrigation scheduling and methods of application of irrigation and management practices for soil and water for crop production. Further, introduction of new high yielding varieties of crops like wheat which is high water demanding in respect of timing and amount of water, because their productivity or growth rate is much faster. But these varieties do not require more water as compared to traditional varieties. It may be understood that fertilizer applications so essential for the new varieties do not increase water requirements of crops but resulted in more production with a given quantity of water, so that water use efficiency increases with fertilizer use. Thus, knowledge of the physical properties of soils in relation to water for efficient management of scarce natural resource is essential to enhance the water productivity of harvested rainwater.

14.2 Soils

Soil mainly consists of finely divided organic matter and minerals (formed due to disintegration of rocks). It holds the plants upright, stores water for plant use, supplies nutrients to the plants and helps aeration. Soils can be classified in many ways *viz.*, based on size particles, (gravel, sand, silt, clay, etc.) and process of formation, and so on.

14.2.1 Soils classification

Soil is classified into the following categories based on their process of formation:

(i) *Residual soils:* The soil formed by disintegration of natural rocks due to the action of air, moisture, frost and vegetation results in residual soils.

(ii) *Alluvial soils:* It is the soil formed by sediment material deposited in bodies of water, deltas and along the banks of the overflowing streams.

(iii) *Aeolian soils:* These soils are formed by deposit through wind action.

(iv) *Glacial soils:* These soils are the products of glacial erosion.

(v) *Colluvial soils:* These are formed by deposition at foothills due to rain wash.

(vi) *Volcanic soil:* These are formed due to volcanic eruptions and are commonly called as volcanic wash.

The soils commonly found in India can be classified into the following groups:

(i) Alluvial soils: These soils include the deltaic alluvium, calcareous alluvial, soils, coastal alluvium and costal sands. This is the largest and most important soil group of India. These soils are formed by transportation of silt in stream and rivers and are deposited in flood plains or along coastal belts.

The main features of the alluvial soils of India are derived from the deposition caused by rivers of the Indus, the Ganges and the Brahmaputra systems. These rivers bring with them the products of weathering of rocks constituting the mountains in various degrees of fineness and deposit them as they traverse the plains. These soils differ from drift sand to barns and from fine silts to stiff clays.

These soils are very fertile and hence large irrigation schemes are feasible areas of such soil. However, the irrigation structures themselves would require strong foundation. These soils are generally deep soil and occur in Indo-Gangetic basin of north India. Alluvial soil of the Gujarat coastal and the Runn of Kutch, exhibit grey and black colours with salt patches and having pH between 7.0 and 9.0. These soils have fairly good water holding capacity and easily manageable.

(ii) Black soils: These soils vary in depth from a thin layer to a thick stratum. The typical soil derived from the Deccan trap is black cotton soil. The black soil contains lime concentration at varying depths. These soils are heavy in texture and their clay content generally ranges from 40 to 60 per cent and these soils may vary from clay to loam. They contain mostly montmarillonitic clay mineral which causes swelling and shrinkage with changing moisture conditions. It is common in Maharashtra, western parts of Madhya Pradesh, parts of Andhra Pradesh, parts of Gujarat and some parts of Tamil Nadu. These soil have a high degree of fertility but some, especially, in the uplands, are rather poor in fertility. These are suitable for the cultivation of rice and sugarcane. Drainage is poor in such soils. Black soils are grouped into: (i) Shallow black soils which have depth of 30 cm or less, (ii) medium black soil which are 30-100 cm in depth, and (iii) deep black soils which have depth more than 1 m.

(iii) Red soils: These are crystalline soils formed due to meteoric weathering of the ancient crystalline and metamorphic rocks. Such soils are found in Tamil Nadu, Karnataka, Goa, south-eastern Maharashtra, eastern Andhra Pradesh, Madhya Pradesh, Orissa, Bihar and some districts of West Bengal and Uttar Pradesh. Many of the so-called red soils of south India are not red. Red soils have also been found under forest vegetation.

(iv) Laterite and lateritic soils: Laterite and lateritic soils occur mostly in south India, but also in Birbhum, Burwan, Bankura and mid Napur district of West Bengal. Laterite is usually radish or yellowish in colour. The formation may be massive or it may be in the form of loose aggregates. It is a formation peculiar to India and some other tropical countries. Laterite rock is composed of a mixture of the hydrated oxides of aluminum and iron with small amounts of manganese oxides. Under the monsoon conditions, the siliceous matter of the rocks is leached away almost completely during weathering. Laterites are found on the hills of Kamataka, Kerala, Madhya Pradesh, the Eastern Ghats of Orissa, Maharashtra, West Bengal, Tamil Nadu and Assam. The laterites soil are generally acidic and have pH between 5.0 and 6.0. Lateric soils are those which are associated with and derived from laterities. These soils have larger proportion of the primary kaolintic minerals than laterites.

(v) Desert soils: A large part of the arid region, belonging to western Rajasthan, Haryana, Punjab, lying between the Indus river and the Aravalli range is affected by desert arid conditions of geologically recent origin. This part is covered with a mantle of blown sand which, combined with the arid climate, results in poor soil development. The Rajasthan desert is a vast sandy plain, including isolated hills or rock outcrops at places. The soil in Rajasthan improves in fertility from west and north-west to east and north-east. These soils are found characteristically in the arid region. Soils are coarse-textured and derived from disintegration of rocks in the adjacent areas and blown in from the coastal region and the Indus valley. These soils are composed of sand to a depth extending beyond 50 cm. Their horizon is weakly developed. They have a fairly high pH and varying amounts of $CaCO_3$.

(vi) Forest soils: These soils contain high percentage of organic and vegetable matter and are also known as humus. Such soils are found in forests and foot hills.

14.2.2 Physical properties of soil

The permeability of soils with respect to air, water and roots are as important soil properties in relation to plant growth of crop. These influence adequate supply of nutrients and water to plant. The permeability of a soil depends on the porosity and the distribution of pore spaces which, in turn, depends on the texture and structure of the soil.

14.2.2.1 Soil texture

The relative proportion of sand, silt and clay determines the soil texture. Most of the soils contain a mixture of sand (particle size ranging from 0.05 to 1.00 mm in diameter), silt (0.002 to 0.05 mm) and clay (smaller than 0.002 mm) in different proportion. The most commonly used classification are those proposed by the United State Department of Agriculture (USDA) and by the International Soil Science Society (ISSS) which are given as under:

Fraction	**Particle diameter**	
	USDA	***ISSS***
Gravel	>2 mm	>2 mm
Very coarse sand	1 to 2 mm	-
Coarse sand	0.5 to 1 mm	0.2 to 2 mm
Medium sand	0.25 to 0.5 mm	-

Fraction	Particle diameter	
	USDA	*ISSS*
Fine sand	0.1 to 0.25 mm	0.02 to 0.2 mm
Very find sand	0.05 to 0.1 mm	-
Silt	0.002 to 0.05 mm	0.002 to 0.02 mm
Clay	<0.002 mm	<0.002 mm

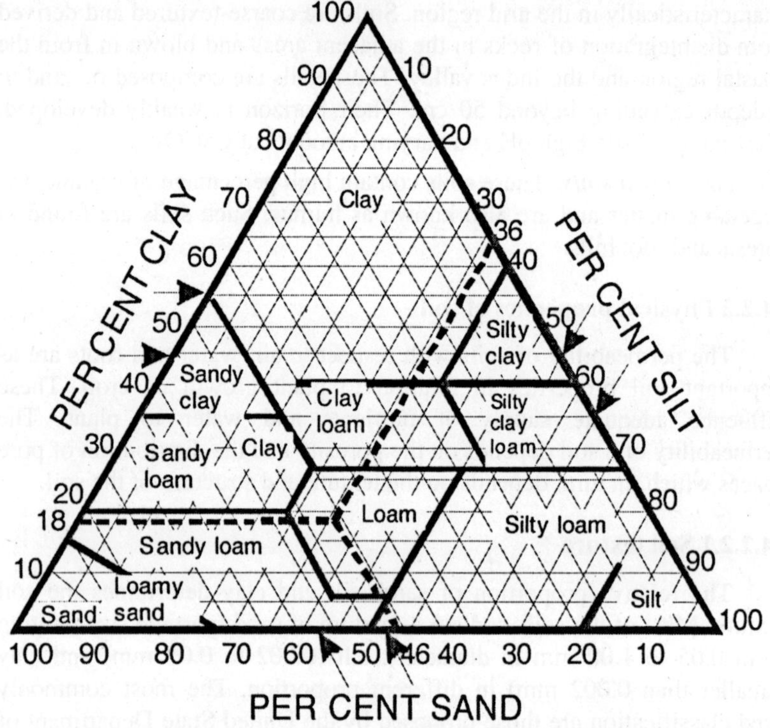

Fig. 14.1: U.S.D.A. soil textural classification chart
(The dotted lines indicate the method of determining the textural class of a soil containing 18% clay, 36% silt and 46% sand. The three lines join at point A which lies within the class 'loam').

If the sand particles dominate in a soil, it is called sand and is a coarse-textured soil. When clay particles dominate (more than 40%) then soil is called as clay soils and is a fine-textured soil. Loam soils (or simply barns) contain about equal amount of sand, silt and clay and are medium-textured soils. The texture of a soil affects the flow of water, aeration of soil and the rate of chemical transformation and all of which are important for plant life. The texture also determines the water holding capacity of the soil.

14.2.2.2 Soil structure

The arrangement of individual soil particles with respect to each other into a pattern is called the soil structure. Volume of space (*i.e.* the pore space) between the soil particles depends on the shape and size distribution of the particles. The pore space in irrigated soils may vary from 35 to 55 per cent. The porosity is the measure of the pore space and is defined as the ratio of the volume of voids (*i.e.* air and water-filled space) to the total volume of soil (including water and air). The pore space directly affects the soil fertility (*i.e.* the productivity value of soil) due to its influence upon the water holding capacity and also on the movement of air, water and roots through the soil. Soils of uniform particle size have large spaces between the particles, whereas soils of varying particle size are closely packed and the space between the particles is less. The particles of a coarse-grained soil function separately but those of fine-grained soils function as granules. Each granule consists of many soil particles. Fine textured soils offer a favourable soil structure permitting retention of water, proper movement of air and penetration of roots which is essential for the growth of crop.

The granules are broken due to excessive irrigation, ploughing or working under too wet (puddling) or too dry conditions. Such working affects the soil structure adversely. The structure of the irrigated soil can be maintained and improved by adoption of following practices:

(i) Ploughing to break hard pan. It increases infiltration rate,

(ii) After ploughing, allowing sufficient time for soil and air to interact before preparing the seed bed or giving pre-sowing irrigation,

(iii) Addition of organic matter: crop residue of the previous crops should be into the soil to form fertilizers, manures, etc.,

(iv) Keeping cultivation and tillage operations to a minimum, and

(v) Adopting a good crop rotation.

Green manures keep the soil fertility high. Crops like sunhamp *dhaincha*, guar, moong, etc. are grown on the fields. When these plants start flowering, ploughing is carried out in the fields so that these plants are buried below the ground surface. Their decomposition removes nutrient deficiencies in the soil.

The tendency of cultivators to grow only one type of crop (due to better returns) should be stopped as this cultivation practice leads to the deficiency of nutrients in the soil which are needed by the crop. If the land is not used for cultivation for one season, the soil recoups its fertility. Alternatively, legume crop may be included in rotation. Rotation of crops (which means growing different crops on the same field by rotation) is also useful in maintaining soil fertility at a satisfactory level.

Soil structure may be classified based on the shape of aggregates into the following:

(i) Simple structure: It is the structure in which natural cleavage planes are absent or indistinct. Simple structure may be a single-grain structure or massive structure. Single grain structure occurs normally only in sands and silts of low organic matter content. In sandy soils, single grain structure facilitates aeration and capillary movement of soil moisture. In other soils, however, single grain structure is undesirable because it results in the absence of large pores needed for aeration. Massive structure is similar to single grain structure, except that it is coherent. Examples of massive structure are dense soil crusts and plough pans.

(ii) Compound structure: In this type of soil structure, natural cleavage planes are distinct. A compound structure may be described according to the relative lengths of their vertical and horizontal axes and by the contour of the edges. Accordingly, they may be grouped into:

(a) Cube-like structure: in this structure, the vertical and horizontal axes are almost of equal length,

(b) Columnar structure: In this structure, vertical axes longer than the horizontal,

(c) Platy structure: in this structure, horizontal axes longer than vertical,

(d) Angular structure: in this structure, the corners and edges are sharp,

(e) Sub-angular structure: It is structure with sharp edges but rounded corners, and

(f) Granular structure: It is structure with both the corners and the edges rounded.

14.2.2.3 Depth of soil

An adequate depth of soil is essential for storing sufficient amount of irrigation water and providing space for root penetration and it cannot be over emphasized. Shallow soils require more frequent irrigations and cause excessive deep percolation losses when shallow soils overlie coarse-textured and highly permeable sands and gravels. While, deep soils would generally require less frequent irrigations, permit the plant roots to penetrate deeper and provide for large storage of irrigation water. As a result, actual water requirement for a given crop (or plant) is more in case of shallow soils as compared to deep soils even though the amount of water actually absorbed by the crop (or plant) may be the same in both types of soils. This is due to the unavoidable water losses at each irrigation of the soil.

14.2.3 Chemical properties of soil

For satisfactory crop yield, soils must have sufficient plant nutrients, such as nitrogen, carbon, hydrogen, iron, oxygen, potassium, phosphorous, sulphur, magnesium, etc. Nitrogen is the most important nutrient among all the nutrients. Nitrogenous matter is supplied to the soil from farmyard manure or from the growing of legume crops as green manures, or from commercial fertilizers. Plants absorb nitrogen in the form of soluble nitrates.

Soils having excess (greater than 0.15 to 0.20 per cent) soluble salts are known saline soils and those having excess of exchangeable sodium (more than 15 per cent or pH greater than 8.5) are called alkaline (or sodic) soils. Excessive amounts of useful plant nutrients such as sodium nitrate and potassium nitrate may become toxic to plants. Saline soils delay or prevent crop germination and also reduce the amount and rate of plant growth because of the high osmotic pressures which develop between the soil-water solution and the plants. These pressures adversely affect the ability of the plant to absorb water.

Alkaline (or sodic) soils tend to have inferior soil structure due to swelling of the soil particles. This changes the permeability of the soil. Bacterial environment is also an important feature of the soil-water-plant relationship. The formation of nitrates from nitrogenous compounds is accelerated due to favourable bacterial activity. Bacterial action also converts organic matter and other chemical compounds into the forms usable by the plants. Bacterial activity is directly affected by the soil moisture, soil structure and soil aeration. Arid soils provide better bacterial environment up to much greater depths as compared to humid climate soils because of their open structure. Besides these, due to low rainfall in arid regions leaching of nutrients (*i.e.* draining away of useful salts) is relatively less and the arid soils are rich in mineral plant food nutrients such as calcium and potassium.

14.2.4 Soil-water relationships

Let us consider that given volume of soil (v) (Fig. 14.2) consists of: (i) volume of solids (V_s), (ii) volume of liquids (water) (V_w), and (iii) volume of gas (air) (V_a). The porosity (n), the volumetric moisture content (w) and the saturation (S) are defined as under:

$$n = \frac{V_v}{V}, \quad w = \frac{V_W}{V}, \quad S = \frac{V_W}{V_v}$$

W_a - Mass of air,

W_w - Mass of water,

W_s - Mass of solid,

W - Total soil mass ($W_a + W_w + W_s$),

V - Total volume ($V_s + V_a + V_w$), and

V_e – Volume of pores.

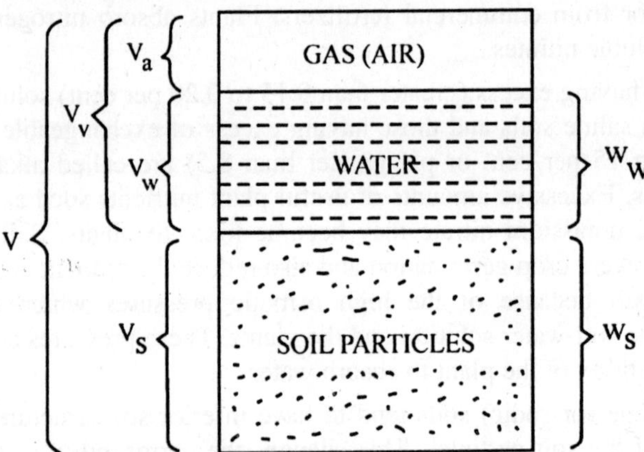

Fig. 14.2: Occupation of space in a soil sample

Volume and mass relationships of soil constituents

A schematic diagram of soil is shown in Fig. 14.3. It may be used to define volume and mass relationships of the three soil phases. The diagram shows the presence of the three phases in relative proportions both in masses and volumes.

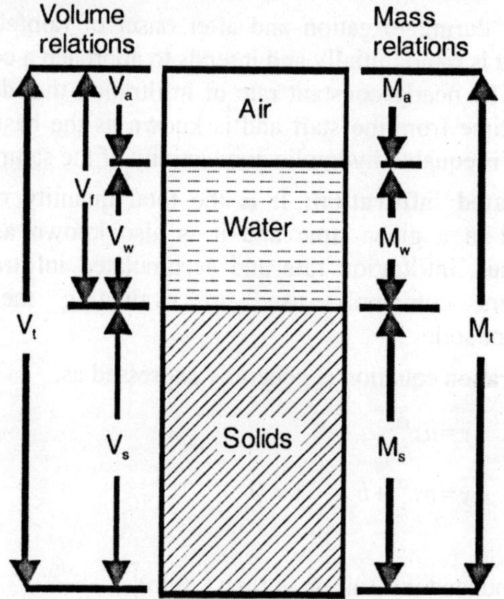

Fig. 14.3: Schematic diagram of the soil as a three-phase system

Let us consider as shown in Fig. 14.3 as,

M_a = Mass of air (negligible)

M_u = Mass of water

M_s = Mass of solid

M_t =Total mass ($M_a + M_w + M_s$)

V_a = Volume of air

V_w = Volume of water

V_f = Volume of pores ($V_a + V_u$)

V_s = Volume of solids

V_t = Total volume ($V_a + V_s$)

14.2.5 Infiltration

It is the movement of water from surface into the soil. Infiltration is one of the important properties of soil which influences irrigation. It not only controls the amount of water entering the soil but also the overland flow.

It is the soil characteristic determining the maximum rate at which water can enter the soil under specific condition. The actual rate at which water is entering the soil any time is known as the infiltration velocity.

This decreases during irrigation and after onset of rainfall event. The decrease of rate is rapid initially and it tends to approach a constant value at saturation. The nearly constant rate of infiltration that develops after elapsed some time from the start and is known as the basic infiltration rate and which is equal to hydraulic conductivity of the saturated soil.

Accumulated infiltration: It is the total quantity of water that enters the soil in a given time and it is also known as cumulative infiltration. Thus, infiltration rate and accumulated infiltration are the two parameters commonly used in evaluating the infiltration characteristics of soil.

The infiltration equation is generally expressed as,

$$y = a t^{\alpha} \qquad \text{....(14.1)}$$

$$y = a t^{\alpha} + b \qquad t \neq 0 \qquad \text{....(14.2)}$$

where,

y = accumulated infiltration in time t, (cm),

t = elapsed time or infiltration opportunity time, (min), and

a, ∝, b = constants (0.0 – 1.0)

The filtration rate at any time t is obtained on differentiating equation (14.1),

$$\frac{dy}{dt} = a \alpha t^{\alpha - 1} \qquad \text{....(14.3)}$$

When data on accumulated infiltration versus time is plotted on an ordinary graph paper, that gives a parabolic curve. While, plotting on a log, log paper these data indicate a linear relationship.

Factors influencing infiltration

Infiltration is a complex process which depends upon the following factors:

(i) soil properties,

(ii) initial soil moisture content,

(iii) previous wetting history,

(iv) permeability and its changes due to surface water movement,

(v) cultivation practices, type of crop being sown,

(vi) organic matter,

(vii) properties of water, and

(viii)climatic effects.

In an initially dry soil, the infiltration rate is high at the beginning of rain (or irrigation), but rapidly decreases with time until a fairly steady state infiltration is attained (Fig. 14.4).

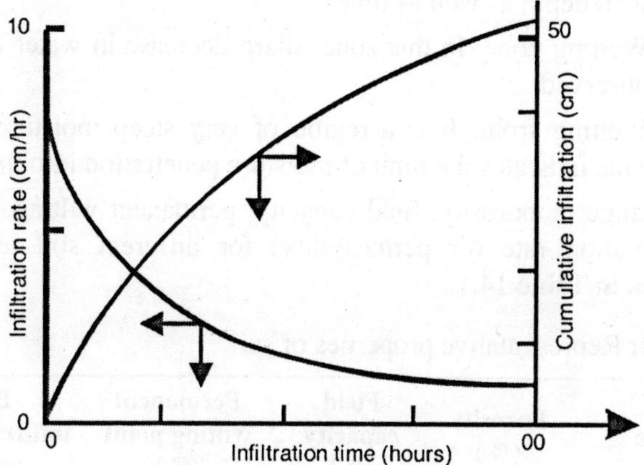

Fig. 14.4: Variation of infiltration rate and cumulative infiltration with time

The moisture profile under ponded infiltration into dry soil can be divided into the following five zones as shown in Fig. 14.5.

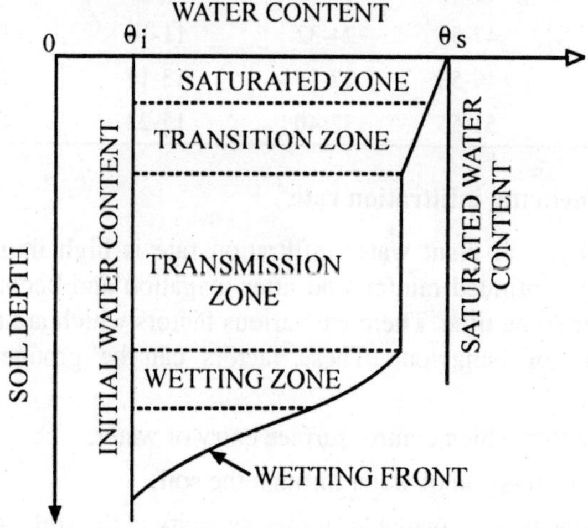

Fig. 14.5: Soil moisture profile during ponded infiltration

(i) Saturated zone: It extends up to about 1.5 cm below the surface and having saturated water content.

(ii) Transition zone: It is about 5 cm thick and is located below the saturated zone. A rapid decrease in water content occurs in this zone.

(iii) Transmission zone: In this zone, water content varies slowly with depth as well as time.

(iv) Wetting zone: In this zone, sharp decrease in water content is observed.

(v) Wetting front: It is a region of very steep moisture content. This indicates the limit of moisture penetration into the soil.

The range of porosity, field capacity, permanent wilting point and basic infiltration rate (or permeability) for different soil textures is summarized in Table 14.1.

Table 14.1: Representative properties of soil

Soil texture	Porosity (%)	Field capacity (%)	Permanent wilting point (%)	Basic infiltration rate (cm/hr)
Sand	32-42	5-10	2-6	2.5-25.0
Sandy loam	40-47	10-18	4-10	1.3-7.6
Loam	43-49	18-25	8-14	0.8-2.0
Clay loam	47-51	24-32	11-16	0.25-1.5
Silty clay	49-53	27-35	13-17	0.03-0.5
Clay	51-55	32-40	15-22	0.01-0.1

Factors influencing infiltration rate

Generally, soil plant water infiltration rate is high initially which limits during continued rainfall and after irrigation and becomes almost constant after some time. There are various factors which are responsible for this kind of behaviour. These factors can be grouped into the following:

(i) Factors which control surface entry of water,

(ii) Transmission of water through the soil,

(iii) Depletion of available storage capacity in the soil, and

(iv) Other factors.

(i) Factors which control surface entry of water: It includes factors such as soil texture, condition of the soil surface, vegetative cover, compaction and organic crop residue, etc.

Mechanical compaction caused by rain drops in fine textured soils (clay) reduces infiltration rate while, it is not significantly influenced in a sandy soil (coarse textured soils). Vegetation is the one of the significant factors for influencing the surface entry of water. Dense vegetation protects the soil from impact. Crop residue and forest litter etc. not only protect the soil from rain impact but also help in the creation of high biotic activity in and beneath such layers and it resulted in increase of infiltration rate. Compaction of soil surface caused by vehicle traffic and livestock would decrease the surface entry of water.

(ii) **Transmission of water through the soil:** It includes factors such as addition of organic matter, tillage and crop management practices. The infiltration rate cannot be increased unless the soil profile has adequate transmission capacity. The transmission of water through the soil is increased by deep ploughing, addition of organic matter and gypsum and also the crop management practices.

(iii) **Depletion of available storage capacity in the soil:** It includes factors such as thickness of the soil horizon, porosity, amount of moisture present which determines the volume of storage available in any horizon. Porosity also determines the resistance to flow. Non-capillary porosity is the most important soil characteristic that influences infiltration. The infiltration rate during the initial period of rainfall or irrigation is greatly influenced by the size, volume and continuity of macro pores (non-capillary) as movement of percolating water is greatly facilitated by such pores. The antecedent soil moisture content has considerable influence on the initial rate and infiltration capacity, both decreasing as the soil moisture content rises. Infiltration rate also influenced by the porosity of the soil which is changed by cultivation or compaction. Cultivation practices increase the porosity of the soil surface by breaking up the surface seals. Since the soil moisture is usually high in winter and spray seasons and low in other seasons. Thus, soil moisture is also responsible for seasonal variation in infiltration capacity. While, the current soil moisture content depends upon the factors such as the time, amount of antecedent infiltration, evaporation rate and transmission of under laying horizons. Thus, storage may directly influence the infiltration rate during the rainfall period.

(iv) **Other factors:** It includes the factors such as temperature, entrapped air, freezing of soil moisture, depth of surface detention and thickness of saturated layers which may affect the infiltration capacity of the soil. Viscosity of water influences the infiltration. This is due to the fact that flow through soil is always laminar and changes in temperature lead to change the viscosity of water. The high rate of infiltration in the tropics is due to low viscosity of warm water.

It is evident from the study that in surface irrigation, increased depth increases initial infiltration slightly but the head has negligible effect after prolonged irrigation. The downward movement of the sheet of water entering the soil compresses the air and hence tends to reduce the infiltration rate as the storm or irrigation progresses. The force causing the flow is large compared to resistance to flow and water will enter the soil surface at a very fast rate. This is one of the reasons of relative values of infiltration capacity at the start of the rainfall or irrigation.

Measurement of infiltration

Infiltration measurements may be made by the following methods:

(i) Cylinder infiltrometer,

(ii) Furrow infiltrometer,

(iii) Inflow-out flow measurement, and

(iv) Volume balance method.

(i) Cylinder infiltrometer: There are many types of infiltrometers. However, only ring type or cylinder infiltrometers are most commonly used. In this method water is artificially applied to a small area and rate of infiltration is determined directly. When water is applied to a plot of land by sprinkling at a uniform rate, the resulting runoff occurring at varying rates is measured and f-value is determined. The area of sample plot may vary from 0.1 to 0.4 m^2. Rainfall simulators may also be used. Cylinder infiltrometers are used to measure infiltration rate in borders with ponding water in the cylinder and noting the fall of ponded water in it. Cylinder ring infiltrometer are of two types: (a) Single cylinder, (b) Double cylinder.

It is the soil characteristics determining the maximum rate at which water can enter the soil under specific condition. The actual rate at which water is entering the soil at any time is known as the infiltration velocity. This is decreases during irrigation after onset of rainfall event. The decrease of rate is rapid initially and it tends to approach a constant value. The nearly constant rate of infiltration that develops after elapsed some time from the start is known as the basic infiltration rate and which is equal to hydraulic conductivity of the saturated soil.

Accumulated infiltration: It is the total quantity of water that enters the soil in a given time and it is also known as cumulative infiltration. Thus, infiltration rate and accumulated infiltration are the two parameters commonly used in evaluating the infiltration characteristics of soil.

The infiltration equation is generally expressed as,

$$y = at^{\alpha} \quad \text{....(14.4)}$$

$$y = at^{\alpha} + b \qquad t \neq 0 \quad \text{....(14.5)}$$

where,

y = accumulated infiltration in time t, (cm),

t = elapsed time or infiltration opportunity time, (min), and

a, α, b = constants (0.0 – 1.0).

The infiltration rate at any time, it is obtained on differentiating equation (14.5),

$$\frac{dy}{dt} = a\alpha t^{\alpha-1} \quad \text{....(14.6)}$$

When data on accumulated infiltration versus time plotted on an ordinary graph paper per gives a parabolic curve. While, plotting on paper, these data indicates a curvilinear relationship.

These methods are discussed in chapter that how infiltration rate is measured. Therefore, rest of the methods will be discussed below.

(ii) Furrow infiltrometer: The measurement of infiltration in furrow poses more complex problem due to its shape as compared to borders. This is due to the fact that infiltration is not uniform throughout the furrow section owing different soil compaction on sides of ridge and the bottom of the furrow. There are following methods which may be used to measure the infiltration rate in furrows: (a) Cylinder infiltrometer, (b) Blocking type furrow infiltrometer, and (c) By-pass furrow infiltrometer.

(a) **Cylinder infiltrometer:** This type infiltrometer is described in sub section of Chapter 8. Cylinders should be installed at the furrow bottom so that it gives more representative data of infiltration as compared to the sides.

(b) **Blocking type furrow infiltrometer:** This infiltrometer is developed by Bondurant (1957). The infiltration in rate is measured directly in the furrow and having approximate similar conditions as that of furrow during irrigation. This infiltrometer consists of two metal plates which are fixed in the furrow to isolate a small length of the furrow. The constant water level can be maintained by a supply tank having constant head device and water level in the tank is measured at frequent intervals. This type infiltrometer gives more accurate measurement of infiltration as compared to cylinder infiltrometers because in furrow infiltrometers effect of side is taken into consideration. This method gives an average infiltration rate over a given furrow section. In this method for determination of accumulated infiltration depth the ratio of the area of the supply tank to the area of infiltration (furrow section) is determined.

(c) **By-pass furrow infiltrometer:** This type of infiltrometer was developed by Shull (1961). It simulates the condition that occurs during the course of irrigation in a furrow. A small segment of furrow section is isolated by the infiltrometer enclosure and water continuous flow through the remaining part of the furrow. This improves the accuracy by reducing the lateral movement of water in the soil and water does not flow within the infiltrometer enclosure. The infiltration rate is determined similar to blocking type furrow infiltrometer method.

(iii) Inflow-out flow measurement, and

(iv) Volume balance method.

Example 14.1: A single cylinder with inside diameter of 30 cm was used for measuring infiltration in a border. Water level in the infiltrometer was kept constant by supplying water from a supply tank having inside diameter of 35 cm. Following observations were recorded.

Elapsed time (min)	0	5	10	15	30	45	60	90	120
Water level in the supply tank (cm)	27.55	26.59	26.12	25.78	25.05	24.46	23.96	23.10	22.44

Fit the Kostiakov type infiltration equation ($d = k\ t^n$) on using least square method.

Solution:

Ratio of the area of supply tank

to the area of cylinder infiltrometer = $\left(\frac{35}{30}\right)^2$ = 1.36

Calculations are tabulated in Table 14.3 Column 4 gives the decrease of water level in supply tank and these values when multiplied by 1.36 give the accumulated infiltration depth as is shown in columns 5. Columns 6 to 9 give calculations for least square method of curve fitting. Fig. 14.6 represents a log-log plot of d versus t.

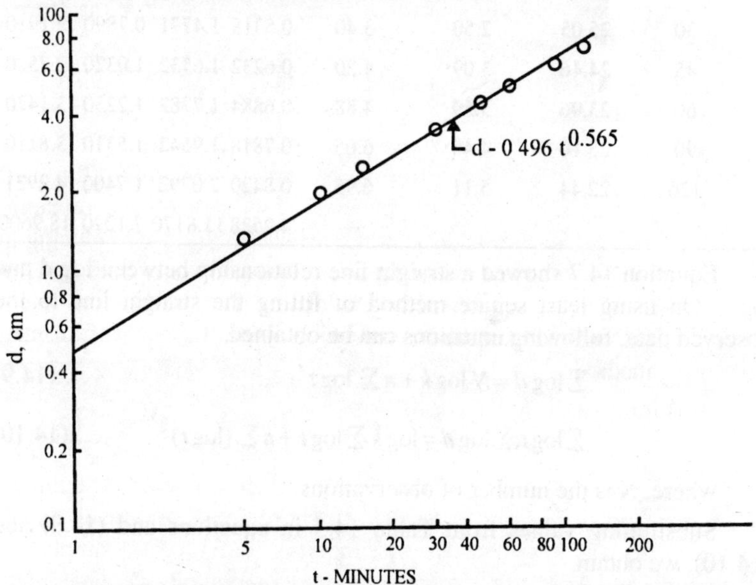

Fig. 14.6: Relation between accumulated infiltration and elapsed time

Method of least square assumes that best line through a set of linear data lies in a position such that the sum of the squares of the differences between the observed and calculated values of d is a minimum. Data are to be fitted in an equation of the form,

$$d = kt^n \qquad \text{....(14.7)}$$

Taking logarithms on both sides of equation [14.7]

$$\log d = \log k + n \log t \qquad ....(14.8)$$

Table 14.2: Calculations for determining infiltration equation for example 14.1. (Initial water level (t=0) (= 27.55 cm)

S. No.	Elapsed time, t (min)	Water level in the supply tank (cm)	Decrease of water level in supply tank (cm)	Accumulated infiltration depth, (d) (cm)	log d	log t	log d × log t	$(\log t)^2$
1	2	3	4	5	6	7	8	9
1	5	26.59	0.96	1.30	0.1139	0.6990	0.0790	0.4899
2	10	26.12	1.43	1.95	0.2900	1.0000	0.2900	1.0000
3	15	25.78	1.77	2.41	0.3820	1.1761	0.4430	1.2820
4	30	25.05	2.50	3.40	0.5315	1.4771	0.7890	2.1910
5	45	24.46	3.09	4.20	0.6232	1.6532	1.0320	2.7520
6	60	23.96	3.59	4.88	0.6884	1.7782	1.2230	3.1420
7	90	23.10	4.45	6.05	0.7818	1.9542	1.5310	3.8110
8	120	22.44	5.11	6.95	0.8420	2.0792	1.7400	4.2921
					4.2528	11.8170	7.1270	18.9600

Equation 14.7 showed a straight line relationship between log d and log t. On using least square method of fitting the straight line to the observed data, following equations can be obtained.

$$\sum \log d = N \log k + n \sum \log t \qquad ....(14.9)$$

$$\sum \log t . \sum \log d = \log k \sum \log t + n \sum (\log t)^2 \qquad ....(14.10)$$

where, N is the number of observations.

Substituting values from Table 14.3 in equations and (14.9) and (14.10), we obtain,

$$4.2528 = 8 \log K + 11.8170 n$$

$$7.1270 = 11.8170 \log K + 18.96 n$$

Solving these equations simultaneously,

$$n = 0.565$$

$$K = 0.496$$

and hence the infiltration equation is

$$d = 0.496 t^{0.565}$$

Example 14.2: Following data were recorded for accumulated depth of water infiltrated at different times while using a single cylinder infiltrometer:

Elapsed time (min)	4	8	12	16	20
Accumulated depth of infiltration, d (cm)	0.97	1.43	1.75	2.05	2.22

Using method of averages, fit the above data into exponential infiltration equation of the form,

$$d = b\,(1 - e^{-rt})$$

Solution:

Equation $d = b\,(1 - e^{-rt})$ can not be brought to a simple straight line form. However the constants b and r can be determined as the procedure described below:

Let us consider that,

$\Delta t_1, \Delta t_2, \ldots\ldots\ldots \Delta t_N$ be N incremental times during which infiltration is measured and for these time intervals, let the incremental depths of water infiltrated be, respectively $\Delta d_1, \Delta d_2, \ldots\ldots\ldots \Delta d_N$, so that,

$$\Delta t_1 = t_2 - t_1$$

$$\Delta t_2 = t_3 - t_2$$

$$\Delta t_N = t_{N+1} - t_N$$

and

$$\Delta d_1 = d_2 - d_1$$

$$\Delta d_2 = d_3 - d_2$$

$$\Delta d_N = d_{N+1} - d_N$$

If the experimental procedure is so designed that accumulated infiltration depths of water are measured for equal increments of elapsed time, as is the case in the present example, then all delta t's would be equal and can be represented by Δt. Thus in the present case Δt has a constant value of 4 min. Then from given equation, we obtain,

$$d_1 = b(1 - e^{-rt_1}) \qquad \text{....(14.11)}$$

and $$d_2 = b(1 - e^{-rt_2}) \qquad \text{....(14.12)}$$

$$d_5 = b(1 - e^{-rt_5}) \qquad \text{....(14.13)}$$

Subtracting equation (14.11) from (14.12) and we obtain,

$$\Delta d_l = b(e^{-rt\Delta} - e^{-rt2})$$

or $$\Delta d_l = e^{-rt\Delta}(b - be^{-r\Delta t}) \qquad ....(14.14)$$

The expression within parenthesis on right hand side of equation (14.14) has a constant value and substituting this expression by M, equation (14.14) can be rewritten as,

$$\Delta d_l = Me^{-rt}{}_1 \qquad ....(14.15)$$

Taking natural logarithm on both the sides

$$1\,n\,\Delta d_l = 1\,n\,M - rt_l$$

Similarly,

$$1\,n\,\Delta d_2 = 1\,n\,M - rt_2$$

$$1\,n\,\Delta d_3 = 1\,n\,M - rt_3$$

and $$1\,n\,\Delta d_4 = 1\,n\,M - rt_4$$

Calculated values of Δ d's are shown in Table 14.4

Substituting values from column 2 and 5 of this table

$$1\,n\,0.46 = 1\,n\,M - 4\,r$$

$$1\,n\,0.32 = 1\,n\,M - 8\,r$$

$$1\,n\,0.30 = 1\,n\,M - 12\,r$$

$$1\,n\,0.17 = 1\,n\,M - 16\,r$$

According to the method of averages, adding up separately first two and last two equations, one gets after simplification,

$$-1.9159 = 2\,1\,n\,M - 12\,r \qquad ....(14.16)$$

$$-2.9760 = 2\,1\,n\,M - 28\,r \qquad ....(14.17)$$

Table 14.3: Calculations for determining infiltration equation for example 14.2

S. No.	Elapsed time, t (min)	Δ t (min)	Accumulated depth of infiltration, d (cm)	Δ d (cm)
1	2	3	4	5
1	4		0.97	
		4		0.46
2	8		1.43	
		4		0.32
3	12		1.75	

S. No.	Elapsed time, t (min)	Δ t (min)	Accumulated depth of infiltration, d (cm)	Δ d (cm)
		4		0.30
4	16		2.05	
		4		0.17
5	20		2.22	

Solving equations (14.16) and (14.17) simultaneously, we obtain,

r = 0..067

and M = 1.74

Substituting the value of M

$$b\,(1\ ^{-}e^{-r\Delta t}) = 1.74 \qquad ....(14.18)$$

On substituting known value of r and Δ t in equation (14.18) and simplifying one gets,

b = 7.6

Hence accumulated infiltration equation is,

$$d = 7.6\,(1 - e^{-0.067})$$

where, d is in cm and t is in mm.

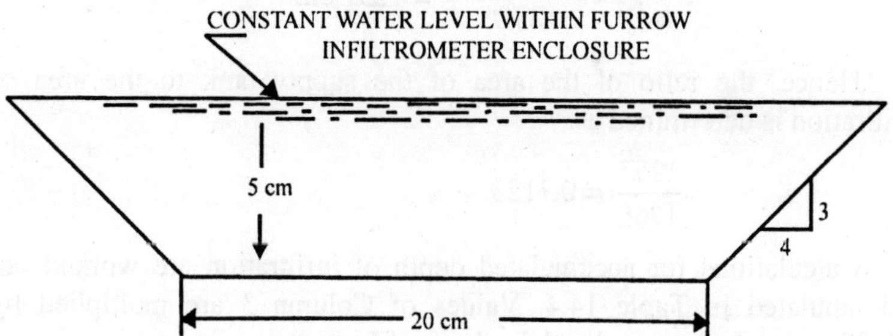

Fig. 14.7: Furrow section for computing accumulated infiltration depths in example 14.3

Example 14.3: A trapezoidal furrow cross-section having 22 cm bottom width and 4 horizontal: 3 vertical side slopes, infiltration was measured by blocking type furrow infiltrometer. The two plates were installed 50 cm apart and the depth of water in the furrow was maintained constant at 5 cm throughout the experiment. Following observations were recorded for the water level in the supply tank of the constant head device having inside diameter of 40 cm.

Elapsed time (min)	0	5	10	15	30	45	60	90	120
Water level in the supply tank (cm)	31.25	29.60	28.60	27.65	25.20	23.03	21.15	17.55	14.20

Determine the accumulated infiltration depths at different times.

Solution:

Furrow section shown in Fig. 14.7 having the top width of water surface, (T) can be determined as,

$$T = 22 + \frac{8}{3} \times 5 = 35.3 \text{ cm}$$

Spacing between the two plates is 50 cm. Area of the top of the water surface is a through which uniform infiltration is assumed to take place is given by,

$$a = 35.3 \times 50$$
$$= 1765 \text{ cm}^2$$

For supply tank of 40 cm dia cross-sectional area is determined as,

$$= \frac{\pi}{4} r^2 = \frac{22 \times 40 \times 40}{4 \times 7} = 1257 \text{ cm}^2$$

Hence, the ratio of the area of the supply tank to the area of infiltration is determined as,

$$= \frac{1257}{1765} = 0.7123$$

Calculations for accumulated depth of infiltration are worked out and tabulated in Table 14.4. Values of Column 3 are multiplied by 0.7123 to get the accumulated depth of infiltration.

Table 14.4: Calculations for accumulated depth

S. No.	Water level in the supply tank (cm)	Decrease of water level in supply tank (cm)	Accumulated infiltration depth, d (cm)
1	2	3	4
0	31.25	0	0
5	29.60	1.65	1.18

S. No.	Water level in the supply tank (cm)	Decrease of water level in supply tank (cm)	Accumulated infiltration depth, d (cm)
10	28.60	2.65	1.89
15	27.65	3.60	2.56
30	25.20	6.05	4.31
45	23.00	8.25	5.88
60	21.15	10.10	7.19
90	17.55	13.70	9.76
120	14.20	17.05	12.14

By using the method of least square or the method of averages for curve fitting between the values of column 1 and 4 of Table 14.4, the infiltration equation can be determined as is illustrated in example 14.2 and 14.3.

Inflow-outflow measurements

The infiltration rate or water intake of the soil can be determined under actual field conditions by measuring the flow into and out of the field with time. There are two general approaches for determination of infiltration. Generally two methods used for determination hydrographs of runoff from natural rainfall on the field.

(i) Analysis of hydrograph of runoff from natural rainfall, and

(ii) Application of artificial water in the enclosed areas.

(i) Analysis of hydrograph of runoff from natural rainfall

This method is suitable for determination of infiltration rate as likely to occur during rainfall rather than during application of irrigation. The volume rates of rainfall and runoff are determined at different elapsed times from runoff hydrograph. The difference of the two gives the volume rate of infiltration at different times. The accuracy of the method depends upon the precision of measurement of rainfall and runoff. Infiltration rate is determined on analysis of hydrograph by any method suggested by Chow (1964) as,

(i) The detention-flow-relationship method,

(ii) The time-condensation method, and

(iii) The block method.

(ii) An application of artificial water in the enclosed area

Water is applied from a rainfall simulator or an irrigation stream at known rate on a small experimental area within the irrigation field.

The infiltration rate of the soil likely to occur during surface irrigation application can be determined by measuring the flow into and out from a small sample plot. Since, the size of stream is known while applying irrigation. This method is more practicable to use in furrows than in borders. The inflow stream size can be measured by an orifice plate, a V-notch or by pre-calibrated siphon tubes. The outflow from downstream end can be measured by a Parshall flume. If length of the furrow and cross-sectional area of .water flow in the furrow are known then infiltration rates at different times can be determined from the known differences of inflow and outflow as,

$$V_i = V - V_r \quad \text{....(14.19)}$$

$$\text{and Infiltration rate} \quad (I) = \frac{V_i}{a} \quad \text{....(14.20)}$$

$$a = T \times i \quad \text{....(14.21)}$$

$$T = b + h.s \quad \text{....(14.22)}$$

Example 14.4: The infiltration rate in a furrow was determined by infiltration measurement in the furrow. The furrow was 100 m long and having uniform trapezoidal section with a base width of 20 cm and side slopes of 4 horizontal to 3 vertical. For a constant inflow rate of 1.0 L/sec, average depth of water flow was estimated to be 6 cm while the outflow recorded at different intervals are given as under:

Clock time (a.m.)	Elapsed time (min)			Outflow, Vr (L/sec)
	Upstream end (x = 0)	Downstream end (x = 100 m)	Average	
8.00	Start	-	-	-
8.20	20	Water front reaches	-	-
8.30	30	10	20	0.66
8.40	40	20	30	0.73
9.00	60	40	50	0.79
9.20	80	60	70	0.82
9.50	110	90	100	0.85
10.40	160	140	150	0.88

Determine the infiltration rate at different time.

Solution:

For given furrow cross-section and depth of flow, the top width of water flow (T) is determined as,

$$T = 20 + \frac{8}{3} \times 6 = 36 \text{ cm}$$

For the furrow length of L= 100 m, the average area C.S. of the top of water surface along which uniform infiltration is assumed to take place is determined as,

$$a = T \times L$$

$$= 36,00,00 \text{ cm}$$

Calculations for infiltration rates are tabulated in Table 14.5.

Table 14.5: Calculations for infiltration rate by inflow outflow measurement of example 14.4 (Inflow = 1 L/sec)

Average elapsed time, t (min)	Outflow, V_r (L/sec)	Volume rate of infiltration, V_i (L/sec)	Infiltration rate, I (cm/min)
1	2	3	4
20	0.66	0.34	0.057
30	0.73	0.27	0.045
50	0.79	0.21	0.035
70	0.82	0.18	0.030
100	0.85	0.15	0.025
150	0.88	0.12	0.020

Volume rates of infiltration, (V_i) as given in column 3 of Table 14.5 are determined from the relationship:

$$V_i = V - V_r$$

where, V is the constant stream size of 1 L/sec and V_r is the outflow at any time corresponding units. Infiltration rates (I) for column 4 of Table 14.5 are obtained from the following relationship:

$$I = \frac{V}{a}$$

From the values of I versus t the infiltration equation can be determined as before.

(iv) Volume balance method

This method gives most accurate measurement of infiltration rate as it represents actual irrigation condition. Finkel and Nir (1960) developed a graphical solution to determine infiltration rate in border irrigation. However, it is not applicable to furrows due to engine, the top width of water flow change with time and distance. The volume balance method gives the most accurate measurements of infiltration rate as compared to other methods in furrow irrigation considering top width is constant (Davis and Fry, 1965 and Smerdan, 1963).

The volume balance method is based on the division of the period of irrigation into convenient increments of time which if made shorter, permit a greater degree of accuracy (Fig. 14.8). At the end of each time increment the total inflow of water (V) into the irrigation channel is equated to the sum of the amount of water that has penetrated the soil and the amount temporary storage above ground. It is expressed as,

$$V = Qt = V_i + V_s \quad \text{....(14.23)}$$

$$V_s = Ax \quad \text{....(14.24)}$$

where,

V = total inflow into the channel in specified time,

Q = rate of water into the irrigation channel,

t = time which has elapsed since the flow began,

V_i = V_i (t) the quantity of water which has penetrated the soil in the channel during the time, t,

V_s = V_s (t), the amount of water stored above the ground surface,

A = average cross section area of water stored in the channel, and

x = the advance length during time, t.

V_i depends upon the rate of infiltration (I) which is assumed to be uniform across the top width but is a function of time, and upon the total wetted length of the channel at each moment. Hence, it is expressed as,

$$V_i = \int_o^x \int_{t_s}^t Q_i \, dt \, ds \quad \text{....(14.25)}$$

where,

t_s = time elapsed from the beginning of the flow until the advancing front reaches a distance s, and

Q_i = T I, where T is average top width of water flow.

The equation (14.25) can be rewritten as,

$$Q_t = \int_0^t \int_{t_s}^t T\,I\,dt\,ds + Ax \qquad \text{....(14.26)}$$

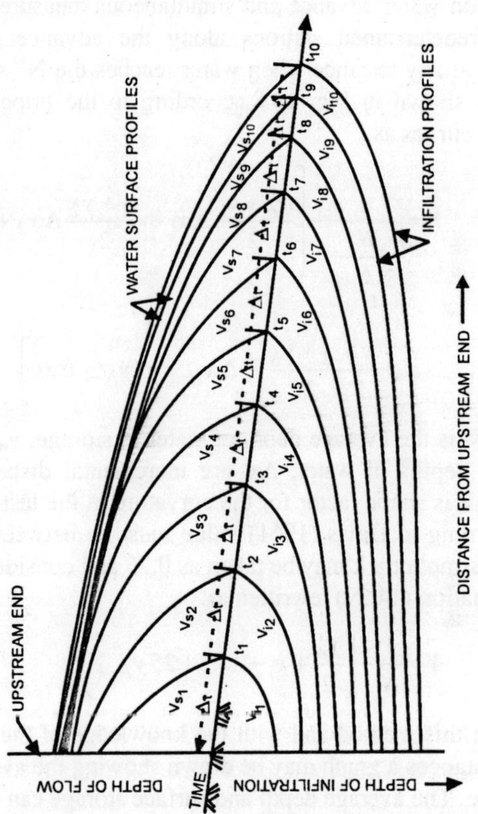

Fig. 14.8: Definition sketch for computing infiltration by volume balance method

Equation (14.26) is not easy to solve but it can be approximated by numerical methods. For a small advance distance Δx, covered during small time increment, Δt equation (14.26) can be written as,

$$Q\,\Delta t = Td\Delta x + A\,\Delta x \qquad \text{.....(14.27)}$$

where, d is the average depth of water infiltrated over Δx during small time increment Δt.

Solving equation (14.27) for d, we obtain,

$$d = \frac{Q\,\Delta t - A\,\Delta x}{T\,\Delta x} \quad .....(14.28)$$

The values of average cross sectional area (A), and average top width (T) for water on ground surface at any instance may be determined from the date on water advance and simultaneous measurement of depth of flow at predetermined stations along the advance distance. The average depth at any instance when water reaches the Nth station may be determined as shown in Fig. 14.8.according to the trapezoidal rule of area under the curves as,

$$\bar{Y} = \frac{1}{\underset{i=1}{\overset{N}{\Sigma}} \Delta x_i} \left[\frac{y_o + y_1}{2} \Delta x_1 + \frac{y_1 + y_2}{2} \Delta x_2 + ... \right.$$

$$\left. ....+ \frac{y_{N-2} + y_{N-1}}{2} \Delta x_{N-1} + k y_{N-1} \Delta xn \right]. \quad ....(14.29)$$

Where, $\bar{y}$ is the average depth of water in storage, y_o, y_1,...y_{N-1} are the measured depths of water, Δx' are incremental distance along the flow path and k is shape factor for the curvature in the last portion of the profile. According to Davis (1961) value must be between 0.5 and 1.0. However, for simplicity k may be taken as 0.75 and considering Δx' to be equal then equation (14.29) rewritten as.

$$\bar{Y} = \frac{1}{N} \left(\frac{y_o}{2} + y_1 + ...1.25\, y_{N-1} \right) \quad ....(14.30)$$

Based on this method and with the knowledge of the advance time for various distances a graph may be drawn showing the average depth of surface storage. The average depth and surface storage can be determined for any time from this graph. The average area and average top width of surface storage may then be determined from channel geometry. Let us consider that channel shape is trapezoidal then,

$$\text{Top width (T)} = B + 2.Z\,\bar{y}$$

$$\text{Average area (A)} = (B + Zy)\;\bar{y}$$

where,

T = average top width,

A = average cross sectional area of water stored in channel,

B = bottom width of the channel, and

Z = side slope.

The average water surface area (A_w) can be determined from channel advance distance and average top width with the following expression.

$$A_w = T.x \quad \text{....(14.31)}$$

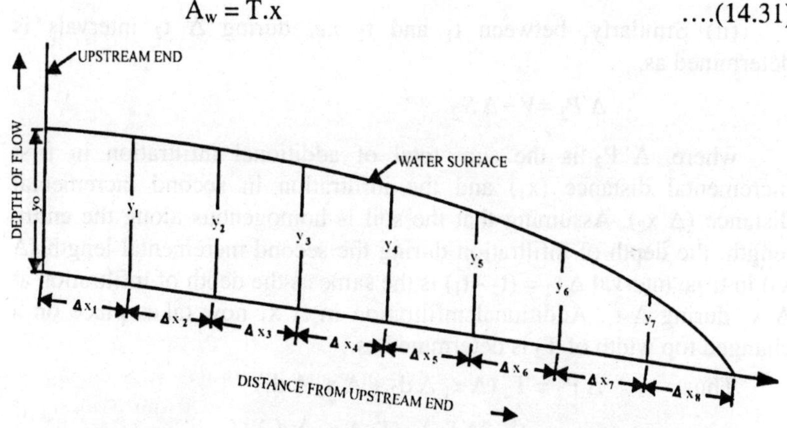

Fig. 14.9: Sketch of computation process for average depth of surface storage

Infiltration rate is assumed to take place uniformly through an area (A_w) at any time. Total time of advance is divided into short equal time intervals and the average top width and average volume of surface storage are determined for different time intervals as discussed above. The advance length for desired time intervals are determined from the pertinent advance equation.

Let us consider, $t_1, t_2 \ldots\ldots t_N$ be the accumulated time periods after equal time increments Δ t so that $t_1 = \Delta t$, $t_2 = 2 \Delta t$ and $t_N = N \Delta t$.

For $\Delta t_1, \Delta t_2, \ldots\ldots \Delta t$, time intervals

$T_1, T_2 \ldots\ldots T_N$ are the average top widths,

$\Delta x_1, \Delta x_2, \ldots\ldots \Delta X_N$ are the incremental distance of advance,

$\Delta S_1, \Delta S_2, \ldots\ldots \Delta S_N$ are the incremental surface storage volumes,

$\Delta P_1, \Delta P_2, \ldots\ldots \Delta P_N$ are the incremental volumes of infiltration,

$\Delta d_1, \Delta d_2, \ldots\ldots \Delta d_N$ are the incremental infiltration depths in the first incremental distance Δx_1, and

$V = Q\Delta t$ is the constant incremental inflow volume.

(i) Increment volume of infiltration between o and t_1 *i.e.* during Δt_1 intervals is determined as,

$$\Delta P_1 = T_1 \Delta x_1 \Delta d_1$$

And also $\Delta P_1 = V - \Delta S_1$

$$\text{Hence, } \Delta d_1 = \frac{V - \Delta S_1}{T_1 \Delta x_1} \quad \text{....(14.32)}$$

(ii) Similarly, between t_1 and t_2 *i.e.* during Δt_2 intervals is determined as,

$$\Delta P_2 = V - \Delta S_2$$

where, ΔP_2 is the sum total of additional infiltration in first incremental distance (x_1) and the infiltration in second incremental distance (Δx_2). Assuming that the soil is homogenous along the entire length, the depth of infiltration during the second incremental length (Δx_2) in time interval $\Delta t_2 = (t_2 - t_1)$ is the same as the depth of infiltration at Δx_1 during Δt_1. Additional infiltration in Δx_1 now takes place on a changed top width of T_2 is determined as,

$$\text{Thus, } \Delta P_2 = T_2 (\Delta x_1 \Delta d_2 + \Delta x_2 \Delta d_1)$$

$$\text{Hence, } \Delta d_2 = \frac{(V - \Delta S_2) - (T_2 \Delta x_2 \Delta d_1)}{T_2 \Delta x_1} \quad \text{....(14.33)}$$

Similarly,

(iii) Between t_2 and t_3 *i.e.* during Δt_3 is obtained as,

$$\Delta P_3 = V - \Delta S_3$$

$$= T_3 \Delta x_1 \Delta d_3 + T_3 \Delta x_2 \Delta d_2 + T_3 \Delta x_3 \Delta d_1$$

$$\text{or } \Delta d_3 = \frac{(V - \Delta S_3) - (T_3 \Delta x_2 \Delta d_2 + T_3 \Delta x_3 \Delta d_1)}{T_3 \Delta x_1} \quad \text{....(14.34)}$$

(iv) In general between interval t_{N-1} and t_N *i.e.* during Δt_N is obtained as,

$$\Delta P_N = V - \Delta S_N$$

$$= T_N \Delta x_1 \Delta d_N + T_N \Delta x_2 \Delta d_N + T_N \Delta x_3 \Delta d_{N-2}$$

$$+ \ldots\ldots\ldots\ldots + T_N \Delta x_N \Delta d_1$$

or we obtain,

$$\Delta d_N = \frac{(V - \Delta S_N) - (T_N \Delta x_2 \Delta d_{N-1} + T_N \Delta x_3 \Delta d_{N-2} + ... + T_N \Delta x_N \Delta d_1)}{T_N \Delta x_1} \quad ....(14.35)$$

The above can be rewritten as,

$$\Delta d_N = \left[(V - \Delta S_N) - T_N \sum_{J=1}^{N-1} \Delta dj \Delta xi \right] \frac{1}{\Delta x_1 T_N} \quad ....(14.36)$$

where, i = N + 1 – J

Equation (14.36) will thus enable the determination for depth of water infiltrated over the first incremental distance Δx_1 during each increment of time. This equation gave the cumulative depth of infiltration within total time may be obtained by successive addition and the infiltration equation may be obtained by least square method of curve fitting or by the method of averages.

When borders, having a uniform top width, the above analysis may be simplified on using the concept of stream size per unit width. Thus, denoting the incremental flow volume per unit width as "V" and the incremental surface storage during any period between t_N and t_{N-1} per unit width as $\Delta S'_{N'}$ equation (14.36) can be written for borders as

$$\Delta d_N = \frac{1}{\Delta x_1} \left[(V' - \Delta S'_N) - \sum_{J=1}^{N-1} \Delta dj \Delta xi \right] \quad ....(14.37)$$

where, i = N + 1 – J

For high degree of accuracy it is advisable to use small time intervals which involve rigorous calculations.

Example 14.5: On a border strip of 6 m width, while A stream size of 14.9 L/sec was applied on a border strip 6 m width. The average depth of surface storage at the end of 4, 8, 12, 16 and 20 minutes was found to be 1.40, 1.47, 1.59, 1.71 and 1.86 cm, respectively. The water front advance relationship is represented by the equation.

$$t_x = 8.71(e^{0.012x} - 1)$$

where t is in min and *x* is in metres.

Determined the depth of infiltration at different time intervals on using volume balance procedure. Following the method of averages for curve fitting, fit the calculated values into Kostiakov-Lewis type equation of the form $d = K t^n$.

Solution:

(i) During the first time interval of 4 min

Distance covered, $(x) = \Delta x_1 = \dfrac{1}{0.012} ln\left(\dfrac{4}{8.71}+1\right) = 30.9\, m$

Total water supplied, $(V) = 4 \times 17.9 \times 60 = 4296$ L

Water x in surface storage, $(\Delta S_1) = 1.40 \times 30.9 \times 6 \times 10$

$= 2595$ L

Volume of water infiltrated, $(\Delta P_1) = V - \Delta S_1 = 1701$ L

Depth of water infiltrated, $(\Delta d_1) = \dfrac{1701}{30.9 \times 6 \times 10} cm = 0.92$ cm

(ii) During the second time interval of 4 min

Cumulative distance, $(x_2) = \dfrac{1}{0.012} ln\left(\dfrac{8}{8.71}+1\right) = 54.2\, m$

Incremental distance, $(\Delta x_2) = 54.2 - 30.9 = 23.3$ m

Water supplied, $(V) = 4296$ L

Cumulative water in surface storage,

$V_{s2} = 54.2 \times 1.47 \times 60 = 4780$ L

Incremental volume of surface storage.

$\Delta S_2 = 4780 - 2595 = 2185$ L

Incremental infiltration volume,

$\Delta P_2 = 4296 - 2185 = 2110$ L

Estimated infiltration volume through Δx_2

$= 23.3 \times 0.92 \times 60 = 1286$ L

Hence infiltration volume through Δx_1

$= 2210 - 1286 = 824$ L

Depth of water infiltrated, $(\Delta d) = \dfrac{824}{30.9 \times 60} = 0.44\, cm$

(iii) Proceeding as above during 3rd time interval of 4 min.

$x_3 = 72.2$ m

$\Delta x_3 = 18.0$ m

$V = 4296$ L

$V_{s3} = 72.2 \times 1.59 \times 60 = 6888$ L

$\Delta S_3 = 6888 - 4780 = 2108$ L

$\Delta P_3 = 4296 - 2108 = 2188$ L

Infiltration through $\Delta x_3 = 18 \times 0.92 \times 60$

$= 994$ L

Infiltration through $\Delta x_2 = 23.3 \times 0.44 \times 60$

$= 615$ L

$$\text{Hence } \Delta d_3 = \frac{2188-994-615}{60\times30.9} = 0.31 \text{ cm}$$

(iv) During 4th time interval of 4 min

$x_4 = 86.8$ m

$\Delta x_4 = 14.6$m

$V = 4296$ L

$V_{s4} = 8910$ L

$\Delta S_4 = 2030$ L

$\Delta P_4 = 2360$ L

Infiltration volume through Δx_4, Δx_3, and Δx_2

$= 60\ (14.6 \times 0.92 + 18 \times 0.44 + 23.3 \times 0.31)$

$= 1718$ L

$$\Delta d_4 = \frac{2360-1718}{60\times30.9} = 0.29 \text{ cm}$$

(v) During 5th time interval of 4 min

$x_5 = 99.2$ m

$\Delta x_5 = 12.4$ m

$V = 4390$ L

$V_{s5} = 11100$ L

$\Delta S_5 = 2190$ L

$\Delta P_5 = 2200$ L

Infiltration volume through Δx_5, Δx_4, Δx_3 and Δx_2

$= 60\ (12.4\times0.97+14.6\times0.44+18\times0.31+23.3\times0.29)$

$= 2021$ L

$$\Delta d_5 = \frac{2200-2021}{60\times30.9} = 0.10 \text{ cm}$$

The calculated values of time and depth of infiltration can be tabulated as given in Table 14.6.

Table 14.6: Values of depth of infiltration at different times

Serial Number	Elapsed time, t min	Δt, min	Δd, cm	Accumulated depth of infiltration, d cm
1	2	3	4	5
1	4			0.92
		4	0.44	
2	8			1.36
		4	0.31	
3	12			1.67
		4	0.29	
4	16			1.96
		4	0.10	
5	20			2.06

Fitting into Kostiakov - Lewis type equation

$$d = Kt^n$$

or log $d = \log K + n \log t$

Substituting in above equation values from column (5) and (2) of Table 14.6 and simplifying, we obtain,

$$0.0132 = \log K + 0.6021\ n$$

$$0.1553 = \log K + 0.9031\ n$$

$$0.2430 = \log K + 1.0792\ n$$

$$0.3118 = \log K + 1.2041\ n$$

$$0.3464 = \log K + 1.3010\ n$$

Adding separately the first 3 and last 3 equations

$$0.385 = 3 \log K + 2.5844\ n \qquad ....(14.38)$$

$$0.9012 = 3 \log K + 35843\ n \qquad ....(14.39)$$

Solving equations (14.38 and 14.39) simultaneously, we get

$$n = 0.5161$$

and $K = 0.4820$

Hence accumulated infiltration equation is

$$d = 0.4820\ t^{0.5161}$$

where, d is in cm and t is in min.

Important terminology pertaining to soil water relationship

Density of solids (G_s): It is defined as ratio of mass of solid to its volume and is expressed as,

$$G_s = \frac{M_s}{V_s\ Cy_w} \qquad \text{....(14.40)}$$

where,

Cy_w = density of water at 4°C

Soil density is the mass per unit volume of the soil particle. The density of dominant soil minerals such as quartz and several of felspars is nearly 2.65 g/cc. Consequently, the particle matter is 2.65 g/cc. Alternatively, it may be said that the true specific gravity of soil is about 2.65. It is also known as real specific gravity or particle density. However, the density of soil organic matter usually ranges from 1.3 and 1.5 g/cc.

Dry bulk density (Gb): It is defined as the ratio of the mass of dried particles to the total volume of soil (including particles and pores) and is expressed as,

$$G_b = \frac{W_s}{V_t} = \frac{W_s}{V_s + V_a + V_w} \qquad \text{....(14.41)}$$

The term dry bulk density and apparent specific gravity are often used synonymously. However, the term apparent specific gravity is a dimensionless quantity, while, bulk density is expressed in g cc or mass per unit volume. 1 g of water fills a volume of 1 cc at normal temperatures, the two terms have equal numerical values. The apparent specific gravity is influenced by factors such as structure, texture and compactness of the soil. These are important soil physical properties which influence the water holding capacity and hydraulic conductivity of soils. For example, when the bulk density of medium to fine-textured subsoils exceeds about 1.7 g/cc then the hydraulic conductivity values will be so low that drainage may become difficult.

Total (wet) bulk density (G_t): It is the mass of moist soil per unit volume and is expressed as,

$$G_t = \frac{W_t}{V_t} = \frac{W_s + W_w}{V_s + V_a + V_w} \qquad \text{....(14.42)}$$

Porosity (n): It is defined as the ratio of the volume of pores (voids) to the total soil volume and is expressed as,

$$n = \frac{V_v}{V_t} = \frac{V_a + V_w}{V_s + V_a + V_w} \quad ....(14.43)$$

Porosity is an index of the relative volume of pores. It is influenced by the textural and structural characteristics of the soil. The porosity of sandy soils usually ranges from 35 to 50 per cent, while that of clay soils from 40 to 60 per cent. The more finely divided are the individual soil particles then greater is the porosity.

Void ratio: It is defined as ratio of the volume of pores to the volume of solid also known as relative porosity and is expressed as,

$$e = \frac{V_a}{V} = \frac{V_a + V_w}{V_s} \quad ....(14.44)$$

This index is advantage as over porosity (n). For given void ratio, the total soil volume changes with volume change of voids, whereas for given porosity the volume of pores may change without change in the volume of solids. Therefore, knowledge of the dynamics of pores due to external load is possible with the information on void ratio. The term 'void ratio' is commonly used in engineering works such as compaction of foundations, embankments, etc. The relationships exist between porosity and void ratio to apparent and true specific gravity as under:

$$G_b = G_s\left(1 - \frac{n}{100}\right) \quad ....(14.45)$$

$$G_s = G_b(1 + e) \quad ....(14.46)$$

Soil wetness

It is defined as the relative water content in the soil. It can be expressed in following ways:

Mass wetness: It is defined as the ratio of mass of water to the mass of soil and is also known as the 'gravimetric water content'. It is expressed as,

$$Mass\ wetness = \frac{M_w}{M_s} \quad ....(14.47)$$

This is a commonly used tern and is often known as soil moisture content. It may be expressed in fraction or, in percentage. In mineral saturated soil, it can range between 0.25 and 0.60 (or 25% to 60%). It depends upon the bulk density.

Volume wetness.: It is the ratio of volume of water to the total soil volume and is expressed on volume basis as,

$$Volume\,wetness = \frac{V_w}{V_t} = \frac{V_w}{V_s + V_f} \qquad ....(14.48)$$

It is also known as volumetric water content. The volume wetness is more preferable over mass wetness as the former is more directly adaptable to the computation of fluxes and water quantities. The volume wetness at saturation varies from 40 and 50 per cent in sandy soils to 60 per cent in clayey soils.

Degree of saturation (S): It is the ratio of the volume of water present to the total pore volume and is expressed as,

$$Degree\,of\,saturation = \frac{V_w}{V_f} = \frac{V_w}{V_a + V_w} \qquad ....(14.49)$$

Where, V_v is the volume of voids (or pore spaces). Thus,

$$V_v = V_w + V_a$$

Obviously, on using equation (14.43), we obtain

$$w = S.n \qquad ....(14.50)$$

Further, consider that weight of water in a wet soil sample is W_w and the dry weight of the soil sample is W_d, then the dry weight moisture fraction, W is expressed as,

$$W = \frac{W_w}{W_d} \qquad ....(14.51)$$

The bulk density (or the bulk specific weight or the bulk unit weight) of a soil mass is the dry weight of the soil per unit bulk volume, *i.e.* W_d/V. The specific weight (or the unit weight) of the soil particles is the ratio of dry weight of the soil particles W_s, to the volume of soil particles V_s, *i.e.* W_d/V_s. Thus,

$$G_b.\gamma_w = \frac{W_d}{V} \; i.e.\, V = \frac{W_d}{G_b.\gamma_w}$$

and $$G_s.\gamma_w = \frac{W_d}{V} \, i.e.\, V_s = \frac{W_d}{G_b.\gamma_w}$$

$$\frac{V_s}{V} = \frac{G_b}{G_s} \qquad ....(14.52)$$

Where, γ_w is the unit weight of water and G_b and G_s are the bulk density or apparent specific gravity of soil and the relative density of soil grains, respectively. Further,

$$1-n=1-\frac{V_v}{V}=\frac{V-V_v}{V}=\frac{V_s}{V}=\frac{G_b}{G_s}$$

$$\therefore G_b = G_s(1-n) \qquad ....(14.53)$$

Also, $$w=\frac{V_w}{V}=\frac{W_w/\gamma_w}{W_d/(G_b \cdot \gamma_w)}=G_b\frac{W_w}{W_d}$$

$$\therefore \quad w = G_b W \qquad ....(14.54)$$

and $\quad w = G_s(1-n)W$

Considering a soil of root zone depth *d* and surface area *A* (*i.e.* bulk volume = *Ad*) then the dry weight of soil

$$W_d = V_s.G_s.\gamma_w$$

$$= Ad(1-n).G_s.\gamma_w$$

Therefore, the dry weight moisture fraction, $W=\frac{W_w}{W_d}$

$$=\frac{V_w.\gamma_w}{Ad(1-n)G_s.\gamma_w} \qquad ....(14.55)$$

Therefore, the volume of water in the soil root zone,

$$V_w = WAd(1-n)G_s \qquad ....(14.56)$$

This volume of water can also be expressed in terms of depth of water which would be obtained when this volume of water is spread over the soil surface area *A*.

$$\therefore \text{Depth of water } (d_w)=\frac{V_w}{A}$$

$$\therefore d_{w^s} = G_s(1-n)Wd \qquad ....(14.57)$$

or $$d_w = w.d \qquad ....(14.58)$$

Example 14.6: A moist soil sample has a volume of 490 cm^3 in the natural state and a weight of 7.94 N. The dry weight of soil is 7.36 N and the relative density of the soil particles is 2.65. Determine the porosity, soil moisture content, volumetric moisture content and degree of saturation.

Solution:

$$G_b = \frac{7.26}{490 \times 10^{-6} \times 9810} = 1.51$$

The porosity, $n = 1 - \frac{G_b}{G_s}$

$$= 1 - \frac{1.55}{2.65} = 43.0 = 43\%$$

The soil moisture fraction, $W = \frac{7.84 - 7.26}{7.26} = .0.079$

$$= 7.9\%$$

The volumetric moisture content = $G_b W$

= 1.51 (.0788)

= 12.06%

Depth of saturation, $S = \frac{w}{n} = \frac{12.06}{43} = .2805 = 28.05\%$

Determination of soil bulk density

An uncompacted soil sample of known volume is obtained from the field for determination of bulk density. Core samplers are used to take undisturbed soil samples. The cylinder of the core sampler has a cutting edge, which is driven into the soil and an uncompacted core obtained within the tube. The samples are carefully trimmed at both ends of the core cylinders. Samples are dried in an oven at 105°C for about 24 hr until all the moisture is driven off and the sample weighed again. The volume of the soil core is the same as the inside volume of the core cylinder. The bulk density is determined as,

$$Bulk\ density = \frac{Weight\ of\ soil\ (gm)}{Volume\ of\ soil\ core\ (cc)} \quad ....(14.59)$$

The bulk density of most of soils ranges from 1.0 g/cc (clays) to 1.8 g/cc (sands).

Basic terms pertaining to soil and water

Some of the terms pertaining to soil and water are discussed as under:

Consumptive use (Cu): It is the quantity of water used by the plant to perform its metabolic activities and that lost due to evaporation and transpiration. Since water used in plant metabolism is negligible for all practical purposes, consumptive use is numerically equal to evapotranspiration. It is generally expressed as depth of water in mm.

Effective rainfall (R_e): It is the precipitation falling during the growing period of the crop which is available to meet the evapotranspiration requirements of the crop. It does not include precipitation which is lost by run off and deep percolation below the root zone of the crop.

Carry-over soil moisture: It is the moisture stored in the root zone depth in the soil before the crop is planted. This moisture also contributes to meet the evapotranspiration needs of the plant.

Wilting coefficient: It is the moisture content in percentage by weight at which plants droop or wilt. The plants usually droop when the rate of transpiration (water lost by plants through leaves) exceeds the rate at which the roots can absorb moisture from the soil and transmit it to the leaves. Plants permanently wilt due to lack of water, when the pull of roots is not great enough to get sufficient water in time to prevent wilting. The amount of moisture in the soil at this instant is known as *permanent wilting percentage* and is held with a force of 15 atmosphere.

Irrigation efficiency ($I\eta$): It is the percentage of applied irrigation water stored in the soil and available for consumptive use by the plant. When the water is measured at the entry to the farm, it is known as *farm irrigation efficiency* and when measured at the field, it is called *field irrigation efficiency.*

Irrigation water requirement (I_R): It is the irrigation water required to be applied for the successful growth of crops. It is obtained by dividing the net irrigation requirement by the irrigation efficiency *viz.*

$$I_R = \frac{N_{IR}}{I\eta}$$

14.3 Root Zone Soil-water

Water serves the following useful functions in the process of plant growth:

(i) Germination of seeds,

(ii) All chemical reactions,

(iii) All biological processes,

(iv) Absorption of plant nutrients through their aqueous solution,

(v) Temperature control, and

(vi) Tillage operations.

Crop growth (or yield) is directly influenced by the soil moisture content. Both excessive water (which results in water logging) and deficient water retard crop growth and reduce the crop yield.

Soil water can be divided into three categories as shown in Fig. 14.10.

(i) Gravitational water,

(ii) Capillary water, and

(iii) Hygroscopic water.

***(i) Gravitational water*:** It is the water held by the soil after drains away the excess water under the influence of gravity is drained away. Soon after irrigation (or rainfall) this water remains in the soil and saturates the soil, thus preventing circulation of air in void spaces.

***(ii) Capillary water*:** It is held within soil pores due to the surface tension forces (1/3 – 31 atm) (against gravity) which act at the liquid-vapour (or water-air) interface. Mostly plants receive this water for their growth and development.

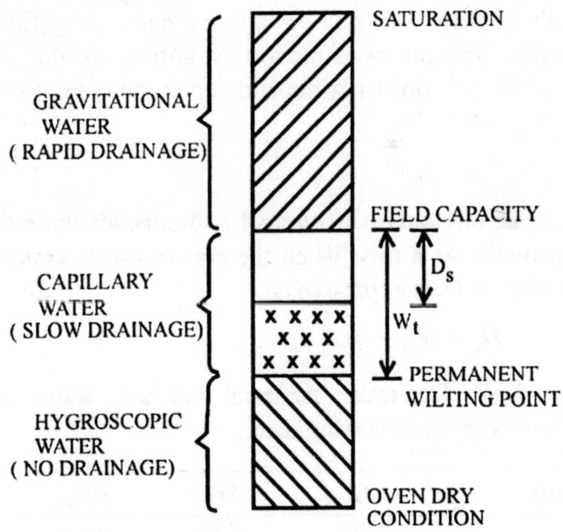

Fig. 14.10: Soil moisture content at different stages

***(iii) Hygroscopic water*:** It is the water held so tightly (tension > 31 atm) as a thin film around each soil particle by the force of adsorption. This water attached to soil particles through loose chemical bonds is known as *hygroscopic water*. This can be removed by heat only and could not be used by plants. However, the plant roots can use a very small fraction of this soil moisture under drought conditions.

When an oven-dry (heated to 105°C for zero per cent moisture content) soil sample is exposed to atmosphere, it takes up some moisture called *hygroscopic moisture.* This can be removed only by heat. If more water is made available, it can be retained as capillary moisture due to surface tension (*i.e.* intermolecular forces). Any water, in excess of maximum capillary moisture, flows down freely and is gravitational water.

The total available moisture (W_t) for a plant (or soil) is determined by the following relation.

$$W_t = (w_{fc} - w_{wp})d$$

where,

d = depth of the root zone (mm).

It is obvious that soil moisture can vary between the field capacity (excess amount would drain away) and the permanent wilting point. However, depending upon the prevailing conditions, soil moisture can be allowed to be depleted below the field capacity (but not below the permanent wilting point in any case), before the next irrigation is applied. This permissible amount of depletion is known as the *management allowed deficit* (D_m). It primarily depends upon the type of crop growing and its stage of growth. Thus,

$$D_m = f_m W_t$$

Where, f_m is, obviously, less than 1 and depends upon the crop and its stage of growth. At a time when the soil moisture content is w, the soil-moisture deficit D_s is expressed as,

$$D_s = (w_{fc} - w).d$$

Example 14.7: Determine the total available water and the soil moisture deficit from the following data.

Soil depth (cm)	G_b	W_{fc}	W_{fp}	W
0-15	1.25	0.25	0.14	0.16
15-30	1.30	0.27	0.13	0.18
30-60	1.35	0.32	0.16	0.23
60-90	1.40	0.32	0.14	0.26
90-120	1.40	0.33	0.16	0.28

Solution:

Depth of soil layers d (mm)	$W_{fc}=G_b W_{fc}$	$W_{wp}=G_b W_{fp}$	$W_t=(W_{fc}-W_{fp})d$	$w=G_b W$	$D_s=(W_{fc}-w)d$ (mm)
150	0.310	0.175	20.625	0.2	15.0
150	0.354	0.169	27.300	0.234	19.5
300	0.432	0.216	64.800	0.3105	32.4
300	0.448	0.196	75.600	0.364	29.4
300	0.462	0.224	71.400	0.392	12.6
		Total	259.725		108.9

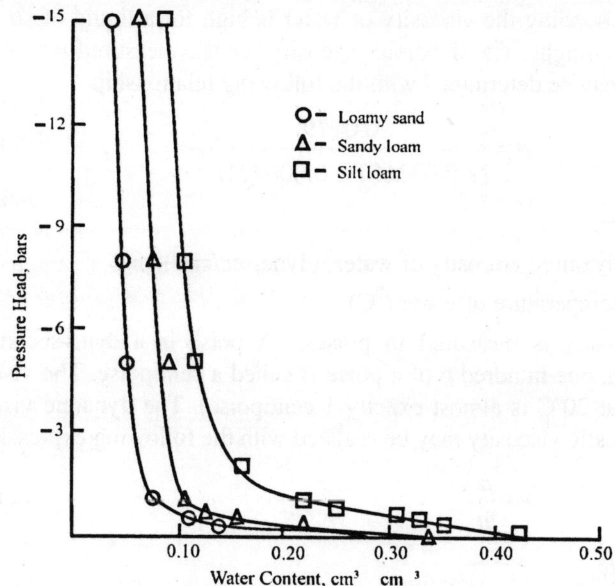

Fig. 14.11: Moisture characteristic curves for soils of different textures [Source: Ghuman and Prihar (1980)]

14.4 Soil Water Retention

Soil is capable of retaining water against the forces of gravity and applied pressures or suctions due to its composition and porous nature. Qualitatively as the amount of applied pressure or suction increases then the amount of water retained by the soil decreases. Consequently, the higher amount of energy is required to extract water from the soil. A plot of soil-moisture convent against suction is known as moisture characteristic curve of the soil (Fig. 14.11). The soils may differ a great deal in their moisture-retention characteristics. Adhesion is the attraction of solid surfaces for water molecules. It is operative only at the solid-liquid interface and hence the film of water established by it is very thin. Cohesion is the attraction of water molecules for each others. This force makes possible a marked thickness of the films of water established by hydration until they attain microscopic size.

Viscosity: Viscosity of a liquid or a gas is its internal friction that brings about resistance to flow. Viscosity is an important factor in infiltration and percolation of water and it also influences the characteristics of fluid flow in open channels and pipes. Viscosity of water, as of most liquids, changes greatly with temperature. Due to hydrogen bonding the viscosity of water is high for a liquid of so low a molecular weight. The dynamic viscosity for the measured temperature of water may be determined with the following relationship:

$$\mu = \frac{0.0179}{1 + 0.03368T + 0.000221T^2} \qquad \text{....(14.60)}$$

where,

μ = dynamic viscosity of water, (dyne-sec/cm^2), and

T = temperature of water (^{0}C)

Viscosity is measured in poises. (A poise is a dyne-second per square cm, one-hundredth of a poise is called a centipoise. The viscosity of water at 20°C is almost exactly 1 centipoise). The dynamic viscosity and kinematic viscosity may be realized with the following expression:

$$\nu = \frac{\mu}{\rho} \qquad \text{....(14.61)}$$

where,

v = kinematic viscosity of water, (cm^2/sec), and

p = mass density of water, (gm/cm^2)

Kinematic viscosity on time and is determined as,

$$v = 0.0022\,t - \frac{1.8}{t}$$

where,

v = kinematic viscosity, (stokes), and

t = seconds

Surface tension: Surface tension is a force pulling inward at the surface of a liquid, tending to make the surface area as small as possible. It is also defined as the force per unit length on the surface of a liquid which opposes the expansion of its surface. Its value for water in air is 72 dynes/sq cm at 25°C. Due to the phenomenon of surface tension liquid molecules are more attracted to each other than to air molecules at an air-water interface. The liquid surface thus behaves like a stretched elastic membrane.

The nature and dimensions of surface tension may be analysed by considering the force that holds a soap bubble together. At an imaginary central plane of a soap bubble, the upper half is held to the lower half by surface tension. Surface tension acts all along the length of the circle that forms contact between the upper half and the lower half of the bubble. Since the surface tension acts on the outside and the inside of the bubble, the total length over which it acts is twice as long as the circumference of the bubble, *i.e.* $4\pi r\sigma$ in which σ is the surface tension and r is the radius of the bubble.

Let us consider that P_b be the pressure on the inside of the bubble and P_a the atmospheric pressure on the outside of the bubble. Hence, the pressure difference (P_b - P_a) acting over the entire area of cross-section of the bubble (πr^2), is the force counteracting the inward pull of surface tension. Equating these two forces, we get,

$$4\pi r\sigma = (P_b - P_a)\pi r^2 \qquad \text{....(14.62)}$$

or
$$P_b - P_a = \frac{4\sigma}{r} \qquad \text{....(14.63)}$$

It is evident from the equation (14.63) that the bigger the radius of the bubble then the smaller is the difference in pressure. Thus, the greater is the curvature of the bubble, greater is the inside pressure within the temperature limits normally encountered in the soil. The surface tension of water does not change appreciably with temperature.

Since g is a constant and σ are also constant for a given liquid at given temperature thus it (equation 14.64) implies that the height of rise is inversely proportional to the radius of the capillary tube. The product δ gh represents the pressure difference (P_0-P) where P_0 is the hydrostatic pressure at the free water surface and P is the pressure on the inside of air-water interface in the capillary. Therefore,

$$P_o - P = \frac{2\sigma}{r} \; or \; P = P_o - \frac{2\sigma}{r} \qquad \text{....(14.64)}$$

It is evident from equation (14.64) that the pressure at the air-water interface is less than that at the free water surface. If the pressure at free water surface at atmospheric pressure is taken as zero, the liquid pressure at the air-water interface is negative and it decreases with decreasing radius of the capillary.

The force of surface tension accounts for most of the water retention in non-shrinking soils while is shrinking soil considerable portion water is retained by high clay content. In these soils, large quantities of water may be retained owing to forces other than surface tension. When a saturated soil loses water without loss of volume then the air enters the system. It results to give rise to air-water interface. The water columns in the soil pores end up in curved surfaces. The size of the pore in which water is held corresponds to the negative pressure below the curved water surface in the pore which balances the extracting force. As more water is lost from the soil and as the remaining water attains a new equilibrium it recedes to pores of smaller size. It implies that decreasing amounts of moisture in soil are associated with increasingly negative pressures. Negative pressure is also called tension or suction to avoid the negative sign. This pressure deficiency, suction or tension is a measure of the unsaturation of soil and denotes a force capable of sucking water into the system.

14.4.1 Capillarity

It is the most important phenomenon of surface tension in soil. Capillarity is defined as the attraction of water into 'hair-like' openings or capillary pores which results in rise of water in small capillary tube or pores. It depend both on cohesion of the liquid and on the adhesion of the liquid to solid wall. The height are rise is determined by the surface tension and the weight of the liquid column.

Let us consider the weight of the column of water elevated to a height (*h*). The vertical component of the surface tension force acting is $2\pi r\sigma \cos^{\theta}$, where, $2\pi r$ line of contact between the liquid and the tube.

σ = surface tension, and

θ, angle of contact.

The weight of the column of water of height (*h*) and diameter (2r) is $g\,\pi r^2 h$.

At equilibrium condition, the vertical component of surface tension is equal to the weight of water column. Thus, equating both these, we get,

$$2\pi r\sigma\cos\theta = g\rho\pi r^2 h$$

$$h = \frac{2\sigma\cos\theta}{g\rho r} \quad \text{....(14.65)}$$

where,

σ = surface tension of water,

r = radius of tube,

ρ = density of water, and

g = acceleration due to gravity, (9.8 gm/sec^2).

If the angle of contact is zero then cosθ is equal 1 and equation (14.65) reduces to,

$$h = \frac{2\sigma}{g\rho r} \quad \text{....(14.66)}$$

The rise of water in a soil is determined when the size of largest effective pore space is known. The relationship between the diameter of soil pores and the height to which water rises above a free water surface is shown in Fig. 14.12.

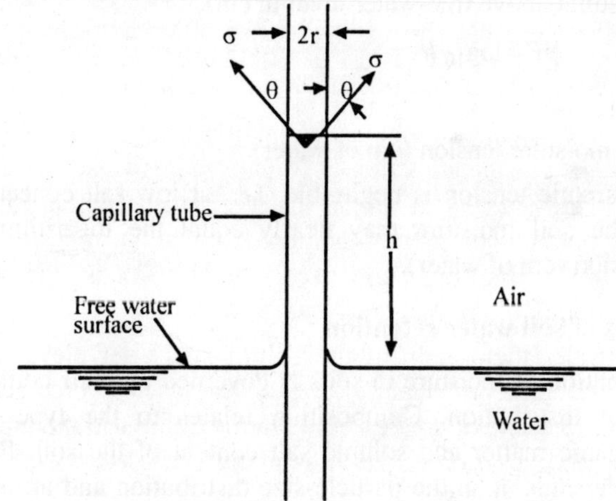

Fig. 14.12: Rise of water in capillary tube

It is evident from the Fig. 14.12 that the larger the diameter of the capillaries then the faster is the rate. While, the rise is so slow there the plants seldom gain from the presence of groundwater in fine texture soils. The capillary rise is much slower in dry range as compared to the wet range.

Soil moisture tension

It is a measure of the tenacity with which water is retained in the soil and shows the force per unit area that must be exerted to remove water from a soil. The tenacity is measured in terms of the potential energy of the water in the soil usually with respect to free water. It is usually expressed in atmospheres, the average air pressure at sea level. (1 atmosphere = 1036 cm of water or 76.39 cm of mercury).

Buckingham (1907) introduced the concept of 'capillary potential' to define the energy with which water is held by soil. This term is not applicable over the entire moisture range. In a wet soil, as long as there is a continuous column of water, it is known as 'hydrostatic potential' while, the intermediate range the term 'capillary potential' is appropriate. In the dry range the term 'hygroscopic potential' would be suitable for the dry range. However, according to Khonke (1968) the term 'soil moisture potential', 'soil moisture suction' and 'soil moisture tension are often used synonymously to cover the entire range of moisture.

pF of soil: Schofield (1935) defined as the logarithm of soil moisture tension and represented by *pF* to logarithm which is an exponential expression of a free-energy difference (based on the height of a water column above free-water level in cm).

$$pF = \log_{10} h \qquad ....(14.67)$$

where,

h = soil moisture tension (cm of water).

if the osmotic tension is negligible, *i.e.*, at low salt concentration, the *pF* of the soil moisture may nearly equal the logarithm of the capillary tension (cm of water).

14.4.2 Forces of soil water retention

The retention of moisture in soils is governed by their composition and pore-size distribution. Composition relates to the type of clay minerals, organic matter and soluble salt content of the soil. Pore-size distribution depends upon the particle-size distribution and arrangement of soil particles. Water is held in the soil due to forces of adsorption at the solid-liquid and surface tension at the liquid-air interfaces.

14.4.3 Surface adsorption

It is the phenomenon in which one substance, in a two-phase system, concentrates close to a surface of the other phase and, thus, exhibits a higher density in that region compared with its density in the rest of phase. These forces keep the soil and water together. In the wet range surface tension is the most important force while, adsorption is the main force in the dry range. Thus, higher the moisture content, the lesser is the attraction of soil for water.

When dry soil is exposed to water vapour then the water molecules are physically attracted to the soil surface and are held tightly. The process of water adsorption on clays is exothermic and the amount of heat liberated (heat of wetting) depends upon the clay content and type of clay mineral and also upon the nature of cations present on the exchange complex. Heat of wetting has been linearly related with cation-exchange capacity of the soil.

The adsorbed water has lower density and a more rigid ice-like structure as compared to free water. The thickness of adsorbed layer is generally believed to be only a few molecular layers.

The electrostatic forces are responsible for adsorption of water on the soil surface. The presence of charges on the clay surface leads to polar adsorption of water with its positive end towards the negative charges on the clay and the negative end away from the surface. If the soil has expanding type of clay minerals then water accommodates itself by pushing apart the clay lattices. This phenomenon is responsible for swelling of clay soils. Such soils absorb and retain water in excess of pore saturation. While draining and drying of such swelling soil after saturation lose considerable quantities of water without entry of air in the system. The loss of water is accompanied by decrease in volume of the soil-water system. When the loss of water exceeds the volume, change in the system, the air begins to enter the soil pores.

The energy tension in a soil depends on the specific surface as well as the structure of the soil and on its solute content. The water retained in the soil through fine capillaries is a function of the surface tension. Salts dissolved in water decrease the free energy of water. Soil water with dissolved salt has lower energy as compared to pure water.

There are two types of interaction which decrease the free energy due to the following reasons:

(i) Solubility of salts, and

(ii) Interaction of water and solid surface.

These factors together decrease the energy of soil water. Thus, the soil water retention and its movement is due to energy effects.

14.4.4 Soil moisture characteristics

It is the relationship between soil moisture content and suction or tension and also known as water retentivity curve of soil. The curve differs widely in different soils (Fig. 14.11). The shape of the curve is a manifestation of the pore size distribution resulting from soil fabric. Soils having more macropores release large amounts of water at low tensions whereas those with higher fraction of micropores release water gradually. It is evident from the Fig. 14.11 that the curves for three sizes of aggregates differ at low tensions and converge into one at tensions beyond 100 cm water. This shows that for these aggregates nearly all the inter-aggregate pores are drained at or below this tension and any water remaining in the system is retained in the intra-aggregate pores. Well-aggregated soil and graded sand yield similar moisture curves except that the former level off at much higher moisture contents as compared to the latter. Moisture-retention curve of a soil also depends upon the direction of change of its moisture content. This phenomenon is called 'hysteresis'. The curve obtained by progressively draining a saturated soil is known as the desorption curve and indicates moisture tension relations of the soil during drying cycle between wetting incidents. While, moisture-tension curve obtained by progressively wetting a dry soil is known as the sorption curve. This curve holds good when soil is wetted by irrigation or rain.

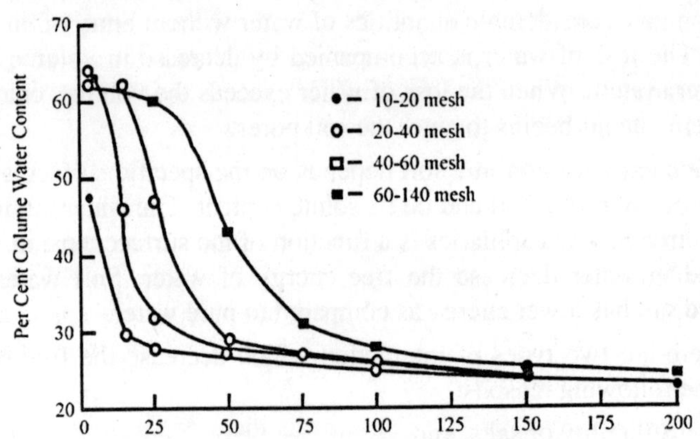

Fig. 14.13: Moisture desorption curves as affected by aggregate size [Source: Sharma and Uehara (1968)]

Moisture characteristic curves are based on a number of retentivity values in the laboratory. These are determined by controlling suction or positive pressures in pressure plate and pressure membrane apparatus at different strips as the soil-water content adjusts to the applied suction.

But under field conditions soil-water content changes and the suction changes accordingly. Suction decreases when soil is wetted by rain or irrigation and increases when soil-water is depleted by drainage, root absorption or evaporation.

14.4.5 Soil moisture constants

Soil moisture moves when subjected to pressure gradients and vapour pressure differences. Thus, soil moisture cannot be said to be constant at any pressure. However, it has been found experimentally that certain moisture contents at particular pressure are of significance in agriculture and these are known as soil moisture 'constants'.

Saturation capacity: When all the pores of the soil are filled with water then it is called as saturation capacity and also known as *maximum water holding capacity.* The tension of water at saturation capacity is almost zero and it is equal to free water surface.

Field capacity: According to Veihmeyer and Henodrickson (1931), it is defined as the amount of water retained in a soil, following saturation, after downward movement has materially ceased. This implies that field capacity is a point of static equilibrium. This situation usually exists one to three days after the soil has been thoroughly saturated by rain or irrigation. At field capacity, the large soil pores are filled with air, the micro pores are filled with water and any further drainage is slow. The field capacity is the upper limit of available moisture range in soil moisture and plant relations. The soil moisture tension at field capacity varies from soil to soil, but in general ranges from 1/10 (sandy soil) to 1/3 (clay soil) atmospheres.

Field capacity is determined by ponding water on the soil surface in an area of about 2 to 5 sq m and permitting it to drain for one to three days, with control of surface evaporation by a polythene sheet or a thick straw mulch. After One to three days the soil is thoroughly wetted then soil samples are collected with an auger from different soil depths at uniform intervals throughout the wetted zone. The moisture content is determined by the gravimetric method.

Moisture equivalent:. It is defined as the amount of water retained by a sample of initially saturated soil material after being subjected to a centrifugal force of 1000 times that of gravity for half an hour.

The moisture equivalent is determined by placing a small sample of soil in the centrifuge and is whirled with a centrifugal force of 1000 times that of gravity for half an hour. The moisture remaining in the sample is determined which gives the value of the moisture equivalent on dry basis.

Permanent wilting percentage: It is the soil moisture content at which plants can no longer obtain enough moisture to meet transpiration requirements; and remain wilted unless water is added to the soil. It is also known as permanent wilting point or wilting co-efficient. At the permanent wilting point the water is held so tightly that roots in contact with the soil cannot remove the water at a sufficiently rapid rate to prevent wilting of the plant leaves. All plants whose root systems thoroughly permeate the soil will wilt at nearly the same soil moisture content when grown in a particular soil in a humid atmosphere.

The moisture tension for permanent wilting point varies from 7 to 32 atmosphere. It depends upon soil texture, the kind and condition of the plants, the amount of soluble salts in the soil solution, and some extent on the climatic environment. In most soils this point has been reported to correspond to retention at 15 bar suction. Although PWP corresponds to 15-bar suction, field crops are known to deplete soil-water to much higher suctions.

The range of soil-moisture content at permanent point is through which plants undergo progressive degrees of permanent or to ultimate wilting. At the permanent wilting point, which is the top of this range, plant growth ceases. Small amounts of water can be removed from the soil by plants after growth ceases, but apparently the water is absorbed only slowly and is enough only to maintain life until more water is available. The moisture content at which the wilting is complete and the plants die (tension >60 bar) is called the ultimate wilting. Though, the difference in the amount of water in the soil between the two points may be small. However, there may be a big difference in tension. For example, Papendick *et al.* (1971) reported that at the time of harvest of dryland wheat, suction in the root-zone ranged from 30 to 45 bar.

According to Philip (1957) the soil-water content at which a plant will appear wilted would depend upon following factors:

(i) Transpiration rate,

(ii) The extent of root system,

(iii) The soil-water transmission rate, and

(iv) The characteristic plant-water potential at which wilting symptoms appear.

Plant recovers from wilting including permanent wilting when soil water content is replenished by irrigation or rain. PWP is considered to be a soil characteristic and its value is generally governed by the amount of clay and type of clay mineral in the soil. Abrol and Khosla (1966) found a significant direct relation between PWP and specific surface area of soil. In certain black-cotton soils, rich in montmorillonite type of clay, its value may be as high as 30% by volume while in certain coarse-textured soils it may be as low as 3-5% by volume.

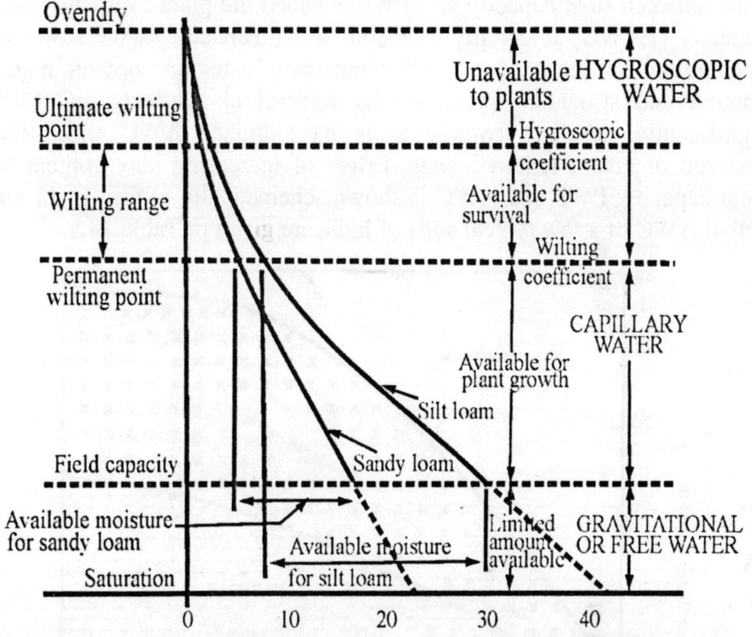

Fig. 14.14: Kinds of water in the soil and difference in available moisture content between a sandy and a silt loam soil. (Note that the silt loam contains more than twice as much readily available water) (Adapted from USDA, SCS Handbook Sec. 45, 1964)

The permanent wilting percentage is determined by growing an indicator plants in container usually in small cans having about 600 g of soil. Sunflower plant is commonly used as the indicator plant. These plants are allowed to wilt and are then placed in a chamber with an approximately saturated atmosphere to test them for permanent wilting. The residual soil moisture content in the container is then determined which gives permanent wilting percentage. The determination of moisture content at 15 atmosphere tension which is the usually assumed value of permanent wilting point in pressure membrane apparatus as suggested by (Richard, 1941).

Available water: It is the water held by the soil between field capacity and permanent wilting point and it is also known as readily available moisture. Water in excess of field capacity drains rapidly and does not stay in soil long enough to be used by plants. Similarly, water used by plants below PWP is generally small.

It is the moisture available for plant use. The storage in the root-zone between field capacity and PWP is called the plant-available-water capacity (PAWC) or simply available water capacity (AWC) of soil. Available-water capacity of soil depends upon its texture, organic matter content and structure. Type of clay mineral also affects the AWC significantly. Medium-textured soils have higher AWC than fine-textured or coarse-textured soils. Effect of increasing clay content on field capacity, PWP and AWC is shown schematically in Fig. 14.15 and actual AWC of a few typical soils of India are given in Table 14.8.

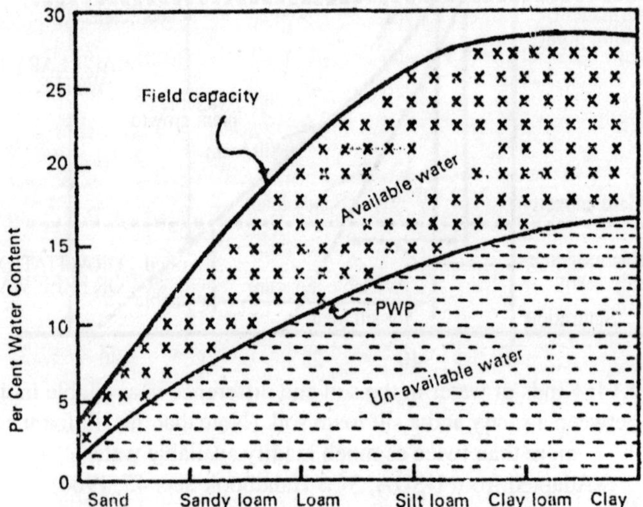

Fig. 14.15: Effect of fineness of soil texture on field capacity, permanent wilting point and available water [Source: Brady (1984)]

Generally, fine-textured soils have a wide range of available moisture as compared to coarse textured soils. While, sandy soils with their larger proportion of non-capillary pore space release most of their water within a narrow range of potential due to the presence of large pores. The three kinds of soil water and the difference in available water between typical sandy loam and silt loam soils. The range of available water holding capacities of different textured soil is summarized in Table 14.7.

Table 14.7: Range of available water holding capacity of soils

Soil type	Percent moisture, based on dry weight of soil		Depth of available water per unit of soil
	Field capacity	**Permanent wilting percentage**	
Find sand	3-5	1-3	2-4
Sandy loam	5-15	3-8	4-11
Silt loam	12-18	6-10	6-13
Clay loam	15-30	7-16	10-18
Clay	25-40	12-20	16-30

(Source: Booher, 1967)

Abrol *et al.* (1968) showed that retention of 0.33 bar suction in low organic matter soils of Haryana was highly positively correlated with per cent silt and per cent 'silt + clay' and negatively correlated with percent sand. Salter *et al.* (1966) found the following relation between AWC and different soil parameters for 20 soils of Great Britain.

AWC (cm/m) = 12.5 - 0.10 (% coarse sand) + 0.103 (% fine sand) + 2.51 (% organic carbon).

Table 14.8: Available water contents of different soils of India as influenced by soil texture

Location	Particle size distribution (%)			Soil texture	Water retained (cm^3 soil)		Available water storage ($cm\ m^{-1}$ soil)
	Sand	**Silt**	**Clay**		**0.33 bar**	**15 bar**	
Alwar	70	23	7	Loamy sand	0.144	0.043	10.1
	55	15	30	Clay loam	0.297	0.120	17.7
Ludhiana (Punjab)	88	7	5	Loamy sand	0.162	0.074	8.8
	79	13	8	Loamy sand	0.232	0.082	15.0
Jabalpur (Madhya Pradesh)	5	35	60	Silty clay	0.444	0.270	17.4
	38	21	41	Clay	0.220	0.120	10.0
Mahboob Nagar (Andhra Pradesh)	18	18	64	Clay	0.474	0.351	12.1
	56	24	20	Silt loam	0.288	0.149	13.9

(Source: Annual Reports, ICAR Project on Improvement of Soil Physical Conditions)

Available-water capacity of a given soil layer in a layer profile is generally greater than the homogeneous profile of the same texture. The AWC is directly related to the depth of root zone.

14.4.6 Terminology

Water intake: Movement of water within soil pertaining to soil water movement. The movement of irrigation or rainwater from the soil surface into and through the soil. It is influenced by several factors such as infiltration and percolation.

Percolation: It is the downward movement of water through saturated or nearly saturated soil in response to the force of gravity. Percolation occurs when water is under pressure or when the tension is smaller than about ½ atmosphere. Percolation rate is synonymous with infiltration rate at saturated conditions.

Inter flow: It is the lateral seepage of water in a relatively pervious soil above a less pervious layer. Such water usually reappears on the surface of the soil at a lower elevation.

Seepage: Seepage is the vertically downward and lateral movements of water into soil or substrata from a source of supply such as a reservoir or irrigation canal. Such water may reappear at the surface as wet pots or seeps or may percolate to join the ground water or may join the subsurface flow to springs or streams. Seepage rate depends on the wetted perimeter of the reservoir or the canal and the capacity of the soil to conduct water both vertically and laterally.

Permeability (k): It is the ability of a formation to transmit water through its pores when subjected to a difference in head.

Coefficient of permeability (k): It is the rate of flow per unit cross sectioned area under unit hydraulic gradients at a specified temperature. Its unit is lpd/m^2 or m/sec. It has dimension of velocity (L/T). The permeability depends upon the grain size distribution, porosity, shape and arrangement of pores, properties of the pore fluid and an entrapped air or gas. It can be expressed as,

$$k = CD^2 \frac{\gamma_w}{\mu} \frac{e^3}{1+e} \quad \text{....(14.68)}$$

where,

C = a constant,

D = effective size of the formation material,

e = void ratio, and

γ_w = unit weight of water at the flow temperature.

Intrinsic or specific permeability (k): It is the property of water bearing medium and is expressed as,

$$K = CD^2 \qquad \text{....(14.69)}$$

Where the constant C summaries the geometrical properties of the porous medium and K has the dimension of L^2.

$$1 \text{ darg} = 0.987 \times 10^{-3} \text{ cm}^2$$

Hydraulic conductivity: It is the effective flow velocity at unit hydraulic gradient and has the dimensions of velocity (LT^{-1}). It is the proportionality factor k in Darcy's law ($v = ki$, in which v is the live flow velocity and i is the hydraulic gradient). The values of k depend on the properties of the fluid as well as it reflects interactions of the fluid with the porous medium, such as swelling of a soil. A soil which has high porosity and coarse open texture has a high hydraulic conductivity value. There are two soils of the same 'total' porosity then the soil with small pores has lower conductivity as compared to the soil with large pores. This is due to the fact of the small pores offer resistance to flow. A soil with pores of many size conducts water faster if the large pores form a continuous path through the profile. Hydraulic conductivity depends almost entirely on structural pores in fine textural soil. If the particles are cemented together to form nearly impermeable layers, it is commonly known as *hardpans*. In other soils, very finely divided or colloidal material expands on absorbing water to form an impervious gelatinous mass that restricts the movement of water.

Hydraulic head: It is the elevation with respect to a reference datum at which water stands in a riser pipe or manometer under a given condition (soil). This will be sum of elevation head, pressure head, and also the velocity head, if the terminal opening of the sensing element is pointed upstream. However, velocity head is negligible due to non-turbulent flow of water in soil. A porous cup may be used for establishing hydraulic contact between the soil water and water in a manometer.

Hydraulic gradient: It is the rate of change of piezometric or hydraulic head with distance.

Hydraulic equilibrium of water in soil: It is the condition for zero flow rate of liquid or film water in the soil. This condition is achieved when the pressure gradient force is just equal and opposite to the gravity force.

14.4.7 Soil water movement

When rainfall occurs or water added at the soil surface enters into the soil and moves down to the lower layers. It continues to move down even when the water supply is cut off or rainfall is stopped at the surface. While, during evaporation the water from lower layers moves towards the soil surface. Similarly, water moves to the plant roots during transpiration and to the drains during drainage of soil. The soil water movement and its rate and director is governed by certain principles as described below.

Movement of soil water caused by difference in potential (ϕ) and it moves from a point of higher potential to that of lower potential. Potential gradients are the driving forces which cause water flow into, through or out of the soil. Total potential of water (ϕ_t) consists of three components as,

$$\phi_t = \phi_h + \phi_z + \phi \qquad \text{....(14.70)}$$

where,

ϕ_h = pressure potential or hydrostatic potential,

ϕ_z = gravitational potential, and

ϕ = osmotic potential due to dissolved salts.

Although temperature and air pressure also influence water potential, their effects are usually small and are ignored. Osmotic potential becomes operative only when the flow occurs through semi-permeable membranes.

Therefore, the movement within the soil occurred only by hydrostatic (ϕ_h) and gravitational (ϕ_z) potential. In terms of the amount of work done in moving a unit amount of water from a standard state to another point against the forces of pressure gravity potential and can be expressed as,

$$\phi = Qgh + QgZ \qquad \text{....(14.71)}$$

where,

Q = density of water,

g = acceleration due to gravity,

h = height of water above the point in question (for saturated soils) and soil water suction for unsaturated soil, and

Z = height above a chosen reference level to the point.

On dividing both sides of equation (14.71) by specific weight of water, we obtain,

$$H = h + Z \qquad \text{....(14.72)}$$

where,

H = hydraulic potential or hydraulic head.

The component 'h' is measured by piezometers in saturated and by tensiometers or other means in unsaturated soils. Hydraulic head (H) has the dimensions of length and is expressed as equivalent height of water column in centimetres or as pressure units, *viz.*, atmospheres or bars.

Movement of water within the soil (saturated)

Movement of water in or within the soil occurs when soil is saturated. In other words, saturated flow in soils occurs when the water is not under tension. All or most of the pores are completely filled with water.

Poiseuille's law: The Poiseuille law expresses the flow of water in a narrow tube,

$$q = \frac{P \pi r^4}{8 l \mu} \qquad \text{.....(14.73)}$$

where,

q = volume of flow per unit time, (cm^3/sec),

P = pressure difference between two ends of the tube of length I (dynes/cm^2),

r = radius of tube, (cm),

l = length of tube, (cm), and

μ = viscosity of liquid, [dyne sec/cm^2 (poises)].

It is evident from the equation (14.73) that the pore size is of outstanding significance. Hence, rate of saturated flow is proportional to fourth power of pore size (radius). Generally, the rate of flow in soils of various textures is in the following order.

Sand > loam > clay

Darcy's law: The water movement in soils is based on Darcy's law, Darcy's law states that the quantity of water passing a unit cross-section of soil is proportional to the gradient or hydraulic head. It is expressed as,

$$q \alpha \left(\frac{h_1 - h_2}{1} \right) a \qquad \text{....(14.74)}$$

or $q = k\,i\,a$

where,

q = volume of flow per unit time, (cm^3/sec),

i = hydraulic gradient, (dimensionless),

a = cross-section of flow area, (cm^2), and

k = hydraulic conductivity, (cm/sec).

The hydraulic gradient is the ratio of difference in hydraulic head, to the distance between the points measured along the path of flow.

$$i = \frac{h_1 - h_2}{l} \qquad \text{....(14.75)}$$

Where, h_1 and h_2 hydraulic heads at points of measurement 1 and 2, respectively, and *l* distance between the points.

Darcy's law is valid for laminar flow. The reynold number (R_n) is defined as,

$$\frac{\rho v d}{\mu}$$

When ρ = density of fluid, *v* = velocity of flow, d = mean diameter of soil particles, and μ = dynamic viscosity of the fluid.

Darcy's law valid as long as the velocity of flow and the size of soil particles are such that the value of R_n is less than 1. This condition is valid for most cases of water flowing through soils and for ground water flow problems involving the flow of water towards wells.

If the hydraulic head is replaced by the difference in total potential (θ) in equation, we obtain velocity of flow as,

$$v = \frac{\kappa \Delta \theta}{l} \qquad \text{.....(14.76)}$$

Where, *l* the distance along the path of greatest change in potential. It is evident from the equation (17.76) that the rate of movement is proportional to the gradient $\frac{\Delta \theta}{\Delta l}$. The negative sign shows that the movement is in the direction of decreasing potential. Equation (14.76) is the generalised expression for soil moisture movement and is analogous to Ohm's law for flow of electricity and to the Fourier's law flow of heat.

The nature of the potential and of the conductivity *k* depends on the soil moisture range in which movement is occurring.

Laplace's equation: This equation is derived based on Darcy's law and the continuity equation to three-dimensional flow of an incompressible fluid (water) through a porous medium (soil) and is expressed as,

$$\frac{\partial^2 \phi}{\partial x^2}+\frac{\partial^2 \phi}{\partial y^2}+\frac{\partial^2 \phi}{\partial z^2}=0 \qquad .....(14.77)$$

It is evident from both the Poisciiille's equation and Darcy's law that the conductivity of water does not depend upon the total amount of pore space. For example, the conductivity of clay, through a porous material, is so low that it makes one of the best confining materials. However, conductivity of sand and gravel depends much upon the sorting and grading of the particles composing the materials which having grains of uniform size being the most permeable. Thus, the value of hydraulic conductivity is not constant but changes with time and management history in most of the soils.

14.4.8 Moisture movement under unsaturated conditions

As water drains from a soil then the larger pores are emptied of water. This shows that contribution of the hydraulic head or the gravitational component to total potential becomes progressively less important and the contribution of the matric potential (Q_m) becomes more important. Thus, for unsaturated flow (equation 14.77) may be governed by the Darcy's law as,

$$v=-k\,\frac{\Delta(Q_m-Q_v}{\Delta l} \qquad .....(14.78)$$

Where, *l* path of greatest change in (Q_m+Q_v).

Therefore, under unsaturated flow conditions is governed by Darcy's law Equation (14.77) with some modifications and qualifications. In unsaturated flow the *k* is considered as a function of water content such as $k(\theta)$, where, θ is the soil moisture content. As the soil moisture content and soil moisture potential decreases, the hydraulic conductivity decreases very rapidly (Fig. 14.16), so that θ_{soil} is 15 bars, *k* is only 10^{-3} of the value at saturation. Philip (1957 a) stated that the rapid decrease in conductivity occurs due to emptying of larger pores first, which greatly decrease the cross-section available for liquid flow.

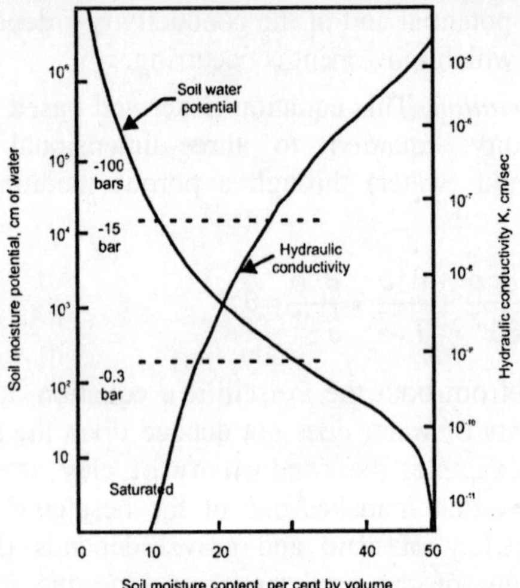

Fig. 14.16: Decrease in hydraulic conductivity k and soil water potential ψ_{soil} with decrease in soil moisture content (Adapted from Philip, 1957 b)

The unsaturated conductivity is a function of soil moisture content as well as number, size and continuity of soil pores. At moisture contents below field capacity, the capillary conductivity is so low that capillary movement is of little or no significance in relation to plant growth. It is evident from different studies that capillary rise from a free water table can be an important source of moisture for plants only when free water is within 60 or 90 cm of the root zone.

The unsaturated flow ceases in sand as compared to fine textured soils at lower tension as the water films lose continuity sooner between the larger particles. The wetter the soil, the greater is the conductivity for water. The range of unsaturated flow in soils of various textures is in the following order,

Sand < loam < clay

This is reverse of the order in saturated flow and unsaturated conductivity in wet range is same or similar as saturated conductivity.

14.4.9 Water vapour movement

Soil water movement in unsaturated soils takes place both in liquid and vapour phases. Water vapour movement is significant only in moist range. While, it is insignificant in high soil water contents (wet range), it increases as void space increases. The continuity of the liquid films is broken and water moves only in the form of vapour at a soil moisture potential of about 15 bar. Diffusion of water vapour occurs due to the vapour pressure gradient as the driving force. The vapour pressure of soil moisture increases with the increase in soil moisture content coupled with temperature and it decreases with the increase in soluble salt content.

14.4.10 Soil moisture

Soil moisture is the ratio of water contained by soil to dry weight of soil. It is expressed in terms of weight basis and volume percentage basis. Measurements of soil moisture are important for scheduling of irrigations and determination of the amount of water to apply in each irrigation. The changes in soil moisture storage with time is important for determination of evapotranspiration. There are also many experimental situations where careful measurement and control of soil moisture is essential for study.

The principal methods of expressing soil moisture are: (i) by the amount of water in a given amount of soil, and (ii) the stress or tension under which the water is held by the soil. The relationship between these two properties throughout the entire moisture range gives a good deal of insight into the physical properties of a soil. This relationship can be shown in the soil moisture characteristic curves (Fig. 14.11)

Soil moisture on weight basis is determined by the following relationship:

$$Soil\ moisture\ content\ (\%) = \frac{W_w - W_d}{W_d}$$

where,

W_w = weight of moist sample, (g), and

W_d = weight of oven dry sample, (g).

The soil moisture on volume basis is determined by the following relationship:

Moisture content, percent by volume = Moisture content percent by weight × Bulk density

The per cent by volume of moisture content obtained by the above relationship is numerically equal to cm of water per metre depth of soil depth.

Methods of soil moisture measurement

There are several methods to measure the soil moisture presently. Out of those some methods are described below:

Gravimetric method

The basic measurements of soil moisture are made on soil samples of known weight or volume. Soil samples are collected with soil auger or sampling tube. The samples are taken from the desired depth at several locations for each soil type. They are collected in small air tight aluminum box. The soil samples are weighed and they are dried in an oven at 105°C for about 24 hr. Then the samples are removed from oven and cooled slowly to room temperature and weighed again. The difference in weight is the amount of moisture in the soil. This method gives most accurate measurement of soil moisture. However, it is used mainly for experimental work. This method consumes more time and is labourious for soil moisture determination.

Example 14.8: An undisturbed soil sample was collected from a field two days after rainfall. The dimensions of the core sampler were 7.5 cm diameter and 15 cm deep. Weight of the core sampling cylinder with moist soil was 2.66 kg and the weight with oven dry soil was 2.51 kg. The weight of the core sampling cylinder was 1.46 kg. Determine the water depth in centimetres per metre depth of soil and available soil moisture.

Solution:

Weight of moist soil = 2.66 - 1.46 = 1.20 kg

Weight of oven dry soil = 2.51 - 1.46 = 1.05 kg

$$\therefore \text{ Moisture content } \frac{1.20-1.05}{1.05}\times 100 = 14.28\%$$

$$\text{Volume of core sampler} = \frac{\pi}{4}d^2\pi = \frac{\pi}{4}\times 7.5\times 7.5\times 15 = 662\,cu\,cm$$

$$\text{Apparent specific gravity} = \frac{\text{Wt. of dry soil in grams}}{\text{Volume soil (sampler) in cu cm}}$$

$$= \frac{1.05\times 1000}{662} = 1.58$$

∴ Moisture content in linear units

= Ap. sp. gr × moisture content

= 1.58 × 14.28 = 22.56 cm per metre depth of soil

Tensiometers

This gives a direct measure of the tenacity with which water is held by soils. Tensiometers measure the matric potential and to determine the soil moisture content.

It consists of a porous ceramic cup filled with water which is buried in the soil at any desired depth and connected to a water filled tube to a manometer or vacuum gauge and install as shown in Fig. 14.17 & 14.18. The scales are generally calibrated in either hundredths of an atmosphere or in centimetres of water. Tensiometers with a mercury manometer are preferred as research tools as it gives greater precision. It is simple and better suited for practical field use.

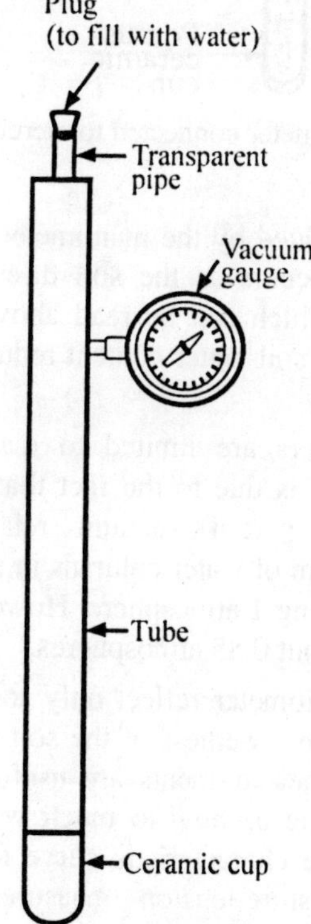

Fig. 14.17: Tensiometer fitted with vacuum guage

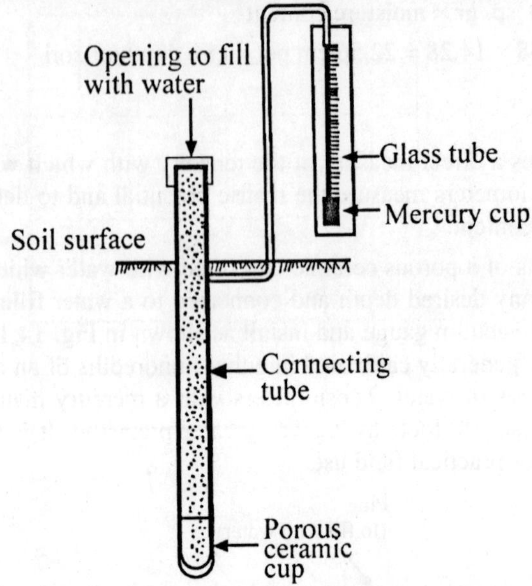

Fig. 14.18: Tensiometer connected to mercury manometer

This pressure is indicated by the manometer or vacuum gauge. An increase in tension that occurs as the soil dries causes the gauge or manometer reading and which can be read above ground, to increase. Conversely, an increase in soil-water content reduces tension and lowers the reading.

Generally, tensiometers are limited to matric suction values of below 1 atmosphere. This is due to the fact that the vacuum gauge or manometer measures a partial vacuum relative to the external atmospheric pressure failure of water columns in macroscopic systems to withstand tensions exceeding 1 atmosphere. However, the useful limit of most of tensiometers is about 0.85 atmospheres.

The readings of tensiometer reflect only soil-moisture tension and does not indicate the relative wetness of the soil surrounding the porous cup. Therefore, tension measurements are useful in deciding when to irrigate and do not indicate as how to much water should be applied. However, special moisture characteristic curve for the particular soil is needed to convert moisture-tension measurements into available-moisture percentages (Fig. 14.19).

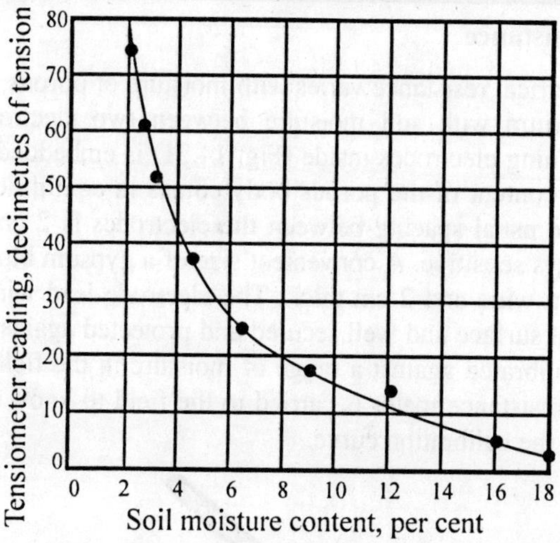

Fig. 14.19: Calibration curve of a tensiometer in a sandy loam soil

Tensiometers are less well suited to use in fine-textured soils in which only a small part of the available moisture is held at a tension of less than 1 atmosphere.

Some experience is required to use a tensiometer and measurements are not reliable due to air entry. Moisture characteristic of common soil types are shown in Fig. 14.20.

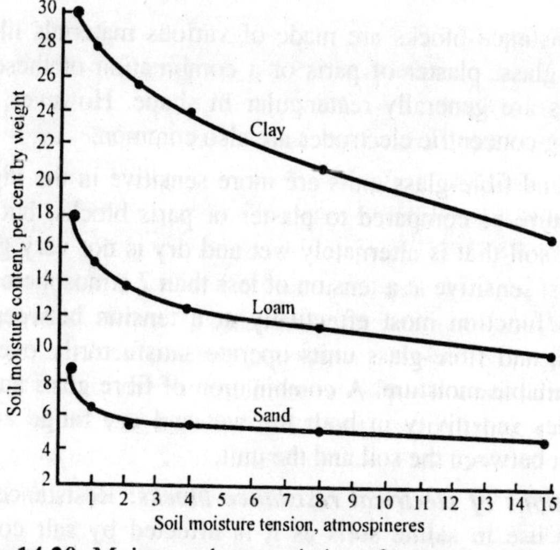

Fig. 14.20: Moisture characteristics of common soil types (Adapted from USDA, SCS Handbook Sec. 15, 1964)

Electrical resistance

The electrical resistance varies with moisture of porous block which is in equilibrium with soil moisture between two electrodes. Porous blocks containing electrodes inside (Fig. 14.21) is embedded in soil then the moisture content of the porous body comes in equilibrium with soil moisture. The usual spacing between the electrodes is 2 cm. A smaller block is always sensitive. A convenient size of a gypsum block is 5.5 cm long, 3.75 cm wide and 2 cm thick. The electrode lead wires are taken out to the soil surface and well secured and protected against rain. These blocks are calibrated against a range of moisture in the field at required depths. The resistance meter is carried to the field to know the moisture content from the calibration curve.

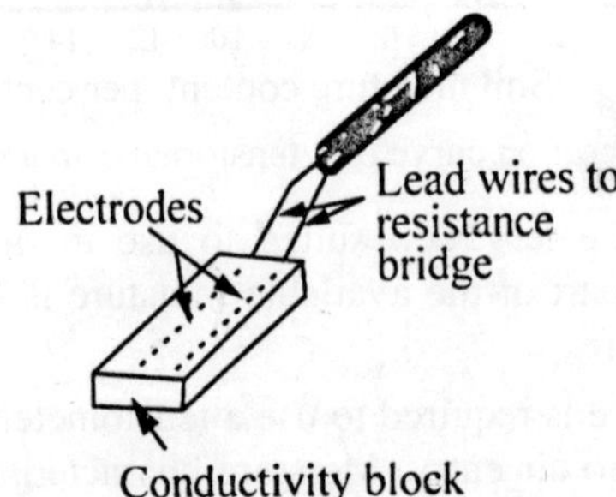

Fig. 14.21: Electrical resistance block

The resistance blocks are made of various materials like gypsum, nylon, fibre glass, plaster of paris or a combination of these materials. These blocks are generally rectangular in shape. However, cylindrical blocks having concentric electrodes are also common.

Nylon and fibre-glass units are more sensitive in the higher ranges of soil moisture as compared to plaster of paris blocks, but often their contact with soil that is alternately wet and dry is not very good. Nylon units are most sensitive at a tension of less than 2 atmospheres. Plaster of paris blocks function most effectively at a tension between 1 and 15 atmospheres, and fibre-glass units operate satisfactorily over the entire range of available moisture. A combination of fibre glass and plaster of paris provides sensitivity in both the wet and dry range and provides good contact between the soil and the unit.

Limitations of electrical resistance blocks: Resistance blocks are restricted to use in saline soils as it is affected by salt concentration. However, gypsum blocks work better even in saline conditions because of buffering action. But nylon and fibre glass blocks cannot be used.

Fibre-glass blocks are more sensitive as compared to plaster of paris. Their readings are also affected by concentrations of fertilizer. In case when the fertilizer is spread in bends than the unit should be placed to one side of the bend. Generally, the resistance blocks should not be used for wet soil moisture range. The measurements of resistance is influenced by change in temperature. Therefore, thermisters should be installed with the blocks to measure the temperature

Neutron moisture meter

The principle of the neutron method is based on the measurement of the number of hydrogen nuclei that are present in a unit volume of soil, their number being a direct function of the number of water molecules contained in that same volume, This measurement is made by inserting a source of fast neutrons and by counting the slow neutrons produced.

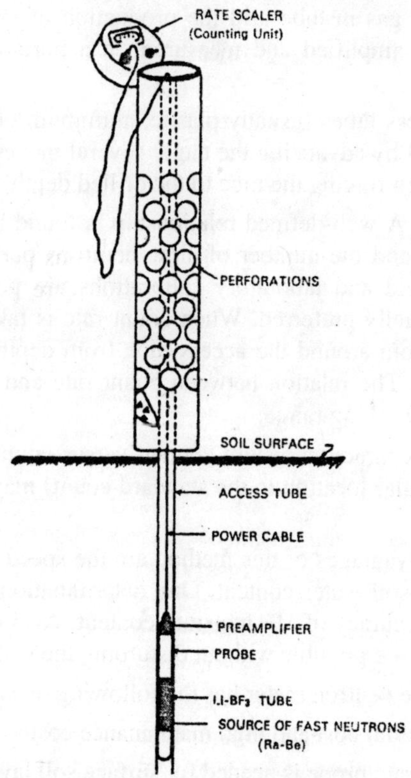

Fig. 14.22: Schematic diagram of neutron moisture meter consisting of the probe unit scaler

The neutron scattering property of hydrogen and slowing of fast neutrons is exploited in neutron moisture meter. Neutrons are slowed down by transfer of energy at each collision. The average number of collisions required to slow a neutron from 2 Me V to thermal energies is 18 for hydrogen. According to Gardner (1965), average number of collisions required for thermalization of neutrons by other elements are 67 for Li, 114 for C, 150 for oxygen and 9A + 6 (A mass number) for nuclei with large mass numbers.

The neutron moisture meter consists of two parts: (i) Probe unit, and (ii) counting unit (Fig. 14.22). Probe contains a fast neutron source usually radium-226 or actinium-227 mixed with beryllium and a slow-neutron detector *e.g.* a tube filled with lithium-enriched boron trifluoride gas. When probe is placed in the soil some of the deflected slow neutrons resulting from collisions with H nuclei will traverse the volume of the detector tube and be captured. Neutron-capture results in transient breakdown of the gas in tube arid the production of short-period pulse which is suitably amplified and measured on a portable counting unit (scaler).

Neutron access tubes (usually plastic, aluminum or steel pipes) are inserted in the soil by advancing the auger several inches beyond the end of the tube and then forcing the tube to the drilled depth.

***Calibration*:** A well-defined relationship is found between the soil-moisture content and the number of slow neutrons per unit time at the detector. Both field and laboratory calibrations are possible, but field calibrations is usually preferred. When count rate is taken, soil samples are also drawn from around the access tube from depths where neutron source is located. The relation between count rate and water content is almost linear for working range.

For higher accuracy, moisture content versus count ratios (the ratio of count at particular location to the standard count) may be used instead of direct count.

The main advantages of this method are the speed and accuracy for determination of soil-water content. One determination takes about one minute with an accuracy of 1% moisture content. Also multiple readings in space and time are possible without disturbing the soil.

However, the neutron meter has the following drawbacks:

(i) High initial cost and high maintenance cost,

(ii) A separate probe is needed for surface soil layer,

(iii) Some inaccuracies may arise in certain soils owing to the presence of Li, B, Cd or other nuclei of high neutron capture or from presence of hydrogen in organic matter and due to this reason it will not work.

Measurement of hydraulic conductivity

Field methods are also available to determine hydraulic conductivity of soil layers below a water-table. Such values are often required for working out the design of drainage system for a given field/area.

Hydraulic conductivity of soil is influenced by size distribution and geometry of its pores. A few continuous large pores can transmit water at faster rate as compared to large number of small pores. Therefore, the hydraulic conductivity of soil is dependent upon its texture, structure, organic matter content, mineralogical composition, exchangeable cations and cropping history.

Table 14.9: Permeability classes based on hydraulic conductivity of soil (After Smith and Browning, 1946)

Permeability class	Hydraulic conductivity (cm hr^{-1})
Extremely slow	< 0.0025
Very slow	0.0025 – 0.025
Slow	0.025 – 0.25
Moderate	0.25 – 2.5
Rapid	2.5 – 25.0
Very rapid	> 25.0

Coarse-textured soils have higher conductivity as compared to fine-textured soils. Similarly, well-aggregated soils transmit water at faster rate as compared to poorly-aggregated soils. Addition of organic matter plays an important role as it promotes soil aggregation. While, puddling aims at reducing hydraulic conductivity of soil by destroying soil structure such that water losses by deep percolation are minimized in puddy cultivation. Soils having montmorillonite type of clay have low hydraulic conductivities because the clay particles swell on wetting and clog the pores and reduce their size. Similarly, if soils having high exchangeable sodium on the exchange complex (alkali or sodic soils) have low hydraulic conductivity because in these soils particles get dispersed on wetting. The low permeability of such soils is increased by replacing sodium on the exchange complex with calcium such as application of calcium sulphate (gypsum) which promotes soil aggregation and porosity.

Hydraulic conductivity of different soils varies widely. According to Smith and Browning (1946) permeability is classified into six classes (Table 14.9) based on hydraulic conductivity.

Unsaturated Hydraulic Conductivity: As the soil-water content (Θ) decreases or the soil water suction (h) increase consequently, the hydraulic conductivity decreases. Thus, the hydraulic conductivity is a function of Θ or h and is often denoted as K (Θ) or K (h). The hydraulic conductivity decreases with decreasing soil-water content due to the following three main reasons.

(i) As the soil-water content decreases then the water is first drained from the larger pores. It is evident that the larger pores transmit water much faster as compared to smaller pores. According to Poiseuille's Law flow rate increases as the square of the size of the pores. For example, if pore size is reduced by one half then flow will be reduced to one fourth.

(ii) As the water drains out of the soil then central part of the pore where the flow is the more ceases to contribute to the flow. Radial distribution of flow velocity in a cylindrical pore obeys the following relation:

$$v = 2v_o \left(1 - \frac{r^2}{r_1^2}\right)$$

where,

v_o = average flow velocity through the pore,

r_1 = radius of pore, and

v = flow velocity at distance r from the centre of the pore

(iii) As the water content decrease then the water-filled porosity decreases and the cross sectional area contributing to the flow decreases.

The value of K (θ) differs widely with soil. Generally, K decreases logarithmically as water content decreases linearly. As θ decrease from field capacity to permanent wilting percentage, K decreases by about 10,000 times. The variation of K with θ for three alluvial soils is shown in Fig. 14.23.

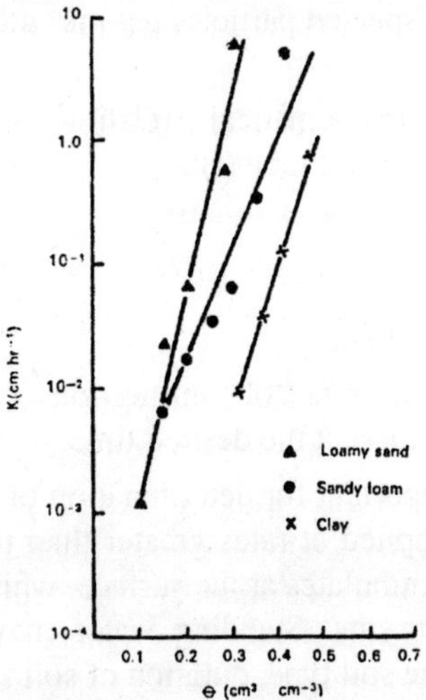

Fig. 17.23: Unsaturated hydraulic conductivity (K) in relation to volumetric soil water content (θ) [Source: Rajpal *et al.* (1977)]

14.5 Dynamics of Field-water

Water in the field is always in moving position and it remains static seldom. It moves from one point to the other. Movement of field water is associated with at least four processes, *viz.* (i) infiltration, (ii) redistribution, (iii) drainage, and (iv) evaporation.

***Infiltration*:** The process of water entry into the soil is referred as 'infiltration' and the water flux into the soil is known as the infiltration rate. The maximum water flux into the soil or the maximum rate at which water can be absorbed by the soil is known infiltration capacity of soil. According to Hillel (1971) the term of infiltration capacity and infiltrability are interchangeable.

When a water head is maintained at the soil surface, the rate of water entry into the soil is more rapid in the beginning and decreases with time. Cumulative infiltration plotted against time yields a curvilinear curve. At long times the infiltration rate tends to approach a constant value, which is known as the terminal infiltration rate or basic infiltration rate. The terminal infiltration rate is approximately equal to the saturated hydraulic conductivity in a homogeneous soil. The infiltration rate decrease with time due to decrease in hydraulic head gradient as the progressive downward movement of water lengthens the wetted zone. Moreover, the soil at the surface gets dispersed by beating action of rain drops or rapid wetting by irrigation (entrapped air ruptures the aggregates). The dispersed particles get into the pores and partly clog them.

The following mathematical relation holds good between cumulative infiltration (I) and time (t):

$$I = a\,t^b$$

where,

a and b are constants.

The instantaneous rate of infiltration can be obtained by differentiating this equation at the desired time.

Infiltrability is important for determination of the hydrological balance of soil as water applied at rates greater than the infiltrability would be lost as runoff or accumulates at the surface, while the runoff results in loss of water and nutrients. Standing water may cause crop damage which depends upon the soil type, duration of soil submergence, crop and its growth stage, and the climatic conditions. Similarly, hydraulic conductivity and infiltrability also depends upon its texture, structure, soil profile and exchangeable cations etc.

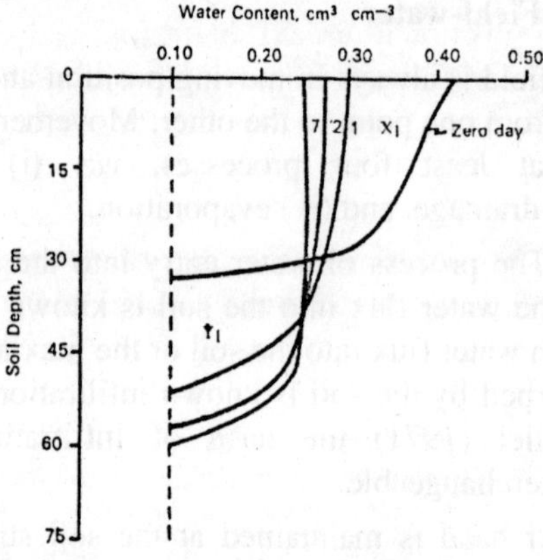

Fig. 14.24: Schematic diagram of water redistribution in soil profile at different times after cessation of water infiltration

Generally, a discontinuity in the texture of soil horizons in the profile decreases the infiltrability. In case a fine-textured layer underlies a coarse layer the infiltration rate decreases due to low permeability of the fine layer. However, if the underlying layer is coarse then the water from the top layer does not move into the coarse soil immediately on reaching the bottom of the fine-textured layer. The advancement of the wetting front stops and the infiltration continues till soil water content at the bottom of the upper layer becomes high enough to permit retention of water in the comparatively large pores of the lower layer. Increase in water content of soil at the bottom of the upper layer decreases the hydraulic gradient across the wetted zone. This leads to decrease the infiltration rate.

Redistribution: When the water supply at the surface is cut off or rainfall stops, the water infiltrated into the soil redistributed itself with time. The upper part of the wetted portion of soil loses water and the lower part of the wetted portion and the contiguous dry layers gain it. If the soil is covered against evaporation then the loss from the upper portion equals to water gain by the lower part (Fig. 14.24).

When the water redistributes in the presence evaporation from the surface the water from the central zone moves in both the directions *viz*., upward and the downward. Distribution of soil water with depth is shown for increasing times after t_0, the time the infiltration ceases. The X_1 area represents the water lost by soil above plane 'd' during the period t_0 to t_1 and the Y_1 area shows the water gained by soil below this plane during the same time interval. The rate of redistribution is initially rapid but decreases with time. Soil-water storage (W) of a layer which is losing water decreases as a function of time (t) and expressed as,

$$W = at^{-b}$$

where,

'a' and 'b' are constants.

Drainage: It is the removal of excess water from the soil so as to ensure adequate soil aeration for crop growth and or maintain proper salt balance in the root-zone. This can be achieved by lowering the water-table in water-logged soils or by not allowing the groundwater to rise beyond a certain threshold limit. During redistribution water moves in the unsaturated soil while, during drainage the water movement occurs in the saturated soil-zone. Adoption of open ditches or tile drains at some depth below the surface lowers the water-table as it induces water to flow from saturated soil to the tile drain. Water in the bottom of the open drain or in the tile is at atmospheric pressure. If we consider the water surface in the drain to be the reference datum in a saturated soil, water at any point above this plane is at a positive pressure equal to the height of the point above the water level in the drain. This hydraulic head gradient produces the driving force for movement of water to the drains (Fig 14.25).

The lines of flow are known as stream lines. The process of water flow to the drain is more complicated than the one directional flow.

According to Luthin (1957) Darcy law is combined with a law of mass conservation which describes the rates of water flow during drainage. The water table is deepest close to the drain and slowest at mid way between two drains.

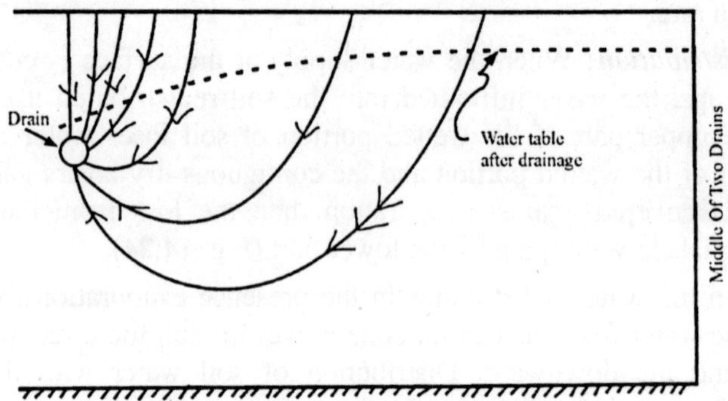

Fig. 14.25: Schematic diagram of pattern of stream lines of water flow to a drain installed in a water-logged land

Darcy Law combined with law of mass conservation describe the rates of water flow during drainage (Luthin, 1957). The water-table is deepest close to the drain and shallowest midway between two drains,

Evaporation: It is process of transformation of water into vapour and its escape into atmosphere. It is expressed as,

$$q = -A \frac{d P_v}{d Z} \quad \text{....(14.79)}$$

Where,

q = evaporative flux,

$\frac{d P_v}{d Z}$ = vapour pressure gradient above the evaporating surface, and

A = transfer coefficient of vapour to the atmosphere.

Evaporation can occur from bare soil, vegetation or free water surface. The following three conditions are essential for evaporation in bare soil (Hillel, 1971):

(i) a continuous supply of heat to convert water into vapour,

(ii) vapour pressure gradient between the bulk in the atmosphere, and the evaporating surface, and

(iii) there must be a continuous supply of water through the soil to its surface.

The soil water potentials reduced due to drying of the surface layers of soil by radiation and wind. If soil in the lower layers has higher water content then potential or hydraulic head gradient is created in the downwards direction which causes the water flows in the upward direction to the surface. The rate of movement of vapour can be determined with the following relation:

$$v = -k(\theta)\frac{dH}{dZ} \qquad \text{....(14.80)}$$

Where,

v = flow rate,

k (θ) = hydraulic conductivity as function of soil-water content,

H = hydraulic head (*i.e.* the sum of gravitational head and hydrostatic head), and

Z = distance in the vertical direction.

Hillel (1971) described several conditions under which evaporation can occur. However, two broad situations of evaporation from soil surface take place are described here *viz.*, (i) in the presence, and (ii) in the absence of a shallow water-table.

(a) In the presence of shallow water table: The evaporation from a soil with a shallow water table increases with evaporativity under isothermal conditions and it attains a steady state in equilibrium with the atmospheric evaporativity. The external conditions govern the suction at the soil surface and more intense the drying the lower will be the soil-water potential at the surface. However, the flux does not increase in direct proportion to the ensuing hydraulic head gradient because as the soil water content decreases, k (θ) decreases very sharply. Therefore, very high rates of evaporation cannot be sustained even from shallow water-table. Sharma and Prihar (1973) proposed the relation between maximum evaporation (Y) with depth of groundwater (X) in sandy loam soil as,

$$X = 149.2 - 150 \log Y$$

where,

X = depth of groundwater (cm), and

Y = maximum evaporation, (mm day^{-1}).

(b) In the absence of water table: The process of evaporation is transient. Evaporation from wet soil can take place in 'three stages.

I stage - One of high constant rate where the soil can supply water fast enough to meet the atmospheric evaporative demand.

II stage - The rate of evaporation decreases with time or with decrease in soil-water content. This stage is called the failing rate stage.

III stage - The third stage is one of reduced evaporation where the evaporation occurs at a low constant rate (Fig. 14.26).

It is evident from Fig. 14.26 that in coarse-textured soil under high evaporativity the first stage is short-lived and the fall in E/E_0 during the second stage is rapid. With increase in fineness of texture and decrease in evaporativity the duration of the first stage gets prolonged and fall in E/E_0 during the second stage becomes more gradual.

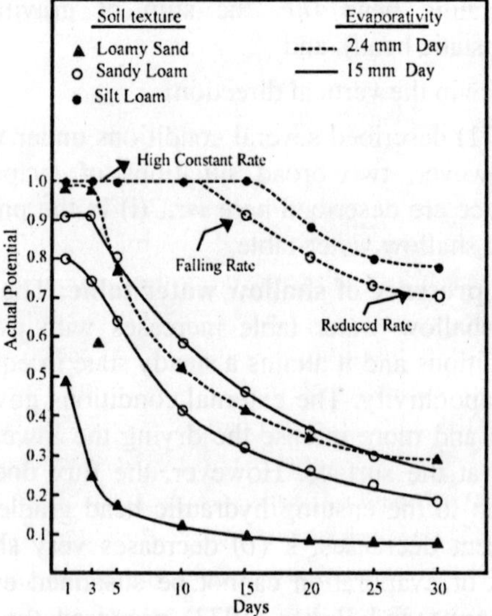

Fig. 14.26: Three stages of evaporation from bare soil as influenced by soil texture and atmospheric evaporativity (Source: Personal communication from Dr. S.K. Jalota, Department of Soils, PAU, Ludhiana)

Rate of water loss is reduced by modification of external conditions in the first stage and by the water transmission properties during the second stage. Potentialities for reducing evaporation losses through cultural practices exist mainly during I and II stage. Shallow tillage creates a dust mulch which acts as barrier to vapour diffusion and use of organic, paper and plastic mulches is often recommended for reducing evaporation losses. The effectiveness of tillage for evaporation reduction depends upon the soil type, ambient environment, time of tillage and tilth and thickness of the tilled layer. Paper and plastic mulches being too costly to be used on field scale. Therefore, straw mulching is generally recommended. The surplus crop residues can be used as surface mulches for reducing evaporation losses from soil inspite of being disposed by burning. Prihar and Arora (1980) found that the amount of evaporation reduction by straw mulch depend upon the rate of mulch, atmospheric evapora-tivity, soil type and time after application of mulch. The amount of evaporation reduction with straw mulch as well as shallow tillage increases up to a point and then starts decreasing. The mulched and unmulched soil may give sence with comparable evaporation losses after long. This is so because with time a dry layer also develops naturally on the surface of untreated soil which reduces the further evaporation losses.

Mulches help to keep the surface layers more moist and to carry-over the moisture in the seed-zone in the event of long dry periods preceding sowing operations.

14.6 Plant-water Relationships

Water constitutes about 80 to 90 per cent of most plant cells and tissues in which there is active metabolism. Therefore, almost every plant process is affected directly or indirectly by water supply. Plant cells forms a continuous liquid phase through the plant from the root hairs to the leaf epidermis. Plant roots take up water and transfer it through the root tissues to the conducting vessels of the plant. These conducting vessels pass it on to the mesophyll tissues of the leaves. Water is evaporated from the leaf through primarily the walls of the sub-stomatal cavities. All these flow are in the liquid phase along a potential gradient. The final transfer of water from the sub-stomatal cavities through the stomata of the leaves is a process of molecular diffusion of water vapour.

The factors influencing the water relations of plants may be grouped into the following:

(i) Soil factors: It includes *viz.*, soil moisture content, texture, structure, density, salinity, fertility, aeration, temperature and drainage.

(ii) Plant factors: It includes type of crop, density and depth of rooting, rate of root growth, aerodynamic roughness of the crop, drought tolerance and varietal effects.

(iii) Weather factors: It includes sunshine, temperature, humidity, wind and rainfall.

(iv) Miscellaneous factors: It includes soil volume and plant spacing, soil fertility, crop and soil management.

14.6.1 Terminology pertaining to crop and water

Critical periods of crop growth: There are specific periods or stages during growing season which mark a particular development in the plant deficiency in available soil moisture content during these periods. Water content adversely affects the ultimate yield of the crop, therefore, these are known *critical periods of crop growth.* For instance, in case of wheat there are following growth stages *viz.*,

Emergence	:	25 days
Crownroot development	:	20 days
Tillering	:	20 days
Jointing	:	20 days
Flowering	:	20 days
Milk	:	15 days
Dough	:	15 days
Total growth period	:	135 days

Of these, emergence, crown root development, tillering and grain filling (which includes flowering, milk and dough stages) are considered as critical periods. It is important that irrigation must be provided during the critical periods in order to remove moisture deficiencies of the soil.

Root zone depth: It is the depth to which roots of a plant penetrate into the soil for development of its plant root system and is also known as effective root zone depth. Adequate soil moisture must be available in the immediate vicinity of the plant roots to enable them to draw water for their growth. Root zone depth mainly depends upon the types of crops and soil. While roots of maize, cotton and wheat penetrate a permeable soil to a depth of 1.20 m, sugarcane can penetrate permeable well-drained soil to a depth of 2 m. It is observed that roots of most crops go much deeper in sandy soils than in clayey soils and root development also varies with crop genotype.

Root zone moisture extraction depth (D): It is the depth in the root zone of the plant from which the plant extracts moisture for its roots development.

It shows the relative amount of moisture extracted from different depths within the crop root zone. The usual extraction pattern in uniform soils with available moisture is shown in Fig. 14.27. It is evident from the Fig. 14.27 that about 40 per cent of the extracted moisture comes from the upper quarter of the root zone, 30 per cent from the second quarter, 20 per cent from the third quarter and 10 per cent from the bottom quarter.

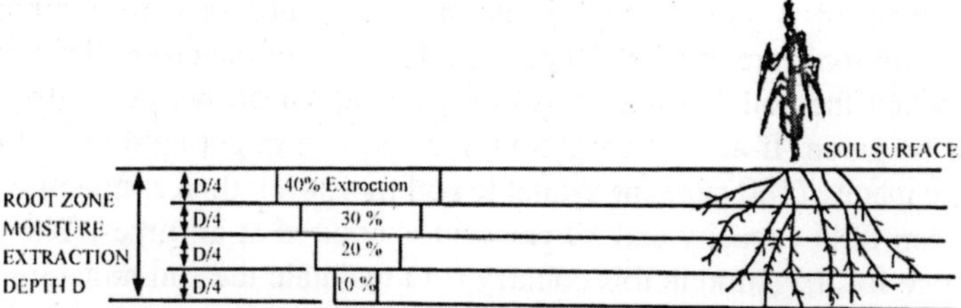

Fig. 14.27: Root zone moisture extraction patter

Values of individual crops fall within a range of ±10 per cent. It may be noted that the root zone moisture extraction depth is numerically equal to the root zone depth. However, for determining irrigation water requirements, the depth of soil is taken as the depth for which roots of an average mature plant are capable of extracting soil moisture to the extent that it would need to be replaced by irrigation, and may not necessarily be the same as the root zone depth of any given plant.

The approximate values of effective root zone depth of some of the common crops grown in deep well drained soils is summarized in Table 14.10.

Table 14.10: Effective root zone depth of some common crops* (grown on very deep, well drained soils)

Rooting characteristics			
Shallow rooted	**Moderately deep rooted**	**Deep rooted**	**Very deep rooted**
Depth of root zone			
60 cm **(1)**	***90 cm*** **(2)**	***120 cm*** **(3)**	***180 cm*** **(4)**
Rice	Wheat	Maize	Sugarcane
Potato	Tobacco	Cotton	Citrus
Cauliflower	Castor	Sorghum	Coffee
Cabbage	Groundnut	Pearl millet	Apple
Lettuce	Muskmelon	Soybean	Grapevine
Onion	Carrots	Sugar beet	Safflower
	Pea	Tomato	Lucerne
	Bean		
	Chilli		

(Source: Gandhi *et al.*, 1970)

Moisture level**:** As crop production suffers due to lack of moisture in the soil, the primary objective of irrigation is to overcome the deficiency by providing moisture at sufficiently low tension within the root zone of a given crop. It is therefore necessary to maintain a minimum moisture level of 50 per cent for most of the crops. This means that when the soil moisture level drops to about 50 per cent, irrigation must be given. If at each irrigation, the soil is brought upto its optimum field capacity (100 per cent available soil moisture), then a moisture level between field capacity and 50 per cent is assured at all times. The usual practice for irrigation in this country is to maintain the soil moisture level during winter season at 50 per cent of available soil moisture level after the first irrigation to the growing crop. During summer season, the rain may bring moisture to 100 per cent over the crop area. In such a case, irrigation should be given only when soil moisture level drops to about 60-70 per cent of the available moisture.

14.7 Water Movement in Soil-Plant-Atmosphere System

The complete path of water from the soil to plant and the atmosphere forms a continuous system which may be analysed by determining the potential difference between soil and atmosphere in contact with root and leaf, respectively. The path of water may be divided into four sequential processes as follows:

(i) the supply of liquid water to the root surface,

(ii) the entry of water into the root,

(iii) the passage of water in the conducting element, and

(iv) the movement of water vapour through and out of the leaves.

The rate of water movement is proportional to the potential gradient and inversely proportional to the resistance to flow at every place. The approximate magnitudes of water potential in the soil-plant-atmosphere system is given in the Table 14.11.

Table 14.11: Approximate magnitudes of water potential in the soil-plant-atmosphere system

Component	Water potential, bars
Soil	- 0.1 -- - 20
Leaf	- 5.0 -- - 50
Atmosphere	- 1000 -- - 2000

(Source: Plaut and Moreshet, 1973)

14.7.1 Terminology pertaining to plant water

***Chemical potential of water*:** It is a measure of the capacity of that substance to work in a system. The chemical potential of water depends on the mean free energy per molecule and the concentration of water molecules.

***Osmotic pressure*:** When a solution is separated from pure solvent by a membrane permeable only to the solvent. This laid to a net flux of solvent into the solution since the chemical potential of the solvent is higher in the pure phase as compared to the solution. This process is called osmosis and the pressure difference which must be applied to the solution to prevent a net flux of solvent is called osmotic pressure. The dehydration is accompanied by an increase in osmotic pressure. However, osmotic pressure is not sufficiently sensitive to be used as an indicator of small changes in water balance.

***Diffusion pressure deficit (DPD)*:** It is defined as the amount by which its diffusion pressure is lower than that of pure water at the same temperature and under atmospheric pressure. The *DPD* of a cell may be regarded as a measure of the pressure with which water would diffuse into the cell when immersed in pure water. The equation for the *DPD* in a cell is expressed as,

$$\text{DPD} = \text{OP} - \text{TP} \qquad \text{....(14.81)}$$

where,

OP = osmotic pressure of the cell contents

TP = turgor pressure within the cell

The *DPD* is a more sensitive indicator of the degree of turgidity.

***Cell water potential*:** It is a measure of the free energy status of water. Under isothermal equilibrium conditions, the various factors involved in cell water relations can be summarized by the following relation:

$$\Psi_{cell} = \Psi_s + \Psi_p + \Psi_m \qquad \text{....(14.82)}$$

where,

Ψ_{cell} = potential of water in the cell

Ψ_s = solutes potential,

Ψ_p = pressure potential, and

Ψ= matric forces

Ψ_s and Ψ_m are both negative while Ψ_p is positive. Ψ_s express the effect of solutes in the cell solution, and Ψ_m express the effect of water-binding colloids and surfaces in the cell. The sum of $\Psi_s + \Psi_p + \Psi_m$ is a negative number, except in fully turgid cells when it becomes zero. In this case the positive pressure potential Ψ_p balances the sum of the negative osmotic and matric potentials (Kramer, 1969). It may be shown that DPD is numerically equal to Ψ_{cell}, but opposite in sign and is expressed as,

$$\Psi_{cell} = - DPD \qquad ....(14.83)$$

The potential of water in a cell is less as compared to pure water. It is negative whereas DPD is positive as it is defined as a deficit.

14.7.2 Indices of plant water stress

Relative leaf water content: Weatherley (1950) proposed that relative turgidity is a good indication of water deficits in plants. It is usually measured in terms of the relative leaf water content (RLWC) and it is the amount of water in the leaf tissue relative to the amount of water held by the tissue at saturation. The RLWC is determined with the following relationship:

$$RLWC = \frac{Fresh\ weight - oven\ dry\ weight}{Fully\ turgid\ weight - oven\ dry\ weight} \times 100 \qquad ....(14.84)$$

Though, the procedure for determination of RLWC is simple but it is not free from errors. Barrs (1986) reviewed several other methods of measuring plant water deficits. The rice leaf discs must float on water for 4 to 6 hr before water deficit is completely removed (Tomar and Ghildyal, 1973c).

The value of RLWC for wilting and other physiological characters depends on type of genotype. Gardner and Ehlig (1965) recorded that plants having 75-85% RLWC exhibited visible wilting. Tomar and Ghildyal (1973a) found that RLWC of rice dropped down to 33% before the crop was permanently wilted. Several important physiological processes are influenced by RLWC of the tissue. Hsiao (1973) recorded that expensive growth was the most sensitive to plant water stress which attributed to the essential role of turgor pressure in sustaining cell enlargement. Decrease in RLWQ also reduces net photosynthesis (Shaw and Laing, 1966) mainly through reduced flow of CO_2 due to stomatal closure (Cox and Boersina, 1967; Turner, 1969) as shown in Fig. 14.28.

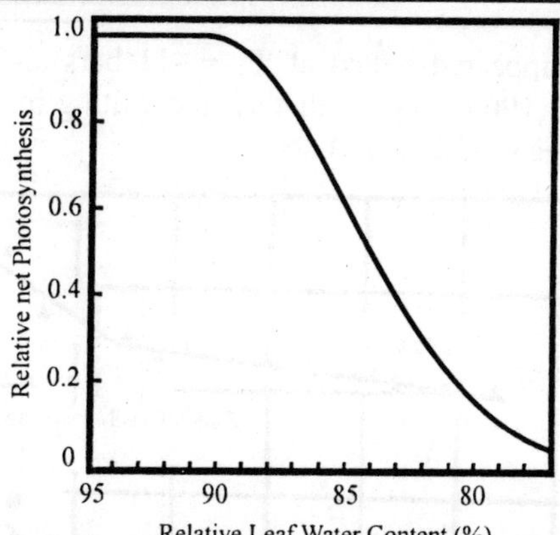

Fig. 14.28: Net photosynthesis as a function of relative leaf water content (RLW) (Source: Shaw and Laing, 1966)

Leaf water potential: It is a measure of energy status of plant water and is a good index of physiological and biochemical phenomena in plant. The leaf water potential (ψ_l) is determined as,

$$\Psi_l = \Psi_m + \Psi_p + \Psi_o \quad \text{....(14.85)}$$

where,

Ψ_m = matric,

Ψ_p = pressure, and

Ψ_o = osmotic potential.

The contribution of Ψ_m to Ψ_l is negligible, therefore, Ψ_m is lumped with ψ_p itself and equation is rewritten as,

$$\Psi_l = \Psi_p + \Psi_o \quad \text{....(14.86)}$$

Varshney (1980) proposed that the term xylem water potential is used to define the tissue water status and osmotic component is ignored as it is very small. Leaf water potential may be determined with thermocouple psychrometer (Tomar and Ghildyal, 1978) and xylem water potential with pressure chamber (Pal and Varade, 1974; Singh *et al.*, 1980). Pressure chamber technique is more rugged and takes only a few minutes for each determination. Hsiao *et al.* (1980) pointed out some common errors that may creep in and should be avoided during determination of Ψ_l.

The leaf water potential decreased sharply as RLWC decreased below 100%, in the intermediate RLWC range the drop in, ψ_l gradual, but below RLWC of 35% the drop in Ψ_l was again very sharp (Fig.

14.29). The plant appeared wilted at $\Psi_l \approx$ -11 bars and Ψ_p = 3 bars. Gardner and Ehlig (1965) showed that visible wilting in several species occurred in the range of -12 to -16 bars.

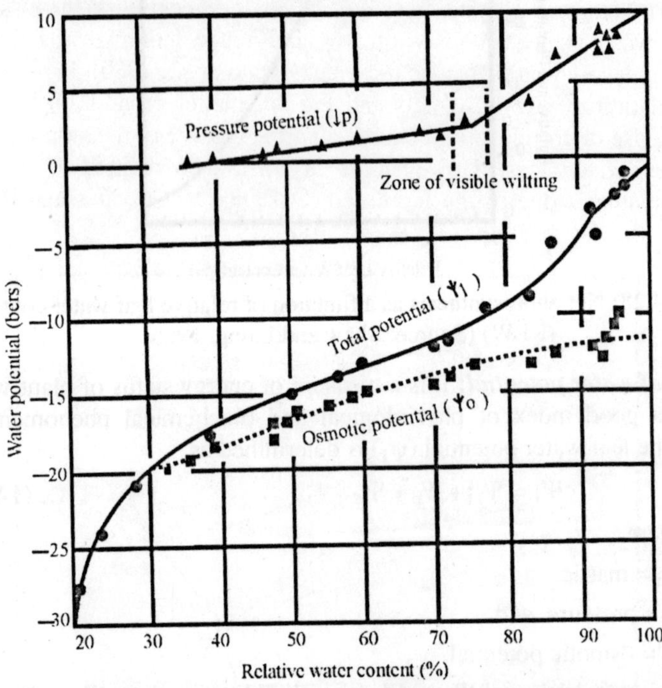

Fig. 14.29: Relationship of plant water potential with relative leaf water content in rice [Source: Tomar and Ghildyal (1973a)]

***Leaf diffusion resistance*:** The stomata are fully open and offer little resistance to the diffusion of carbon dioxide and water vapour into and out of the leaf in the presence of daylight and absence of water deficiency. Thus, the transpiration proceeds at potential rates. The stomatal apertures are regulated by guard cells and tissue water status. Plant water deficit does not seem to affect the stomatal apertures up to a point. However, the deficit exceeds a certain threshold value then stomata begin to close partially and may close completely when stress reaches critical level. The reduction in the stomatal apertures increases the resistance to diffusion of water vapour out of the leaf. Consequently, the transpiration slows down. Several researchers (Szeica and Long, 1969; Van Bavel, 1967; Sandhu and Horton, 1977) have reported that leaf diffusion resistance (LDR) is closely related to plant water status (Fig. 14.30). Varshney (1980) observed that the LDR was 3.8 sec cm^{-1} at

0900 hr while, delaying irrigation by 6 days after the normal irrigation date than in well-watered maize. The LDR increased to 32.4 sec cm^{-1} at 1400 hr.

The stomata operate as safety valves in checking water loss when tissue water status becomes too low and hence minimize the development of severe water stress in the plant (Hsiao *et al.*, 1980). However, the assimilation of CO_2 also falls and the dry matter production is slowed down. Use of an-transpirant causes a partial closure of the stomata which reduces water loss. However, it is also accompanied by reduced photosynthesis due to the increased resistance to CO_2 passage into the leaf.

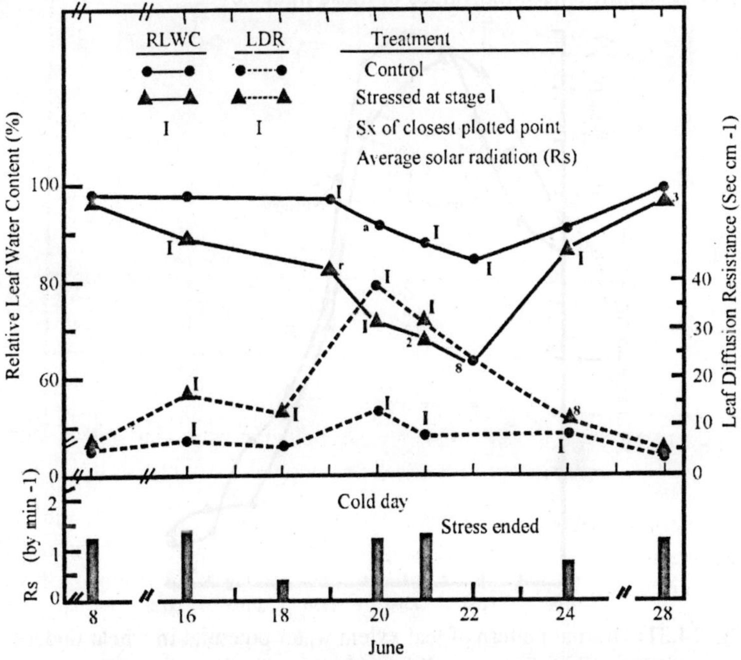

Fig. 14.30: Variations in relative leaf water content and leaf diffusion resistance as influenced by stress at boot stage in oats [Source: Sandhu and Horton (1977)]

A crop water stress index (CWSI): It is computed on the basis of linear regression of TC-TA with VAD under water stressed on non-stressed condition. Inessence it is ratio of difference TC-TA at given VAD with linear of TC-TA to that of difference of upper limit of TC-TA with lower limit of TC-TA (Idso *et al.*, 1977; 1981 a,b and Jackson *et al.*, 1981).

Crop water stress has also been estimated using spectral characteristics and micro-waves (Chopra and Katra, 1988) and chlorophyll fluorescence (Seemann, *et al.*, 1986).

14.7.3 Diurnal changes in plant water stress

All the three parameters of plant water deficit such as RLWC, ψ_l and LDR, exhibit large diurnal changes because these, are affected by soil moisture status in the root-zone and environmental conditions. Beside these, the age of the plant and position of the sampled leaves on the plant also influences the values of these indices.

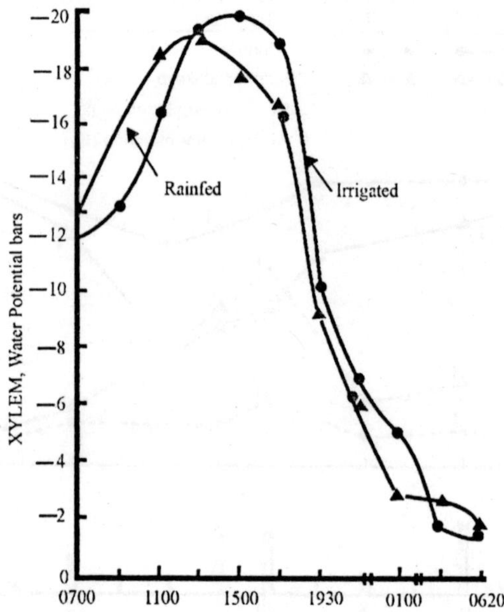

Fig. 14.31: Diurnal pattern of leaf xylem water potential in wheat under varying soil moisture conditions [Source: Singh et al. (1980)]

Transpiration during night is almost negligible. The predawn values of ψ_l seem to be in equilibrium with soil-water potential. But as the day advances, the stomata open up and evaporative demand increases, the ψ_l decreases till early afternoon when the cycle reverses (Fig. 14.31). Singh *et al.* (1980) recorded that changes in ψ_l are sharper under high soil-water status. The changes are quicker in the morning and evening hours and relatively small during major part of the day.

Hsiao *et al.* (1980) proposed that if crop growth and CO_2 assimilation are of major concern then ψ_1 must be taken at mid-day when it is the minimum. They also pointed out that if ψ_1 reaches a value corresponding to visible wilting at mid-day and the CO_2 assimilation may only, be slightly depressed for one or two hours that day. But if mid-day ψ_1 is five or six bars lower than the threshold ψ_1, the assimilation and dry matter production are greatly reduced.

14.7.4 Water uptake by plants

Water moves in the soil-water plant-atmosphere system due to the difference in the energy status of water from one region of the system to the other. It always moves from a region of higher potential to that of a lower potential. The greater the potential gradient then faster is the rate of water movement. The other factor that influences the rate of movement of water in the system is the conductance of the pathway. The drop in the energy status of leaf water caused by loss of water on its exposure to radiation, wind and vapour-deficient atmospheres. It is transmitted through the stem to the root in contact with soil water. This creates the gradient needed, for flow of water from the soil to the leaf through/root and stem of the plant. Flow of water in soil-plant-atmosphere system is generally expressed in terms of resistance which is the reciprocal of conductance. The flow of water is expressed as,

$$q = \frac{\psi_s - \psi_1}{R_t} \qquad \text{....(14.87)}$$

where,

q = rate of water uptake,

ψ_s = water potential in soil,

ψ_1 = water potentials in leaf, and

R_t = total resistance offered to water flow by the system

The total resistance from soil to leaf has three components arranged in series. Therefore, equation 14.87 can be rewritten as,

$$q = \frac{\psi_s - \psi_1}{R_{soil} + R_{root} + R_{shoot}} \qquad \text{....(14.88)}$$

When the water content of leaf *is* not changing then rate of water uptake is equal to the rate of transpiration (T). Therfore, substituting T for q and rearranging the terms, we obtain,

$$\psi_1 = \psi_s - T\,(R_{soil} + R_{root} + R_{shoot}) \qquad \text{....(14.89)}$$

It indicates that ψ_l does not have a unique relation with ψ_s. The conductance of soil decreases or the resistance increases sharply with decrease in soil wetness. Additionally, the soil resistance is influenced by root length density (cm root cm^{-3} soil). The lower the root density, greater is the soil resistance to water flow.

Gajri (1980) found that root density had a large influence on ψ_l, and by inference on water uptake, at a given ψ_s.

14.7.5 Evapotranspiration losses

Water losses front the cropped fields consisting of evaporation from the soil surface and transpiration from the plant surfaces are collectively termed as evapotranspiration (ET). Evaporation from soil surface constitutes a variable fraction of the total ET loss depending upon the extent and architecture of crop canopy and the frequency of soil wetting by rain and/or irrigation. Peters and Russel (1959) observed that in a widely spaced maize crop planted in large lysimeters, direct evaporation from soil constituted approximately 50% of the evapotranspiration. In thickly growing crops which receive water at long intervals, the evaporation losses may be a rather small fraction of the total ET. Under field conditions it is difficult to measure separately the evaporation and transpiration losses.

Under non-limiting soil water conditions ratio of actual to potential transpiration T_a/T_P was found to be linearly related to square root of leaf area index (LAI) of sorghum between 0.1 and 2.7 as,

$$\frac{T_a}{T_p} = -0.21 + 0.71(LAI)^{0.5} \quad ....(14.90)$$

For LAI greater than 2.7 the relative transpiration was independent of leaf area. Under limiting soil-water conditions, however, the relative transpiration was significantly affected by average soil-water content of the root zone, rooting density and LAI (Kumar, 1981). The relative transpiration is not only a function of soil-water status of the root-zone but also of LAI, rooting density and evaporative demand.

14.7.6 Water stress and plant growth

As plant-water decreases it causes loss of turgidity is the closing of stomata. Since stomata are the pathways for the diffusion of CO_2 into the leaf, the closure of stomata reduces the rate of photo-synthesis along with a reduction in transpiration rate. Thus, the plant-water status has direct relation with plant growth. A decrease of 80% in the net photosynthetic rate (NPR) of oats was observed when a severe water stress was imposed at boot and grain-formation stage (Sandhu and Horton, 1977).

Table 14.12: Effect of root density on leaf water potential (at 1000 to 1100 hr) of full-grown wheat on sandy-loam soil

Available soil water fraction in 0-180 cm profile	Root weight in 180-cm profile (mg cm^{-2} surface)	Leaf water potential (bars)
0.77	4.9	-17.1
0.75	9.0	-15.5
0.73	11.5	-13.6
0.66	12.1	-13.3
LSD (.05)	-	1.4

(Source: Gajri and Prihar, 1985)

Ritchie (1976) summarized the effects of soil-water deficits on plant growth and observed that during vegetative growth a decrease in soil-water availability was first reflected in decreased expansive growth without any decrease in photosynthesis or transpiration. This reduces the sink size of active leaves for the storage of photosynthates which are then preferentially transported to root and root growth is stimulated at the expense of shoot growth. If plants grown under field conditions where large soil volumes are available for rooting it showed a higher degree of adaptation to soil-water deficits. An adapted plants reduced their water requirements through senescence as the soil water decreased.

References:

Abrol, I.P. and Khosla, B.K. 1966. Surface area-a rapid measure of wilting point of soils. *Nature*, Lond. 212: 1382.

Abrol, I.P., Khosla, B.K. and Bhumbla, D.R. 1968. Relationship of texture to some important soil moisture constants. Geoderma 2: 33-9.

Arkley, R.J. 1963. Relationship between plant growth and transpiration. *Hilgardia* 34: 559-84.

Aronovici, V. S. 1955. Model study of ring infiltrometers performance under low initial soil moisture, *Soil Science Soc.* Amer. Proc., 19: 1-6.

Arora, V.K. and Prihar, S.S. 1983. Regression models of dryland wheat yields from water supplies in Ustifluvent in Punjab, India. *Fld Crops Res.* 6: 41-50.

Barrs, H.D. 1968. Determination of Water deficits in plant tissues. In *Water Deficits and Plant Growth.* pp. 236-347. Kozlowski T.T. (ed) Academic Press, New York.

Bassett, D.L, and Tinney, E.R. 1960. Water surface configuration and velocity of advance in hydraulic laboratory tests. Proc. of the ARS-SCS Workshop on Hydraulics of Surface Irrigation, ARS 4143 11-18, U.S. Dept. of Agri.

Bayer, L.D., Gardner, W.H. and Gardner, W.R. 1976. *Soil Physics.* Wiley Eastern Ltd, New Delhi.

Bondurant, J.A. 1957. Developing a furrow infiltrometer. *Agricultural Engineering*. 38, No. 8, 602-604.

Bouyoucos, C.J. 1953. More durable plaster of paris moisture blocks. *Soil Sci.* 76: 447-51.

Brady, N.C. 1984. The Nature and Properties of Soils. pp. 103. Macmillan Publishing Co., N. York.

Briggs, L.J. and Shantz, H.L. 1913. The water requirements of plants. I. Investigations in great plants in 1910 and 1911. *U.S. Dep. Agric. Bur. Pl. Indus. Bull.* 284.

Burgy, R.H. and Luthin, J.N. 1957. Discussion on the paper "A test of the single and double ring type of infiltrometers", Trans. Amer. Geophysical Union. Vol. 38. No.20.

Chow, V.T. 1964. Hand book of applied hydrology. New York, McGraw Hill Book Company.

Cope, F. and Trickett, E.S. 1965. Measuring soil moisture. *Soils Fert.* 28: 201-08.

Cox, L.M. and Boersma, L. 1967. Transpiration as a function of soil temperature and soil water stress. *Pl. Physiol.* 42: 550-56.

Criddle, W.D., Davis, S., Pair, C.H. and Shockley, D.G. 1956. Methods for evaluating irrigation systems. Agricultural Hand Book No. 82, p. 24. U.S. Dept. of Agri. Soil Conservation Service.

Davis, J.R. 1961. Estimating rate of advance for irrigation furrows. Trans. Amer. *Soc. of Agri. Eng.*, 4, No. .1, 53-54, 57.

Davis, J.R. and Fry, A.W. 1963. Measurement of infiltration rates j irrigated furrows. Trans. Amer. *Soc. of Agri. Eng.*, 6, No. 4, 318-319.

Day, P.R. 1965. Particle fractionation and particle size analysis. In Black, C.A. (Editor-in-Chief). Methods of Soil Analysis. pp. 545-66. Part 1, Monograph No. 9, ASA Inc., Publisher, Madison, Wisconsin, U.S.A.

Denmead, O.T. and Shaw, R.H. 1962. Availability of soil water to plants as affected by soil moisture content and meteorological conditions. *Agron. J.* 54:385-90.

Dewit, C.T. 1958. *Transpiration and crop Yields.* Institute of Biology and Chemical Research on Field Crops and Herbage. Wageningen, Netherlands, Mededeling 59.

Finkel, H.J. and Nir, D. 1960. Determining infiltration rates in an irrigation border. *J. of Geophysical Research*, 65, No. 7, 2125-2131.

Fok, Yu-Si. and Hansen, V.E. 1966. One dimensional infiltration into homogenous soil, 3. Irrigation and Drainage Division, Proc. Amer. *Soc. of Civil Eng.* 92, No. JR 3, 35-48.

Gajri, P.R. and Prihar, S.S. 1985. Rooting, water use and yield relations in wheat on loamy sand and sandy loam soils. *Fld Crops Res.* 12: 115-32.

Gardner, W. and Widtsoe, J.A. 1921. The movement of soil moisture. Soil Science. 11: 215-232.

Gardner, W.H. 1965. Water content. In *Methods of Soil Analysis.* Vol. 1. ed. (C.A. Black). Am. Soc. Agron. Madison, Wisconsin, U.S.A.

Gardner, W.R. and Ehlig, C.F. 1965. Physical aspects of internal water relations of plant leaves. *Pl. Physiol.* 40: 705-13.

Ghuman, B.S. and Prihar, S.S. 1980. Chloride displacement by water in homogeneous columns of three soils. J. Soil Sci. Soc. Am. 44: 17-21.

Haise. H.R., Phelan, J.T., Lawson, L. S. and Shockley, D.G. 1956. The use of cylinder infiltrometers to determine the intake characteristics of irrigated soils. ARSSCS, ARS. 417, p 10. U.S. Dept. of Agri.

Hanks, R.J. 1974. Model for predicting plant yield as influenced by water use. *Agron. J.* 66: 660-65.

Hanks, R.J. and Rasmussen, V.P. 1982. Predicting crop production as related tq plant water stress. *Adv. Agron.* 35: 193-215.

Hansen, V.E. 1955. Infiltration and soil water movement during irrigation. *Soil Science.* 79, No. 2, 93-105.

Hansen, V.E. 1960. Mathematical relationship expressing the hydraulics of surface irrigation. Proc. of the ARS-SCS Workshop on Hydraulics of Surface Irrigation. ARS-41-43: 29-35, U. S. Dept, of Agri.

Hillel, D. 1971. Soil and Water, Physical Principles and Processes. Acad. Press New York.

Horton, R. E. 1940. Approach towards a physical interpretation of infiltration capacity. Proc. *Soil Science Soc. Amer.* 5; 399-417.

Hsiao, T.C. 1973. Plant responses to water stress. *Ann. Rev. Pl. Physiol.* 24: 519-70.

Hsiao, T.C., O' Toole, J.C. and Tomar, V.S. 1980. Water stress as a constraint to crop production in tropics. In *Priorities for Alleviating*

Soil-Related Constraints to Food Production in the Tropics. Int. Rice Res. Inst., Los Banos, Philippines.

Idso, S.B., Jackson, R.D. and Reginato, R.J. 1977. Remote sensing of crop yield. *Science*, 196-19.

Idso, S.B., Jackson, R.D. Indder, P.J. Jr., Reginato, R.J. and Hatfield. 1981a. Normalizing the stress degree day parameter for environmental variability. *Agric. Meteorol.* 24: 44-45.

Idso, S.B., Reginato, R.J., Reincosty, S.C. and Hotfield, J.L. 1981b. Determining soil induced plant water potential depression in alfalfa by means of infra red thermometry. *Agronomy J.*, 73: 826-30.

Jackson, R.D., Idso, S.B., Reginato, R.J. and Pinter, P.J. Jr. 1881. Canopy temperature as a crop water stress indicator water resources res. 17: 1133-1138.

Jensen, M.E. 1968. Water consumption by agricultural plants. In *Water Deficits and Plant Growth II.* Kozlowski T.T. (ed) Academic Press, New York.

Kumar, Ramesh. 1981. Effect of Soil Water Content, Rooting Density and (LA!) on Warier Uptake by Wheat. M.Sc. Thesis, Punjab Agric. Univ., Ludhiana.

Lal, R. 1969. Hydraulics of water flow in irrigation borders and furrows. Unpublished Ph. D. thesis, Indian Institute of Technology, Kharagpur.

Lehane, J.J. and Staple, W..I. 1965. Influence of soil texture, depth of soil moisture storage and rainfall distribution on wheat yields in south-western Saskatchewan. *Can. J. Soil Sci.* 45: 207-19.

Lewis M. R. 1937. The rate of infiltration of water in irrigation practice. Trans. Amer Geophysical Union 18: 361-368.

Lewis, M. R. and Mime, W. R. 1938. Analysis of Border irrigation. *Agricultural Engineering.* 19, No. 6, 267-272.

Luthin, J.N. (ed). 1957. Drainage of Agricultural Lands. Monograph 7, Am: Soc. Agron., Madison, Wisconsin, USA.

Minhas, B.S. Parikh, K.S. and Srinivasan, T.N. 1974. Towards the structure of a production function for wheat yield with dated inputs of irrigation water. *Water Resour. Res,* 10:383-93.

Pal, D. and Varade, S.B. 1974. Fabrication and performance of pressure-bomb apparatus for the measurement of leaf-water potential. *Indian J. Agric. Sci.* 44: 259-61.

Papadakis, J.A. 1941. A rapid method for determining soil moisture. *Soil Sci.* 51: 279-81.

Papendick, R.L., Cochren, V.L. and Woody, W.M. 1971. Soil water potential and water content profiles with wheat under spring and summer rainfall. *Agron. J.* 63: 73 1-34.

Penman, H.L. 1948. Natural evaporation from open water, bare soil and grass. *Proc. Roy. Soc. Lond.* 193: 120-45.

Penman, H.L. 1956, Evaporation: An introductory survey. *Neth. J. Agric. Sci.* 4: 9-29.

Peter, D,B. and Russel, M.B. 1959. Relative water losses by evaporation and transpiration in field corn. *Proc. Soil Sci. Soc. Am.* 23: 170-73.

Philip, J. R. 1954. An infiltration equation with physical significance. *Soil Science.* 77: 153-157.

Philip, J. R. The theory of infiltration. The infiltration equation and its solution. *Soil Science.* 83:345-357.

Philip, J.R. 1957. The physical principles of soil Water movement during the irrigation cycle. 3rd Congr. Int. Comm. Irrig Drain. Quest. pp. 125-28, 154.

Prihar, S.S. and Arora, V.K. 1980. Crop Response to Mulching with Crop Residues. PAU Deptt of Soil Bull. 33 pp.

Prihar, S.S. and Sandhu, B.S. 1967. Studies on immersion method of soil moisture determination. *J. Res.* 4: 157-61.

Prihar, S.S. and Sandhu, B.S. 1968. A rapid method of soil moisture determination. *Soil Sci.* 105: 142-46.

Prihar, S.S., Khera, K.L., Sandhu, K.S. and Sandhu, B.S. 1976. Comparison of irrigation schedules based on pan evaporation and growth stages in winter wheat. *Agron. J.* 68: 650-53.

Rajpal, Jagan Nath and Khanna, S.S. 1977. Experimental verification of the recently proposed calculation method for two parametric diffusivity-water content function. *J. Indian Soc. Soil Sci.* 25: 95-106.

Ram, M. 1963. Surface irrigation. Trans. 5th Congress of international Commission on Irrigation and Drainage, Tokyo. 4: 16.465-16.483.

Rasmussen, V.P. and Hanks, R.J. 1978. Spring wheat model for limited wiater conditions. *Agron. J.* 70: 940-44.

Richards, L.A., Gardner, W.R. and Ogata, G. 1956. Physical processes determining water loss from soil. Proc. *Soil Sci. Soc. Am.* 20: 310-14.

Ritchie, J.T., Burnett, E., and Henderson, R.C. 1972. Dryland evaporative flux in sub-humid climate. III. Soil water influence. *Agron. J.* 64: 168-73.

Ritchie, J.T. 1976. Soil water deficits effects on plant processes. *Agron. Abstr.* 11 pp

Sahay, J. 1960. Infiltration rate measurement. Unpublished B. Tech. Thesis, Kharagpur, Indian Institute of Technology.

Salter, P.J., Berry, G. and Williams, J.B. 1966. The influence of textured on moisture characteristics of soil. III. Quantitative relationship between particle size, composition and available-water capacity. *J. Soil Sci*. 17: 93-8.

Sandhu, B.S. and Horton, M.L. 1977. Response of oats to water deficits. I. Physiological characteristics. *Agron. J.* 69: 357-60.

Schiff, L. 1957. Discussion on the paper "A test of single and double ring type of infiltrometer". Trans. Amer. Geophysical Union. Vol. 38, No. 20.

Seemann, J.R., Johan, W. Paston, J.S. Berry, J.A. 1986. Temperature and leaf osmatic potential of factors in the accumulation of photosynthesis to high temperature in desert plants. *Plant Physiols*. 80: 916-30.

Sharma, D. R. and Prihar, S.S. 1973. Effect of depth and salinity of ground water on evaporation and soil salinization. *Indian .J. agric. Sci.* 43: 45-49.

Sharma, M.L. and Uehara, G. 1968. Influence of soil structure on water relations in low humic latosols water retention. Proc. *Soil Sci. Soc. Am.* 32: 765-70.

Shaw, R.H. and Laing, D.R. 1966. Moisture Stress and plant response. In *Plant Environment and Efficient Water Use.* pp.73-94. Pierre W.H. *et al.(ed).* Am. Soc. Agron. Wisconsin, U.S.A.

Shull, H. A. 1961. Bypass furrow infiltrometer. trans. Amer. Soc. of *Agri. Eng*. 4, No. 1, 15-17.

Singh, N.T., Brar, G.S. and Aggarwal, G.C. 1980. Effect of Soil moisture regime on xylen water potential of three wheat varieties. *Symp. Indian Soc. Soil Sci.,* Karnal (India).

Skinner, R.J. 1975. Moisture determination in soil. In Soil Physical Conditions and Crop Production. Ministry of Agriculture. *Fisheries & Food Tech. Bull.* 29: Her Majesty's Stationery Office, London.

Smerdon, E. T. 1963. Subsurface water distribution in surface irrigation. J. Irrigation and Drainage Division. Proc. *Amer. Soc. of Civil Eng*. 89, No. IR-1, 1-15.

Smith, R.M. and Browning, D.R. 1946. Some suggested laboratory standards of sub-soil permeability. *Proc. Soil Sci. Soc. Am.* 11: 21-26.

Sur, H.S. and Singh, N.T. 1973. Water-retention and transmission characteristics of soils of the Pilot Project Patiala. *J. Res.* PAU, Ludhiana 10: 190-98.

Szeica, G. and Long, I.F. 1969. Surface resistance of crop canopies. *Water Resour. Res.* 5: 622-33.

Tisdall, A. L. 1951. Antecedent soil moisture and its relation to infiltration. *Australian Journal of Agricultural Research*. 2: 342-348.

Tomar, V.S. and Ghildyal, B.P. 1973a. Short note on the wilting phenomenon in plants. *Agron. J. 65:* 514-15.

Tomar, V.S. and Ghildyal, B.P. 1973b. Internal leaf water status and transport of water in rice plant. *Agron. J.* 65: 861-65.

Tomar, V.S. and Ghildyal, B.P. 1973c. Determination of water status of the rice plant. *Indian J. Agric. Sci* 43: 754-59.

Turner, N.C. 1969. Stomatal resistance to transpiration in three contrasting canopies. *Crop Sci.* 9: 303-07.

Van Bavel, C.H.M. 1967. Changes in canopy resistance to water loss from alfalfa induced by soil water depletion. *Agric. Meteorol.* 4: 165-76.

Varshney, M. L. 1960. Laboratory studies on one-dimensional movement of water in unsaturated soils. Unpublished M. Tech. Thesis. Indian Institute of Technology Kharagpur.

Varshney, O.P. 1980. Effect of Soil Water Stress on Water Potential and Photosynthetic Activity in Wheat and Maize Leaves. Ph.D. Thesis, Punjab Agric. Univ., Ludhiana.

Veihmeyer, F.J. and Hendrickson, A.H. 1931. The moisture equivalent as a measure of the field capacity of soils. *Soil Sci.* 32: 18 1-93.

Warrington, G.R. 1970. Transpiation and Dry Matter Response to Atmospheric Humidity, Matric Suction and Fertility. Ph.D. dissertation, Utah State Univ. Logan, Utah.

Weatherley, P.E. 1950 Studies in the water relations of the cotton plant. I. The field measurements of water deficits in leaves. *New Phytol.* 49: 81-97.

Wilde, S.A. and Spyridakis, D.M. 1962 Determination of soil moisture by immersion method. *Soil Sci.* 94: 132-33.

Wotsan, K. K. 1959. A note on the field use of a theoretically derived infiltration equation. *J. Geophysical Research*. 64: 1611-1615.

EXERCISES

14.1. Classify the soils.

14.2. Differentiate the following

(a) Soil texture & Structure

(b) Infiltration & accumulated Infiltration

(c) Soil & mass wetness

(d) Gravitational & capillary water

(e) Saturated capacity & field capacity

(f) Permeability & Hydraulic conductivity.

(g) Percolation & seepage

14.3. Explain soil water relationships

14.4. Discuss the factors affecting infiltration.

14.5. Describe different methods for measurement of infiltration rate.

14.6. Define the following

(a) Consumptive use

(b) Carry over soil moisture

(c) Wilting coefficient

(d) Dynamic viscosity

(e) Moisture equivalent

(f) Root zone depth

14.7. Explain capillarity

14.8. Explain soil water movement

14.9. State the Darcy's law & it limitations.

14.10. State the Poiseuille's low.

14.11. Discuss the water vapory movement.

14.12. Describe the method for measurement of soil moisture

14.13. Discuss the use of tensiometer & it limitations.

14.14. Describe the method for measurement for Hydraulic conductivity.

14.15. Explain dynamics of field soil water.

14.16. Explain plant water relationships.

14.17. Discuss the water movement in soil-plant atmospheres system.

14.18. Explain the different indices plant water stress.

14.19. What is the evapotranspiration losses?

14.20. Explain water uptake by plants.

14.21. Determine the soil moisture content, by gravimetric method. The following data are given.

Weight of moist soil, including box = 180.04 gm

Weight of oven dry soil, including box = 157.84 gm

Weight of moisture box alone = 42.2 gm

14.22. A tube well is having a capacity of 4500 L/hr operates 18 hours each day during the irrigation season. How much area can be irrigated if the irrigation interval is 20 days and depth of irrigation is 6 cm?

14.23. Determine the size of the irrigation stream required to irrigate an area of 20 ha in a region having no effective rainfall and where the peak consumptive use is 5 mm/day. The irrigation system is operated for 12 hr each day. The field irrigation efficiency is 50 per cent.

14.24. An area of 2 hectare was irrigate in 20 hr with a stream of 40 L/sec. Depth of root zone was 1 m and available moisture holding capacity is 16 cm/m. Irrigation was given at 50% of available moisture was depleted. Water application efficiency was 60%. Determine the water storage efficiency.

14.25. Fit the Kostiakov equation Values of depth to infiltration at different times is given as under:

S. No.	Elapsed time, t mm	Accumulated depth of infiltration, d cm
(1)	(2)	(3)
1	4	0.97
2	8	1.43
3	12	1.75
4	16	2.05
5	20	2.22

■■■■

Chapter 15

WATER REQUIREMENT ESTIMATION FOR CROPS

15.1 Introduction

The estimation of the water requirement (WR) for different crops is one of the basic needs for scheduling irrigation crop planning on a farm and for the design and planning of any irrigation project based on water demand and water availability. Water requirement may be defined as the quantity of water, regardless of its source, required by a crop or diversified pattern of crops in a given period of time for its normal growth under field conditions at a place. Water is needed mainly to meet the demands of evapo-transpiration (ET) and the metabolic activities of the plant, both together known as consumptive use (C_u, or U). Since the water used in the metabolic activities of the plant is negligible, ET is practically considered equal to C_u. Water requirement, includes the losses due to ET (or C_u) plus the losses during the application of irrigation water (unavoidable losses) and the quantity of water required for special operations such as land preparation, transplanting, leaching, etc. Therefore, water requirement is expressed by the following relation:

$$WR = IR + ER + S \quad ....(15.1)$$

where,

IR = Irrigation requirement,

S = Soil profile contribution, and

ER = Effective rainfall.

The field irrigation requirement of a crop refers to the water requirement of crops exclusive of effective rainfall and contribution from soil profile, and it may be expressed as:

$$IR = WR - (ER + S) \quad ....(15.2)$$

15.2 Important Definitions

Consumptive use: It is defined as the quantity of water lost in evaporation, transpiration and that used by the crop in metabolic activities.

Evaportanspiration: It is the total water lost due to transpiration from a crop and evaporation from the soil for a particular area during a specified time.

Water requirement: It is defined as the water needed for raising a crop in a given period. It includes consumptive use and other economically unavoidable losses and that applied for special operations, such as land preparation, transplanting, leaching of salts etc.

Irrigation requirement: It is defined as the quantity of water, required to be applied for successful growth of crops, exclusive of precipitation, groundwater, contribution and other natural resources.

Critical period: It is the period during which yield of crop is affected adversely due to inadequate moisture.

Effective rainfall: It is that part of the precipitation which forms part of the water requirement of crops.

Duty of water(s): It is relationship between the irrigation water and the area of crop that matures fully with the amount of water is known as duty of water. It is generally expressed as area per unit rate of flow.

Delta(s): It is the depth of water required excluding rainfall for the full crop period.

Potential evaportanspiration (PET)

It is defined as the evaportanspiration that occurs when the ground is completely covered by actively growing vegetation and there is no limitation of the soil moisture. It may be considered to be the upper limit of evaportanspiration for a crop in a given climate.

Seasonal consumptive use (U)

The total amount of water used in evaporation and transpiration by a crop during the entire growing season is called seasonal consumptive use. It is expressed as depth of water in cm or volume in hectare-cm per hectare. Seasonal consumptive use values are required to evaluate and determine seasonal irrigation water supplies.

Peak period consumptive use

The average daily water use rate during the few days of the highest consumptive use of the season is called the peak period use rate. This is the design rate to be used in planning an irrigation system. The peak-use period for various crops in a given area may occur at different times in the crop season.

Reference transpiration (ET_0)

It is evapotranspiration from the reference crop as a hypothetical crop with an assumed height of 0.12 m with a surface resistance of 70 sm^{-1} and an albedo of 0.23 closely resembling the evaporation from an extensive surface of uniform green grass (Allen *et al.,* 1988).

Evaporation (E)

Evaporation is the process during which a liquid changes into a gas. The process of evaporation of water in nature is one of the basic components of the hydrological cycle by which water changes to vapour through the absorption of heat energy. This is the only form of moisture transfer from land and oceans into the atmosphere.

Transpiration (T)

Transpiration is the process by which water vapour leaves the living plant body and enters the atmosphere. It involves continuous movement of water from the soil into the roots, through the stem and out through the leaves to the atmosphere. The process includes cuticular transpiration or direct evaporation into the atmosphere from moist membranes through the cuticle, and stomatal transpiration or outward diffusion of water vapour previously evaporated from imbibed membranes into intercellular spaces within the plant into the atmosphere through the stomata.

Transpiration ratio

The effectiveness of the plants for use of water is generally expressed in terms of its transpiration ratio. It is defined as the amount of water transpired by a crop in its growth to produce a unit weight of dry matter.

Evaportanspiration (ET) or consumptive use (Cu or U)

The water use by crops includes evaporation and transpiration into one term evaportanspiration (ET), as it is difficult to separate these two losses in cropped fields. The term consumptive use is used to designate the losses due to evaportanspiration and the water that is used by the plant for its metabolic activities. Since the water used in the actual metabolic processes is insignificant (about 1% of ET or less) the term consumptive use is generally taken equivalent to ET. It thus includes all water consumed by plants plus the water evaporated from bare land and water surfaces in the area occupied by the crop. The relative amount of direct evaporation from the land and water surfaces and transpiration depends usually on the extent of ground cover.

Methods for estimation of evapotranspiration

There are following methods for estimation of evapo-transpiration:

(i) Direct measurement of evapotranspiration,

(ii) From evaporation data, and

(iii) Estimation from climatological data

15.3 Direct Measurement of Evapotranspiration

There are following methods for direct measurement of evapotranspiration -

(i) Lysimeter, (ii) field experimental plots, (iii) soil moisture depletion studies, and (iv) water balance method.

These methods yield very reliable values of ET provided elaborate installations and precise measurements are made. The methods are, however, costly, laborious and time consuming.

(i) Lysimeter

Lysimeters are large tanks filled with soil and supported on some type of weighing mechanism. Estimation of reference evapotranspiration (ET_0) involves the growing of crops in large lysimeters installed in cropped fields and measuring their water loss and gains. The soil and crop conditions in these lysimeters should be as close to the surrounding field conditions as possible. From irrigation point of view, lysimeters are set up to enable the operator to measure the water balance, *i.e.* water added, water retained by the soil, and water lost through evaporation, evapotranspiration and deep percolation. These measurements involve weighing which may be made with scales or by floating the lysimeters in water or a suitable heavy liquid in which case the change in liquid displacement is computed against water loss front the lysimeter. This technique is useful for measurement of field water losses provided the lysimeters are of appropriate size and necessary precautions are taken in the installations. For short crops, the lysimeter should be at least one cubic metre in volume (Mcllroy, 1957) and for tall crops *viz.*, sugarcane it should be four cubic metres or more in volume (Chang, 1968). The lysimeters must be surrounded by a fairly large guard area of the same crop with similar moisture conditions in order to minimize the clothsline effect. If the data recorded are to bear close relation to the crop water losses in the surrounding field. Thornthwaite (1954) suggested that a guard area of 4 ha may not be large enough (Chang, 1968). Water is applied in measured amounts to the lysimeter, as irrigation is applied to the surrounding cropped area. The overflow and deep percolation, if any, are measured. The water received either through irrigation or precipitation, excluding overflow and deep percolation, constitutes the water used by the crop in the process of evapotranspiration. Reference evapo-transpiration rate is determined from weight changes of the lysimeter, using the water balance equation.

$$ET_0 = \text{Weight change} + \text{water added} - \text{percolation} \qquad ....(15.3)$$

Though lysimeters provide the means of precise and direct measurement of the amount of water supplied to and lost by the crop however, it encounters a number of problems. The major limitations are the reproduction of physical conditions such as temperature, water table, soil texture and density etc., within the lysimeter comparable to those outside in the field. To ensure proper drainage, the bottom of an isolated soil column will often require the artificial application of a moisture suction equivalent to that present at the same depth in the natural soil (Coleman, 1946). In some cases the soil temperature in lysimeter is raised to such an extent that the air conditioning of the whole system becomes a necessity (Chang, 1968).

Types of lysimeter

Lysimeters are generally classified as weighing type and drainage type (Fig. 15.1). In drainage type lysimeters, the inflows and drainage are measured, but changes in storage within the soil are not measured. Weighing type lysimeters: It includes an evaluation of each element in the water balance equation. The changes in storage are measured by weighing the soil block. Weighing is accomplished either manometrically or by a suitable scale system.

Example 15.1: A weighing type lysimeter was installed at Dryland Farming Research Station, Arjia, Bhilwara in wheat field to measure the evapotranspiration of the wheat crop. The manometer reading on a particular day showed that the lysimeter lost 5.5 kg weight during 24 h period in a day. The lysimeter had an area of 1 m^2. Determine the average evapotranspiration per hour on that day.

Solution

Loss of water by evapotranspiration = Loss of lysimeter weight = 4.5 kg in day of 24 h = 5.5 x 1000 = 5500 cm^3

$$\text{Evapotranspiration per hour per cm}^2 \text{ area} = \frac{\textit{Water loss by evapotranspiration}(cm^3)}{\textit{Lysimeter area }(cm^3)\times \textit{period}(h)}$$

$$= \frac{55}{10\times 24} = \frac{55}{240}$$

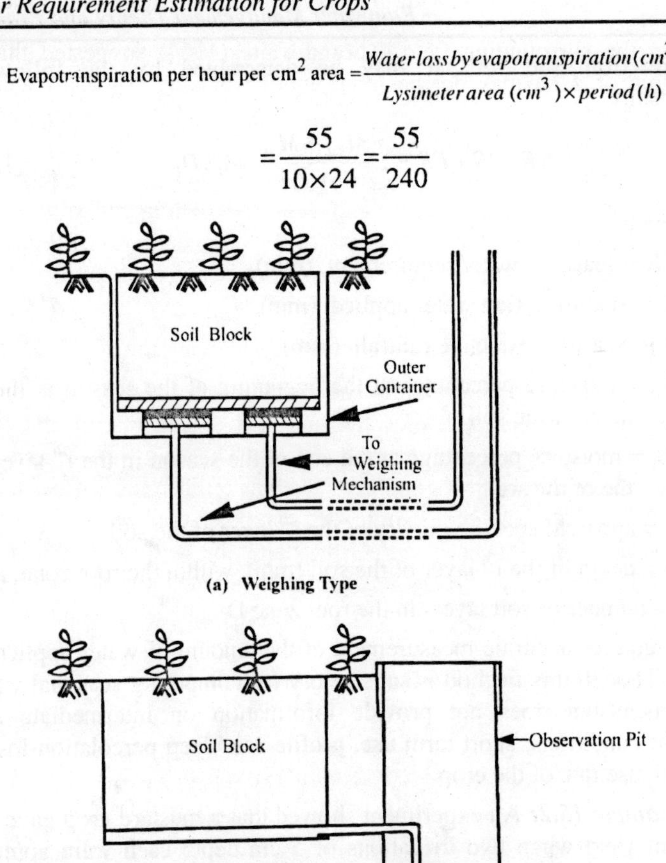

Fig. 15.1: Lysimeters

(ii) Field experimental plots

This method involves the measurements of water supplies to the field and changes in soil moisture contents of field plots. It is sometimes more dependable for determination of seasonal water requirement of crops as compared to measurements with small tanks or lysimeters and also not free from limitations. The main limitation of this method accurate measurement of soil moisture. The seasonal water requirements are estimated by adding measured quantities of irrigation water, the effective rainfall received during the season and the contribution of moisture from the soil. It may be determined by the following relationship:

$$WR = IR + ER + \sum_{i=l}^{n} \frac{M_{bt} - M_{at}}{100} . A_t . D_t \qquad ....(15.4)$$

where,

WR = seasonal water requirement, (mm),

IR = total irrigation water applied, (mm),

ER = seasonal effective rainfall, (mm),

M_{bt} = moisture percentage at the beginning of the season in the ith layer of the soil,

M_{et} = moisture percentage at the end of the season in the ith layer of the of the soil,

A_i = apparent specific gravity of the ith layer of the soil,

D_i = depth of the ith layer of the soil (mm), within the root zone, and

n = number of soil layers in the root zone D

It requires accurate measurement of the amount of water applied to a field. Though this method is satisfactory for computing seasonal water requirement but does not provide information on intermediate soil moisture conditions, short term use, profile use, deep percolation losses and peak use rate of the crop.

Example 15.2: An experiment showed that a mustard crop gave the optimum yield when two irrigations of 5 cm depth each were applied. The crop received an effective rainfall of 8.5 cm during the crop period. The average soil water content at sowing and at harvesting of the crop was 21.32 and 14.25%, respectively. The root zone depth of the crop was 75 cm and bulk density of the soil was 1.49 g/cm^3. Estimate the water requirement of the crop.

Solution:

On using equation (15.1)

Irrigation applied = 5 x 2 = 10 cm

Rainfall received 8.5 cm

Soil water use during the crop season $= \frac{2132 - 14.25}{100} \times 1.49 \times 75$

$= 7.90$ cm

On putting values in equation (15.1), we obtain, the water requirement of the crop (WR) = 10.0 + 8.5 + 7.9 = 26.4 cm.

Example 15.3: Estimate the available soil water holding capacity of a soil in centimetres in the 60-cm soil profile from the following data:

Soil depth (cm)	Field capacity (%)	PWP (%)	Bulk density (g/cm^3)
0-15	25.1	10.8	1.51
15-30	24.8	11.1	1.52
30-45	24.4	11.4	1.54
45-60	23.9	11.3	1.55

Solution:

Available soil water on using following equation holding capacity (ASW)

$$= \sum_{i=l}^{n} \frac{FC_i - PWP_i}{100} \times As_i \times D_i$$

where,

FCi = field capacity of the ith soil layer

PWP_i = permanent wilting point of the ith soil layer

As_i = bulk density of the soil of the ith layer

D_i = depth soil (cm) of the ith soil layer

ASW = available soil water holding capacity, (cm)

On putting values of different layers of soils, we obtain,

$$0-15\,cm = \frac{25.1-10.8}{100} \times 1.51 \times 15 = 3.24\,cm$$

$$15-30\,cm = \frac{24.8-11.1}{100} \times 1.52 \times 15 = 3.12\,cm$$

$$30-45\,cm = \frac{24.4-11.4}{100} \times 1.54 \times 15 = 3.00\,cm$$

$$45-60\,cm = \frac{23.9-11.3}{100} \times 1.55 \times 15 = 2.92\,cm$$

Then, total ASW is obtained by summarizing up above all these. Thus, the total ASW 3.24 + 3.12 + 3.00 + 2.92 = 12.28 cm

(iii) Soil moisture depletion

This method usually used to determine the consumptive use of irrigated field crops grown on the fairly uniform soil and the depth to the ground water is such that it will not influence the soil moisture fluctuation within the root zone. It involves measurement of soil moisture at different depths at a number of times throughout the crop growth period. Greater the number of measurements, more is the information obtained. Consumptive use (C_u) is determined from the change in soil water content in successive samples by the following relationship:

$$u = \sum_{i=l}^{n} \frac{M_{it} - M_{2t}}{100} \times A_i \times D_i \qquad \text{....(15.5)}$$

where,

u = water use from the root zone for a successive sampling periods or within one irrigation cycle, (mm),

n = number of soil layers sampled in the root zone depth, (D),

M_{it} = soil moisture percentage at the time of first sampling in the ith layer,

M_{2t} = soil moisture percentage at the time of second sampling in the ith layer,

A_t = apparent specific gravity of the ith layer of the soil, and

D_t = depth of the ith layer of the soil (mm).

Seasonal consumptive use ($C_u = \Sigma$ u) is estimated by summing of consumptive use values of each sampling interval. A correction is made by adding PET value (s) for accelerated water loss for the interval (s) just after irrigation (s) and before soil moisture sampling.

Example 18.4: Soil samples were drawn from a chickpea field two days after an application of irrigation when the earliest soil sampling could be made and just before the neat irrigation applied 25 days after. The soil water content was estimated by the thermo-gravimetric method. The soil data are given in the following table. Determine the (a) daily consumptive use during the period between two soil sampling and (b) soil water deficit in the crop root zone of 50 cm.

Soil depth (cm)	Soil water content (%) 2 days after irrigation	Soil water content (%) before the following irrigation	Field capacity (%)	Bulk density (g/cm^3)
0-25	25.5	14.1	25.8	1.49
25-50	25.4	14.6	26.0	1.52

Solution:

On using equation (15.5)

$$0-25\,cm=\frac{25.1-14.1}{100}\times 1.49\times 20=5.10\,cm$$

$$25-50\,cm=\frac{25.4-14.6}{100}\times 1.52\times 25=4.92\,cm$$

Therefore, the total water used = 5.10 + 4.92 = 10.02 cm in 25 days. Average daily consumptive use by the crop = 10.02/25 = 0.40 cm or 4 mm.

Soil water deficit in the two different layers before the application of irrigation is:

$$0-25\,cm=\frac{25.8-14.1}{100}\times 1.49\times 25=5.23\,cm$$

$$25-50\,cm=\frac{26.0-14.6}{100}\times 1.52\times 25=5.20\,cm$$

Thus, total soil water deficit in root zone = 5.23 + 5.20 = 10.43 cm.

(iv) Water balance method

It is also called as inflow-outflow methods and is suitable for large areas (watersheds) over long periods. It may be expressed by the following hydrological equation:

Precipitation = Evapotranspiration + surface runoff + sub-surface drainage + change in soil water contents

$$W = P + I - R - D - ET_0 \qquad \text{....(15.6)}$$

where,

W = change in water content in the root zone during certain period, (mm),

P = Precipitation during the period, (mm),

I = Irrigation, (mm),

R = Runoff, (mm),

D = deep percolation losses, (mm), and

ET_0 = evapotranspiration, (mm).

This method required adequate measurement of all factors, except evapotranspiration. The value of evapotranspiration is computed from the measured data.

18.4 Estimating Evapotranspiration from Evaporation Data

The close relationship exists between the rate of consumptive use by crops and the rate of evaporation from a well located evaporation pan. Evaporimeters incorporate the effects of all climatic factors. These are more accurate in estimating short-term fluctuations in ET than empirical formulae that depend on fewer of the climatic factors involved. The standard U.S. Weather Bureau Class A pan evaporimeter or the sunken screen pan evaporimeter (Sharma and Dastane, 1968) may be used for the measurement of evaporation.

USWB Class-A pan evaporimeter

This evaporimeter is the most widely used evaporation pan (Fig. 15.2). It is made of 20 gauge galvanized iron sheet and diameter of 120 cm and 25 cm in depth. It is painted white and exposed on a wooden frame in order that air may circulate beneath the pan. It is filled with water to a fixed depth of about 20 cm. The water surface level is measured daily by means of a hook gauge in a stilling well, and evaporation is computed as the difference between observed levels adjusted for any precipitation measured in a standard rain gauge. Water is added each day to bring the level to a fixed point in the stilling well. A measuring cylinder is used for measurement of add water in the pans. This pan have higher rates of evaporation than large free water surface, and a factor of about 0.7 is usually adopted for converting the observed evaporation rate to those of large water surface areas. This coefficient called as pan coefficient.

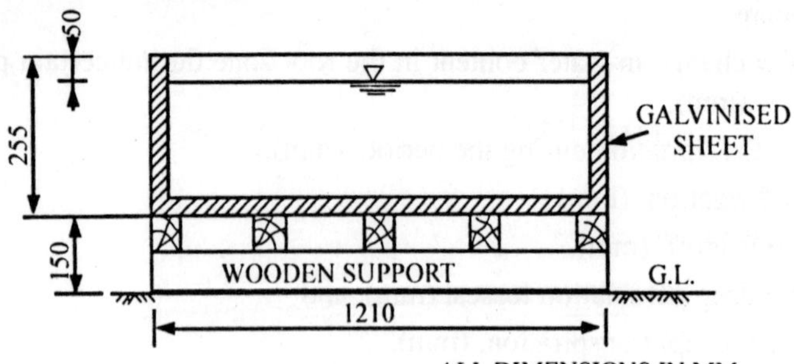

Fig. 15.2: US Class A evaporation pan

Table 15.1: Values of crop factor to compute consumptive use from USWB Pan evaporation data (measured or computed).

Stage of crop growth in % of growth season	Wheat (Ludhiana)	Wheat (Pune)	Maize (Ludhiana)	Maize (Alabama, USA)	Sorghum (Alabama, USA)	Cotton (Pune)	Sugar cane (Hawaii)	Rice (los Banos, Phillippines)	Groundnut (Israel)	Small vegetables (Hargreaves data)
0	0.14	0.30	0.40	0.40	0.42	0.22	0.34	1.00	0.30	0.25
5	0.17	0.40	0.42	0.40	0.44	0.22	0.37	1.02	0.30	0.28
10	0.23	0.51	0.47	0.43	0.46	0.23	0.40	1.03	0.32	0.30
15	0.33	0.62	0.54	0.46	0.48	0.24	0.44	1.05	0.35	0.38
20	0.45	0.73	0.63	0.52	0.50	0.26	0.50	1.07	0.40	0.45
25	0.60	0.84	0.75	0.59	0.52	0.35	0.60	1.09	0.49	0.50
30	0.72	0.92	0.85	0.67	0.56	0.58	0.72	1.11	0.60	0.55
35	0.81	0.96	0.96	0.76	0.59	0.80	0.86	1.13	0.69	0.58
40	0.88	1.10	1.04	0.85	0.64	0.95	0.93	1.16	0.78	0.60
45	0.90	1.10	1.07	0.93	0.71	1.03	0.98	1.18	0.85	0.63
50	0.91	1.00	1.09	1.02	0.79	1.08	1.02	1.20	0.90	0.65
55	0.90	0.91	1.10	1.09	0.92	1.08	1.05	1.21	0.94	0.65
60	0.89	0.80	1.11	1.14	1.01	1.07	1.07	1.22	0.96	0.65
65	0.86	0.65	1.10	1.19	1.07	1.05	1.10	1.22	0.95	0.63
70	0.83	0.51	1.07	1.21	1.09	1.00	1.13	1.21	0.94	0.60
75	0.80	0.40	1.04	1.32	1.09	0.93	1.16	1.19	0.91	0.58
80	0.76	0.30	1.00	1.22	1.05	0.85	1.19	1.16	0.86	0.55
85	0.71	0.20	0.97	1.16	0.99	0.73	1.20	1.10	0.79	0.50
90	0.65	0.12	0.89	1.06	0.91	0.62	1.20	1.03	0.72	0.45
95	0.58	0.10	0.81	0.95	0.82	0.50	1.19	0.96	0.64	0.38
100	0.51	0.10	0.70	0.75	0.70	0.40	1.19	0.86	0.55	0.30
Seasonal	0.61	0.61	0.86	0.86	0.75	0.68	0.89	1.10	0.68	0.50

(Source: Water Management Division, Dept. of Agr., Irrig. Govt. of India, 1971)

Evapotranspiration is equal to pan evaporation multiplied by the crop coefficient factor. The value of crop coefficient factor for any crop depends on its foliage characteristics, stage of growth, climate and geographical location. Generally, values of consumptive use are low during the early stages of crop growth and increase as the plant approaches grand growth and again declines with maturity. The values of crop coefficient should be determined for a particular crop for particular location experimentally. The value of crop coefficient for some of the common crops at selected locations for different stages of the growth is given in Table 15.1.

Sunken screen pan evaporimeter

Sharma and Dastane (1968) developed at the Indian Agricultural Research Institute, New Delhi, provides a simple device to make reasonable estimates of consumptive use. The ratio between evapotranspiration and evaporation from USWB Class A pan (ET/E) is about 0.5 to 1.3 after establishment of the crop. However, the same ratio in a sunken screen pan evaporimeter was observed to be 0.95 to 1.05. In other words, the evaporation value obtained from it closely approximated the evapotranspiration. This should be extensively tested under varying climatic conditions and for different crops.

This consists of three parts, *viz.*, an evaporation pan, a stilling well and a connecting tube as shown in Fig. 15.3. The evaporimeter is made of 20 gauge galvanized iron sheet having 60 cm diameter and 45 cm depth. It is painted white and fitted with a screen of 1/24 or 6/20 mesh which is held tight over the pan by bending it at the brim and pressing hard. The stilling well as 15 cm diameter and 45 cm depth and is also fitted with a screen cover of the same mesh as that of the evaporation pan. It has a pointer attached to the side wall and bent upward in the centre at right angle. The evaporimeter is installed by digging a pit of suitable size, placing the pan and back filling the earth with due compaction. The top edge of the pan protrudes 10 cm over the soil surface. It is necessary to avoid runoff from the surrounding area entering into the pan. The water level is maintained at the same height as the soil level outside. Thus, the tip of the pointer, free water surface in the pan and the soil surface are at the same level.

The water level in the pan is brought in level with the pointed tip and the pan is set at work. Observations of falling water level are recorded at suitable intervals (24 hr). This is accomplished by adding water with a measuring cylinder and recording the quantity of water added to bring the water level back to the pointer tip. The volume of water (ml), added is converted into depth (mm) by dividing with the area of the pan plus that of the stilling well.

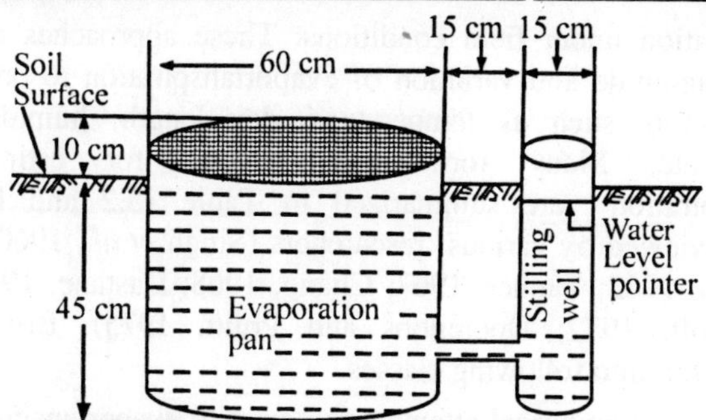

Fig. 15.3: Cross-sectional view of the sunken screen pan evaporimeter (Source: Sharma and Dastane, 1968)

It is installed in duplicate to enable leakage detection. The minimum distance between two evporimeters is kept 3 m. Clean tap water is used in the evaporimeter. The pan is leaned occasionally and painted white once in a year and checked scrupulously for leakage. It is installed under natural conditions in a field which does not provide obstruction to wind and is aligned perpendicular to the main direction of wind to avoid mutual interference.

Piche evaporimeter

It consists of a graduated tube of 1.5 cm diameter and 30 cm length with one end open. The open end is covered by a drier paper held in place by a metallic device. The tube is filled with water and turned upside down. The water slowly wets the paper and evaporates. Water loss is read on the tube and it gives as a measurement of evaporation. Owing to the smaller size of Piche evaporimeter, the rates of evaporation are in excess of the rates of water used by the crops. Piche values are also higher than those obtained by the USWB Class A evaporation pan. Measurement of evaporation with Piche evaporimeters are usually poorly correlated with evapotranspiration and evaporation from other pans. It overestimates wind effects and grossly underestimates radiation effects. The evaporating surface of the Piche evaporimeter is often subject to contamination by dust, oil and other foreign materials.

15.5 Estimating Evapotranspiration from Climatological Data

Evapotranspiration is predicted on the basis of climatological data because difficulty arises in obtaining accurate direct measurements of pan evaporation under field conditions. These approaches are related with the magnitude and variation of evaportanspiration to one or more climatic factors such as temperature, day length, humidity, wind, sunshine, etc. Many formulae developed for estimation of evaportanspiration are summarized in Table 15.2 and have been critically reviewed by various researchers (Singh *et al.*, 1960; Israelsen and Hansen, 1962; Tanner, 1967; Chang, 1968; Dastane, 1972; Taylor and Aschroft, 1972; Doorenbos and Pruitt, 1975). Broadly these approaches fall into following classes.

(i) Purely empirical attempts to correlate evaportanspiration with one or more climatic factors, and

(ii) The application of a more theoretical approach.

Table 15.2: Variables included in the different formula to estimate crop water requirements

Formulae	Variables														Results
	Temperature	Air humidity	Dry-wet bulb	Day light hrs.	Sunshine hrs./ Cloud cover	Radiation	Wind velocity	Evaporimeter	Crop data	Crop factor	Soil factor	Correction factor	Precipitation	Barograph	
Makkink 1957, Holland	X					X									ETP (grass), monthly.
Blaney-Criddle. 1964, USA	X			X						X					CU crop, monthly.
Jensen-Haise, 1963. USA	X			(X)	(X)	X				X		(X)			ETP crop and ET crop
Penman, 1948-56, UK	X	X		(X)	(X)	X	X	X		(X)					E_0 or ET crop.
Bouchet	X	X						X				X			ETP crop.
Halkais, 1955, USA												X			CU crop.
Lowry-Johnson, 1942, USA	X											X			ET of valley, entire growing season.
Thornthwaite, 1955, USA	X				(X)							X			ET crop, monthly, under 60-70% avail, soil water.

Formulae	Temperature	Air humidity	Dry-wet bulb	Day light hrs.	Sunshine hrs./ Cloud cover	Radiation	Wind velocity	Evaporimeter	Crop data	Crop factor	Soil factor	Correction factor	Precipitation	Barograph	Results
	Variables														
Turc-Langbein, 1954 France	X												X		E_0 or ET crop, annual river basin.
Sarov, USSR	X											X			ET for optimum product,
Haude, 1952, Germany	X		X	(X)			(X)					X			ET crop.
Skvortsov, 1950, USSR			X									X			ET crop.
Blaney-Morin, 1942, USA	X	X		X						X					ET crop, monthly.
Prescott, 1949, USA	X	X	X								X				ETP crop.
Hatstead, 1951, USA	X	X	X	X											ETP crop.
Rohwer, 1931, USA	X	X					X				X			X	ET crop.
Ivanov, 1957, USSR	X	X													ET crop under optimum water conditions.
Kostiakov, USSR	X	X													ET crop.
Turc, 1954, France	X	X		X	X	(X)									ET crop.
Hargreaves, 1956 USA	X	X		X						X					E_0 (A pan) or ET crop.
Turc 1953, France	X					X				X	X			X	ET crop.
Christianson, 1966, USA	X	X			X	X	X					X			E_0 (A pan).
Thoruthwaite-Mather	X				(X)						X	X	X		ET and water balance,

Formulae	Temperature	Air humidity	Dry-wet bulb	Day light hrs.	Sunshine hrs./ Cloud cover	Radiation	Wind velocity	Evaporimeter	Crop data	Crop factor	Soil factor	Correction factor	Precipitation	Barograph	Results
	Variables														
Munson, 1960, USA	X		X		X	X				X		X			PE index, Cu.
Walker	X			(X)	X	(X)						X			E_0 (A pan) or ET crop.
Olivier, 1961, UK	(X)	(X)	X		(X)	X					X		X		Basic water requirements for crop/land Unit.
Rijtema, 1957, Holland	X	X		(X)	(X)	X	X		X						ETP crop.
Mcllory, 1961. Australia	X	X	(X)	(X)	(X)	X	X			X					ETP crop.
Linacre, 1967, Australia	X	X		X	X	X	X		X						ET crop.
Van Bavel, 1956. USA	X			X	X	(X)						X			ETP crop.
Sverdrup, 1952,USA	X	(X)	X				X					X			ET per unit surface.

() when essential parameters are unavailable. (Source: Doorenbos and Pruitt, 1975)

Table 15.3: Maximum possible duration of sunlight in the northern hemisphere expressed in units of 30 days of 12 hours each

N	J	F	M	A	M	J	J	A	S	O	N	D
0	1.04	0.94	1.04	1.01	1.01	1.04	1.04	1.04	1.01	1.04	1.01	1.04
5	1.02	0.93	1.03	1.02	1.06	1.03	1.06	1.05	1.01	1.03	0.99	1.02
10	1.00	0.91	1.03	1.03	1.08	1.06	1.08	1.07	1.02	1.02	2.98	0.99
15	0.97	0.91	1.03	1.04	1.11	1.08	1.12	1.08	1.02	1.01	0.95	0.97
20	0.95	0.90	1.03	1.05	1.12	1.11	1.14	1.11	1.02	1.00	0.93	0.94
25	0.93	0.89	1.03	1.06	1.15	1.14	1.17	1.12	1.02	0.99	0.91	0.91
26	0.92	0.88	1.03	1.06	1.15	1.15	1.17	1.12	1.02	0.99	0.91	0.91
27	0.92	0.88	1.03	1.07	1.16	1.15	1.18	1.13	1.02	0.99	0.90	0.90

N	J	F	M	A	M	J	J	A	S	O	N	D
28	0.91	0.88	1.03	1.07	1.16	1.16	1.18	1.13	1.02	0.98	0.90	0.90
29	0.91	0.87	1.03	1.07	1.17	1.16	1.19	1.13	1.03	0.98	0.90	0.89
30	0.90	0.87	1.03	1.08	1.18	1.17	1.20	1.14	1.03	0.98	0.89	0.88
31	0.90	0.87	1.33	1.08	1.18	1.18	1.20	1.14	1.03	0.98	0.89	0.88
32	0.89	0.86	1.03	1.08	1.19	1.21	1.21	1.15	1.03	0.98	0.88	0.87
33	0.88	0.86	1.03	1.19	1.19	1.20	1.22	1.15	1.03	0.97	0.88	0.86
34	0.88	0.85	1.03	1.09	1.20	1.20	1.22	1.16	1.03	0.97	0.87	0.86
35	0.87	0.85	1.03	1.09	1.21	1.21	1.23	1.16	1.03	0.97	0.86	0.85
36	0.87	0.85	1.03	1.10	1.21	1.22	1.24	1.16	1.03	0.97	0.86	0.84
37	0.86	0.84	1.03	1.10	1.22	1.23	1.25	1.17	1.03	0.97	0.85	0.83
38	0.85	0.84	1.03	1.10	1.23	1.24	1.25	1.17	1.04	0.96	0.84	0.83
39	0.85	0.84	1.03	1.11	1.23	1.24	1.26	1.18	1.04	0.96	0.84	0.82
40	0.84	0.83	1.03	1.11	1.24	1.25	1.27	1.18	1.04	0.96	0.83	0.81
41	0.83	0.83	1.03	1.11	1.25	1.26	1.27	1.19	1.04	0.96	0.82	0.80
42	0.82	0.83	1.03	1.12	1.26	1.27	1.28	1.19	1.04	0.95	0.82	0.79
43	0.81	0.82	1.02	1.12	1.26	1.28	1.29	1.20	1.04	0.59	0.81	0.77
44	0.81	0.82	1.02	1.13	1.27	1.29	1.30	1.20	1.04	0.95	0.80	0.76
45	0.80	0.81	1.02	1.13	1.28	1.29	1.31	1.21	1.04	0.94	0.97	0.75
46	0.97	0.81	1.02	1.13	1.29	1.31	1.32	1.22	1.04	0.94	0.79	0.74
47	0.77	0.80	1.02	1.14	1.30	1.32	1.33	1.22	1.04	0.93	0.78	0.73
48	0.76	0.80	1.02	1.14	1.31	1.33	1.34	1.23	1.05	0.93	0.77	0.72
49	0.75	0.79	1.02	1.14	1.32	1.34	1.35	1.24	1.05	0.93	0.76	0.71
50	0.74	0.78	1.02	1.15	1.33	1.36	1.37	1.25	1.06	0.92	0.76	0.70

To find the maximum possible duration of sunlight for any month, multiply 2 x 30 x coefficient. To find the mean daily maximum duration of bright sunshine hours (iv) for any month multiply the coefficient by 12 x 30 and divide by the number of days in the month. For southern latitudes read the table commencing from July instead of January.

Thornthwaite Method

According Thornthwaite (1948) method the potential evapotranspiration (unadjusted) is obtained by the following relation:

$$PET = 16.0\left(\frac{10T}{I}\right)^{e} \; mm/month \qquad ....(15.7)$$

where,

T= mean monthly temperature, (^{0}C),

I = 'heat index' for the 12 months in a year, where

$I = \Sigma i \; (T/5)^{1\text{-}514}$ for each month in the year

$a = 6.75x10^{-7} I^{3} - 7.71 \times 10^{-5} I^{2} + 1.792 \times 10^{-2} I + 0.49239.$

Steps followed for determination of PET by this method are as under:

(i) Find the mean monthly temperature for each month of the year T.

(ii) Calculate = $(T/5)^{1\text{-}514}$ for each month

(iii) Calculate $I = \Sigma i$

(iv) Using the value of I calculate value of a

(v) Calculate PET (unadjusted) by using Eq. 15.7

The potential evapotranspiration (unadjusted) calculated with the above formula is corrected for the duration of sunlight by multiplying with the factors given in Table 15.3 for the month and latitude.

The actual evapotranspiration is calculated by multiplying the potential evaportanspiration with a factor whose value depends upon the particular crop and region. Table 15.4 gives such factors for some crops.

Table 15.4: Normal seasonal consumptive-use coefficient for some irrigated crops

Crop	Length of growing season or period in months	Consumptive use coefficients (seasonal k)
Beans	3	0.65
Corn	4	0.75
Cotton	7	0.70
Potatoes	3	0.70
Rice	3-4	1.00
Small grain	3	0.75
Sorghum	5	0.70

(Source : Withers and Vipond, 1974)

The most commonly used formulae for determination of evapotranspiration are those proposed by Blancy-Criddle, Thornthwaite, Penman and Christiansen and Pennan Montheith etc.

Blaney-Criddle formulae

According to Blaney and Criddle (1950) the amount of water consumptively used by crops during their growing seasons is closely related with mean monthly temperature and daylight hours. The consumptive is determined by relationship developed by Blaney and Criddle (1950) as,

$$U = K.F. = \Sigma k.f = \Sigma u = \Sigma \frac{k.t.P}{100} \quad ....(15.8)$$

where,

U =seasonal consumptive use of water by the crop for a given period (inches),

u = monthly consumptive use, (inches),

K =empirical seasonal consumptive use crop coefficient for the growing season.

F = sum of the monthly consumptive use factors (f) for the growing season,

k = empirical consumptive use crop coefficient for the month=u/f;

$$f = \frac{t \times p}{100},$$

t = mean monthly temperature (^{0}F), and

p = monthly daylight hours expressed as percentage of daylight hours of the year (Table 18.3)

The following steps are involved for calculations of consumptive use:

(i) Calculate the mean monthly temperature in ^{0}F.

(ii) Calculate p from Table (18.3) as,

$$P = \frac{Possible\ sunshine\ hours\ for\ the\ particular\ month}{Possible\ hours\ for\ the\ whole\ year} \times 100$$

(i) Calculate F as Σpt

(ii) Select suitable value for k obtain kf

The mean monthly temperature of a locality is obtained from the weather data. In the absence of a more reliable information based on local experimental data, the values of consumptive use crop coefficients (k) for common irrigated crops is given in Table 15.4 may be used. The procedure for estimation of PET by this method is shown in Example 15.7.

Example 15.7: Estimate the consumptive use of water by wheat crop sown on 1 November and harvested on 20 March at Sriniketan, West Bengal (23°39') using the Blaney-Criddle formula from the following data:

Month	**Mean monthly air temperature ^{0}C**	**Corresponding mean monthly air temperature ^{0}F**	**Percent monthly day light hours**	**Assumed crop coefficient (k)**
November	23.26	73.87	7.45	0.65
December	19.34	66.81	7.49	0.60
January	18.51	65.3	7.55	0.50
February	21.30	70.34	7.16	0.70
March	26.68	80.02	8.40	0.75

Solution:

$$\text{November: } u = \frac{ktp}{100} = \frac{0.65 + 73.87 \times 7.45}{100} = 3.57\,in$$

Similarly,

$$\text{December: } u = \frac{0.60 + 66.81 \times 7.49}{100} = 3.00\,in$$

$$\text{January: } u = \frac{0.50 + 65.3 \times 7.55}{100} = 2.47\,in$$

$$\text{February: } u = \frac{0.70 + 70.34 \times 7.16}{100} = 3.53\,in$$

$$\text{March: } u = \frac{0.75 + 80.02 \times 8.40}{100} = 5.04\,in$$

the consumptive use for 20 days in March = 3.25 in

Therefore, the seasonal consumptive use (u) = 3.57 + 3.00 + 2.47 + 3.53 + 3.25 = 15.82 in or 40.18 cm

Modified Blaney – Criddle Method

This method generally gives sufficiently accurate estimates of seasonal consumptive use owing to the inclusion of locally developed crop coefficient factor (K). Basically, it considers that the consumptive use is dependent only on temperature and day length which is not fully true. Crop water requirements have been found to vary widely between climates having similar air temperature but different humidity and wind conditions. Therefore, the consumptive use crop coefficient (K) will vary not only with crop but also with climatic conditions. Doorenbos and Pruitt (1975) have rejected the use of crop coefficient (K) normally applied in the original Blaney- Criddle approach, because (1) the original crop coefficients (K) are heavily dependent on local conditions, and wide variation of K values reported in literature make the selection of this value rather difficult, (2) the relationship between Blaney-Criddle f-values and PET can be adequately described for a wide range of temperatures for areas having minor variations in relative humidity, sunshine and wind velocity, and (3) once PET has been determined by any standard method, one Set of crop factors (k_c) can be used to determine crop ET. Thus, Doorenbos and Pruitt (1975) suggested the following relationships for determination of 'f' factor in Blaney-Criddle formula.:

$$f = p\,(0.46\,t + 8.13),\ \text{using } t \text{ in } {}^0\text{C} \qquad \ldots(15.9)$$

where,

t = the mean of daily maximum and minimum temperature in °C or over the month considered, and

p = the mean daily percentage of annual day time hours for a given month and latitude (Table 15.3).

Table 15.5: Monthly percentage (p) of the day- time hours of the year for different latitudes

Latitude (degrees)	Jan.	Feb.	Mar.	Apr.	May	June	July	Aug.	Sept.	Oct.	Nov.	Dec.
1	2	3	4	5	6	7	8	9	10	11	12	13
						North						
60	4.67	5.70	8.05	9.66	11.72	12.39	12.33	10.72	8.57	7.00	5.04	4.15
59	4.81	5.78	8.05	9.60	11.61	12.23	12.21	10.60	8.56	7.07	5.09	4.31
58	4.99	5.85	8.06	9.55	11.44	12.00	12.00	10.56	8.56	7.13	5.13	4.55
57	5.14	5.93	8.07	9.51	11.32	11.77	11.87	10.47	8.54	7.19	5.27	4.69

Latitude (degrees)	Jan.	Feb.	Mar.	Apr.	May	June	July	Aug.	Sept.	Oct.	Nov.	Dec.
1	2	3	4	5	6	7	8	9	10	11	12	13
56	5.29	6.00	8.10	9.45	11.20	11.67	11.69	10.40	8.52	7.25	5.54	4.89
55	5.39	6.06	8.12	9.41	11.11	11.53	11.59	10.32	8.51	7.30	5.62	5.01
54	5.53	6.12	8.15	9.36	11.00	11.40	11.43	10.27	8.50	7.33	5.74	5.17
53	5.64	6.19	8.16	9.32	10.88	11.31	11.34	10.19	8.52	7.38	5.83	5.31
52	5.75	6.23	8.17	9.28	10.81	11.13	11.22	10.15	8.49	7.40	5.94	5.43
51	5.87	6.25	8.21	9.26	10.76	11.07	11.13	10.05	8.48	7.41	5.97	5.46
50	5.98	6.32	8.25	9.25	10.69	10.93	10.99	10.00	8.44	7.43	6.07	5.65
48	6.13	6.42	8.22	9.15	10.50	10.72	10.83	9.92	8.45	7.56	6.24	5.86
46	6.30	6.50	8.24	9.09	10.37	10.54	10.66	9.82	8.44	7.61	6.38	6.05
44	6.45	6.59	8.25	9.04	10.22	10.38	10.50	9.73	8.43	7.67	6.51	6.23
42	6.60	6.66	8.28	8.97	10.10	10.21	10.37	9.64	8.42	7.73	6.63	6.39
40	6.73	6.73	8.30	8.92	9.99	10.08	10.34	9.56	8.41	7.78	6.73	6.53
38	6.87	6.79	8.34	8.90	9.92	9.95	10.10	9.47	8.38	7.80	6.82	6.66
36	6.99	6.86	8.35	8.85	9.31	9.83	9.99	9.40	8.36	7.85	6.92	6.79
34	7.10	6.91	8.36	8.80	9.72	9.70	9.88	9.33	8.36	7.90	7.02	6.92
32	7.20	6.97	8.37	8.72	9.63	9.60	9.77	9.28	8.34	7.93	7.11	7.05
30	7.30	7.03	8.38	8.72	9.53	9.49	9.67	9.22	8.34	7.99	7.19	7.14
28	7.40	7.02	8.39	8.68	9.46	9.38	9.58	9.16	8.32	8.02	7.27	7.27
26	7.49	7.12	8.40	8.64	9.37	9.30	9.49	9.10	8.32	8.06	7.36	7.35
24	7.58	7.17	8.40	8.60	9.30	9.19	9.41	9.05	8.31	8.10	7.43	7.46
22	7.76	7.22	8.41	8.57	9.22	9.12	9.31	9.00	8.30	8.13	7.50	7.56
20	7.73	7.26	8.20	8.52	9.14	9.02	9.25	8.95	8.30	8.19	7.58	7.88
18	7.88	7.26	8.40	8.46	9.06	8.99	9.20	8.81	8.29	8.24	7.67	7.89
16	7.94	7.30	8.42	8.45	8.98	8.98	9.07	8.80	8.28	8.24	7.72	7.90
14	7.08	7.39	8.43	8.44	8.90	8.73	8.99	8.79	85.28	8.28	7.85	8.04
12	8.08	7.40	8.44	8.43	8.84	8.64	8.90	8.78	8.27	8.28	7.85	8.05
10	8.11	7.40	8.44	8.43	8.91	8.57	8.84	8.74	8.26	8.29	7.89	8.08
8	8.13	7.41	8.45	8.39	8.75	8.51	8.77	8.70	8.25	8.31	7.89	8.11
6	8.19	7.49	8.45	8.39	8.73	8.48	8.75	8.69	8.25	8.41	7.95	8.19
4	8.20	7.58	8.46	8.33	5.65	8.40	8.67	8.63	8.21	8.43	7.95	8.20

Latitude (degrees)	Jan.	Feb.	Mar.	Apr.	May	June	July	Aug.	Sept.	Oct.	Nov.	Dec.
1	2	3	4	5	6	7	8	9	10	11	12	13
2	8.43	7.62	8.47	8.22	8.51	8.25	8.52	8.50	8.20	8.45	8.16	8.42
0	8.49	7.67	8.49	8.22	8.49	8.22	8.49	8.49	8.19	8.49	8.22	8.49
						South						
0	8.49	7.67	8.49	8.22	8.49	8.22	8.49	8.49	8.19	8.49	8.22	8.49
2	8.55	7.71	8.49	8.19	8.44	8.17	8.43	8.44	8.19	8.52	8.27	8.55
4	8.64	7.76	8.50	8.17	8.39	8.08	8.20	8.41	8.19	8.56	8.33	8.65
6	8.71	7.81	8.50	8.12	8.30	8.00	8.19	8.37	8.18	8.59	8.38	8.74
8	8.79	8.84	8.51	8.11	8.24	7.91	8.13	8.32	8.18	8.62	8.47	8.84
10	8.85	7.86	8.52	8.09	8.18	7.84	8.11	8.28	8.18	8.65	8.52	8.90
12	8.91	7.91	8.53	8.06	8.15	7.79	8.08	8.26	8.17	8.67	8.58	8.95
14	8.97	7.97	8.54	8.03	8.07	7.70	7.08	8.19	8.165	8.69	8.65	9.01
16	9.09	8.02	8.56	7.98	7.96	7.57	7.94	8.14	8.14	8.76	8.72	9.17
18	9.18	8.06	8.57	7.93	7.99	7.50	7.88	8.90	8.14	8.80	8.80	9.24
20	9.25	8.09	8.58	7.92	7.83	7.41	7.73	8.05	8.13	8.83	8.85	9.32
22	9.36	8.12	8.58	7.89	7.74	7.30	7.76	8.03	8.13	8.86	8.90	9.38
24	9.44	8.17	8.59	7.87	7.60	7.24	7.58	7.99	8.12	8.89	8.96	9.47
26	9.52	8.28	8.00	7.81	7.56	7.07	7.49	7.87	8.11	8.94	9.10	9.61
28	9.61	8.31	8.61	7.79	7.49	6.99	7.40	7.85	8.10	8.97	9.19	9.74
30	9.69	8.33	8.63	7.75	7.43	6.94	7.30	7.80	8.09	9.00	9.24	9.80
32	9.76	8.36	8.64	7.70	7.39	6.85	7.20	7.73	8.08	9.04	9.31	9.87
34	9.88	8.41	8.65	7.68	7.30	6.73	7.10	7.69	8.06	9.07	9.38	9.99
36	10.06	8.53	8.67	7.61	7.10	6.59	6.99	7.59	8.06	9.15	9.51	10.21
38	10.14	8.61	8.68	7.59	7.03	6.46	6.87	7.51	8.05	9.19	9.60	10.34
40	10.24	8.65	8.70	7.54	6.96	6.33	6.73	7.46	8.04	9.23	9.69	10.42
42	10.39	8.72	8.71	7.49	6.85	6.20	6.650	7.39	8.01	9.27	9.79	10.57
44	10.52	8.81	8.72	7.44	6.73	6.04	6.45	7.30	8.00	9.34	9.91	10.72
46	10.68	8.88	8.73	7.39	6.61	5.87	6.30	7.21	7.98	9.41	10.03	10.90
48	10.85	8.98	8.76	7.32	6.45	5.69	6.13	7.12	7.96	9.47	10.17	11.09
50	11.03	9.06	8.77	7.25	6.31	5.48	5.98	7.03	7.95	9.53	10.32	11.30

(Source: U.S.D.A., A.R.S., 1962)

Table 15.6: Monthly consumptive use crop coefficients (K) for use in Blaney-Criddle formulae

	Month											
Crop	**Jan.**	**Feb.**	**Mar.**	**Apr.**	**May**	**June**	**July**	**Aug.**	**Sept.**	**Oct.**	**Nov.**	**Dec.**
Rice				0.85	1.00	1.15	1.30	1.25	1.10	0.90		
Maize				0.50	0.60	0.70	0,80	0.80	0.60	0.50		
Wheat	0.50	0.70	0.75	0.70						0.70	0.65	0.60
Sugarcane	0.75	0.80	0.85	0.85	0.90	0.95	1.00	1 00	0.95	0.90	0.85	0.75
Cotton				0.50	0.60	0.75	0.90	0.85	0.75	0.55	0.50	0.50
Vegetables	0.50	0.55	0.60	0.65	0.70	0.75	0.80	0.80	0.70	0.60	0.55	0.50
Berseem	0.50	0.70	0.80	0.90	1.00				0.60	0.65	0.70	0.60
Citrus	0.50	0.55	0.55	0.60	0.60	0.65	0.70	0.70	0.65	0.60	0.55	0.55

(Source: Dastane, 1972)

ET_0 values are now determined using the relationship between f and ET_0 (Fig. 15.4).

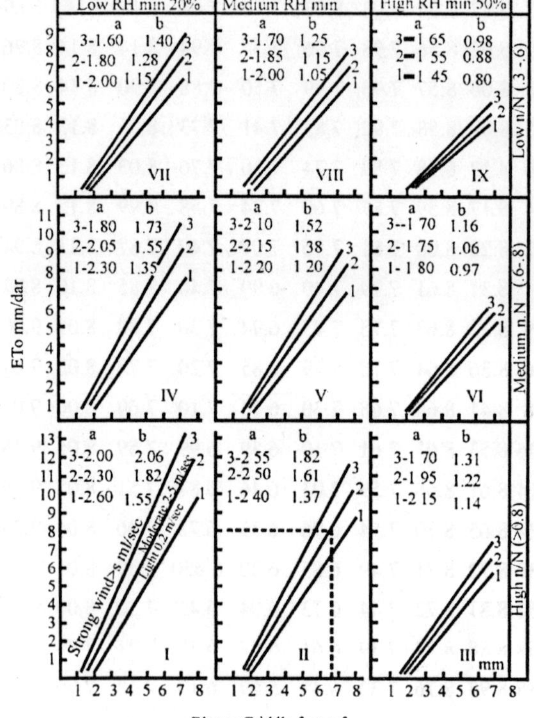

Fig. 15.4: Graphs for modified Blaney Criddle

These figures are developed using values of day time minimum, humidity (RH min), ratio of actual maximum possible sunshine hours (n/N), and day time wind condition (u) at 2 meters height. After determining ET_0, ET crop is determined on using the following relation:

$$ET\ (crops) = kc \times ET_0 \qquad \text{....(15.10)}$$

where,

kc = crop coefficient

Since f is expressed in mm per day, ET_0 and ET (crop) are also expressed in mm/day and represent the mean daily values for the period considered (usually for one month)

They have further pointed out that the use of Blaney-Criddle method to determine mean daily PET should normally be applied for periods no shorter than one month and for each calendar month for each year of record instead of using mean temperature of several years. According to them the formula is not suitable for use in the following regions:

(i) equatorial regions where temperatures remain fairly constant, but other weather parameters change

(ii) small islands where air temperature is affected by the surrounding sea temperature showing little response to seasonal change in radiation

(iii) high altitudes where day time radiation is practically independent of night temperature, and

(iv) climates with a high variability in sunshine hours during transition months

Example 15.8: Calculation of the consumptive use of water (C_u) for sugarcane crop at Delhi (28.4° N) using Blaney-Criddle formulae

Month	**Mean monthly air temperature °F (t)**	**Monthly crop coefficient (K) from table 15.4**	**Percent day light hrs (p) from table 15.3**	**Monthly consumptive use (inches)** $u = \frac{k.t.p}{100}$ **(3x2x4)**
January	55.7	0.75	7.38	3.08
February	60.4	0.80	7.02	3.39
March	69.3	0.85	8.39	4.94
April	80.6	0.85	8.69	5.95

Month	Mean monthly air temperature °F (t)	Monthly crop coefficient (K) from table 15.4	Percent day light hrs (p) from table 15.3	Monthly consumptive use (inches) $u = \frac{k.t.p}{100}$ (3x2x4)
May	88.0	0.90	9.48	7.51
June	92.3	0.95	9.41	8.25
July	87.1	1.00	9.60	8.36
August	84.2	1.00	9.18	7.73
September	82.8	0.95	8.33	6.55
October	76.6	0.90	8.01	5.52
November	66.0	0.85	7.25	4.07
December	56.7	0.75	7.24	3.08
		$U = \sum_{i=1}^{12} u$ = 68.43 inches, or = 173.8 cm		

Table 15.7: Correction factors for length and number of days in a month for use in the Thornthwaite formula (Mean possible duration of sunlight expressed in units of 30 days of 12 hr each)

Latitude (degrees)	Jan.	Feb.	Mar.	Apr.	May	June	July	Aug.	Sept.	Oct.	Nov.	Dec.
1	2	3	4	5	6	7	8	9	10	11	12	13
						North						
0	1.04	0.94	1.04	1.01	1.04	1.01	1.04	1.04	1.01	1.04	1.01	1.04
5	1.02	0.93	1.03	1.02	1.06	1.03	1.06	1.05	1.01	1.03	0.99	1.02
10	1.00	0.91	1.03	1.03	1.08	1.06	1.08	1.07	1.02	1.02	0.98	0.99
15	0.97	0.91	1.03	1.04	1.11	1.08	1.12	1.08	1.02	1.01	0.95	0.97
20	0.95	0.90	1.03	1.05	1.13	1.11	1.14	1.11	1.02	1.00	0.93	0.94
25	0.93	0.89	1.03	1.06	1.15	1.14	1.17	1.12	1.02	0.99	0.91	0.91
26	0.92	0.88	1.03	1.06	1.15	1.14	1.17	1.12	1.02	0.99	0.91	0.91
27	0.92	0.88	1.03	1.07	1.16	1.15	1.18	1.13	1.02	0.98	0.90	0.90
28	0.91	0.88	1.03	1.07	1.16	1.16	1.18	1.13	1.02	0.98	0.90	0.90
29	0.91	0.87	1.03	1.07	1.17	1.16	1.19	1.13	1.03	0.98	0.90	0.89
30	0.90	0.87	1.03	1.08	1.18	1.17	1.20	1.14	1.03	0.98	0.89	0.88
31	0.90	0.87	1.03	1.08	1.18	1.18	1.20	1.14	1.03	0.98	0.89	0.88

Latitude (degrees)	Jan.	Feb.	Mar.	Apr.	May	June	July	Aug.	Sept.	Oct.	Nov.	Dec.
1	2	3	4	5	6	7	8	9	10	11	12	13
32	0.89	0.86	1.03	1.08	1.19	1.19	1.21	1.15	1.03	0.98	0.88	0.87
33	0.88	0.86	1.03	1.09	1.19	1.20	1.22	1.15	1.03	0.97	0.88	0.86
34	0.88	0.85	1.03	1.09	1.20	1.20	1.22	1.16	1.03	0.97	0.87	0.86
35	0.87	0.85	1.03	1.09	1.21	1.21	1.23	1.16	1.03	0.97	0.86	0.85
36	0.87	0.85	1.03	1.10	1.21	1.22	1.24	1.16	1.03	0.97	0.86	0.84
37	0.86	0.84	1.03	1.10	1.22	1.23	1.25	1.17	1.03	0.97	0.85	0.83
38	0.85	0.84	1.03	1.10	1.23	1.24	1.25	1.17	1.04	0.96	0.84	0.83
39	0.85	0.84	1.03	1.11	1.23	1.24	1.26	1.18	1.04	0.96	0.84	0.82
40	0.84	2.83	1.03	1.11	1.24	1.25	1.27	1.18	1.04	0.96	0.83	0.81
41	0.83	0.83	1.03	1.11	1.25	1.26	1.27	1.19	1.04	0.965	0.82	0.80
42	0.82	0.83	1.03	1.12	1.26	1.27	1.28	1.19	1.04	0.95	0.82	0.79
43	0.81	0.82	1.03	1.12	1.26	1.28	1.29	1.20	1.04	0.95	0.81	0.77
44	0.81	0.82	1.02	1.13	1.27	1.29	1.30	1.20	1.04	0.95	0.80	0.76
45	0.80	0.81	1.02	1.13	1.28	1.29	1.31	1.21	1.04	0.94	0.79	0.75
46	0.79	0.81	1.02	1.13	1.29	1.31	1.32	1.22	1.04	0.94	0.79	0.74
47	0.77	0.80	1.02	1.14	1.30	1.32	1.33	1.22	1.04	0.93	0.78	0.73
48	0.76	0.80	1.02	1.14	1.31	1.33	1.34	1.23	1.05	0.93	0.77	0.72
49	0.75	0.79	1.02	1.14	1.32	1.34	1.35	1.24	1.05	0.93	0.76	0.71
50	0.74	0.78	1.02	1.15	1.33	1.36	1.37	1.25	1.06	0.92	0.76	0.70
						South						
5	1.06	0.95	1.04	1.00	1.02	0.99	1.02	1.03	1.00	1.05	1.03	1.06
10	1.08	0.97	1.05	0.99	1.01	0.96	1.00	1.01	1.00	1.06	1.05	1.10
15	1.12	0.98	1.05	0.98	0.98	0.94	0.97	1.00	1.00	1.07	1.07	1.12
20	1.14	1.00	1.05	0.97	0.96	0.91	0.95	0.99	1.00	1.08	1.09	1.15
25	1.17	1.01	1.05	0.96	0.94	0.88	0.93	0.98	1.00	1.10	1.11	1.11
30	1.20	1.03	1.06	0.95	0.92	0.85	0.90	0.96	1.00	1.12	1.14	1.21
35	1.23	1.04	1.06	0.94	0.89	0.82	0.87	0.94	1.00	1.13	1.17	1.25
40	1.27	1.06	1.07	0.93	0.86	0.78	0.84	0.92	1.00	1.15	1.20	1.29
42	1.28	1.07	1.07	0.92	0.85	0.76	0.82	0.92	1.00	1.16	1.22	1.31
44	1.30	1.08	1.07	0.92	0.83	0.74	0.81	0.91	0.99	1.17	123	1.33

Latitude (degrees)	Jan.	Feb.	Mar.	Apr.	May	June	July	Aug.	Sept.	Oct.	Nov.	Dec.
1	2	3	4	5	6	7	8	9	10	11	12	13
46	1.32	1.10	1.07	0.91	0.82	0.72	0.79	0.90	0.99	1.17	1.25	135
48	1.34	1.11	1.08	0,90	0.80	0,70	9,76	0.89	0.99	1,18	1.27	1.37
50	1.37	1.12	1.08	0.89	0.77	0.67	0.74	0.88	0.99	1.19	1.29	1.41

(Source: Thornthwaite, 1948)

Table 15.8: Monthly values of 'i' corresponding to the mean monthly temperature in °C

T °C	0.0	0.1	0.2	0.3	0.4	0.5	0.6	0.7	0.8	0.9
0			0.01	0.01	0.02	0.03	0.04	0.05	0.06	0.07
1	0.09	0.10	0.12	0.13	0.15	0.16	0.18	0.20	0.21	0.23
2	0.25	0.27	0.29	0.31	0.33	0.35	0.37	0.39	0.42	0.44
3	0.46	0.48	0.51	0.53	0.56	0.58	0.61	0.63	0.66	0.69
4	0.71	0.74	0.77	0.80	0.82	0.85	0.88	0.91	0.94	0.97
5	1.00	1.03	1.06	1.09	1.12	1.16	1.19	1.22	1.25	1.29
6	1.32	1.35	1.39	1.42	1.45	1.49	1.52	1.56	1.59	1.63
7	1.66	1.70	1.74	1.77	1.81	1.85	1.89	1.92	1.96	2.00
8	2.04	2.08	2.12	2.15	2.19	2.23	2.27	2.31	2.35	2.39
9	2.44	2.48	2.52	2.56	2.60	2.64	2.69	2.73	2.77	2.81
10	2.86	2.90	2.94	2.c9	3.03	3.08	3.12	3.16	3.21	3.25
11	3.30	3.34	3.39	3.44	3.48	3.53	3,59	3.62	3.67	3.72
12	3.76	3.81	3.86	3.91	3.96	4.00	4.05	4.10	4.15	4.20
13	4.25	4.30	4.35	4.40	4.45	4.50	4.55	4.60	4.65	4.70
14	4.75	4.81	4.86	4.91	4.96	5.01	5.07	5.12	5.17	5.22
15	5.28	5.33	5.38	5.44	5.49	5.55	5.60	5.65	5.71	5.76
16	5.82	5.87	5.93	5.98	6.04	6.10	6.15	6.21	6.26	6.32
17	6.38	6.44	6.49	6.55	6.61	6.66	6.72	6.78	6.84	6.90
18	6.95	7.01	7.07	7.13	7.19	7.25	7.31	7.37	7.43	7.49
19	7.55	7.61	7.67	7.73	7.79	7.85	7.91	7.97	8.03	8.10
20	8.16	8.22	8.28	8.34	8.41	8.47	8.53	8.59	8.66	8.72

T °C	0.0	0.1	0.2	0.3	0.4	0.5	0.6	0.7	0.8	0.9
21	8.78	8.85	8.91	8.97	9.04	9.10	9.17	9.23	9.29	9.36
22	9.42	9.49	9.55	9.62	9.68	9.75	9.82	9.88	9.95	10.01
23	10.08	10.15	10.25	10.28	10.35	10.41	10.48	10.55	10.62	10.68
24	10.75	10.82	10.89	10.95	11.02	11.09	11.16	11.23	11.30	11.37
25	11.44	11.50	11.57	11.64	11.71	11.78	11.85	11.92	11.99	12.06
26	12.13	12.21	12.28	12.35	12.42	12.49	12.56	12.63	12.70	12.78
27	12.85	12.92	12.99	13.07	13.14	13.21	13.28	13.36	13.43	13.50
28	13.58	13.65	13.72	13.80	13.87	13.94	14.02	14.09	14.17	14.24
29	14.32	14.39	14.47	14.54	14.62	14.69	14.77	14.87	14.92	14.99
30	15.07	15.15	15.22	15.30	15.38	15.45	15.53	15.61	15.68	15.76
31	15.84	15.92	15.99	165.07	16.15	16.23	16.30	16.38	16.46	16.54
32	16.62	16.70	16.78	16.85	16.93	17.01	17.09	17.17	17.25	17.33
33	17.41	17.49	17.57	17.65	17.73	17.81	17.89	17.97	18.05	18.13
34	18.22	18.30	18.38	18.46	18.54	18.62	18.70	18.79	18.87	18.95
35	19.03	19.11	19.20	19.28	19.36	19.45	19.53	19.61	19.69	19.78
36	19.86	19.95	20.03	20.11	20.20	20.28	20.36	20.45	20.53	20.62
37	20.70	20.79	20.87	20.96	21.04	21.13	21.21	21.30	21.38	21.47
38	21.56	21.64	21.73	21.81	21.90	21.99	22.07	22.16	22.25	22.33
39	22.42	22.51	22.59	22.68	22.77	22.86	22.95	23.03	23.12	23.21
40	23.30									

(Source: Thornthwaite, 1948)

Table 15.9: Extra-terrestrial radiation Q_Δ expressed in equivalent evaporation in mm/day

J	F	M	A	M	J	J	A	S	O	N	D
3.8	6.1	9.4	12.7	15.8	17.1	16.4	14.1	10.9	7.4	4.5	3.2
4.3	6.6	9.8	13.0	15.9	17.2	16.5	14.3	11.2	7.8	5.0	3.7
4.9	7.1	10.2	13.3	16.0	17.2	16.6	14.5	11.5	8.3	5.5	4.3
5.3	7.6	10.6	13.7	16.1	17.2	16.6	14.7	11.9	8.7	6.0	4.7
5.9	8.1	11.0	14.0	16.2	17.3	16.7	15.0	12.2	9.1	6.5	5.2
6.4	8.6	11.4	14.3	16.4	17.3	16.7	15.2	12.5	9.6	7.0	5.7

J	F	M	A	M	J	J	A	S	O	N	D
6.9	9.0	11.8	14.5	16.4	17.2	16.7	15.3	12.8	10.0	7.5	6.1
7.4	9.4	12.1	14.7	16.4	17.2	16.7	15.4	13.1	10.6	8.0	6.6
7.9	9.8	12.4	14.8	16.5	17.1	16.8	15.5	13.4	10.8	8.5	7.2
8.3	10.2	12.8	15.0	16.5	17.0	16.8	15.6	13.6	11.2	9.0	7.8
8.8	10.7	13.1	15.2	16.5	17.0	16.8	15.7	13.9	11.6	9.5	8.3
9.3	11.1	13.4	15.3	16.5	16.8	16.7	15.7	14.1	12.0	9.9	8.8
9.8	11.5	13.7	15.3	16.4	16.7	16.6	15.7	14.3	12.3	10.3	9.3
10.2	11.9	13.9	15.4	16.4	16.6	16.5	15.8	14.5	12.6	10.7	9.7
10.7	12.3	14.2	15.5	16.3	16.4	16.4	15.8	14.6	13.0	11.1	10.2
11.2	12.7	14.4	15.6	16.3	16.4	16.4	15.9	14.8	13.3	11.6	10.7
11.6	13.0	14.6	15.6	16.1	16.1	16.1	15.8	14.9	13.6	12.0	11.1
12.0	13.3	14.7	15.6	16.0	15.9	15.9	15.7	15.0	13.9	12.4	11.6
12.4	13.6	14.9	15.7	15.8	15.7	15.7	15.7	15.1	14.1	12.8	12.0
12.8	13.9	15.1	15.7	15.7	15.5	15.5	15.6	15.2	14.4	13.3	12.5
13.2	14.2	15.3	15.7	15.5	15.5	15.3	15.5	15.3	14.7	13.6	12.9
13.6	14.5	15.3	15.6	15.3	15.0	15.1	15.4	15.3	14.8	13.9	13.3
13.9	14.8	15.4	15.4	15.1	14.7	14.9	15.52	15.3	15.0	14.2	13.7
14.3	15.0	15.5	15.5	14.9	14.4	14.6	15.1	15.3	15.1	14.5	14.1
14.7	15.3	15.6	15.3	14.6	14.2	14.3	14.9	15.3	15.3	14.8	14.4
15.0	15.5	15.7	15.3	14.4	13.9	14.1	14.8	15.3	15.4	15.1	14.8

Lat	J	F	M	A	M	J	J	A	S	O	N	D
50^0	17.5	14.7	10.9	7.0	4.2	3.1	3.5	5.5	8.9	12.9	16.5	18.2
48^0	17.6	14.9	11.2	7.5	4.7	3.5	4.0	6.0	9.3	13.2	16.6	18.2
46^0	17.7	15.1	11.5	7.9	5.2	4.0	4.4	6.5	9.7	13.4	16.7	18.3
44^0	17.8	15.3	11.9	8.4	5.7	4.4	4.9	6.9	10.2	13.7	16.7	18.3
42^0	17.8	15.5	12.2	8.8	6.1	4.9	5.4	7.4	10.6	14.0	16.8	18.3
40^0	17.9	15.7	12.5	9.2	6.6	5.3	5.9	7.9	11.0	14.2	16.9	18.3

Lat	J	F	M	A	M	J	J	A	S	O	N	D
38^0	17.9	15.8	12.8	9.6	7.1	5.8	6.3	8.3	11.4	14.4	17.0	18.3
36^0	17.9	16.0	13.2	10.1	7.5	6.3	6.8	8.8	11.7	14.6	17.0	18.2
34^0	17.8	16.1	13.5	10.5	8.0	6.8	7.2	9.2	12.0	14.9	17.1	18.8
32^0	17.8	16.2	13.8	10.9	8.5	7.3	7.7	9.6	12.4	15.1	17.2	18.1
30^0	17.8	16.4	14.0	11.3	8.9	7.8	8.1	10.1	12.7	15.3	17.3	18.1
28^0	17.7	16.4	14.3	11.6	9.3	8.2	8.6	10.4	13.0	15.4	17.2	17.9
26^0	17.6	16.4	14.4	12.0	9.7	8.7	9.1	10.9	13.2	15.5	17.2	17.8
24^0	17.5	16.5	14.6	12.3	10.2	9.1	9.5	11.2	13.4	15.6	17.1	17.7
22^0	17.4	16.5	14.8	12.6	10.6	9.6	10.0	11.6	13.7	15.7	17.0	17.5
20^0	17.3	16.5	15.0	13.0	11.0	10.0	10.4	12.0	13.9	15.8	17.0	17.4
18^0	17.1	16.5	15.1	13.2	11.4	10.4	10.8	12.3	14.1	15.8	16.8	17.1
16^0	16.9	16.4	15.2	13.5	11.7	10.8	11.2	12.6	14.3	15.8	16.7	16.8
14^0	16.7	16.4	15.3	13.7	12.1	11.2	11.6	12.9	14.5	15.8	16.5	16.6
12^0	16.6	16.3	15.4	14.0	12.5	11.6	12.0	13.2	14.7	15.8	16.4	16.5
10^0	16.4	16.3	15.5	14.2	12.8	12.0	12.4	13.5	14.8	15.9	16.2	16.2
8^0	16.1	16.1	15.5	14.4	13.1	12.4	12.7	13.7	14.9	15.8	16.0	16.0
6^0	15.8	16.0	15.6	14.7	13.4	12.8	13.1	14.0	15.0	15.7	15.8	15.7
4^0	15.5	15.8	15.6	14.9	13.8	13.2	13.4	14.3	15.1	15.6	15.5	15.4
2^0	15.3	15.7	15.7	15.1	14.1	13.5	13.7	14.5	15.2	15.5	15.3	15.1
0^0	15.0	15.5	15.7	15.3	14.4	13.9	14.1	14.8	15.3	15.4	15.1	14.8

Table 15.10: Mean daily maximum duration of bright sunshine hours N for different months and latitudes

Northern Lats. / Southern Lats.	Jan. / July	Feb. / Aug.	Mar. / Sept.	Apr. / Oct.	May / Nov.	June / Dec.	July / Jan.	Aug. / Feb.	Sept. / Mar.	Oct. / Apr.	Nov. / May	Dec. / June
50^0	8.5	10.1	11.8	13.8	15.4	16.3	15.9	14.5	12.7	10.8	9.1	8.1
48^0	8.8	10.2	11.8	13.6	15.2	16.0	15.6	14.3	12.6	10.9	9.3	8.3
46^0	9.1	10.4	11.9	13.5	14.9	15.7	15.4	14.2	12.6	10.9	9.5	8.7

Northern Lats. Southern Lats.	Jan. July	Feb. Aug.	Mar. Sept.	Apr. Oct.	May Nov.	June Dec.	July Jan.	Aug. Feb.	Sept. Mar.	Oct. Apr.	Nov. May	Dec. June
44^0	9.3	10.5	11.9	13.4	14.7	15.4	15.2	14.0	12.6	11.0	9.7	8.9
42^0	9.4	10.6	11.9	13.4	14.6	15.2	14.9	13.9	12.9	11.1	9.8	9.1
40^0	9.6	10.7	11.9	13.3	14.4	15.0	14.7	13.7	12.5	11.2	10.0	9.3
35^0	10.1	11.0	11.9	13.1	14.0	14.5	14.3	13.5	12.4	11.3	10.3	9.8
30^0	10.4	11.1	12.0	12.9	13.6	14.0	13.9	13.2	12.4	11.5	10.6	10.2
25^0	10.7	11.3	12.0	12.7	13.3	13.7	13.5	13.0	12.3	11.6	10.9	10.6
20^0	10.0	11.5	12.0	12.6	13.1	13.3	13.2	12.8	12.3	11.7	11.2	10.9
15^0	11.3	11.6	12.0	12.5	12.8	13.0	12.9	12.6	12.2	11.8	11.4	11.2
10^0	11.6	11.8	12.0	12.3	12.6	12.7	12.6	12.4	12.1	11.8	11.6	11.5
5^0	11.8	11.9	12.0	12.2	12.3	12.4	12.3	12.3	12.1	12.0	11.9	11.8
0^0	12.1	12.1	12.1	12.1	12.1	12.1	12.1	12.1	12.1	12.1	12.1	12.1

(Source: Doorenbos and Pruitt, 1975)

Table 15.11: Values of a T_a^4 for various temperatures when computing evaportanspiration by the Penman method (after Criddle)

Temperature (^{0}K)	σT_a^4 (mm water/day)
270	10.73
275	11.51
280	12.40
285	13.20
290	14.26
295	15.30
300	16.34
305	17.46
310	18.60
315	19.85
320	21.15
325	22.50

(Source: Israelsen and Hansen, 1962)

Note: Heat of vaporization was assumed to be constant at 590 cal/gm of water.

Penman Method

Penman (1948) developed an equation for evaporation from open water surface taking into consideration several climatic parameters based on combination of energy balance and sink strength. The developed equation for estimation of evaporation is expressed as,

$$E_0 = \frac{\Delta Q_n + \gamma E_a}{\Delta + \gamma} \quad ....(15.11)$$

where,

E = evaporation from open water surface (mm/day)

Δ = slope of saturation vapour pressure *vs.* temperature curve (de_a/dT_a) at the mean air temperature T_{as} mm Hg per °C (Fig. 15.6),

e_a = saturation vapour pressure of the evaporating surface (ea) in mm Hg at mean air temperature Ta (Fig. 15.7). (Here ea is considered equal to ea by assuming zero temperature gradient between surface (s) and air (a) temperature).

T_a = mean air temperature in ^{0}K = 273 + ^{0}C,

Q_n = net radiation (mm of water) = Q_Δ (1–r) $(0.18 + 0.55\frac{n}{N})$ - $\sigma T_a^4(0.55–0.092)\sqrt{e_d})$ $(0.10+0.90\frac{n}{N})$,

r = reflection coefficient of evaporating surface, 0.06 for open in water surface,

Q_Δ = Angot's value of mean monthly extra-terrestrial radiation in mm of water/day (Table 15.9),

$\frac{n}{N}$ = ratio between actual and possible hours of bright sunshine (Table 15.10 for N),

σ = Stefan-Boltzman constant (Table 15.11 for $\sigma T^4{}_{a)}$),

e_d = saturation vapour pressure of the atmosphere, in mm Hg, at dew point temperature = $\frac{\text{RH(mean)}}{100} \times e_a$, where RH is mean relative humidity,

Υ = psychrometfic constant or the ratio of specific heat of air to the latent heat of evaporation of water (0.49 [or 0°C and mm Hg),

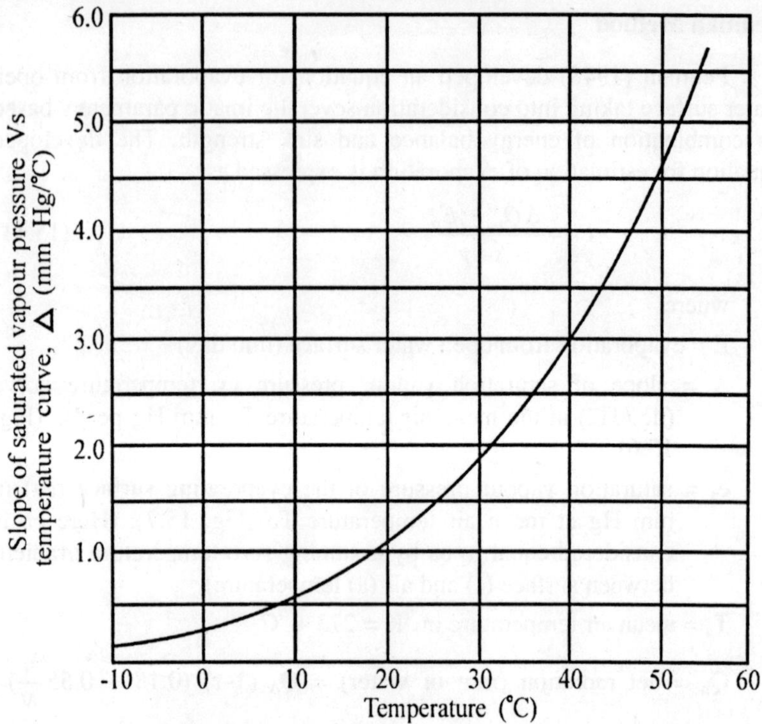

Fig. 15.6: Slope of saturation vapour pressure *v/s* temperature curve

E_a = an aerodynamic component where, e_s is considered equal to e_a = 0.35 (e_a - e_d) (1+0.0098 u_2), and

u_2 = wind speed in miles/day at 2 metres height = $u_1 \left(\frac{\log 6.6}{\log h}\right)$

where, u_1 is wind speed in miles per day at any other height h in feet.

The Penman equation determines evaporation from a free water surface, the results must be modified to obtain evapotranspiration (ET) estimates for crops. Therefore, for estimation of ET_0 (crop), it should be multiplied with the crop coefficient (k). The values of K for short grass are given by Penman as 0.6 from November to February, 07 from March, April, September and October and 0.8 from May to August. For accurate estimates of ET, the values of K may be locally determined. Penman formula can be used for direct estimation of PET by using an appropriate value of reflection coefficient (r) for fresh green vegetation, which is taken as 0.25 for most crops. For more accuracy plant resistance to transpiration and wind functions also need to be included but this leads to many complexities.

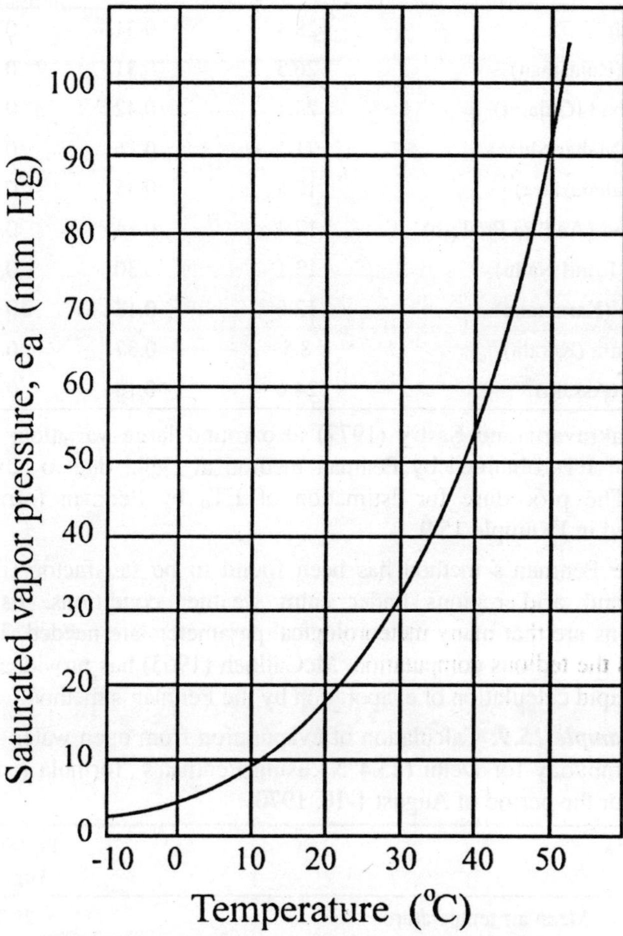

Fig. 15.7: Saturated vapour pressure *v/s* temperature

Several other constants used in Penman's equation also need to be evaluated with reference to the climatic conditions of a place. The values of 'a' and 'b' in the equation, $Q = Q_\Delta$ (a + b n/N) for instance varied from place to place (Doorenbos and Pruitt, 1975). Gangopadhyay *et al.* (1970) has worked out these values for different places in India which are shown in Table 15.12.

Table 15.12: Values of constants, 'a' and 'b' in Penman formula for different places of India

Place	Latitude $_0$N	Value of 'a'	Value of 'b'
New Delhi	28.4	0.31	0.46
Jodhpur (Rajasthan)	26.3	0.31	0.49
Ahmedabad (Gujarat)	23.1	0.42	0.30
Nagpur (Maharashtra)	21.1	0.16	0.68
Pune (Maharashtra)	18.5	0.35	0.40
Hyderabad (Andhra Pradesh)	17.4	0.14	0.55
Madras (Tamil Nadu)	13.1	0.30	0.44
Banglore (Karnataka)	13.0	0.18	0.62
Trivendrum (Kerala)	8.5	0.37	0.38
Shillong (Assam)	24.6	0.18	0.66

Chakravarti and Sastry (1977) also found large variations in the values of ET_0 obtained by Penman method at Delhi due to advective effect. The procedure for estimation of ET_0 by Penman formula is illustrated in Example 15.9.

The Penman's method has been found to be satisfactory in both humid and arid regions under calm weather conditions. Its main limitations are that many meteorological parameters are needed. Thus, it involves the tedious computation. McCulloch (1965) has provided tables for the rapid calculation of evaporation by the Penman's method.

Example 15.9: Calculation of evaporation from open water surface (E_0) in mm/day for Delhi (23.4°N) using Penman's formula (equation 15.11) for the period of August 1-10, 1970

A.	**Data**	**Period 1-10 Aug. 1970**
1.	Mean air temperature	28.7°C
2.	Mean relative humidity (%)	88.5
3.	Mean sunshine hours (n)	3.43
4.	Possible sunshine hours (N from Table 15.8)	13.1
5.	Value of the ratio n/N	0.26
6.	Wind speed at 3 m height in miles/day (u_1) = 90.02 miles/day and at 2 m height = u_1 x 0.93 = u_2 = 83.7	83.7

7.	Extra-terrestrial radiation in mm/ day (Q_Δ from table 15.7)	15.7
8.	Reflection coefficient	0.06

B. Solving expression Q_Δ (1 - r) (a + b . n/N)

9.	(1 - r) =(1 - 0.06) =	0.94
10.	(a + b . n/N) = (0.18 + 0.55 x 0.26) =	0.32
11.	Item No.7 x Item No.9 x Item No. 10 =	4.72

C. Solving expression σT_a^4 (0.56 - 0.092 $\sqrt{e_a}$) (0.10 + 0.90 n/N)

12.	Vapour pressure: (i) e_a (from Fig. 15.7 =	29.2
	(ii) e_a (recorded) =	25.5
	(iii) $\sqrt{e_d}$ =	5.04
13.	σT_a^4 (from Table 15.9) =	16.72
14.	$(0.56 - 0.092\sqrt{e_d})$ = (0.55 - 0.092 x 5.04) =	0.10
15.	(0.10 + 0.90 . n/N) (0.10 + 0.90 x 0.26) =	0.23
16.	Item No. 13 x Item No. 14 x Item No. 15 =	0.38

D. Solving for Q_a = Item No. 11 - Item No. 16 = (4.72 - 0.38) = 4.34

E. Solving for $E_a = 0.35 (e_a - e_d) (1 + 0.0098\ u_2)$
= 0.35 (29.2 - 25.5) (1 + 0.0098 x 83.7) = 2.37

F. Solving for $E_0 = \dfrac{\Delta Q_n + \gamma E_a}{\Delta + \gamma}$

$$= \frac{1.65 \times 4.34 + 0.49 \times 2.37}{1.65 + 0.49} = 3.89 \text{ mm/day}$$

(For Δ value, see Fig. 15.6)

G.	(i) PET = E_0 x K for August = 3.89 x 0.8 =	3.11 mm/day
	(ii) PET by using 0.25 as the value of 'r' for fresh green vegetation in equation (15.8) =	3.15 mm/day
	(iii) PET by using in equation (18.8) the value of r = 0.25 and the values of 'a' and 'b' in above item 10 as 0.31 and 0.46 respectively for Delhi given by Gangopadhyay *et al.* (1970) =	4.24 mm/day

Modified Penman formula

This method also requires more climatic parameters based intensive studies in the world than the other methods. Doorenbos and Pruitt (1975) modified Penman method for accurate for estimation of the reference crop ET and gave tables to facilitate the necessary computations. The equation for estimation of the reference crop evaporation is given as under:

$$ET_0^* = W \cdot R_n + (1 - W) \cdot f(u) \cdot (e_a - e_d) \qquad ....(15.12)$$

radiation term + aerodynamic term

where,

ET_0^* = the reference crop evapotranspiration (mm/day (not adjusted),

e_a = saturation vapour pressure (mbar) at the mean air temperature in ^{0}C (Table 15.10),

e_a = mean actual vapour pressure of the air in mbar = ea x $\frac{RH\ Mean}{100}$, where, RH = relative humidity. This can also be found out from dry and wet bulb temperatures or dew point temperature (see original reference),

f (u) = a wind related function (Table 15.11), and estimated as, f (u) $= 0.27\ \Delta + \frac{42}{100}$ where, 42 is the mean win velocity (km/dag) at 2 m height,

(1-W) = a temperature and elevation related weighting factor for the effect of wind and humidity on ET_0 (Table 15.13),

W = a temperature and elevation related weighting factor for the effect of radiation on ET_0 (Table 15.14), and

R_n = net radiation (same as $Q_n = R_{ns} - R_{nt}$)

R_n is estimated as follows,

R_{ns} = the net incoming shortwave solar radiation = R_Δ (1 - a) (0.25+0.50 n/N) in which R_Δ is same as Q_Δ or extra-terrestrial radiation expressed in equivalent evaporation in mm/day, given in Table 15.7; n/N is the same as explained in equation (15.8), and a is same as 'r' or reflection coefficient; the value of which is taken as 0.25 for most crops (Table 15.14 gives conversion factors for RA to R, for a given reflection of 25% and ratios of n/N), and

R_{nl} = the net long wave radiation f(t) . f(ed) . f(n/N), the values of which are given in Tables 15.15, 15.16 and 15.17, respectively.

For accurate estimation of ET_0, the unadjusted reference crop evapotranspiration, ET is adjusted for day and night-time weather conditions with the help of Fig. 15.6. This adjustment and the revised wind function term are the important variations made by Doorenbos and Pruitt to the original Penman method. The modified Penman relationship gives very satisfactory values of crop ET.

Table 15.13: Saturation vapour pressure e_a in mbar as a function of mean air temperature t in 0^0C

Temp. 0^0C	0	1	2	3	4	5	6	7	8	9	10	11	12	13	14
E_a mbar	6.1	6.6	7.1	7.6	8.1	8.7	9.4	10.0	10.7	11.5	12.3	13.1	14.0	15.0	16.1

Temp. 0^0C	15	16	17	18	19	20	21	22	23	24	25	26	27	28
E_a mbar	17.0	18.2	19.4	20.6	22.0	23.4	24.9	26.4	28.1	29.8	31.7	33.6	35.7	37.8

Temp. 0^0C	29	30	31	32	33	34	35	36	37	38	39
E_a mbar	40.1	42.4	44.9	47.6	50.3	53.2	56.2	59.4	62.8	66.3	69.9

Table 15.14: Values of wind function, f (u) = 0.27 $\left(1+\frac{u_2}{100}\right)$ for wind run at a 2m height* in km/day

Wind km/day	0	10	20	30	40	50	60	70	80	90
	-	0.30	032	0.35	0.38	0.41	0.43	0.46	0.49	0.51
100	0.54	0.57	0.59	0.62	0.65	0.67	0.70	0.73	0.76	0.78
200	0.81	0.84	0.86	0.89	0 92	094	0.97	1.00	1.03	1.05
300	1.08	1.11	1.13	1.16	1.19	1.21	1.24	1.27	1.30	1.32
400	1.35	1.38	1.40	1.43	1.46	1.39	1.51	1.54	1.57	1.59
500	1.62	1.65	1.67	1.70	1.73	1.76	1.78	1.81	1.84	1.80
000	1.89	1.92	1.94	1.97	2.00	2.02	2.05	2.08	2.11	2.15
700	2.16	2.19	2.21	2.24	2.27	2.29	2.32	2.35	2.38	2.40
800	2.43	2.46	2.48	2.51	2.54	2.56	2.59	2.62	2.65	2.65

900	2.70

*** where wind data are not collected at 2 m height the appropriate corrections for wind measurements taken at different heights are given below:**

Measurement height, m	0.5	1.0	1.5	2.0	3.0	4.0	5.0	6.0	10.0
Correction factor	1.35	1.15	1.06	1.00	0.93	0.88	0.85	0.83	0.77

Table 15.15: Values of weighting factor (1 - W) for the effect of wind and humidity on ET at different temperatures and altitudes

Temp. ^{0}C		2	28	30	32	34	36	38	40
(1 - W) at altitude	m								
	0	.25	.23	.22	.20	.18	.17	.16	.15
	500	.24	.22	.21	.19	.18	.16	.15	.14
	1000	.23	.21	.20	.18	.17	.15	.14	.13
	2030	.21	.19	.18	.16	.15	.14	.13	.12
	3000	.19	.18	.16	.15	.14	.13	.12	.11
	4000	.18	.16	.15	.14	.13	.12	.11	.10

Temp. ^{0}C		2	28	30	32	34	36	38	40
(1 - W) at altitude	m								
	0	.25	.23	.22	.20	.18	.17	.16	.15
	500	.24	.22	.21	.19	.18	.16	.15	.14
	1000	.23	.21	.20	.18	.17	.15	.14	.13
	2030	.21	.19	.18	.16	.15	.14	.13	.12
	3000	.19	.18	.16	.15	.14	.13	.12	.11
	4000	.18	.16	.15	.14	.13	.12	.11	.10

W = Δ/(Δ + ϒ) where Δ is the rate of change of the saturation vapour pressure with temperature and ϒ is the psychrometric constant.

Table 15.16: Values of weighting factor W for the effect of radiation on ET_0 at different temperatures and altitudes

Altitude m	Temperature °C																			
	2	**4**	**6**	**8**	**10**	**12**	**14**	**16**	**18**	**20**	**22**	**24**	**26**	**28**	**30**	**32**	**34**	**36**	**38**	**40**
	Values of W																			
0	.43	.46	.49	.52	.55	.58	.61	.64	.66	.68	.71	.73	.75	.77	.78	.80	.82	.83	.84	.85
500	.44	.48	.51	.54	.57	.60	.62	.65	.67	.70	.72	.74	.76	.78	.79	.81	.82	.84	.85	.86
1000	.46	.49	.52	.55	.58	.61	.64	.66	.69	.71	.73	.75	.77	.79	.80	.82	.83	.85	.86	.87
2030	.49	.52	.55	.58	.61	.64	.66	.69	.71	.73	.75	.77	.79	.81	.82	.84	.85	.86	.87	.88
3000	.52	.55	.58	.61	.64	.66	.69	.71	.73	.75	.77	.79	.81	.82	.84	.85	.86	.87	.88	.89
4000	.54	.58	.61	.64	.66	.69	.71	.73	.75	.77	.79	.81	.82	.84	.85	.86	.87	.89	.90	.90

Table 15.17: Conversion factor for extra-terrestrial radiation to net solar radiation for a given reflection α of 25% and different ratios of actual to maximum sunshine hours $(1 - \alpha)(0.25 + 0.50\ n/N)$

n/N	0.0	.05	.10	.15	.20	.25	.30	.35	.40	.45	.50	.55	.60	.65	.70	.75	.80	.85	.90	.95	1.00
$(1 - \alpha)(0.25 + 0.50\ n/N)$	.19	.21	.22	.24	.26	.28	.30	.32	.34	.36	.37	.39	.41	.43	.45	.47	.49	.51	.52	.54	.56

Table 15.18: Correction for temperature f (t) on long wave radiation R_{nl}

T °C	**0**	**2**	**4**	**6**	**8**	**10**	**12**	**14**	**16**	**18**	**20**	**22**	**24**	**26**	**28**	**30**	**32**	**34**
$f(t) = \sigma TK^4$	11.0	11.4	11.7	12.0	12.4	12.7	13.1	13.5	13.8	14.2	14.6	15.0	15.4	15.9	16.3	16.7	17.2	18.1

Table 15.19: Correction for vapour pressure f (e_d) on long wave radiation R_{nl}

e_d mbar	6	8	10	12	14	15	18	20	22	24	26	28	30	32	34	36	38	40
$f(e_d) = 0.34 - 0.444\sqrt{e_d}$	.21	.22	.20	.19	.18	.16	.15	.14	.13	.12	.12	.11	.10	.09	.08	.08	.07	.06

Table 15.20: Correction for the ratio actual maximum bright sunshine hours f (n/N) on long wave radiation R_{nl}

n/N	0.0	.05	.10	.15	.20	.25	.30	.35	.40	.45	.50	.55	.60	.65	.70	.75	.80	.85	.90	.95	1.00
f (n/N) = 0.1 + 0.9 n/N	.10	.15	.19	.24	.28	.33	.37	.42	.46	.51	.55	.60	.64	.69	.73	.78	.82	.87	.91	.96	1.0

Steps for estimation of ET_0 on using Fig. 15.8

ET ~ 1.2 ET_0*

Important in areas with moderate to high radiation (>8 mm/day) during summer months, which consistently have low humidity, winds during much of the day (>4 m/sec) and calm night-time conditions with high night humidity values approaching 100%. Curve in Fig. 15.8 applies for day-night wind ratios of about 3 to 4 with very high RHmax.

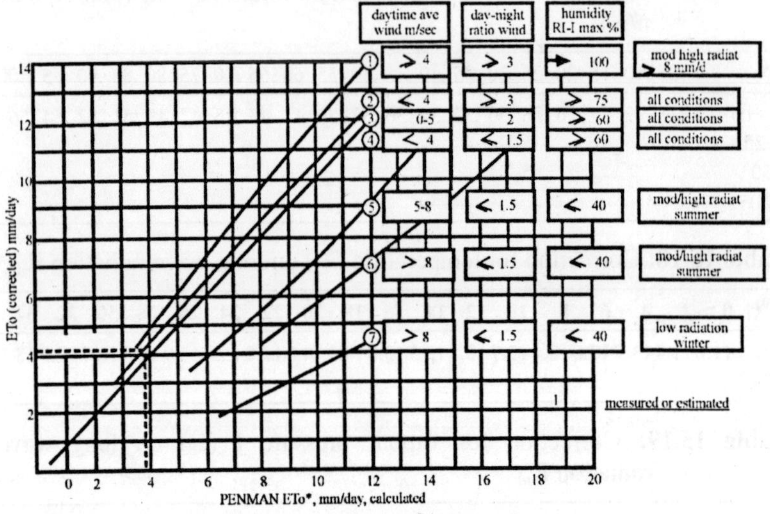

Fig. 15.8: Correction on calculated ET_0^0 (Penman) for day and night time, wind and humidity conditions

ET_0 ~ 1.05 to 1.10 ET_0*

Applies in areas where daytime winds are <4 m/sec and nights are very calm and humid, *i.e.* RHmax > 75%. The daily wind distribution should give a day- night ratio of 3 or more. To adjust $ET0_0$*, Curve b in Fig. 18.8 should be used.

$ET_0 \sim ET_0$*

Daytime wind speeds of about twice those during the night-time are generally found and maximum relative humidity >60%. No adjustment of Penman ET_0* is required curved c.

$ET_0 \sim 0.75$ to $0.95\ ET_0$*

ET_0 will be slightly over predicted in areas with moderate to high radiation and with wind speeds <4 m/sec but about equal during the day and night. When relative humidity during night-time is > 60% curve d in Fig. 15.8 applies.

$ET_0 \sim 0.65$ to $0.80\ ET_0$*

Relatively rare but applies for spring, summer and autumn conditions with moderate to high radiation and when wind speeds during the day and night are between 5 and 8 m/sec with maximum relative humidity < 40%. Curve e in Fig 15.5 should be used.

$ET_0 \sim 0.55$ to $0.65\ ET_0$*

Also rare, but applies during spring, summer and autumn with moderate to high radiation and when wind speeds during day and night are > 8 m/sec arid with relative humidity in day and night < 40%. See curve f in Fig. 15.8.

$ET_0 \sim 0.30$ to $0.35\ ET_0$*

Very rare, but will apply under very strong winds of > 8 m/sec during both day and night, while relative humidity, day and night, is low (< 40%). Curve g in Fig. 15.8 will apply but only when radiation is low, *i.e.* during late autumn and winter (R_{ns} <4 mm/day).

Christiansen method

Christiansen (1968) developed a revised equation estimation of pan evaporation from climatic data when reliable measured pan evaporation data are not available for estimation of evapotranspiration. It is expressed as,

$$E_0 = K_{ev} \,.\, R \,.\, C_t \,.\, C_{10} \,.\, C_h \,.\, C_3 .\, C_6 \,.\, C_m \qquad \text{....(15.13)}$$

where,

E_0 = computed pan evaporation equivalent to Class A pan evaporation,

K_{ev} = dimensionless empirically developed constant, the value of which is given by Christiansen as 0.473,

R = extra-terrestrial radiation in the same evaporation units as E_v (Table 15.18) and C_t, C_w, C_h and C_6 coefficients for temperature, wind velocity, relative humidity, percent of possible sunshine and elevation, respectively and C_m monthly coefficient or factor by which all the basic formulae would have to be multiplied to obtain the measured evaporation and averaged to obtain mean monthly values C_m. The values of C_m mostly range between 0.90 to 1.10 and vary from latitude to latitude.

Each coefficient of an equation (15-13)_is expressed by a linear, second degree or possibly a third or fourth degree equation. There are coefficients are dimensionless. The standard values were approximately the mean values for the data analyzed. The value of the constant K_{ev} depends on the selection of standard values for each climatic factor. A expression, second degree equation of the coefficients.

$$C\infty = A + BX + CX^2 \qquad ....(15.14)$$

where, X represents a dimensionless parameter of a climatic or other factor which is the ratio of the climatic or other factor to a standard value of the factor for which the coefficient is unity. The simplified and non-dimensionless equations for different coefficients are given as under:

$$C_t = 0.393 + 0.02796\ T_0 + 0.0001189\ T_0^2$$

where,

T_0 = mean monthly temperature (^{0}C) from Table 15.19

$$C_w = 0.708 + 0.00339\ W - 0.0000039\ W^2$$

where,

W = wind velocity in km/day at 0.6 m height from Table 15.19. The equivalent wind velocities at 2 m height are also given in Table 15.19. Wind velocity observed at other heights may be converted to desired height by using factors given in Table 15.12.

$$C_{hn} = 1.25 - 0.87\ H_n + 0.75\ H_n^2 - 0.85\ H_n^4$$

where,

H_n = mean monthly relative humidity at noon expressed decimally, *i.e.* 40% 0.40 (Table 15.19)

or $C_{hm} = 1.25 — 0.37\ H_m - 0.60\ H_m^5$

where,

H_m = mean relative humidity for the month expressed decimally (Table 15.19). The approximate relationship between H_n and H_m is given as $H_n = 0.40\ H_m + 0.60\ H_m^2$

$$C_0 = 0.542 + 0.80\,S - 0.78\,S^2 + 0.62\,S^3$$

where,

S = mean sunshine percentage expressed decimally, *i.e.*, S = 80% = 0.80 (Table 15.19)

$$C_0 = 0.970 + 0.0000984E$$

where,

E = elevation (m) from Table 15.20

The coefficients in dimensionless forms are as below:

For mean temperature (^{0}C) for $T_0 = 20^0$

$$C_t = 0.393 + 0.5592\left(\frac{T_e}{T_o}\right) + 0.04756\left(\frac{T_e}{T_o}\right)^2$$

For mean wind velocities above the evaporative pan (at 0.6 m height), for W_o = 60 miles/day or 96.56 km/day

$$C_w = 0.708 + 0.3276\left(\frac{W}{W_o}\right) - 0.036\left(\frac{W}{W_o}\right)^2$$

For mean relative humidity at noon or average humidity for 11 hr and 17 hr, for H_o = 40% or 0.40

$$C_{hn} = 1.250 - 0.348\left(\frac{H}{H_o}\right) + 0.120\left(\frac{H}{H_o}\right)^2 - 0.0218\left(\frac{H}{H_o}\right)^4$$

or, where mean humidity is available,

$$C_{hm} = 1.250 - 0.212\left(\frac{H_m}{H_{mo}}\right) - 0.038\left(\frac{H_m}{H_{mo}}\right)^5$$

in which H_{mo} = 57.4% or 0.574

For mean sunshine percentage, for S_o = 80% or 0.80

$$C_8 = 0.542 + 0.64\left(\frac{S}{S_o}\right) - 0.4992\left(\frac{S}{S_o}\right)^3 + 0.3174\left(\frac{S}{S_o}\right)^3$$

For elevation, for E_o = 305 m,

$$C_o = 0.970 + 0.030\left(\frac{E}{E_o}\right)$$

Table 15.21: Mean monthly values of extra-terrestrial radiation expressed as equivalent evaporation at 20ºC, in mm per month of water*

Latitude (degrees)	Jan.	Feb. 28 days	Feb. 29 days	Mar.	Apr.	May	June	July	Aug.	Sept.	Oct.	Nov.	Dec.
1	2	3	4	5	6	7	8	9	10	11	12	13	14
North 60	45	99	102	217	337	458	496	486	391	260	144	58	30
50	117	167	173	286	384	483	503	501	430	317	221	131	97
45	154	201	208	318	404	491	505	505	445	343	257	168	132
40	191	234	242	348	421	497	505	508	458	367	291	204	169
35	229	265	275	374	435	500	503	508	468	388	322	240	207
30	267	295	306	398	446	500	498	505	474	407	350	274	245
25	304	323	334	419	455	498	490	500	478	422	376	307	283
20	339	348	361	437	460	491	479	491	478	435	400	339	320
15	372	371	384	451	461	482	465	479	475	442	422	368	356
10	402	391	405	462	459	469	448	464	468	448	441	395	390
5	429	408	423	470	453	454	428	446	458	450	457	420	421
Equator	453	422	437	473	445	435	406	424	445	449	468	441	450
South 5	474	434	449	473	433	413	382	401	429	443	474	459	474
10	493	442	458	469	417	389	354	374	409	436	478	475	496
15	509	446	462	462	399	362	325	345	386	424	478	487	514
20	521	449	465	452	378	333	294	375	361	409	475	497	527
25	531	449	465	439	359	302	261	282	335	392	470	502	538
30	537	446	462	422	327	269	227	248	305	371	462	504	548
35	541	438	454	402	297	234	192	213	274	248	450	503	553
40	539	428	443	379	266	198	157	177	241	322	435	499	555
50	530	397	411	322	203	129	91	106	160	262	392	484	550

***Christiansen computed the values in inches from data of Napier Shaw. These values are multiplied by 25.4 and reported here in mm.**

(Source: Christiansen, 1968)

Table 15.22: Coefficients for temperature, wind velocity, relative humidity and sunshine percentage

Mean Temperature		Wind velocity (km/day)			Relative humidity				Sunshine	
		Anemometer height								
°C	Coefficient C_t	0.6 m W	2.0 m W	Coefficient C_w	Noon H_n (%)	Coefficient C_{hn}	Mean H_m (%)	Coefficient C_{hm}	S (%)	Coefficient C_s
1	2	3	4	5	6	7	8	9	10	11
0	0.393	0	0	0.708	10	1.170	10	1.213	0	0.542
1	0.421	8	27	0.735	15	1.130	15	1.194	4	0.573
2	0.448	16	50	0.762	20	1.105	20	1.176	8	0.601
3	0.478	24	71	0.788	22	1.093	22	1.168	10	0.615
4	0.507	32	87	0.814	24	1.082	24	1.161	12	0.628
5	0.536	40	103	0.839	26	1.071	26	1.153	16	0.653
6	0.565	48	119	0.863	28	1.060	28	1.145	20	0.676
7	0.595	56	132	0.887	30	1.050	30	1.138	22	0.687
8	0.624	64	145	0.911	32	1.039	32	1.130	24	0.698
9	0.654	72	158	0.934	34	1.030	34	1.121	26	0.708
10	0.684	80	169	0.956	36	1.020	36	1.113	28	0.718
11	0.715	89	180	0.978	38	1.010	38	1.105	30	0.729
12	0.746	97	192	1.000	40	1.000	40	1.096	32	0.738
13	0.777	105	201	1.021	42	0.990	42	1.087	34	0.748
14	0.808	113	213	1.042	44	0.981	44	1.077	36	0.758
15	0.839	121	222	1.062	46	0.970	46	1.067	38	0.767
16	0.871	129	232	1.081	48	0.960	48	1.057	40	0.777
17	0.903	137	240	1.100	50	0.949	50	1.046	42	0.786
18	0.935	145	250	1.119	52	0.938	52	1.035	44	0.797
19	0.967	153	258	1.137	54	0.927	54	1.023	46	0.805
20	1.000	161	267	1.155	56	0.914	56	1.010	48	0.815
21	1.033	169		1.172	58	0.902	58	0.996	50	0.824
22	1.066	177	288	1.188	60	0.888	60	0.981	52	0.834
23	1.099	185		1.204	62	0.873	62	0.966	54	0.844

Mean Temperature		Wind velocity (km/day)				Relative humidity			Sun-shine	
		Anemometer height			Noon					
°C	Coeffi-cient C_t	0.6 m W	2.0 m W	Coeffi-cient C_w	H_n (%)	Coeffi-cient C_{hn}	Mean H_m (%)	Coeffi-cient C_{hm}	S (%)	Coefficient C_s
1	2	3	4	5	6	7	8	9	10	11
24	1.133	193	299	1.220	64	0.858	64	0.949	56	0.854
25	1.167	201		1.235	66	0.841	66	0.931	58	0.865
26	1.201	209	316	1.250	68	0.823	68	0.912	60	0.875
27	1.235	217		1.264	70	0.804	70	0.890	62	0.886
28	1.270	225	332	1.277	72	0.784	72	0.868	64	0.897
29	1.304	233		1.290	74	0.762	74	0.843	66	0.908
30	1.339	241	345	1.303	76	0.738	76	0.817	68	0.920
31	1.375	250		1.315	78	0.713	78	0.788	70	0.932
32	1.410	258	364	1.327	80	0.686	80	0.757	72	0.945
33	1.446	266		1.338	82	0.657	82	0.724	74	0.958
34	1.432	274	380	1.348	84	0.625	84	0.688	76	0.971
35	1.518	282		1.358	86	0.592	86	0.650	78	0.985
36	1.555	290	396	1.368	88	0.555	88	0.608	80	1.000
37	1.591	298		1.377	90	0.517	90	0.563	82	1.015
38	1.628	306	411	1.386	95	0.408	95	0.434	84	1.031
39	1.666	314		1.391	100	0.280	100	0.280	86	1.047
40	1.703	322	427	1.401					88	1.064
41	1.741								90	1.082
42	1.779								92	1.100
43	1.817								94	1.119
44	1.855								96	1.139
									98	1.160
									100	1.181

(Source: Christiansen, 1968)

Table 15.23: Coefficient of elevation, C_e

Elevation E (metres)	Coefficient C_s	Elevation E (metres)	Coefficient C_s
0	0.970	1067	1.075
30	0.973	1097	1.078
61	0.976	1128	1.081
91	0.979	1158	1.084
122	0.982	1189	1.087
152	0.985	1219	1.090
183	0.988	1250	1.093
213	0.991	1280	1.096
244	0.994	1311	1.099
274	0.997	1341	1.102
305	1.000	1372	1.105
335	1.003	1402	1.108
366	1.006	1433	1.111
396	1.009	1463	1.114
427	1.012	1494	1.117
457	1.015	1524	1.120
488	1.018	1554	1.123
518	1.021	1585	1.126
549	1.024	1615	1.129
579	1.027	1646	1.132
610	1.030	1676	1.135
640	1.033	1707	1.138
671	1.036	1737	1.141
701	1.039	1768	1.144
732	1.042	1798	1.147
762	1.045	1829	1.150
792	1.048	1981	1.165
823	1.051	2134	1.180
853	1.054	2286	1.195
884	1.057	2438	1.210

Elevation E (metres)	Coefficient C_s	Elevation E (metres)	Coefficient C_s
914	1.060	2581	1.225
945	1.063	2743	1.240
975	1.066	2896	1.255
1006	1.069	3048	1.270
1036	1.072	3200	1.285

(Source: Christiansen, 1968)

To estimate evaporation, each coefficient may be determined and plotted as a function of the climatic factor or better tabulated for a full range of values as shown in Tables 15.19 and 15.20. Similarly the values of R may also be tabulated for each month as a function of latitude as in Table 15.18. To estimate of the pan evaporation for a given month, mean values of each factor are tabulated. From the tables, the values of coefficients are determined. When data for one or more of the climatic factors is missing, the user may either omit the coefficient by considering it unity or estimate the probable value of the missing factor, if it is believed appreciably different from the standard value.

Christiansen's equation may not yield desirable results when the mean values of the climatic factors of a place differ significantly from the standard values. Thus the standard values of the climatic factors must be worked out for the local data for reliable estimates.

The values of PET estimated by different methods based on climatological parameters, have been described earlier. These need to be adjusted for actual crop ET since under natural field condition, Estimation of ET_0 from PET, suitable crop coefficients (K_c) should be evolved for different crops, soils and climatic conditions and also for different stages of growth for the same crop. The ET/PET ratios usually range between about 0.2 during the early establishment stage of crop growth and about 1.0 when the crop develops maximum canopy and root system. The ratios again get abruptly reduced as the crop approaches senescene and maturity stages. This ratio may exceed 1.0 in areas prone to advection since the same is not reflected in PET estimated by most of the formulae. The correct estimation of ET from PET, therefore, depends upon the accuracy of the K_e values adopted. As far as possible K_e values should be verified from the direct methods available for ET determinations Doorenbos and Pruitt (1975) have elaborated the procedure for selection of appropriate values of K_c where direct values of K_0 are not easily available.

Once the ET is determined or estimated, the values of water requirement of crops could be determined by adding to ET the application losses and the water required for special needs as shown in equation 15.1. Application losses comprised of the ineffective rainfall and losses during the application of irrigation water. Water required for special needs could be directly measured. Water requirement for crops could also be determined by measuring the irrigation water, the effective rainfall and the soil profile contribution as shown in equation (15.1).

Example 15.10: Computation of pan evaporation by using Christiansen's formula from the climatological data of Delhi (28.4 0N and 228 m altitude)

Months	Extra-terrestrial radiation R (mm/month) from Table 15.18	Mean monthly temperature 0 °C	Ct (Table 15.19)	Mean relative humidity (%) H_m	C_{hm} (Table 15.19)	Sunshine hours: Maximum possible N (Table 15.19)	Sunshine hours: Mean observed (n)	Sunshine hours: Percentage (n/N) x 100	C_s (Table 15.19)	Wind velocity at 2 m (km/day*) W	C_w (Table 15.19)	Elevation (E metres) C_e for 228 m (Table 15.20)	Estimated pan evaporation Ev (mm/month) = 0.473 RCtChm CsCwC0Cm** = 0.473 (2x4x6x10x12x13)
Jan.	279	12.6	0.765	84.9	0.670	10.4	8.2	78.8	0.991	98.2	0.830	0.992	54
Feb.	304	15.8	0.865	82.1	0.724	11.1	8.6	77.5	0.981	118.3	0.862	0.992	75
Mar.	405	20.7	1.023	71.3	0.876	12.0	8.5	70.8	0.937	122.8	0.870	0.992	135
Apr.	449	27.0	1.235	48.4	1.054	12.9	9.1	70.5	0.935	149.5	0.919	0.992	229
May	499	31.1	1.379	41.1	1.091	13.6	9.9	72.0	0.945	187.5	0.992	0.992	320
June	495	33.5	1.464	52.0	1.035	14.0	7.1	50.7	0.828	243.3	1.106	0.992	313
July	503	30.6	1.361	78.2	0.785	13.9	6.4	46.0	0.805	196.4	1.010	0.992	199
Aug.	475	29.0	1.304	86.3	0.644	13.2	6.0	45.1	0.801	136.2	0.895	0.992	130
Sept.	412	28.2	1.277	81.7	0.729	12.4	7.6	61.2	0.882	118.3	0.862	0.992	133
Oct.	358	28.8	1.160	75.2	0.827	11.5	9.3	80.8	1.006	96.0	0.827	0.992	130
Nov.	285	18.9	0.964	76.7	0.807	10.7	9.1	85.0	1.039	87.0	0.813	0.992	85
Dec.	257	13.7	0.799	83.3	0.701	10.3	8.2	79.6	0.994	98.2	0.832	0.992	54

* The wind velocity at 3 m height was converted to that at 2 m height by multiplying with a factor 0.93 as given in the foot note of Table 15.11.

** The value of C_m was assumed as 1.0

(Source : Michael, Hukari and Singh, 1977)

Effective rainfall

It is defined as the proportion of total rainfall which is useful or utilizable for productive purposes. it has been interpreted different by the workers in different disciplines.

In the field of dryland agriculture, when the land is left fallow, effective rainfall is that which can be conserved for the following crop. An individual farmer considers that effective rainfall is that quantity which is useful in raising crops planted on his land. Water which moves out of the field by runoff or by deep percolation beyond the root zone of his crop is ineffective.

According to Food and Agricultural Organization (FAO), the effective rainfall has defined the effective rainfall as that of part the total annual or seasonal rainfall which is useful directly and/or indirectly for crop production at the site where it falls, but without pumping (Dastane, 1974). Therefore, it includes rain intercepted by living or dry vegetation, that is lost by evaporation from the soil surface (during seed bed preparation), the precipitation lost by evapotranspiration during growth and that fraction which contributes to leaching and facilitates other cultural operations, either before or after sowing, without any harm to yield and quality of the principal crop. Thus, the ineffective rainfall is that portion of the total annual or season rainfall which is lost by surface runoff, deep percolation beyond root zone, the moisture remaining in the soil after the harvest of the crops and is not useful for the next crop. The period considered is from the start of the first tillage operation till the harvest and not only the actual crop growing period.

Measurement of effective rainfall

Different methods and criteria have been employed to measure the effective rainfall in different countries (Dastane, 1974). The evaluation of effective rainfall involves measurements of rainfall and/or irrigation, losses by surface runoff, percolation beyond the root zone and soil moisture use by the crops. Information is needed for the rooting depth of crop plants. Different components discussed above could be measured directly or indirectly and either individually or in an integrated way field lysimeters are advocated for the direct and precise measurements.

Ramdas (1960) method: In this method, a small percolating filled in soil type cylinder is used for direct measurement of percolation losses in the field. However, this does not provide day- to-day information of effective rainfall and the soil conditions in the container may vary from that in the field. Cylinders of proper length are used with a suitable number of replications. This method can yield useful results.

Stanhill (1958) method: This method is used for measurement of effective rainfall. An integrating gauge consisting of a rainfall receiver which is connected to a water collecting reservoir and this in turn is connected to an evaporating surface (plate) to represent crop evapotranspiration. The reservoir is also provided with an overflow device to measure surface runoff and can easily be set up in the field and moisture balance can be recorded at any time. However, main difficulty lies in the selection of evaporating plate. specially its size and porosity. The plate also gets dusted under field conditions.

Dastane *et al.* (1966) method: In this method, a set of three cylindrical containers or drums, say A, B and C is used. The diameter of these drums about 60cm diameter and 125 cm high filled with soil. These drums are embedded about 3/4th in the soil to assess ET, percolation, and ineffective rainfall in the rice field. Drum 'A' is with bottom intact while drums 'B' and 'C' are bottomless. The drum 'C' is fitted with an overflow device at a suitable height to measure runoff. These drums are installed in the field at a convenient distance from each other and are replicated suitably. Rice seedlings are transplanted in all the three drums at the same time when planting is done in the surrounding field. Daily differences in water levels in different drums are recorded. Drum 'A' gives ET values. The difference in the readings of drums 'A' and 'B' gives the value of ET+percolation losses. Drum 'C' is used for measuring runoff losses. This technique is well suited for rice fields and is simple and cheap. The drums without bottom may be installed without disturbing the soil as far as possible.

Estimation of effective rainfall

A number of empirical formulae can also be used for determination of effective rainfall. The formula developed under a given set of conditions may not be applicable to different conditions elsewhere. U.S. Bureau of Reclamation (Stamm, 1967) developed a rough guide for estimating effective rainfall. It is used for arid and semi-arid regions in which the mean seasonal precipitation of the five driest consecutive years is used as the basis. The effectiveness of increments of monthly rainfall range from more than or 90% of the first 25 mm to 0% for precipitation greater than 150 mm. The same has been modified by Doorenbos and Pruitt (1975), to account for the effectiveness of rainfall as given below:

Monthly rainfall increment, mm	25	50	75	100	125	150	>150
Effective %	90	85	75	50	30	10	0

A more detailed evapotranspiration/precipitation ratio method has been developed to determine effective rainfall based on extensive field data by Soil Conservation Service of U.S.D.A. (1969).

Factors influencing effective rainfall

The proportion of effective rainfall out of the total rainfall received is influenced by several factors such as the amount, intensity and duration of rainfall, other climatic parameters, land and soil characteri-stics, ground waters position from the surface as well as the nature and management of crops. These factors may either act singly or collectively and may interact with each other in their effect in determining the effective rainfall.

Rainfall characteristics: The greater quantities and high intensities of rainfall reduce the effective rainfall fraction and increases runoff and reduced infiltration due to limited infiltration and moisture holdng capacity. Similarly, uneven distribution of rainfall decreases effective rainfall whereas an even spread increases it. A well distributed rainfall in frequent light showers is more conducive to crop growth as compared to heavy down-pours. The intensity frequency and amount of rainfall are high during July and August and which reduces effective rainfall in India. The most of the rainfall is effective due to its low intensity, frequency and amount during November to April.

Land slope: It influences the time available for the rain water to infiltrate into the soil (infiltration opportunity time). Water stays longer on flat and levelled land and has a longer opportunity time as compared to sloping, rolling and undulating lands which influences opportunity time of rain water for uniform infiltration and thereby effective rainfall fraction.

Characteristics of the soil: The soil properties with influence infiltra-tion, and moisture retention, release and movement influence the effective rainfall. The greater infiltration rates and hydraulic conductivity of the soil increase the infiltration and reduce runoff. The effective rainfall increases with increased water holding capacity of soil. The amount of water held and retained by a soil depends upon its depth, texture, structure and organic matter content. The finer is the texture than greater is the storage capacity. The moisture content of the soil at the time of occurrence of the rain (initial moisture content), effects the effective rainfall considerably. The higher the moisture content, the infiltration rate is which results into higher surface runoff. Thus, effective rainfall is reduced. The effective rainfall is lower in irrigated areas as compared to unirrigated areas. Rainfed areas often have greater deficit of moisture in the soil.

Groundwater characteristics: The effective rainfall is greater is deep water table than shallow water table. Water moves upwards in the soil by capillarity, causes in reduction of contents moisture and enhance the amount of effective rainfall.

Management practices: Any management practice which influences runoff, infiltration, hydraulic conductivity or evapo-transpiration also influences the effective rainfall. Bunding, terracing, contour tillage, ridging and mulching reduce runoff and increase effective rainfall by reducing surface runoff.

Crop characteristics: The effective rainfall is directly proportional to the rate of water uptake by the plant. Thus crop characteristics influencing the rate of water uptake are the degree of ground cover, rooting depth and stage of growth which also influences it. The deep-rooted crops increase the effective rainfall in a given area. Rainfall is ineffective for most crops just before harvesting.

Carry-over soil moisture: It is the moisture stored in the crop root zone depths between cropping seasons or before the crop is planted. This moisture help to meet the water needs of the crop. The rain occurring just prior to sowing may be equivalent to one full irrigation and this is generally deducted when determining seasonal irrigation requirements. An additional benefit of excess pre-sowing season rain is the leaching of salts accumulated in the root zone in the summer season. Carry-over of soil water is of great importance in rainfed farming areas.

Groundwater contribution: Water moves upward in the soil by capillarity and reduces the soil moisture deficit. Lower the water table than lesser are the contribution of groundwater in effective rainfall.

Crop coefficients

The reference crop evapotranspiration (ET_0) is determined for estimation of crop water requirements. For determination of the evapotranspiration values for particular crop, following relation is used:

$$\text{Evapotranspiration} = \text{Crop coefficient} \times ET_0$$

Crop coefficients are affected by the crop characteristics, time of planting or sowing, stage of crop development and climatic conditions including frequency of rain or irrigation. For determination of crop coefficients, the crop development is considered in four stages as,

1. Initial stage: Germination and early growth when the soil surface is not covered by the crop.
2. Crop development stage: From the end of the initial stage to attainment of effective full ground cover.
3. Mid-season stage: From attainment of effective full ground cover to time of start of maturity.
4. Late season stage: From end of mid-season stage until full maturity or harvest.

K_e values is determined on the following procedure:

1. The sowing dates, growing season and length of crop development stages should be determined from local information.
2. K_e value for initial stage is determined on using Fig. 15.9 after determination of irrigation and rainfed frequency.

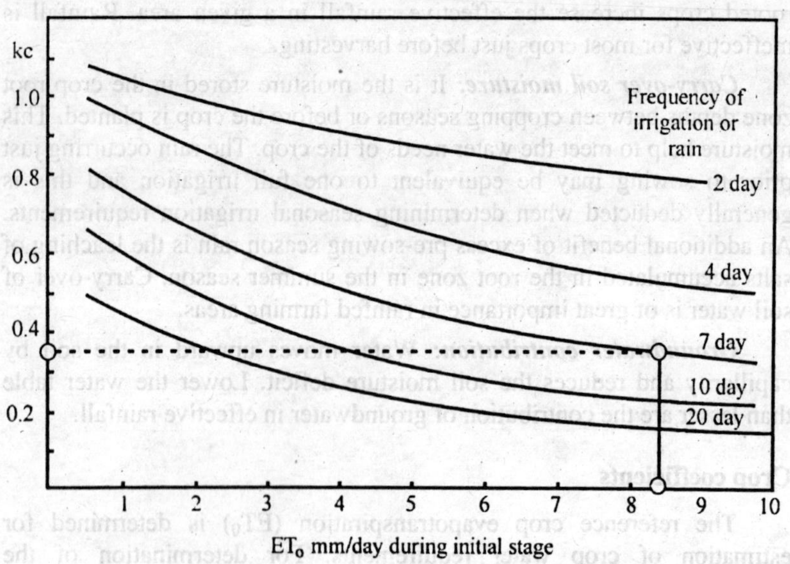

Fig. 15.9: Diagram of crop coefficients

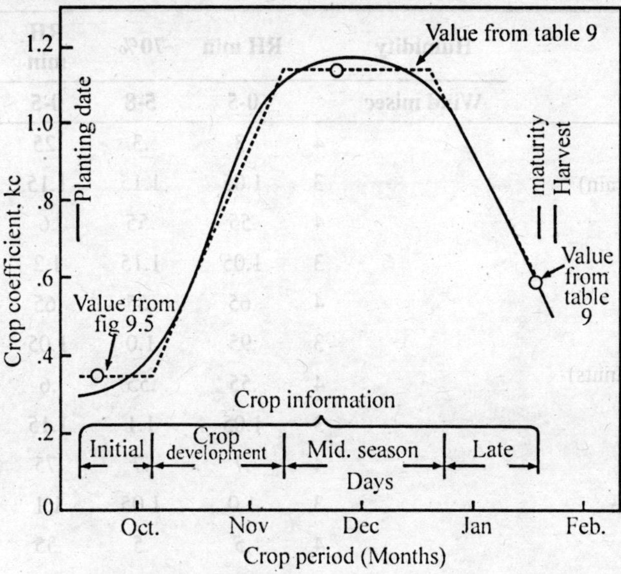

Fig. 15.10: Calculation of crop coefficients

3. K_c values for the mid .season as well as late season are selected on using values in Table 15.24.
4. The selected values of K_c are plotted as shown in Fig. 15.10. From the initial stage to mid season and from mid season to maturity. A straight line relation is assumed. Thus, the K values for any stage of the crop period are determined.

Table 15.24: Crop coefficient K_c for field and vegetable crops for different stages of crop growth and prevailing climatic conditions

Crop	Humidity		RH min	70%	RH min	70%
	Wind misec		**0-5**	**5-8**	**0-5**	**5-8**
	Crop stage					
All field crops	Initial crop dev.	1	Use fig. 15.9			
		2	by interpolation			
Barley		3	1.05	1.1	1.15	1.2
		4	.25	.25	.2	.2
Beans (green)		3	.95	.95	1.0	1.05
		4	.85	.85	.9	.9

Crop	Humidity	RH min	70%	RH min	70%
	Wind misec	0-5	5-8	0-5	5-8
Pulses	4	.3	.3	.25	.25
Corn (grain)	3	1.05	1.15	1.15	1.25
(maize)	4	.55	.55	.6	.6
Cotton	3	1.05	1.15	1.2	1.25
	4	.65	.65	.65	.7
Peanuts	3	.95	1.0	1.05	1.1
(Groundnuts)	4	.55	.55	.6	.6
Potato	3	1.05	1.1	1.15	1.2
	4	.7	.7	.75	.75
Sorghum	3	1.0	1.05	1.1	1.15
	4	.5	.5	.55	.55
Tomato	3	1.05	1.1	1.2	1.25
	4	.6	.6	.65	.65
Wheat	3	1.05	1.1	1.15	1.2
	4	.25	.25	.2	.2

Example 15.11: Calculate the rates of seasonal consumptive use and periodic consumptive use during the two irrigation intervals and during the maturation period of a mustard crop of 120 days duration from the following data:

(i) Mean soil Water contents at sowing and at harvesting in the root zone are 20.5 and 14.5 per cent respectively.

(ii) Mean bulk density of the soil in the root zone is 1.50 g/cm^3.

(iii) Mean soil water contents immediately before and 2 days after the first irrigation of 6cm depth applied at 40 days after sowing are 15.1 and 19.9 per cent, respectively.

(iv) Mean soil water contents immediately before and after the second irrigation of 6 cm depth applied at 75 days after sowing are 14.3 and 19.0 per cent, respectively.

(v) The effective rainfall received by the crop during the first and second irrigation intervals and during the season are 4.3 cm, 4.8 cm and 13.1 cm, respectively.

(vi) Potential evapotranspiration during the two-day period immediately after the first and second irrigations are 3.0 mm and 3.5 mm per day respectively.

(vii) Rooting depth of the crop is 75 cm.

Solution:

Consumptive use during the 40-day period before the first irrigation

$$=\frac{205-15.1}{100}\times 1.75\times 75=6.1\,cm$$

Consumptive use during the 33-day period between 42-75 days

$$=\frac{19.9-14.3}{100}\times 1.5\times 75=6.3\,cm$$

Consumptive use during the last 43-day period after second irrigation till the harvest of the crop (77-120 days)

$$=\frac{19.0-14.5}{100}\times 1.5\times 75=5.1\,cm$$

The PET during two 2-day periods each time immediately after first an second irrigations = 3.0 x 2 + 3.5 x 2 = 13.0 mm or, 1.3 cm

Therefore, Seasonal consumptive use = Soil water use + PET + Effective rainfall

= (6.1 + 6.3 + 5.1) + 1.3 + 13.1 = 31.9 cm

Consumptive use per day during the 40-day period before the first irrigation

$$=\frac{\text{Water use}+\text{effective rainfall,cm}}{\text{Period,days}}=\frac{6.1+4.3}{40}=0.26\,\text{cm day}$$

Similarly, the consumptive use rate per day during the 33-day period before second irrigation

$$=\frac{6.3+4.8}{33}=0.34\,cm\ day$$

and, the consumptive use rate per day during the 43-day period before till harvest of the crop

$$=\frac{5.1-4.0}{43}=0.21\,cm\ day$$

Example 15.12: Calculate the potential evaportanspiration by wheat sown on 1 November and harvested on 20 March at Sriniketan, West Bengal (23°39' N) using the Thornthwaite formula from the following data:

Month	Mean monthly air temperature °C	i value	Correction factor	Assumed crop coefficient (k)
January	18.51	7.25	0.94	0.50
February	21.30	8.97	0.89	0.70
March	26.68	12.63	1.03	0.75
April	30.34	15.26		
May	30.88	15.75		
June	27.72	13.38		
July	29.29	14.53		
August	28.67	14.06		
September	28.70	14.09		
October	26.91	12.79		
November	23.26	10.27	0.92	0.65
December	19.34	7.69	0.92	0.60

Solution:

I = 146.6

a = .000000675 x 146.6 x 146.6 x 146.6 – 0.0000771 x 146.6 x 146.6 + 0.01792 x 146.6 + 0.49239 = 2.1297 – 1.6586 + 2.6283 + 0.49239 = 3.59179

Month	10t	10t/I	$e=1.6(10t/I)^a$	Correction factor	PET (cm)	Cu (cm)
November	232.6	1.586	8.3862	0.92	7.72	5.01
December	193.4	1.319	4.3253	0.92	3.98	2.39
January	185.1	1.262	3.6907	0.94	3.47	1.74
February	213.0	1.452	6.1075	0.89	5.44	3.81
March	266.8	1.819	13.7210	1.03	14.13	10.60

The PET value for 20 days in March = 14.13 x 20/31 = 9.12 cm

Consumptive use (Cu) for 20 days in March = 10.60 x 20131 = 6.84 cm

Therefore, the PET during the season = 7.72+3.98+3.47+5.44+9.12
= 29.73 cm

Seasonal consumptive use = 5.01 + 2.39 + 1.74 + 3.81 + 6.84
= 19.79 cm

Example 15.13: Compute the reference crop ET_0 in mm/day for Pantnagar (29^0 N and 283.39 m altitude) using the modified Penman formula from the following data:

Given data for December 1981

1.	Mean air temperature	14.41 ^{0}C
2.	Mean relative humidity	72%
3.	Mean sunshine hours (n)	5.81 h/day
4.	Possible sunshine hours (n)	10.2 h/day
5.	Value of the ratio n/N	0.57
6.	Extra-terrestrial radiation mm/day	8.6
7.	Wind speed at 3 m height in km/day (u_1)	129.6
	u_1= 129.6, therefore, the wind speed at 2 m height (u_1 x 0.93)	120.53
8.	Reflection coefficient	0.25

Solution:

(a) Solving the aerodynamic term, (I - W) f(u) (e_a - e_d)

9.	e mbar	16.14 mbar
10.	e_d mbar = $\dfrac{e_a \times RH\ mean}{100} = \dfrac{72 \times 16.14}{100}$	11.62 mbar
11.	(e_a - e_d) = 16.14 – 11.62	4.52
12.	f(u)	0.85
13.	(I - W)	0.38
14.	0.38 x 0.85 x 4.5	1.46

(b) Solving for radiation term, W . R_n = W(R_{ns} - R_{nl})

15.	R_{ns} = R_A x factor B = 8.6 x 0.39 =	3.35
16.	R_{nl} = f(s) f (e_d) f (n/N) = 13.5 x 0.19 x 0.60	1.54
17.	R_n = (R_{ns} - R_{nl}) = (3.35 - 1.54) = 1.81	

18.	W	0.62
19.	W . R_n = 1.81 x 0.62	1.12
(c)	ET_0 (unadjusted) = Item 14 + Item 19 = 1.46 + 1.12	2.58
(d)	Factor for adjustment of ET_0	0.92
(e)	Adjusted ET_0 (= 2.58 x 0.92)	2.37 mm/day

Example of finding out the adjustment factor, C

Pantnagar, December Rs. = 5.81 mm/day, RH max 90%

U day 6 m/s, U night 3 m/s.

U day/U night = 2.0

Then, the C value = 0.92

(Source: Misra and Ahmed, 1987)

Example 15.14: Estimate the ET_0 by the radiation method from the following data at Sriniketan West Bengal:

(i) Month - November

(ii) T mean - 23.2^{0}C

(iii) n - 7.7 h/day

(iv) Day time wind velocity - low

(v) RH mean - high

(vi) Place - Sriniketan, West Bengal (23.4^{0}N)

Solution:

RA for 23.4°N for November = 10.8 mm/day

N for 23.4°N for November = 10.7 h

R_s = (0.25 + 0.50 n/N) RA = (0.25 + 0.50 x 7.7/10.7) x 10.8 = 6.59 mm/day

W for 23.2°C, 79 m altitude = 0.72

W . R_s = 0.72 x 6.59 = 4.74 mm

ET_0 = C(W . R_s) 0.70 x 4.74 = 3.3 mm/day

[C value for RH and wind velocity low = 0.70]

For RH high, wind velocity low from Fig. 18.4.

ET_0 = 3.4 mm/day

Example 15.15: Estimate the duty of water for a mustard crop of 120 day duration when its water requirement is 40 cm.

Solution:

Total volume of water available during the crop period of 120 days fro a continuous flow of one cumec of water = 1 m^3 x 60 x 60 x 24 x 120 = 10368000 m^3

Water requirement of the crop = 40 ha-cm x 100 m^3 = 4000 m^3

[1 ha-cm of water = 100 m^3 of water]

$$\text{Therefore, the duty of water} = \frac{10368000}{4000} = 2592 \text{ ha.}$$

References

Alten, R.G., Pereira, L.S., Raes, D. and Smith, M. 1998. Crop Evapotranspiration Guidelines for Computing Crop Water Requirements. Irrigation and Drain, Paper, No. 56 FAO, Rome, Italy 300 pp.

Bandyopadhyay, P.K. and Mallick, S. 2003. Actual evapo-transpiration and crop coefficient of wheat (*Tr**iricum aesrivum*) under varying moisture levels of humid tropical canal command, Agricultural Water Management, 59(1), 33-47.

Blaney, H.F. and Criddle, W.D. 1950. Determining Water Requirements in Irrigated Areas from Climatological and Irrigation Data. *USDA Soil Conservation Service Tech.* Paper 96.

Bowen, I.S. 1926. The Ratio of Heat Losses by Conduction and by Evaporation from any Water Surface. *Phys Rev*. 27, 779-787.

Dastane, N.G. 1970. Water management research in 1970. *Indian J. Agron* 15: 415-16.

Dastane, N.G. 1974. Effective rainfall in irrigated agriculture. Irrigation and Drainage paper 25. Food and Agriculture Organization of the United Nations, Rome, 62 p.

Doorenbos, J. and Pruitt, W.O. 1977. Crop water requirements, Irrigation and Drainage Paper No. 24 (rev). FAO, Rome, Italy. 144 p.

Doorenbos, J. and W.O. Pruitt 1975. *Crop Water Requirements*, Food and Agriculture Organization of the United Nations, Rome.

Downey, L.A. 1971. Visual irrigation guide for maize (*Zea mays* L). *Agron J.* 63: 887-89.

Grassi, C.J. 1964. Estimation of evapotranspiration from climatological formulas. M.S. Thesis College of Engg. Utah State Univ. Logan: 101.

Haise, H.R. and Kelley, O.J. 1946. Relation of moisture tension to heat transfer and electrical resistance in plaster of Paris blocks *Soil Sci.* 61:411-22.

Jensen, M.E. 1972. Programming irrigation for greater efficiency In *optimising* the soil physical environment towards of greater crop yields pp. 133-71 (ed. Hillel, D), Acad. Press, New York.

Jensen, M.E., Robb D.C.N. and Franzoy, C.E. 1970. Scheduling irrigation using climate-crop soil data, *J. Irrig. Drain, Div. Am. Soc. Civil Engg.* 96:25-38.

Jension, M.E. and Haise, H.R. 1963. Estimating evapotranspiration from solar radiation. *J. irrig. Drain. Div. Amer. Soc. Civ. Engg.* 89:15-41.

Mavi, H.S. and Chaursia, R. 1980. A comparison of open pan evaporation with computed values under Punjab Conditions. National Seminar on Agro-Climatology, Coimbatore: 70-78.

Michael, A.M. 1978. Irrigation: Theory and Practice. Vikas Pub. Stephens, J.C. and Stewart, F.H. 1963. A comparison of procedures for computing evaporation and evapo-transpiration Pub. No. 62. Inst. Assn. Sci. Hyderabad L.V.G.C. Barkeley. 123-133.

Michael, A.M. 2002. Irrigation Theory and Practices. Vikas Publishing House Pvt. Lid, New Delhi, India.

Michael, A.M., S.B. Hukkeri and N.P. Singh 1977. Estimating water requirement of crops (in) Water Requirement and irrigation Management of Crops in India, IA, R.I. Monograph No.4. Water Technology Centre, IARI, New Delhi.

Miller, D.E. and Hang, A.N. 1980. Deficit high frequency irrigation of sugarbeet with the line-source technique *J. Soil Sci. Soc. Am.* 44:1295-98.

Pal, A.M. and S.B. Hukkeri 1979. Manual on irrigation wager management, Govt. of India, Ministry of Agriculture and Irrigation, New Delhi.

Palmer, J.H. 1967. Diurnal variation in leaf and boll temperatures of irrigated cotton grown under two soil moisture regimes in a semi-arid climate. *Agric. Met.* 4: 39-54.

Papadakis, 3. 1965. Potential evapotranspiration Bureau of Airs: 54.

Parhi, P.K. Mabra, S.K. and Singh, R. 2008. Scheduling of irrigation for wheat in a semi-Humid Tropical and canal command using water balance approach *J. Indian Water Soc.* 28 (1): 1-10.

Penman, H.L. 1948. Natural Evaporation from Open Water, Bare Soil, and Grass. Roy Soc., London, Proc. Ser. 4. 193, 120-146.

Penman, H.L. 1948. Natural evaporation from open water, bare soil and grass. Proc. Roy. Soc. Lond. 193: 120- 120-45.

Penman, H.L. 1956. Evaporation: An Introduction *Nethrl J. Agric. Sci.* 4:9-29.

Penman, H.L. 1963. Vegetation and Hydrology, Tech. Communication No. 53. Commonwealth Bur. of Soils. England.

Prihar, S.S., Cheri, K.L., Sandhu, K.S. and Sandhu, B.S. 1976. Comparison of irrigation schedules based on pan evaporation and growth stages of winter wheat *Agron. J.* 60, 650-653.

Prihar, S.S., Gajri, P.R. and Narang, R.S. 1974. Scheduling irrigation to wheat using open pan evaporation *Indian J. agric. Sci.* 44: 567-71.

Prihar, S.S., Sandhu, B.S. and Singh, N.T. 1975. A critical appraisal of research on irrigation scheduling to crops. Proc. Second World Congr. Water Resources International Water Resources Association, New Delhi. 1:345-51.

Singh, C.B., Khera, K.L., Sandhu, B.S. and Aujia, T.S. 2004. Optimizing irrigation schedule to spring sunflower on loamy and sandy loam soils. *Indian J. Soil. Cons.* 32(3):221-224.

Singh, M. Gandhi, R.T. and Raheja, P.C. 1960. A critical review of the methods used to determine water requirements of crops and suggestion for planning future irrigation experiments in India. *India J. Agric. Sci.* 4:272-85.

Singh, P. 1991. Evapotranspiration evaporation and crop yield relationship for wheat in humid climate. *J. Inst. Eng.* 77, 65-71.

Slatyer, R.O. 1967. Plant water Relationships Acad, Press, New York.

Stanhill, G. 1961. A. comparison of methods of calculating potential evapotranspiration form climatic data. *Israel J. Agric Res.* 11: 157-71.

Thornthwaite, C.W. 1948. An Approach toward a Rational Classification of Climate. Geogr. Rev. 38. 55-94.

Thornthwaite, C.W. 1948. An approach towards a rational classification of climate. Geographical review. 38:55-94.

Thornthwaite, C.W. 1954. A re-examination of the concept and measurement of potential evapotranspiration. In the measurement of potential Evapo-transpiration (Ed.) J.R. Mather Seabrook, New Jersey, USA, pp. 200-2009.

Withers. Bruce and Stanley Vipond 1974. Irrigation: Design and Practice, B.T. Batsford Ltd., London.

Yiraew, M. and Brown, P. 1990. Predicting daily evapotranspiration from short-term values *J. Irrig. Drain. Eng.* 115, 387-398.

EXERCISES

15.1 Distinguish the following:

(a) Transpiration and evaporation

(b) Seasonal consumptive use and peak period consumptive use

(c) Crop water requirements and water requirements

(d) Evapotranspiration and potential evapotranspiration

15.2 Write short note on the following:

(a) Measurement of evapotranspiration

(b) Lysimeter

(c) Effective rainfall

15.3 Describe the different methods for determination of evapotranspiration from climatological data.

15.4 State the Blaney-Criddle method for determination of evapotranspiration and its limitations.

15.5 Describe the method of estimation evapotranspiration Penman methods and its merit and demerit.

15.6 What is the water requirement of crops and why it is needed?

15.7 What is the meaning of consumptive use? On what factors does it depend? How would one determine the consumptive use for a given crop?

15.8 The consumptive use for a given crop is 90 mm. Determine the field irrigation requirement if the effective rainfall and the irrigation efficiency in the area are 15 mm and 60 per cent, respectively.

15.9 Determine the consumptive use and irrigation requirement for small grains to be grown with the help of Balaney-Criddle formula. Given that value of x is 0-65 months growing season are 21.5°5C, 23.35°C and 24.55°C, respectively. The percentage of day time hours for the same period are 8.7, 9.3 and 9.5. The monthly average rainfall is 3.30 cm, 4.85 cm and 5.41 cm, respectively. Assume water application efficiency as 60 per cent. **(Ans. 33.3 cm)**

15.10 Determine the potential evapotranspiration by Penman's method for potato crop for the month of November having the following monthly average conditions.

(a) Air temperature = 25°C

(b) Relative humidity = 72 per cent

(c) Radiation rate = 987 cal cm^{-2} day^{-1}

(d) Percent actual sun shine hours to the maximum possible = 70 per cent

(e) Reflection coefficient = 5 per cent

(f) Wind velocity at 2 m, height = 100/cm/day

(Ans. 12.50 cm/month)

■■■■

Chapter 16

IRRIGATION SCHEDULING

16.1 Introduction

Proper harvested rainwater should be applied at the time of actual need of the crop with just enough water to wet the effective root zone soil. The interval between two irrigations should be as wide as possible to save irrigation water without affecting adversely the growth and yield. The principal aim is to obtain maximum crop yield by making the most efficient and economic use of water. It is desirable to apply irrigation neither more than the crop requirement nor economise water so much to force water stress to crop plants affecting the yield severely. Irrigation applied before the time of actual crop need encourages only losses of water through higher evapotranspiration and deep percolation. On the other hand, delayed irrigation causes plant water stress that depresses the growth activities and yield.

Scheduling irrigation

An ideal irrigation schedule must indicate when irrigation water is to be applied and the quantity of water to be applied. Several approaches for scheduling irrigation have been used by scientists and farmers throughout the world, each one has its own advantages and disadvantages.

Time of irrigation application is usually governed by two major factors namely (1) water need of crops and (2) availability of irrigation water. Water need of crops is, however the prime factor to decide the time of irrigation.

16.2 Water Needs of Crop

Crop plants require water to meet the transpiration loss, build up body tissue and to carry on biochemical and physiological activities within the body. Transpiration, which is considered as a vital physiological activity of plants occurs continuously as long as the water supply is maintained. Also, a continuous evaporation occurs from the moist soil surface in crop field. After irrigation the evapotranspiration begins at a peak rate drawing water from the moist soil below and continues till there is available water in soil. This causes a continuous decline in soil water content. Rate of evapotranspiration decreases continuously sometime after completion of irrigation with reduction in available soil water below the field capacity. There are following soil water regime:

(i) ***Lowest point of soil water regime***: It is the stage reached within a few days after irrigation when the rate at which soil water is available for extraction by crop plants becomes equal to the normal consumptive use rate.

(ii) ***Optimum soil water regime***: It is the range of available soil water in which plants do not suffer from water stress and all the plant activities occur at an optimal rate. Field capacity is the uppermost limit of optimum soil water regime for crops other than rice.

A soil water deficit below optimum soil water regime causes water stress in plants causing decline in growth and yield, as the rate of availability of soil water falls short of the normal consumptive use rate. Irrigation is, therefore, needed when this lowest limit optimum water regime is reached and it is considered as the most appropriate time for irrigation.

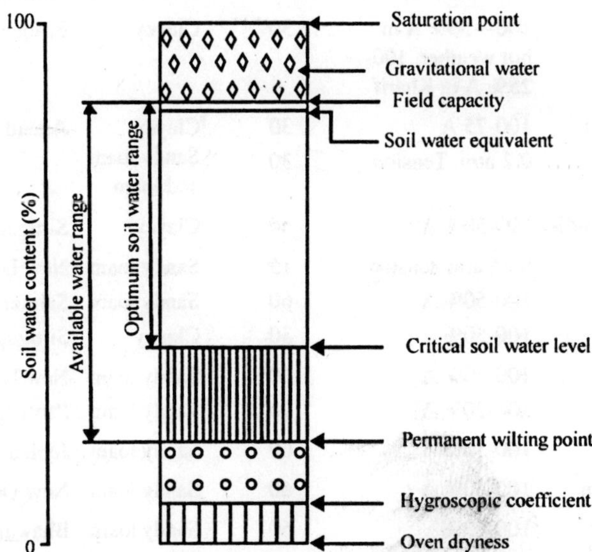

Fig. 16.1: Schematic diagram of soil water status and critical level of soil water

Critical level of soil water: It is the stage of available soil water below which water stress begins to cause serious fall in crop growth and subsequently the yield. This level of available soil water coincides with the lowest level of optimum available soil water regime. The optimum available soil water regime and the critical level of soil water is represented in Fig. 16.1.

Every crop has a characteristic optimum water regime and is summarized in Table 16.1. The water regime varies with the crop cultivars abilities to extract water from different soil layers.

Table 16.1: Optimum soil water regime for different crops in India

Crop	Optimum soil water regime	Soil depth for soil water measurement (cm)	Soil type	Place
1	2	3	4	5
Rice	Submergence to 4 cm water height	-	All	Bikramganj, Mandya, Kharagpur, New Delhi, Siruguppa, Bhawanisagar
Sorghum	100—50% A in hot weather, 100-25% A in Kharif	30	Clayey	Sirugupa
Pearl millet	100-75 A	30	Clayey	Anand
	0.2 atm. Tension	30	Sandy loam and loam	
Crowfoot millet	100-50% A	30	Clayey	Siruguppa
Maize	0.65 atm. tension	15	Sandy loam	New Delhi
	100-50% A	60	Sandy loam	Sabour
	100-50%	30	Clayey	Siruguppa
Wheat	100-50% A	60	Sandy loam	New Delhi
	100-20% A	60	Sandy loam	Pantnagar
Barley	100-50% A	60	Sandy loam	Jobner
Green gram	100-50% A	60	Sandy loam	New Delhi
Groundnut	100-60%	60	Sandy loam to loam	Bhawanisagar

Crop	Optimum soil water regime	Soil depth for soil water measurement (cm)	Soil type	Place
1	2	3	4	5
Cotton	100-20%	30	Clay loam to clayey	Bhawanisagar
Sugarcane	100-75% A at formative stage and 100-25% A at other stages	30 at formative stage and 90 other stages	Loam	Anakapalle
Tobacco	0.3 atm. tension	18	Sandy loam	New Delhi
Jute	100-50% A	30	Sandy loam	Barrackpore
Berseem	0.25 atm. Tension 0.25 atm. Tension 100-75% A	22 30 30	Sandy loam Sandy loam Sandy loam	New Delhi Hissar Jhansi
Green fodder of sorghum, lucerne and cluster bean	100-75% A	30	Sandy loam	Jhansi
Potato	0.30 atm. Tension	15	Sandy loam	New Delhi Kharagpur
Onion	0.65 atm. tension	8	Sandy loam	New Delhi
Tomato	100-60% A	120	Sandy loam to loam	New Delhi
Root vegetables radish, turnip, beet root	0.20 atm. tension	18	Sandy loam	New Delhi
Sugarbeet	0.4 atm. tension 0.25 atm. tension	60 25	Loam Sandy loam	Padegaon

(Source: Dakshinamurti *et al.*, 1971)

16.2.1 Surface distribution of water

The water occurred below the soil surface may be divided into two zones, *viz.*, the zone of saturation and the zone of aeration as shown in Fig. 16.2.

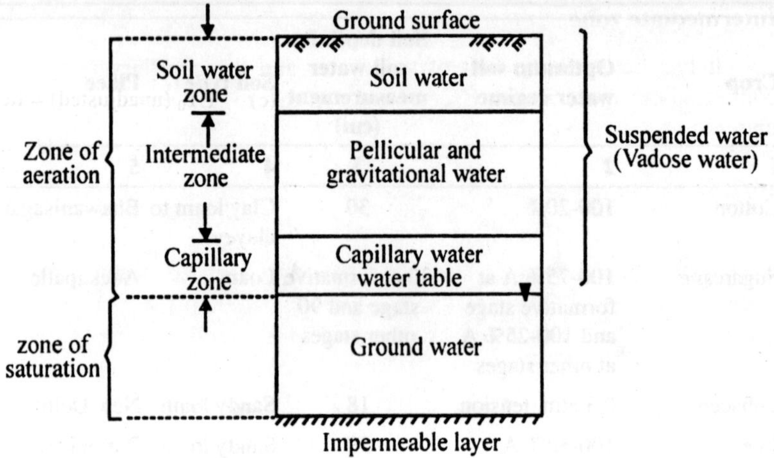

Fig. 16.2: Classification of subsurface water

Zone of saturation

In this zone all the .pores are completely filled or saturated with water. The saturated zone is bound at the bottom by impermeable strata. The upper surface of the zone of saturation (when there is no overlying impermeable strata) is known as the water table. Water occurring in the zone of saturation is commonly referred as groundwater.

Zone of aeration

It is occurring above the zone of saturation it consists of pores occupied partially by water and partially by air. This zone may be subdivided into three belts, *viz.*, (1) the belt of soil Water, (2) the intermediate belt, and (3) the capillary fringe. The soil water zone is immediately below the surface and extends through the root zone of the crops or other vegetation. It is the region from which plants extract the moisture necessary for their growth.

Capillary fringe

It refers to the zone immediately above the zone of saturation. Water in this zone is suspended by capillary forces similar to those which cause water to rise in a capillary tube. The thickness of the capillary fringe depends upon the texture of the soil. If pores are narrower then the water will rise higher.

Intermediate zone

It lies between the belt of soil water and the capillary fringe and connects both these zones. Excess water in this region moves downward under the force of gravity.

16.2.2 Availability of irrigation water

Irrigation water is often in short supply in most rainfed areas and therefore it demands a careful and economic use. Economy of water helps to bring more areas under protective irrigation and leads to enhance crop production under limited water supply or harvested rainwater. In areas where water is scarce, farmers are not able to apply normal irrigation to crops and are forced to skip some irrigations.

Therefore one should decide priority stages of crops when irrigations are to be applied and the stages when one can afford to miss irrigation. The critical stages of water need of crops receive the foremost attention. It is essential to simultaneously consider and weigh the relative importance of the various stages for irrigation and the availability of water. A preferential status of crop stages according to their relative importance to yield should be considered for irrigation in rainfed areas where water scarcity is the common feature of water scarcity.

16.2.3 Critical stages of water need

During the life cycle of a crop plant, there are some critical stages in the life cycle when the plant badly needed water. Water stress beyond a certain limit during crop growth period in different stages of crop causes a definite set back to growth processes and the yield is adversely affected. These stages are known as the critical stages of water requirement. Irrigation planning should be such that these stages do not usually coincide with the periods of peak consumptive use by different crops. It will not be correct to consider that crops at these critical stages require more water as their water needs are utmost. Critical stages of water requirement are usually the turning points in plant life cycle. This can be represented by a sigmoid or S-shaped growth curve (Fig. 16.3).

The curve indicates two most important points of change in the growth rate under: (i) The point of inflection, (ii) the point of deflection. The point of inflection indicates a sudden spurt in vegetative growth and the point of deflection represents the slowing down of vegetative growth and initiation of the reproductive phase. Crops demand for adequate water at these stages and cannot afford to sustain water stress without serious reduction in growth and yield. These two stages of crop life are, therefore, considered as the moat critical stages of water requirement.

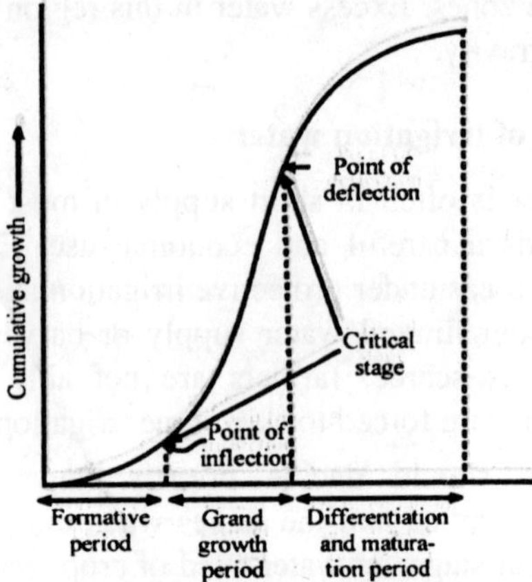

Fig. 16.3: Characteristics of S-shaped growth curve showing critical stages of water need in crop plants

When crop plants are young and delicate, they are not able to sustain water stress and demand a liberal supply of water. Again, with the start of grand growth period, crop puts up a faster rate of vegetative growth and transpiration and biochemical activities in plants occur at a higher rate. This leads to tremendous increase in water need of the crop and the supply of water should be adequate to maintain the normal rate of active growth and evapotranspiration. Water stress at the sensitive stages causes a serious retardation in growth process that ultimately depresses the yield. The sensitive stages differ from one crop to the other (Table 16.2).

Table 16.2: Moisture sensitive stages of important crops for irrigation scheduling

Crop	Important moisture sensitive stages
Rice	Panicle initiation, flowering
Wheat	Crown-root initiation, jointing, milking
Sorghum	Seedling, flowering
Maize	Silking, tasseling, grain filling
Pearl millet	Flowering, panicle initiation
Finger millet	Panicle initiation, flowering
Groundnut	Rapid flowering, pegging, early pod formation

Crop	Important moisture sensitive stages
Red gram, Green gram, blackgram	Flowering and pod formation
Sugarcane	Formative stage
Sesame	Blooming stage to maturity
Sunflower	Two weeks before flowering to two weeks after flowering
Safflower	From rosette to flowering
Soybean	Blooming and seed formation
Cotton	Flowering and boll development
Tobacco	Transplanting to full bloom
Chillies	Flowering
Potato	Tuber initiation to tuber maturity
Onion	Bulb formation to maturity
Tomato	From the commencement of fruit set
Peas	Flowering and pod development
Cabbage	Head formation until becoming firm
Carrot	Root enlargement
Citrus	Fruit setting and enlargement stage

Water stress at these stages causes low tillering, branching, pegging, tuber bulking, inadequate flowering and in extreme case, flower drops, poor setting of grains or fruits, bad filling of grains or serious fruit drops depending on the type of crops and it results into reduction of crop yield. In dwarf wheat crown root initiation stage is the most important critical stage of water need as crown roots fail to develop in dry soils. Lack of adequate water at this stage reduces tillering and affects the yield adversely. It is true that crops require adequate water supply throughout their life cycle for best growth and yield. Only in the latter stages of crop maturity, water supply is reduced or cut-off to obtain uniform and quicker crop maturity. Crops may be allowed to stand water stress to some extent during certain periods of life except at the critical stages to save some water under situations of water scarcity.

16.3 Determination of Critical Periods of Water Need

To make a judicious use of rainwater for irrigation particularly when the water supply is limited specially in dryland areas. Rainwater harvested, stored should be used judiciously to improve the water productivity. Thus, it is essential to determine the critical periods of water need of crops. For this purpose, a crop is subjected to predetermined water stress at different stages of growth and then the corresponding yield reductions are considered. It is then related to the yield of crop that has not been subjected to any water stress and irrigated according to the normal schedule. Another to decide the critical periods is to miss irrigations at different stages of the crop and then relating the corresponding yield reductions with the yield from control plot which is irrigated normally. Periods at which yield reductions are significant are considered as the critical periods of water need in the life of the crop or critical stages vis-a-vis critical period of water need.

The effect of water shortage in different stages of rice can be studied from Table 16.3. It is evident from the Table 16.3, that water shortage during tiller initiation, primordial growth and flowering stages reduced the crop yield considerably whereas, the water shortage during maximum tillering caused only a slight reduction. These appeared to be no adverse effect of water shortage during maturity period.

Table 16.3: Effect of water shortage at different stages of rice on yield

Period of water stress		**Grain yield (t/ha)**			
Crop age in days	**Stage**	**N.P. 130**		**Taichung Native 1**	
		Absolute	**Relative**	**Absolute**	**Relative**
Never		6.7	10.0	7.4	10.0
0 to 20	Tiller initiation	4.9	7.3	6.1	8.3
20 to 40	Maximum tillering	6.3	9.3	6.9	9.4
40 to 60	Primordial growth	5.3	7.9	6.1	8.3
60 to 80	Flowering	5.1	7.5	6.1	8.3
80 to 100	Maturity	6.7	10.0	7.3	9.9

(Source: Gautam and Dastane, 1970)

16.4 Criteria for Irrigation Scheduling

Since rainwater is scarce resource in dryland areas. Therefore, an emphasis should be laid on making the most efficient and economic use of rainwater for crop production. There are two situations *viz.*, water harvested from rain is adequate and when harvested rainwater is short. When adequate harvested water is available on demand, farmers often irrigate their crops earlier to the time of actual need in their eagerness to obtain good growth and high yield of crops. Their attention is to produce more yields per unit area of land without consideration of the amount of water used. This leads to wastage of valuable rainwater and it may sometimes cause damage to crops and lands owing to over irrigation which results into salinity. On the other hand, a delay in irrigation for lack of proper knowledge may force water stress to crop causing decline in yield. The optimum scheduling of irrigation under this situation should be such that crop needs to avoid both over and under-irrigation and to ensure a high rainwater use efficiency. Since adequate rainwater is not available in most places, attention must be given to produce the maximum yield per unit of water used by rational distribution of water among different crops over the growing seasons for maximum returns and high water productivity.

A thorough knowledge of the soil-water-plant-atmosphere relationships is essential for efficient scheduling of irrigation since irrigation needs of crops are decided by the evaporative demand of the ambient atmosphere, soil water status and plant characteristics. The criteria for scheduling irrigation may be grouped into following categories:

(i) Plant criteria,
(ii) Criteria based on soil water status,
(iii) Climatological criteria,
(iv) Combination approach, and
(v) Others

16.4.1 Plant criteria

Plants show up certain characteristic changes in their constitution, appearance and growth behaviour with changes in available soil water and atmospheric condition. These changes in plants are often valuable indicators for application of irrigation to crops. Different plant criteria considered to schedule irrigation are the general appearance of crop plants, plant water potential and water content of plant tissues, growth, critical periods of water need, indicator plant, stomatal opening, leaf diffusion resistance and plant temperature. There are following plant criteria for scheduling irrigation in crops:

(i) Plant appearance
(ii) Plant water potential
(iii) Plant growth

(i) Plant appearance

Water stress changes some characteristics which usually occur in the general appearance of plants. It may change in the normal colour of plant or distortions of plants such as wilting or drooping of plants and curling or rolling of leaves. Some crops are very sensitive to soil-water changes and develop scarcity symptoms easily, while others do not. Some plants such as wheat change their normal deep green colour to lighter green and then to yellow when water stress occurs. Changes in colour appear first in the lower leaves. On the other hand, deep green and light green colours in alfalfa are indicative of water stress and adequate supply of water respectively. Water stress is also shown by rolling of leaves in wheat or rice or by temporary wilting of plants as with sunflower and sugarbeet during the hottest part of the day. Fruit plants do not easily show up water stress by changes in appearance until serious retardation in growth takes place. This technique is however quite simple and rapid, but suffers from many deficiencies. Change in colour may be misleading since nutritional disorder, insect damage, disease attack and varietal character cause variable changes in foliage colour.

(ii) Plant water potential and water content

Some crops show strong correlation between the water content of leaf or leaf sheath and the available soil water *viz.*, sugarcane. Water content of sugarcane leaf sheaths decreases with decrease in available soil water and this has been taken as advantage for determining the time of irrigation in sugarcane in Hawaii (Clement, 1960). The relative leaf water content (RLWC) and leaf water potential change with variations in soil water availability or owing to lag between water absorption by plants and evaporative demand of the atmosphere. Namken (1965) found that irrigation to cotton should be given when its RLWC dropped below 72% in afternoon over a long period. Boyer (1970) and Sandhu and Horton (1977) reported adverse physiological and growth phenomena specific to plant species with fall in the RLWC and water potential below certain critical limits. Shardakov (1957) and Potinov (1959) suggested the use of plant water potential as a guide for irrigation in cotton and wheat. However, it required sophisticated equipment, intricate measuring devices which are costly and lack of proper standardization of these instruments deter the use of this technique on a large scale.

(iii) Plant growth

Water stress changes cell elonganation in plants. Subsequently retardation in growth of height or inter-nodal length occurs. Timing of irrigation can be set as and when the normal growth rate is observed to decline. This is possible continuous measurement of plant growth is maintained as is done in sugarcane in Hawaii (Clement, 1960). Namken *et al.* (1971) found radial changes in stem as a guide to irrigate cotton. This technique offers difficulties owing to unavailability and high costs of equipment, inadequate standardization of the method, difficulties in selection of proper growth parameter and precise growth measurement and so on. The major drawback of this approach for scheduling irrigation is that the plants suffer before they show any retardation in growth processes.

(iv) Critical crop stages

Irrigation scheduling may be decided based on some growth stages at which moisture stress leads to irrecoverable yield losses. This is the period during which the development stage is shown as critical period or water stress period (Table 16.2). If the sufficient rainwater is available, irrigation is scheduled whenever soil moisture is depleted to critical moisture level, say, 25 or 50 per cent of available soil moisture under limited supply condition. Then irrigation is scheduled at moisture sensitive stages and irrigation is skipped at non-sensitive stages.

The crown root initiation, tillering, flowering, milk and dough stages in wheat, branching, flowering and pod development stages in mustard, pegging, pod setting and pod development stages in groundnut and tasselling, silking and grain filling stages in maize are very specific and can be easily identified by a common farmer for scheduling irrigation. Application of irrigation at these critical stages, crop yield will not affected adversely. Majumdar (1998) suggested that scheduling of irrigation based on these critical stages is most convenient for ordinary Indian farmers who may need, at the most with some guidance or education initially. However, it may have a drawback if stages are not so well defined.

(v) Indicator plant

There are some plants sensitive to soil-water deficit. They may be used for detecting irrigation scheduling in crops that do not show symptoms of water stress easily or exhibit the same when they have already suffered seriously. Sunflower plants may be used as in indicator plants in onion crop. An indicator plant for irrigation should be such that it should show the water stress before the crop has suffered from water stress. When an indicator plant is grown in a crop field, care should be taken not to shade the plant by crop plants.

(vi) Stomatal aperture

Opening of stomata in plants is regulated by soil-water availability. Stomata remain fully open when the supply of water is adequate, when they start closing with initiation of water stress in soils to restrict the transpiration. Slatyer and Shmueli (1967) recommended that the stomatal aperture is an indicator of water deficit in plants because it influences both photosynthesis and transpiration by its carbon dioxide and water vapour transport. Water deficit in plants is directly related to availability of soil water and it may be used for irrigation in crops.

(vii) Leaf diffusion resistance

Leaf resistance to vapour diffusion is primarily governed by the degree of stomatal closure under sufficient day light condition. It is regulated mainly by the leaf water deficit. Several researches reported a close relationship between leaf diffusion resistance (LDR) and plant water stress. Sandhu and Horton (1977) found that on cloudy days LDR is a sensitive index of internal water balance in the mild to moderate stress range and holds a promise for scheduling irrigation. Similarly, Sij *et al.* (1972) found that LDR of sorghum leaves followed the plant water status and suggested that advice known as diffusion porometer is used to measure LDR. It may be a useful tool in irrigation scheduling. The LDR is a sensitive index of plant water stress in sugarcane and can be used as a basis for scheduling irrigation as suggested by Prihar and Sandhu (1987).

(viii) Plant temperature

Solar radiation received on earth surface heats up leaf tissues besides causing evapotranspiration and heating up the ambient air. Increase in water deficit rises with plant canopy temperature. Many scientists have shown that leaf or canopy temperature is a sensitive index of plant water status in soybean, oats, barley, alfalfa, wheat, sorghum and corn (Wiegand and Namken, 1966; Horton *et al.*, 1970; Carlson et al, 1972; Miller *et al.*, 1971; Sandhu and Horton, 1978; Idso *et al.*, 1981). During hotter part of the day under semi-arid climate, fully exposed leaves of oats subjected to mild and moderate stress when 1°C to 4°C warmer than non-stressed leaves (Sandhu and Horton, 1978). It was also found that unstressed sugarcane and mungbean crops were cooler by 8°C and 9°C, respectively. Then the difference between the stressed and unstressed canopy temperatures is a better index of water deficit than the difference between plant canopy and air temperatures.

Canopy temperature (Tc) is measured with infrared thermometer. It also simultaneously measures air temperature (Ta) and displays Tc-Ta values. The difference between canopy temperature and air temperature (Tc-Ta) may be used for scheduling irrigation. In case transpiration is normal, due to its cooling effect, canopy temperature as compared to air temperature. The negative values of Tc - Ta reveal that plants have sufficient amount of water and irrigation is not required. When Tc-Ta values are zero or positive, then it indicates stress and irrigation is scheduled. Stress-degree days (SDD) is useful for irrigation scheduling and are summed in a manner that is analogous to growing-degree days.

$$SDD = \Sigma\,(Tc\text{-}Ta) \qquad \text{....(16.1)}$$

Canopy temperature should be measured during mid-day when air temperature is maximum. Yield reduction is maximum when the total number of cumulative SDD' exceeds 10 to 15 between two consecutive irrigations.

16.4.2 Criteria based on soil water status

Irrigation scheduling based on soil water content is the most accurate and dependable method. Determination of available soil water is more important than determining the total water content of soils. For this purpose, knowledge of the optimum water regime of crops and the available water holding capacity of soils is essential. Irrigation is applied when the soil water content reaches the lowest point of optimum soil water regime. The optimum water regime for a crop in a region is determined experimentally by correlating yields with the water contents of soils. The optimum soil water regimes of different crops are presented in Table 16.1. When limited irrigation is limited water then it may be applied at any other pre-decided level of soil water below the optimum soil water regime.

Determination of soil water from time to time becomes essential to know when the soil water reaches the level at which the irrigation is best applied. Different methods are used to determine the soil water and farmers may choose any one of the methods according to their needs, accuracy wanted and facilities available for determining soil water.

The criteria used based on soil water status to schedule irrigation to crops is discussed as under:

Soil moisture depletion

The available soil moisture in the root zone is a good criteria for scheduling irrigation. When the soil moisture in a specified root zone is depleted to a particular level (which is different for different crops), it is to be replenished by irrigation. For crops like maize, wheat etc., scheduling irrigation may be performed at 25 per cent depletion of available soil moisture is adequate. For drought resistant crops *viz.*, sorghum, pearl millet, finger millet, cotton etc., scheduling irrigation at 50 per cent depletion of available soil moisture is sufficient. Irrigation water can be applied at a predetermined soil water tension at a specified depth as it is considered that plant response to irrigation is better related with soil water suction than with soil water content. These approaches are reliable, but cannot be recommended to farmers because it is difficult to measure soil water Content or soil moisture tension and use of equipments. Soil moisture is approximately determined by farmers with feel and appearance method and irrigation is scheduled accordingly.

Moisture content can be roughly estimated by taking the soil from root zone into the hand making it into a ball. It requires a lot of experience to estimate the soil moisture by this method and rough guidance is summarized in Table 16.4.

Depth-interval of irrigation

Irrigation scheduling on how much depth of water to be applied or interval between irrigations was not found very effective owing to various limitations. Since the water retentive capacity of soils varies widely with type of soil and its physical parameters, and root zones of crops vary with types of crops and their rooting characteristics at different growth stages, the depth and interval of irrigation require modifications in different soils and at various crop growing periods. An arbitrarily fixed depth or interval of irrigation has misleading effects on crop growth and yield.

Critical level of soil water

It is the level of available water below which the crop growth and yield decline drastically. Thus, is the lowest level of the available soil water. This level once determined experimentally for various crops in different soil types and soil conditions may be used for irrigation scheduling. This approach has been widely recommended for adoption. A periodical determination of soil water content is made to know the time when the soil water is likely to reach the critical level. This criterion is similar to the concept of available soil water depletion for deciding the time of irrigation. The depth of irrigation however, needs revision upwards every time with increasing vegetative growth and rooting depth of an actively growing crop.

Table 16.4: Guide for judging the amount of available soil moisture range

Available soil moisture range	Coarse texture (loamy sand)	Moderately coarse (sandy loam)	Medium texture (loam and silt loam)	Fine texture (clay loam and silty clay loam)
Field capacity 75 per cent to 100 per cent	On squeezing, no free water appears on soil, but wet outline is left on hand Tends to stick together slightly, sometimes forms a very weak ball under pressure	Similar symptoms for all types of soils Forms weak ball, breaks easily, does not slick	Forms a ball, is very pliable, slicks readily	Easily ribbons out between fingers, has slick feeling
50 per cent to 75 per cent	Appears to be dry, do not form a ball with pressure	Tends to form a ball under pressure, but seldom holds together	Forms a ball somewhat plastic Sometimes slick slightly with pressure	Forms a ball, ribbons out between thumb and forefinger
25 per cent to 50 per cent	As above, but ball is formed by squeezing very firmly	Appears to be dry, do not form a ball unless squeezed very firmly	Somewhat crumbly but holds together with pressure	Somewhat pliable, forms ball under pressure
0 to 25 per cent	Dry, loose, single grained, flows through fingers	Dry, loose, flows through lingers	Powdery, dry, sometimes slightly crusted but easily broken down into powdery conditions	Hard, baked, cracked, sometimes has loose crumbs on surface

(i) Soil water tension

Many scientists have used this criterion for irrigation scheduling in crops in various parts of the world. Miller and Duley (1925) were the earliest to use this approach for scheduling irrigation to maize. Richards (1942) and Richards and Marsh (1961) used tensiometer techniques for irrigation. Tayler (1952) suggested the use of mean soil water tension of the entire root zone for deciding the time of irrigation and found the same more dependable than the soil water tension measured at shallow depth of soil. The tensiometer has been considered as a useful device for measuring soil moisture tension and it is used for scheduling irrigation to orchard and vegetable crops particularly on coarse textured soils where more of the available water is held at low tensions. Taylor (1965) prepared a guide that included soil water tension as a criterion for irrigating crops on different soils under different climatoligical conditions.

Irrigation may be scheduled based on soil moisture tension. Tensiometers are installed at specified depth in the root zone. When the soil moisture tension reaches a specified value (0.5, 0.75 or 1.0 bars etc.) then irrigation is applied. These are generally used for irrigating orchards, especially in coarse textured soils. However, tensiometers do not give any information on the amount of irrigation water to be applied at each irrigation.

The use of tensiometer for scheduling irrigation did not find much favour with common farmers as this device presents certain difficulties in its use.

The water content is calibrated from the soil-tension curve. There exists a time lag in tension-equilibrium between the porous cup and the surrounding soil that makes the tensiometer. It indicates that it shows energy status of soil water earlier to the existence of the actual soil water content.

(ii) Electrical resistance

The electrical resistance varies inversely with the water content in soils and is also tried to schedule irrigation. For this purpose, resistance blocks made of gypsum, nylon, nylon-resin etc. were used. Crops were irrigated when the electrical resistance reached a certain value. The value could be decided experimentally for crops by using the resistance blocks. Haise and Kelley (1946) and Taylor (1952) suggested the use of gypsum blocks for irrigation scheduling to crops that are capable of withstanding higher tensions. This method has following drawbacks and did not become popular:

(i) Resistance blocks cannot be used at low tension at which following most of the available water is held by soils,

(ii) Difficulty of deciding the depth of irrigation as resistance blocks do not directly show the prevailing soil water drawbacks content and

(iii) The existence of a time-lag in tension-equilibrium between the porous block and the surrounding soil which causes showing up the energy status of soil water earlier.

16.4.3 Climatological approach

Evapotranspiration mainly depends on climate. The amount of water lost by evapotranspiration is estimated from climatological parameter and when ET reaches a particular level, irrigation is scheduled. The amount of irrigation given is either equal to ET or fraction of ET. Different methods are used in climatological approach *viz.*, IW/CPE ratio method and can evaporimeter method. The approach of irrigation scheduling to crops is complicated for an ordinary farmer.

(i) Empirical formulae

In this method, meteorological data are used for estimating the evapo-transpiration and consumptive use for controlling irrigation amount. For this purpose, different empirical formulae have been developed on using different meteorological parameters *viz.*, Penman (1948) and Thornthwaite (1948); Blaney-Criddle (1950); Christiansen (1968) and Penman Monteith method. These formulae are used for estimating potential evapotranspiration and then the estimated evapotranspiration is used for irrigation scheduling by water budget method. The daily evapotranspiration loss is deducted from the soil water reserve in root zone soil after irrigation and a balance is worked out. When the balance shows that the soil water is depleted to a predetermined level, say, the lower level of optimum soil water regime then irrigation is applied to replenish the water lost through evapotranspiration. The adoption of empirical formulae for irrigation control demands required the knowledge of water holding capacity of soil and a continuous record of rainfall and other meteorological parameters. This approach of irrigation scheduling to crops is complicated for an ordinary farmer.

(ii) Evaporimeter

Evaporimeters *viz.*, Class-A Open Pan Evaporimeter, Sunken Screen Open Pan Evaporimeter and atmometer may be used for measurement of evaporation loss.

The evaporation loss is used to determine the consumptive use by crops by multiplying the evaporation values with crop coefficient values. The coefficient varies from 0.6 to 0.8 for most crops at their different stages. Irrigation is applied when crops consume the available soil water to a certain limit. It is calculated on the basis of consumptive use rate as determined by evaporimeters. Sunken screen evaporimeter value can be used from the period of 25 per cent ground coverage by crops till their maturity (Gautam and Dastane, 1970). With pan evaporimeter, the irrigation is schedules to apply when a certain amount of water gets evaporated from the pan. The values of pan evaporation for this purpose are found for various crops at their different growth stages under different soil and climatic conditions.

(iii) Can evaporimetry

In this method small cans of one litre capacity (14.3 cm height and 10 cm diameter) are used to indicate evaporation from the cropped field. These cans are painted white and covered with 6/20 size mesh. An indicator pointer is fixed at 1.5 cm below the brim. When irrigation is given bringing the soil to field capacity, the can is filled up with water to pointer level and kept at the crop height. Evaporation from can is directly related to crop evapotranspiration. Irrigation is scheduled when the water level in the can falls to a predetermined level (equal to the amount of water to be applied at each irrigation) and can is filled again to the pointer level.

(iv) IW/CPE approach

Prihar *et al.* (1974) suggested the use of IW/CPE ratio as a practical basis for irrigation scheduling. This approach is based on the close and direct relationship of crop evapotranspiration with pan evaporation. When irrigation is applied, water is lost from the soil through evapotranspiration in the same way as the evaporation occurs from an open pan evaporimeter.

In this approach, a known amount of irrigation water (IW) is applied when cumulative pan evaporation (CPE) reaches a predetermined level. The amount of water given at each irrigation ranges from 4 to 6 cm, the most common being 5 cm of irrigation. Scheduling irrigation at an IW/CPE ratio of 1.0 with 5 cm of irrigation water means 5 cm of irrigation water is applied when the cumulative pan evaporation reaches 5 cm. Generally, irrigation scheduled at 0.75 to 0.8 ratio with 5 cm of irrigation water.

Example 16.1: Determine cumulative evaporation required for scheduling irrigation at 0.5, 0.6, 0.75 and 0.8 with 4 cm of irrigation water in wheat.

Solution:

Cumulative pan evaporation at IW/CPE ratio of 0.5 = IW/CPE =

$$\frac{5}{CPE} = 0.5$$

$$\text{CPE x } 0.5 = 5$$

$$\text{CPE} = \frac{5}{0.5} = \frac{50}{5} = 10\,cm$$

∴ Irrigation of 5 cm is given when cumulative pan evaporation 10 cm.

Similarly,

$$\text{CPE at 0.6 ratio} = \frac{5}{0.6} = 8.33\,cm$$

$$\text{CPE at 0.75 ratio} = \frac{5}{0.75} = 6.66\,cm$$

$$\text{CPE at 0.8 ratio} = \frac{5}{0.8} = 6.25\,cm$$

In IW/CPE ratio approach, irrigation can also be schedule at level of CPE by varying amount of irrigation water.

Example 16.2: Determine the amount of water for each irrigation for scheduling irrigation at 0.5 and 0.8 IW/CPE ratio with 8 cm of CPE.

Solution:

Amount of water to be given at IW/CPE ratio of 0.5

$$= \frac{IW}{8} = 0.5$$

$$\text{IW} = 0.5 \text{ x } 8 = 4 \text{ cm}$$

Amount of water to be given at IW/CPE ratio of 0.8

$$= \frac{IW}{8} = 0.8$$

$$IW = 8 \times 0.8 = 64 \text{ cm}$$

By changing either the amount of irrigation water or cumulative pan evaporation, irrigation frequency can be changed. Since evaporation is directly related to ET of crop, irrigation is scheduled based on the need of water of the crop. However, the equipment to measure pan evaporation and irrigation water is not available with the farmers.

16.5 Combination Approach

Doorenbos and Kassam (1979) developed combination approach using soil moisture depletion approach and climatologjcal approach for sufficient and deficit irrigation water conditions. Similar to soil moisture depletion approach, total available soil water (Sa) is defined as the depth of water in mm/m between field capacity and permanent wilting point. The proportion of the total available soil water that can be depleted without causing actual evapotranspiration (ETa) to become less than potential evapotranspiration (ETm) is known as fractions (P) of the total available soil water. The value of the fraction (P) depends on the crop, the magnitude of ETm and the soil. Some crops, such as vegetables continuously need relatively moist soil to maintain ETa while, as other such as cotton and sorghum can deplete soil water much further before ETa falls below ETm. Crops can be grouped according to the fraction (P) to which the available soil water (Sa) can be depleted while maintaining ETa equal to ETm (Table 16.4). The value of P varies with growth period and in general, it is greater during the ripening period due to low evapotranspiration at this period. For conditions where ETm is high, P is smaller, the soil is comparatively wet when ETa becomes less than ETm under high ETm conditions as compared to when ETm is low. Consequently, the fraction (P) varies with the level of ETm (Table 16.5).

Table 16.5: Crop groups according to fraction of soil water depletion (P)

Group number	Crops
1	Onion, potato
2	Cabbage, grape, pea, tomato
3	Groundnut, sunflower, wheat
4	Cotton, maize, safflower, sorghum, soybean, sugarcane. tobacco

Adopted from 'Yield Response to Water'; FAO. Irrigation and Drainage, Paper No. 33.

Table 16.6: Soil water depletion fraction (P) for different crop groups and maximum evapotranspiration rates (ETm)

Crop group	ETm (mm/day)							
	2	**3**	**4**	**5**	**6**	**7**	**8**	**9**
1	0.50	0.43	0.35	0.30	0.25	0.23	0.20	0.20
2	0.60	0.58	0.48	0.40	0.35	0.32	0.28	0.25
3	0.80	0.70	0.60	0.50	0.45	0.43	0.38	0.35
4	0.88	0.80	0.70	0.60	0.55	0.50	0.45	0.43

(Source: FAO, Irrigation and Drainage Paper No. 33

For scheduling irrigation based on the combination approach, the information needed is Sa and P which are related soil properties and ETm which related atmospheric condition. The available soil moisture (Sa) is multiplied with root zone depth of the soil (D) to obtain the total amount of water present in the root zone (D. Sa). Then fraction of available water is calculated by multiplying P and D. Sa. Irrigation interval obtained by dividing P.D. Sa by ETm. This approach is suitable to temperate regions where ETm is low.

Example 16.3: ETm for maize in July is 8 mm/day. Total available moisture is 150 mm/m depth of soil. The root zone depth is 0.6 m. Determine the irrigation interval for the month of July.

Solution:

Fraction P for maize at 8 mm ETm from Table 16.6 = 0.45

Total available soil water (Sa) = 150 mm/m

Total available soil water in root zone depth = D.Sa = 150×0.6

= 90 mm

Fraction P in 90 mm = $90 \times 0.45 = 40.5$ mm

$$\text{Irrigation interval} = \frac{P.D.Sa}{ETm} = \frac{40.5}{8} = 5.6 \text{ days}$$

Thus, irrigation is to be scheduled at 6 days interval to maintain actual evapotranspiration at maximum evapotranspiration level.

16.5.1 Other techniques

There are following other techniques for irrigation scheduling:

(i) Soil cum sand mini plot technique

(ii) Sowing high seed rate

(iii) Remote sensing

The following methods can be recommended to farmers for scheduling irrigation:

(i) Soil-cum-sand mini-plot technique

(ii) Increased plant population

(iii) Can evaporimeter

(i) Soil-cum-sand miniplot technique

In this technique, one-cubic-metre pit is dug in the middle of the field. About five per cent of sand by volume is added to the dug soil, mixed well and the pit is filled up in the natural order. Crops are grown as usual in the entire area of the field including the pit area. The plants in the pit show willing symptoms earlier than the other plants in the remaining area. Irrigation is scheduled as soon as wilting symptoms appear on the plants in the pit.

(ii) Sowing high seed rate

In this technique, one square-metre plot is selected than crop is grown with four times thicker than normal seed rate. Because of high plant density plants show wilting symptoms earlier as compared to the rest of the crop area indicating the need for scheduling of irrigation.

(iii) Remote sensing

This method is used for a single crop when it is grown on a large area irrigation scheduling can be done with the help of remote sensing data. Reflectance of solar radiation by the plants with sufficient amount of water is different from that of stressed plants. This principle can be used for scheduling irrigation.

16.6 Frequency and Interval of Irrigation

The terms, frequency of irrigation and interval of irrigation are closely related and are often interchangeable. Higher the frequency of irrigation, lower is the interval between two irrigations in a given period. While, with lower frequency the interval between two irrigations increases. The *interval of irrigation* indicates the time gap, usually expressed in days, between two subsequent irrigations. The total amount of water required by a crop for producing an optimum yield is termed as *delta of water* and it is synonymous with water requirement of crop or depth of water.

The principal objectives of rainwater irrigation management are to make the most effective use of scare water coupled with higher crop productivity and crop production, to prevent losses of water and to save water as much as possible to irrigate new areas. Fewer irrigations at longer intervals encourage saving of water. Losses of water occur in the irrigation practice and the losses may take place in conveyance channels and in fields every time when irrigation is applied. The loss may get aggravated due to inadequate knowledge and experience of the farmers in water application to crops. With higher frequency of irrigation, surface soils remain moist for longer periods leading to higher evapotranspiration losses. Thus, frequency of irrigations should be as low as possible to avoid water loss from evaporation. Of course, frequent irrigations with smaller depths of water each time are often more conducive to higher yields than heavier irrigations at long intervals. Thus, the delta of water remaining the same within a certain limit (Dastane, 1969) and is shown in Fig. 16.4.

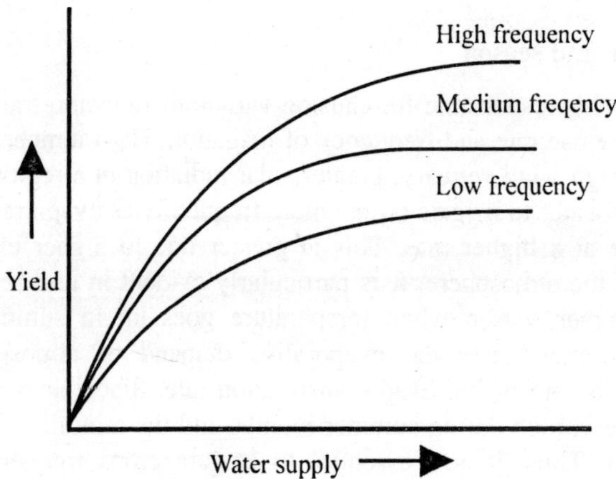

Fig. 16.4: Schematic representation of yield-irrigation frequency relationship

Immediately after irrigation when the soil is wet, evapotranspiration occurs at a potential rate. It starts declining some days after irrigation as the surface soil dries up. Dry and loose soil surface helps to reduce evaporation. Since soil water declines progressively owing to continuous evapotranspiration, the rate of evapotranspiration also declines progressively with the advance of time after irrigation. Therefore, the longer the interval between irrigations, the greater is the saving of water. Besides, a longer interval between two irrigations cuts down the number of irrigations during the growing season. However, care should be taken not to cause any water stress beyond a certain limit by making the irrigation interval unduly long unless compelled to do so for under water scarcity condition. Irrigation is usually advised at the lowest limit of the optimum water regime. Thus, the interval between two irrigations should normally be the time taken by crops to reduce the soil water from field capacity to the lowest level of optimum soil water regime.

16.6.1 Factors affecting frequency of irrigation

The two main factors namely, water need of crops and the availability of irrigation decide the irrigation frequency. Once these two factors are known, the frequency of irrigation is influenced by the following factors:

(a) Climate and season,

(b) Soil characteristics,

(c) Crop characteristics, and

(d) Crop and water management practices.

(a) Climate and season

Climate is responsible for causing variations in evapo-transpiration, consumptive use rate and frequency of irrigation. High temperature, low humidity, high wind velocity, greater solar radiation in a region increase the need of water to irrigate crops more frequently as evapotranspiration takes place at a higher rate. This is greater due to higher evaporative demand of the atmosphere. It is particularly evident in arid regions and during summer season when temperature goes up in summer which increases tremendously the evaporative demand of atmosphere and which results into higher evapotranspiration rate. Since air remains dry, temperature is high during summer months and the rainfall is low in the arid region. Thus, it is a essential to irrigate crops frequently. With greater evapotranspiration, frequent replenishment of soil water becomes essential to maintain the optimum growth of crops. On the other hand, during the rainy season, high rainfall and greater relative humidity reduce the irrigation requirement of crops and irrigations may be applied at longer interval. While, in humid areas where rainfall is higher, the irrigation requirements of crops are lower and irrigations are applied less frequently. During winter season the crop water demand is much less due to the lower temperature and evaporative demand of the atmosphere. Therefore, the irrigation requirement during winter season is lower and irrigations may be applied at longer interval.

(b) Soil characteristics

Water retentive capacity of soil is considered as the most important factor deciding the frequency and interval of irrigation. The soil characteristics *viz.*, soil texture, structure, aggregates and organic matter content influence the water retentivity of soils. A soil with greater water retentive capacity serves as a bigger water reservoir for crops and can supply water for longer duration. Consequently, frequency of irrigation is lower and interval of irrigation is longer in heavier soils and in soils with crumb structure, good organic matter content and low content of soluble salts. While, the frequency is higher in porous sandy soils with coarse texture, poor structure and low organic matter content. Retention of greater amount of available water is considered more important as compared to total quantity of water retained by a soil.

Depth of soil is another factor that influences the frequency of irrigation. A shallow soil cannot hold enough water to meet the crop requirement for a longer period. Thus, frequent irrigations are required with smaller depth of water each time. Irrigations at longer interval is applied to deep soil that has a greeter water storage capacity. Such a soil can supply water for longer duration particularly when the root system is quite deep and extensive.

(c) Crop characteristics

Crops are having different characteristics and vary in their water need or use of water, sensitivity to water stress, water extraction capacity and optimum water regime. Thus, frequency of irrigation varies with crop. A crop having higher consumptive use rate consumes the soil water quickly and requires more frequent replenishment of soil water. Rice crop thus requires frequent irrigations as compared to wheat crop. Crops like vegetables, onion, tobacco and sugarbeet that require a higher level of water to be maintained in the soil and need frequent irrigations as compared to cereals oil seed crops *viz.*, wheat, *bajra,* maize, pearl millet and mustard.

Many crops have different varieties that are either sensitive or tolerant to drought conditions. Varieties sensitive to drought conditions require frequent irrigations as compared to tolerant varieties. The dwarf wheat varieties are usually more sensitive to water stress as compared to the tall wheat varieties. Thus, dwarf wheat variety need frequent irrigations.

Shallow or deep, fibrous or tapering, vertically or laterally extensive root systems decide the frequency of irrigation. When the root system is shallow and fibrous, crops are not able to utilize water from deeper soil layers and are frequently irrigated with smaller depth of water to wet only the upper soil layers. Crops with deeper and extensive root system command a greater depth of soil and water reserve and require irrigations at longer interval. Sometimes, they may also get water from water table which is not deep enough. Shallower water table reduces the irrigation requirements and help to increase the interval between irrigations. Besides, the concentration and relative proportion of the root mass in different soil layers decides the water extraction capacity. The extraction capacity of crops from different depths of soils is shown in Fig. 16.5. Maximum quantity of water is extracted from the upper 25 per cent of the effective rooting depth and the extraction is least from the last 25 per cent of the rooting depth which is about one-fourth of that from the upper section. It is evident from the water extraction pattern that a higher frequency with smaller depth of irrigation each time is preferable for crops with shallow root system that extract most of their water need from the upper soil layers.

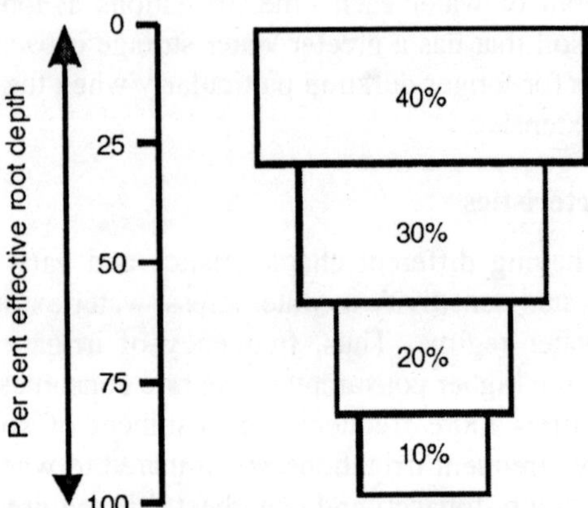

Fig. 16.5: Design water extraction pattern from effective rooting depths

Frequency of irrigation also varies with stages of crop growth. The consumptive use rate, sensitivity to water stress and rooting characteristics of crops change at different crop growth stages. When a crop is young and delicate it needs frequent irrigations. Plants at this stage are very susceptible to water stress, because their root system is not well developed, deep and extensive enough to extract water from deeper soil layers. Subsequently, the consumptive use rate gradually increases and at the same time the root system also develops then irrigations can be applied at longer interval, as roots are able to draw water from greater volume of soils or from greater depth of soil. When a crop approaches maturity, the demand for water greatly declines because of steep fall in consumptive use rate. Besides this, the well-developed root system can also draw water from deeper layers. Therefore, the irrigation requirement of crop declines with approach of maturity and crops are irrigated at longer intervals.

(d) Crop and water management practices

In-situ soil moisture conservation practices *viz.*, deep summer ploughing, sowing method, land treatments, crop residues, mulching and crop cultural practices like weeding and hoeing help to reduce runoff from the field in rainy season and the evaporation loss and conserve more soil water for crop use. Thus, there is a reduction in irrigation requirement of crops. Method of irrigation, depth of water applied each time and the water distribution efficiency influence the frequency of irrigation. Sprinkler irrigation recommended in porous and lighter soils demands frequent irrigations which cause more evaporation loss as the surface soil remains moist for a longer period. Therefore, a smaller depth of irrigation is applied each time. While, irrigations are applied at longer interval when the surface irrigation is adopted in soils that have higher water retentive capacity and are not too porous.

Some important definitions

Irrigation period

It is the time that can be allowed for applying one irrigation to a given design crop area during the peak consumptive use period of the crop and usually expressed in days. It is a function of the peak-period consumptive use rate. It is considered for designing the irrigation system capacity and equipment. The irrigation system must be so designed that the irrigation period is not greater than the irrigation interval. It is determined with the following relation:

Irrigation period (days) =

$$\frac{\text{Net depth of soil depletion in the design crop area just before irrigation, cm}}{\text{Peak period consumptive use rate of crop cm}} \quad \text{....(16.2)}$$

Different crops and varieties have different types of roots having different characteristics. Thus, irrigation period is different.

Design irrigation frequency

It is the period between two irrigations that is necessary to irrigate the design crop area during the period of peak consumptive use of the crop to be irrigated and usually expressed in days. The design irrigation frequency is used to decide the capacity of the irrigation system to be able to supply the required water to crops in the area. Irrigation frequency depends on how fast soil water is extracted when crop is transpiring at its maximum rate. The average consumptive use rate during this period is used for planning the system. The design irrigation frequency may be determined by the following relationship,

Design irrigation =

$$\frac{(F_c - M_b) \times A_s \times D / 100}{\text{Peak period of consumptive use of crop, cm}} \quad \text{....(16.3)}$$

where,

F_c = field capacity, (percent),

M_b = soil water content just before irrigation, (percent),

A_s = apparent specific gravity of soil or bulk density of soil, (dimensionless), and

D = depth of crop root zone, (cm).

Depth of irrigation

It refers to the depth to which the irrigation water applied would cover an area. Depth of irrigation is a function of the water retentive capacity of the root- zone soil and the extent of soil water depletion at the time of irrigation. For example, a 10-centimetre depth of irrigation to a hectare of land represents the volume of water which when allowed to stand without any loss and infiltration into the soil would stand over one hectare area to a depth of 10 cm. The net depth of irrigation is determined as the amount of water required to bring the soil water content just before an irrigation to field capacity in the crop root zone soil. The water content of soil just before application of irrigation must be known to estimate the net depth of water required to be applied. It is determined by the following relation:

$$d = \sum_{i=1}^{n} \frac{F_{c_i} - M_{b_i}}{100} \times A_{s_i} \times D_i \quad \text{....(16.4)}$$

where,

d = net depth of water to be applied or net irrigation (cm),

F_{c_i} = field capacity of the i-th layer of soil, (percent),

M_{b_i} = water content of the i-th layer of soil just before irrigation, (percent),

A_{s_i} = parent specific gravity of i-th layer of soil, (g/cm^3),

D_i = depth of i-th layer of soil in the root zone, (cm) and

n = number of soil layers in the root zone, (D), n.

Therefore, irrigation should consist of the quantity of water required to replenish the soil water depleted in the crop root zone just before irrigation applied. It should neither be lighter nor heavier than actual depth of water needed. A lighter irrigation leads to under-irrigation and lower soil remaining dry. While, a heavier one results into over-irrigation causing loss of water through deep percolation. Roots do not grow in dry soil. Thus, in case of under irrigation, the root development gets restricted and shallow rooting takes place due to the lower soils remain dry. Roots of an actively growing plant move deeper provided the soil below the growing point of roots is moist. Therefore, the depth of irrigation should be increased with growth of crop to wet the extending root zone. Shallow irrigation causes the roots already grown in deeper layers to gradually die due to water deficit. While, when a heavy irrigation is applied, some amount of water percolates down beyond the root zone and gets wasted.

Generally, the soil zone that accounts for 90 per cent of the root mass needs be wetted by irrigation when the crop is fully grown, but for an actively growing crop the soil little below the actively growing roots should be made moist. The depth of irrigation required for different soil types when soil water is depleted to 50 per cent of availability are shown in Table 16.7.

Table 16.7: Irrigation depths required for different soils at 50 per cent soil water depletion

Soil class	**Depth of irrigation (mm) per metre depth of soil**
Sandy soil	40
Sandy loam soil	60
Loam soil	80
Clay loam soil	100
Clay soil	125

[Source: Gautam and Dastane (1970)]

Factors affecting depth of irrigation

Depth of irrigation is inversely related to the frequency of irrigation. When the frequency of irrigation is more, the depth of irrigation applied each time is usually smaller provided the delta of irrigation remains the same. Factors that govern the frequency of irrigation, influence also the depth of irrigation. There are following factors mainly concerned in modifying the depth:

(a) Depth of effective root zone soil,

(b) Water retentive capacity of soil,

(c) Degree of soil water depletion, and

(d) Apparent specific gravity of soil.

Depth of soil to be irrigated regulates the depth of irrigation. For deeper roots, more water is required to wet the soil layers up to which roots have developed. Since roots function as the principal water absorbing organ of plants, application of smaller irrigation continuing keeps the lower soils dry and it would lead either to shallower development of roots or the deeper roots that have grown earlier will die gradually. Therefore, a greater depth of water should be applied each time to an actively growing crop so that roots grow deeper. A smaller depth of water is applied to crops that have shallower root system. The depth of water is also smaller when the crop is young and it roots are shallower. On application of higher amount of water, some water will percolate down beyond the root zone and get wasted. With growth of the crop and its root system, the depth of irrigation should be increased.

Depth of irrigation is a function of the water retentive capacity of root zone soil. Soils of heavier texture with a greater amount of capillary pores can retain more water in their pore spaces than lighter soils with more of non- capillary pores. While, soils with crumb structure, higher organic matter content and water-stable aggregates retain more water. The depth of irrigation is more in soils with a greater water retentive capacity as compared to soils with a lower water retentive capacity.

The consumptive use of crop decides the depth of irrigation. Soil water gets depleted continuously after an application of irrigation till the next irrigation. The amount of soil water depleted from the field capacity level in the effective root zone is to be replenished to bring back the soil water content to the field capacity to continue the normal crop growth.

The apparent specific gravity of soil is determined to calculate the net depth of water to be applied to the crop. The value of bulk density is important since it gives the weights of soil solids involved in water retention. The depth of irrigation increases with increase in bulk density of soil. The apparent specific gravity is the bulk density without considering the units of the bulk density. Tillage operation, soil texture, structure, organic matter content of the soil influence the bulk density of soil and consequently affect the depth of irrigation.

References

Blaney, H.F. and Criddle, W.D. 1950. Determining water requirements in irrigated areas from climatological and irrigation data. *US Dep. Agr Soil Conserv. Bull. Tech.* Pap. 96, p. 48.

Boyer, J.S. 1970. Leaf enlargement and metabolic rates in corn soybean and sunflower at various leaf-water potentials. *Plant Physiol.* 46: 233-235.

Burman, R.D. and Painter, L.T. 1964. Influence of soil moisture on leaf colour and foliage volume of beans grown under greenhouse condition. *Agron. J.* 56: 420-423.

Carlson, R.B., Yarger, D.N. and Shaw, R.H. 1972. Environmental influences on the leaf temperature of two soybean varieties grown under controlled irrigation. *Agron. J.* 64: 224-229.

Christiansen, H.F. 1968. Pan evaporation and evapotranspiration from climatic data. *J. Irrig. Drain. Div. ASCE, 94,* IR.-2. *Proc. paper No. 5988,* pp. 243-265.

Clements, H.F. 1960. Factors affecting germination of sugarcane, *Howaii Plant. Rec.* 44*:* 117.

Dakshinamurti, C., Michael, A.M. and Dastane, N.G. 1971. Water resources and their optimum utilization in agriculture. *Proc. Sym. Water Resources, Indian last. Ed., Bangalore,* C_{6-1}, C_{6-14}.

Dastane, N.G. 1969. New concepts, practices and techniques in the field of water use and management. *Proc. ICAR Sym. Water Management,* Hissar, pp. 109-133.

Gautam, O.P. and Dastane, N.G. 1970. Agronomic practices and water-use patterns for higher crop yields. *Agriculture Year Book-New Vistas in Crop Yields.* ICAR, New Delhi.

Haise, H.R. and Kelley, O.J. 1946. Relation of moisture tension to heat transfer and electrical resistance in plaster of Paris blocks. *Soil Sci.* 61: 411-422.

Idso, S.B., Reginato, R.J., Reicosky, D.C. and Hatfield, J.L. 1981. Determining soil induced plant-water potential depressions in alfalfa by means of infrared thermometry. *Agron. J.* 73: 826-830.

Majumdar, D.K. 1988. Advancing frontiers of agricultural sciences and technology for enhanced crop production. *Proc. 75th Indian Sci. Cong. Part II: Section X: Presidential Address.* Indian Sci. Cong. Assoc., Calcutta, pp. 1-34.

Majumdar, D.K. 1990. Building a sustainable national edible oil security system. *Agri. Situation in India* 44(11): 899-908.

Majumdar, D.K. and Mandal, Muralidhar 1984. Effects of irrigation based on pan evaporation and nitrogen levels on the yield and water use in wheat. *Indian J. Agric. Sci.* 54 (7): 613-614.

Miller, A.A., Jensen, R.E., Baner, A. and Norum, E.B. 1971. nfluence of atmospheric and soil environmental parameters on diurnal fluctuations on the water status of barley. *Agric. Meteorol.* 8: 93-105.

Miller, D E. and Duley, F.L. 1925. The effect of a varying moisture supply on the development and composition of the maize plant at different periods of growth. *Res. Bull. Missouri Agric. Expt. Stn.* 76.

Namken, L.N. 1965. Relative turgidity techniques for scheduling cotton irrigation. *Agron. J.* 57: 38-41.

Namken, L.N., Bartholic, J.F. and Rankles, J.R. 1971. Water stress and stem radial construction of cotton plants *(Gossypium hirsutum* L.) under field conditions. *Agron. J.* 63: 623-627.

Penman, H.L. 1948. Natural evaporation from open water, bare soil and grass. *Proc. Royal. Soc.* A193: 120-145.

Potinov, N.S. 1959. The physiology of irrigated wheat (Russian). *Izd. Akad. Nauk.* SSSR, p. 554.

Prihar, S.S. and Sandhu, B.S. 1987. *Irrigation of Field Crops-Principle and Practices.* ICAR, New Delhi.

Prihar, S.S., Gajri, P.R. and Narang, R.S. 1974. Scheduling irrigation to wheat using open pan evaporation. *Indian J. Agric. Sci.* 44: 567-571.

Prihar, S.S., Khera, K.L., Sandhu, K.S. and Sandhu, B.S. 1976. Comparison of irrigation schedules based on pan evaporation and growth stages in winter wheat. *Agron. J.* 68: 650-653.

Raheja, P.C. 1961. Water Requirement of Indian Field Crops. *ICAR Res. Series No. 28,* ICAR, New Delhi, p. 25.

Richards, L.A. 1942. Soil moisture tensiometer materials and construction. *Soil Sci.* 53: 241-248.

Richards, S.J. and Marsh, A.W. 1961. Irrigation based on soil suction measurements. *Proc. Soil Sci. Soc. Amer* 25: 65-69.

Ritchie, J.T., Burnet, E. and Henderson, R.C. 1972. Dryland evaporative flux. III. Soil water Influence. *Agron. J.* 64: 168-173.

Robins, J.S. and Domingo, C.E. 1956. Moisture deficit in relation to the growth and development of dry beans. *Agron. J.* 48: 67-70.

Sandhu, B.S. and Horton, M.L. 1977. Response of oats to water stress. 1. Physiological characteristics. *Agron. J.* 69: 357-360.

Sandhu, B.S. and Horton, M.L. 1978. Temperature response of oats to water tress in the field. *Agric. Mel.* 19: 329-336.

Shardakov, V.S. 1957. Principle for determining the water periods of cotton plant in relation to the magnitude of suction force of leaves (Russian). In "Physiological Questions in Cotton and Grasses". *Akad. Nauk Uzbek* SSR (Tashkant) 1: 5-32.

Sij, J.W., Kanemasu, E.T. and Tiare, I.D. 1972. Stomatal resistance, net photosynthesis and transpiration in PMA-treated sorghum-A field study. *Crop Sci.* 12: 733-735.

Slatyer, R.O. and Shmueli, E. 1967. Measurement of internal water status and transpiration. [In Irrigation of Agricultural Lands, Ed. Hagan, R.M., Haise, H.E. and Edminster, T.W. 1967. Number 11 in the series Agronomy, *Amer Soc. Agmn.*, Madison, Wisconsin, USA, pp. 337-353.

Taylor, S.A. 1952. Use of mean soil moisture tension to evaluate the effect of soil moisture and crop yields. *Soil Sci.* 74: 433-436.

Taylor, S.A. 1965. Managing irrigation water on the farm. *Amer Soc. Agri. Engg. Trans.* 8: 433-436.

Thornthwaite, C.W. 1948. An approach towards a rational classification of climate. *Geogr. Rev.* 28: 55-94.

Wiegard C.L. and Namken, L.N. 1966. Influence of plant moisture stress, solar radiation and air temperature on cotton leaf temperature. *Agron. J.* 58: 582-586.

Yadav, Rajiv 1995. Studies on growth and response of rapeseed to levels of irrigation and plant population. *M.Sc. Thesis, Visva-Bharati,* Palli Siksha Bhavana, Sriniketan, West Bengal.

EXERCISES

16.1 Explain irrigation scheduling and critical stages of water need.

16.2 Discuss the surface distribution of water.

16.3 Describe the criteria for irrigation scheduling in crops.

16.4 Define the relative leaf water content.

16.5 Explain the plant criteria for irrigation scheduling in crops.

16.6 Describe the climatological approach for irrigation scheduling.

16.7 Determine the amount for each irrigation for scheduling irrigation at 0.5 and 0.8 IW/CPE with 10 cm of CPE. **(Ans. 5 and 8 cm)**

16.8 Determine the irrigation interval for maize during the month of July. Given that ETm for maize in July 8 mm/day. Total available moisture is 140 mm/m depth of soil. The root zone depth is 0.5 m. **(Ans. 2.8 days)**

16.9 Describe the following:
- (a) Soil-cum-sand mini plot technique
- (b) Sowing high seed rate
- (c) Plant appearance
- (d) Critical crop stages
- (e) Indicator plant
- (f) Electrical resistance

16.10 Differentiate the following:
- (a) Irrigation frequency and irrigation interval
- (b) Depth of irrigation and crop root zone

16.11 Describe the factors affecting frequency of irrigation.

16.12 Discuss factors affecting depth of irrigation.

■■■■■

Chapter 17

PLANNING OPERATION OF WATER RECYCLING SYSTEM

17.1 Introduction

The growing demand of food production needs efficient operation and management of water recycling system. Most of the approaches followed aim at meeting the crop demand with available water to get maximum production. Scheduling of water delivery is one among them and it is a core activity that has more influence on the performance of the system compared to other irrigation activities (Chambers, 1988). Irrigation scheduling deals with two questions, when and how much to irrigate. Before cultivating a crop, the availability of water in the reservoir, chances of rainfall during the crop growing periods, groundwater potential etc., are to be assessed. In developing countries like India, scientific approach is needed for efficient utilization of the scarce natural resources. In water deficit recycling systems, attempts can be made to utilize rainfall to the maximum extent for crop growth.

Water is harvested in water harvesting structures during rainfall season and recycled during the dryspell period in *kharif* season or after the recession of rainfall. The performance of the water delivery system is judged in terms of equity, reliability, stability and predictability after the structural improvement carried out. Management information system (MIS) should be developed in such a way to improve its performance and to manage the system throughout the season for community rainwater recycling system. The MIS coupled with the communication system streamlined the operation and management of the system effectively (Santhi and Pundarikanthan, 1999). A multi-criteria scheduling procedure for rotational water distribution system may be used which incorporated the man power management aspect along with other irrigation performance indicators (Santhi and Pundarikanthan, 2000).

17.2 Need for Planning of Operation

The availability of utilizable water is limited while need of water in various sectors is growing day by day. Irrigation assumes special importance due to the fact that it alone consumes lion share which accounts for 80 per cent of available water. Thus, there is need for planning of efficient management of scarce harvested rainwater resource and it can not be over emphasized. The objective of recycling harvested rainwater is to supply and apply the right amount of water at the right place and right time to enhance productivity of both crops and water for livelihood security of farmers.

The performance of any water recycling system is judged by the following indicators:

(a) *A proper time and area irrigated against planned to be irrigated*: It is the ratio actual area irrigated to the area planned to be irrigated in any water recycling system,

(b) *Water logging*: A land is said to be water logged when the pores of the soil within the root zone of plants get saturated and normal growth of the plant is adversely affected due to insufficient air circulation. Poorly planned and managed recycling system is the main cause of the water logging whenever it occurs,

(c) *Tail-end deprivation in case of community tanks or large reservoirs*: It is present in almost every community or large reservoir and is reflected in terms of water supply, irrigation intensity, crops grown, cultivation practices, yield and incomes. Tail-end deprivation of water supply and irrigation is usually more acute problems in new ones, and

(d) *Yield*: Crop yields and farmers income usually falls from head reach to tail-end reach. However, it can be higher in the middle reaches if the head reach is waterlogged and tailed-end reach as if groundwater is used extensively.

17.3 Causes of Poor Operation of Water Recycling System

There are following causes for poor operations of any recycling system:

(i) Lack of technical skill in planning, implementing and monitoring system,

(ii) Poor man-management,

(iii) Technical deficiencies in the physical system,

(iv) Lack of interaction between engineering and agricultural experts,

(v) Higher conveyance losses,

(vi) Lack of poor coordination between different departments,

(vii) In proper assessment of personal and other facilities,

(viii) Lack of communication facilities in the command,

(ix) Poor extension services,

(x) Lack of farmers participation in the water management,

(xi) Absence of consummative use of groundwater and surface water, and

(xii) Problems related to the land settlement

17.4 Important Terminology Pertaining to Community or Large Water Harvesting System

Canal: It is defined as an artificial channel constructed on the ground to carry water from a river or another canal or a reservoir to the fields.

Permanent canal: It is continuous source of water supply. Such canals are also called perennial canals.

Inundation canal: These canal draws its supplies from a river only during the high stage of the river.

Irrigation canal: These canals carry water from its source to agricultural fields.

Navigation canal: These are used for transport of goods are known as navigation canals.

Power canal: These are used to carry water for generation of hydroelectricity.

Feeder canal: These canals are used to feed two or more canals.

Main canal: It supplies directly from the reservoir through the head regulator and acts as a feeder canal supplying water to branch canals and major distributaries. Usually, direct irrigation is not carried out from the main canal.

Branch canals (also called 'branches'): These canals take their supplies from the main canal. Branch canals generally carry a discharge higher than 5 cumecs and act as feeder canals for major and minor distributaries. Large branches are rarely used for direct irrigation. However, outlets are provided on smaller branches for direct irrigation.

Major distributaries (also called 'distributaries' or rajbaha): These canals carry 0.25 to 5 cumecs of discharge and take their supplies generally from the branch canal and sometimes from the main canal. The distributaries feed either watercourses through outlets or minor distributaries.

Minor distributaries (also called 'minors'): These are small canals which carry a discharge less than 0.25 cumec and feed the watercourses for irrigation. They generally take their supplies from major distributaries or branch canals and rarely from the main canals.

Watercourse: It is a small channel which takes its supplies from an irrigation channel (generally distributaries) through an outlet and carries water to the various parts of the area to be irrigated through the outlet. It is maintained by the farmers.

17.4.1 Important terms pertaining to irrigation

Gross commanded area (GCA): It is defined as the gross irrigable area which can be commanded by canal irrigation. Such area is usually enclosed on the boundary by drainage across which no further irrigation can reach. GCA includes unculturable area, *e.g.*, ponds, small drainage, forests, alkaline and barren land, roads, houses, etc.

Culturable Commanded area (CCA): It is the culturable area commanded by canal irrigation on which crops can be satisfactorily grown. CCA is determined by subtracting all the unculturable land from the gross commanded area as expressed,

$$CCA = \text{GCA - Unculturable area of GCA} \quad ....(17.1)$$

Intensity of irrigation: It is the percentage of CCA proposed to be irrigated annually. As irrigation is done according to the crop season *viz.*, *winter* or summer, the precise definition of intensity of irrigation is the percentage CCA irrigated at a time in one crop season.

Crop ratio: It is the ratio of the areas under the crops of summer and winter. It is also known as summer-winter ratio. Crop ratio for Punjab, Haryana is 1:1.5; for Uttar Pradesh 1:1.2 and for Madhya Pradesh, Maharashtra, Orissa, West Bengal an Andhra Pradesh 1:1.

Overlap allowance: It crops of one season extend into another season. In such situations, overlapping exists in both seasons require water simultaneously, thus creating an extra demand of water. Generally, to cater this need and overlap allowance of 5 per cent of canal discharge is provided. This indicates that the canal discharge is increased by 5 per cent for this period of overlap.

Capacity factor: It is the ratio of the average discharge of a canal at a point to the authorised full supply discharge (AFSD) or the capacity of the canal at that point. Generally, capacity factor varies from 0.5 to 0.9.

Delta (Δ): It is the depth of water that would result over a given area when irrigated from a given discharge for a specific length of time. It may be also defined the depth of water required to bring a crop to maturity during its period of growth. It is generally expressed in cm. Average values of delta for some of the important crops grown in different states of the India are presented in Table 17.1.

Table 17.1: Average delta for different crops

	Season	Crop	Delta (cm)
(a)	*Kharif*	Sugarcane	140-200
		Rice	120-180
		Cotton	50-75
		Maize	45-60
(b)	*Rabi*	Tobacco	60-65
		Gram	40-60
		Wheat	35-60
		Fodder	30-50

Base Perid (B). It is the time in days between the first irrigation of a crop at the time of sowing to its last irrigation before harvesting.

Duty (D): It is a relationship between the quantity of water and the area over which this quantity of water is applied to the crop being grown. It may be also defined as the number of hectares under a particular crop brought to the state of maturity by a constant supply of 1 cumec of water flowing continuously for the base period.

Duty of canal water may be high or low depending upon the location of the control points. High duty indicates that a small amount of water can mature comparatively over large area under a crop while, low duty indicates that large amount of water matures comparatively small area under a crop. It occurs when converse is true. When the duty of canal water is determined at the head works, it is called *gross duty* which is low duty. However, when the duty is determined at an outlet called *outlet factor* and it is high. This occurs due the fact that considerable water losses take place in the course of conveyance from the head works to the field channel. Thus, the water available for irrigation at the tail end is much less and the area to be cultivated much larger.

The relation between duty (D), base period (B) and delta (Δ) is derived as follows with the definition of duty and delta,

1 m^3/s of water for *B* days gives depth Δ metres over *D* hec. of land

Thus, 1 m^3/s for 1 day gives depth Δ metre over D/B ha.

Hence, 1 m^3/s for 1 day gives $\frac{\Delta \times D}{B}$ ha.m. of water ....(17.2)

Now 1 m^3/s of water flowing for a day measures 1×60×60×24 = 86,400 cubic metres of water = 86,400 $m^2 \times$ m.

$$1 \text{ ha} = 10{,}000 \text{ m}^2$$

$$1 \text{ m}^3/\text{s} = \frac{86{,}400}{10{,}000} = 8.64 \text{ ha.m.} \quad ....(17.3)$$

On equating equation (17.2) and (17.3) we obtain,

$$\frac{\Delta \times D}{B} = 86.64$$

or $$D = \frac{8.64\,B}{\Delta} \quad ....(17.4)$$

Paleo: It is known as the first irrigation before sowing of the crop in order to add sufficient water to unsaturated zone of the soil for the initial growth of the crop including germination.

Kor watering: It is the first irrigation after a crop has grown to a. few centimeters or the second irrigation from the beginning. Generally, Kor watering requires maximum depth of water known as *Kor depth* and is therefore important for designing irrigation channels. The period of Kor water is called *Kor period.* This Kor period for a crop in a given region depends on the duration during which rainfall is likely to occur then smaller is the water requirement. Accordingly, the Kor period is least in humid region as compared to dryer region.

Example 17.1: A water course has GCA of 1400 hectares and CCA is 80 percent of GCA. The intensity of irrigation for wheat crop is 45 per cent and for rice crop 30 per cent. Wheat has a Kor period of 28 days while rice has a Kor period of 20 days. Determine the outlet discharge neglecting losses. Assume Kor depths for wheat and rice as 15 cm. and 22 cm, respectively.

Solution:

$$\text{GCA} = 1400 \text{ ha.}$$

$$\text{CCA} = 1400 \times \frac{80}{100} = 1120 \text{ ha.}$$

$$\text{Area irrigated under wheat} = 1120 \times \frac{45}{100} = 554 \text{ ha.}$$

$$\text{Area irrigated under rice} = 1120 \times \frac{30}{100} = 336 \text{ ha.}$$

Eq. (17.3) is rewritten as,

$$\text{Duty or outlet factor} = \frac{8.64 \times \textit{Kor period in days}}{\textit{Kor depth in metres}}$$

$$\text{Duty for wheat} = \frac{8.64 \times 28}{0.15} = 1613 \text{ ha/m}^3\text{/s}$$

$$\text{Duty for rice} = \frac{8.64 \times 20}{0.22} = 785 \text{ ha/m}^3\text{/s}$$

$$\text{Discharge at outlet} = \frac{\textit{Area irrigated}}{\textit{duty}}$$

$$\text{Outlet discharge for wheat} = \frac{554}{1613} = 0.34 \text{ m}^3\text{/s}$$

$$\text{Outlet discharge for rice} = \frac{336}{785} = 0.43 \text{ m}^3\text{/s}$$

17.5 Dynamics of Water Harvesting Structures

The basic dynamics of any water harvesting reservoir system can be described by the following flow equation at any time period (*t*).

$$S(t) = S(t\text{-}1) + In(t) = [IR(t) + Spill(t)] - L(t)\ (1) \qquad \text{....(17.5)}$$

where,

$S(t)$ = storage volume of reservoir at the end of period t,

$IR(t)$ = gross irrigation water release volume from the reservoir during the period *t*,

$In(t)$ = total net inflow volume to reservoir during the period *t*,

$Spill\ (t)$ = spilled release volume from reservoir during the period *t, and*

$L(t)$ = total loss due to evaporation from the reservoir during the period *t*.

Simple bounds on variables

$$S\,min \leq S(t) \leq S\,max \qquad t = 1,2,3,.....36 \qquad \text{....(17.6)}$$

$$IR(t) \le Demand(t) \quad \text{....(17.7)}$$

S min is the dead storage volume and *S max (t)* is the maximum storage volume.

The release into the canals should be less than or equal to the irrigation demand.

The computational process of water balance equation is carried out as follows on using Microsoft Excel:

Balance storage = End storage + Inflow

Storage available = Balance storage - Evaporation loss-S_{min}

if. Storage available ≥ Demand. Release = Demand

else, Release Storage available

Total storage = Storage available - Release + S_{min}

Spill = Total storage – Reservoir capacity

(if total storage ≥ Reservoir capacity, else spill = zero)

Deficit = Demand – Release

End storage = Total storage – Spill

17.6 Main Objectives of Operation and Criteria of Good Performance

17.6.1 Main objectives

The objective of any water recycling systems is the timely delivery of the irrigation water necessary to satisfy crop water requirements. This objective is achieved by three-tier system. It consists of three main activities as follows:

- Planning the operation (preparation of irrigation plans),
- Implementation of the Plan (actual water distribution), and
- Monitoring of the operation (collection of data related water use and preparation of the corresponding reports) to execute the plan.

To undertake these activities different kinds of personnel are required with specific qualifications depending on the type of water distribution system and other local characteristics.

17.6.2 Criteria of good performance of water recycling system

The criteria for judging the performance of recycling systems in community tank or large reservoir can be vastly different for different groups of people depending upon their concerns (Table 17.2). Chambers (1984) suggested the most common criteria generally accepted for judging the performance of an irrigation system are productivity, equity and stability which together contribute to the objective of well-being.

Table 17.2: Performance criteria of good water recycling system (large)

Type of person	Possible first criterion performance for good system
Landless Labourer Farmer	Increased labour demand, days of working and wages. Delivery to his or her farm of an adequate, convenient, predictable and timely water supply for preferred farming practices.
Irrigation engineer	Efficient delivery of water from headworks to outlet.
Agricultural engineer	Efficient delivery and field application of irrigation water and from the outlet to the root zone of the crop.
Agronomist	Creation and maintenance of the optimum moisture regime and plant growth and, in particular, maximizing production of that part of the plant which is the harvestable product.
Agricultural economist	High and stable farm production and incomes.
General economist	A high internal rate of return.
Political economist	Equitable distribution of benefits, especially to disadvantaged groups.
Sociologist	Participation of irrigators in management.

(Source: Chambers, 1984)

Productivity

It is defined as the ratio of total crop yield (t)/total area irrigated (ha) or total crop yield/total water delivered in the irrigated area. It may be also defined as the ratio of output and input. The productivity can be determined at the farm gate, at the outlet or at upstream points in the system including the point of diversion or storage (Pant, 1984). Typical points of input and output measurements for different professionals are depicted in Fig. 17.1 (Chambers, 1984). Improved water supply influences the adoption of high-yielding agriculture practices by farmers which justifies the productivity criterion of performance.

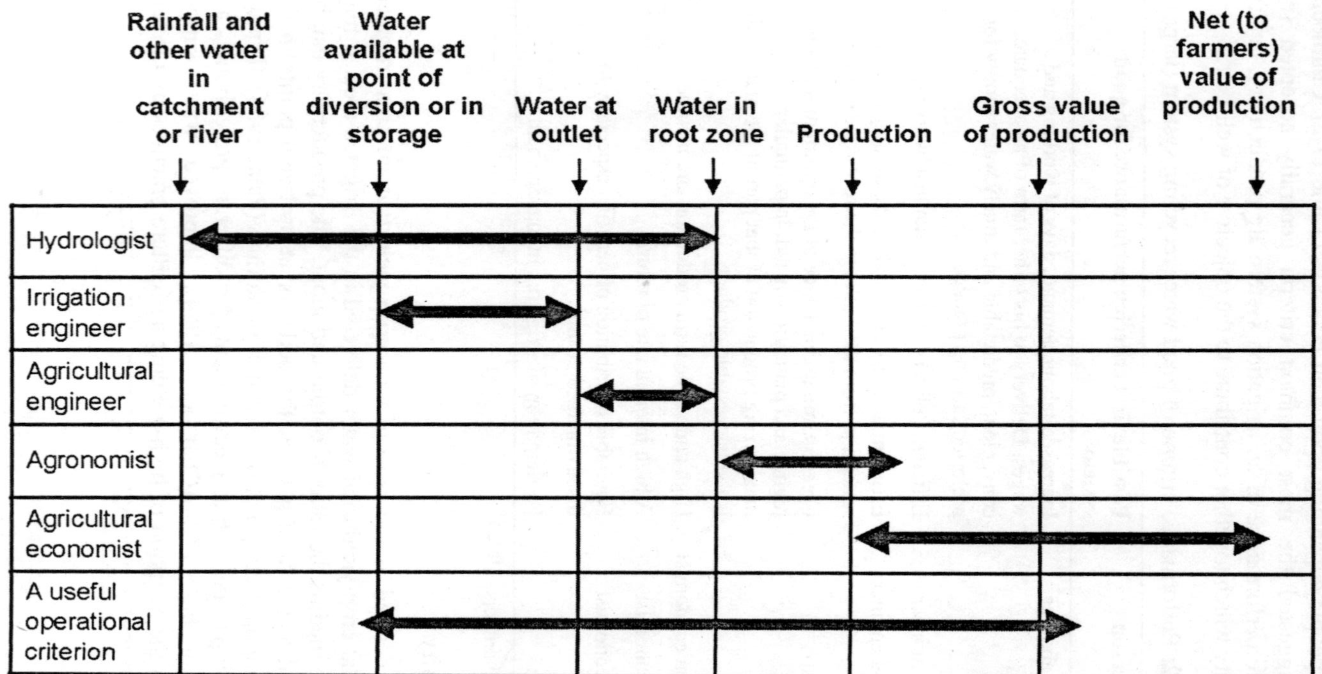

Fig. 17.1: Typical points of input and output measurements of productivity for different professions

Equity

Chambers (1984) defined equity as equality, fairness and even-handed dealing in matters of allocation and appropriation of irrigation water in community tank or large reservoirs. There are several ways to decide the equality of supplies to different farmers. Out of this there are two methods of prior appropriation and of proportionate equality. In the first method, first exploits a resource establishes a right to continue to do so. Thus, head reach farmers and early comers to water recycling system establish their right over irrigation water, even though the tailed-end farmers or late comers get less water or no water. The second method implies that the supply of water is in proportion to the size of the land-holding as in the *warabandi* system of north-west India (Mahalotra, 1982). However, Chambers (1984) found that water distribution is far from meeting even the criterion of water proportional to land and, hence, both these methods have been criticised for their inequity in India.

Stability

Stability of productivity as well as equity is important in any community tank or large reservoirs. Stability can either be short-term stability or long-term sustainability.

Short-term: It refers to the variations in productivity and equity between irrigation seasons and is a function of climate, water supply, storage and control, system management and other factors such as pests, diseases and availability of labour and other inputs. It is also known inter-seasonal stability. It can be measured by comparing performance between seasons.

Long-term sustainability: It is the "environmental stability" and "du*rabi*lity" and refers to the prevention or minimising of adverse effect of physical changes such as water-logging, leaching of nutrients from soils, salinity, erosion, silting, the 'mining' of ground water and infestations with weeds (Chambers, 1984). Sustainability can be measured in terms of ground water levels, salinity, erosion, or silting through inspecting works and by measuring long-term trends in productivity and equity.

17.7 Planning Operation of Recycling System

The basic objective of planning is to match supply with demand as closely as possible. The planning is a complex and laborious exercise when "Irrigation Plans" is prepared for large water recycling systems or just a simple meeting where farmers are informed about the amount of water available and the time delivery schedule. The complexity or simplicity of the planning process varies from situation to situation depending on the scope for manipulating supplies to meet demand. However, planning is essential in all cases and performed in two-way communication with the farmers. As water is a scarce natural resource in water recycling system specially under dryland ecosystem. Thus, the importance of this planning process cannot be overemphasized.

The preparation of an Irrigation or Crop Plan involves following main steps:

(i) Estimation of future water supply,

(ii) Estimation of water demand of the expected cropping pattern, and

(iii) Matching supply and demand.

17.7.1 Estimation of future water supply

The future water supply may be a fairly determined. The water supply depends upon the several factors such as the characteristics of the dry and wet season, the type of water storage utilized, the reliability of climatic data and the effective rainfall during the irrigation season.

There are following conditions for easier determination of the water supply:

(i) Pumping or diversion from a stream or a river with an average flow much greater than the one pumped. Thus, it clearly indicates that supply is greater than the area to be irrigated.

(ii) Pumping from fairly, abundant aquifers,

(iii) Dam storage where the season for filling the reservoir does not coincide with the irrigation season during which hardly any water contribution can be expected.

In all above conditions, the available water resources can be determined precisely at the beginning of the irrigation season.

However, there are other cases in which a certain degree of uncertainty exists about the availability of the water to be recycled. In such situation, water availability is determined on the basis of conservative estimates such as taking into consideration of rainfall at probability of 75 or 80 percent. The mathematical models based reliable climatic data can also be a very useful tool to forecast the water available for different levels of risks. However, it is most important in these cases to develop an alternative plans which can be adopted according to changing climatic conditions. A dynamic programme is essential so that plan can be modified according to changes in the weather technically rather than fixed plan. However, the bottleneck problems are the rapid communication of this information to the farmer so that farmers can adjust the irrigation or crop plan accordingly to the changing conditions.

17.7.2 Estimation of water demand

The water demand is basically determined by the expected cropping pattern and taking into consideration of irrigation efficiencies at the farm and project level.

1. *Cropping pattern*: The cropping pattern can be known by the land and revenue records of the command area of the water recycling system. However, the difficulty of foreseeing the expected cropping pattern in any water recycling system varies according to the degree of freedom allowed to farmers in their choice of crops and the timing of their cultivation activities and market price of the out of the crops. Alternatively, demand can be controlled by means of differential control over the water supply pattern in following ways:
   (i) Permitting certain highly water-consuming crops to be grown only in certain designated areas which will be the only ones to receive sufficient water;
   (ii) Legal limits can be placed on the areas to be covered by certain crops (with fines being exacted for contravention of the rules), and
   (iii) There can be a free choice of cropping, when response to market demand will be the main determining factor.
2. *Irrigation efficiencies*: To determine the actual demand, the efficiency of the water distribution system and of application must be known. However, this is usually not available in all the water harvesting and recycling system. Because it is time-consuming and required qualified staff. This again stresses the need for some specialized unit to take care of this important function, but the important thing is that they are carried out by someone. In the absence of reliable field data on efficiencies, empirical data can be used as given in Table 17.3 and 17.4.

Table 17.3: Water application efficiency in different irrigation method and soil type

Irrigation method	Application practices	Water application efficiency (Ea × 100) Soil texture	
		Heavy	Light
Sprinkler	Day time application, moderately strong wind Night application	60 70	60 70
Localized		80	80
Basin	Poorly levelled and shaped	60	60
	Well levelled and shaped	75	75
Furrow, border	Poorly graded and sized	55	40
	Well graded and sized	65	50

(Source: FAO, 1984)

Table 17.4: Conveyance losses in canals

Material	Loss in cumecs per million square metres of wetted perimeter (or water surface)
Impervious clay loam	0.88 to 1.24
Medium clay loam under laid with hard pan at depth of not over 0.60 to 0.90 m below bed	1.24 to 1.76
Ordinary clay loam, silty soil or lava ash loam	1.76 to 2.65
Gravelly or sandy clay loam, cemented gravel, sand and clay	2.65 to 3.53
Sandy loam	3.53 to 5.29
Loose sand	5.29 to 6.17
Gravel sand	7.06 to 8.82
Porous gravel soil	8.82 to 10.58
Gravels	10.58 to 21.17

17.7.3 Matching supply and demand

Once the supply and demand has been determined then the most important part is to decide on water distribution practices or other measures which allow the closest possible matching of supply and demand. Although the spectrum of possible situations is very large. However, three main situation can be clearly distinguished as under:

(i) Water recycling system with water supply greater or equal to demand,

(ii) Water recycling system with a moderate water deficit, and

(iii) Water recycling system with a large water deficit.

(i) Water recycling system with water supply greater or equal to demand: It is easier to operate the water recycling system in this system due to abundant water supply and it is most favourable from the management point of view. However, such type of water recycling systems are less efficient in terms of returns per units of water. This kind of situation often occur during the construction period of very large irrigation reservoir lesser demand than supply till their completion.

However, in well designed water harvesting and recycling system, supply and demand should be equal. Further, it is not only important to check the seasonal volumes needed but also to check if the demand in the peak month can be met. If the supply is less as compared to the peak month demand under such situations the usual corrective measure is to advance the sowing dates of some of the crops in order to avoid coincidence of peak demands.

(ii) Water recycling system with moderate water deficit: When water deficiency is order of 10-20% of the available supply is known as water recycling system with water supply moderate water deficit. This can either be a periodic situation found only in "dry" years or recurrent every year. This occurs due to accepted risk in the designing of a water harvesting and recycling system in the first case, while latter can be caused due to designs associated more with social factor than technical ones and other factors such as changes in planned cropping pattern, overestimation of the potential water supply, technical deficiencies of the system etc. may lead to this kind of situation. This systems offer the best opportunities for maximizing the returns from the available water.

In this situation, suitable water distribution practices and methods, combined with some of the measures described in subsequent section can be of immense helpful in matching the supply and demand.

(iii) Water recycling system with a large water deficit: When the water deficit is greater than 50 per cent of the available supply called as water recycling system with large water deficit. However, this system is not designed to irrigate the whole command area at cropping intensities of 100-200 per cent. Thus, 'demand' may be fulfilled if 40-60 per cent intensities are achieved.

These water recycling systems are designed with the following objectives:

(a) Important social objective of benefiting as many people as possible.

(b) Other reason may be the result of an under estimation of crop water requirements.

(c) Maximizing government revenue.

Whatever may be the reason, these systems have frequently yielded less than expected. Production per hectare is low and large areas are sometimes abandoned due to salinity problems. However, there are some reasons for this situation.

- No financial support to the farmers to make efficient use of available water,
- Faulty irrigation practices and to control salinity and also efficient water use,
- No preventative measures (drainage system) will be provided to prevent build-up of soil salinization, and
- The water distribution network is much longer than it should have been; therefore it gives greater water losses.

These problems do not only occur in water recycling system with large water deficiencies. These problems are likely to have more damaging effects under very limited supply.

17.8 Measures to Match Supply and Demand

Several measures and water distribution practices can be helpful to bridge the gap between supply and demand. The measures of water deficit can be grouped into the following categories:

(i) Measures related to the cropping pattern,

(ii) Measures related to water distribution practices, and

(iii) Measures related to water fees

These measures can be recommended as a sole or in integrated manner to reduce the demand.

17.8.1 Measures related to the cropping pattern

There are following three main measures that can be recommended to reduce water demand:

(i) Changing the sowing time,

(ii) Introduction of low water requirement crops, and

(iii) Reducing the irrigated area.

All these measures are the most effective to reduce water deficit but they are also the most difficult to implement. It requires that the management of the irrigation scheme be invested with the necessary authority to introduce such changes in the cropping pattern. Otherwise a long dialogue may take place between the management and the farmers to convince them of the necessity for this measure which may not always end with a positive decision.

17.8.1.1 Changing sowing time

By changing sowing time and other cultivation practices can reduce large water deficit during the peak demand of water recycling system. Hence, water supplies are very carefully changed according to the stage of crop development and farmers should also know that to obtain enough water at periods of peak requirements such as for land preparation, etc. Farmers should keep a pre-planned timetable for the same. Thus, careful planning leads to allow controlled staggering of cultivation practices between different sections of the same recycling system.

This leads to a more rational use available machinery and man-power. Crop calendar adjustment is the adjustment of either advancing or postponing the starting dates of crop cultivation from existing practice and consider the rainfall contributions towards crop water requirement. Thus, the water availability is maximized by considering the effective rainfall, evaporation loss from the system and evapotranspiration.

The moisture availability index (MAI) is used to assess the moisture availability during crop growing period as suggested by Thornthwaite Mather (1955). The moisture availability index (MAI) is expressed as,

$$\text{MAI} = \left[\frac{R}{ET} - 1\right] \times 100 \quad \text{....(17.8)}$$

where,

MAI = Moisture availability index, (%),

R = Rainfall, (mm), and

ET = Evapotranspiration of reference crop, (mm).

This index helps to know the importance of level of irrigation in the cultivation of any crop in a given area. However, MAI alone cannot be sufficient to ascertain the irrigation release from the rainwater harvesting structures or recycling system for the crop and the crop water requirement, Hence, the performance of the recycling system needs to be evaluated by some other means. Adequacy indicator is used as proposed by Molden & Gates (1990) for evaluating the performance. It is expressed as,

$$\text{Adequacy index (AI)} = \frac{I}{n}\sum_{i=1}^{n}\frac{D_i - D_{fi}}{D_i} \qquad \text{....(17.9)}$$

where,

D_i = Demand for the period i,

D_{fi} = Deficit for the period i, and

n = Number of periods in crop period (100 days at ten days per period).

The starting of cultivation period of crops play a vital role in efficient functioning of any system. Before starting the cultivation of crop, the water storage either in the tank or in the reservoir should be adequate in quantity. It is evident from the results of study carried out by Ramakrishnan *et al.* (2009) that cultivation in the third period of November month (21st - 30th November) recorded the higher performance indicator for groundnut cultivation and domestic water supply among the four options in Sathanur reservoir of Tamil Nadu (Table 17.5).

Table 17.5: Comparison of average performance indicator for different options of starting cultivation of groundnut

S. No.	Option/policy	PI for release to groundnut	PI for domestic water supply	PI for tank supply	PI for old command area release
1	First period of December (1-10 Dec.)	0.84	0.83	0.85	0.61
2	Second period of December (11-20 Dec.)	0.79	0.83	0.86	0.62
3	Third period of November (21-30 Nov.)	0.89	0.84	0.85	0.61
4	Total command area of 29.43% irrigation in dry crop grown in third period of November	0.80	0.63	0.85	0.61

17.8.1.2 Introduction of low water requirement crops

It is an effective measure to reduce water demand such as clover for alfalfa, sorghum for maize, etc. To achieve this, both crops have similar characteristics or end purposes. Introducing crops should be financially attractive to the farmer. The adoption of this measure depends upon the communication, farmer's choice and income.

17.8.1.3 Reducing the irrigated area

It is the most expedient way of reducing the demand, but it is difficult to implement. This can be achieved by reducing the water allocation in turn rather than physically reducing the irrigated area. This results in reduction in the area irrigated by the farmer. However, the farmer often decides to take a risk and irrigates a much larger area than technically feasible. The water deficit still can be reduced with adoption of suitable levelled land and using good water management practices. The risk taken by the farmer might yield high returns but unfortunately this rarely is achieved.

17.8.1.4 Irrigation application methods

Most of the surface irrigation methods yield reasonably high field application efficiency provided that the land has been prepared properly and due care has been taken during irrigation. However, introduction of pressurized irrigation methods such as the sprinkler and the drip irrigation method seem to be more promising than others in most of the conditions. These methods require much higher initial investment, energy for generating pressure and silt-free water. The sprinkler irrigation may result in considerable saving of money as well as water when land preparation and greater percolation losses are considered. The average cost of a sprinkler irrigation system may be approximately ₹ 15,000 per hectare and can possibly be recovered in about 2 years' time. The drip irrigation method is highly efficient and better suited for fruit crops, vegetables and cash crops like sugarcane, cotton, groundnut, etc. Singh (1989) summarized the comparison of different irrigation methods (Table 17.6). It clearly indicates that sprinkler and drip irrigation systems are superior over the surface methods (*i.e.* flooding check basin, border strip and furrow methods).

Table 17.6: Comparison of suitability of irrigation methods

Site factor	Surface Method	Sprinkler Method	Drip Method
Soil	Uniform with moderate to low infiltration	All	All
Topography	Level to moderate slopes	Level to rolling	All
Crops	All	Tall crops limit type of system	High Value
Water supply	Large streams	Small stream nearly continuous	Small stream continuous and clean
Water quality	All but very salty	Salty water may harm plants	All
Efficiency	50-80%	70-80%	80-90%
Labour requirement	High	Low to moderate	Low to high
Capital requirement	Low to high	Moderate to high	High
Energy requirement	Low	Moderate to high	Low to moderate
Management skill	Moderate	Moderate	High
Weather	All	Poor in windy conditions	All

In India, sprinkler as well as drip irrigation has considerable scope to meet the ever increasing demand of water for livelihood security of increasing population. This is possible only by saving irrigation water through micro irrigation system and extension of irrigation facilities. The micro irrigation system will not save water but water can also be used for expansion of irrigated area and expansion of industries. Sprinklers must find useful application in undulating sandy terrains of Rajasthan, Gujarat, Haryana and Punjab state of India. Drip irrigation is ideally suited for fruit orchards, vegetable crops and some cash crops. Singh (1989) found that water saved due to the introduction of these two methods in favourable regions may be enough to increase the irrigation potential for additional 5 million hectares.

17.8.1.5 Other measures for reducing the cropped area

Other measures for reducing the cropped area are as follows:

(i) Eliminating the areas furthest from the distribution point. It is quite common in small village tank when the farmers may have fragmented holdings some near the tank, others farther away. In these cases to reduce the water demand, water should be applied in the nearest areas. This also reduces the seepage and evaporation losses.

(ii) Delivery of water to preferable section of command area: Delivery of water in certain sections of the command area only, with the rotation from season to season. This is only feasible when irrigation is supplementary and farmers are able to grow rainfed crops in the concerned season.

17.8.2 Measures related to water distribution practices

There are only two measures that can be recommended to reduce the water deficiency:

(i) Reducing the water allocation but keeping the same water distribution method, and

(ii) Changing the water distribution method to a more efficient one.

17.8.2.1 Reducing the water allocation

It can be achieved in the following different ways:

(i) Allocation of water to preferential or high value crops,

(ii) Deficit irrigation or decreasing the amount of water per irrigation, and

(iii) Reducing the number of irrigation or extending irrigation intervals.

(i) *Allocation of water to preferential or high value crops*: This is the most common where high value crops such as fruit trees, nursery produce, vegetables are grown near to less valuable ones. The valuable crops must receive their necessary water and whatever is left can be utilized for the other crops. This kind of measure is recommended when the interests of the farmers and the management are the same.

(ii) *Deficit irrigation or decreasing the amount of water per irrigation*: This can be applied in proportion to the water deficit provided that no possible effects on the crop yield or minimize the effects on production. The amount of water is decreased in such a way that the effect on crop production is minimized. The first alternative is the most commonly adopted measure because of its simplicity. However, the second one offers much better possibilities. The deficit irrigation based on new developments resulting from studies made on crop yield response to water (FAO, 1979). This method can be extremely effective than one single crop is growing, but its effectiveness decreases with the number of crops grown. This is difficult due to fact that the intervals that fit one crop well may not necessarily do so for others.

With adoption these measures, moderate water deficit can be reduced. While, large reductions in flow in canals will often hamper their operation *e.g.* as the canals are designed to run above 75 percent capacity. Thus it will not distribute water proportionally if run below this limit.

(iii) *Reducing the number of irrigation or extending irrigation interval*: It is the measure most commonly used to cope with water deficits. For example, if any water recycling system is capable to provide 8 irrigations in particular year but during the next year the supply is only half of this amount. Thus, the system will deliver 4 irrigations at double intervals of time. Although this measure is most common, as in other cases and also the possibility of giving the 4 irrigations so that the crop can make best use of the water.

17.8.2.2 Changing the water distribution method

Among the different water distribution methods, some of the methods are more efficient from water management point of view. The possibilities of changing the water distribution method are very limited as a certain method is normally linked with a specific technical design. This shows that changing the method means changing the physical system to some extent.

17.8.3 Measures related to water fees

Increasing the water fees lead to decrease the amount of water used. However, this measure should be exercised with great care and only where the water supply is measured. In this system, prices are associated with volume of water received by the farmers. Another important requirement is that the farmer should have some knowledge of soil-plant-water relationship. The increase in water fee should be in substantial manner so that farmers can easily adopted this system.

17.9 Water Delivery System

The actual distribution of water has different characteristics depending on the water distribution method to be used. There are following water distribution methods:

1. On-demand,
2. Semi-demand,
3. Canal rotation and free demand,
4. Continuous flow, and
5. Rotational system

17.9.1 On-demand

In this system, water is available to the farmer at any time that the intake or hydrant is opened. Therefore, the amounts to be used are not limited but water consumption is usually metered and paid for by cubic metre. This system is generally designed with high-level technology. The degree of human intervention is minimal since they operate on automatic principle, *i.e.* when the water level or pressure drops in a canal or pipe due to the opening of an inlet, the level or pressure is immediately reinstated by an automatic device which calls for a greater supply provided by automatic gates or valves. The efficiency of this system is very high (up to 90 per cent) particularly when using pipes for the distribution.

The great merit of this method is that it allows the farmer to use the water when it is most necessary for the crops. The main demerit is high costs and it needs a high level technology in the construction and maintenance of the systems. Thus, these systems are more suitable for developed countries than developing ones and unsuitable for the undeveloped countries.

17.9.2 Semi-demand

In this system, water is made available to the farmer within a few days (generally 2-7 days) of his request. The amount is often limited to a certain volume per hectare. Farmers made a request for water to the management system and be informed according to availability of water and time of schedule. The amount to be supplied to the farmer is usually fixed in relation to the number of hectares. A known amount of water is an indispensable requirement for such a system. This will help in the measurement to calculate the time and flow needed for each request in order to prepare the water distribution programme.

To avoid possible excess use of water by the farmers some restrictions may be imposed, such as the number of irrigations per year is limited, or a certain time (7-10 days) must elapse between two requests for water.

The advantage of this system is that its easier to operate, when the need arises (peak month) or during exceptionally dry years, it can also function on a fixed rotation.

The only disadvantage of this distribution system is that it is having low efficiency at times of low demand. The opening and closing of canals for a few farmers leads considerable water loss which resulted into low efficiency.

17.9.3 Canal rotation and free demand

In canal rotation system, secondary canals receive water by turns such as every 7 days, and once the canal has water farmers can take the needed of amount of water at the time they wish.

This system is recommended particularly when there is mixed control of management. The public administration undertakes the operation of the main and secondary canals, and the farmers either take the water freely from those canals or distribute it themselves from a smaller canal. The duration of the turns is generally the result of experience in the area. When the number of crops grown in the recycling system or region is fairly large, there is not much opportunity for rationalizing the duration of the turns. However, if the number of crops is limited or some crops clearly prevail then the duration of the turns can be determined in a rational basis.

This type of system can be handled easily by public organizations but it implies large operational losses. The low efficiency of the system is mainly due to disassociation of demand and supply. Canals are filled every turn irrespective of the demand. Some adjustments can be made in the canal water levels according to the demand pattern but still operational water losses are bound to be high. The system can easily be improved by properly organizing the farmers along the watercourses and regulating the flows according to the requests received from the groups of farmers on each watercourse.

17.9.4 Continuous flow

In this system, the farmer receives a small but continuous flow that compensates the daily crop evapotranspiration through out the season.

Continuous flow is perhaps the simplest water distribution system. However, it is the least efficient. This is due to delivery from field to field and it results in large evaporation losses. Under this system water losses by deep percolation and excessive runoff are high.

Where water is scarce, the continuous flow system can lead to considerable social unrest because farmers on higher land get the water needed while those lower down get very little or nothing.

Although continuous submergence of paddy fields has certain advantages (saving of labour, prevention of crop damage, better control of weeds, etc.), the system is more and more restricted to areas where water is plentiful and its efficient use is therefore less important.

17.9.5 Rotational system

In rotational system, secondary canals receive water by turns and the individual farmers within a given canal area receive the water at a pre-set schedule of time and generally in a limited quantity. All canals receive water by turns and farmers on the tertiary canals or watercourses receive water at a pre-set schedule of time and in the allowed quantity. This system is an improvement of the canal rotation when the rotation is not only of the main canals receiving water but also of the farms. It is a highly efficient system from the operational point of view and socially fair since it gives an equal chance to everyone.

The water distribution method is normally linked with the design of the conveyance system. However, some exceptions therefore, there is chance to change the water delivery system once a water distribution method has been selected. For selection of the water delivery method, the important factors including social, technical and economic characteristics must be taken into consideration. Each of the water delivery methods has its own merits and demerits which may, or may not, suit local conditions.

Another promising system of water delivery is through water cooperatives which purchase water in bulk and then distribute it among their members (farmers). One example of this distribution is the Mohini Water Cooperative Society near Surat which is considered as the first successful irrigation cooperative in Gujarat.

17.9.5.1 *Warabandi*

It is the rotational delivery system of water and is the most feasible and water delivery system and offers many advantages. *Warabandi* is an integrated management system from source (community tank or reservoir) down to the farm gate, *i.e. nakka*. It is also defined as a system of equitable water distribution by turns according to a predetermined schedule specifying the day, time of duration of supply to each farmer in proportion to land holdings in the outlet command. The water from the source is carried by the main canal which feeds two or more branch canals (which operate by rotation) and may not carry the total required supply in the *warabandi* system (Fig. 17.2).

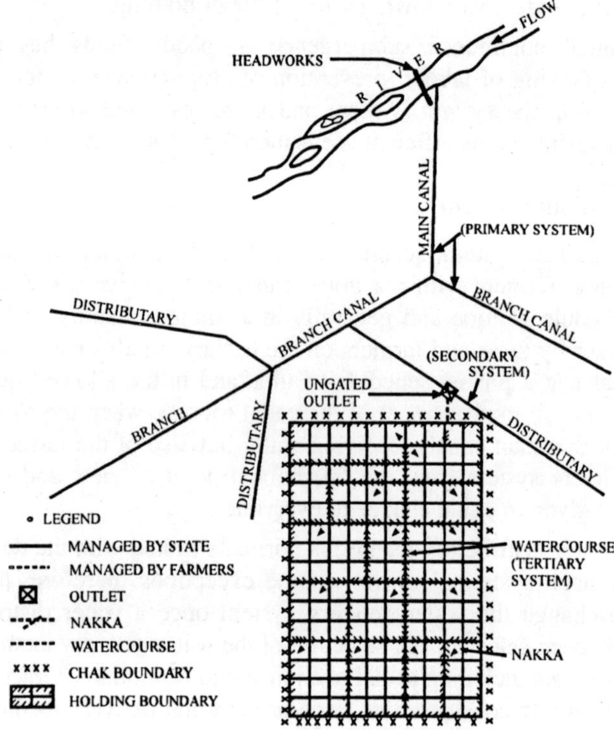

Fig. 17.2: Typical *warabandi* distribution system

This is the primary distribution system which runs throughout the irrigation season with varying supply. The secondary distribution system consists of a large number of distributaries which too run by rotation but carry full supply. The distributaries supply water to the watercourses through outlets. These watercourses run full supply when the supplying distributary is running. Water is then allocated to various fields (or farms) situated along the waterly course by a time roster. This is the tertiary distribution system.

The main objective of this system is to attain high efficiency of water use by imposing water scarcity on every user. The system ensures equitable distribution and safeguards the interest of the farmer whose field is located at the tail end of the conveyance system.

In this system, each unit of culturable commanded area is allocated a certain rate of flow of water, termed water allowance, whose value is generally a compromise between demand and supply. The carrying capacity of distributaries and watercourses is designed on the basis of water allowance. Whenever distributaries run, they are expected to carry their full supply. The outlets to watercourses are so planned and constructed that all the watercourses on a distributary withdraw their authorised share of water simultaneously.

To check the dangers of water-logging and salinity, no distributary is allowed to operate all the days during any crop season. The ratio of the operating period of a distributary and the crop period is called the *capacity factor* of the distributary. For example, the capacity factors for summer and winter are 0.8 and 0.72, respectively for the Bhakra project. It implies that each distributary would receive its full supply for a period of about 144 and 129 days, respectively in a crop season of 180 days.

The ratio of the resultant irrigated area to the culturable command area is known as intensity of irrigation. Its value is 62 per cent for the Bhakra project. The intensity of irrigation is an index of the actual performance of the irrigation system.

The design of distributaries and watercourses is related to the culturable command area (which is fixed) instead of the variable cropping pattern in warabandi system. The total amount of water available at the source has its own limitations and it may not always be possible to expand or augment the supply to keep pace with the ever-increasing demand of the cropping pattern. Therefore, there is an obvious advantage of relating the design to the culturable command area instead of the needs of the cropping pattern.

It is evident from the results of studies that the *warabandi* system, although efficient and acceptable to the farmers, does not result in equitable distribution primarily due to the losses in the water courses. These losses may cause about 25-40 per cent reduction in the share of the farmers at the tail-end reach. Hence, some weightage should be given to the time allotment for tail-enders to compensate for these losses while determining the schedule of operation.

17.9.5.2 Determination of roster of turn

The cycle of turns on a watercourse or its branch starts from its head, proceeds downstream and ends at its tail. Before a farmer receives his share of water, some time is spent in filling up the empty watercourse between the point of taking over and the beginning of his holding. This time is known as *bharai* (*i.e.* filling time) and is debited to a common pool time of 168 hrs and credited to the account of the concerned farmer.

The supply in the watercourse has to be stopped when the tail-end farmer is having his turn. The water filled in the watercourse during the common pool time *(bharai)* can be discharged only into the field of the tail-end farmer. Thus, the total time spent on the filling of the watercourse should be recovered from him in lieu of this. But he does not receive this water at a constant rate. This leads to efficient field application beyond supply limit. Therefore, the tail-end farmer is compensated for it and is allowed a certain discount on the recovery of the *bharai* time. This value of *bharai* is known as *jharai* (*i.e.* emptying time). Thus, the correct determination of *jharai* time is an unresolved problem and its present values are favourable to the tail-end farmer.

After allowing for *bharai* and *jharai,* the flow time (FT) for a unit area and for an individual farmer is determined by the following relationship:

$$\text{(FT) for unit area} = \frac{168-(\textit{Total bharai}-\textit{Total jharai})}{\textit{Total area}} \qquad \text{....(17.10)}$$

$$\text{FT for a farmer} = (\text{FT for unit area} \times \text{area of farmers field}) + (\text{his } \textit{bharai} - \text{his } \textit{jharai}) \qquad \text{....(17.11)}$$

Thus, *bharai* is generally zero in the case of the last farmer and *jharai* is zero for all except the last farmer. It evident from Equations 17.10 & 17.11, that the losses in watercourse are not taken into consideration.

17.9.5.3 Night irrigation

Another important issue is the operation of the system in night irrigation. The canal water continues to flow at nights as well and is either badly used or wasted. Darkness, cold, fear, normal working hours and desire for sleep discourage the irrigation staff, farmers and labourers to work at night. About 47 per cent of the 24 hours day is the time of darkness (between sunset to sunrise with an allowance of 20 minute twilight period each after the sunset and before the sunrise) in India.

Chambers (1984) found that about 40 per cent of the canal irrigation water on medium and major systems (*i.e.* with commands over 2000 ha) is either applied in night irrigation or sent into drains at night. He further estimates that, except in north-west India where *warabandi* is practised at night, 25 per cent of the canal irrigation water is wasted at night and much of the 15 per cent which is applied at night is used inefficiently. According to Chambers (1984) farmers usually dislike night irrigation due to the following reasons:

(i) Loss of sleep and disruption in the normal sleeping duration,

(ii) Discomfort due to cold nights and difficulty in moving around in sticky soils and mud,

(iii) Danger and fear of snakes, scorpions, accidents, violence including murder and other problems related to law and order,

(iv) Inefficient application of water due to darkness, and

(v) Higher costs due to higher night wages, non-availability of family labour, especially women, old people and those very young to work at night and need of firewood, beverages and lighting.

In addition to above reasons, field conditions, type of crop and its stage of growth may add to the difficulties of irrigating at night. Sloping lands with difficult soils and standing crops are difficult to be irrigated at night. Paddy, trees and crops which are low and wide apart or are in early stages of growth can be irrigated with relative ease. High and dense crops and crops in the later stages of growth are relatively difficult to be irrigated at night. The difficulties in night irrigation may influence a farmer's choice of crop if he has to rely on only water received at night.

Chambers (1984) observed that night irrigation is preferred by farmers due to the following reasons:

(i) In warmer regions, farmers find it more comfortable to irrigate at night,

(ii) Part-time farmers having other work during the day would prefer night irrigation, and

(iii) Tail-end farmers may get relatively more adequate and reliable supply during night.

In addition to the above characteristics of night irrigation, the following features must also be noted:

(i) It is during the night that illicit appropriation of water (breaching of bunds, removing checks, blocking streams, opening pipes, pumping out of channels etc. to secure water for an individual or a group) takes place.

(ii) During nights, evaporation losses are much less. This saving is, however, more than offset by wastage.

(iii) Most physical damages such as erosion in steep minors, watercourses and field channels due to relatively larger flows on account of lesser withdrawals upstream, and aggravation of water-logging and associated soil salinity and alkalinity problems seem to occur at night.

It is, therefore, obvious that night irrigation poses many problems. This situation can be improved either by reducing irrigation at night or improving it (Chambers, 1984).

Night irrigation can be improved through the following main ways as suggested by Chambers (1984):

(i) Making flows predictable and manageable: For this purpose, the *warabandi* system has obvious advantages. Each farmer knows the time of his turn and does not need to recruit labour to capture and guard his supply.

(ii) Improving convenience and efficiency by measures such as (a) good lighting, (b) organisation of groups of farmers for night irrigation for mutual benefits of shared labour and common protection from water raiders, etc., (c) installing and maintaining structures, channels, fields and water application methods requiring minimum observation and adjustment, (d) shaping fields for easier water application, and (e) providing ways and means for easy movement from residences to the fields and within the fields.

(iii) Choosing easy crops as paddy, trees and crops which are low and widely spaced.

(iv) Zoning for night flow which means using day flows for more difficult crops and soils and night flows for relatively easier crops and soils.

(v) Phasing night irrigation for shorter, warmer and moon-lit nights would obviously be more convenient for farmers.

17.10 Monitoring the Operation

It is an extremely important activity and aims with the following purposes:

(i) Short-term: It means of management control-comparing the actual pattern of water distribution with what it should have been and helps to identify reasons for divergences, and

(ii) Long-term: It is amassing informations on water supply, demand and performance etc. in past seasons as a ride to planning and implementation in forthcoming seasons.

Monitoring will also perform function such as preparation of the annual Irrigation or crop plan and preparation of the annual report covering the irrigation season to only relevant information for the purpose of performance evaluation and future planning.

A monitoring and planning unit could also be extremely useful in recycling systems affected by drainage problems, in order to monitor the behaviour of the system. Recycling systems affected by problems of periodic increases in sediment in the water or poor quality could also benefit greatly from the services of the monitoring and planning unit.

17.11 Maintenance

The irrigation network is perhaps the most costly element of any water harvesting structures and is designed to last a long time. Maintenance is required due to silt deposition, weed infestation, malfunctioning of structures and other undesirable situations which make it practically impossible to control the flow in these canals. This results in inability to deliver the necessary water and distribute it equitably. It is not surprising that farmers working in those recycling systems sometimes feel frustrated because they know the potential benefits of irrigation but cannot get to their expectations.

17.11.1 Reasons for poor maintenance

There are following most important reasons for poor maintenance:

(i) Insufficient funds made available to the management,

(ii) Lack of interest by the farmers in participating or collaborating in the maintenance work,

(iii) Poor organization of the work, and

(iv) Lack of interest in the participation.

17.11.2 Main functions of maintenance operation

The maintenance service is entrusted with the overall responsibility for keeping the irrigation and drainage systems working in a satisfactory manner, within the limitations imposed in initial design.

The maintenance of recycling system performed with the following functions:

(i) Planning and maintenance activities,

(ii) Implementing the maintenance activities planned and those unforeseen, and

(iii) Monitoring the above mentioned activities.

The optimum cycle of maintenance is the time that can safely elapse between two consecutive maintenance operations of a constructed element (canal, road, drain, etc.) without that element failing and disrupting the efficient operation of the whole. A certain degree of malfunctioning (10-20 per cent reduction in absolute efficiency with respect to initial design) is normally acceptable between the two consecutive maintenance operations.

17.11.3 Types of maintenance

There are following three main types of maintenance:

(i) *Routine or normal maintenance*: It includes all work necessary to keep the irrigation system functioning satisfactorily and is normally done annually.

(ii) *Special maintenance*: It includes repairs of damage caused by major disasters, such as floods, earthquakes and typhoon. The unforeseeable nature of such natural phenomena make it very difficult to take specific preventive action. Hence, safeguards shall be installed in particularly prone areas. For this, a "special reserve fund" or budget allocation should be established for repair work.

(iii) *Deferred maintenance*: It includes any work necessary to regain the lost flow capacity in canals, reservoirs art structures when compared to the initial design. It often includes large modifications to the canal system and structures arising from important changes (cropping patterns, drainage problems, etc.) that have occurred in an irrigation scheme. In practice, it is difficult to differentiate between so-called 'deferred maintenance' and a ‘rehabilitation programme’. The difference is mainly of a financial nature, because ‘deferred maintenance’ is normally undertaken with funds from the national budget allocated to operation and maintenance while rehabilitation programmes are considered as an investment and the funds come from a different source (loans, national development banks, etc.).

The lined water recycling requires little maintenance while, unlined requires more and regular maintenance. The maintenance is required due to the following causes:

(a) Silting,
(b) Weed infestation,
(c) Water infiltration, and
(d) Erosion and drainage network.

17.11.3.1 Silting

Excessive sedimentation is perhaps the most common problem affecting the performance to earth canals/channels. It reduces the storage capacity of water harvesting structures and carrying capacity of canals. According to Malik (1978) there are following causes of silting:

(i) Excessive silt entry at the main canal intake,
(ii) Disproportionate withdrawal by branches,
(iii) Prolonged heading up at control points,
(iv) Drifting sand,
(v) Inadequate transport capacity of channels,
(vi) Re-entry of excavated material by rain and wind action,
(vii) Malfunctioning of intakes,
(viii) Haphazard sediment excavation,
(ix) Excessive weed growth, and
(x) Wrong channel regulation.

Causes 1 to 5 indicate defective design, 6 to 9 inefficient maintenance, while 10 denotes improper canal or channel operation. Corrective measures for defective design are difficult to implement since they require major physical changes and need heavy investments. However, the effects of defective design can be reduced by proper maintenance. Incorrect operation is also a major cause, of silting. Canals carrying a heavy load of material in suspension should not be allowed to run at less than three quarters of their capacity since at lower capacities the velocity decreases inducing silting.

In order to determine the amount of silt to be removed, a topographical survey is made by taking the sections every 50 or 100 m according to the required degree of precision. With the data obtained the area of the sections is calculated. The volume of sediment (Vp) for each stretch of length l, will be determined as,

$$Vp = \frac{1}{2}(A_1 + A_2)\, l \quad \text{....(17.12)}$$

where A_1 and A_2 are the transversal section of the sediment and l the length of the section, and therefore for a canal of constant section:

$$Vp = \{\frac{1}{2}(A_1 + A_n) + A_2 + A_3 + \ldots\ldots. + A_{n-1}\}\, l \quad \text{....(17.13)}$$

Once all the measurements of the maintenance are complete, they should be grouped according to activity in order to study the manpower and machinery requirements.

Silt clearance

Silt is still removed manually in many parts of the world, provided that the water levels in the canals can lowered sufficiently or, even better, the canals dried for several days. This method is quite effective, although the actual organization of the work can be a problem.

However, where water-borne diseases are known to be prevalent, the use of labour should be restricted to those canals that can be dried completely for several days, otherwise mechanical means should be recommended.

Productivity of labour is generally low due to the muddy conditions in which they often work. Although some effort has been made to increase productivity by developing more appropriate tools like dredging scoops, specially designed digging hoes and forks, traditional tools (head baskets and shovels) are still used and productivity remains low. The output of manual maintenance varies widely from 2 to 8 m^3/manday, depending on several factors such as working conditions, tools, lifting and hauling distance.

Several types of machines are utilized for desilting and canal reshaping. The productivity depends largely on how well suited the machine is to the particular work. As already mentioned, only large water harvesting and recycling system are likely to have specialized machinery for each type of maintenance work. The different types of machineries are available for silt clearance and their selection is mainly influenced by its reach and working conditions such as wet or dry, accessibility, amount and type of work, weed infestation, etc.

17.11.3.2 Weed infestation

Weed infestation may reduce the flow of canal water which leads to greater water losses. There are two types of weeds which reduces the carrying capacity of canal as,

(i) *Earth weeds*: They root in the soil and their habitat is not the water; they proliferate on the canal slopes and in the banks, benefiting from favourable soil moisture conditions,

(ii) *Aquatic weeds*: They can either root in the water or the earth but their habitat is in the water. Robson (1976) classified them as under:

- *Emergent plants*: These are plants growing in the water and whose foliage emerges above the surface such as common read (*Phragmites communis*),
- *Floating leaved plant*: There are two sub-groups with floating leaves: in one, the plants are rooted in the mud and their leaves float flat on the surface, in the other, plants are not rooted but free-floating on the surface such as water hyacinth,
- *Submerged plants*: This group consists of plants whose foliage is totally submerged; a number of them produce flowers which emerge above the surface; one or two plants are free-floating, but most are rooted in the mud,
- *Algae*: This group consists of a variety of algae of various forms, including unicellular algae and the large filamentous forms.

A combination of labour and machinery is required particularly in lined canals, where silt is removed and accumulated at certain points by labour and then removed by mechanical means.

Measures to control weed infestation in canal

Weeds are normally removed by cutting, mowing or dredging. Emergent and submerged weed are best cut at near the base the stem, leaving roots and rhizomes undisturbed. Weeds need cutting at regular intervals throughout the season and the interval varies according to the environmental conditions and species. In temperate regions during the summer, it is usually not more than four weeks. The stage of growth at which the plant is cut affects the rate of growth; cutting at an early growth stage tends to be more effective as compared to later one.

Four main methods are commonly used to control canal vegetation: (a) manual, (b) mechanical, (c) chemical, and (d) biological. The selection of method depends mainly on the availability of labour, the predominant weed species, the environmental and economic conditions.

(a) *Manual*: The method of manual removal of silt may be applied to weed control. However, since it requires a little more skill, the selection of an appropriate hand tool is more important and will lead to relatively high productivity. Table 17.7 illustrates some of the characteristics of these hand tools and their related productivity (FAO Irrigation & Drainage Paper 33).

Table 17.7: Hand tools for control of canal vegetation

Type	**Use**	**Dimension of canal**	**Productivity**
Scythe	Submerged weeds + grass and reeds on banks	Small canals up to 0.6-0.8 m depth	15-25 m^2/hour
Sickle grasshook	Submerged weeds + grass and reeds on banks	Small canals up to 0.75-1.25 m depth	8-12 m^2/hour
Rake, fork	Removing cut material, lifting floating weeds, removing algae	-	Highly variable depending on weed species and vegetation density
Chain knives and chain scythes	Submerged weeds + grass and reeds on banks	Upto 6 m width	4-60 m^2/hour (two or three labourers)

(b) *Mechanical*: There are many types of machines available for weed control purposes. It is preferable to adapt a regular farm tractor for use with different attachments for removing weeds. Rubber wheeled 40-60 hp farm tractors can travel along channel banks with a maximum slope of 1:3 percent using the PTO (power take-off) for the implement. This will not require specialized operator neither training nor equipments. Hydraulically operated machines are used for slope steeper than 1:3. Tractor-dragged chains are more effective on steeper slopes. Two 60 hp tractors can easily remove weeds in 6 m wide channel. Mower and cutter attachments are used mainly for grass and reeds on banks of watercourses, while buckets are more suitable for emergent and submerged weeds. Launches can be used for mowing and cutting emergent weeds on banks, if the size of the channel permits their passage. Boats with cutting attachments are used for large waterways of shallow depth (up to 2.5 m) and where floating vegetation prevails. According to FAO Irrigation and Drainage Paper 33, details and productivity of the different types of machinery are summarized in Table 17.8.

(c) *Chemical*: Chemicals have been developed which can control weeds effectively and safely, but adequate precautions are required. They also offer an economic system of weed control in certain circumstances. However, the use of herbicides should be limited because of their possible adverse effects on the environment. It is known that some herbicides may affect the quality of water to the extent that it becomes harmful to humans, animals and crops. Therefore, they should be selected with care. Where their use may prove hazardous, it may be necessary to limit or even prohibit them. Table 17.9 indicates a summary of some available herbicides and their use (FAO Irrigation and Drainage Paper 33).

Table 17.8: Special characteristics of power moved tools for control of canal weeds

Type	Attachment	Use	Outreach	Productivity	Remarks
Tractor powered attachments or trailers	Oscillating grass cutter bar	Emergent weeds	3-6 m	1-2.5 km/hr[1]	Machines can only work on one side of the canal at a time (crossing places needed to complete operation). Industrial tractors (70-150 hp) increase outreach to 8 m.
	Rotary cutter	Emergent weeds	- do -	1-2.5 km/hr[1]	
	Flail mower	Emergent weeds	- do -	1-2.5 km/hr[1]	
	Chain, harrow	Submerged + emergent	Width 2 m	0.5-3 km/hr	
Hydraulic excavator with attachments	Oscillating grass cutter bar	Emergent weeds	Horizontal outreach: 6-12 m, depth: 3-5 m	800-1200 m/day[1]	Wheeled machines 1 metre less effective outreach than tracked machines
	Rotary cutter	Emergent weeds	- do -	0.1 ha/hr	Can work on both sides of canal at sane time
	Flail mower	Emergent weeds	- do -	0.1 ha/hr	
	Mud bucket	All types	- do -	400-600 m/day[2]	

Type	Attachment	Use	Outreach	Productivity	Remarks
	Weed cutting bucket	All types	- do -	600-800 m/day[2]	
Dragline	Weed rake Mud bucket	Submerged + algae + emergent	9-21 m (drag-line throw)	500 m/day[1]	Max. channel width depends on access and configuration of channel
Large boats (10-15 hp)	Oscillating rigid knives Modified (T type) grass cutters	Bank weeds + floating leaved plants + emergent weeds + algae	Width: 6-10 m	1-4 km/hr with a width of cut of 1.5-2.8 m	Max. depth of cut: 1.5-2.8 m; mm. depth of cut: 0.5-1.0 m; max. river flow velocity: 2.5 km/hr.
Small boats (4-5 hp)	Oscillating rigid knives grass cutter	- do -	Width: 5-6 m	1.4 km/hr with a width of cut of 1.0-1.8 m	Max. depth of cut: 1.0-1.8 m; mm depth of cut: 0.5-1.75 m; max. river flow velocity: 2.5 km/hr

Adapted from Robson (1976) and other sources

[1] Covering a swathe of 1.5-2 m.

[2] Refers to the complete cleaning of medium sized canals (5-6 width)

Table 17.9: Selected herbicides for control of aquatic weeds

Type	Use	Dose	Safety interval[1]	Remarks[2]
Aromatic solvents	Submerged weeds	40-80 ppmv[3]	-	Inject under water, Toxic to fish, distasteful in water
Acrolein	Submerged weeds	4-7 ppmv	-	Inject under water, Very toxic to aquatic fauna
Amitrole-T	Floating + emergent weeds	1-1.5 kg/ha	-	Spray on foliage. In USA can only be used in drainage canals
2, 4-D amine salt	Broad leaved plants near water	2-4 kg/ha	3 weeks	Spray on foliage, especially for water hyacinth
2, 4 5-T	Ditch bank control (woody vegetation)	4 kg/ha	-	Spray on plants

Type	Use	Dose	Safety interval[1]	Remarks[2]
Dalapon	Aquatic grasses + cat-tail	10-25 kg/ha	5 weeks	Spray on foliage, repeat applications. Safe to fish
Diquat	Submerged + floating weeds	1-1.15 kg/ha	10 days	Spray on foliage. Considered safe at recommended dosage
Dichlobenil	Submerged + some floating weeds	1 mg/l	4 weeks	Spray on foliage. No adverse effect on wild life. At high conce-ntrations can be toxic to fish
Copper sul-phate (CSP)	Algae	1.0 ppmw[4]	-	Toxic to fish, distasteful in water when suggested dose exceeded
Sodium arsenite	Submerged + floating weeds	4 ppmw	-	Inject or spray, highly toxic (0.02 g may kill a man) to mammals. Its use is discouraged
Endothall	Submerged weeds	1.5-4 ppmw	7-25 days	Inject underwater. Some of the salts (long chain type) are toxic to fish and mammals

Thomson (1976), Unesco (1974) and other sources

[1] Before use of water for irrigation

[2] Time to spray one hectare: Hand operated back sprayer, 10 litres capacity -9 hours

Back mounted power sprayer - 6.5-7 hours
Tractor mounted boom sprayer - 4-5 hours

[3] ppmv - parts per million by volume ppmw - parts per million by weight

(d) Biological: Biological control may become more important in the future in view of the adverse effect of other methods (high costs, danger to the environment). The most common method is the introduction of an animal, fish or insect which feeds especially on the problem plant. Recently, attention has been given to the use of the grams carp (*Ctenopharyngodon idella*) for the control of submerged weeds. In Some cases, it appears to be an economical and effective way of weed control.

Sometimes weed growth can be prevented by fluctuations in the water regime, for example, holding a canal dry for 3-6 days is the most effective to algae control. A system was developed to drain the canals and let them dry for five days; this prevented weed growth for six months (Bhakra Irrigation System in India). However, the success of this method depends upon the weed species.

17.11.3.3 Water infiltration

Water leaks through canal banks can be caused by burrowing small crabs and water rats or by rotting plants and roots which were not removed from the canal bank seat during construction. Ants are also known to be a problem even in concrete- lined canals. These leaks can be repaired by following the path of the leak through the bank either by hand digging or hydraulic backhoe if available and once the path has been found, the trench must be carefully backfilled and compacted. Canal leaks, if not repaired in time, results in major breaches in banks causing far greater inconvenience and most costly repairs.

Water seepage through porous soils may also be a major problem. Seepage through banks can be considerably reduced by trenching them and burying a plastic membrane or thick slurry made from the excavated material. The trench is backfilled with sand after the barrier has been interred.

17.11.3.4 Erosion of banks

Canal banks can be eroded by heavy rainfall or wind, improper canal operation, stock grazing or passage by drinking animals, and the transit of vehicles. Heavy rainfall or wind can cause serious damage to unprotected banks. Seeding of grasses on the canal bank is a cheap and effective protective measure. Short growing varieties such as *Agropyron riparium* (streambank wheat-grass), *Psathyrostachys juncea* (Russian wildrye), *Festuca ovina* (sheep fescue) and *Phleum bertolonii* (dwarf timothy) give good results.

Abrupt and rapid shutting of canal water may also contribute to erosion of the banks. The practice of leaving, a canal empty during the rainy season will contribute considerably to erosion of canal slopes.

Cattle and sheep damage the channel bank in different ways (Swales, 1976). Cattle tend to push the moist bank material it the water-line into the waterway when they drink. Whereas, sheep graze the banks bare thereby allowing wind and rain to wash away the bank material.

Erosion of canals can be repaired by mechanical means or manually by re-building the worn canal banks. However, care should be taken to construct a proper joint between the old and the new part, otherwise the canal will deteriorate at the same place.

The most effective measures are of a preventive nature: such as seeding grass discussed earlier, fencing the canal, and construction of special places for animal watering and bathing.

17.12 Assessment of Water Charges

These are the sum paid by the farmer as his contribution to government investment for the storage and distribution of water and to cover the expenses related to the operation, maintenance and administration of the scheme.

Community tanks or large water harvesting structures (reservoirs) involve huge expenditure for their construction. The operation and maintenance of these structures also require finances. With the introduction of these water harvesting structures in an area, which increases the water resources for irrigation and the farmers are immensely benefited. Therefore, it is only appropriate that farmers should be suitably charged for the supply of irrigation water.

The assessment of irrigation water charges can be made in one of the following ways:

(i) Assessment on area basis,

(ii) Volumetric assessment,

(iii) Assessment based on outlet capacity,

(iv) Permanent assessment, and

(v) Consolidated assessment.

(i) *Assessment on area basis*: In this, water charges are fixed per unit area of land irrigated for each of the crops grown. The rates of water charges depends upon the cash value of crop, water requirement of crop and the time of water demand with respect to the available supplies in the source. Since the water charges are not related to the actual quantity of water used, the farmers (particularly those whose holdings are in the head reaches of the canal) tend to over irrigate their land. This results in uneconomical use of available irrigation water and low water productivity besides depriving the cultivators in the tail reaches of the canal of their due share of irrigation water. However, this method of assessment, being simple and convenient, is generally used for almost all irrigation projects in India.

(ii) *Volumetric assessment*: In this method, the charges are in proportion to the actual amount of water received by the cultivator. Therefore, this method requires installation of water meters at all the outlets of the irrigation system. Alternatively, modular outlets may be provided to supply a specified discharge of water. This method results in economical use of irrigation water and enhances water productivity. It is an ideal method of assessment. However, this method has several drawbacks described below:

1. This method requires the installation and maintenance of suitable devices for water measurement,
2. These devices require adequate head at the outlet,
3. Further, there is a possibility of water theft by cutting of banks or siphoning over the bank through a flexible hose pipe, and
4. The distribution of charges among the farmers, whose holdings are served by a common outlet, may be difficult. Thus, the method has not been adopted in India because of above drawbacks.

(iii) *Assessment based on outlet capacity*: In this method, water charges based on outlet capacity. It is a simple method and is workable if the outlets are rigid or semi-modular and the channel may run within their modular range.

(iv) *Permanent assessment*: This method is suitable in some regions where artificial irrigation is not essential. Irrigation water is supplied provided to meet the water demand only during the drought years. Every farmer of such a region has to pay a fixed amount. The farmers have to pay these charges even for the years for which they do not take any water. A farmer has also to pay a tax on the land owned by him.

(v) *Consolidated assessment*: In this method, both the land revenue and the water charges are combined and the cultivators are accordingly charged.

Water rates may also be classified as,

(i) Monomial and binomial rates, and

(ii) Increasing or decreasing rates.

(i) Monomial and binomial rates

(a) Monomial rates

It is a single sum paid for the water received. Sometime this payment covers only O&M expenses, while others cover part of the repayment of the investments. This type of billing is normally associated with payment per unit of land or harvested crop.

(b) Binomial rates

It is made up of two components such as one fixed rate (constant for a number of years) and a variable rate (changes vary from year to years). This type of billing is normally associated with the volumetric method of payment. The fixed rate is a quota paid annually to repay part or all of the investments made in the irrigation system and to cover some fixed costs. Once repayment of the investment is completed this quota is normally cancelled, although in some cases it is continued as a smaller amount to cover part of the fixed operating cost. The variable rate means to cover the variable expenses incurred in operating and maintenance of recycling systems. Therefore, it changes from year to year with charge in operation and maintenance cost.

Binomial rates are most preferable as it is a more precise method of billing. However, it is not always fully understood by the farmers.

(ii) Increasing or decreasing rate

The volumetric methods have variable rates according to the consumption (block rating) in some countries. The rate is fixed upto a certain volume which is constant. If this is exceeded a new rate is then additional volume of water is given. In this way the rate is changing with change of volume of water delivered to the farmers. Thus, the greater the consumption, higher is the rates so as to discourage users from applying too much water. There are a few cases when the reverse has been assayed, especially during the first years of a new irrigation scheme when farmers are sometimes fearful of using too much water.

Although method of increasing water rates can be justified in extreme conditions of water scarcity, its application must be carefully considered especially where the level of the farmers' education is low and they might not understand the principle of the method.

17.13 Canal Outlets

An *outlet* is a masonry structure through which water is admitted from the distributary into a watercourse. It is used to transfer canal into the watercourses. Outlet is constructed at the junction of the watercourse and the distributary. These are used in community and large water harvesting structures. It also serves the purpose of water measurement, supplied from the distributary.

The discharge though an outlet is usually less than 0.085 cumec. It plays a vital role in the *warabandi* system of water distribution. Thus, an outlet acts as a head regulator for the field channel.

The main objective of an outlet is to provide ample supply of water to the fields whenever needed. If the total available supply is insufficient, the outlets must ensure equitable distribution under deficit water supply. The efficiency of an irrigation system depends upon the proper functioning of canal outlets. Therefore, an outlet should satisfy the following recruitments for proper functioning:

(i) The outlets must be strong and simple with no moving parts which would require periodic attention and maintenance,

(ii) The outlets should be tamper-proof and if there is any interference in the functioning of the outlet, it should be easily detectable,

(iii) The cost of outlets should be less as a large number of them have to be installed in an recycling system,

(iv) The outlet should be able to draw sediment in proportion to the amount of water withdrawn so that there is no silting or scouring problem in the distributary downstream of the outlet, and

(v) The outlets should be able to function efficiently even at low heads.

The selection of type of an outlet and its design are governed by several factors such as water distribution policy, water distribution method, method of water assessment, sources of supply and the working of the distributary channel.

Water may be distributed on the basis of either the actual area irrigated in the previous year or the actual culturable command area. The discharge from the outlet should be capable of being varied in the first case while, discharge remain fixed in the second case. The method of water distribution may be such that each cultivator successively receives water for a duration in proportion to his area. Or, alternatively, all the cultivators share the outlet discharge simultaneously. The first system is better as it results in less loss of water. The outlet capacity is decided based on the method of water distribution.

If the water assessment is made on volume basis then the outlet discharge should remain constant and not change with variation in the water levels of the distributary and the watercourse. While, if water charges are decided on the basis of area then the variation in the outlet capacity with water levels of the distributary and the watercourse is relatively immaterial.

17.13.1 Types of outlets

There are following three types of canal outlets commonly used:

(i) Non-modular outlets,

(ii) Semi-modular outlets, and

(iii) Modular outlets.

(i) *Non-modular outlets*: In these outlets, discharge capacity depends upon the difference of water levels in the distributary and the watercourse. The non-modular outlet is usually in the form of a submerged pipe outlet or a masonry slice which is fixed in the canal bank at right angles to the direction of flow in the distributary (Fig. 17.3). The discharge through non-modular outlets fluctuates over a wide range with variations in the water levels of either the distributary or the watercourse. Such an outlet is controlled by a shutter at its upstream end. The loss of head in a non-modular outlet is less than that in a modular outlet. Thus, non-modular outlets are very suitable for low head conditions. However, in these outlets, the discharge may vary even when the water level in the distributary remains constant. Hence, it is very difficult to ensure equitable distribution of water at all outlets at times of keen demand of water.

(ii) *Semi-modular outlets*: The discharge in this outlet (semi-modular or flexible outlet) depends only on the water level in the distributary and is unaffected by the water level in the watercourse provided that a minimum working head required for its working is available. The simplest type of semi-modular outlet is a pipe outlet discharging freely into the atmosphere and other types of semi-modular outlets are Kennedy's gauge outlet (Fig. 17.4), open flame outlet, orifice-semi-modular. A semi-module is more suitable for achieving equitable distribution of water at all outlets of a distributary. The only disadvantage of a semi-modular outlet is that it involves comparatively greater loss of head.

(iii) *Modular outlets*: In these outlets, discharge is independent of the water levels in the distributary and the watercourse, within reasonable working limits. These outlets may or may not have moving parts. However, Gibb's rigid modular outlet and Khanna's rigid do not have any moving parts (Fig. 17.5). These are also called rigid modules. Modular outlets with moving parts are not simple to design and construct and are, hence, expensive.

A modular outlet, supplies fixed discharge therefore, farmers can plan his irrigation accordingly. However, if supply is in excess or deficient in the distributary then the tail-end reach of the distributary may either get flooded or may be deprived of water. This is due to the factor that the modular outlet would not adjust its discharge level in the distributary. But if an outlet is to be provided in a branch canal which is likely to run with large fluctuations in discharge, a modular outlet would be an ideal selection. The outlet would be set at a level low enough to permit it to draw its due share when the branch is running with low supplies. When the branch has to carry excess supplies to meet the demands of the distributaries, the discharge through the modular outlet would not be affected and the excess supplies would reach upto the desired distributaries. Similarly, if an outlet is desired to be located upstream of a regulator or a raised crest fall, a modular outlet would be best suitable for the purpose.

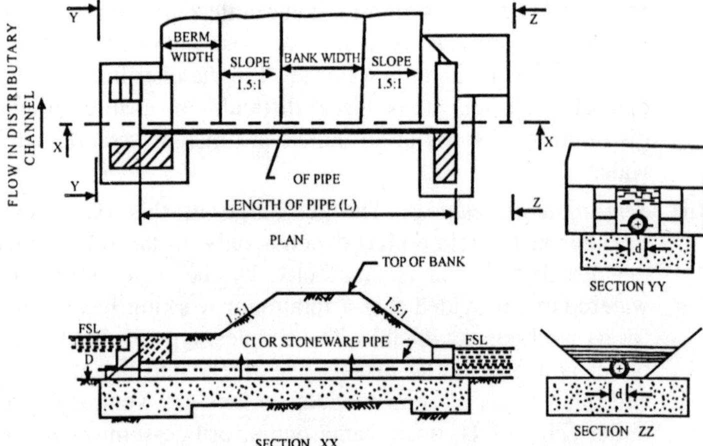

Fig. 17.3: Pipe outlet (non-modular outlet)

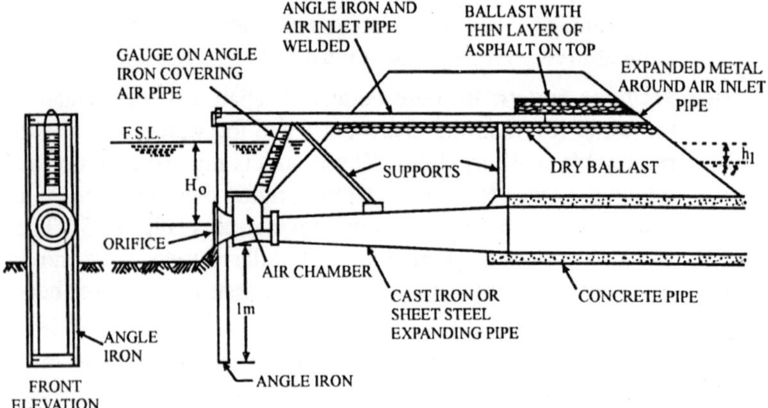

Fig. 17.4: Kennedy's gauge outlet (semi-modular outlet)

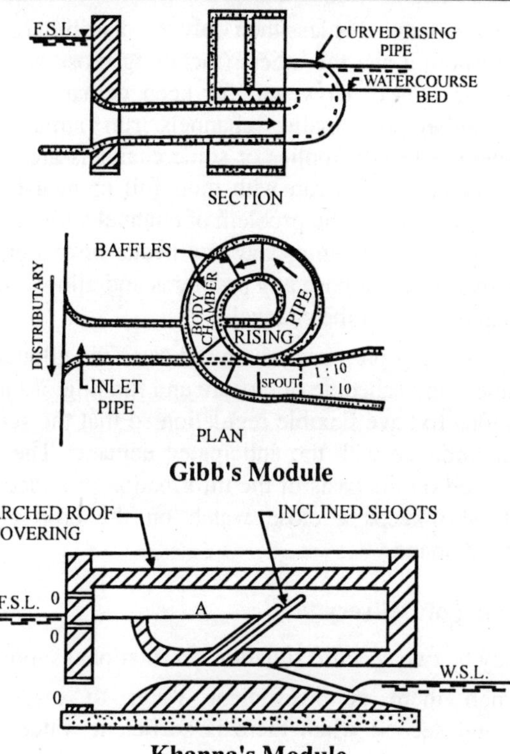

Fig. 17.5: Modular outlets

Canal regulation

The method of distribution of available supplies in water harvesting structures (community tank or large reservoirs) is termed *canal regulation*.

If a significant demand for water in any command area of a canal exists then the canal has to be kept flowing. However, canal can be closed, if the water demand falls below a specified quantity and it is reopened when the water demand exceeds the specified minimum quantity. Normally, there always exists a demand in some part of the command area of any major canal in large water harvesting structures (large reservoirs). These canals can be closed only for a very small period (say three to four weeks in a year). These canals run almost continuously and carry discharges much less than their full capacity, when either demand is less or available supplies are insufficient.

In case, when demand is less then only the distributaries which need water are kept running and the others (including those which have very little demand) are closed. However, for keen demand but insufficient supplies then either all smaller channels run simultaneously and continuously with reduced supplies or some channels are closed turn by turn and the remaining ones run with their full or near-full capacities. However, alternative posses the problem of channel silting, weed growth, increased seepage, water-logging and low heads on outlets. While, the second alternative does not pose any problems and allows sufficient time for inspection and repair of the channel.

A roster is usually prepared which indicates the allotted supplies to different channels and schedule of closure and running of these channels. It is advantageous to have flexible regulation so that the supplies can be allocated in accordance with the anticipated demand. The allocation of supplies is decided on the basis of the information provided by the canal revenue staff who keeps a close watch on the crop condition and irrigation water demand.

17.14 Scheduling of Delivery

There may be two specific situations in rotational supply:

First when there is insufficient water to meet crop water requirements and second when there is sufficient water to meet crop water requirements.

The first situation is quite innovative and is the result of recent findings in the study of the relationship between water and crop yield. While, second situation has been selected to illustrate some of the practical problems associated with rotational distribution when based on technical calculations.

17.14.1 Rotational supply with limited water

In this situation, the main objective is to modify a rotational supply so that crop yield reductions can be kept to the minimum.

The most commonly adopted solution to reduce the quantity of water given at each irrigation in proportion to the following relationship:

$$\frac{\text{available water}}{\text{crop water requirements}} \quad \text{....(17.14)}$$

or, to extend the interval between two consecutive irrigations in proportion to the following relationship:

$$\frac{\text{crop water requirements}}{\text{available water}} \qquad ....(17.15)$$

However, these practices are not fully satisfactory since they are not in accordance with the physiological development of the plant. It is known that water stress at certain periods of crop growth adversely affects the crop yields while at other times the effect is much less significant. Therefore, water savings should be made predominantly during those periods when the plant is less sensitive to water stress and reduced to the minimum during the sensitive periods. The critical periods are different for each crop and have variable lengths of time depending on the climatic characteristics.

Theory of water saving

Many researchers developed water-yield relationships during recent years. Although researchers continue to develop the relations, but some of the results already obtained offer a satisfactory level of reliability for planning purposes. Crop yield is estimated as function of the water supply or on using yield response to water as suggested by FAO Irrigation and Drainage paper No. 33.

One simple way indicates that there is a linear relationship between relative yield decreases (1-Ya/Ym) and relative evapotranspiration deficit (1-ETa/ETm) for each characteristic growth period such as establishment (0), vegetative (1), flowering (2), grain formation (3) and ripening (4). This constant is called "*yield response factor*" (ky) and its value is given by the following relationship:

$$ky = \frac{\left(1 - \frac{Ya}{Ym}\right)}{\left(1 - \frac{ETa}{ETm}\right)} \qquad ....(17.16)$$

where,

Ya = actual yield,

Ym = maximum yield,

ky = yield response factor,

ETa = actual evapotranspiration, and

ETm = maximum evapotranspiration.

The values of ky *for* each growing period and crops may be used from Table 17.10. Once these values are known equation (17.16) can be utilized to predict yields for given deficits of evapotranspiration (which reflect the water deficits).

Equation (17.16) may be used for two or more growing periods as,

$$\text{Growing period 1: } (1-\frac{Ya_1}{Ym}) = ky_1 = (1-\frac{ETa_1}{ETm_1}) \quad \text{....(17.17)}$$

$$\text{Growing period 2: } (1-\frac{Ya_2}{Ym}) = ky_2 = (1-\frac{ETa_2}{ETm_2}) \quad \text{....(17.18)}$$

On using Equations (17.17 & 17.18), we obtain,

$$\left(\frac{Ym-Ya_1}{Ym-Ya_2}\right) = \frac{ky_1}{ky_2}.\frac{ETm_2}{ETm_1}.\frac{ETm_1-ETa_1}{ETm_2-ETa_2} \quad \text{....(17.19)}$$

Assume that the tolerable yields decrease ($Ym_1 - Ya_1$) and ($Ym_2 - Ya_2$) are equally distributed among the main periods, equation (17.19) can be rewritten as,

$$1 = \frac{ky_1}{ky_2}.\frac{ETm_2}{ETm_1}.\frac{ETm_1-ETa_1}{ETm_2-ETa_2} \quad \text{....(17.20)}$$

or

$$\frac{ETm_2-ETa_2}{ETm_1-ETa_1} = \frac{ky_1}{ky_2}.\frac{ETm_2}{ETm_1} \quad \text{....(17.21)}$$

Since the evapotranspiration in each period (ETm_1, ETm_2) is proportional to the water supply in those periods (V_1, V_2), Equation (17.21) can be expressed as,

$$\frac{V_2-Va_2}{V_1-Va_1} = \frac{ky_1}{ky_2}.\frac{V_1}{V_2} \quad \text{....(17.22)}$$

where,

V_1 and V_2 = water applied when full supply available for growing periods 1 and 2,

Va_1 and Va_2 = water applied with limited supply for growing periods 1 and 2,

$V_1 - Va_1$ = water deficit in growing period 1 = WS_1, and

$V_2 - Va_2$ = water deficit in growing period 2 = WS_2.

Equation (17.22) can be rewritten as,

$$\frac{WS_1}{WS_2} = \frac{ky_2}{ky_1}.\frac{V_1}{V_2} \quad \text{....(17.23)}$$

Table 17.10: Yield response factor (ky)

Crop	Vegetative period (1)			Flowering period (2)	Grain formation (3)	Ripening (4)	Total growing period
	Early (la)	Late (lb)	Total				
Alfalfa			0.7-1.1				0.7-1.1
Banana							1.2-1.35
Bean			0.2	1.1	0.75	0.2	1.15
Cabbage	0.2				0.45	0.6	0.95
Citrus							0.8-1.1
Cotton			0.2	0.5		0.25	0.85
Grape							0.85
Groundnut			0.2	0.8	0.6	0.2	0.7
Maize			0.4	1.5	0.5	0.2	1.25
Onion			0.45		0.8	0.3	1.1
Pea	0.2			0.9	0.7	0.2	1.15
Pepper							1.1
Potato	0.45	0.8			0.7	0.2	1.1
Safflower		0.3		0.55	0.6		0.8
Sorghum			0.2	0.55	0.45	0.2	0.9
Soybean			0.2	0.8	1.0		0.85
Sugar beet							
Beet							0.6-1.0
Sugar							0.7-1.1
Sugarcane			0.75		0.5	0.1	1.2
Sunflower	0.25	0.5		1.0	0.8		0.95
Tobacco	0.2	1.0					0.9
Tomato			0.4	1.1	0.8	0.4	1.05
Watermelon	0.45	0.7		0.8	0.8	0.3	1.1
Wheat							
Winter			0.2	0.6	0.5		1.0
spring			0.2	0.65	0.55		1.15

(Source: FAO Irrigation & Drainage paper No. 33)

Determination of the rotational supply

The rotational supply which minimizes reductions in crop yield can be determined on using equation (20.23). The procedure involves the following steps for determination of supply as suggested by FAO Irrigation & Drainage paper No. 33.

(i) Basic data,

(ii) Critical periods for water deficits,

(iii) Amounts of water to be saved,

(iv) Available water per month,

(v) Determination of extended irrigation interval, and

(vi) Determination of the depth of irrigation

(i) *Basic data*: It includes,

- Net irrigation requirements,
- Available annual water supply,
- Cropping pattern, and
- Irrigation interval under full supply (it can also be calculated if soil moisture parameters are known).

(ii) *Critical periods for water deficits*: These are obtained from Table 17.10. High ky values indicate a sensitive period for water shortage and savings should be reduced in that period otherwise a considerable yield reduction will occur.

(iii) *Amounts of water to be saved*: It is the difference between needed supply and available supply. The saving should not be greater than 50 percent of the full supply in any case because under such conditions equation (17.20) may not be valid. The allocation of the savings over the periods can be determined on using equation (17.23), whereas the savings per month can be determined by the following relationship:

$$\frac{WS\ month}{WS\ period} = \frac{V\ month}{V\ period} \qquad \text{....(17.24)}$$

where,

WS month = water to be saved in a given month,

WS period = water to be saved in the chosen growing period,

V month = water applied in the month under full supply, and

V period = water applied in the chosen period under full supply.

(iv) *Available water per month*: It is the difference between the water applied under full supply (V) and water savings (WS) under limited supply.

(v) *Determination of extended irrigation interval*: It consists in extending the irrigation interval but keeping the depth of irrigation constant. The extended interval can be determined taking into account the interval under full supply, which is determined through the traditional procedures and the ratio V/Va or:

$$i \text{ (limited supply)} = I \text{ (full supply)} . V/Va \qquad ....(17.25)$$

After determining the optimum irrigation interval, some practical changes may be introduced in scheduling in order to facilitate the operation such as the introduction of a fixed interval throughout a part of the irrigation season.

(vi) *Determination of the depth of irrigation*: It consists in keeping the initial (fully supply) interval but reducing the depth of irrigation. Although this is not a very practical measure, it is done in some instances. The depth of each irrigation is determined by the following relationship:

$$d = Va \frac{\text{Interval (under full supply)}}{\text{30 (number of days per month)}} \qquad ....(17.26)$$

Example 17.2: A water recycling system can irrigate 10,000 ha, under full supply but the area could be increased if water were available. The only crop grown is sugarcane. Average yield (full supply) is 90 t/year. The net crop water requirements are 21 500 m^3/ha/year and the gross irrigation requirements, with a 50 percent overall efficiency, are 43 400 m^3/ha/year. It is decided to save 20% of the crop water requirements and utilize the saved water to extend the irrigated area. Water saved to extend the irrigation area. Determine the following:

(a) Production increased,

(b) Amount of water saved, and

(c) Increased area of sugarcane.

a. Amount of water to be saved:

at the farm level $\frac{20}{100} \times 21500 = 4300\, m^3 / ha$

at project level $10000 \times 4300 \times \frac{1}{0.5} = 86000000\, m^3 / year$

b. Expected reduction in yield due to water shortage:

The reduction in yield can be determined on using Equation (17.16) as,

$$1-\frac{ETa}{ETm}=ky\,(1-\frac{Ya}{Ym)}$$

where,

ETa = 21500 – 4300 = 17200 m^3,

ETm = 21500 m^3,

ky = 1.25, and

Ym = 90 t

Putting the values in the Equation (17.16), we obtain:

$$1-\frac{17200}{21500}=1.25\,(1-\frac{Ya}{90)}$$

$$=\frac{0.2}{1.25}=90-Ya$$

$$0.16=90-Ya$$

$$Ya=89.84\,t$$

c. Incremental irrigated area and yield:

The water saved will allow irrigation of:

$$86000000\times\frac{1}{17200}\times\frac{1}{2}=2500\,ha$$

The sugarcane production arising from this additional area would be:

2500 ha x 89.84 t/ha = 224600 t

d. Total production:

10000 ha x 89.84 t/ha + 2246000 = 920860 t

whereas the production under full supply conditions was only:

10 000 ha x 90 t/ha = 900000 t

Two important objectives have been obtained: on one hand the total agricultural production has been increased, and on the other, the irrigated area has been increased by 25 percent and consequently benefit increased from it. If the intention is to maximize income and beneficiaries, the above methodology can be a useful tool. However, it should be stated that when water is scarce, designing an irrigation scheme based on full supply conditions will maximize neither income nor potential beneficiaries.

The suggested procedure can be improved by using some more sophisticated mathematical calculations or the trial and error method, if the economic optimum is desired.

17.14.2 Rotational scheduling under full water supply

The rotational supply is determined according to the technical principles involved. However, such calculations generally lead to water distribution patterns which are often impractical and therefore rarely applied. Simplification of the irrigation scheduling is often necessary but oversimplification can lead to wasting large amounts of water. Therefore, an example is illustrated.

Example 17.3: Determine the rotational schedule for a farm with the following characteristics:

a_1) irrigable area: 3.7 ha

a_2) cropping pattern by type of soil:

tomato	1.5 ha in soil A1
tomato	0.8 ha in soil A2
maize	0.25 ha in soil A1
pepper	0.50 ha in soil A1
pepper	0.50 ha in soil A2
beans	0.15 ha in soil A1
Total	3.7 ha

a_3) hydraulic characteristics of soils:

soil A1: W1 = water holding capacity = 93 mm/m

depth of soil = 0.8 m^1

soil A2: W1 = water holding capacity = 87 mm/m

depth of soil = 0.8 m^1

fraction of available soil water = p = 0.65

a_4) ET crop (in mm):

	Tomato	Maize	Pepper	Beans
Oct.	65	66	65	31
Nov.	91	85	92	92

Dec.	124	124	125	103
Jan.	52	125	50	34
	332	400	232	260

Solution:

(i) The depth of irrigation to be applied is determined as,

$$d = p \cdot w \cdot d \qquad \text{....(17.27)}$$

where,

p = fraction of available soil water,
W = water holding capacity, (mm/m),
D = rooting depth, (m), and
d = depth of irrigation, (mm).

On putting the values given in Equation (17.27), we get:

Soil A_1 $d_1 = 0.65 \times 93 \times 0.8^3 = 48$ mm

Soil A_2 $d_2 = 0.65 \times 87 \times 0.8^3 = 45$ mm

(ii) The irrigation intervals are determined with the following relationship:

$$\text{Irrigation interval } (I) = \frac{d}{ET\,crop} \qquad \text{....(17.28)}$$

where, ETcrop is evapotranspiration (mm/day).

The irrigation intervals for tomato in soil A_1 are determined as follows:

$$\text{ETcrop (Oct)} = \frac{65}{30} = 2.17 \text{ mm/day}$$

$$I_1 = \frac{48}{2.17} = 22\, days$$

For the next interval:

$$\text{ETcrop (Nov)} = \frac{91}{30} = 3.03 \text{ mm/day}$$

The application of Equation (17.28) to the following month gives a discontinuity in the intervals because October cannot be divided into equal intervals by the 22 day period. Therefore, in order to have a continuous irrigation schedule, the remaining days in October must also be considered:

2.17 mm/day x (30 – 22) days + 3.03 x δ days = 48

δ = 10.1 days

therefore I_2 = 8 + 10.1 = 18 days

The next interval in all within November, *i.e.*:

$$I_3 = \frac{48}{3.03} = 15.8 = 16\, days$$

The next interval is in part of November and December:

$$(30 - 10 - 16) \times 3.03 + \delta \frac{124}{30} = 48$$

$$\delta = 8.60 = 9 \text{ days}$$

$$I_4 = 4 + 9 = 13\, days$$

17.15 Farmer's Participation

The farmer forms the target group for all irrigation management such as water user association or water *panchayat*. Thus, it is a good source of first hand communication right from the investigation to the operational stage of any water harvesting structure and recycling system. Therefore, it is reognised that the participation of the farmer at all stages of the water recycling system would be very beneficial. Yoganarasimhan (1989) has suggested farmer's participation in the following three phases:

Phase 1 - Initiation

- Publicising the proposal, without restriction as a promotional measure to gain popular support,
- Extending invitations for information meeting,
- Scheduling public hearing,
- Warranting the critical assessment of objections raised as well as of individual alternatives of supplementary proposals,
- Facilitating discussions and debates on proposal components and details,
- Arranging for secret ballots to be cast in preparation of final decision,
- Facilitating the nomination and election of farmers' representatives to join working groups,
- Drafting the statutes of organisation which may only be adopted by majority vote through secret ballot,

- Imparting courage, inspiration, and resolution to farmers, and
- Creating an atmosphere of trust and confidence in the feasibility of local participation.

Phase 2 - Continuation

- Establishing formal organisations of farmers,
- Recognising the roles of organization,
- Forming joint committees charged with proposing and reviewing.
- Conducting training courses,
- Inviting contributions from farmers in the process of implementation, management, operation and maintenance, and in situations where repairs are required,
- Creating a group atmosphere of collaboration and cooperation, and
- Strengthening the task-oriented leadership based on majority group consensus and support.

Phase 3 - Perpetuation

- Assisting farmers to maintain their level of participation and increase its intensity,
- Relating their duties to improve their socio-economic status,
- Monitoring the performance of programmes to generate feedback into assistance and instruction, and
- Inducing farmers to actively participate in the management through unreserved recognition of their achievements.

References

Abernethy, C. 1986. Performance Measurement in Canal Water Management - A Discussion, Overseas Development Institute, London.

Adiseshiah M.S. 1985. 'Some Thoughts on Agriculture in the Seventh Plan', *17th Lal Bhadur Shastri Memorial Lecture* Indian Agricultural Research Institute, New Delhi.

Ali, S.H. (Chairman), 1982. *Report of the Commission for Irrigation Utilization* Govt. of Andhra Pradesh, Hyderabad.

Anonymous 1981. *Sixth Five-Year Plan 1980-85,* Planning Commission, Govt. of India.

Anonymous 2002. "National Water Policy", Ministry of Water Resources, Govt. of India. New Delhi, pp. 10.

Anonymous, 1981. 'Cropping Pattern Studies for Gomti-Kalyani Doab,' *Ford Foundation Project,* Research Group, WRDTC, University of Roorkee, Roorkee, India, Dec.

Anonymous, 1983. Resource Analysis and Plan for Efficient Watet Management: A Case Study of Mahi Right Bank Canal Command Area, Gujarat, Water Technology Centre, Indian Agricultural Research Centre, New Delhi.

Anonymous. 1983. 'Diagnostic Analysis of Farm Irrigation Systems on the Gambhiri Irrigation Project,' Rajasthan, India, *Water Management Synthesis Report no. 17,* Colorado State University, USA.

Ashok Raj, P.C. 1979. Onset of effective monsoon and coastal dryspells; A computer based forecasting technique. Indian Agricultural Research Institute. Research Bulletin No. 11.

Bottrall, A. 1981. 'Comparative Study of the Management and Organisation of Irrigation Projects,' *World Bank Staff Working paper no. 458,* Washington.

Chambers, R (Ed.), 1988. Managing canal irrigation: practical analysis from South Asia, Oxford Publishing. New Delhi, pp. 20-25.

Chambers, R. 1988. *Managing Canal Irrigation,* Oxford and IBH Publishing Co. Pvt. Ltd., New Delhi.

Chauhan, H.S. 1989. 'Drip Irrigation Status and Prospects in India', *National Seminar on New Perspectives in Water Management* held at Indore.

Chawla, A.S. 1989. 'Conjunctive Use of Surface and Ground Waters', *National Seminar on New perspectives in Water Management* held at Indore.

Dahe, P.D., and Srivastava, D.K. 2002. "Multireservoir multiyield model with allowable deficit in annual yield", *J. Water Resour Plang. Mgmt.*, ASCE, Vol.128(6), pp. 406-414.

Dandy, G.C., Connartv, M.C., and Loucks, D.P. 2002. "Comparison of Yield Assessment of Multiple Reservoir System", *J. Water Resour. Plang. Mgmt.* ASCE, Vol.123(6), pp. 350-358.

Dooronbos, J. and Pruitt, W.D. 1977. Crop water requirements. Irrigation and Drainage Paper No. 24, FAO, Rome: 179.

FAO 1992. "CROPWAT, a computer program for irrigation planning and management" by M. Smith, FAO Irrigation and Drainage Paper No. 26. Rome.

FAO, 1984. Irrigation and drainage paper 4 Rev.: Irrigation Practices and Water Management, Oxford and IBH Publishing Co. Pvt. Ltd., New Delhi, pp. 63.

FAO. 1992. CROPWAT, a computer programme for irrigation planning and management by M. Smith. FAO Irrigation and Drainage paper No. 26. Rome.

Frederick, L.H. 1984. *World Bank Irrigation Experience,* Overseas Development Institute, London.

Garg, O.N. and Gupta, D.K. 1988. "Economics of Canal Lining - A Systems Study", *Seminar on Canal Lining* held at Roorkee.

Ghosh, R. 1989. 'Waterlogging - Problem and Remedies', *National Seminar on New Perspectives in Water Management* held at Indore.

Jayaraman, T.K. *et al.,* 1983. 'Diagnostic Analysis of Farm Irrigation System on the Mahi-Kadana Irrigation Project', Gujarat, India, *Water Management Synthesis Report 18,* Colorado State University, Colorado.

Jhajharia, D., and Srivastava, D.K. 2008. Optimization Study of reservoirs in Sabri Basin", Paper presented in International Conference, ICWRDM, 23-24 October 2008, BITS Pilani, Dist. Jhunjhunu, Rajasthan, India.

LINDO 2002 Release 6.1 (15 Jan 02) © 2002 by LINDO systems Inc., Chicago, USA.

Malhotra, S.P. 1982. 'The *Warabandi* System and Its Infrastructure', *Publication No. 157,* CBIP, New Delhi.

Molden, D.J., Gates, T.K. 1990. Performance measures for evaluation of irrigation water delivery system. *Journal of Irrigation Drainage Engg.*, ASCE 116(6), 804-823.

Munot, J.S., Purohit R.C. and Singh, Jaspal, 1988. Optimal Irrigation Scheduling for Wheat Crop of Udaipur Region,' *J. of Indian Water Resources Society.*

Padhi, G.N. 1983. 'Pricing Policy for Irrigation Water', *Ph.D. Thesis* WRDTC, University of Roorkee.

Padhi, G.N. and Suryanshi, A.R. 1982. 'Evaluation of Water Distribution Method of Left Salawa Distributary of Upper Ganga Canal System,' *IWRS Workshop on Water Distribution Practices.*

Pant, N. (Editor), 1984. *Productivity and Equity in Irrigation Systems,* Ashish Publishing House, New Delhi.

Pant, N. 1981. Some Aspects of Irrigation Administration (A Case Study of Kosi Project), Naya Prakashan, Calcutta.

Pradhan, P.C., Behera, B.D. and Senapati, P.C. 1999. Forecasting Dry and Wet spells in Ghumusal Udayagiri Block of Orissa for strategic crop planning. *Indian J. Dryland Agric. Res. & Dev.*, 14(2): 68-76.

PWD. 1990. Draft report on revised operational rule for Sathanur Dam and its canals. National Water Management Project, Public Works Department, Govt. of Tamilnadu. India. 1-20.

Santhi, C. and Pundarikanthan, N.V. 1997. Performance evaluation of the water delivery system of Sathanur irrigation project. *Journal of Indian Water Resources Society*. 17(3), 8-12.

Santhi, C. and Pundarikanthan, N.V. 1999. Application of management information system in improving irrigation system performance in a developing country. *Water International*. 24(3), 229-239.

Santhi, C. and Pundarikanthan, N.V. 2000. A new planning model for canal scheduling of rotational irrigation. *Journal of Agriculture Water Management*. 43, 327-343.

Santhi, C., Neelakantan, T.R., Pundarikanthan, N.V. and Srinivasa Prasad, P. 2002. Distributary scheduling of rotational irrigation system with partial canal gate-openings, Journal of Indian Water Resources Society. 22(2), 47-56.

Sarma, P.B.S. and Rao, V.V. 1992. Identification of critical dryspell for irrigation planning. *Water and Energy International*, Vol. 49, 29-36.

Sarma, P.B.S., Rao, N.H. and Michael, A.M. 1983. Water Resources Utilization and Management, WTC report on "Resource Analysis and Plan Efficient Water Management: Case Study on Mahi Right Bank Canal Area, Gujarat", Water Technology Centre, Indian Agricultural Research Institute, New Delhi.

Seckler, D. and Nobe, K. 1983. 'The Critical Role of the Management Factor in Developing Countries,' *Paper presented for the International School of Economic Development,* Colorado State University.

Shivanappan, R.K., Kandaswamy, P. and Selvaraj, K.V. 1984. 'Water Management Practices in Parambikulam Aliyar Project Command Area,' Tamil Nadu, *Proc. of National Symposium on Strategies for irrigation Development,* held at Bihar College of Engineering in Patna, Mangalik Prakashan, Saharanpur.

Singh, Bharat 1989. 'Agriculture and Irrigation', *National Seminar on New Perspectives in Water Management* held at Indore.

Singh, K.K. (Editor), 1981. *'Warabandi* for Irrigated Agriculture in India,' *CBIP Publication no 146.*

Singh, R.V. 1989. 'Increasing Irrigation Efficiencies through Improved Methods of Water Application,' *National Seminar on New Perspectives in Water Management* held at Indore.

Sivanappan, R.K., Kandaswamy, P. and Selvaraj, K.V. 1984. 'Water Management Practices in Parambikulam Aliyar Project Command Area', Tamil Nadu, *Proc. of National Symposium on Strategies for Irrigation Development,* held at Bihar College of Engineering in Patna, Mangalik Prakashan, Saharanpur.

Swaminathan, M.S. 1982. *Science and integrated Rural Development,* Concept Publishing Company, New Delhi.

Thornthwaite, C.W. and Mather, J.R. 1955. The Water Balance, Climatology, Vol. 1. I Drexel Institute of Technology, Centerton, New Jersey.

Yoganarsimhan, G.N. 1989. 'Some Musings on People's Participation in Irrigation Water Management,' *J. of Indian Water Resources Society.*

EXERCISES

17.1 Describe the common criteria for judging the performance of a water recycling system.

17.2 Discuss the inadequacies of present-day canal irrigation management in India, and suggest suitable methods of improvement.

17.3 Discuss the causes of poor operation of recycling system.

17.4 Define the following:

(a) Inundation canal

(b) Water course

(c) Gross command area

(d) Crop ratio

(e) Delta

17.5 Derive the relationship between duty, base period and delta.

17.6 Explain the criteria for good performance of recycling system.

17.7 Describe the planning operation of recycling system.

17.8 Describe the measures to bridge the gap between supply & demand.

17.9 Explain the water delivery system.

17.10 Explain wara bandi.

17.11 Describe different types of maintenance.

17.12 Discuss the assessment of water charges.

17.13 Describe different types of outlets.

17.14 Explain schedule of rotational delivery.

■■■■

Chapter 18

WATER RECYCLING

18.1 Introduction

Water harvesting is the collection of surface runoff for its productive uses in farm ponds or nadies, anicuts and community tanks including small, medium and major reservoirs. The source of water may be from roof, fields (cultivated or uncultivated) or intermittent or perennial streams or rivers. Productive use includes, domestic water supply, livestock water, water for irrigation of crops, fodder or trees and used for fish farming.

Agriculture production is not only severely affected by critical dryspells during crop growth period but also from unscientific use of the available harvested rainwater. Adoption of improved methods of application of irrigation water with surface and micro-irrigation techniques would help in managing the scarce available natural water resource to enhance the water productivity.

18.2 Effects of Life Saving Irrigation

Response of life irrigation from harvested rainwater can be applied during dryspell period to increase production during summer season and also raising winter crops.

The best thing that can be done in case of terminal drought is to irrigate the crop and enhance the productivity by avoiding crop failure. As crop failure is common feature of rainfed crops due to terminal drought. Thus some examples of recycling rainwater are cited to illustrate the response of life saving irrigation. Under severe drought at Arjia, one protective irrigation increased yields by 377 over 1004 kg ha^{-1} in maize. During the experimental period of 7 seasons, there was a total failure of crop in one year and drought occurred in 3 seasons at Arjia. At Agra, the protective irrigation increased the yield by 696 kg over 633 kg ha^{-1} in pearlmillet. During the experimental period of 7 seasons, the drought occurred in 3 seasons at Agra (Table 18.1). In moderate drought at Varanasi, the protective irrigation to rice increased yields by 320 kg over 1190 kg ha^{-1}. During the experimental period of 7 seasons drought occurred in 3 seasons at Varanasi. One supplemental irrigation gave an advantage usually to the tune of 200 kg grain ha^{-1} cm^{-1} water applied.

Table 18.1: Response to supplemental irrigation

System (seasons)	Protective irrigation (Mq/ha)	No irrigation (Mq/ha)	Percent increase
Post rainy season sorghum (7)	1.07	0.62	158
Late rainy season green chillies (7)	1.00	0.64	157

(Source: Sankar *et al.*, 1998)

Another way to overcome the mid season drought is to give a protective irrigation with harvested rainwater. At Rajkot (severe drought), a protective irrigation increased groundnut haulm fold by 62 kg/ha. During the experimental period of 6 seasons, drought occurred in 3 seasons, crop failed completely in one, year while in other 2 seasons yields were almost equivalent to seed rate. In moderate drought, at semi arid Vertic soils in maize. Out of 7 years experimental period protective irrigation was not required in any year less drought prone areas, a life saving irrigation to rice increased yields by 698 kg over 1040 kg ha^{-1}; during the experimental period of 4 seasons, drought occurred only in one year.

In rice-wheat system under sub humid vertisols, performance of a farm pond was evaluated over seven years. On an average, at total of 3.55 ha-m rainfall was received per annum. In that, 8 percent was used for irrigation while 4 percent was lost through seepage and evaporation losses. Rest of 88 percent was runoff. An irrigation of 0.04 mm was given over 2.4 ha in the post- rainy season as pre-sowing irrigation for wheat. The irrigated area was 72 percent of catchment. In an eight-year period, irrigations in 3 seasons for pre-sowing, one season each at crown root initiation stage, flowering and filling stages (two times), and tillering were given. In two seasons, wheat could not be grown due to early cessation of monsoon. Rice needed irrigation in two years. In case of supplemental irrigation, higher sustainable value index was obtained when the limited water was applied to post-rainy season crops.

Water Use Efficiency (WUE) in response to irrigation techniques

It is evident from the results of the study that irrigation to furrow after three unirrigated rows (3 RUN) gave the higher WUE as compared to all other irrigation practices to less application of water which was reflected by the reduction in wetted area. The WUE was 14.1, 27.0, 25.5, 46.4 and 52.9 per cent higher in treatment 3-RUN as compared to 2-RUN, AFA, AF, EF and B irrigation practices, respectively. The WUE in irrigation to furrow after 2-RUN irrigated rows was 13.3, 15.1, 37.69.4 and 47.4 per cent higher as compared to AF, AM, EF and B irrigation practices, respectively. There was not much variation in WUE between irrigation practices of AF and AFA.

Though the irrigation practice did not significantly influence the growth and yield of cotton, their influence on water use parameters were noteworthy. The modified irrigation practice (3-RUN) saved considerable amount of irrigation water applied by reducing the wetted area of soil surface and irrigation time since the entire root zone is not used for water storage in modified irrigation practice.

Though the WUE of irrigation to furrow 2-RUN were more than AF and A.FA practices, these two irrigation practices after 2 un-irrigated runs produced 145 and 189 kg/ha lesser cotton seed yield, respectively as compared to AFA irrigation practice. Thus, the results of the study indicate that the irrigation practice AF and or AFA should be adopted for maximizing the yield per unit area and 2-RUN for maximizing the production utilizing the maximum area in the case of limited water supply.

Water use efficiency or water productivity can be enhanced through use of improved irrigation techniques either by surface. The more command area can be pressurized irrigation and also increase the yield.

Table 18.2: Seed cotton yield and water use efficiency in cotton, as affected by irrigation methods

Irrigation Methods	Years						Mean
	1993	1994	1995	1995	1996	1997	
	Ludhiana			Bathinda			
	Seed cotton yield, kg/ha						
B	1291	1463	952	2069	1459	798	1339
EF	1255	1656	999	1944	1500	986	1390
AFA	1385	1721	1064	1999	1573	1161	1484
2-RUN	1258	1415	984	1882	1525	971	1339
3-RUN	1160	1373	945	1858	1517	918	1295
AF	1365	1675	1033	2139	1451	1119	1464
CD (0.05)	NS	NS	NS	NS	NS	192	

	Water use efficiency (kg/ha-cm) of irrigation water applied						
B	26.1	38.0	22.5	85.5	54.0	25.2	43.6
EF	29.8	51.1	27.4	97.7	66.1	38.2	51.7
AFA	48.5	72.3	43.3	121.2	82.8	54.3	70.4
2-RUN	52.1	76.6	52.3	162.2	94.1	59.9	82.9
3-RUN	59.3	86.2	62.2	189.6	106.8	75.2	96.5
AF	47.4	72.5	44.5	137.1	78.4	51.6	71.9

B = Irrigation to border
EF = litigation to each furrow
AFA = Irrigation to alternate furrow irrigated alternately
AF = Irrigation to same alternate furrow throughout
2-RUN = Irrigation to furrow after 2-unirrigated rows
3-RUN = Irrigation to furrow after 3-unirrigated rows

18.3 Methods of Irrigation

According to Walker and S.K. Ogerbee (1987) any irrigation system consists of the following four sub-system:

(i) The water supply subsystems which may include diversion from rivers water harvesting structures *viz.*, surface ponds, nadi or farm pond etc. and water from wells.

(ii) The water delivery or conveyance sub-systems which will include canals and water distribution field channel and different structures required to control erosion and runoff.

(iii) The irrigation application methods or use of subsystems can be one of the three main types namely, (a) surface irrigation, (b) subsurface irrigation, (c) pressured irrigation system *viz.*, sprinkler and drip irrigation.

(iv) The water removal system *i.e.* the drainage system.

For small water harvesting structures, the drainage system is not required. Because of small quantity of harvested rainwater is available which will not create problem of drainage. The first two systems are discussed earlier in chapter.

Irrigation techniques

Irrigation water can be applied to crop of lands broadly in three general ways:

(i) Surface irrigation which includes the following:

(a) Uncontrolled (or wild or free) flooding method,
(b) Border strip method,
(c) Check method,
(d) Basin method, and
(e) Furrow method.

(ii) Subsurface irrigation

(iii) Pressurized irrigation methods *viz.*, sprinkler and drip irrigation

Selection of irrigation methods

Each of the above methods have some merits and demerits, and the choice of the method depends on the following factors as suggested by Walker and Skogerboe (1987).

(i) Size, shape and slope of the field,
(ii) Soil characteristics,
(iii) Nature and availability of the water supply subsystem,
(iv) Type of crops being grown,
(v) Initial development costs and availability of funds,
(vi) Preferences and past experience of the farmers, and
(vii) Purpose of irrigation either maximization of production or water productivity

18.3.1 Surface irrigation

Although the application irrigation with surface method is the oldest and most common method of irrigation it does not result in high levels of performance due to application efficiency and water efficiency. This is mainly because-of uncertain infiltration rates which are affected by year-to-year changes in the cropping pattern, cultivation practices, climatic factors and many other factors. These factors resulted in correct estimation of irrigation efficiency in surface irrigation methods.

(a) Uncontrolled flooding

In this method water is applied to the cropland without any preparation of land and without any levees or bunds to guide or restrict the flow of water on the field, the method is known as 'uncontrolled', 'wild' or 'free' flooding. Irrigation water is brought to field ditches and then admitted at one end of the field. Thus entire field is flooded without any control.

Generally, uncontrolled flooding the entire field results in excess irrigation at the inlet region of the field and insufficient irrigation at the outlet end. Application efficiency is also low because of either deep percolation (in case of longer duration of flooding) or flowing away of water (in case of shorter flooding duration) from the field. The application efficiency upon the depth of flooding, the rate of intake of water into the soil, and the size of the stream and topography of the field.

This method is suitable when water is available in large quantities, the land surface is irregular or levelled and the crop being grown is unaffected because of excess water. The advantage of this method is that it required the low initial cost of land preparation and required less labour for controlling the water. This is offset by the disadvantage of greater loss of water due to deep percolation and runoff.

(b) Border strip method

For application of irrigation by this method the farm as divided into a number of strips which can be 3-20 metres wide and 100-400 metres long. These strips are separated by low levees (or borders). The strips are level between levees but slope along the length according to natural slope of the field. For efficient application of irrigation, the slope should be between 0.2 and 04 per cent. But, slopes may be as flat as 0.1 per cent and as steep as 8 per cent for application of irrigation by this method. In case of steep slopes, care should be taken to prevent erosion of soil.

The slope also depends upon the type of soils. Clay loam and clayey soils require much flatter slopes (around 0.2%) of the border strips because of low infiltration rate. Medium soils may have slopes ranging from 0.2 to 0.4%. Sandy soils require large slopes ranging from 0.25 to 0.6%.

Water is diverted to these strips from supply channels and it flows slowly from upstream end to towards the downstream end and in the process it wets and irrigates the soil. The water supply is stopped when it recedes from the upstream end to the downstream end. In this process it wets and irrigates the cropland in the border strip.

The border strip method is well recommended to soils of moderately low to moderately high intake rates and low erodibility. This method is suitable for all types of crops except those which require prolonged flooding *viz.*, rice. It is difficult to maintain because of the slope. The main disadvantage of this method is that it requires preparation of land involving high initial cost.

(c) Check method

For application of irrigation by check method the entire field is divided into a number of small and almost levelled plots (compartments or 'Kiaries') surrounded by levees. Water is diverted from the water supply channels in the field to these plots by turn. This method is suitable for a wide range of soils ranging from very permeable to heavy soils. The farmer very efficiently controls over the distribution of water in different areas of farm. Loss of water through deep percolation (near the supply ditch) and surface runoff can be reduced and adequate irrigation of the entire farm can be achieved by this method. Thus, method gives application efficiency compared to earlier methods. This method requires more labours for allowing and closing the supplies to the levelled plots. Besides, this there is some loss of cultivable area which is occupied by the levees. Sometimes, levees are made sufficiently wide so that some 'row' crops can be grown over the levee surface to utilize the levees areas.

(d) Basin method

Generally, this method is used to irrigate orchards. In this method, one basin is made to irrigate one tree. However, where conditions are favourable, two or more trees can be included in one basin in the initial stage due to plant mortality.

(e) Furrow method

In this method, small furrows are construed small channels along the primary direction of the movement of water between the crops rows and letting the water flow through these small furrow. It is also known as 'furrow', 'creases' or 'corrugations'. When is desired into the furrows from field channels then water infiltrates through the wetted perimeter of the furrows and moves vertically and then laterally to saturate the soil.

Furrow lengths may vary from 10 metres to 500 metres. Although, the furrow length of 100 metres to 200 metres are the desirable lengths and more common. Very long furrows may result in excessive deep percolation losses and soil erosion near the upstream end of the field. Preferable furrows slope ranges between 0.5 and 3.0 per cent. Most of the soils have been satisfactorily irrigated with furrow slope ranging from 3 to 6 per cent (Hanson, 1979). For steep slopes, care should be taken to control erosion. Spacing of furrows for row crops (such as maize, potatoes, sugarbeet, etc.) is guided by the required spacing of the plant rows. The furrow stream should be small enough to prevent the flowing water from coming in direct contact with the plant. The furrow depth of 20 to 30 cm are satisfactory for soils of low permeability. For more permeable soils, furrow may be kept 8 to 12 cm deep. The size of stream also influences the furrow spacing.

Water is diverted to furrow from supply channels through small openings made in earthen banks. Alternatively, a small-diameter pipe of light weight plastic or rubber can be used to siphon water from the supply channels to the furrows without disturbing the banks of the earthen ditch.

The essential condition for furrow irrigation is the wetting of only about half to one-fifth of the field surface. This reduces the evaporation loss considerably. Puddling is also done in heavy soils to lessened water losses and it is possible to start cultivation soon after irrigation. Furrows provide better on-farm water management capabilities among surface irrigation methods and variable and severe topographical conditions. Furrows can be easily changed. This method gives very high-irrigation efficiency as compared to other surface irrigation method. For example this can be achieved by changing number of furrows according to change in supply.

There are following disadvantages of furrow irrigation:

(i) Possibility of increased salinity between furrows,

(ii) Loss of water at the downstream end unless end dikes are used,

(iii) The necessity of one extra tillage work, *viz.*, furrow construction,

(iv) Possibility of increased erosion, and

(v) Furrow irrigation requires more labour than any other surface irrigation method.

18.3.2 Subsurface irrigation

It (or simply sub-irrigation) is the practice of applying water to soils directly under the soil surface. According to Hansen (1979) the following conditions favour sub-irrigation:

(i) Impervious subsoil at a depth of 2 m or more,

(ii) A very permeable subsoil,

(iii) A permeable loam or sandy loam surface soil,

(iv) Uniform topographic conditions, and

(v) Moderate ground slopes.

In sub-irrigation, water is distributed in a series of perforated pipes. The ditches are of about 0.6 to 0.9 m deep and 0.3 m wide having vertical side. These ditches are spaced 45 to 90 m apart in the field.

It is favourable for the production of cash crops (*i.e.* high-priced crops) on small areas. A perforated pipe distribution is placed in the soil well below the soil surface for application irrigation. This application of irrigation water is known as artificial sub-irrigation. Soils which permit free lateral movement of water, rapid capillary movement in the root zone soil and very slow downward movement of water in the subsoil are very suitable for artificial sub-irrigation. The main disadvantage of this method is that it requires very high cost. However, the water consumption is as low as one-third of the surface irrigation methods. The method also improves yield.

18.3.3 Pressurized irrigation techniques

Pressurized irrigation with sprinklers, introduced about 50 years ago, contributed much to modernizing agriculture and increasing water use efficiency. However, from the standpoint of agriculture in arid and semiarid regions, the most important development has been the introduction of drip irrigation. Drip irrigation was developed in Israel and introduced into Israeli agriculture less than 35 years ago. Since then it has been disseminated all over the world with great success.

Therefore, the need is to enhance the efficiency of the existing irrigation projects and use of water saving technologies to bring additional areas under irrigation with same available water. Pressurized irrigation system uses pipe network to transport water and water is applied either by sprinkling action or by dripping very near to the plant to irrigate desired area. Pressurized irrigation works on concept of applying low irrigation volume at frequent intervals. This helps in avoiding the loss of water due to deep percolation and enhancing irrigation efficiency (Table 18.3).

Considering these facts the paper present an overview of pressurized irrigation systems with respect to its development, present state of the art, adoption, applicability, advantages, limitations, emerging concepts, strategies and its significance in dryland agriculture in context to Indian conditions.

Table 18.3: Irrigation and Application efficiencies of various irrigation systems

S. No.	System	Irrigation efficiency	Application efficiency
1.	Surface methods of irrigation	70-90%	50-95%
2.	Sprinkler irrigation systems	70-90%	60-95%
3.	Micro irrigation systems	80-90%	70-95%

Based on the availability of water, the crop planning and application of assessed critical water and fertilization combination will be beneficial for achieving the optimal production per unit of water and nutrient. Results of an experiments indicated that a minimum of 29% water savings can be achieved under the without significantly reducing the yield. No significant difference was obtained in yield of potato fertilized by conventional method or via fertigation. However, 31% of nitrogen saving was achieved without significantly reducing the yield.

Pressurized irrigation technologies can be defined as any irrigation system using pressurized or gravity head, pipe distribution at the farm or field level and can be grouped into the following:

a) Sprinklers,

b) Micro irrigation and

c) Piped distribution for surface irrigation.

These irrigation systems offer higher water use efficiencies. There are following major potential benefits of pressurized irrigation systems:

1. Improved conveyance and application efficiency,
2. Improved control over timing and depth of irrigation,
3. Reduced labour, fertilizer and energy requirement,
4. Effective irrigation of desired quantity and duration,
5. Facilitate use on even undulating terrain,
6. Provide better control and use on small discharges,
7. Facilitate use of poor quality waters.

Pressurized irrigation systems further can be classified based on nozzle type, provide a wide range of options to growers ranging from large-scale irrigation machines such as centre pivots and linear move systems, to drag hose sprinkler systems, mini sprinklers and drip systems.

18.3.3.1 Sprinkler irrigation

It is the method of application of irrigation water to the crop land or soil surface in the form of a spray, which is somewhat similar to rain. This method started in the beginning of this century for nurseries and orchards. It is popular in developed countries and is gaining popularity in developing countries too due to water savings, because light irrigation can be applied with this method.

Sprinklers have been used on all types of soils on lands of different topography and slopes, and for many crops. According to Hansen (1979), the following conditions are favourable for sprinkler irrigation:

(i) Very permeable soils which do not permit good distribution of water by surface method *viz.*, sand, soil and sandy loam soils,

(ii) Lands which have steep slopes and easily erodible soils,

(iii) When irrigation channels are too small to distribute water efficiently by surface irrigation, and

(iv) Lands with shallow soils and undulating topography which will not require proper levelling as required in surface methods of irrigation.

Sprinklers which facilitate a spray pattern, apply water to the entire irrigated area. Depending upon the nozzle type, pipe network system and portability, sprinkler irrigation system has several classifications. In India mainly hand moved portable or semi portable sprinkler systems are in use.

These types of systems mainly comprises of

a) Water source and pumping unit,

b) Filtration and fertigation units (as per the need),

c) Pipe networks,

d) Risers,

e) Accessories consisting of flow regulating valves and fittings and

f) Sprinkler nozzles.

Main lines: Main lines may be permanent or portable depending upon the field conditions. Permanent lines are made of steel, PVC and asbestos cement. Portable lines are made of light weight aluminium pipes.

Lateral lines: are usually portable. Quick-coupled aluminium pipes are best for portable laterals. Usual lengths are 5, 6 or 12 m. Each length has quick couplings.

Sprinkler heads: Sprinkler heads most important component of a sprinkler system. The sprinkler heads through nozzles direct the jet stream into the air.

Debris screens: Debris screens are usually needed when surface water is used as the source of irrigation. Its function is to keep the system free of trash that might plug the sprinkler nozzles. When heavy loads are expected two or more screens of progressively finer mesh can be used.

Desilting basins: Desilting basins are required to trap sand or suspended silts when the water comes from streams, open ditches or well water having silt. Sometimes desilting basins and debris screens are built as a combination structure.

Booster pumps: Booster pumps are used when a sprinkler irrigation system is used with an existing pumping system installed in a well and the pump capacity is insufficient to force the water through sprinklers.

Take off-valves: Take off-valves are needed to control pressures in the lateral lines.

Flow control-valves: Flow control-valves are used to regulate pressure and discharge of individual sprinkler when unevenness of the ground causes an unequal distribution of pressure along the lateral.

Types of sprinkler system

Sprinkler system is classified based on the following:

(i) Classification based on the arrangement for spraying water criteria.

(ii) Classification based on portability.

(i) Classification based on the arrangement for spraying water criteria

1. *Rotating head system:* Along the length of the lateral pipe, risers are fixed at uniform intervals with small size nozzles placed on them.
2. *Perforated pipe system:* Holes are perforated in the lateral irrigation pipes. It is usually designed for low operating pressure of 0.5 to 2.5 kg/cm^2 and is used for irrigating lawns, gardens, vegetable fields and other plots with crop height not exceeding 40 to 60 cm.

(ii) Classification based on portability

Sprinkler system is classified into the following groups based on portability.

1. *Portable system:* In this sprinkler system, it consists of portable main lines, laterals and pumping units.
2. *Semi-portable system:* In this system, except location of water source and pumping plant all other parts are portable.
3. *Semi-permanent system:* It consists of portable lateral lines, permanent main lines, sub-mains, stationary water source and pumping plant.
4. *Solid* set *system:* In this system, laterals are positioned in the field early in the crop season and remain for the season.
5. *Permanent system:* It consists of permanently laid mains, sub mains and laterals and a stationary water source and pumping plant.

Uniformity of coverage

In sprinkler irrigation system, uniformity of coverage is achieved with the following:

- Tie irrigation efficiency of sprinkler will depend upon the degree of uniformity of water application.
- The water spray distribution characteristics of sprinkler and their spacing will regulate the uniformity of water application.
- With lower pressure, the drops are larger and water from the nozzle falls in a ring away from the sprinkler.
- With higher pressure, water from the nozzle breaks up into fine drops and fall very near the sprinkler.
- High wind velocity distorts the application pattern.
- Under favorable condition, the depth of water applied, surrounding the sprinkler, decreases as the distance from the sprinkler increases.
- To obtain uniformity of water application, the wetted circle of the adjacent sprinklers should be overlapping.

Uniformity coefficient

Christiansen (1942) proposed a measurable index for the evaluation of the uniformity of sprinkler irrigation. This coefficient is based on the sum of the absolute deviations of each observation from the mean observation. It is expressed by:

$$Cu = 100[1.0 - (\Sigma X/mn)] \quad ....(18.1)$$

where,

m = average value of all observations, mm

n = total no. of observation points, and

X = numerical deviation of individual observations from the average application rate, mm

A uniformity coefficient of 85% or more is considered to be satisfactory.

Design of sprinkler irrigation system

One of the major concerns in the design of sprinkler irrigation system is ensuring high efficiencies in its performance uniformity of water distribution and economy to suit the conditions of a particular site. Uniformity of water from sprinklers varies depending upon pressure, wind, rotation of sprinkler, spacing and many other factors.

The step-by-step procedure in the planning and design of a sprinkler irrigation system is enumerated below:

1. Inventory of resources and conditions

(i) Map of the area
(ii) Water supply - source, availability and dependability
(iii) Climatic conditions
(iv) Depth of irrigation
(v) Irrigation interval
(vi) Water application rate (suggested guidelines in Table 18.7)
(vii) Sprinkler spacing (suggested guidelines in Table 18.8)
(viii) Power source

2. Types of systems and layout

To provide the optimum application rate with the greatest degree of uniformity of distribution, the location of the pumping unit, the orientation of mains and laterals, sprinkler spacing, operating pressure and nozzle sizes are determined.

(i) Location and nature of water supply

- The well may be located at a high corner or, more likely, at the center of the farm to minimize the distance the water must be pumped.

Table 18.4: Suggested maximum application rates (cm/hr) for sprinklers for average soils, slope and tilth

Soil texture and profile	Slope			
	0-5%	**5-8%**	**8-12%**	**12-16%**
1. Coarse sandy soil to 2m	5.0	3.7	2.5	1.3
2. Coarse sandy soil over more compact soils	2.7	2.5	2.0	1.0
3. Light sandy loams to 2 m	2.5	2.0	1.5	1.0
4. Light sandy loams over more compact soils	2.0	1.3	1.0	0.8
5. Silt loams to 2 m	1.3	1.0	0.8	0.5
6. Silt loams over more compact soils	0.8	0.6	0.4	0.3
7. Heavy textured clays or clay loams	0.4	0.3	0.2	0.1

Table 18.5: Maximum spacing of sprinklers under windy conditions

S. No.	Average wind speed	Spacing
1.	No wind	65% of the diameter of the water spread area of a sprinkler
2.	0 - 6.5 km hr.	60% of the diameter cfthe water spread area of a sprinkler
3.	6.5 – 13 km hr.	50% of the diameter of the water spread area of a sprinkler
4.	Above 13 km hr.	30% of the diameter of the water spread area of a sprinkler

(ii) Orientation of laterals

- The main should be located in the general direction of the steepest slope, with the laterals at right angles and as close as is practical to the contour.
- The arrangement selected should provide for a minimal investment in irrigation pipe, have a lower labour requirement.
- Any system in which the laterals are moved should be planned for successive irrigations in strict rotations, so that

the interval between irrigations is the same for each portion of a field.

- The possible arrangements for mains, laterals and sprinklers should provide for a minimal investment in irrigation pipe, have a lower labour requirement and provide for an application of water over the total area in the required period of time.

3. Sprinkler selection and spacing

The selection of the sprinkler depends mainly on the diameter of coverage required, pressure available and sprinkler discharge. The required discharge of an individual sprinkler is determined by:

$$Q = (S_i \times S_m \times I) / 360 \qquad \text{....(18.2)}$$

where,

Q = required discharge of individual sprinkler (L/sec),

S_i = spacing of sprinklers along the laterals, (m),

S_m = spacing of laterals along the main, (m) and

I = optimum application rate, (cm/hr),

Height of sprinkler riser pipes:

The minimum height of riser is 30 cm when the riser pipe is of 2.5 cm diameter and 15 cm when it is of 1.8 to 2 cm diameter.

4. Capacity of the sprinkler system

$$Q = 2780\,[(A \times D) / (F \times H \times E)] \qquad \text{....(18.3)}$$

where,

Q = discharge capacity of the pump, (L/sec),

A = area to be irrigated, (ha),

D = net depth of water application, (cm),

F = no. of days allowed for the completion of one irrigation,

H = no. of actual operating hours per day, and

E = water application efficiency, (percent)

5. Hydraulic design of sprinkler systems

Discharge of sprinkler nozzle: The discharge of a sprinkler nozzle may be computed from the formula:

$$Q = C.a.\sqrt{(2.g.h)} \qquad ....(18.4)$$

where,

C = co-efficient of discharge,

a = cross-sectional area of nozzle, (m),

h = pressure head at the nozzle, (m) and

g = acceleration due to gravity, (m/sec^2).

Design of sprinkler laterals

- Total pressure variation in the laterals, should not be more than 20% of the higher pressure.
- Friction loss in the lateral line should be computed where flow decreases along its length. First, using Hazen-William formula, the friction loss with full flow to the end of line is determined. Then from standard table 'F' factor for the appropriate number of sprinklers, is multiplied to obtain the appropriate friction loss in the lateral.
- The design capacity for sprinklers on a lateral is based on the average operating pressure. The average head, H_a, for design in a sprinkler line can be expressed by

$$H_a = H_o + \tfrac{1}{4} H_f \qquad ....(18.5)$$

In which H_o is the pressure at the sprinkler on the farthest end. H_f, friction loss, in the laterals (should be within 20% of the average pressure).

- Where the lateral is on nearly level land or on the contour, the head H_n at the main is

$$H_n = H_o + H_f \qquad ....(18.6)$$

By solving for Ho in equation (18.5), by substituting in equation (18.6).

$$H_n = H_a + 3/4\, H_f + \tfrac{1}{2} H_e + H_n \qquad ....(18.7)$$

where,

H_e = max difference in elevation between the first and last sprinklers on the D lateral, m

H_n = the riser height, m

Main line pipe

The selection should be based on economic considerations. Main line friction loss of about 3 m for small systems and 12 m for large systems may be allowed.

Pumps and power unit

In selecting a suitable pump, it is necessary to determine the maximum total head against which the pump is working. This may be determined by:

$$H_t = H_n + H_m + H_j + H_s \quad \text{....(18.8)}$$

where,

H_t = total design head against which the pump is working

H_n = maximum head required at the main to operate the sprinklers on the lateral at the required average pressure, including the riser height

H_m = maximum friction loss in the main and in the suction line, m

H_j = elevation difference between the pump and the junction of the lateral and the main, m and

H_s = elevation difference between the pump and the source of water after draw down m.

Cost estimation for sprinkler system

After complete design of the sprinkler system, an estimate of cost is required.

The following items are considered for estimation of operating cost of sprinkler system.

- Well
- Pump house
- Power
- Electrical installation
- Engine
- Pump unit, including foot .valve, non-return valve, suction and discharge pipes.
- Main line and sub-main
- Lateral lines
- Sprinklers and risers

- Special equipments
- Installation charges
- Other costs and charges

Performance evaluation of sprinkler

For evaluation of performance water samples were collected after the water was discharged by the micro-sprinklers. Utilizing the depth distribution data of a single nozzle, overlapped patterns of various micro-sprinkler spacing were obtained. Different uniformity coefficients developed by Christiansen, Wilcox-Swailes, Hart and Merriam-Keller were computed and compared for each of the different overlapped patterns.

Christiansen (1941) introduced the most commonly used measure of uniformity called Christiansen's Uniformity. It is known as Coefficient (UCC) and is expressed as,

$$UCC = 100[1 - \frac{\sum_{i=1}^{N} (x_i - x)}{Nx}] \quad ....(18.9)$$

where, Σ (x_i - x) is the sum of the absolute deviations of individual observations, x_i from the mean of the observations, x; and N is the number of observations.

Wilcox-Swailes (1947) proposed replacement of the 2nd term in the parenthesis of Eq. (18.9) by the coefficient of variation (s/x), Wilcox-Swailes Uniformity Coefficient (C_{ws}) is thus expressed as,

$$C_{ws} = 100(1 - \frac{S}{\bar{x}}) \quad ....(18.10)$$

where,

S = standard deviation, and

$\bar{x}$ = mean

Hart (1961) developed Hawaiian Sugar Planters Association (HSPA) Uniformity Coefficient (UCH) and expressed as,

$$USH = 100(1 - 0.798\frac{S}{\bar{x}}) \quad ....(18.11)$$

This uniformity is based on normal (Gaussian) distribution. If water application depths are assumed to be normally distributed, the

Christiansen Uniformity Coefficient (UCC) and HSPA Uniformity Coefficient (UCH) will have the same value.

Merriam and Keller (1978) suggested a new parameter as Distribution Uniformity (DU) similar to USDA Pattern Uniformity. Coefficient indicating the uniformity of infiltration throughout the field. The distribution uniformity (DU) is determined with the following relationship:

$$DU = \frac{\text{Avg. depth of water caught in lower quarter}}{\text{Avg. depth of water}} \times 100 \quad \text{....(18.12)}$$

Single nozzle performance evaluation micro-sprinklers

Single nozzle test was conducted on five commercially available micro-sprinklers at four pressure levels and four riser heights to study the performance parameters such as average application rate, mean application depth (D_a), effective maximum depth (D_{xe}), absolute maximum depth (D_{xa}) and coefficient of variation (C). The application rate varied from 11.82 to 0.67 mm/hr at operating pressures 0.5 and 2.0 kg/cm. The range of mean depth at varying pressures and riser heights varied from 1.01 to 8.05 mm/hr. The CV varied from 68 to 282%. In general, the average application rate and mean depth decreased with increase of pressure due to increase in area of coverage at all the riser heights for all the micro-sprinklers. Effective and absolute maximum depth increased with increase in nozzle size. The coefficient of variation increased with increase in pressure.

The criteria to evaluate performance characteristics will have to be developed and modified for a given condition. Only a few of the criteria have been picked up and their values are determined for specific common use. The earlier criteria were developed by Merriam and Keller (1978) as distribution characteristic (DC) which was defined as the ratio of the area which received more than half of the average application to the total wetted area. High DC values indicated adequately irrigated area to be a large fraction of total area. With the development of a variety of micro-sprayers and micro-spinners, DC alone was not found sufficient to characterise emitter performance.

Single nozzle micro-sprinkler study was conducted at the University Campus, Pantnagar (UP). Due to lack of testing standards for micro-sprinklers, the American Society of Agricultural Engineers standard ASAE S 330.1 "Procedure for sprinkler distribution testing for research purpose" (ASAE, 1996) was used as a guideline for the test.

The amount of water collected in the catch cans was then converted into precipitation rate (mm/hr) for further analysis. Manjunatha *et al.* 1998 summarized the various performance parameters determined for different micro-sprinkler which are described below.

Average application rate (AAR)

This parameter was determined by methods described by Merriam and Keller (1978) and Boman (1989), later referred as Keller and Boman method. According to Keller, the radius at which the approximate average application rate occurs for each concentric quarter of the area, can be computed by using effective radius as obtained by multiplying with 0.40, 0.60, 0.78 and 0.93 for inner, second, third and outer quarter, respectively. The average application rate received over total wetted area was the sum of the rates at quarter points divided by four. But, according to Boman method, the average application rate was calculated as the ratio of the total precipitation rate and the total number of catch cans receiving water.

Mean application depth (D)

It is defined as the average depth of water caught in the catch cans located within a distance of effective radius from the micro-sprinkler.

Coefficient of variation (CV)

Post *et al.* (1986) proposed coefficient of variation, CV a performance parameter in order to better characterize micro-sprinkler performance. The CV is the ratio of mean application depth (D_a) to the standard deviation.

Effective maximum depth (D_{xe})

It is defined as the average depth of water caught in the five per cent of the catch cans that had greatest water depths.

Evaluation of different uniformity coefficients for various micro-sprinklers

In order to evaluate the performance of five makes of commercially available micro-sprinklers of different nozzle size (1.50, 1.60, 1.73, 1.92 and 2.34 mm) operated at different pressures (1.0, 1.5 and 2.0 kg/cm^2) at a riser height of 35 cm, experiments were conducted to measure the depth distribution pattern from a single leg test. Overlapped patterns of various sprinkler spacing (from 3m x 3m to 8m x 8m) are obtained by utilizing the depth distribution data of a single nozzle. Different uniformity coefficients developed by Christiansen, Hart. Wilcox-Swailes and distribution uniformity were computed for each of the different overlapped patterns. The uniformity in water distribution varied greatly from one make to the other under the various operating conditions, the highest being 95% and the lowest being 9%. In general, as the operating pressure of micro-sprinkler increases, all the uniformity coefficients increases and uniformity decrease with increase in sprinkler spacing.

One of the major concerns in the design of micro-sprinkler irrigation system is ensuring uniformity of water distribution. Christiansen (1941) concluded that uniformity of distribution of water from sprinklers varies depending upon pressure, wind, rotation of sprinkler, spacing and many other factors. A nearly uniform application is possible with proper sprinkler patterns and with optimum spacing of sprinklers. Kerr *et al.,* 1980 concluded that only 2 of 15 uniformity coefficient (UCC) values were acceptable (UCC 70%) at 25 per cent overlap and 7 of 15 were acceptable at 50 per cent overlap. The main objective of this study is to evaluate the performance of different micro-sprinklers based on different uniformity coefficients for different sprinkler spacing operating at various pressures.

One of the basic requirements of these systems is to apply water as uniformly as possible to fill the root zone of the crop with water. The size of water droplets after breaking is of 0.5 and 4.0 mm in size as size of droplet also affects the uniformity pattern. Smaller droplets tend to fall nearer the sprinkler whereas the larger ones fall towards periphery of wetted diameter. The common uniformity measures are Christiansen's Uniformity Coefficient (UC) and Distribution Uniformity (DU). Christiansen's Uniformity Coefficient reflects how equal (or unequal) the application rates are throughout the field, whereas, Distribution Uniformity is based on the low quarter of irrigated area and it does not tell how big or severe the dry spot is. A lower value of uniformity coefficients is indicative of large variation in application rates, while higher values indicate uniform application of water within the wetted area.

Normally, DU values are lower than the UC values for the same set of data. In case of sprinklers uniformity coefficient values of less than 70% are not acceptable. In case of high value crop and where chemicals are to be applied with irrigation water, the uniformities should be high (e.g. DU greater than 80%, or UC greater than 87%).

Advantages and limitations of sprinkler system

There are following advantages of the systems:

(i) Provides good control of water application, which can reduce vegetative growth and improve the product quality,

(ii) Reduces effects of wind-blown sandy soils on young crops,

(iii) Provides a high degree of flexibility in design and operation, and can be automated to save labour and energy expenditure,

(iv) Nutrients can be applied through the irrigation system (fertigation),

(v) Can be used to cool the crop during heat waves, to protect against frost and to manage plant health,

(vi) Has the ability to keep soil soft for emerging seedlings with frequent short irrigations,

(vii) Both proper drainage, sprinklers can be used efficiently to flush accumulated salts down the profile, and

(viii) Promotes rapid germination and crop establishment.

There are following limitations with sprinkler systems:

(i) High capital cost,

(ii) Distribution uniformity can deteriorate over time,

(iii) Prone to wind effects on uniformity of' water distribution,

(iv) May create a humid environment in the crop canopy which, can lead to disease, and

(v) Can cause crop damage where saline irrigation water is applied.

18.3.3.2 Drip irrigation

The evaporation losses depend upon the size of the drop and weather condition *viz.*, wind and temperature. It is application of irrigation near the plant through (also known as drip irrigation) an extensive network of small diameter pipes. The water outlet device in the pipe is called an 'emitter' or dripper which discharges only few litres of water per hour. Soil capillary forces cause the emitted water to spread laterally and vertically.

(1) Valves to control the flow, (2) discharge meter, (3) pressure controls and (4) filters (to remove all debris, sand and clay to reduce clogging of the emitters) are usually located at the source of the water and are together referred to as the 'control head'. Lateral lines are

generally flexible PVC pipes of 12 to 32 mm in diameter and the emitters are inserted into the lateral lines at a predetermined spacing chosen to fit crop and soil conditions, (6) the emitters must produce a relatively, small flow and yield almost constant discharge without causing clogging of the emitters.

Flow is controlled manually or set to automatically deliver either (i) desired amount of water for a predetermined time, or (ii) water whenever soil moisture decreases to a predetermined amount.

Drip irrigation apply water efficient water application in the root zone of small trees and widely spaced plants without wetting the soil where no roots exist. In arid regions, the irrigation efficiency of 90 per cent can be achieved with application of very good management it may approach the ideal value of 100 per cent. The main reasons for the high efficiency of drip irrigation are its capability to produce and maintain continuously high soil moisture content in the root zone and the reduction of weed growth (due to limited wet surface area), which compets with the crop for water and nutrients. Insect, disease and fungus problems are also reduced with reduction in the wetting of the soil surface.

Drip irrigation results in better quality and greater crop yields, by maintaining a nearly constant soil moisture content in the root zone. Fruits which contain considerable moisture at the time of harvesting (such as tomatoes, grapes, berries, etc.) respond very well to drip irrigation. However, this method is not well (from practical as well as economic considerations) for closely growing crops such as wheat and other cereal grains due to high cost involved.

One of the main problems of drip irrigation is the clogging of small conduits and openings in the emitters due to sand and clay particles, debris, chemical precipitates and organic growth. In this method, only a part of the soil is wetted and, hence, it must be ensured that the root growth is not restricted. Another problem of encounter is on account of the dissolved salt left in the soil as the water is used by the plants. If the rainwater flushes the salts near the surface down into the root zone, severe damage to the crop may result. In such situations, application of water by sprinkler or surface irrigation may become necessary.

Design of drip irrigation system

The following general informations are required for designing a drip irrigation system:

1. Water source – The source of water is usually a well or a tank storing harvested runoff. The distance and depth of water source effect the pressure head required.
2. Types of crops - Different crops require different plant spacing and k irrigation requirements. The general layout of the system and especially the emitters will depend on the type of crop.
3. Topographic condition - It is necessary to know the general land slope to determine the size and location of main and submain lines.
4. Soils. The soil characteristics such as infiltration capacity, texture and structure, and bulk density must be known when selecting emitter type, determining spacing, and setting up irrigation schedules.
5. Climatic records - The climatic records will show when and how often in various seasons of the year.

The flow conditions in the submains and the laterals of a drip irrigation system can be considered as steady and spatially varied with lateral outflows. The flow from submains into laterals or the outflow of each emitter from a lateral is controlled by the pressure distribution along the submain and laterally lines. The pressure distribution along a drip irrigation line-submain or lateral-is controlled by the energy drop through friction and the energy gain or loss due to either down or up slopes.

If the pressure distribution along a lateral line can be determined, uniform irrigation can be achieved by adjusting the size of emitters, or the length and size of the micro-tube, or slightly adjusting the spacing between emitters. If the design allows a certain variation of emitter outflow along the lateral line, a single type of emitter can be used, eliminating the troubles of adjustments. The degree of variation of emitter outflow can be shown by using the uniformity coefficient equation of Christiansen.

The variation of discharge from emitters along a lateral line is a function of the total length and inlet pressure, emitter spacing, and total flow rate. This creates a design problem to select the right combination of length and pressure in order to achieve an acceptable, non-uniform pattern of irrigation.

Basic Hydraulics

A drip irrigation system is a combination of different sizes of plastic pipes and emitters, which are usually considered as smooth pipes.

Following empirical equations used for estimating the loss of head by friction in a drip irrigation line as,

1. Blassius equation

$$f = \frac{0.316}{(R_n)^{1/4}} \qquad ....(18.13)$$

where,

f = friction coefficient, and

R_n = Reynolds number

2. Williams and Hazen equation (using C= 150)

$$\Delta H = 15.27\left(\frac{Q^{1.852}}{D^{4.871}}\right)L \qquad ....(18.14)$$

where,

ΔH = energy drop by friction, (m),

Q = total discharge in the pipe, (L/sec), and

L = length of the pipe section, (m).

Equation (18.14) is frequently used to compute the energy drop of a main line section.

Since the discharge in the line decreases with respect to the length, the total energy drop will be less than the one given in equation (18.14). The total energy drop, ΔH, for lateral or submain can be determined by introducing an *F*-value as a reduction coefficient (Howell and HileI, 1974) or determined by integration (Wu and Gitlin, 1975). The total energy drop by friction for lateral or submain can be estimated as,

$$\Delta H = 5.35\left(\frac{Q^{1.852}}{D^{4.871}}\right)L \qquad ....(18.15)$$

where,

ΔH = total energy drop by friction at the end of lateral (or submain), (m),

Q = total discharge at the inlet of lateral (or submain), (litres/sec),

D = inside diameter of lateral (or submain), (cm),

L = the total length of lateral (or submain), (m), and

The discharge in the lateral (or submain) decreases with respect to its length, the energy gradient line will not be a straight line but results into the exponential type curve. The shape of the energy gradient line can be expressed by a dimensionless energy gradient line as,

$$R_t = 1 - (1 - i)^{2.852} \qquad \text{....(18.16)}$$

$$R_t = \Delta Hi / \Delta H, \text{ and } i = l / L$$

where,

R_t = the energy drop ratio,

ΔH = total energy drop determined by equation (18.15),

ΔH_i = total energy drop expressed in metres at a length ratio i,

L = total length, (m),

l = *a* giver length measured from the head end of the line, (m)

Equation (18.16) used to determine the energy drop pattern along a lateral (or submain) when the total energy drop, ΔH, is determined and known. If a drip irrigation line, lateral or submain, is laid on a level ground the pressure variation along the line will be indicated by the energy gradient line. If a drip irrigation line is laid on up or down slopes the pressure variation will be affected by the slopes. The pressure variation (the change of pressure with respect to length) can be determined as a linear combination of energy slope and line slope by assuming that the change of velocity head in the line is small and hence negligible. This can be expressed as suggested by Wu and Gitlin, (1974) and is given as,

$$\frac{dh}{dl} = S_f \pm s_o \qquad \text{....(18.17)}$$

where,

S_f = energy slope (the slope of energy gradient line), and

s_o = *the* line slope.

Plus (+) sign is for downslope and minus (-) sign for upsIope.

Equation (18.17) can be used for both uniform and non-uniform slopes,

Hydraulically, the pressure variation along a lateral line will cause an emitter flow variation along the lateral and a pressure variation along a submain will cause a lateral line flow variation (into each lateral line) along a sub- main. For the most commonly used emitters and also assuming turbulent flow in the laterals, the emitter flow (or lateral line flow for submain) and the pressure head can be expressed as,

$$q_i = c\sqrt{h_i} \quad \text{....(18.18)}$$

where,

q_i = emitter flow (or into laterals for submain) at the ith section,

c = a coefficient,

h_i = pressure head at the ith section

The pressure variation and emitter flow (or lateral line flow) variation are related and can be expressed as,

$$q_{var} = 1-(1-h_{var})^{0.5} \quad \text{....(18.19)}$$

The emitter flow variation (or lateral flow for submain), q_{var} is defined as,

$$q_{var} = \frac{q_{max} - q_{min}}{q_{max}} \quad \text{....(18.20)}$$

where,

q_{max} = the maximum emitter flow, and

q_{min} = the minimum emitter flow along the lateral line (or maximum and minimum lateral flow in submain)

The pressure variation h_{var} is defined as,

$$h_{var} = \frac{h_{max} - h_{min}}{h_{max}} \quad \text{....(18.21)}$$

where, h_{max} and h_{min} are the maximum and minimum pressure heads, respectively, along the lateral (or submain).

Criteria for designing drip irrigation system is generally based on an emitter flow variation of less than 20 per cent (about 40 per cent pressure variation) for lateral line; and a lateral flow variation of less than five per cent (about ten per cent pressure variation) for a submain.

Design charts and design procedure

Design charts have been developed on the basis of basic hydraulics of drip irrigation lines and a computer simulation, for lateral, submain and main line design has also been developed (Wu *1975,* Wu and Gitlin, 1974). The design charts and their design procedures are described as under:

1. Lateral line design charts for uniform slopes

The lateral line design charts for the two commonly used sizes, namely, 12 mm and 16 mm pipes are shown in Figs. 18.1 and 18.2, respectively.

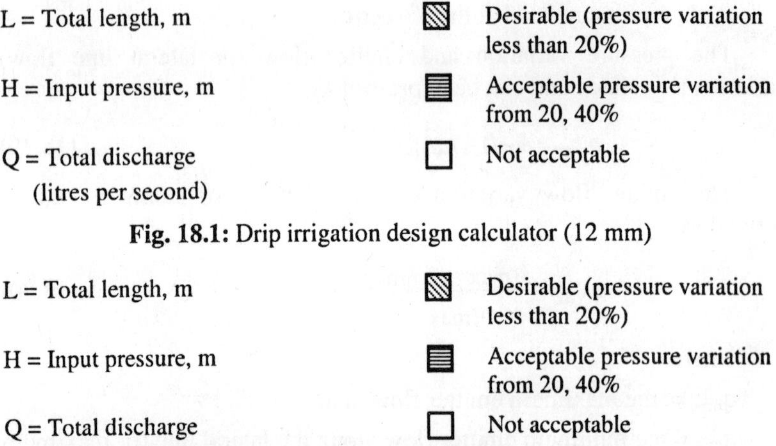

Fig. 18.1: Drip irrigation design calculator (12 mm)

Fig. 18.2: Drip irrigation design calculator (16 mm)

The design procedure is consists of following steps:

Step 1. Establish the lateral length *(L)* and operating pressure head *(H),* ratio *L/H* and total discharge (*Q*) in iitres per second *(LPS).*

Step 2. Move vertically from *L/H* to the given total discharge *(LPS)* line in quadrant II; then establish a horizontal line into quadrant I.

Step 3. Move horizontally from *L/H* to the per cent slope line in quadrant IV; then establish a vertical line into quadrant I.

Step 4. The intersection point of these two lines in quadrant I determines the acceptability of the design.

Desirable: Pressure variation less than 20 per cent, or emitter flow variation less than 10 per cent.

Acceptable: Pressure variation 20 to 40 per cent, or emitter flow variation about 10 to 20 per cent.

Not recommended: Pressure variation greater than 40 per cent or emitter flow variation larger than 20 per cent.

Example 18.1: A lateral line length in a vegetable field is 50 m and the slope of the lateral line is one per cent downslope. There are 150 emitters spaced equally in the lateral line. The emitter flow is four litres per hour at an operating pressure head of 10 m. Design the lateral line size.

Solution:

Given

Lateral length L	=	50 m
Operating pressure H	=	10 m
Number of emitters	=	150
Total discharge Q	=	600 L/hr
	=	0.167 L/sec

Design procedure

Step 1. Determine L/H = 5.

Step 2. From Fig. 18.1, move vertically from *L/H* = 5 to the total discharge line, *Q* = 0.167 L/sec; then establish a horizontal line into quadrant I.

Step 3. Move horizontally from *L/H* = 5 to the 1 per cent downslope line in quadrant I; then establish a vertical line into quadrant I.

Step 4. The intersection point is in the acceptable range, so the 12 mm lateral line is selected.

2. Lateral design for non-uniform slopes

Wu and Gitlin (1975) developed a simple design chart for lateral line design on non-uniform slopes (Fig. 18.3). The chart is dimensionless and can be used for both metric and FPS units. The design procedure involved following steps:

Step 1. Divide the non-uniform slope profile into several sections in which each section can be considered as a uniform slope; determine the slope of each section; calculate the energy gain (or loss) in each section by its slope; and determine the total energy gain by slopes for any length along the line (ΔH_i).

Step 2. Plot the non-uniform slope pattern in a dimensionless form l/L *vs* $(\Delta H_i'/L)$ in quadrant I.

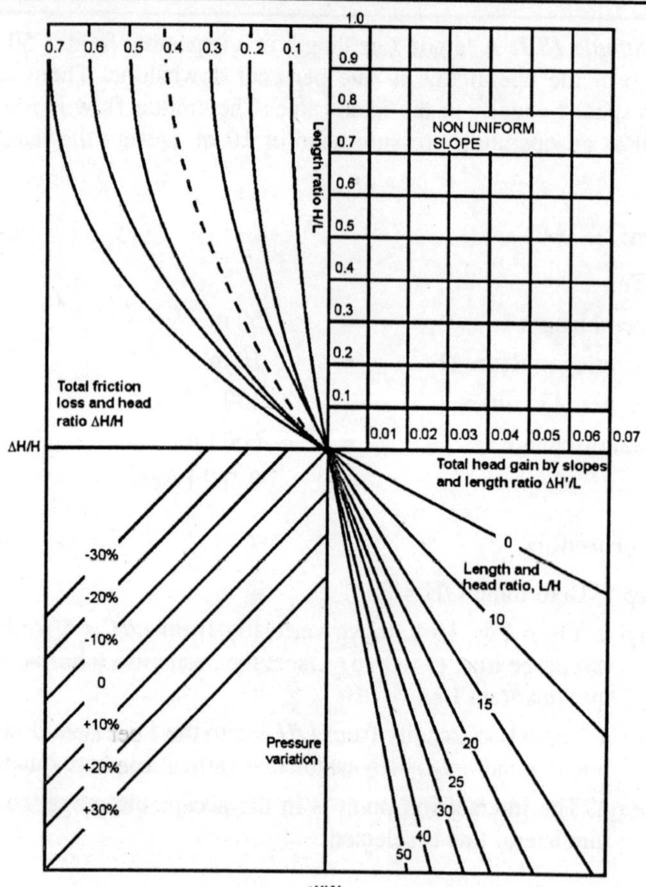

Fig. 18.3: Design chart for non-uniform slopes (with design example 18.2).

Step 3. Determine the total energy drop, ΔH, by friction, using Equation 18.31 and calculate the total energy drop (ΔH) and operating pressure head (H) ratio, ($\Delta H/H$).

Step 4. Determine the lateral length (L) operating pressure head (H) ratio L/H.

Step 5. Select a point from the non-uniform slope profile in quadrant I, usually, points between two slopes and at the middle of a section.

Step 6. From the point selected in step 5 draw a vertical line downward to the determined L/H in quadrant IV and established a horizontal line into quadrant III.

Step 7. Also from that point (in step 5) draw a horizontal Line to the determined $\Delta H/H$ in quadrant ii and then establish a vertical line into quadrant III.

Step 8. The location of the intersection point of these two lines *in* quadrant III will show the pressure variation from the operating pressure.

Step 9. Repeat the same procedure for several other points in the dimensionless non-uniform slopes profile and check the pressure variations of these points from the operating pressure.

Example 18.2: A 120 m long 16 mm diameter lateral line is laid on a non-uniform slope field. The non-uniform situation can be expressed as follows:

0-30 m	3% down
30-60 m	2% down
60-90 m	0%
90-120 m	3% down

The operating pressure head for drip irrigation is 10 m and the total discharge for the lateral line (120 m long) is 0.13 L/sec. Check the pressure variation.

Solution:

Given

$L = 120$ m

$H =$ I0 m

$D = 16$ m

(a) Determine

$\Delta H = 1.5$ m (using Equation 21.15)

$L/H = 12$

$\Delta H/H = 0.15$

(b) Non-uniform slopes

1/L	ΔH_i (m)	ΔH_i'lL
0.25	0.9	0.0075
0.50	1.5	0.0125
0.75	1.5	0.0125
1.00	2.4	0.0200

The values of 1/*L vs.* ΔH_i'*lL* are plotted in quadrant I.

(c) Follow the procedure given in the previous section to check the pressure variation for the four points at length ratios of 0.25, 0.50, 0.75 and 1.0. The results are shown in Fig. 18.3. It may be observed in quadrant III (Fig. 18.3) that the pressure variation along the lateral line is less than ten per cent. Hence, the design is acceptable.

3. Simplified design chats for the submain

Design charts similar to those shown in Figs. 18.3 and 18.4 can be developed for submain design if the charts are made for larger pipe sizes. The non-uniform slope design chart can also be used for the submain. However, since submain lengths are relatively short, a uniform slope can usually be assumed.

Wu and Gitlin (1974, 1975) were developed two simplified charts for submain designing under following conditions:

(i) one for submain slopes less than 0.5 per cent

(ii) for submain slopes equal to or larger than 0.5 per cent.

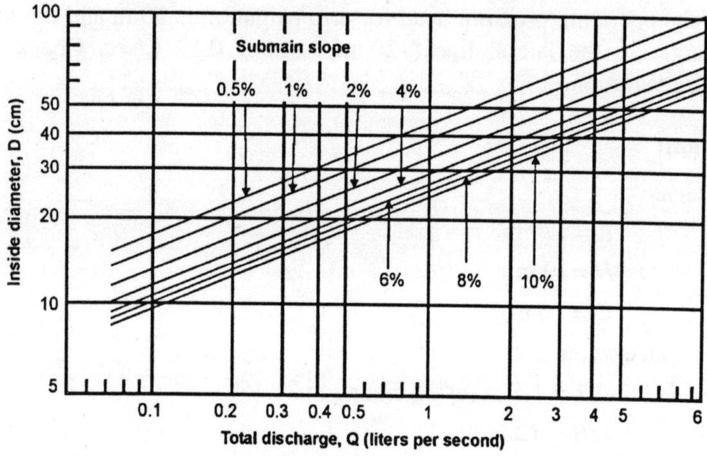

Fig. 18.4: Submain design chart-slope equal to or larger than 0.5%

The two simplified submain design charts are shown in Figs. 18.4 and 18.5. Since the submain length is usually relatively short, ranging from 20 to 60 m. and the operating pressure head is about 10 m, the length and head ratio *L/H* will be only equal to or less than six. Under this situation the two simplified submain design charts will give a lateral flow variation less than five per cent. The design procedure involves following steps:

Step 1. Determine the total discharge Q, for the submain.

Step 2. Determine the length and pressure head ratio, *L/H*.

Step 3. Determine the submain slope. If the slope is equal to or larger than 0.5 per cent, use Fig. 18.4 to design the submain size.

Step 4. If the slope is less than 0.5 per cent, use Fig. 18.5 to design the submain size.

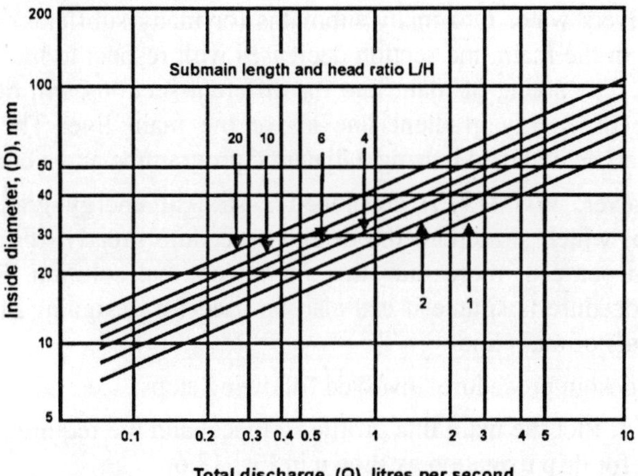

Fig. 18.5: Submain design chart-slope less than 0.5% and allowable pressure variation 10%

Example 18.3: A submain, 10 m long, is installed in a vegetable field. The lateral line spacing is 1 m and there are 10 lateral lines connected to the submain. The submain is laid on a level situation (zero slope). Design the submain size.

Solution

Given

Submain length, $L = 10$ m

Operating pressure, $H = 10$ m (Example 18.1)

Total discharge for each lateral 0.167 L/sec.

Design procedure

1. Total discharge. $Q = 1.67$ L/sec.
2. Submain length and pressure head ratio, $L/H = 1$.
3. Submain slope = 0.
4. Using Fig. 18.5, the submain size is designed as 25 mm.

4. Main line design

Design a main line is not a problem if it is a simple drip irrigation system in which the main line delivers water for only one or two subfields. The design of main line is simply a pipe design; the main line size is determined based on the total discharge required in the main and the length and the allowable energy drop by using Equation 18.16. When a line delivers water into many submains (or many subfields), the total discharge in the main line section decreases with respect to the length of main line. The sizing of main line for different sections will depend on the shape of energy gradient line above the main line. The optimal solution can be obtained b using a dynamic programme and a computer.

However, WU (1975) developed a straight energy gradient line (Fig. 18.6) which simplified the design procedure greatly. It costs only about two per cent more than that of the optimal solution. Since the design procedure is simple it can also be used for designing alternative field layouts of main lines.

The design procedure involved following steps:

Step 1. Plot the main line profiles (slopes) and the required pressure for drip irrigation as shown in Fig. 18.6.

Step 2. Draw a straight energy gradient line from the available operating pressure to the required pressure profile (Fig. 18.6) so that the energy gradient line is above the required pressure profile.

Step 3. Determine the energy slope which is the slope of the straight energy gradient line, $\Delta H/L$.

Step 4. Determine the required discharge in each main line section.

Step 5. Design the main line size by using the nomograph (Fig. 18.7), based on the energy slope (determined in step 3) and total discharge (determined in step 4) for each mainline section.

Example 18.4: A drip irrigation system is to be designed for a 20 hectare papaya field. The field is rectangular in shape and is divided into 0.4 hectare sub-plots which are controlled by submains connected from the main line. The main line is laid in the centre of the field with 25 sub-plots on both sides of the main line. Each subplot is about 130 m long and 30 m wide. Each main line section is 30 m long. The design capacity is 2 L/sec for each sub-plot. There is a total of 24 main line stations and at the end of each section there are outlets to supply 4 L/sec for irrigating both sides. The main line slopes, the required operating pressure 7 m), and input pressure (17.5 m) are shown in Fig. 18.7. Design the main line.

Solution:

Design procedure:

1. Plot the main line profile as shown in Fig. 18.6.
2. Plot the required pressure head (7 m) along the main line profile as shown in Fig. 18.6.
3. Determine the energy slope. From Fig. 18.6 the slope of the straight energy gradient is determined as 1 per cent.
4. Design the main line size. The main line size is determined from the nomograph (Fig. 18.7), using 1% energy slope. The results are shown in Table 18.6.

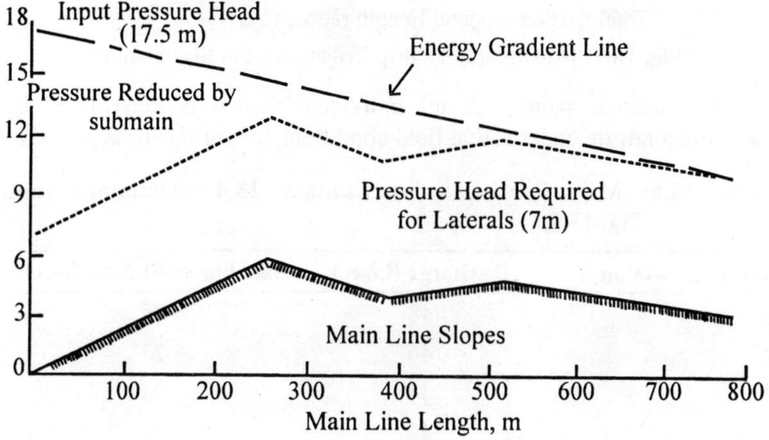

Fig. 18.6: Main line layout and energy gradient line in Example 18.4.

Evaluation of drip irrigation systems

Drip irrigation research can be applied for the following studies and objectives:

1. Optimal soil moisture for crop growth
2. Water use by different crops under drip irrigation
3. Improved crop yield and quality
4. Interaction of irrigation and application of fertilizer on crop growth
5. Unsaturated soil water movement

6. Irrigation water quality and the movement and distribution of salt in the root zone
7. Irrigation system design and management

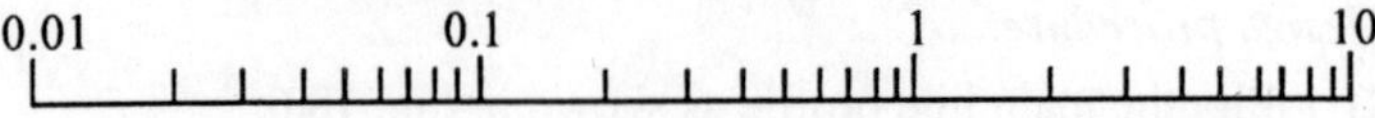

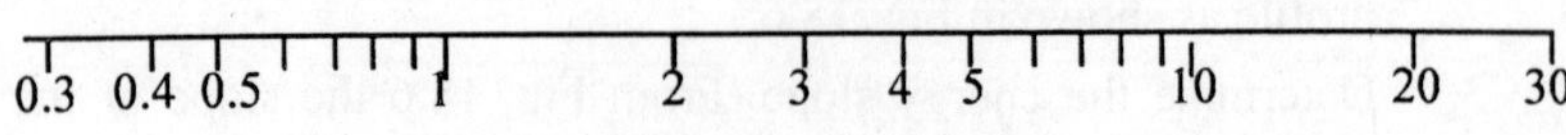

Pile size, D (Inside diameter, cm)
(Plastic Pipe)c

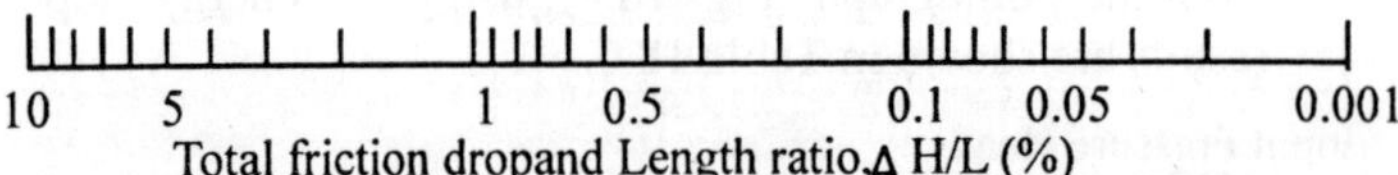

Total friction dropand Length ratio,Δ H/L (%)

Fig. 18.7: Nomograph for drip irrigation main line design.

The wetting pattern of an individual nozzle is determined by operating a nozzle under actual field conditions for various time periods.

Table 18.6: Main line sizes for Example 18.4, determined from Fig. 18.8.

Main line section	Discharge (L/sec)	Main line size inside dia, cm
0*	48	25
1	46	25
2	44	25
3	42	20
4	40	20
5	38	20
6	36	20
7	34	20
8	32	20
9	30	20
10	28	20
11	26	20
12	24	20

Main line section	Discharge (L/sec)	Main line size inside dia, cm
13	22	15
14	20	15
15	18	15
16	16	15
17	14	15
18	12	15
19	10	12.5
20	8	12.5
21	6	10
22	4	10
23	2	7.5

*There is an outlet at the entrance of section 1 for irrigating the sub-plots on both sides of the section.

The moisture penetration pattern in the soil is determined by measuring the soil moisture *in-situ.* The moisture measuring instruments should be installed at different depths and preferably in a radial pattern, away from the drip nozzle. The effective area of coverage of the nozzle will determine the spacing of the nozzles and the number of nozzles required to wet the root spread area of the crop.

While evaluating crop response under the drip method of irrigation, the crop parameters such as growth, maturity, produce, and disease and weed infestation are recorded. Detailed account of salt movement in the soil is maintained by periodical soil sampling and testing. Soil samples are taken from different soil depths and locations away from the nozzle. Graphical representation of salt concentration is maintained to evaluate the movement and concentration of salts.

Factors affections water productivity

It can be concluded from the results of various studies that the water productivity of harvested rainwater can be enhanced with the following key factors:

(i) Growing of low water requirement and high value crops,

(ii) Weed control

(iii) Irrigation through improved methods (surface or micro irrigation system)

(iv) Use of mulches

(v) Wind break or shelter belt

(vi) Use of antitransparents

Above these factors not only save the water through reduction of evaporation, seepage losses, deep percolation losses and evapotranspiration but bring more command areas under irrigation with saved water which results into increase in productivity of both crops and water and also save labour inputs which reduces cost of production.

Except irrigation application techniques all other factors discussed in detail given earlier and subsequent chapters. Therefore, this chapter mainly deals with irrigation techniques.

18.4 Micro irrigation

Micro-irrigation refers to low-pressure irrigation systems that spray, mist, sprinkle or drip. The water discharge patterns differ because emission devices are designed for specific applications due to agronomic or horticultural requirements. Micro-irrigation components include pipes, tubes, water emitting devices, flow control equipment, installation tools, fittings and accessories. With the help of the small emitting devices micro irrigation delivers water onto the soil surface very near the plant or below the soil surface directly into the plant root zone. Types of micro irrigation differ based on the emission devices. Of various form of micro irrigation, drip irrigation system is most common of all in India. Drip irrigation system which is a permanently installed system comprises of mains and sub mains pipes of PVC buried beneath the ground and laterals of LDPE or LLDPE on the surface. Nozzles namely drippers, micro tubes etc. are attached on the laterals depending upon the plant spacing which delivers water. In other case laterals are also buried beneath the surface, this type of drip system is known as sub-surface drip system and mostly adopted for closely spaced crops such as sugarcane, cotton etc.

Emission devices

The water is delivered to plants through emitters in a micro-irrigation system where lateral pipes are laid on the surface. It is made up of plastic and facilitates small but precise discharge. The quantity of water delivered from these emitters is usually expressed in litre per hours (lph) gallons per hour (gph). These emitters dissipate water pressure through the use of long-paths, small orifices or diaphragms. Some emitters are pressure compensating having capability to discharge water at a constant rate over a range of pressures. Emission devices deliver water in three different modes: drip, bubbler and micro-sprinkler. In drip mode, water is applied as droplets or trickles, in bubbler mode, water 'bubbles out' from the emitters. While in micro-sprinkler mode water is sprinkled, sprayed, or misted. These emitters are available in several discharge increments as some emitters are adapted to apply water to closely spaced crops planted in rows (one or more than one plants) whereas others are used to irrigate several plants at once.

Types of emission sources

Line source - Line source can be described as a system that seeps water uniformly along the full length of the line. These are often referred to as soaker or porous hose lines.

Point surface – The point source drip irrigation systems can be defined as emitter attached on laterals at fixed spacing. With the help of attached emitters the water is delivered to plants at specific points.

Bubbler irrigation - Bubblers typically apply water on a "per plant" basis. Bubblers are very similar to the point source external emitters in shape but differ in performance.

Micro-sprinkler irrigation – Micro-sprinklers are emitters commonly known as sprinkler or spray heads. They operate by throwing water through the air, usually in predetermined patterns.

Micro irrigation system components

There are following basic system components of micro irrigation system.

(i) Control head consisting of pump and power unit, a backflow prevention device if chemicals arc used with water, a filter,

(ii) A water distribution network system consisting of main, submain and lateral lines, and

(iii) Emission devices for controlling the volume of water and pressure in the system.

Advantages of micro irrigation

There are following advantages of micro irrigation over other irrigation methods.

(i) Water is discharged uniformly along the lateral (pipe fitted with drippers) even on moderately sloping terrain. The invention of compensated drippers enables uniform irrigation of steeper slopes and long distances.

(ii) Fertilizers can be supplied to the plant via the drippers together with the water (fertigation).

(iii) Water and fertilizers are delivered directly to the root system rather than to the total area of the field, thereby economizing on both water and fertilizers.

(iv) The quantity of water delivered can be optimized to fit different soil types, avoiding percolation of water beyond the root zone.

(v) The emergence of weeds is minimized.

(vi) Exploitation of poor quality water (saline water or effluents) is made possible.

(vii) Drippers with a given discharge of water (liters per hour) can be installed at any spacing to accommodate the needs of any crop.

(viii) Drip irrigation is the most efficient method as regards water saving. Since the drippers emit the water directly to the soil adjacent to the root system, which absorbs the water immediately, evaporation to the air is minimal. This effect is especially important under the conditions prevailing in arid zones.

(ix) In irrigation by sprinklers or by surface methods, evaporation is enhanced by winds; by comparison, in drip irrigation the impact of winds is minimal.

(x) Drip irrigation, unlike sprinkler irrigation, makes it possible to utilize saline water by eliminating direct contact between the water and the leaves, thus avoiding burns.

(xi) Drip irrigation causes salts to be continuously washed away from the root system, avoiding salt accumulation in the immediate vicinity of the roots when irrigating salinized soils or irrigating with saline water.

(xii) Drip irrigation allows the use of sewage water because the water is delivered directly to the ground, minimizing health risks.

(xiii) High-quality drip irrigation equipment can last for fifteen to twenty years if handled properly.

Micro irrigation is becoming popular for vegetable and horticultural crops in many areas of the country. Spray or spinner emitters (micro sprinklers) are often preferred over drip system since they provide larger diameter welting pattern. This characteristic is especially desirable in areas with coarse textured soils where lateral movement of soil water is limited.

18.4.1 Recycling through microirrigation

Furrow/border irrigation is seen as being wasteful. Low water use efficiency is problem since total water loss through conveyance under unlined irrigation system is 71%, broken up as 15% in canal, 7% in distributaries, 22% in water courses and 27% in field losses (Navalawala, 1991). The most efficient and economically feasible method of irrigation is therefore, drip where 50-70% water saving with 10-15% increase in crop yield could be effected (Sivanappan *et al.*, 1987). It is however practised only in limited regions in 0.228 million hectare due to several technological constraints, though about 27 million ha area in the country is suitable for drip irrigation (Samuel, 2000).

Singandhupe and Verma (2002) summarized the importance of micro irrigation system and its application in different crops is described below.

Sugarcane

Importance of drip irrigation in sugarcane (*Saccharum officinarum* L.) has been widely accepted by most of the farmers. In drip system, irrigation water, applied through small emitters comes in direct contact with root systems and frequent application of water keeps soil profile at field capacity or just below field capacity water content and hence the cyclical effect of moisture stress which occurs in furrow is avoided. On the other hand due to easy availability of water in drip irrigation, root system is proliferated at surface layer and sugarcane being a tall and sturdy plant, can not withstand under high wind speed and hence lodging of the .crop occurs. To avoid such adverse effect on quality of sugarcane due to ridging, earthing up is generally provided when the crop is six month old. The experiment, conducted at Rahuri in Maharashtra state (longitude 74° and latitude 19°24', altitude 7 m above sea level) during January-February, 1982 in clay soil (vertisol), recorded maximum sugarcane yield of 146.4 t ha1 (19.9% higher yield) with 29.5% saving of irrigation water in drip, applied daily @ 0.6 pan evaporation (PE) during 0-2 months, 0.8 PE during 3-5 months, and 0.8 PE thereafter at two days interval as compared to furrow irrigation at 75 mm cumulative pan evaporation (CPE) with earthing up in which sugarcane yield was 122.11 ha^{-1} and irrigation water applied was 231 cm. To reduce the cost of laterals and emitters, changes in plant geometry is essential for economic consideration since the installation cost of drip systems is very high, which is not easily acceptable to farmers. Accordingly, during 1988-92 (Jan.-Feb. planting), further study was made in same soil with paired row and normal planting method and recorded the highest cane yield (137.5 t ha^{-1}) in drip under pit method with 64% saving of water over normal planting with furrow, where irrigation was applied at 75 mm CPE (Table 18.7).

Table 18.7: Effect of crop geometry and Irrigation system on yield of sugarcane (1988-92)

Treatment	Yield (t ha^{-1})	Irrigation water (cm)	Effective rainfall (cm)	Total water (cm)	Increase in yield (%)	Saving of Water (%)
Drip						
Paired row	129.4 (108.1)	79.88 (82.26)	20.03 (18.60)	99.91 (100.86)	12.43 (12.24)	56.92 (58.67)
Pit method	137.5 (112.30)	58.37 (62.10)	20.03 (18.60)	78.04 (80.70)	19.42 (19.65)	68.45 (68.95)
Furrow system						
Alternate furrow	94.4 (82.9)	123.03 (130.1)	30.49 (25.83)	153.51 (155.93)	-18.02 (-13.97)	33.50 (34.95)
Paired row furrow	112.6 (94.3)	137.15 (149.2)	30.49 (25.83)	167.64 (174.95)	-2.17 (-2.04)	25.86 (2544)
Normal furrow	115.1 (96.3)	185.00 (200.00)	30.49 (25.83)	215.49 (225.83)		

(Figures in parentheses indicate yield of ratoon sugarcane and related parameters)

In paired row planting with alternate day irrigation through drip, the yield and water saving was higher by 12.43 and 56.9%, respectively. Application of irrigation water at 75% CPE in alternate furrow and paired row furrow recorded low cane yield due to higher moisture stress. In this case, irrigation was applied in alternate furrow by skipping one furrow. In case of ratoon sugarcane, similar trend was observed but the yield was reduced by 12.2 to 18.3% over main crop under different treatment levels. Shan *et al.* (1977) made field trials with three forms of drip irrigation and compared with surface irrigation. Best results were obtained with one drip tube along every cane row which was significantly better than having one tube in the narrower inter-row of cane, grown at alternate 0.97 m and 2.26 m spacing (as against 1.62 m spacing for an other trials) and both were better than use of tube in alternate inter-row of equal row spacing. In drip irrigation, fertigation is highly effective as plant nutrients are directly placed near root zone and additional manpower is not required for fertilizer application. Considering the importance of fertigation, field experiment was conducted at same location (Rahuri) during 1993-94 and 1994-95 with three nitrogen (250, 187.5, 125 kg N ha^{-1}) and three irrigation levels (irrigation at 1, 2 and 3 day interval). These treatments were compared with furrow method where irrigation was applied at 75 mm CPE and nitrogen @ 250 kg ha^{-1} was applied in four splits up to six month crop age as top dressing. Irrigation applied at 2 day interval recorded maximum yield (157.3 t ha^{-1}) with 43.2% saving of water as compared to furrow method in which the sugarcane yield was 131.41 ha^{-1} and irrigation water applied was 169 cm. Further delaying irrigation by 3 and 4 days through drip reduced cane yield marginally by 2.9% and 5.5%, respectively. Application of nitrogen @ 250 kg ha^{-1} in four splits through drip gave maximum yield of 158.7 t ha^{-1} followed by 147.11 ha^{-1} at 187.5 kg N ha^{-1} and 136.71 ha^{-1} at 125 kg N ha^{-1}. In furrow method, top dressing of applied nitrogen is not as much effective as drip irrigation since available soil moisture in profile is inadequate due to longer irrigation interval. Further disadvantage is that in re-watering; the mineralized nitrogen moves in to deeper layer and becomes in-effective. But in drip method, continuous supply of adequate water increases the availability of applied nitrogen and there is no chance of leaching loss. Considering the importance of fertigation, frequency of fertilizer application increased, to 20, 10 and 4 splits at weekly, fortnightly and monthly interval, respectively in both paired row planting (0.7 m), four row planting (0.9 m) in drip irrigation and was compared with normal planting in furrow method (ridge and furrow of 1.0 m) with 4 splits in 1995-96. The irrigation was applied at 75 mm CPE. Only one dose of fertilizer (250 kg N ha^{-1}) was applied. Application of nitrogen in 20 splits at weekly interval up to six month crop age, produced maximum yield of 172.51 ha^{-1} in four row planting pattern followed by 171.11 ha^{-1} in two row planting pattern. However, in furrow irrigation, sugarcane yield was only 138.0 t ha^{-1}. Planting of sugarcane in two and four rows saved 54.5% and 48.7% water, respectively over furrow method in which the amount of water applied was 198.0 cm. During 1989-91 (two years) at Parbhani (Maharashtra) in clay soil, irrigation @ 40, 60 and 80% pan evaporation was applied at 2, 3 and 4 day interval and maximum sugarcane yield of 97.91 ha^{-1} was recorded at 2 day interval with 80% pan evaporation. Delayed irrigation interval to 3 and 4 day reduced sugarcane yield by 9.6 and 11.5%, respectively over 2 day interval. At Bhavanisagar (Tamil Nadu), performance of July planted sugarcane was evaluated in sandy loam soil with three levels of irrigation and nitrogen during 1995-96. Irrigation through drip at 75% pan evaporation on alternate day with 80% and 60% wetted area produced same yield (167.6 t ha^{-1}) but was higher by 33.6% over furrow method in which irrigation was scheduled at IW/CPE 0.75 with 5 cm depth of water. With regard to saving of water, irrigation at 60% wetted area saved 36.8% water and hence 60% wetted. area at 75% pan evaporation was acceptable, in ill drained clay soil, at Navsari, (Gujarat), response of solid water soluble and liquid fertilizer with different levels (50, 75 and 100% of the recommended dose *i.e.* 300:125:125 kg NPK ha^{-1}) was tested through drip in sugarcane under paired row planting technique (60-120-60 cm) during 1994-95 and compared with furrow method with only one recommended dose of fertilizer. Sugarcane yield in drip method with alternate day irrigation was 7.3 percent higher (106.71 ha^{-1}) than furrow method (99.41 ha^{-1}) wherein irrigation was applied at IW/CPE 0.75 (1W = 8 cm). The saving in irrigation water was also higher by 42.5% (Table 18.8).

Table 18.8: Effect of split application of fertilizer through drip on yield of sugarcane during 1994-95 at Navsari (Gujarat)

Treatment	Yield (t ha^{-1})	Irrigation water (cm)	Increase in yield (%)	Saving of water (%)
Fertilizer through drip at 50% rate				
Weekly	103.6	65.5	4.2	42.5
Fortnightly	108.7	65.5	9.4	42.5
Monthly	110.4	65.5	11.1	42.3
Mean	107.6			
At 75% rate				
Weekly	115.6	65.5	16.3	42.5
Fortnightly	109.9	65.5	10.6	42.5
Monthly	105.9	65.5	6.5	42.5
Mean	110.5			
At 100% rate				

Treatment	Yield (t ha^{-1})	Irrigation water (cm)	Increase in yield (%)	Saving of water (%)
Weekly	105.2	65.6	3.8	42.5
Fortnightly	102.1	655	2.7	42.5
Monthly	99.1	65.6	-7.1	42.5
Mean	102.1			
Furrow method at 100%rate	106.7	114.0		

Cotton

At Rahuri, a comparative study was made with drip, furrow and pitcher method of irrigation with and without mulch (sugarcane trash/wheat straw @ 5 t ha^{-1}) in cotton (*Gossypium hirsutum*) during 1981-84. In drip method, irrigation was given at 0.8 pan evaporation daily, in pitcher 1.0 liter day^{-1} plant^{-1} and in furrow method at 75 mm CPE with 7 cm depth of water Application of irrigation water through drip produced maximum seed cotton yield of 2.911 ha^{-1} in mulch and 2.76 t ha^{-1} in without mulch. Thus drip irrigation produced 38.2 and 21.6% higher yield and saved 51.4% irrigation water as compared to furrow method wherein 88 cm water was applied. In pitcher, though the saving of water was 66%, the reduction in yield was 28.5 and 17.8% under without and with mulch, respectively. Hence pitcher planting is not feasible to grow deep rooted cotton crop. At Parbhani, two planting geometry (90 x 90 cm as normal planting and 60 x 60 x 120 cm as paired row planting) was compared with furrow irrigation with normal planting Irrigation in furrow method was given at TW/CPE 0.9 in alternate furrow and in drip method at 0.4, 0.6, 0.8 and 1.0 ETc on alternate day Normal planting and paired row planting were not effective, however, irrigation through drip at 0.8 ETc recorded maximum seed cotton yield (2.70 t ha^{-1}) which was 16.4% higher than furrow method but there was negligible saving of water.

Banana (*Musa Paradisiaca* L.)

This crop is very succulent, water loving and requires huge quantity of irrigation water under conventional method of irrigation. Hence drip irrigation is highly feasible to economize substantial quantity of irrigation water. At Rahuri (Maharashtra), experiments on irrigation schedule through drip and flood method (check basin) were taken during 1986-92 (three crops) with four wetted area *i.e.* 40, 50, 60 and 70% in drip method and three cumulative pan evaporation value, *i.e.* 50, 75 and 100 mm in check basin. Planting of banana sucker (cv basrai dwarf) was done in crop geometry of 1.5 in x 1.35 in. The irrigation in drip system was given on alternate day with 0.6, 0.8, 0.9 and 1.0 pan evaporation during 0-2, 3-5, 6-8, 9-12 and 13 month onward, respectively. However, in case of check basin method, irrigation was applied at 50, 75 and 100 mm CPE with 6 cm depth up to earthing up and 8 cm thereafter. The result showed that irrigation at 70% wetted area. on alternate day produced maximum fruit yield of 62.36 t ha^{-1} followed by 60.46 t ha^{-1} at 60% wetted area with 42.2% and 49.8% saving of water as compared to check basin method (55.08 t ha^{-1}, when irrigation was scheduled at 75 mm CPE. The irrigation requirement at 75cm CPE was 262 cm. To economize cost of drip system by changing crop geometry without any adverse effect on yield, experiment was conducted in medium deep clay soil during 1993-95 at Parbhani (Maharashtra) with three planting geometry (normal planting 1.5 x 1.5m, square planting 0.9 x 0.9 x 2.1 in, rectangular planting 0.9 x 1.5 x 2.1 in) with three drip irrigation levels (0.6, 0.8 and 1.0 ETc). These treatments were compared with check basin method at IW/CPE 1.0 (1W = 7 cm). Fruit yield in case of normal planting and rectangular planting were comparable (44.6 and 45.5 t ha^{-1}) but square planting had very low yield (37.4 t ha^{-1}) because of higher competition for sunshine, exchange of air etc (Table 18.9).

Table 18.9: Effect of crop geometry and irrigation schedule in drip Irrigation on fruit yield of banana during 1993-95 at Parbhani (Maharashtra)

Treatment	Fruit Yield (t ha^{-1})	Water applied (cm)	Increase in yield (%)	Saving of water (%)	Seasonal water use (cm)
Normal planting					
(1.5x15m)					
0.6 Etc	30.3	153.8	-40.6	52.3	255.6
0.8 Etc	46.4	205.1	-9.0	37.6	303.5
1.0 Etc	57.1	256.3	12.0	22.1	339.3
Mean	44.6				
Square planting					
(0.9 x 0.9) x 2.1 m					
0.6 Etc	30.9	153.8	-39.4	53.2	245.6
0.8 Etc	41.1	205.1	-19.4	37.6	280.0

Treatment	Fruit Yield (t ha^{-1})	Water applied (cm)	Increase in yield (%)	Saving of water (%)	Seasonal water use (cm)
1.0 Etc	40.1	256.3	-21.4	22.1	325.0
Mean	37.4				
Rectangular planting					
(0.9 x 15) x 2.1 m		.		.	
0.6 Etc	46.7	153.8	-8.4	53,2	215.3
0.8 Etc	42.7	205.1	-16.3	37.6	293.4
1.0 Etc	47.0	256.3	-7:8	22.1	335.2
Mean	45.5				
Check basin	.				
(Normal planting)	51.0	329.0			360.6

With regards to beneficial effect of irrigation level on fruit yield, maximum yield of 57.1 t ha^{-1} was recorded at 1.0 ETc under normal planting. The saving of irrigation water at this level was 22.1% over check basin method (321 cm). Hedge and Srinivas (1990) conducted field experiment during 1985-88 to evaluate (the performance of banana cv. Robusta under drip and basin irrigation with different levels of evaporation replenishment (20, 40, 60, 80, 100 and 120% of USAB class A pan). Banana fruit yield increased significantly in both plant and ratoon crops with increasing evaporation replenishment up to 80% in check basin and up to 60% in drip method. At Water Technology Centre for Eastern Region, Bhubaneswar, banana crop was irrigated through drip with six levels *i.e.* IW/CPE 0.5, 0.65, 0.8, 1.0, 1.25 and 1.5 and irrigation at IW/CPE 0.8 were found to the best treatment in which the maximum fruits was 120.2 per plant (Anonymous, 1999).

Vegetable crops

At Rahuri (Maharashtra) in clay soil, drip irrigation @ 80% of pan evaporation was compared with furrow irrigation at 5 and 10 day interval with mulch and without mulch during wet season of 1981-83. in case of drip irrigation, yield of Okra (*Abelmoschus esculentus* (L) Moench) was increased by 11.9% with saving of water by 45.2% as compared to furrow method where irrigation was applied at 5 and 10 days interval The experiment on summer brinjal (*Solanuni melongena* L) was conducted for two years (1983- 85) with four levels of irrigation (60, 75, 90 and 105 mm CPE) in ridge and furrow method; two levels in drip (daily and alternate day) and one level in pitcher irrigation. The result showed that irrigation through drip on alternate day produced maximum yield of 29.72 t ha^{-1} but was statistically at par with daily irrigation. However, in case of ridge and furrow, irrigation at 75 mm CPE was quite economical as compared to irrigation at 90 and 105 mm CPE. Hence irrigation on alternate day through drip is economical, in case of bitter gourd (*Momordicae harantia* L.), four levels of irrigation (20, 40, 60 and 80 mm CPE with 6 cm depth of water) and one level in drip method on alternate day were compared during 1984-86 with paired row planting technique (0.60-1.4 x 0.8 m) in both methods. The result showed that irrigation through drip n alternate day produced maximum yield of 21.471 ha^{-1} with 52.8% saving in water as compared to ridge and furrow method when irrigation was applied at 60 mm CPE. Ridge gourd (*Luffa acutangula Ruxb.*) is also profitable crop of the western part of Maharashtra and irrigation through drip on alternate day produced 20.061 ha' fruit yield with 59°A, saving in irrigation water over furrow method in which irrigation was applied at 75 mm CPE and fruit yield and water requirement were 17.13 1 ha^{-1} and 42 cm, respectively.

Fruit crops

Pomegranate

Pomegranate (*Punica granatum* U.) requires irrigation water from January to July for harvest of good quality fruit in pre-rainy season in Maharashtra and frequent application of water through drip is the only suitable method to achieve good quality fruits. The experiment was conducted at Rahuri during January-August 1980 on July 1975 planted pomegranate cv. Ganesh. It was observed that every day drip irrigation 100% pan evaporation with 30% wetted area and crop geometry of 3 x 3 m with mulch (sugarcane trash @ 5 t ha^{-1}) produced 7.06 t ha^{-1} with 63% saving of water over check basin without mulch in which die amount of irrigation water applied was 108 cm In check basin, irrigation was applied at 6-7 day interval with 4 cm depth of water each time. Application of mulch provided proper hydro thermal regime particularly during summer season where temperature and evaporation rate were maximum and thus sugarcane trash saved 13.3% and 33.3% water in check basin and drip irrigation, respectively over control *i.e.* without mulch.

Mango

In high rainfall region (annual rainfall 3600 ram) particularly during June to September, major portion of rain water goes in to Arabian Sea due to steep slope (1-30%) and crop suffers due to inadequate water during post monsoon period considering these problems, experiment was conducted at Dapoli (Maharashtra) during 1984-85 in mango cv Alfanso (*Magnifera indica* L). The irrigation was applied through drip @ 15, 30, 45, and 601 tree^{-1} week^{-1} and compared with hand watering, applied @ 601 tree^{-1} week^{-1}. The result showed that application of irrigation water @ 601 tree^{-1} week^{-1} through drip increased fruit yield by 152.7% over hand watering in which number of fruits per plant was 36.4 as against 60.5 in the highest irrigation level under drip system.

Acid Lime

At Madurai (Tamil Nadu), performance of drip irrigation with four levels (100, 75, 50 and 25% pan evaporation) with check basin (irrigation at IW/CPE = 0.6 with LW =6 cm) was tested in sandy loam soil during 1991-92 to 1992-93. The acid lime was planted in spacing of 3.19 x 3.19 m. The result showed that application of water 100% pan evaporation produced 20.8% more fruit yield without any saving of water as against check basin method in which the fruit yield was 65.2 kg tree^{-1} year^{-1} and water applied was 9.45 m^3. However, further reduction in irrigation water at 50% pan evaporation, saved 49.7% water with 1.8% more fruit yield.

Oilseed crops

Due to high evaporative demand during hot season, water requirement of summer groundnut is very high. Considering the importance of such precious resources, the experiment was conducted at Navsari (Gujrat) during 1995-96 on summer groundnut (*Arachis hypogaea* L.) with various levels of irrigation in drip and was compared with flood irrigation with flat bed and ridge and furrow method, in flood method, irrigation was applied at 1W /CPE 0.7 and in drip method at 30, 50, 70 and 90% pan evaporation through micro tube. It was observed that application of irrigation through drip at 90% pan evaporation produced maximum pod yield of 2.411 ha^{-1} with 10.2% saving in water as compared to flat bed method in which the pod yield was 1.741 ha^{-1} and irrigation water applied was 60 cm.

Adoption of drip irrigation in cash crops, horticultural crops and vegetable crops is highly economical since the saving of irrigation water and increase in yield is substantial. Changing crop geometry by pairing crop row saves 50% lateral and dripper cost since drip irrigation system is not economical to marginal fanners due to higher installation cost. Application of fertilizer through conventional method is not effective but drip irrigation system is quite effective since fertilizer is applied through drip is directly coming in root contact and used efficiently. Similarly frequent application of fertilizer through drip increases crop yield and also saves substantial quantity of fertilizer. Split application of phosphorus and potassium is quite possible in drip method unlike basal dose in furrow method.

18.5 Some Emerging Concepts and Innovations

To make pressurized irrigation affordable for even small farmers' variety of low-cost drip systems that cut the cost of conventional systems by two thirds have been developed. The cost of materials is reported to be reduced by making a lateral line movable so that it could be used on 4-10 rows of plants rather than one. The cost was further reduced by replacing emitters with micro tubes or baffled holes, which are cheaper. They also make it possible to use simpler lower cost filters. Some low cost innovations are:

Bucket kits consist of a simple household bucket attached to a pole at shoulder height, which supplies a drip line with 26 micro tubes, each of which waters four plants. A starter bucket kit costs Rs. 125/- in India, including the bucket, and irrigates a 20-25 sq. meter kitchen garden to feed a family of 6, using 2-4 buckets of water a day.

The drum kit uses a 200-liter drum instead of a bucket, and uses five lateral lines to irrigate a 125 square meter plot for a cost of Rs. 1250/-. It can be expanded in 125 square meter increments at a cost of Rs. 700/- for each expansion unit.

Shiftable drip systems: The shiftable drip system reduces capital cost by using increased labour. This is done by making lateral lines shift able, so that each line is capable of irrigating ten rows of plants instead of one. Water drips out of baffled holes or curled micro-tubes instead of more expensive emitters.

Larger low-cost drip systems: For 1000 to 10,000 square meters cost Rs. 31250/-per hectare for crops like cotton. One lateral line can irrigate four rows by using micro tubes. This does not include the cost of a pressure pump.

A LEPA type micro sprinkler has been attempted by AICRP on water management at Agronomic Research Station, Chalakudy of Kerala Agricultural University, India to provide irrigation in vegetable, horticultural, ornamental plants and lawns. The rotating sprinkler head is made of a small piece of 12mm/8mm diameter good quality LDPE (Low Density Polyethylene) pipe, plugged at both ends by end caps. The length of the unit is 8 cm for 12 mm and 5 cm for 8 mm pipe. Holes of 1 mm diameter are drilled on opposite sides of the pipe and pivoted at the centre to stand on a stake. The discharge rate is reported to be 30 to 45 lph at an operating pressure ranging from 0.3-1.0 kg/cm^2 for different pipe diameter. The wetted diameter was 210-230 cm when nozzle rotates at a height of 30-90 cm above the ground. This system does not have any sophisticated components, is free from clogging, provides better sub-surface uniformity, avoids any losses due to percolation or runoff, facilitate the use of fertilizers and herbicides and most importantly it has considerably low initial investment and operational cost.

Family drip system: A new, modified drip irrigation system has recently been developed for small-time farmers. Known as the "family drip system" or the "gravitation drip system" it is designed for use where both the water supply and the financial means are very limited. A simple container - such as a barrel or a hand-built container coated with a plastic layer is filled with water and positioned about one meter above the ground. By opening a valve, the water flows by gravitation into the drip system. The container can be refilled using a manual pump, and fertilizer may be added to the water, as in the conventional drip system.

18.6 Quality of Irrigation Water

Needs of a healthy human environment places high demands on the quality irrigation water. Irrigation water must not have direct or indirect undesirable effects on the health of man, animals and plants. The irrigation water must not age the soil and not endanger the quality of surface and ground waters with which it comes into contact. The presence of toxic substance in irrigation water may threaten the vegetation besides degrading the suitability of soil or future cultivation. Generally, surface water, groundwater and suitably treated waste waters are used for irrigation purposes. In examining quality of irrigation water, attention is focused on the physical, chemical and biological properties.

The effect of undissolved substances in irrigation water on the soil fertility depends on the size, quantity and nutrient content as well as the type of soil. Fine-grained soil particles in irrigation water may improve the fertility of light soils, but may adversely affect the permeability and aeration characteristics of heavier soil. The undissolved substances may settle in the irrigation systems and thus result either in the reduction of their capacity or even failure of some installations such as pumping plants. The use of water having unpleasant odour for irrigation may unfavourably affect the farmers.

The quality of irrigation water depends mainly on the type and content of dissolved salts. The problem arises when the total salt content of irrigation water is so high that the salts accumulate in the root zone. The plant then draws water and nutrient from the saline soil with great difficulty which affects the plant growth. The presence of toxic salts is harmful to the plants.

The suitability of irrigation water from biological considerations is usually decided by the type and extent of biological animation.

Special care should be taken when waste water is to be used for irrigation. Waste waters for irrigation can be classified as municipal wastes, industrial wastes and agricultural wastes. Waste waters selected for irrigation must be suitable and their use must be permissible from the sanitary and agricultural considerations as well as from the point of view of smooth operation of the irrigation system. Fields should be located in the immediate vicinity of the waste water resources. The requirement of a minimum distance of about 200 m between the irrigated area and residential buildings must be observed. Also, at a wind velocity of more than 3.5 m/s, sprinkler irrigation should not be used.

Land and water are most natural resources. Efficient use of these two is necessary for maximum benefit. When the availability is not evenly distributed in crop production, the one that is in short supply becomes a constraint for the maximum exploitation of the other. However, water is a very important resource in crop production of lion share (85 per cent) is consumed in agriculture sector. Therefore, use of harvested rainwater should be made efficiently to enhance the productivity.

18.7 Multiple use of harvested rainwater

Sustainable management of natural resources requires optimal management of land and water resources in conjunction with human and livestock resources, However, among all these resources, the water is the critical one as the efficient utilization of other resources is dependent upon temporal and spatial availability of water resource. In the past, the water systems have been evaluated in terms of their ability to provide water for crop production and valued in terms of the 'crop per drop' produced. However, low prices of wheat, rice and other major crops produced on irrigation systems usually do not lead to substantial increase in overall economic and livelihood scenario of the people involved. With increasing emphasis on sustainable development of natural resources, it is logical to integrate appropriate farming practices to enhance water productivity through multiple use of water leading to a shift from 'crop per drop' to a more holistic concept of 'food per drop'. The multiple use of water has now become indispensable to make the cost of water affordable in view of rising cost of water resource creation.

This has also become important in view of increasing cost of water resource development and growing demand of different sectors, *viz.*, industry and municipalities. Thus, the usage of water only for a single purpose can no longer sustain the high cost of creation and management as well as to meet the demand of different sectors.

Unfortunately irrigation systems including large water infrastructures and harvesting systems of water are recognized as common pool resources supplying water for agricultural production, but potential of utilizing water for other uses is often overlooked. The design of existing systems as well as those in the process do not account for any additional use of water. Similarly the participatory management policies mainly emphasize user's involvement in only irrigation and domestic water supply. While farmers organizations are being promoted, their membership and structure which is limited to agricultural land holders do not take into account the multiple uses or users of water. These uses have challenging opportunities for increasing water productivity at various scales by integrating crops. Horticulture, fishery, livestock, small scale industries, domestic needs etc. In rural and pert- urban areas, water is used or could be used for productive purposes such as grossing vegetable and fruits, livestock, poultry, fishery besides domestic needs. However, intensification of multiple use has implications for downstream flow both in terms of quality and quantity. Developing technologies and strategies that accommodate different user groups without hampering the rights of downstream users remain. A major challenge for improving the overall productivity, as well as equity of water use. Although it is being done at research farm and small farmers level.

This chapter deals with available technologies, different constraints in upscaling of technologies and strategies for refining design and implementation to harness the huge socio-economic potential of multiple use system of water.

Assessment of technologies and potential of multiple use of water

The work done on multiple uses of water in India is very limited, Srivastava and Satpathy (2004) and Sikka and Bhatnagar (2004) have summarized the work on multiple use of water in India in almost all domains of water. *i.e.* harvested rainwater, canal water after outlet and in the network itself pumped water from tubewell before it is delivered to the fields, seasonal or perennial waterlogged areas in vast inland tracts and saline coastal belt of eastern India. Srivastava *et al.* (2004) for Orissa and Samra *et al.* (2002) for Haryana conditions have reported that multiple use of water for aquaculture, horticulture etc. in runoff recycling based irrigation system increases the returns very significantly. Srivastava *et al.* (2003) have reported that the tanks constructed primarily for irrigation on community as well as private lands could be used for multiple use. However, due to poaching, aquaculture and papaya failed in community ponds but in private ponds, the farmers were able to generate a sizable additional income by growing fish. vegetables and pigeonpea on embankment.

Srivastava *et al.* (2004) has reported that it is feasible to change from surface irrigation system to hybrid pressurized system to increase the irrigation efficiency of the flow based minor irrigation system in plateau by adding a service reservoir as an adjunct. This service reservoir can be used for multiple use and it was found that the annual cost of the pond was recovered by the returns from papaya on embankment and fish. The B:C ratio of the system was 2.55 but if ducks and intensive vegetable on outward slopes could also be integrated, the total cost of the system inclusive of drip and sprinkler can be recovered from multiple use of reservoir itself.

WTC and TNAU, Coimbatore explored the possibility of multiple use of water by installing the cages in the Lower Bhawani Project (LBP) irrigation canal. Although results were not found encouraging for the cage fish culture (grow out fishes) in running water system from economical point of view but raising upto 5 cm fingerlings at the end of the canal supply period has potential in view of very good market demand. The net return from rearing of fish fly to fingerlings in a canal of 1 km length (100 cages) was worked out to be Rs. 10,000/- (Mayilswami *et al.*, 2004).

Exploring the feasibility at ICAR-Research Complex for Eastern Region, Patna, Bhatnagar and Sikka (2003) found fish yield of 11.0 t ha^{-1} with weekly water exchange by integrating secondary reservoir or small tank in the tubewell based irrigation system. The integration of horticultural crops *viz.*, tomato, banana, lemon and guava planted on the bunds has given good results. Duckery, animal husbandry, piggery and poultry can easily integrate with the system using the secondary reservoir as a central unit. It was found as an effective mechanism for better regulation of water as well as enhancing water productivity and livelihood opportunity.

Sinhababu and Venkateswarlu (1996) developed an adaptable production technology of rice-fish integrated farming system for rainfed lowlands of eastern India. The system involves construction of dikes and refuge/trench, with refuge/trench being used for growing rice and fish and dikes for growing vegetables and fruit crops. The profit over investment was 132%. In Mekong Delta of southern Vietnam, farmers implement such system of fish trenches connected to the adjacent rice fields (Prein 2002).

The multiple use technology is not limited to the research farm alone. Some enterprising farmers like S. Darshan Singh in Punjab and Shri Radhakant Sahoo in Orissa, have taken innovative initiatives for multiple use of water using tubewell water and seasonal waterlogged area, respectively, which has increased the water productivity and their income manifold (Sahoo *et al.,* 2003).

In coastal areas of eastern India, WTCER, Bhubaneswar properly designed and constructed subsurface water harvesting structures (SSWHS) to mitigate the early drought in monsoon season and to provide irrigation during post-monsoon and summer season for chronically waterlogged area having saline groundwater. Integration of aquaculture in the system improved the water productivity which varied from Rs. 15.84 m^{-3} to Rs. 80.43 m^{-3} with an average of Rs. 36.20 m^{-3} while taking total income from all the structures. If only the benefit is considered, the water productivity ranged from Rs. 9.00 m^{-3} to Rs. 53.70 m^{-3} with an average of Rs. 23.26 m^{-3} (Sahoo *et al.*, 2003; Sahoo and Verma 2003; Sahoo *et al.,* 2003).

It is evident from the above that there has been a scope of technologies for multiple use of water for different domains of water *i.e.* rainwater harvesting systems, canal commands, canal itself, rainfed lowlands, tubewell commands. The uses of water were either non-consumptive or low consumptive. Aquaculture was a common use in all the systems as it is a non-consumptive and most productive use of water. The area of bunds required to facilitate aquaculture is utilized for growing horticultural crops irrigated by stored water. If security environment permits, the water body may be used for duck rearing which is again a non-consumptive use. These technologies have been found quite attractive at research farm, but they are not transferred at farmers' field at expected pace.

Constraints in design and implementation

There is immense potential of multiple water use system in all water domains. However, the challenge lies in upscaling these technologies for their adoption on a larger scale taking in to account the socio-economic conditions. Based on past experiences, farmers' perception and available literature, some of the constraints in adoption of these technologies on a large scale are as it (Srivastava, 2007):

1. Social impediments such as religious feelings, poaching especially at remotely located ponds, theft, local conflicts in case of community water bodies or open access water bodies and water rights.
2. Perenniality of water availability in the streams and water bodies.
3. Mortality of fish fry due to change in environment, intrusion of snakes and carnivorous fishes.
4. Loss of natural fish feed (planktons) due to excessive water flow through fish pond
5. Lack of capital investment and resources in developing the system, provision of aeration and mixing specially for resource poor and socially disadvantaged section of rural population.
6. Lack of awareness, technical know-how and Institutional support.
7. Lack of multiple use systems and technologies for different bio-physical, agro-ecological and socio-economic conditions and crop-livestock-fish-enterprise combinations.

Innovations need to overcome constraints

To overcome above-mentioned constraints, few innovations in design of multiple use systems and their implementation are required. The technologies for multiple use systems are more or less standardized under research farm conditions for different water domains. However, their designs require fine tuning and large scale field testing under different agro-ecological, socio-economic and management level conditions before they are recommended for wide-spread adoption.

Based on experience of authors in the field, few of them for each water domain, *i.e.* water harvesting system, canal water, groundwater and waterlogged area due to water congestion have been identified as follows:

Water harvesting systems (WHS)

According to Srivastava (2007), the WHS can be divided in three categories depending upon their location and use as follows:

(i) Near the homestead which can have good management practices,

(ii) Far off from homestead which are under poor management practices and suffer from poaching, and

(iii) WHSs located in highly permeable soils where water does not last beyond monsoon season. Different design parameters and utilization strategies are required for optimal utilization of these water bodies

(i) For WHSs: which are located near homestead and can be managed properly, a combination of aquaculture-horticulture (both fruits and vegetables) on embankment-duck raising rudimentary livestock raising + irrigated agriculture can be adopted. A major constraint in these tanks is the drying up of water by February-March, reducing the growth period of fish and therefore, overall productivity. For enhancing the growth period, following options can be studied for their efficiency also as suggested by Srivastava (2007):

(a) Enhancing the capacity of tank by increasing the depth to have sufficient water during summer season even after accounting for the seepage losses. This can be done only in high rainfall areas and the economics of creating additional capacity for aquaculture has to be worked out. Its implication in overall design of multi tank-based irrigation system will also have to be worked out.

(b) A small adjunct reservoir (lined) can be constructed along with the main tank. The frylings can be stocked in main tank and may be allowed to grow upto December, By December, the size of frylings may be 10-15 cm., *i.e.* fingerlings stage. These fingerlings can be shifted to small tank for high density stocking where they remain as stunted till next July. In July, these fingerlings can be restocked in main tank along with fresh stock of frylings. In December, the grown up fish can be harvested for table purpose and fingerling can be harvested for stocking in small adjunct tank, and thus the cycle continues (Fig. 18.8).

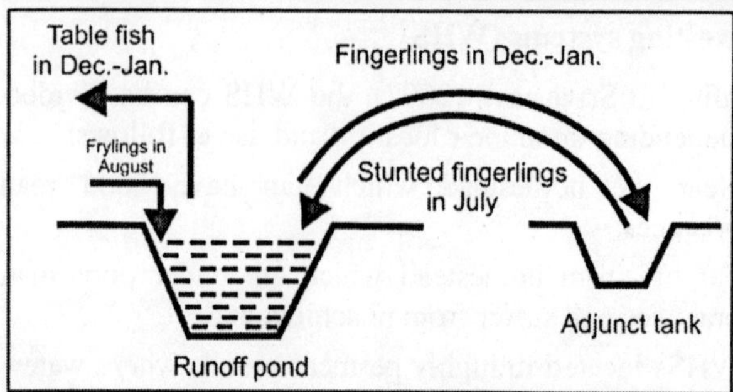

Fig. 18.8: Schematic diagram of adjunct tank system for intensive aquaculture (Source: Srivastava, 2007)

(i) In multi-tank and multi-well based irrigation system (Srivastava *et al.,* 2004), the yield of the open dugwell can be used for maintaining the water level in pond for aquaculture by taking care of seepage and percolation loss. However, this can be possible only when electricity is available for pumping. The overall economics of use of water from well for continued aquaculture and duck rearing may be compared with summer crop cultivation.

(ii) WHSs, which are located away from homestead are prone to poaching, can be put up to multiple use system by exploring the aquatic crops which will spread over the water surface. The product of these crops should not be prone to poaching. Makhana *(Euryle ferox)* and water chestnut (*Trapa bispinosa*) are two such crops which are suitable for western India. Similarly other crops can also be explored for other agro-ecological conditions. Aquaculture with specific spacing of fishes can be integrated with these crops. A suitable system both in terms of design as well as cultivation practices are to be developed.

(iii) In plateau as well as in hilly areas, the land near the homestead have soils of coarse texture and thus have high seepage rates. Therefore, unlined tanks are not feasible. WTCER, Bhubaneswar has developed design of LDPE film lined tanks. A multiple use system comprising of aquaculture + duck rearing ÷ rudimentary livestock + horticulture + domestic water use + vegetable activation on a limited scale can provide better livelihood to farm families. This aspect requires research on proper designing and its evaluation (Fig. 18.9).

Canal water or community water harvesting system

The multiple use of water especially aquaculture in canals has tremendous potential in vast area available.

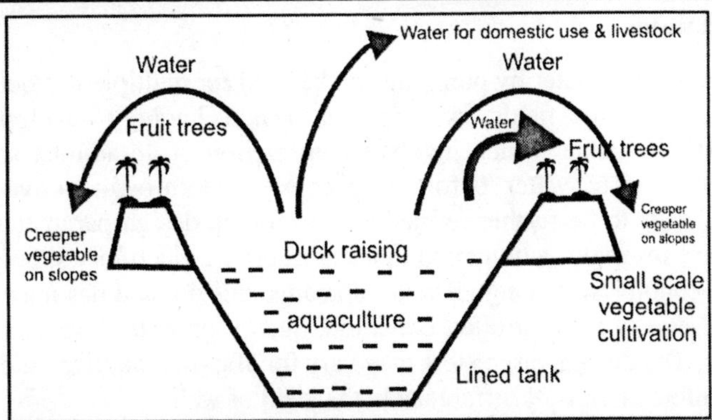

Fig. 18.9: Schematic diagram of multiple use system based on lined tank on top areas of watershed in hills and plateau areas (Source: Srivasatava, 2007)

However, the limitations of water availability, flow rate, and the water velocity, the canal sections will have to be identified where aquaculture can be practiced. Central Institute of Freshwater Aquaculture, Bhubaneswar has developed design of floating cage for controlled aquaculture in large water bodies. There is need of refining this technology for its use in canal system under varying conditions of flow and water availability. Some modification in canal sections may also be required to ensure water availability for longer period. The effect of these interventions on canal hydraulics requires detailed study to avoid any adverse effect on overall canal management. This technology has a very significant social impact as it can serve as a tool of increasing the number of stakeholders in canal system by involving landless people in the enterprise. A large-scale adoption of this system can alleviate the stigma on canal irrigation that it serves only land owners and increases the disparity between land owners and landless.

The canal water can also be put for multiple uses below the outlet especially in plateau areas, which have sufficient slope to ensure pipe flow through gravity. An adjunct service reservoir connected to the outlet and then to a network of pipes for conveyance of water can ensure highly improved irrigation efficiency. Srivastava *et al.* (2006) have shown that the conveyance efficiency below the outlet is 75% with unlined field channels and a pipe network may improve it to 95-100%. The cost of pipe network and service reservoir can easily be recovered through multiple use of reservoir by integrating aquaculture, duck rearing and horticultural crops. Thus multiple use planning can ensure not only additional returns but also improve irrigation efficiency.

Ground water

The ground water by pumping can be used for multiple use before it is conveyed to the fields for irrigation. The technology developed at ICAR-RCER, Patna, which involves construction of an adjunct service reservoir to store water before it goes to field through conveyance system, needs to be further refined by developing design parameters for the service reservoir, automation in terms of erratic electricity supply and overall economics, Among all water domains, this domain has maximum potential due to its controlled condition, quality of water and exchange facilities. The design parameters may vary for alluvium aquifers and hard rock aquifers in view of difference in the yield of well.

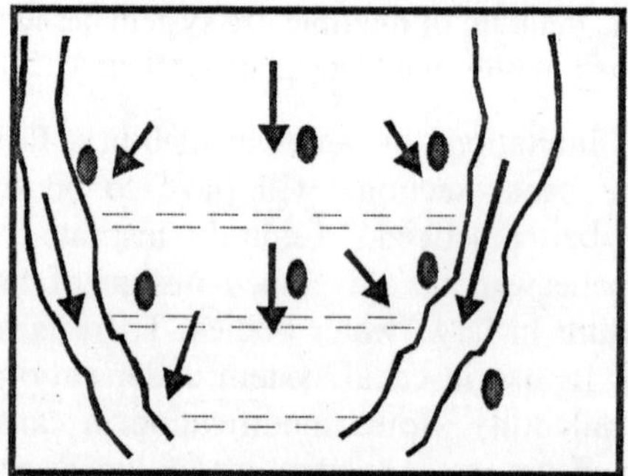

Fig. 18.10: Schematic diagram of reclamation of waterlogged land between two rivers in mouth of delta (Source: Srivastava, 2007)

Waterlogged lowlands

Rainfed lowlands in India are the most poorly exploited land mass. They are monocropped and productivity is low (1 to 1.5 t ha^{-1}). To achieve higher productivity from this underutilized otherwise high potential area, the multiple use should he linked reclamation to make the reclamation economical. According to Srivastava (2007), most of the waterlogged lowlands can be divided in two parts: (i) saucer shaped areas having very high water congestion and (ii) water congested area lying between two rivers. A possible option for reclamation of such lands has been given in Fig. 18.10 and Fig. 18.11. If suitable multiple use systems could be designed which are compatible to such systems, the reclamation will be economical and the productivity can be increased manifold.

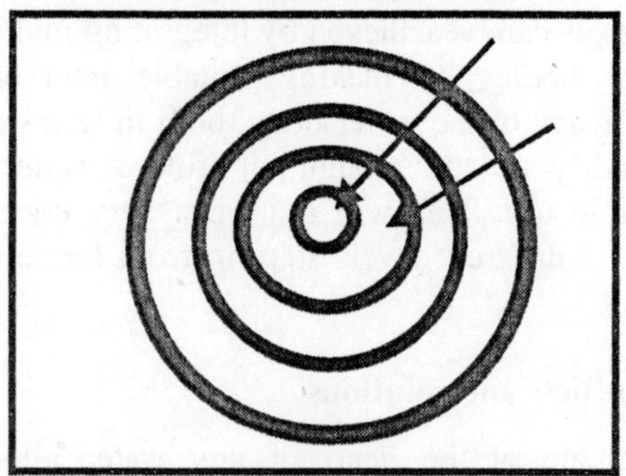

Fig. 18.11: Schematic diagram of reclamation of saucer shaped waterlogged area (Source: Srivastava, 2007)

The implementation of multiple use system in all domains will require integration of both consumptive and non-consumptive uses of water to achieve maximum water-use- efficiency in terms of production and net returns.

The strategy for managing this vital resource would be two fold as suggested by Srivastava (2007):

(i) to get maximum optimal returns from created water to use same water resource,

(ii) to use same water to meet the demand of different sectors in complementary manner.

The first part is the integration of water uses for multiple purposes both consumptive and non-consumptive to achieve maximum water-use-efficiency in terms of both production and net returns. The second part is designing a system where emission outflow from one user serves the demand of another user with or without any minor treatment.

The technologies of integration of water uses for multiple purposes will increase farm productivity production without any net increase in water consumption and enable diversification in higher value crops, including aquatic species, utilization of otherwise wasted on-farm resources, reduction in net environmental impacts of semi intensive farming practices, ensuring diversification of risk, self-employment, flow of income throughout the year and to satisfy the needs of various sectors with limited water availability.

The implementation of the multiple use system will require efforts on two counts, first to integrate them in existing resources and second to modify the design of future projects to make them suitable for multiple use.

The above goals can be achieved by integrating multiple use system in existing water bodies by making suitable interventions without jeoparding the efficacy of the water body, both in terms of quantity and quality, and modifying the design of future water resources to incorporate multiple use. This will require a very concerted effort of capacity building at different levels starting from farmers to the design engineers.

Water rights, conflicts and solutions

Water rights are at the heart of any water allocation system. Reallocation has to take care of not only efficiency but also fundamental issue of equity. The delineation of water rights is further complicated when we take into consideration multiple uses as well as multiple users. The demands of multiple users may be in conflict both in terms of quantity and time. Each case requires a separate understanding of the problem and probably an innovative solution, because just as water is a fluid and dynamic resource, the water rights are also fluid and dynamic, rarely a single consistent system. Therefore, water rights scenario needs to be tackled to take care of conflicting interests both in terms of quantity and quality of downstream water.

References

Anonymous, 1999. Annual Report 1998-99, Water Technology Centre for Eastern Region, Bhubaneswar, Orissa (India) 99-100.

Aujla, M.S., Singh, C.J., Kumar, and Sandhu, B.S. 1992. Efficiency of different methods of irrigation in Indian rape (*Brassica nepus*) *Journal of Arid Environments*. 22: 369-374.

Aujla, M.S., Singh, C.J., Vashist, K.K. and Sandhu, B.S. 1991. Evaluation of methods for irrigation of cotton in a canal irrigated area of South-West Punjab, India Arid Soil Research and Rehabilitation. 5: 225-234.

Bhatnagar, P.R. and Sikka, A.K. 2003. Enhancing water productivity through multiple uses. *Ext. Booklet No. E-26 PAT-14,* ICAR-RCER, Patna.

Bodman, G.B. and Coleman, E.A. 'Moisture and Energy conditions during Downward Entry of Water into Soils', *Proc. of American Society of Soil Science,* 1943.

Booher, L. J. 1967. *Surface irrigation.* Food and Agriculture Organization of the United Nations, Rome. 161 p.

Criddle, W. D., Davis, *S.,* Pair, C. H. and Shockley, *G.* 1956. *Methods of evaluating irrigation systems.* Agriculture Handbook No. 82, *Soil Cons.* Service, U. S. Dept. of Agriculture, Washington, D. C. 24 p.

Dadhwal. V.K. 1999. Remote Sensing Applications for Agriculture-Retrospective and Perspective. Proc. ISRS National Symposium on RS applications for Natural Resources- Retrospective and Perspective. Bangalore, 11-22pp.

Davis, J. R. 1961. Estimating rate of advance for irrigation furrows. Trans. Am. Soc. Agr. Engrs. Vol. 4, No. 1.

Deshmukh, M. T. 1970. *Hydraulics of flow in irrigation furrows.* M.Sc. thesis, Division of Agrl. Engg., Post Graduate School, Indian Agricultural Research Institute, New Delhi. p. 157.

Dhindwal, A.S., Agarwal, M.C., Jaiswal, C.S., Ram Deo, Prabhaker, J. and Aujla, M.S. 1999. Status of Research on Agricultural Water Management in Northern Region. All India Project for Research on Water Management Directorate of Water Management Research, WALMI Complex, Phulwari Sharif, Patna (Bihar).

Dwivedi, R.S. and Sreenivas, K. 1998. Delineation of salt-affected soils and waterlogged areas in the Indo-Gangetic plains using IRS IC LISS Ill data. *Int. J. of Remote Sensing*, 19(14), 2739-2751.

Dwivedi. R.S. & Ravi Sankar, T., Venkataratnam, L., Karale, R.L., Gawande, S.P., Sheshagiri Rao, K.V., Senchanduri, Bhaumic, K.R. & Mukherjee, K.K. 1992. The management of shifting cultivation areas - A Remote Sensing Approach. Asian & Pacific Rem. Sens. J. 4 (2).

Ehlers, M. and Amer, S. 1994. Geoinformatics: An integrated approach to acquisition, Proceeding and Production of Geo-data. In Proceedings EGIS '91, Brussels, pages 306-312, 1991.

FAO. 1993. Sustainable development of drylands and combating deserti. Cation Rome.

Fry, A. W. and Gray, A. S. 1970. Sprinkler irrigation handbook. Rain Bird Sprinkler Mfg. Corp., Glendora, California. 42 p.

Goldberg, D. and Shmueli, M. 1970. Drip irrigation-a method used under arid and desert conditions of water and soil salinity. Trans. Am. Soc. of Agrl. Engrs. Vol. 13, No. 1, pp. 38-41.

Grimes, D.W., V.K. Walhood and W.L. Dickens. 1968. Alternate furrow irrigation for San Joaquin Valley Cotton. California Agriculture. 22: 4-6.

Groot, Richard. 1989. Meeting educational requirements in Geomatics. ITC Journal, 1989 (1): 1-4.

Hall, W. A. 1960. Design of irrigation border checks. Agricultural Engineering, Vol. 41: pp. 439-442.

Hansen, V.E., Israelsen, O.W. and Stringham, G.E., *Irrigation Principles and Practices,* 4th ed., John Wiley & Sons, 1979.

Hedge, D. M. and Srinivas, K. 1990. Growth, productivity and water use of banana under drip and basin irrigation in relation to evaporation replenishment. *Indian J. Agron.* 35(1-2): 106-112.

Howell, T.A. and Hiler, F.A. 1974. Designing of trickle irrigation laterals for uniformity. Jl. of the irrigation and Drainage Division, ASCE, Vol. 100, No. IR 4, Proc. Paper 10983, pp. 443-454.

Howell, T.A. and Hiler, F.A. 1974. *Trickle irrigation lateral design.* Transactions. American Society of Agricultural Engineers, Vol. 17, No. 5, pp. 902-908.

Imperial Chemical Industries of Australia and New Zealand Ltd., 1970. *Polyplot hydraulic design for multi-nozzle irrigation system.* ICIANZ Ltd. Nicholson Street, Melbourne, Australia.

Israelson, O. W. and Hansen, V. F. 1962. *Irrigation principles and practices.* 3rd ed. John Wiley and Sons, Inc., New York. 447 p.

Kataria, D. P. 1970. *Hydraulics of flow in check bczsin irrigation.* M.Sc. thesis, Division of Agrl. Engg., Post Graduate School, Indian Agricultural Research Institute, New Delhi. 135 p.

Lewis, M. R. and Milne, W. F. 1938. *Analysis of border irrigation. Agricultural Engineering*, Vol. 19, pp. 267-272.

Longnecker, D.E., E.L. Thaxton Jr. and P.J. Lyerly. 1969. Variable row spacing of irrigated cotton as a means for reducing production costs and conserving water. *Agronomy Journal.* 61: 101-104.

Mayilswami, C., Palanisami, K. Rangasami, M.V. and Kumar, V. 2004. *In: Resource Conserving Technolgies for Social Upliftment* (Eds. V.N. Sharda, K.S. Dadhwal, S.K. Dhyani, G.P. Juyal, P. Dogra, A. Raizada and O.P.S. Khola), IASWC, Dehradun, pp. 548-554.

Michael, A. M. 1968. *Evaluation of hydraulic characteristics influencing border irrigation design.* Ph.D. thesis, Dept. of Agril. Engg, Indian Institute of Technology, Kharagpur. 420 p.

Michael, A. M. and Ojha, T. P. 1966. *Principles of agricultural engineering,* Vol. II, Jam Brothers, Jodhpur. 534 p.

Michael, A. M. and Pandya, A. C. 1971. Hydraulic resistance relationship in irrigation borders. Jl. of Agricultural Engineering Research, 16(1), pp. 72-80.

Michael, A. M. and Pandya, A. C. 1971. Water front advance in irrigation borders. Jl. of Agricultural Engineering Research, 16(1), pp. 62-71.

Michael, A. M., Reeser, R. M. and Knierim, G. C. 1965. How to *improve irrigation farming in India.* University of Udaipur, College of Agriculture, Jobner, Extn. Bul. No. 2. 60 p.

Michael, A. M., Shri Mohan and Swaminathan, K. R. 1972. *Design and evaluation of irrigation methods.* I.A.R.I. Monograph No. 1, Water Technology Centre, Indian Agricultural Research Institute, New Delhi. 208 p.

Molenaar, A. 1960. *Irrigation by sprinkler.* Food and Agriculture Organization of the United Nations, Rome. 91 p.

Moncada, A. 1970. Evaluation of sprinkler performance from can catches set in different patterns. M.Sc. thesis, Utah State University, Logan. 215 p.

Nagarajan, S. 1983. Plant disease epidomology, Oxford and IBH Publishing Co., New Delhi.

Navalawala, B. N. 1991. Water logging and its related issues in India. *J. Irrigation and Power,* Jan. 1991, 55-64.

Navalawala, B.N. 1999. Improving management of irrigation resources. Yojana, Jan. 1999, 81.

Oldeman, L. R. Hakkeling, R. T. A., and Sombroek, W. G. 1990. *Global assessment of soil degradation (GLASOD)* (Wageningen. Nairobi: ISRIC/UNEP).

Philip, J. R. and Farrel, D. A. 1964. *General solution of the infiltration-advance problem in irrigation hydraulics.* Jl. of Geophysical Research, Vol. 69, No. 4, pp. 621-631.

Pillsbury, A.F. 1968. *Sprinkler irrigation.* Food and Agriculture Organisation of the United Nations, Rome. 179 p.

Prein, M. 2002. Integration of aquaculture into crop-animal systems in Asia. *Agricultural Systems.* 71: 127-146.

Premier Equipment Pvt. Ltd. 1971. *Operating and instruction manual for Premier Irrigation systems.* Pub. No. T 26, Issue No. 1. 35 p.

Proceedings of the national seminar on "Water management for sustainable agriculture problems and perspectives for the 21st centaury", IARI, New Delhi, April, 15-17.

Punmia, B.C. and Pandey, B.B.L. *Irrigation and Power Engineering,* Standard Publishers & Distributors, New Delhi, 1981.

Saha, S. K. 2000. Crop yield modeling using Remote Sensing and GIS - future prospect Proc. international Symposium on Geoinformatics beyond 2000, IIRS, Dehradun.

Sahoo, N. and Verma, H.N. 2003. Development and management of small water resource for irrigation and fish production in coastal belt. *In Proc of Tenth National Water Convention* pp . 411-421.

Sahoo, N., Verma, H.N., Jena, S.K. and Tripathy, V.K. 2003. Downsizing water resource development technology for cyclone affected coastal poor farmers. *In Proc. of International Seminar on Downsizing Technology for Rural Development (ISDTRD, 2003).* Regional Research Laboratory, Bhubaneswar, India. pp. 811-816.

Sahoo, N., Verma, H.N., Jena, S.K., Sethi, R.R. and Tripathy, V.K. 2003. Micro water resources development and management in coastal waterlogged area. A case study *In Proc. of International Conference on Water and Environment 2003.* pp 205 – 217. (ISBN 81-7764-548-X).

Samra, J.S., Agnihotri, Y., Mittal, S.P., Bansal, R.C., Aggarwal, R.K. and Yadav, R.P. 2002. *Environment Impact Assessment of Participatory Watershed Management in Shivaliks-Relmajra.* Bulletin No. 1-46/C-10. CSWCRTI, Chandigarh.

Samuel, J. 2000. Micro irrigation for fruit crops in India. In Proceeding of International conference on Micro and Sprinkler irrigation systems held at Jalgaon (Maharashtra) during February 8-10, 2000.

Schwab, G.O., Frevert, R. K., Edminster, T. W. and Barnes, K. K, 1966. *Soil and water conservation engineering.* 2nd ed. John Wiley and Sons, Inc., New York. 683 p.

Shan, L.L.P., Hardy, M. and Chan P. (1977). Drip irrigation of sugarcane. Proceedings 16th Congress, International Society of Sugarcane Technologists, 1977, 1219-1225.

Sikka, A.K. and Bhatnagar, P.R. 2004. Enhancing water productivity through multiple uses at different scales. *In: Resource Conserving Technologies for Social Upliftment.* (Eds. V.N. Sharda, K.S. Dadhwal, S.K. Dhyani, G.P. Juyal, P. Dogra, A. Raizada and O.P.S. Kolay), IASWC, Dehradun, pp. 411-420.

Sinhababu, D.P. and Venkateswarlu, B. 1996. *Rice-fish in Rainfed Lowlands.* Central Rice Research Institute, Cuttack, India.

Sivanappan, R.K., Padmakumari, O and Kumar, V. 1987 Drip Irrigation. Keerthi Publishing House Pvt. Ltd. Coimbatore, Tamil Nadu (India), 1-42.

Srivastava, R.C. 2007. Multiple use of water – A review of some suggestions for implementation. *Indian J. Soil Cons.* 35(3): 198-203.

Srivastava, R.C. and Satpathy, K.K. 2004. Getting more from the water and watersheds. Multiple uses of resources. *In: Watershed Management Challenges, Improving Productivity Resources and Livelihoods.* (Eds.

B.R. Sharma, J.S. Samra, C. A. Scott and S.P. Wani). IWMI. Srilanka, pp. 82-94.

Srivastava, R.C., Kannan, K. Mohanty, S., Sahoo, N., Mohanty, R.K., Nanda, P., Das, M. and Verma, H.N. 2003. Micro-level Water Resource Development through Rainwater Management for Drought Mitigation in Sub-humid Plateau Areas of Eastern India. Publication No. 15. WTCER, Bhubaneswar, India.

Srivastava, R.C., Mohanty, S., Singandhupe, R.B., Biswal, A.K., Ray, L.I.P. and Sahoo, D. 2006. Studies on canal water based pressurized irrigation system in a minor irrigation command. *J. Agril. Engg.* 43(4): 28-35.

Srivastava, R.C., Mohanty, S., Singandhupe, R.B., Mohanty, R.K., Bahera, M.S., Ray, L.I.P. and Sahoo, D. 2004. Feasibility of canal water based pressurized irrigation system in high rainfall areas of eastern India. *In: Proc. Emerging Technologies in Agricultural Engineering*. Anamaya Publishers, New Delhi. pp. 261-266.

Srivastava, R.C., Singandhupe, R.B. and Mohanty, R.K. 2004. Integrated farming approach for runoff recycling systems in humid plateau areas of eastern India. *Agricultural Water Management* 64(3): 197-212.

Stoner, E.R. and Baumgardner. M.F. 1981. Characteristics variation in reflectance of surface soils. *Soil Sci. Soc. Am. J.* Vol. 45: 1161-1165.

Suresh Kumar, S.B. Dhaimodkar and Pande, L.M. 2002. The assessment of potential land use change in irrigation command using remote sensing and GIS. *Journal of Indian Society of Remote Sensing*, Vol. 30, No.3: 157-166.

United States Department of Agriculture, 1960. *SCS National Engineering Handbook, Section 15 Irrigation,* Chapter II, *Sprinkler irrigation.* USDA, SCS, Washington, D.C. 83 p.

Walker, W.R. and Skogerboe, G.V. *Surface Irrigation. Theory and Practice,* Prentice Hall Inc., USA, 1987.

Williams, G. S. and Hazen, A. 1960. *Hydraulic tables,* 3rd ed., John Wiley and Sons, Inc., New York.

Wu, I Pai and Gitlin, H. M. 1973. *Hydraulics and uniformity for drip irrigation.* j. of the Irrigation and Drainage Division, ASCE, Vol. 99, No. IR 3, Proc. Paper 9786, pp. 157-168.

Wu, I Pai and Gitlin, H. M. 1974. *Drip irrigation design, based oil uniformity.* Transactions, American Society of Agricultural Engineers, Vol. 17, No. 3, pp. 157-168.

Wu, I Pai and Gitlin, H. M. 1975. *Drip irrigation design on non-uniform slopes.* Paper presented at the 1975 Winter Meeting of American Society of Agril. Engineers, Chicago, Illinois.

Wu, I Pai and Gitlin, H. M. 1975. *Energy gradient line* for *drip irrigation laterals. Jl. of the Irrigation and Drainage Division*, ASCE, Vol. 101, No. IR 4, Technical Notes, Proc. Paper 11750. pp. 323-326.

Wu, I Pai. 1975. Design of drip irrigation mainlines. Jl. of the Irrigation and Drainage Division, ASCE, Vol. 101, No. IR 4, Proc. Paper 11803, pp. 265-278.

EXERCISES

18.1. Discuss the effect of irrigation techniques on water use efficiency.

18.2. Explain pressurized irrigation system and its role in dryland eco-system.

18.3. What is uniformity coefficient?

18.4. Define the following:
- (i) Effective maximum depth
- (ii) Mean application depth
- (iii) Coefficient of variation

18.5. Describe micro irrigation system.

18.6. Inlist the factors affecting water productivity.

18.7. Describe different types of surface irrigation method.

18.8. What condition sub-surface irrigation method is used?

18.9. Discuss multiple uses of harvested rainwater.

18.10. Explain water harvesting system.

18.11. What is the strategy for managing vital resources?

18.12. Describe important physical and chemical properties of soil which are important from considerations of irrigation.

18.13. What is the meaning of consumptive use? On what factors does it depend? How would one calculate the consumptive use for a given crop?

18.14. The field capacity and the permanent wilting point for a given soil are 35 and 15 per cent, respectively. Determine the storage capacity of soil within the root zone of the soil which may be taken as 80 cm. At a given time the soil moisture in the field is 20 per cent and a farmer applies 25.0 cm of water. What part of this water would be wasted? Assume porosity of soil as 40 per cent and relative density as 2.65.

18.15. Determine the frequency of irrigation at the following stage of growth of crop.

Field capacity of soil 30%

Permanent wilting point 10%

Effective root zone depth 0.75 m

Consumptive use 12 mm/day

Apparent specific gravity of soil 1.6

(including the effect of porosity)

18.16. The consumptive use for a given crop is 90 mm. Determine the field irrigation requirement if the effective rainfall and the irrigation efficiency in the area are 15 mm and 60 per cent, respectively.

18.17. The following data wee obtained in determining the soil moisture content at successive depths in the root zone prior to applying irrigation water:

Depth of sampling	Weight of moist soil sample	Weight of oven dried soil sample
0-25 cm	1.35 N	1.27 N
25-50 cm	1.36 N	1.26 N
50-75 cm	1.23 N	1.15 N
75-100 cm	1.11 N	1.02 N

The bulk specific gravity of the soil in the root zone is 1.5. The available moisture holding capacity of the soil is 17.8 cm/m depth of soil.

Determine (i) the moisture content at different depths in the root zone, (ii) the moisture content in the entire root zone prior to irrigation, and (iii) depth of water to be applied to bring the moisture content to the field capacity.

18.18. Describe various methods of irrigation mentioning their advantages disadvantages and applicability to different field conditions.

■■■■

Chapter 19

RECYCLING OF INDUSTRIAL EFFLUENTS

19.1 Introduction

Water recycling is an age old concept. By knowingly or unknowingly people have been reusing the harvested water in their house garden and crop fields during the scarcity periods. Now-a- days, it is compulsion to efficient reuse of the wastewater because of whole hydrological cycle is under drastic change due to unwise exploitation of water resources and uneven distribution of the same and also due to rapid industrialization. River, pond and other water bodies have under gone degradation of water quality that is also one of the major compelling demands. It is the need of the day to reuse the treated wastewater. In many parts of the country the wastewater is already an important element. The treated wastewater is employed in several ways such as, agricultural irrigation, landscape irrigation, industrial recycling, groundwater recharge, recreational use, non-potable use and potable use. Agricultural irrigation and landscape irrigation cause surface and groundwater problem. In industries, recycling and reuse cause scaling, corrosion, biological growth and fouling. In groundwater, recharge cause organic chemicals deposition and thereby toxic effect. Rest recreational, non-potable and potable use are having infection problem. The major problem of the treated wastewater is a problem of pathogens, bacterial growth, fouling, aesthetic view and public acceptance. Instead of all these demerits, it has a lot of merits also and presently, having pressing demand for reuse application. We need that treatment system which can correct all these demerits and make the treated water for the proper reuse. As the population increases, its need of water also increases. Only surface and groundwater are not sufficient to fulfill the need, hence, treated wastewater may be one of the sources of water for the various purposes. (Ramakrishna, 2002) During past 15 years, a wide variety of technologies have been developed for the treatment of wastewater for reuse. In this chapter we discuss the reuse of wastewater and conservation of water in industries.

19.2 Historic Perspective

Early development in the field of water reuse are with the historic practice of land application for the disposal of wastewater with advent of sewerage systems in the nineteenth century, domestic wastewater was used at sewage farms and by 1900 there were numerous sewage farms in Europe and in the United State. While these sewage firms were used primarily for waste disposal, incidental use *was* made of the water for crop production or other beneficial uses (Veatch, 1938). Ranjan and Sabuman (2004) summarized the reuse of water at different locations in the world (Table 19.1).

Table 19.1: Wastewater reuse application at different locations

Year	Location	Water reuse application
1912-1985	Golden Gate Park, San Francisco, CA	Watering lawns and supplying ornamental lakes.
1926	Grand Canyon National park, AZ	Toilet flushing, lawn sprinkling, cooling water and boiler feed water
1929	City of Pomona, CA	Irrigation of lawns and gardens.
1942	City of Baltimore, MD	Metals cooling and steel processing at Bethlehem steel company.
1960	City of Colorado springs, CO	Landscape irrigation for golf courses, parks, cemeteries, and freeways
1961	Irvine ranch water district, CA	Irrigation and domestic water including toilet flushing in high-rise buildings.
1962	Country sanitation District of Los Angeles country, CA	Groundwater recharge using spreading basins at the Montebello fore bay
1976	Orange country water District CA	Groundwater recharge by direct injection in to the aquifers at the water factory-2 I
1977	City of St. Petersburg FL	Irrigation of parks, golf course schoolyards, residential lawns and cooling tower makeup water.
1985	City of EL Paso, TX	Groundwater recharge by direct injection into the Hueco bolson aquifers, and power plant cooling water
1987	Monterey Regional water pollution control agency, Monterey, CA	Agricultural irrigation of food crops eaten uncooked including artichoke, celery, broccoli, lettuce and cauliflower.

Outside of United States Mexico, Tunisia, City of Windhock (Namibia), Australia. Israel, Japan, and Spain are also started reuses of treated wastewater.

19.3 Need for reuse of Industrial Effluents

During the past few decades, Indian industries have registered a quantum jump which has contributed to high economic growth but simultaneously it has also given rise to severe environmental pollution. Consequently, ambient air and water quality are seriously affected which are far lower in comparison to the international standards. The problem is worse in the case of water pollution. It is found that one-third of the total water pollution comes in the form of effluent discharge, solid waste and other hazardous wastes. The rapid industrialization in developing countries had led to tremendous pressure on the available finite water resources. This phenomenon is quit evident in the areas where water intensive industrial units like textile dyeing, leather tanning, paper and pulp processing, etc. thrive rapidly. Among the various kinds of pollution caused by industrialization, the problem of water pollution due to industrial effluents has attained greater dimensions in India. Indiscriminately discharged industrial effluents naturally affect the water quality as well as natural ecosystem.

The availability of water is unevenly spread across the country and there is an urgent need to utilize the alternate sources of water for agriculture in order to meet the demand. Increasing industrialization and urbanization in India has brought in its wake the major problem of safe disposal of wastewater. Good quality of water is inadequate for normal living benefits and is getting polluted due to industrial discharge including those of paper, textile, tannery, distilleries and other industries. Hence, efforts should be made in future for proper treatment and safe disposal of industrial effluents and use in agricultural land in order to increase our food grain production and enhancing environmental quality (Yadav *et al.*, 2006).

India receives on average annual rainfall equivalent of about 4000 cubic km^3. This is unevenly distributed across different parts of the country and most of the rainfall is confined to the monsoon season, from June to September while, small proportion of rains is received due to south-east monsoon and cyclonic rains in different parts of the country. Thus, while India is considered to be rich in terms of annual rainfall and total water resources, water is spatially and temporally very unevenly distributed. Based on per capita water availability, some river basin fall in the category of water scarce and water stressed regions, and several others suffer from absolute scarcity. Though water resource availability is estimated to be 1085.9 billion m^3, annual average utilizable per capita water resources vary considerably from as high as 3020 in the Narmada basin to as low as about 180 m^3 and less in the Sabarmati basin, as against a desired availability of 1700 m^3 per capita per year. The estimated per capita water availability has also declined from 6008 m^3 a year in 1947 to 2266 m^3 in 1997, as per TERI's Green India 2047 study (TERI, 1998). This declining figure gives a broad indication of the growing water scarcity in the country in the last 50 years since independence (Singhal, 2003). These problems can be overcome by means of proper utilization of wastewater because the wastewater acts as source of plant nutrients, irrigation and thus help in meeting the food grain demand. Though agricultural sector is still the single largest user of groundwater, there have been growing demands for this resource from other sectors, the process which got accelerated due to rapid urbanization, fast industrial expansion and increasing demographic pressure.

There are following objectives for utilization of agriculture industrial effluents for rapid urbanization and increasing population is putting pressure on scare water resources.

- Water pollution abatement due to no discharge or reduced discharge of pollutants in receiving water such as in the touristic coastal waters.
- Increasingly stringent water pollution control requirements and as a result, availability of high quality effluents for various beneficial uses (Dahl. Stephen, 2003), such as irrigation to crops for possible nutrient values.
- Public policy and ethics encouraging water conservation and reuse.
- One of the water demand and drought management strategies in overall water resources management.
- Wastewater also acts as a resource when it is properly treated and used for the various purposes.
- To prevent adverse effect on environment through the wastewater.
- Safe disposal of industrial effluents.

Reuse application of waste water is shown in flow diagram (Fig. 19.1).

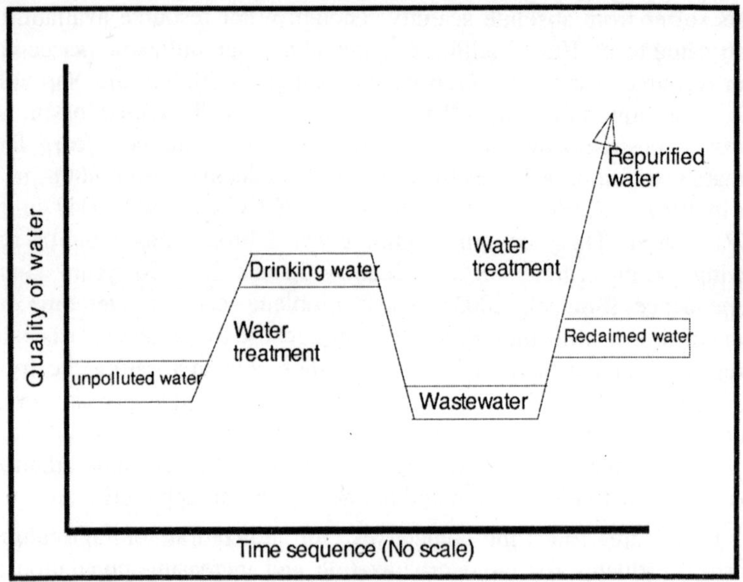

Fig. 19.1: Flow diagram of reuse of application of waste water (Source: Asano Takshi, 2000)

The quality changes during municipal use of water in a time sequence. Through the process of water treatment, drinking water is produced, which has an improved water quality meeting applicable standard for drinking water.

19.4 Environmental Impacts of Industrial Uses of Water

All kinds of pollution including industrial pollution cause severe impact on environment. There is now a global awakening on pollution control though recovery of reusable matter from waste effluents.

Pollution control

Control of pollution at source to the extent possible with due regard to techno-economic feasibility would necessitate recycling of wastes at every stage of production. In case of metal processing industries, the waste effluent contains recoverable metals, which it not extracted are lost in aqueous solution, resulting in pollution. Taking nickel, zinc etc. as examples, there are various techniques for metal extraction from waste effluent *e.g.* precipitation, electrolysis and ion exchange. In case of manufacture of organic chemicals, recovery of organic chemicals is possible and is being practised even in this country.

In a large number of cases, the wastewater disposed off in streams is composed of varying degree of pollution. The cleaner streams can be easily treated and used for purposes or which water quality does not have to be brought to a high level of purity. Thus, the cost of water, which can be recycled or re-used, can be kept lower than the marginal cost of one unit of water. Water needed for flushing toilets, watering lawns, air conditioning and cooling can often be drawn from treated effluents.

It should also be recognized that the phenomenon of dilution and dispersion in the ambient water body is an integral part of the water quality management and the management programme should take advantage of its availability provided that certain basic minimal pollution control devices are installed at the polluting sources. The industry specific effluent standards which are being evolved at the National level are to be recognized as "Minimal National Standards" (MINAS) and envisage treatment of all wastes to certain minimum standards regardless of the type of wastewater and locations. No State Board is required to relax the MINAS but if the quality criteria of the ambient water reaches at some warrant stricter effluent-quality, the State Boards shall prescribe that and thus would make the MINAS altered to suit the location. This model is effective in halting the obvious pollution immediately and envisages a steady progress in meeting the water quality objectives. It also provides a fair degree of flexibility to the Regulatory Authority for control of water pollution.

The water cess rates are too low to discourage excessive water use in industries, because of disparity in production costs of the industries with effluent treatment plants and those without it in view of the capital and O&M costs. Parity in the competitive market can be achieved only by resorting to heavy penalty to the defaulters.

Provisions for environmental protection

Various steps have been taken by the Government for prevention and/or minimization of industrial pollution. A number of Acts have been enacted such as the Water (Prevention and Control of Pollution) Act-1974: Water (Prevention and Control of Pollution) Cess Act -1977: The Environment (Protection) Act-1986 etc.

The Water (Prevention and Control of Pollution) Act, 1974, is the first enactment, by the Parliament in this direction. This is also the first specific and comprehensive legislation institutionalizing simultaneously the regulatory agencies for controlling water Pollution. The Central Pollution Control Board at the Centre and the Pollution Control Boards in States came into being in terms of this Act. Another related legislation enacted was the Water (Prevention and Control of Pollution) Cess Act, 1977 in order to conserve this vital natural resource and to augment the finance of these regulatory agencies.

Environmental Clearance for 29 categories of projects was made mandatory by the Environmental Impact Assessment (EIA) Notification issued in January 1994 under the Environmental (Protection) Act, 1986. Environmental clearance is granted by the Impact Assessment Agency in the Ministry of Environment & Forests.

The EIA Notification provides for two stage clearance for site specific projects like mining including prospecting, pit-head thermal power plants, river valley projects and ports and harbours: Site clearance is given in the first stage and the final environmental clearance in the second stage. An amendment to the EIA Notification introduced in April 1997 has made public hearing mandatory for all the cases where the environmental clearance is required. The views of the affected public, NGO's and all the concerned sector are taken into account during the public hearings. The environmental clearance is not accorded to the project without a positive response during the public hearings. The project is considered and recommended or rejected by the Expert Advisory Committee constituted under the Ministry of Environment & Forests.

It is thus seen that an elaborate procedure has been formulated to ensure that minimal water pollution is caused by various industries. However, inspite of adequate regulatory mechanisms, substantial pollution is caused in various water bodies by the discharge of effluents from the industries. This is partly due to inadequate effort in implementation and partly due to the shortage of controlling/implementing mechanism. In order to minimize pollution of water bodies, a strict implementation of various legislation and procedures is necessary. Simultaneously, sensitization awareness should also be created among the public and concerned agencies and the industrial sector about the maintenance of proper quality of surface and ground water resources. Another effective method of reducing such pollution is the recycling and reuse of water which would reduce the amount of effluents into the water bodies. However, this is stage in our country and a lot of research is required to be carried out to find effective and economical ways of recycle and reuse.

19.5 Characteristics of Different Industrial Effluents

Mainly paper and pulp industries, distilleries, sugar, tannery, textile, food processing, meat processing, coir, printing, chemical and fertilizer industry effluents are the major potential sources of effluent discharge. Vadu *et al.,* (2006) summarized the characteristics of various industrial effluents (Table 19.2) discharged from industries of varying nature and households. Industrial effluents are an important source of organic matter and plant nutrients and have potential value as a complementary source for commercial fertilizers and water source for irrigation is given Table 19.3 (Solaimalai *et al.,* 2002).

Table 19.2: Characteristics of different industrial effluents

Parameters	Units	Distiflery[1]	Textile[#]	Tannery[##]	Pulp and paper+	Sago[+]	Sewage[**]
pH		8.9	9.8-11.8	7.0-9.0	8.4	5.2-6.0	8.00
EC	dSm^{-1}	27.5	8.0-10.0	-	1.6	2.6-3.4	1.15
HOD	$mg\ l^{-1}$	40000	680-840	1000-3000	375	-	2108
COD	$mg\ l^{-1}$	102000	1.60-1790	2500-5000	1439	-	1720
TS	$mg\ l^{-1}$	92505	512-7180	.	1818	2850-3700	3900
TDS	$mg\ l^{-1}$	5975	-	11000-15000	1044	2600-3700	-
TSS	$mg\ l^{-1}$	86530	155-195O	2000-4000	774	200-400	-
Nitrogen	$mg\ l^{-1}$	1988	45-50	-	20 (ammonical)	15-20	196
Phosphorus	$mg\ l^{-1}$	428 (Phosphate)	20-25 (Phosphate)	-	3	2-2.8	2.24
Potassium	$mg\ l^{-1}$	10010	1.2	-	12	80-120	-
Chloride	$mg\ l^{-1}$	10400	100-570	6000-9500	776	8-9.6 (CI)	8.60 (CI)
Sulphate	$mg\ l^{-1}$	2980	660-1600	2400-4000	384	1.2-2.1	-
Sodium	$mg\ l^{-1}$	640	78.7	-	575	1-1.5	-
Chromium	$mg\ l^{-1}$	-	-	7.2-26.2	-	-	-
Oil and grease	$mg\ l^{-1}$	-	-	1.5-4.0	-	-	-

TSS-Total suspended solids, TDS-Total dissolved soilids, TS-Total solids,
[1]Devarajan *et al.,* (1993a), [#]Ramachandran (1994), [##]Mahimairaja *et al.,* (1998),
+Pushapavalli (1990), [+]Senthilvelan (1994), **Ramanathan *et al.,* (1977)

Paper factory effluent has high load of COD and BOD and significant concentration of soluble organic matter. Tannery effluent contained large amounts of N, P, Ca, Na and Cr etc. The dyeing factory effluent is alkaline and contains high amounts of total suspended and dissolved solids and is rich in dissolved oxygen. The spent wash from distilleries and breweries is highly acidic and contains solid that are mostly organic, high in K (7-15% of the total solids) and found to contain wood reaction products, spend cooking and bleaching chemicals, chlorine compounds, fibers, solid inorganic compounds and sodium compound. Phosphatic fertilizer factory effluents registered low pH and contained high amounts of dissolved solids, fluorides and sulphates. Dairy effluents are slightly alkaline or less saline in nature and contain higher amounts of organic matter and plant nutrients (NH_4-N, Cl_2, PO_4, NO_3-N and SO_4). The rayon factory effluents have high salt and heavy metals like Cu and Zn (Solaimalai *et al.,* 2002).

Table 19.3: Irrigation and nutrient potential of some major industrial wastewater in India

Source	Volume (10^6 m^3 $year^{-1}$)	Nutrient contributed (10^3 t $year^{-1}$)		
		N	P_2O_5	K_2O
Sugar	170	1.43	218	-
Distillery	6.87	69.38	11.33	274.8
Dairy	165	18.15	13.36	14.35
Pulp & paper	341.6	-	0.88	7.99
Fertilizers	57.7	7.34	4.42	-

19.6 Effluent Utilization

Large and regularly produced industrial effluents are recycled or reused for different purposes after proper treatment. Its common uses are in agriculture for irrigation as well as crop fertilization, in aquaculture for fish production, application for forest land, in land reclamation and other related uses. These effluents supply major and minor nutrients required for plant growth as well as supply water requirements for crop growth in subsequent sections as summarized by Yadav *et al.* (2006).

(i) Distillery effluents

Sugar industry is mushrooming as a result of which large amounts of molasses is produced. This by-product is used as a feedstock in alcohol industries. During the process of distillation, huge quantities of distillery wastes known as spent wash are produced and there is disposal problem of this waste. In distillery industry, for every litre of alcohol produced, about 15 litres of spent wash are released as wastewater.

Uses : The manurial value of this effluent can profitably be used as supplement to the fertilizer when used along with irrigation water after proper dilution. This will solve the twin problem of disposal and also substitute some quantity of fertilizer (Zalawadia and Raman, 1994). The effluent when diluted suitably (1:10 to 1:50) reduced its BOD and EC amenable for irrigation to crops. The nutrient of soil and yield of crops were increased and there was no adverse effect on soil physical, chemical and biological properties (Augustine Selvaseelan *et al.,* 2001).

Soil sodicity is characterized by high pH, water soluble and exchangeable sodium. The basic principle in the reclamation of sodic soils is to replace the Na ion from soil exchangeable sites by cations like H^+, Ca^{++} and Mg^{++} ions. Since the distillery spentwash is highly acidic and contains fairly good amount of Ca and Mg and other essential plant nutrients, this could be used as an organic amendment to improve the soil properties, particularly in reclamation of non -saline sodic soils. The potential of distillery spent wash to supply H^+ (since distillery effluent is highly acidic), Ca^{++} and Mg^{++} has been used as a basic principle to reclaim the sodic soil (i. e. to replace Na^+ ions in the soil exchangeable complex). Rajukannu *et al.* (1996) found that application of distillery spent wash @ 5 lakh lit ha^{-1} to the air-dried non-saline sodic soil followed by two or three leachings with water, could effectively reclaim sodic soil and suggested a fair period of 60 days from the days of its application and transplanting of rice seedlings for the establishment of crops.

Devarajan *et al.* (1993) found that sugarcane Co-8021 gave a normal yield with 50 times diluted spent wash and also found that the sugarcane yield was not affected with one time application. Devarajan and Oblisami (1995) reported that the distillery spent wash application in rice increased the availability of N, P, K, Ca, Mg, micronutrients and organic matter contents in soil. The maximum grain yield was recorded in rice variety ADT 42 due to 75 times diluted distillery spent wash treatment, which was at par with 100 times diluted spent wash applied plot (Chinnusamy *et at.,* 2001). Joshi *et al.* (1996) reported that effluent treatment at 20% dilution with 50% NP application resulted to the best yield (5.3 t ha^{-1}) in case of maize, saving 50% N and P and 100% K, respectively. Yield of crops like sugarcane, rice, cotton, maize, sorghum, bajra and redgram was significantly increased with the diluted (1: 10 to

1: 50) effluent application (Zalawadia and Raman, 1994). Rajannan *et al.* (1998) found that irrigation with 50 times diluted spent wash recorded higher juice purity, brix and yield of sugarcane CoC. 771 and also found that distillery spent wash at 40 to 50 times dilutions increased the yield of sugarcane, banana, oil seed crops and rice.

(ii) Tannery effluents

Tanning industry is one of the major sources of pollution as it disposes large quantities of effluents rich in chromium and sodium into rivers and on land (Naidu and Laughlin, 1993). For producing one kg of finished leather, the tanning industry lets out 30-35 litres of effluents with high load of chromium, sodium, calcium, ammonium, magnesium, sulphates, chlorides and bicarbonates with organics like fats, dyes and acids (Singaram, 1995).

Uses : To manage the soil already affected, soil amendments *viz.*, composted coir pith, pressmud, gypsum and FYM @ 10 t ha^{-1} are recommended and grain and straw yields in three different locations are compared with control. Among the amendments, Singaram (1995) reported that coir pith performed better than other amendments. Coir pith kept the soil moist, prevented the capillary rise of salts from sub-surface layers and facilitated greater leaching of salts from the root zone.

Effect of tannery effluent water dilutions with well water irrigation with and without trash mulch on millable cane production revealed that maximum population of 1,32,000 millable canes ha^{-1} was recorded by well water irrigation and it was followed by 1,17,000 ha^{-1} in effluent dilution with well water in 25: 75% combination. Cane length and yield were also in similar trend (Kathiresan *et al.,* 1998).

Non-edible crops like flower crops may be suitable for effluent irrigation since the risk of heavy metals entering into the food chain is largely avoided. Growth and marketable flower yield of crops were significantly higher with 50% effluent followed by 100% effluent irrigation than well water irrigation. The crops *viz.*, Gundumalli, Jathimalli and Tuberose showed a high degree of tolerance towards sodicity caused by effluent irrigation, whereas Nerium was found relatively sensitive towards effluent irrigation. Root crops accumulated, more chromium and sodium followed by leaves and flowers suggesting that Cr was less mobile within plant (Mahimairaja *et al.,* 1999). Reuse of tannery effluent for commercial tree plantation is an effective alternative method of land disposal (Sakthivel *et al.,* 1999).

Irrigation with tannery effluent impaired soil productivity and inhibited biological activities of soil microbes when used at high concentration, however, when this effluent was diluted with water and used for irrigation, it acted as a growth adjustment and simultaneously helped in disposal of a part of the waste water (Augustine Selvaseelan *et al.,* 2001). Kumaran *et al.,* (2003) reported that eucalyptus, casuarinas, pungam and neem trees were well established in tannery effluent affected areas and teak establishment was poor. Grain yield of maize increased with increase in dilution of tannery effluent upto 50 times and applications of amendments especially composted coir pith at the rate of 25 t ha^{-1} (Thangavel *et al.,* 2003).

(iii) Paper and pulp mill effluent

Paper and pulp industry is one of the largest consumers of water and most of the water used is discharged as a waste. A quantity of 90, 000 gallons of water is required to produce one tonne of paper and the discharge of effluent is upto 80,000 gallons (Dubey and Joshi, 1980. Liquid wastes such as brown stock wash, caustic extractions and combined chlorination of hypochlorite wastewater are produced in large quantities during the manufacturing of paper in the Kraft paper mill. The effluent is alkaline and rich in organic and inorganic compounds. Disposal of this waste water has been a problem as it contains high amounts of suspended solids, high BOD, COD and lignin. At the same time, agriculture is in need of water for irrigating lands.

Uses : Disposal of lignin bearing coloured paper mill waste water through crop irrigation seems to be an appropriate approach as soil has capacity to retain the lignin and coloured bodies through its physico-biochemical reactions besides providing water for agriculture (Solaimalai *ct al.,* 2002).

It contains high amount of exchangeable sodium, which develops alkalinity in soils. However, when such effluents were used with additional nitrogen and gypsum, it gave increased yields of wheat and rice and lowered soil sodicity. The application of pulp and paper mill effluent to sugarcane tended to increase soil enzyme activities like amylase, invertase, cellulase, dehydrogenase and phosphatase. More microbial activity in the rhizosphere was observed in the plants irrigated with 50-75% effluent concentration. The retention of moisture necessary for growing sugarcane in sandy soils applied with irrigation with effluent water was attributed to humification and mineralization of organic matter (Augustine Selvaseelan *et al.,* 2001). Paper mill effluent irrigation also increased the beneficial organisms like Azospirillum, Azotobacter and

Rhizobium (Dhevagi *et al.,* 2003). Math Srinivaschari (2000) reported that DTPA extractable Zn, Cu was the highest when soil grown with rice ADT 36 was irrigated with undiluted effluent irrigation followed by 1:1 diluted effluent and good quality water. Among the soil amendments normal NPK application and gypsum was found to be comparable with each other and recorded higher soil available Zn and Cu over the rice husk ash in kar and pishanam seasons, respectively. Hence, the application of rice husks ash process was significant in preventing the micronutrients from reaching toxic levels in soils.

Different organic amendments with effluent irrigation to totmato decreased the bulk density of the soil and slightly increased the pH and organic carbon content. Number of fruiting clusters, fruit weight and fruit yield also increased due to the application of amendments such as bioearth, pressmud and farm yard manure alongwith effluent irrigation. Better germination, shoot and root growth and vigour index of seedlings at 75, 50 and 25% of treated effluent water were observed in maize, blackgram, greengram and soybean (Dhevagi, 1997). Gomathi and Oblisami (1992) stated that germination percent decreased at 100% concentration of the effluent and only 75% of the seeds germinated at 25 and 50% concentration. The length of the root, shoot and vigour index of the tree species tested was considerably decreased with increasing concentration. Narasimharao and Narasimharao (1992) found that paper mill effluent water could be safely used for irrigation of rice and cotton on alluvial soil having loamy to sandy loam texture as these crops are tolerant to salinity and are able to give economic yields in these areas. But irrigation of tobacco and chilli with this paper mill effluent water led to poor quality produce and reduced crop yield. In rice, the grain yield was increased due to effluent irrigation by 25% than that of well water irrigation. Effluent tended to decrease the calcium uptake of rice in neutral soil and increased the Na uptake in acid soil. The soil amendments had a favourable effect on the P and K uptake of rice in acid soil. The use of 50% diluted clarified effluent was found superior to raw effluent irrigation more particularly in neutral soil (Velu *et al.,* 1998). The adverse effect of effluent from paper and pulp industry would be alleviated by restoring to the application of N, P, K along with organic and inorganic amendments such as pressmud, farm yard manure and gypsum.

(iv) Textile effluents

Textile industries use large quantity of water during the manufacturing of fabric. The average quantity of water used to process 1 kg fabric is 150-200 litres. Much of these waters will come out as wastewater. In the process of boiling, bleaching and drying, various chemicals such as soda ash, caustic soda, HCl, H_2SO_4, sodium peroxide, hazardous polluting organic chemicals apart from high concentrations of various cations, anions and heavy metals.

Uses : Application of diluted effluent improved the groundnut germination and increased the chlorophyll, carbohydrate and protein contents and the biologically treated effluent enhanced the yield and quality of many cereals and pulses. This effluent when passed through fly ash can be safely used for afforestation (Augustine Selvaseelan *et al.*, 2001). Increase in soil pH, organic carbon, calcium and sodium concentration and decrease in water holding capacity due to addition of textile effluents have been reported. The soil flooded with dyeing factory effluents contained more N, P and K content when compared to other soil (Solaimalai *et al.*, 2002).

Jothimani *et al.*, (1998) studied the response of *green* gram (CO 4) to biotreated dye factory effluent (bio-fungal, bio-bacterial, chemical, chemo+bio-fungal and chemo+bio-bacterial) in clayey loam soil and reported that chemo and bio treated effluent recorded significantly higher symbiosis, vigour index, number of pods per plant, root and shoot length. Vijayakumari (2003) reported that dilution of effluent and water (25: 75) resulted in significantly higher germination percent, vigour index, shoot and root length of soybean crop.

(v) Sago effluent

Sago industry is an important agro-industry that processes tapioca into starch. During this process, 1 tonne of tapioca results in the formation of 30,000 to 40,000 litres of wastewater. The colour of the effluent varies between light yellow to grey. Because of the starch content in the wastewater, the microbial activity is more in the stagnated wastewater resulting in unpleasant odour (Balagopal *et al.*, 1977). The sago factory effluent contains appreciable quantities of suspended and dissolved solids and is acidic in nature with high electrical conductivity. It also contains high amounts of phosphorus and potassium and moderate level of nitrogen with very low quantities of trace elements.

Uses : Applications of sago factory effluent bear a negative impact on the physical and chemical properties of soil. The wastewater irrigated soils have a lower apparent and absolute specific gravity, low water holding capacity and has higher porosity. As it contains higher nutrient status, effluent irrigated soils show increased concentrations of essential plant nutrients. However, long term application may turn the soil as acidic (Solaimalai *et al.,* 2002).

(vi) Fertilizer industry effluents

Discharge of fertilizer factory effluents into water bodies causes eutrophication problems which leads to many disturbances in the human health. Hence, in order to avoid such problems, these wastewaters which are rich in nutrients can be effectively utilized in crop production after proper dilution with good quality water.

Uses : Application of fertilizer factory effluent to the soil resulted in significant increase in nutrient contents of the soil. Irrigation with 2.5 and 5% effluents of fertilizer factory wastewater enhanced the growth ad development of rice (Augustine Selvaseelan *et al.*, 2001).

(vii) Sewage effluent

Sewage includes wastewater from residents, commercial buildings and institutions such as schools and hospitals. The volume of sewage effluent varies from country to country depending upon the standards of living, human habits and availability of water supplies. Per day per capita wastewater produced varies from 50 to 300 litres (Solaimalai *et al.,* 2002).

Uses : Ramanathan *et al.* (1977) reported that application of sewage effluents resulted in a considerable increase in hydraulic conductivity value of 6.27 cc h^{-1} non-capillary porosity 16.20%, total porosity 58.20% and aggregate stability 50.50% as compared to well water irrigated field. Application of sewage effluent resulted in marked increase, in soil organic carbon content (1.02 and 0.97%) compared to ordinary water (0.72 to 0.65%) during the first year as well as in the second year experiment, respectively, and also increased the green fodder and dry matter yield of BN 2 hybrid grass (Malarvizhi and Rajamannar, 2001). Chandrasekaran and Rajkannan (2003) found that water holding capacity of soils was found to be increase by the continuous application of sewage irrigation.

(viii) Other industrial effluents

In addition to utilizing the wastewaters released from paper, tannery, textile, sago distillery and fertilizers, the effluents of zinc smelter, photofilm, viscose factory and dairy plant effluents can also be used for irrigation.

The yield of maize-gram and blackgram-wheat cropping sequences in summer and winter remained unaltered when zinc smelter effluent was used for irrigation (Totawat, 1991). Zinc smelter effluent, when diluted with well water, had no adverse effect on germination of common summer crops like maize, sorghum and blackgram and winter crops like wheat, barley and mustard (Totawat and Subhas Chauvan, 1992).

Application of diluted photoflim factory effluent slightly inhibited the seed germination, seedling development, reduced the chlorophyll content and carbohydrate and protein synthesis in crops like maize, sorghum, etc. (Manonmani *et al.,* 1992).

The effect of various concentrations of dairy effluent on seed germination of wheat, gram, oat, berseem, pearl millet and sorghum was studied and found that germination percentage was maximum with 25% effluent concentration (Gautam *et al.,* 1992).

19.7 Remediation techniques

The question of how to treat toxic plumes and waste deposits is one of the most complicated question of the present day, as it has legal, technical, economical and many scientific aspects. The factors that were identified as major impediments to groundwater restoration include; contaminant factors such as non-aqueous phase liquids (NAPLs), high sorption potential, and continued leaching from source areas; hydrogeologic factors such as heterogeneities, low permeability units, and fractures; and design factors such as-pumping rates, recovery well locations, and screened intervals. This all open up a broad range of considerations and activities including research and development into technologies and systems directed towards the prevention of danger. As a result, researchers have been challenged to develop new and innovative remediation techniques to clean up the contaminated sites faster, more effectively, economically and with a greater degree of safety.

During the past decade, ground water scientists and engineers have devised a number of methods to contain and/or remediate groundwater contamination. The groundwater pollution remediation technologies can be generally classified into on site techniques and in-situ techniques. The commonly used on site technique is pump and treat method. The in-situ technologies, which directly remediate contaminated groundwater, are still not well developed, primarily because site characterization is complex, expensive and inaccurate. Moreover, contaminant plume is often difficult to define exactly. In this chapter, the onsite techniques of pump and treat for pollution remediation and containment is described briefly (Sidhardhan and Thamarai, 2004).

(i) Pump-and-treat remediation systems

Almost all hazardous waste sites pose some form of threat to their local groundwater environment. Cleaning contaminated groundwater is a lengthy process, and most clean up efforts are expensive. Pump-and-treat technology is one of the technologies utilized for remediating contaminated groundwater (Mercer *et al.,* 1990). In this method, the contaminated water is extracted and treated using carbon adsorption canisters or using above ground bioreactors. The treated water is either injected back into the aquifer or disposed of above ground. Unfortunately, in many instances, pump-and-treat methods have failed to remove contaminants at the rates once expected. Although pumping at first tends to remove high concentration levels from an aquifer, the concentrations quickly drop to lower levels (Haley *et al.,* 1992). This is because, the contaminant mass fraction that was absorbed to the soil matrix always tend to be released at a much slower rate, thus limiting the overall removal process. Similarly, mass transfer limitations associated with the contaminant sources such as trapped LNAPL or DNAPL pools also pose limitations to pump-and-treat treatment systems.

(ii) Bioremediation systems

More recently, several novel biotechnologies have been attempted to simulate native microbial population (both aerobic and anaerobic) for degrading ground-water contaminants. Bioremediation technologies are more promising because they are engineered to use microbially-mediated degradation processes to detoxify and degrade environmental contaminants in-situ.

Although subsurface bioremediation is viewed as an innovative technology, wastewater treatment engineers have been successfully using microbes to degrade complex wastes for over hundred years, using either attached or fixed-film reactors. However, unlike wastewater treatment systems in which conditions are optimized to ensure homogeneity and uniform mixing, subsurface bioremediation systems should frequently address multiphase, heterogeneous environment with mixed contaminant present at various concentration levels. In addition, whenever a bioremediation option is proposed for a site, a question arises as to whether the treatment system will utilize indigenous microbial specie or species of engineered microbes that would be artificially introduced.

Both these approaches have their own relative merits and demerits, but the use of indigenous microbes seems to be a better choice for soil and ground-water treatment.

While introduced microbes often can be engineered to outperform the native microbes in a bioreactor, it will be hard to control the conditions in a subsurface to favour the growth of an introduced microbial species. Therefore, most bioremediation efforts are aimed towards optimizing the growth conditions for an in situ population by supplying required nutrients (primarily nitrogen and phosphorous), electron donor (acetate or lactate) or acceptor (oxygen, nitrate, or sulphate), and by providing optimal pH conditions.

During the past two decades, tremendous amount of research efforts have been dedicated for studying the mechanisms by which microbes degrade and metabolise hazardous chemicals. Early success was related to petroleum hydrocarbon remediation sites where several aerobic organisms were discovered to degrade common petroleum contaminants. Field observations also showed depleted levels of oxygen near source areas, supporting the hypothesis that aerobes are actively utilizing these contaminants.

Several innovative ideas utilizing various methods for injecting oxygen-saturated water, hydrogen peroxide, pure oxygen, oxygen release compounds, and/or air were attempted to deliver oxygen to the subsurface to facilitate aerobic degradation of petroleum compounds. However, further research in this area indicated that petroleum compounds and other organic materials can naturally degrade not only under aerobic conditions but also under anaerobic conditions, utilizing other electron acceptors such as nitrate, iron, manganese and sulphate, and via methanogenesis (Border *et al.,* 1995). For example, Christensen *et al.* (1994) lists several cases for natural attenuation of landfill leachate plumes in ground-water aquifers.

The review shows that most leachate plumes are relatively narrow and do not, in terms of width, exceed the width of' the landfill, diverse microbial communities have been identified within leachate plumes and believed to be responsible for degrading contaminants via both aerobic and anaerobic pathways using various electron acceptors A detailed field study of fuel-hydrocarbon plumes at two U.S. Air Force sites demonstrated that natural processes are degrading fuel hydrocarbons under both aerobic and anaerobic conditions (Wiedemeir *et al.,* 1995).

A recent survey over 200-petroleum hydrocarbon plumes concluded that most groundwater plumes originating from petroleum-related sources will naturally attenuate with time, if the sources are removed (Rice *et al.,* 1995). However, if the plume poses immediate threat to any downstream receptors then an active remediation strategy might have to be used to control plume migration. Similar natural attenuation arguments for managing chlorinated solvent plumes (and other chemical and metal pollutants) are currently being debated in the ground-water remediation literature; but the evidences for proving natural attenuation of chlorinated solvent contaminants are not as convincing as the evidences available for petroleum contaminants. It seems that the degradation of chlorinated solvent chemicals in natural environments may not be as ubiquitous as petroleum hydrocarbon chemicals, and in most cases the degradation process will be limited by the availability of a electron donor to support dechlorination reactions (Wiedemeir *et al.,* 1997).

(iii) Source clean up alternatives

Pump-and-treatment and/or other remediation technologies that primarily address remediation of dissolved ground-water pumps can only be effective if the source is completely removed. Often, continued release of the contaminants trapped in the partially saturated or unsaturated region above the water table (known as the vadose zone) may act as a prolonged source and render the treatment process ineffective.

Several clean up technologies have been developed for removing adsorbed contaminants or trapped LNAPL or DNAPL sources in the unsaturated soil zone. Soil vapour extraction technology is the most promising and proven in situ technology useful for source removal activities (Johnson *et al.,* 1990). In a basic vapour extraction system, wells with blowers or vacuum pumps are used to remove contaminant vapors from the vadose zone and thereby speeding up the volatilization process. When operated properly, soil venting can be a cost-effective remediation process for vadose zone soils contaminated with gasoline, solvents, and other volatile chemicals.

Other conventional source control measures such as slurry wall, sheet pile cutoff walls, grouting, geomembranes, surface seals, and surface drainage methods can also be used to contain the source area. Domenico and Schwartz (1990) provide a good overview of these standard geotechnical methods. In situ techniques such as vitrification, biogenic precipitation, and soil heating are also available for treating radioactive metal and other chemical pollutants.

19.7 Planning for Reuse of Wastewater

In the planning and implementing wastewater reclamation and reuse, the reuse application will usually govern the wastewater treatment needed and degree of reliability require for the treatment process and operations. Because wastewater reclamation entails the provision of the continuous supply of water to consistent quality, the reliability of the treatment processes and operation must be evaluated in the planning stage. Specific reuse categories and treatment technology that may be applicable will depend on the location and type of wastewater management employed. World wide the most common use of reclaimed wastewater has been for agricultural irrigation, landscape, irrigation, Industrial recycling and reuse in groundwater recharge, recreational/ environmental use, non-potable urban use and potable reuse.

The categories of wastewater and their issues are shown in Table 19.4.

Table 19.4: Wastewater categories and their issues/constraints

Wastewater categories	Issues/constraints
Agriculture irrigation Crop irrigation Commercial nurseries **Landscape irrigation** Parks School yards Freeways medians Golf courses Cemeteries Greenbelts Residential	Surface and ground water pollution if not managed properly. Marketability of crop and public acceptance. Effect of water quality, particularly salts, on soils and crops. Public health concerns related to pathogens (bacteria, viruses, and parasites). Use for control of area including buffer zone. May result in higher user costs.
Industrial recycling and reuse Cooling water Boiler feed Process water Heavy construction	Constituents in reclaimed wastewater related to scaling, corrosion, biological growth, and fouling. Public health concerns, particularly aerosol transmission of pathogens in cooling water.
Groundwater recharge Groundwater replenishment Salt water intrusion control Subsidence control	Organic chemicals in reclaimed wastewater and their toxicological effect. Total dissolved solids, nitiates and pathogens in reclaimed wastewater.

Wastewater categories	Issues/constraints
Recreational/environmental uses Lakes and ponds Marsh enhancement Stream flow augmentation Fisheries Snowmaking	Health concerns of bacterial and viruses. Eutophication due to nitrogen (N) and phosphorus (P) in receiving water. Toxicity to aquatic life.
Non-potable urban uses Fire protection Air conditioning Toilet flushing	Public health concerns on pathogens transmitted by aerosols. Effect of water quality on scaling, corrosion, biological growth and fouling. Cross-connection.
Potable reuse Blending in water supply Pipe to pipe water supply	Constituents in reclaimed wastewater, especially trace reservoir organic chemicals and their toxicological effects. Aesthetics and public acceptance. Health concerns about pathogen transmission, particularly viruses.

(Tchobanoglous and Burton 2000)

The exclusive use of water in industry for heat transfer is largely limited to cooling and boiler operations. Cooling water- mostly from indirectly cooling can represent up to 90% of total water consumption in industries depending on such application. Due to their high water demand and relatively low water quality requirement compared to high technology applications, cooling tower system are ideal for water recycling.

The quality of water demanded for boiler operation is much higher than for cooling because of the highly deleterious effects namely scaling and corrosion of the solution chemistry under the conditions at which boilers operate. The higher water quality required impose a higher processing cost for recycling but the end product water is none the less require only for physical heat transfer. This is to be distinguished for the case where the water forms a solvent for some chemical or biochemical reaction such as cleaning or disinfection.

In the swimming pools, where water clarity and microbial quality is key, these operate at a make up rate of around 10% with dilution required principally to suppress accumulation of dissolved solids generated from the reaction between the chlorine disinfectant and pollutants arising from bathers. The key economic driver for recycling is the requirement to operate pool at around 30°C with a residual free chlorine level of around 1 mg/l. A 100% make up would thus incur huge costs in freshwater heating and wastewater chemical reduction of the residual chlorine. It is far more economical, therefore, to invest in additional process plant for maintaining pool water clarity (usually by sand filtration) and reducing the organic load (by adsorption on activated carbon and/or by chemical oxidation with the chlorine or ozone) so as to reduce chlorination by-product formation (Judd, Dr Simon, 1999).

Water used for irrigation can vary greatly in quality depending upon the type and quantity of dissolved salts. The consequence of evapotranspiration is salt accumulation in the soil profile. The physical and mechanical properties of soil, such as degree of dispersion of the soil particles. Stability of aggregates, soil structure and permeability are sensitive to the types of exchangeable cations present in soil. Four categories of potential management problems associated with quality are

(1) Salinity

TDS => 2100 mg/I

(2) Specific ion toxicity

SAR = > 9

Cl => 350 mg/I.

B => 3mg/I

(3) Water Infiltration rate

(4) Miscellaneous

N => 30mg/l.

HCO_3 => 500mg/I.

pH = 6.5 - 8.4

(Metcalf and Eddy, 2003)

Possible recycled water grade and treatment is given in Table 19.5 and linking recycled water application and grade is given in Table 19.6.

Table 19.5: Possible recycled water grade and treatment

Grade	Description	FC/100 ml	Treatment
5 star	Potable		Advanced multi-barrier treatment processes effective against microbiological and chemical pollutants.
A 4 star	Open access	<1	Secondary + membrane filtration + disinfection Secondary + coagulation + filtration + disinfection
A 3.5 star	High contact	<10	Secondary + filtration + disinfection
B 3 star	Medium contact	<100	Secondary + disinfection

Grade	Description	FC/100 ml	Treatment
C 2 star	Low contact	<1000	Secondary + disinfection
			Advance primary + filtration + disinfection
			Up flow anaecrobic sludge blanket + disinfection
D 1 star	Restricted	<10000	Secondary + maturation ponds
			Oxidation pond system

Table 19.6: Linking recycled water application and grade- examples on the draft Australian water recycling guidelines

Grade					
Restricted D (1 star) Fe <10000	**Low contact C (2 star) Fe<1000**	**Medium contact B (3 star)**	**High contract A *(3.5* star) Fe<10**	**Open access A+ (4 star) Fe<1**	**Application**
Not allowed	Not allowed	Water quality met post mixing	Allowed	Allowed	**Surface waters/ Impoundments** Surface water protected water supply Primary contact Recreation
Water quality met post mixing	Allowed	Allowed	Allowed	Allowed	Surface water- secondary water supply Secondary contact recreation
Allowed	Allowed	Allowed	Allowed	Allowed	Surface water - restricted access Ornamental water bodies
Not Allowed	Not Allowed	Not Allowed	Water quality Met post mixing	Allowed	**Ground water** Ground water potable aquifer Direct injection
Not Allowed	Water quality Met post mixing	Allowed	Allowed	Allowed	Retention >12 month Groundwater potable aquifer Percolation>3m and Retention >12 months

Grade					
Restricted D (1 star) Fe <10000	**Low contact C (2 star) Fe<1000**	**Medium contact B (3 star)**	**High contract A *(3.5* star) Fe<10**	**Open access A+ (4 star) Fe<1**	**Application**
Allowed with hold 4 hrs	Allowed	Allowed	Allowed	Allowed	Non food crops Silviculture, turf farms Fodder, fiber and seed crops
Not Allowed	Not Allowed	Allowed except pigs	Allowed	Allowed	**Pasture animals and fodder** Stock water
Not Allowed	Allowed with hold 5 days	Allowed	Allowed	Allowed	Pasture fodder for daily cattle and pigs
Not Allowed	Allowed	Allowed	Allowed	Allowed	Pasture and fodder for beef cattle, sheep Dairy wash down waster
Not Allowed	Not Allowed	Not Allowed	Allowed	Allowed	**Food crops** Food eaten raw including salad vegetable And root crops RW contacts edible portion
Not Allowed	Allowed	Allowed	Allowed	Allowed	Food cooked is processed before eating orchards - No RW contact on edible portion
Not Allowed	Not Allowed	Not Allowed	Allowed	Allowed	**Urban and residential** Residential gardens, car washing pavement washing, toilet flushing
Not Allowed	Allowed With hold 4 hours	Allowed	Allowed	Allowed	Sporting field, golf course, parklands open space, landscaping fire protection
Not Allowed	Allowed	Allowed	Allowed	Allowed	**Commercial and industrial** Open system, minimal aerosols
Not Allowed	Allowed	Allowed	Allowed	Allowed	Road making, soil compaction, concrete mixing, dust suppression

(Source: John Anderson, water 21, August 2001, Magazine of the International Water Association)

How safe is the wastewater reuse?

Despite a tong history of wastewater reclamation and reuse in many parts of the world question of safety of wastewater reuse is still difficult to define and acceptable health risk have been hotly debated. The important factors are:

- Presence of the chemicals
- Presence of the pathogens

When treated municipal wastewater effluents are uses in the urban environment results in possibility of direct contact with man, considerable health concerns may be justified. These health concerns are specifically directed in the industrialized country with high health standard, to the presence of coterie pathogens because of their low dose infectivity, long-term survival in the environment, difficulty of monitoring, and low removal efficiency in the conventional wastewater treatment.

Planning for reuse

The planning of the wastewater should he in the proper manner because it may cause many adverse effects on the user and plant operation. Through the planning wastewater should be properly treated and make it fit for the reuse should be properly treated (Peavy Rowe, and Tchobanoglous, 1985). In this planning each and every part has it's own importance but economical and water quality aspects are the most important.

The wastewater reclamation and reuse facilities plant should include the following analyses:

- Assessment of wastewater treatment and disposal needs.
- Assessment of water supply and demand.
- Assessment of water supply benefits based on wastewater reuse potential.
- Analysis of reclaimed wastewater market.
- Engineering and economical analyses of alternatives.
- Implementation plan with financial analysis.

Current Management Scheme is shown in Fig. 19.2

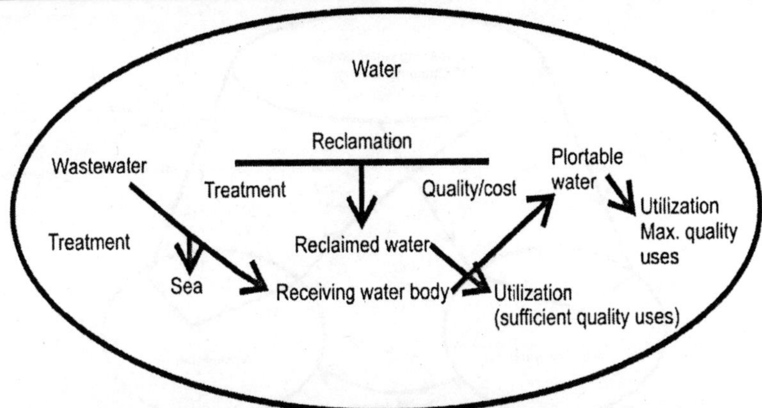

Fig. 19.2: Schematic diagram for current management scheme

Past Management Scheme is shown in Fig. 19.3

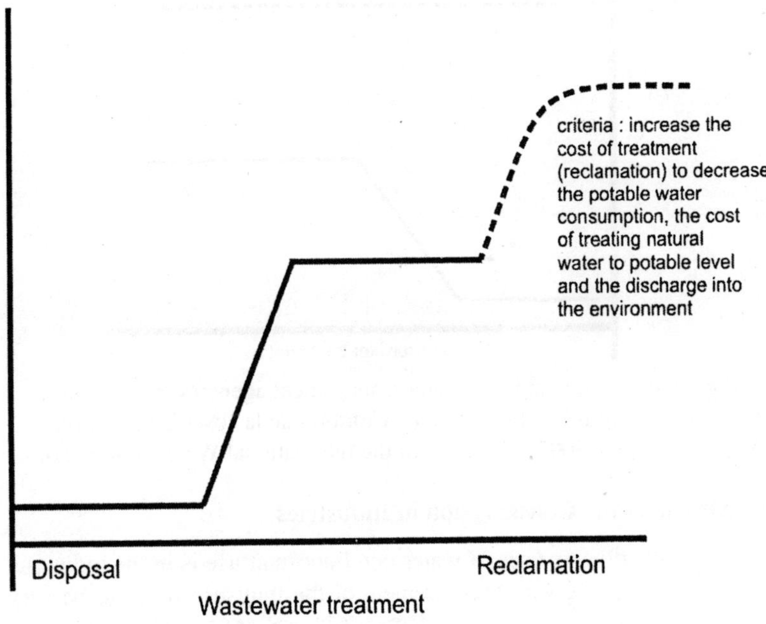

Fig. 19.3: Past management scheme

Present and past water management agencies in Catalonia are shown in Fig. 19.4.

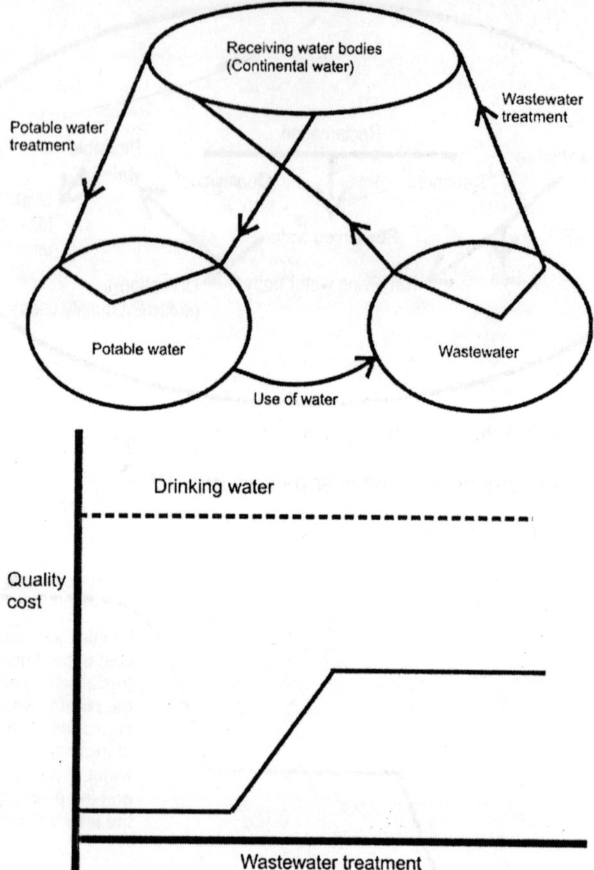

Fig. 19.4: Present and past water management agencies in Catalonia (Source: Sala lluis and Serra Manel, Consorci de la costa Brava, Girona, Water 21 August 2002. Magazine of the International Water Association)

19.8 Measures for Conservation in Industries

Amongst diverse uses of water one important use is in the industrial section. In India, the water requirement of the industries used to be very small compared to other sectoral demands in earlier years. With the rapid growth in population coupled with overall economic development due to industrialization, the requirement for industrial sector is increasing at a fast rate. Projected water requirement various sectors of Indian industries and wastewater generation is given in Table 19.7. When industrial demand is concentrated in specific locations, heavy point loads are created on available water resources. Thus water availability is a major factor in industrial location as non-availability of quality water and scarce situations have forced some industries to shut down. Industries require water for various purposes most of which are non-consumptive, thus making reuse through recycling and other conservation measures possible. The amount of water consumed for any product, therefore varies widely depending upon the process used, plant efficiency, technology employed, the degree to which water is re-circulated and other factors. There are no fixed norms for water demand for industries but rather a range of values determined by the technology used selection of plant and process, practice in providing maximum recycling to reduce demand and pollution.

Table 19.7: Projected water requirement for various sectors of Indian industries

S. N	Category of Industry	Water Requirement m^3/tonnes/yr/yr	Waste Water Generation m^3/tonne	Year 1997		Year 2000	Year 2010	Year 2025	Year 2050
				Production in 1000 tonnes	Water Req. Mm^3/yr	Water Req. Mm^3/yr	Water Req. Mm^3/yr	Water Req. Mm^3/yr	Water Req. Mm^3/yr
1	2	3	4	5	6	7	8		
1.	Integrated Iron & Steel	22	15	119.390	2626.58	3829.1	5837.70	6012.60	12035.1
2.	Smelters	82.5	65	174	11.7	16.76	24.14	32.31	44.35
3.	Petrochemicals & Refinery	17.0	13.0	1.255	21.34	23.47	30.61	37.76	55.61
4.	Chemicals-Causic Soda	5.5	1.0	1,459.30	8.03	8.65	10.20	12.67	19.07
5.	Textile & Jute	200	15&i	24,730.20	4863.02	8153.72	19018.78	36701.40	4693.50
6.	Cement	4.5	NA	76.00	418.00	599.50	985.50	1777.50	3370.50
7.	Fertilizer	16.7	3.5	11,155.30	147.51	220.11	630.96	1105.09	1192.75
8.	Leather Products	30	28.0	912.5	30.12	1244.6	65.74	93.08	147.83
9.	Rubber	6.6	5.0	403.24	2.66	3.04	4.30	6.41	9.54
10.	Food Processing	6.8	NA	96.450	848.76	991.76	3441.82	9442.76	12298.3

S. N	Category of Industry	Water Requirement m³/tonnes/yr/yr	Waste Water Generation m³/tonne	Year 1997		Year 2000	Year 2010	Year 2025	Year 2050
				Production in 1000 tonnes	Water Req. Mm³/yr	Water Req. Mm³/yr	Water Req. Mm³/yr	Water Req. Mm³/yr	Water Req. Mm³/yr
1	2	3	4	5	6	7	8		
11.	Inorganic Chemicals	200	4	2.200	96.80	165.44	1600.00	3346.00	615.20
12.	Sugar	2.2	0.4	14.500	36.21	46.20	71.17	334.40	636.90
13.	Pharmaceuticals	25	NA	4,190.00	104.75	124	209.25	276.15	429.15
14.	Distillery (Req. per 1000 litres)	22	12	1,790.80	3939.76	6357.2	66.31	318.00	5203.92
15.	Pesticides	6.5	5.0	86	3.78	6.67	1.99	4.82	8.38
16.	Paper & Pulp	200	175	3,100	852.50	1259.5	207.00	10240.0	19490.0
17.	General Engineering	2.2	175.0	5,779	1271.38	1433.2	23.66	27.82	55.63
	Total				15282.9	24482.9	32229.13	69768.77	102535.7

Industrial effluents constitute a major source of waste water discharges, which would contain different kinds of toxic pollutants. Treatment of industrial waste water and recycling are essential to meet standards of effluent water quality. With the quality of water becoming poor, availability of fresh water being scarce and statutory environmental regulations becoming more stringent, optimization in use of water demands a closer monitoring by industrial sector.

The requirement of water for industrial use depend upon many factors *viz.*, use of appropriate technology requiring least quantity of water, quality of raw material and its source, sources of water *i.e.* freshwater/groundwater/river, disposal of effluent, policy of re-circulation etc. Some of the industries requiring considerable quantity of water for their production can be broadly classified as small scale industries, chemicals and petrochemicals, sugar, steel, paper, fertilizer, textile, food processing, coal building, non-ferrous metals, cement etc.

In the small-scale sector (SSI), the water intensive industries belong to the sectors namely the food items, hosiery and garments, paper product and printing, chemical and chemical products, iron and steel industry etc. Water consumption varies depending upon the nature, however, in the changing global industrial scenario; the growth rate in SSI sector has been estimated to reduce. The requirement of water in chemical sector by pharmaceutical and pesticides industry is insignificant due to recycling. However, the major water intensive industries are dyes, caustic soda, and soda ash.

In sugar industry water is required for cooling purpose and processing. The requirement varies according to size of the plant, technology being followed and the recycling process. Modernizing the plants and updating the processing technology can reduce the future demand/consumption of water in sugar production.

Water consumption in steel industries varies depending upon the process and production technology used. In view of continuous increase in production, the water demand would also increase in future. In paper industry water is an integral part in paper making and transporting fibre from raw material however, the water demand depend upon the type of raw material used. In India, the paper and pulp industries extensively use intra plant recycle and reuse of waste water from washing, screening and bleaching. The water consumption is likely to reduce due to conservation measures.

In the textile industry, the processing sector is one of the major consumers of water. Estimating the water requirement for textile industry is complex, as the industry is diversified with heterogeneous structure of production. Since water is going to be scarce in coming years, the strategy for reduction in use of water is necessary. The water demand in fertilizer plants widely varies depending on the type of plant-product and he process technology employed. A major portion of the water is used for cooling purposes and boiler feed water. The future plants should be based on latest technology having zero effluent.

Water consumption in coal mining sector is for coal washeries, washing and cleaning of mining buildings and equipment, dust suppression in mines, drinking water, fire fighting etc. In some groups of mines, the water consumption is very high. The food processing industry is also a fast growing sector and besides the organized sector, there are large numbers of units in unorganized sector. The requirement of water varies in food processing industry from unit to unit depending upon the product, technology used etc.

The non-ferrous metal industry in the country is confined to production of aluminum, zinc, lead and copper. The water demand depends upon location and size of mines/smelter, grade of metal contained in the ore and technology used. Future demand of water would get reduced due to recycling process of water in this sector.

In building sector, the brick industry is predominantly in the unorganized sector. There is acute housing shortage in the country at present aid therefore the future growth in demand of bricks for housing is expected to be high. Similarly, cement which is manufactured through wet/dry processing, is also required for construction activities. The cement industry is power intensive industry and has installed captive power units, as adequate power is not available. Hence, the water requirement is both for manufacture of cement and for the captive power plant. Thus the water demand in this sector will also be high in future.

There are following measures for conservation of water in industries:

(i) Economy in industrial use of water,

(ii) Recycling and reuse of water in industry,

(iii) Treatment of industrial wastewater,

(iv) Water quality norms for various industries, and

(v) Adoption of new technology

(i) Economy in industrial use of water

For effecting economy in industrial use of water, the tariff rates have to be prescribed such that the industry is compelled to look into technological interventions leading to reduced use per unit production, which in the end will further reduce generation of pollution load and consequent cost of treatment. National/ international benchmarks for water consumption and water reuse in each of identified 17 categories of industries should be fixed and efforts made to achieve these benchmarks within a specified time frame, say by the year 2010. Water cess or pollution tax is generally levied and collected by the pollution control boards. Differential rates of cess are charged depending upon the. use of water, the rate being higher here pollution load is expected to be more resulting in higher cost of pollution abatement and as such, its effectiveness as a deterrent for economizing use of water needs to be looked into.

(ii) Recycling and reuse of water in industries

As the demand in industries is going to increase, the technological development in processing and methods of reusing water are expected to reduce the demand of fresh water. Recycling is defined as the internal use of wastewater by the original user prior to discharge to a treatment system or other point of disposal. Wastewater is recovered, treated or untreated and then recycled fox repetitive use by the same user. The term reuse applies to waste waters that are discharged and then withdrawn by a user other than the discharger. Reclaimed waters from wastewater after treatment are generally used for "agricultural" irrigation, cooling water, algal cultivation and pisciculture, apart from other industrial uses. India though predominantly rural has still a large urban population. The urban centres are also the nuclei of industrial growth. The wastes (effluents), if reused within the industry with/without treatment as permissible would help in minimizing fresh water requirements while achieving reduction in wastewater volume for final treatment before discharge, deriving economy at both ends.

In most cases industrial water uses are non-consumptive making reuse possible through recycling and conservation measures. Recycling of wastewater also helps in recovery of certain commercially viable by-products. Process industries are major users of water and can recycle or reuse water wastes for lesser duty purpose. Cascade concept is adopted for reusing water discarded from a process requiring higher purity to a process requiring higher purity to a process requiring lower purity. If required, a simple treatment process may be interposed between the processes. Water reuse is more economical if included at the design stage by modification of the existing system. In most of the inland towns, in arid and semi arid areas, where suitable lands are generally available in the neighborhood for development of irrigation, treated effluent can economically be used for industrial effluents are used for irrigation. The water recovered from effluents is mainly used for less important uses like gardening and cleaning. It is necessary to improve the production technology, and low or no waste technology needs to be adopted. Though, such technologies may be costly at the initial stage it would prove economical in the long run.

(iii) Treatment of industrial waste water

The rapid industrialization calls for wastewater treatment and its disposal to be so planned and sited so as to protect people, the quality of water (both surface and ground) and environment from adverse impacts. The industrial units must set up effluent treatment plants fpr treating the wastewater to the desired standard before releasing to water bodies. Effective checks and monitoring should be placed in position and deterrent punitive measure be taken against defaulting units. These should be open for inspection by the State. Pollution Control Board for action. For small units located in various industrial estates common effluent treatment plants be set up and the industry should share the capital and O&M cost of the plants.

The hazardous waste from industries is any residue that may cause significant hazards to human health or the environment unless adequately handled, stored, transported, treated and disposed. The major groups of industries that generate hazardous wastes include inorganic and organic chemicals, petroleum refineries, iron and steel, non-ferrous metals (smelting and refining), leather tanning and finishing, and metal finishing. The major problem in respect of hazardous waste management in India is the absence of a systematic assessment of the quantity and pollution potential of hazardous waste. Preliminary data gathered by Ministry of Environment & Forests, CPCB and institutions like National Environment Engineering Research Institute and National Productivity Council suggest that the disposal methods currently practised for hazardous wastes management are not environmentally compatible.

Toxic wastes are also very harmful. These are of two types, namely solid toxic wastes and toxic effluents. Specialized treatment facilities exist in the country for selected categories of solid toxic wastes. Detailed guidelines are available for handling and disposal of radioactive wastes to conform to national as well as international standards. Special care is taken to ensure that leakages do not contaminate soil surface and/or ground water resources even in the long run. Toxic effluents, however, are not segregated in the industries and are often discharged mixed with other effluents. Generally, waste water of industries such as sugar, distilleries, dairies, tanneries etc. can he treated by biological methods such as stabilization ponds, activated sludge process trickling filtration aerated lagoons etc. Other industrial wastes such as pulp and paper synthetic fibre etc. have to undergo simple physico-chemical methods-of treatment, but the industries discharging toxic wastes such as-electro-plating, metallurgical, caustic chlorine etc. may require more elaborate techniques.

The CPCB has notified wastewater generation standards for various categories of industries under Environmental (Protection) Rules, 1986. The above Act stipulates that every industry should have its own treatment plant, yet in practice this dictum is seldom followed because of the shortage of energy and costly operation involved. However, according to a recent survey conducted by the CPCB, although about 85 percent of the-total number of large and medium water polluting industries in the country have effluent treatment systems, their performance, in most cases, is found to he far from satisfactory.

(iv) Water quality norms for various industries

Water quality norms for various industries are diverse depending on the nature of the industry and the individual processes using water within that industry. The water used ranges between any water to water of extreme purity. Factors like total hardness and hardness due to calcium and magnesium are important requirements. Although temperature is not of much consideration, it is important for water used for cooling in industry. As a flatter of fact, many of the industrial processes do not need any good quality and highly treated water. However, it may need water free from bicarbonates, chlorides, sulphates, iron etc. in order to prevent scale formation, colorings of fabric etc. As such, many of the present industries are switching over to reuse of' recycled wastewater to meet their water requirements.

Water for industrial needs is very much affected by its availability, its quality for different production processes and level of technology cost of treatment, to bring it to acceptable standards, wastewater disposal facilities, discharge regulations and possibilities of reuse. In our country, it is a general tendency to reconsider the first three factors whereas the other factors are neglected. As already pointed out above, generally the requirement of water for industries is estimated on the basis of per unit of product and extrapolated to the future period. This ignores the important facts, namely (i) substitution or reuse possibilities and (ii) impact of technology changes on water utilization rate and water quality requirements. Research is being carried out all over the world to re-use the wastewater including intensive efforts for improving the technology of water supply and use, substitution of dry process for wet process, substitution of non water using goods and services and geographic redistribution of activities designed to minimize the cost of water to the user or to the community.

(v) Adoption of new technology

Adoption of new technologies of treatment is an important conservation measure. Some technologies like up flow, anaerobic sludge blanket, biogas generation from waste, use of oxygen, bleaching to replace chlorine in pulp and paper industry, use of cartridges for treatment of electroplating wastes etc. are well known. They aim at resource conservation, waste minimization and recovery and reuse of wastes to the extent possible. They improve the financial viability of the industry and reduce possible conflict with public interest. Experience in the industrialized nations has proved that anti-pollution technology has also been cost effective in terms of health, property and environmental damage.

Once pollution is replaced as a cost and a symptom of inefficiency in production, it will motivate investments in clean technology, thus reducing waste and pollution. This may require financial incentives such as tax break, low interest loans, investment and depreciation allowances. Proper raw material, water and energy pricing policies can make water minimization more economical. Further, through disincentives such as waste charges and non-compliance charges, etc, the industries can be pushed to adopt clean technologies.

19.9 Strategies for Conservation

As observed from the water demand scenario, there would be a quantum jump in water consumption level in industrial sector from the year 2010 onwards. It is, therefore, essential that liberal use of water be avoided in coming years. Water use management is necessary to conserve precious natural resources in which both public and private sectors have important roles to play. The water efficiency programmes reduce water demand, brings savings in water and wastewater treatment, lessen environment impact and sustain water quality. There is need for laying down certain strategies in view of scarcity of fresh water sources as well as due to water pollution hazards. Some of the strategies having bearing on water conservation and management in relation to industrial sector as given below, may be taken into cognizance while framing any future policies.

Industries should be classified based on their fresh water requirement as highly intensive, moderately intensive etc. Water availability should be one of the criteria to be kept in view while deciding the location of new highly or moderately water intensive industries.

- The Ministry of Industry, in consultation with the Ministry of Water Resources, based on studies should fix up upper limits of water requirements for the production of various industrial commodities.
- Major changes are required in the system of prices and other economic incentives affecting water use and allocation. Although, price instruments exist, the levels of existing incentives and the form in which they are applied result in minimal and in some cases negative impact. There is an urgent need to bridge these incentives gaps in incentives.

- Technology improvement, reuse and other forms of demand management need to be given much more emphasis rather than the traditional supply side oriented approaches in the past.
- Water use efficiency in India in the industrial sector is quite poor as compared to use by similar industries in other countries. Substantial water and often cost savings, can be obtained by introducing water saving by treatment of waste water and changing industrial processes.
- The existing system of subsidies and tax structure on investment in pollution control, water conservation and water recycling technologies should be reviewed. Particular attention needs to be paid to introduce a significant and punitive variable tax on the act of pollution.
- Tariff rates have to be prescribed such that the industry feels compelled to look into technological interventions leading to reduced use per unit production. For effecting maximum conservation, production processes have to be modified, to have lesser generation of effluent water. Waste utilization technologies/clean production technologies with emphasis on waste minimization should be encouraged for adoption.
- Instead of allowing location of hazardous industries and insisting on zero effluent condition in semi-arid and arid areas, industrial zoning be done in a manner that in such areas water intensive industries are not permitted, especially those releasing toxic effluents. Hazardous waste treatment and disposal need to be so planned and sited as to protect people and environment from adverse impacts.
- Selection and zoning of industries associated with potential risks, especially those releasing toxic waste need a thorough analysis and planning before they are set up in any water basin. The concerned State Government and other local bodies should have a coordinated approach in selecting and locating industries of a specific nature with respect to their water requirement and facilities for waste water disposal. A policy for zoning the water basins according to the types of industries, quantity of water consumed/discharged need to be laid down.
- Minimum National Standards evolved by the CPCB based on minimum treatment concept have to be strictly followed while clearing proposals for locating industries in polluted stretches of water bodies. Apart from obtaining clearance from

SPCB/CPCB, clearance from the concerned Ministry dealing with water resources be made mandatory for discharging effluents in the drainage system.

- The CETP (Common. Effluent Treatment Plant) Schemes have already been implemented in some states. The investment in CETP is shared by Central Government, State Government and by the industrial units. The CETP Schemes should be encouraged in other industrial areas and clusters of industries particularly in SSI Sector.
- States may also have to come out with their industry policy based on availability of Water in certain sectors in industry.
- There should be a comprehensive and periodic water audit of industries for effective management strategies.
- Last but not the least an awareness has to be created in the concerned agencies and the industrial sector about the maintenance of proper quality of surface and ground water resources.

The amount of industrial water consumption is usually a small portion of the water withdrawal but is the second main user not only in the world but in our country as well. The effect of industry on water resources includes not only the increased water use and great quantities of waste discharge to the hydrologic system but also changed characteristics of the resource, whether surface or groundwater. Therefore, the approaches for water resources efficiency, conservation and reduction or elimination of waste discharges should be most realistic and effective for the present and near future. These should be acceptable to the users, ecologically admissible and economically profitable on physiographic features and the water use.

19.10 Strategies for Recycling Industrial Effluents in Agriculture

Some of the Management techniques for recycling industrial effluents for irrigation purposes are discussed earlier and other described as under:

1. Gypsum application

Due to its low solubility (0.25-0.30%) and cost, gypsum is suitable for correcting a favourable Ca: Na ratio or Ca: Mg ratio.

2. Leaching requirement

It is the fraction of irrigation water that must be leached through the root zone of the crop to keep the salinity of the soil below a specific limit.

3. Crop selection

To overcome the toxicity and other hazardous problems, selection of suitable crops is important. Crops like barley, cotton and sugarbeet are moderately tolerant. Crops like safflower, wheat, soybean, etc. can be grown if irrigation water is saline. Crops like datepalm, sugarbeet and berseem, etc. can be grown if irrigation water is having more of boron.

4. Sowing crop on the side of the ridges is advisable.

5. Fertilizer application

Additional dose of inorganic N does not change salinity tolerance and the response is rather reduced. Addition of phosphorus reduces the toxicity of chlorine.

6. Method of irrigation

Poor quality water is not suitable for use in sprinkler irrigation. Less application of saline water with frequent interval results in restricted salt build up. Any flood irrigation system involves complete coverage of the soil surface with treated effluent and is normally not an efficient method of irrigation. This system will also contaminate vegetable crops growing near the ground and root crops and will expose farm workers to the effluent more than any other method. Thus, from both the health and water conservation points of view, border irrigation with wastewater is not satisfactory. Furrow irrigation, on the other hand, does not wet the entire soil surface. This method can reduce crop contamination, since plants are grown on the ridges, but complete health protection cannot be generated.

7. Drainage

In order to maintain a satisfactory salt balance in the root zone, excellent drainage of soil is most essential.

8. Overaged seedlings

Transplantation of rice with overaged seedlings in better establishment in salt affected soil results in better establishment.

9. Mulching

Mulching with locally available plant materials helps in reducing salt problems by reducing evaporation and by increasing infiltration. Salts are leached into lower layers even with rainfall by application of mulches.

10. Crop rotation

Inclusion of crops such as rice in the rotation reduces salinity.

11. Dilution of wastewater with fresh water and addition of gypsum and also possing through fly ash.

References

Anderson John, 2001. "Prospect for International guideline for water recycling", Water 21, August 2001, Magazine of the international water Association.

Asano Takashi, 2000. "Groundwater Recharge with reclaimed municipal wastewater and Waters of Impaired quality", Wastewater Reclamation and Reuse, Arab school on Science & Technology, New age International (P) limited.

Asano, Takshi, 2000. "Reclaimed Wastewater as a Water Resource", *New* age International (P) limited, Arab School on Science & Technology, pp 2-19.

Augustine Selvaseelan, A., Maheshwari, M., Rajannan, G. and Thiagarajan, T. M. 2001. Agricultural utilization of industrial wastewater. In: *Agriculture, S.* Kannaiyan and C. Ramasamy (eds.). Tamil Nadu Agric. Univ., Coimbatore. pp. 100-13.

Balagopal, C., Maini, S. B. and Hrishi, N. 1977. Microbial treatment of starch factory effluences and the production of single cell protein. *J. Root Crops* 3: 47-50.

Chandrasekaran, N. and Rajkannan, B. 2003. Continuous applications of sewage effluents on soil physical properties. *Madras agric. J. 90:211-15.*

Chinnusamy, C., Annadurai, K., Jayanthi, C., Veeraputhiran, R. and Karunanithi 2001. Organic amendments and distillery effluent on soil fertility and productivity of rice. In National Seminar on Use of Poor Quality Water and Sugar Industrial Effluents in Agriculture. ADAC & RI, Trichy, Feb. 5. pp. 84.

Dahl, Stephen, 2003. "A Matter Approach to the Design and Operation of a Wastewater Reuse Scheme".

Devarajan, L. and Oblisami, G. 1995. Effect of distillery effluent on soil fertility status, yield and quality of rice. *Madras agric. J.* 82: 664-65.

Devarajan, L., Rajannan, G., Ramanathan, 0. and Oblisami, G. 1993. Sugarcane cultivation with distillery effluent. *SISTA Sugar J.* 20: 23-25.

Devarajan, L., Rajannan, G., Ramanathan, G. and Oblisami, G. 1993a. Studies on the effluent of alcohol, distillery effluent on soil fertility status, yield and quality of crops produced in scheme report submitted to Sakthi Sugar Ltd., Sakthi Nagar. pp. 96-234.

Dhevagi, P. 1997. Studies on the impact of irrigation of treated effluent from bagasse based paper mill on agro-ecosystem. Ph. D. thesis, Tamil Nadu Agric. Univ., Coimbatore.

Dhevagi, P., Maheshwari, M., Ilamuragan, K. and Oblisami, G. 2003. Influence of paper mill effluent on soil microflora. *J. Eco Toxicol. Environ. Monit.* 13: 103-09.

Dubey, R. P. and Joshi, R. C. 1980. Studies on the removal of odour from pulp mill wastewater. *Indian Pulp Pap.* 34: 13-20.

Eldho. T.I. 2003. Remediation of Contaminated Groundwater - Scope of Onsite Pump and Treat Technology - IE Journal -CV Vol.84.

Eldho.T.I. 2001. Ground water Contamination - The challenge of pollution control and protection. *Journal of Indian Water Works Association.*

Gautam, D. D., Kumar, K. and Bishnoi, S. 1992. Effect of dairy effluent on seed germination of some rabi and kharif crop plants. *J. Environ. Biol.* 13: 7-12.

Gomathi, V. and Oblisami, G. 1922. Effect of pulp and paper mill effluent on germination of tree crops. *Indian I Environ. Health 34:* 326-28.

Jackson, Rihert & ORI), Llizabeth, 2002. "Grey Water Reuse- Benefit or Liability, The UK Perspective", Water 21, August 2001, Magazine of the International Water Association, pp 3 8-39.

Jimencz. Blanca, Chavez Alma, Harrior, J Eugenio 2002. "Impact and Potential of Reused water in the Mezqital valley", Water 21, August 2001, Magazine of the International Water Association, pp 34-36.

Joshi, H. C., Pathak, H., Chaudhari, A. and Kaira, Naveen 1996. Distillery effluent as a source of plant nutrients-Prospects and problems. *Fertd. News* 41: 41-47.

Jothimani, P., Prabakaran, J., Rajannan, G. and Ramasamy, K. 1998. Response of greengram CO 4 biotreated dye factory effluent. In: Water World '98-National Seminar on Application of Treated Effluent for Irrigation. REC, Trichy. pp. 87.

Judd, Dr Simon, 1999. "Industry's Reuse Challenge", Water 21, August 2001, Magazine of the International Water Association. Sept - Oct, pp 48-49.

Kansal, B. D. 1994. Effects of domestic and industrial effluents on agricultural productivity. In: *Management of Agricultural Pollution in India,* G. S. Dhalial and B. W. Kansal (eds.). Common Wealth Publishers, New Delhi. pp. 157-73.

Kathiresan, G., Selvaraj, N., Manickam, G. and Sridhar, P. 1998. Effect of tannery effluent dilutions on growth and yield of sugarcane. In: Water World '98-National Seminar on Application of Treated Effluent for Irrigation. REC, Trichy. pp. 32-37.

Kumaran, S., Balasubramanian, V. and Balasubramanian, A. 2003. Screening of suitable tree species for the areas affected with tannery effluent. *Crop Res,* 25: 192-94.

Mahimairaja, S., Anandh Kumar, Saravanan, K, Ramasamy, K. and Naidu, R. 1999. Effect of tannery effluent irrigation on .crop, soil and ground water. Paper presented in International Workshop on Environmental Impacts of Metals. Tamil Nadu Agric. Univ., Coimbatore, 6-7 July. pp. 47-48.

Mahimairaja, S., Sakthivel, S., Divakaran, J., Naidu, R. and Ramasamy, K. 1998. Extent and severity of contamination around tanning industries in Vellore district. Paper presented during the workshop or "Towards better management of chromiurr rich tanning wastes". Tamil Nadu Agric Univ., Coimbatore. pp. 52.

Malarvizhi, P. and Rajamannar, A. 2001. So. fertility management in fodder' cultivated area through sewage water irrigation *Madras agric. J.* 88: 472-77.

Manonmani, K., Murugeswaran, R. and Swaminathan, K. 1992. Effect of photofilm factory effluent on seed germination and seedling development of some crop plants. *J. Ecobiol.* 4: 99-105.

Math Srinivaschari, Dhakshinamoorthy, M. and Arunachalam, G. 2000. Accumulation and availability of Cu, Mn and Fe in soils polluted with paper mill waste water. *Madras agric. J.* 87: 237-40.

Mathkour, Fauez Al 2000. "Treated Sewage Effluent for Landscape Irrigation in Kuwait", Wastewater Reclamation and Reuse, Arab school on Science & Technology, New age International (P) limited, pp 227-242.

Metcalf and Eddy, 2003. "Wastewater engineering Treatment and reuse", forth edition, Tata McGraw Hill Publishing Company limited, New Delhi.

Mujeriego Rafael, Carbo Mrio Sala Lluis and Tiirret Joseph, 2000. "Agronomic and public Health Assessment of Reclaimed Water Quality for Landscape Irrigation", Wastewater Reclamation and Reuse, Arab school on Science & Technology, New age International (P) limited.

Naidu, R. and McLaughlin, M. J. 1993. Heavy metal contamination of agricultural land and ground water in Tamil Nadu, India and in Australia. Report of ACIAR. pp. 26.

Nandy Tapas, Kaul, S.N. 2000. Sewage Treatment Plant in India Current Practices and Recommendation for Improvement, Environment Pollution Control Journal, vol. 3 No. 5, pp 5-11.

Narasimharao, P. and Narasimharao, Y. 1992. Quality of effluents water discharged from paper board industry and its effect on alluvial soil and crops. *Indian J. agric. Sci.* 62: 9-12.

Ogoshi, Masashi, Suzuki, Yutaka and Asano, Takashi 2002. "Nonpotable Urban Water reuse - A case of Japans Water Recycling", Water 21, August 2001, Magazine of the international water Association, pp 31-32.

Poonam Tyagi, Buddhi.D, Rubina Chotidhary and Sawhney R.L. 2000. Degradation of Ground Water Quality due to Heavy metals in Industrial Areas of India - A Review. *Indian Journal of Environmental Protection*, Vol. 20, No.3.

Prabhakar Clement.T 1998. Remediation of Contaminated Ground Water - Lessons from U.S.A. *Journal of Indian Water Works Association.*

Pushpavahli, R. 1990. Studies on the characterization of pulp and paper mill effluent and its effect on soil profile characteristics and on germination, yield and juice quality of sugarcane. M. Sc. (Agri.) thesis, Tamil Nadu Agric. Univ., Coimbatore.

Rajannan, G., Devarajan, L., Prabakaran, J. and Oblisami, G. 1998. Assessment of impact on the use of distillery effluent as irrigation water to banana crop. In: Water World '98- National Seminar on Application of Treated Effluent for irrigation. REC, Trichy. p. 24.

Rajukannu, K., Manickam, T. S., Shanmugam, K., Chandrasekharan, A. and Gladis, R. 1996. Distillery spent wash-Development of technology for using it as an amendment for reclamation of sodic soils. In: Proc. National Symp. on Use of Distillery and Sugar Industry Wastes in Agriculture. ADAC & RI (TNAU), Tiruchirappalli. pp. 40-45.

Ramachandran, K. 1994. Studies on the effect of dyeing factory effluent on soils and adjoining groundwater. M.Sc. (Agri.) thesis, Tamil Nadu Agric. Univ., Coimbatore.

Ramakrishna, C.R. 2002. Proceedings of "Modern Trends in Water Resources Development and Environmental management", Department of Civil Engineering Vellore, Institute of Technology, Vellore, pp C1-C4.

Ramanathan, G., Loganathan, S., Krishnamoorthy, K. K. and Balaraman, K. 1977. The effect of sewage irrigations on soil characteristics. *Madras agric. J.* 64: 194-96.

Rowe, Peavy, and Tehohanoglous, 1985. "Environment Engineering", McGraw Hill, International Edition.

Sakthivel, S., Divakaran, J., Ramasamy, K. and Naidu, R. 1999. Treated tannery effluent irrigation for tree plantations and its impact on soil and ground water. Paper presented in International Workshop on Environmental Impacts of Metals. Tamil Nadu Agric. Univ., Coimbatore, 6-7 July. pp. 46-47.

Sala Lluis, Mujeriego Rafael, Serrqa Manel & Asano, Takashi 2002. "Reuse & Conservation, Spain sets the example", (Aug 2002) Water 21 Magazine of International water Association.

Senthilvelan, S. P. 1994. Effect of continuous application of sago factory water on physico-chemical properties of soils and certain crops. M.Sc. thesis, Tamil Nadu Agnic. Univ., Coimbatore.

Sincero, A.P., Gregorian A, Sincero, 1999. "Environmental engineering A Design Approach", Prentice Hall of India Private Limited New Delhi.

Singaram, P. 1995. Effect of industrial and sewage effluents on soil and crop production. In *Agricultural Inputs arid Environment,* S.P. Palaniappan (ed). Scientific Publishers, Jodhpur. pp. 357-88.

Singhal, S. 2003. India 2025-Environment. Research Projects on India2025 conducted by Centre for Policy Research, New Delhi, July 2003. http://planningcommission.nic.in/reports/ sereport/ser/vision2025/ env 2025.doc. pp. 6-7.

Sl Wakeel, Atta, F. El- Nashar Baumy M. and Al - Hamad O.A, "Reuse of Sewage Effluent for Irrigation", pp 244-258.

Solaimalai, A., Baskar, M. and Ramesh, P.T. 2002. Environmentally safe disposal of waste waters. In: *Environment, Pollution and Management,* A. Kumar, C. Bohra and L. K. Singh (eds.). A.P.H. Publishing Corporation, New Delhi. pp. 15-73.

Tchobanoglous, G. 2000. Appropriate Technologies for wastewater treatment and reuse", Arab school on Science & Technology, New age International (P) limited.

TERI 1998. Looking back to think ahead: GREEN India 2047. Tata Energy Research Institute, New Delhi. pp. 346. Thangavel, P., Rajannan, G.

and Balagurunathan, R. (2003). Effect of tannery effluent on soil properties and yield of maize. *J. Ecobio.* 15: 161-68.

Totawat, K. L. 1991. Effect of effluent from zinc smelter on metallic cations in soil and crops in cropping sequence. *J. Indian Soc. Soil Sci.* 39: 542-48.

Totawat, K. L. and Subhash Chauvan 1992. Genotypic tolerance of crops to smelter effluent. *Indian J. Environ. Hith.* 34: 28-32.

Velu, V., Arumugam, A. and Arunachalam, G. 1998. Impact of paper mill effluent irrigation on the yield and nutrient uptake in rice. In: Water World 98-National Seminar on Application of Treated Effluent for Irrigation. REC, Trichy. pp. 18.

Vijayakumari, B. 2003. Impact of textile dyeing effluent on growth of soybean. *J. Ecotoxicol. Environ. Mont* 13: 59-64.

Vishnu ram, B, G, Venkatasubramani, GS. Sunder, M, Lenin, 2001. Quality of Groundwater - "A Challenge of the New Millennium Proceeding of Modem in Water Resources Development and environmental Management". VIT, Vellore, pp. C5-C8.

Zalawadia, N. M. and Raman, S. 1994. Effect of distillery wastewater with graded fertilizer levels on sorghum yield and soil properties. *J. Indian Soc. Soil Sci.* 42: 575-79.

EXERCISES

19.1. Define the recycling of industrial effluents?

19.2. Describe historic perspective of industrial influents and its reuse.

19.3. What is the need of recycling of industrial effluents?

19.4. Describe the environmental impacts of industrial use of water.

19.5. Describe the characteristics of different industrial effluents.

19.6. Differentiate the following:

(a) Distillery effluents and textile effluents

(b) Sewage effluent and fertilizer industry effluents

(c) Pump and treat remediation system and bio-remediation systems.

19.7. Describe the strategies for recycling of industrial effluents in agriculture.

19.8. Describe the remediation of industrial effluents.

19.9. Explain the planning for reuse of waste water.

19.10. Describe the measures for conservation of water in industries.

19.11. Discuss the strategies for conservation water in industries.

■■■■

Chapter 20

WATER RECYCLING IN SALTY LANDS

20.1 Introduction

Presence of certain chemical substances known as hazard salts beyond permissible limits either in soil or in harvested rainwater will affect crop growth and ultimately little crop yield and also water productivity. Most of the salt affected soils are found in the arid and semi-arid ecosystem tracts of the world. Some soils have the salts as original deposits. Recycling water obtained from surface runoff may be of satisfactory quality otherwise salt build-up in the root profiles could occur over a period of time. This will occur due to the fact that during crop growth period water evaporates increases the salts concentration in the rootzone. High water tables at some locations could bring the salts to the surface due to high rainfall or seepage from channels, canals etc. If irrigation is applied from groundwater which contains large amounts of salts, it enhances the salt build-up in the soil profiles. Salts could also he brought in through sub-surface inflows.

20.2 Salt Affected Soils

Extensive areas of land in the arid and semi-arid regions of India have gone out of cultivation due to rise of water table and accumulation of salts. Excessive irrigation and poor water management are the main causes of water logging and salt build-up. Thus, development of soil salinity is a challenge to the scientists, engineers and researchers to make the irrigated agriculture permanent for 'livelihood' food security to the millions of the peoples. It is evident that accumulation of salts in soils leads to unfavaourable soil-water-air relationships, consequently decreases crop production and results into land going out of cultivation.

High concentrations of certain salts when present in the soil and water then hampers the efficient management of soil and rainwater management and consequently crop production and water productivity effected.

Increase in concentration of soluble salts increases the solute suction which thus reduces the availability of soil water to plants. Some of them like sodium carbonate, soluble borates etc. could be toxic to plants. These salts are harmful indirectly to plants due to rise in the soil pH by sodium carbonate, which makes nutrients *viz.*, phosphates, manganese and zinc becomes unavailable to plants. High concentrations of exchangeable sodium produce a soil structure unbreakable which reduces permeability, aeration, infiltration rate and soil workability.

Salts found in the soil solution and are also linked with the clay particles. There is a continuous interchange or exchange of salts as ions between the soil solutions and clay complex, which establish an equilibrium situation. The salts found in the soil solution which are known as soluble salts. Which can be removed from the soil by leaching and subsequent drainage. The salts retain by clay are known as exchangeable salts. These salts should be exchanged before they could be removed by leaching. The soluble salts that occur in soils consist mostly of certain proportions of the cations of sodium, calcium, magnesium and potassium and the anions of chloride, sulphate, bicarboante, carbonate and nitrate.

20.2.1 Important terminology pertaining to salt problems

Exchangeable sodium percentage (ESP): It is the degree of saturation of the soil exchange complex with sodium and may be determined as,

$$ESP = \frac{\text{Exchangeable sodium (milli equivalents per 100 gm soil)}}{\text{Cation - exchange capacity (milli equivalents per 100 gm soil)}} \times 100$$

Cation exchange capacity (CEC): It is the capacity of soil to absorb total quantity of cations by cation exchange and usually expressed as milli-equivalents per 100 grams.

Sodium adsorption ratio (SAR): It is defined as the ratio of soil extracts and irrigation water, used to express the relative activity of sodium ions in exchange reaction with soil. It is expressed as,

$$SAR = \frac{Na^{+}}{\sqrt{(Ca^{++} + Mg^{++})/2}} \times 100 \qquad (20.1)$$

The ionic concentrations are expressed in milli-equivalents per litre.

Leaching: It is the process of removal of soluble materials (salts) by the passage of water through soil.

Leaching requirement: It may be defined as the fraction of water entering the soil that must pass through the rootzone in order to prevent soil salinity from exceeding a specified value. Leaching requirement is used primarily under steady-state or long term average conditions.

Example 20.1: Analysis of 5 g soil sample reveled that the total concentration of exchangeable cations was 1.35 meq and that of exchangeable sodium was 0.9 meq. Express the cation exchange capacity in meq/100 gm soil and exchangeable sodium in percent.

Solution:

In 5 g soil the total exchangeable cations = 1.35 meq.

∴ In 100 g soil the total exchangeable cations

$$=\frac{1.35\times100}{5}$$

Hence, CFC = 27 meq/100 gm soil

$$\text{Exchange Na percent, } ESP=\frac{\text{Exchange Na}}{\text{Total exchange cations}}$$

$$=\frac{0.9\times100}{1.35}=67$$

$$\text{or exchangeable Na/100 gm soil } =\frac{0.9\times100}{5}=18\text{ meq.}$$

Gypsum requirement (GR): It is determined by Schoonover's method. In this method a saturated solution of gypsum is prepared by adding 5 gm gypsum in one litre of water and shacking mechanically for 10 minutes. The suspension is then filtered. 5 gm soil is weighed and 100 ml solution of saturated gypsum is added. The mixture is shaken for 5 minutes on a mechanical shaker or 30 minutes by hand. It is filtered and Ca+Mg are determined by standard versenate solution.

GR (Mg/100 gm soil) = [Ca concentration of added gypsum solution (Meq/l) – (Ca+Mg) concentration of filtrate (Meq/l)] × 2

It is essential that the soil should be free from soluble salts for GR estimation.

Example 20.2: A quantity of 100 ml of gypsum solution having 35 meq/l concentration as calcium, on reacting with 5 g of an alkali soil revealed 30.5 meq/l of Ca+Mg concentration in filtrate. Determine the gypsum requirement in meq/100 g soil.

Solution:

Weight of soil = 5 g

Volume of gypsum solution added = 100 ml

Concentration of (Ca+Mg) after shaking with soil = 30.5 meq/l

Gypsum requirement (meq/100 g soil) = 2(35-30.5)

= 9 meq/100 g soil

Gypsum requirement = 9 meq Ca/100 gm soil

If 1 meq Ca/100 g soil = 1.72 metric tones per hectare for 15 cm on depth soil

Then for 9 meq/100 g soil is as,

= 9 × 1.72 metric tones/ha for 15 cm depth of soil

= 15.48 metric tones/ha for 15 cm depth of soil

20.2.2 Salt build up and salt balance

Salt building in irrigated soils: There are following main causes for the development of salinity and alkalinity in soils:

(i) Use of saline irrigation water,

(ii) Deposition of salts on soil surface from high sub-soil water table,

(iii) Seepage from the canals,

(iv) Arid climate,

(v) Poor drainage, and

(vi) Backwater flow or intrusion of sea water in coastal areas.

All these factors, either single or in association with other factors are responsible for the development of saline and alkali soils.

Salt accumulation: Saline or alkali soils are seldom formed in-situ by residual weathering of primary rocks. However, hydrological conditions contribute substantially to the development of salinity and alkalinity. The hydrological processes release the salt constituents from the primary and secondary minerals to the surrounding waters. There are two specific reasons as,

(i) Salt concentration is gradually increased due to high evaporative demand in water sources in arid regions, and

(ii) Salts released are transported to lower layers and ultimately by streams to the ocean in humid regions. However, the salts deposited by irrigation water are leached down to the lower layers of the soils by rainfall. It is contradicfom in arid regions the released salts may not be transported a away because of insufficient rainfall.

20.3 Classification of Salt Affected Soils

Salt affected soils are classified based on the soluble salt concentration and exchangeable sodium percentage as these two are the important parameters influencing plant growth and soil condition. Since these parameters are dependent on the soil moisture content. Thus, a common moisture content *viz.*, saturation on a disturbed soil sample in the laboratory is used for their determination. A soil sample is brought to saturation by stirring in distilled water until a characteristic end point is reached, where the soil glistens and flows slightly without free water standing on the surface. The solution (known as saturation extract) is extracted by placing the sample in a funnel with filter paper and applying suction. The salty soil is classified based on three indices such as, (i) EC, (ii) ESP and (iii) pH.

The total soluble salt concentration is estimated by measuring the electrical conductivity of the saturation extract. An electrical conductivity meter is used for measuring.

(i) Electrical conductivity (EC) – It is the total salt content of soil solutions or irrigation waters at 25°C and expressed dsm^{-1}.

The units of EC is dsm^{-1} which is the reciprocal of resistivity and it is expressed in ohms. The resistivity is the resistance in ohms of a conductor which is 1 cm long and has a cross-sectional area of 1 cm^2. As the standard unit of EC *i.e.* mhos/cm is a large unit so far as the solutions under consideration, it is customary to choose a smaller sub unit for expressing the data accurately. According to international system EC is expressed as dsm^{-1} 1 dsm^{-1} = 1 mhos/cms:

Parts per million (PPM) – It is numerically equivalent to milligrams of the substance per litre. Milliequivalent per litre is the milliequivalent weight of anion or compound in a litre of solution. EC is determined from million equivalent per liters (meq/l) by the following relationship:

$$meq/1 = 10 \times EC \times 1O^3 \qquad ....(20.2)$$

EC for irrigation waters and soil extracts varies from 0.1 to 5.0 milli-mhos per cm.

Electrical conductivity (EC) is determined from parts per million (PPM) by the following relationship:

$$ppm = 0.64 \times EC \times 10^6 \qquad ....(20.3)$$

For irrigation waters in the range 100 to 5000 micromhos per cm. Meq/1 are converted to ppm by multiplying meq/1 for each ion by its equivalent weight and adding them.

Equivalent weight: It indicates combining capacity of an clement or compound with hydrogen. It is expressed as the weight in grams of hydrogen. Equivalent weight is determined by dividing the atomic weight with valency of the ion or compound. The milli-equivalent weight is one thousandth part of the equivalent weight. The equivalent weights of some ions and salts relevant to problems of soil and water are given in Table 20.1.

Table 20.1: Equivalent weights of some common constituents

Name	Equivalent weight	Name	Equivalent weight
(i) Cations		(iii) Salts	
Calcium (Ca^{++})	20.04	Calcium chloride ($CaCl_2$)	55.50
Magnesium (Mg^{++})	12.16	Calcium sulphate ($CaSO_4$)	68.07
Potassium (K^+)	39.10	Calcium carbonate ($CaCO_3$)	50.04
Sodium (Na^+)	23.00	Magnesium chloride (MgCl)	47.62
(ii) Anions		Magnesium carbonate ($MgCO_3$)	42.16
Carbonate (CO_3^{--})	30.00	Potassium chloride (KCI)	74.56
Bicarbonate (HCO_3^-)	61.01	Sodium chloride (NaC1)	58.45
Chloride (Cl^-)	35.46	Sodium carbonate (Na_2CO_3)	53.00
Sulphate (SO_4^{--})	48.03	Sodium bicarbonate ($NaHCO_3$)	84.01

Example 20.3**:** 9 g of sodium chloride has been dissolved in 2 litres of water. Determine the salt concentration in forms of ppm, mhos, millimhos, micromhos/cm, and meq/litre.

Solution:

Amount of salt = 9 gm = 9 x 1000

= 9000 milligrams

Amount of water = 2 litres

$$\therefore \text{Salt concentration in mg/1} = \frac{9000}{2} = 4500 \text{ ppm}$$

$$EC = \frac{ppm}{640} = \frac{4500}{640} = 7.03 \text{ millimhos/cm}$$

$$= 7.03 \times 1000$$

$$= 7030 \text{ micromhos/cm}$$

$$= \frac{7.030}{1000} = .00703 \text{ mhos/cm}$$

$$\text{meq/l} = \frac{\text{Concentration of salt in mg/l}}{\text{Equivalent weight}} = \frac{4500}{58.45} = 76.99$$

(i) Exchangeable solution (ESC) – It is the ratio of exchange sodium content to cation exchangeable capacity and is expressed as,

$$ESP = \frac{\text{Exchangeable sodium content (meq/per 100 gm soil)}}{\text{Cation Exchange Capacity (meq per 100 gm soil)}} \times 100 \quad ....(20.4)$$

Every soil has certain capacity to adsorb the positively charged constituents of dissolved salts (cations), such as calcium magnesium, sodium and potassium. This is known as the cation exchange capacity. The different adsorbed cations can be exchanged for one another, *viz.*, calcium by sodium depending upon their relative concentrations in the soil solution.

Exchangeable sodium is the amount of adsorbed sodium in the soil, expressed in per cent of the exchange capacity, in milli-equivalents per 100 g of soil.

When the exchangeable sodium percentage becomes high (exceeds 10 to 20 per cent of the exchange capacity), the soil develops poor physical conditions (like low permeability to water and air, hard when dry and sticky when wet, etc.), high pH values (above 8.5) and associated nutritional disorders, resulting in reduction of crop growth. The poor physical conditions develop from the action of exchangeable sodium, in dispersing the soil aggregates which results into clogging of the soil pores.

Determination of ESP is time consuming, therefore, it can be determined quickly with sodium adsorption ratio (SAR). Although it is less reliable estimate. ESP is determined with following relationship:

$$ESP = \frac{100\,(-0.0126 + 0.01475\,SAR)}{1 + (-0.0126 + 0.01475\,SAR} \quad ....(20.5)$$

Soluble sodium can be assessed on using a flame photometer and soluble calcium and magnesium by titration. Concentrations, thus determined in the saturation extract and these are expressed in milliequivalents/ litre.

(ii) pH – It is the soil reaction or is the negative logarithm of the hydrogen ion activity. It is used as an index of acidity or alkalinity of soils. It is determined using a pH meter.

The above different parameters determined with standard procedure as suggested by above are given in USSL Staff (1969). A general classification, of the salt affected soils based on above indices is given in Table 20.2.

Table 20.2: Classification of salt affected soils

Classification	EC of the Saturation Extract dsm^{-1} at 25^0C	ESP value	pH value	Problems
Saline soils	>4	<15	Less than 8.5	High salt concentration
Alkali soils or non-saline Sodic	<4	>15	Usually between 8.5 to 10	pH & soil structure
Saline-Alkali soils or Saline Sodic	>4	>15	Not higher than 8.5	High salt & sodium content

Saline soil: It is the soil which having electrical conductivity of the saturation extract (ECe) more than 4 millimhos/cm at 25°C, exchange sodium percentagable (ESP) is less than 15 and the pH of the saturated soil paste is less than 8.5.

Alkali soil: It is the soil which has exchangeable sodium percentage (ESP) more than 15 and EC of the saturation extract less than 4 dsm^{-1}. The pH value of such soils is more than 8.5 and ranges from 8.5 to 10.

Saline-alkali soil: It is the soil which has EC of hte saturation extract greater than 4 dsm^{-1} at 25°C and ESP value greater than 15. The pH of the saturated soil paste is greater than 8.5.

However, saline and saline alkali soils are better structured than alkali soils. Other factors particularly the texture has modifying influence. The exchangeable sodium is influenced by the texture, type of clay minerals, potassium content and organic matter content. High ESP would be more problematic in fine textured soils as compared to in coarse textured soils.

20.4 Quality of Irrigation Waters

Generally, all irrigation waters contain some soluble salts in them whatever may be source either surface or groundwater etc. The quality of harvested water varied with watershed characteristics and same was compared with well and river water for irrigation (Table 20.3). However, it is evident from the results that the quality of the harvested water was on par with that of river water and could be safely used for supplemental irrigation.

Table 20.3: Quality of runoff water vis-à-vis ground and river water in the region

Source of water	Chemical constituents (ppm)						
	CO_3	HCO_3	Cl^-	SO_2^-	Ca^{++}	Mg^{++}	B^{+++}
Sreedharagadda (Well No. 1)	Nil	729.6	1014.0	1598.4	210.4	257.7	2.80
Sreedharagadda (Well No. 2)	Nil	605.7	710.0	1708.8	176.8	191.9	3.80
Somasamudram (Well No. 1)	Nil	729.6	1149.0	1502.4	277.8	212.0	3.30
Somasamudram (Well No. 2)	Nil	715.5	1081.0	1598.4	25.2	40.4	2.06
Tungabhadra river	Nil	84.0	19.0	Nil	16.6	4.7	0.035
Farm pond	Nil	178.7	101.5	Nil	16.8	5.0	0.52

(Source: Hayavadana Rao and Chittaranjan 1967)

There are two conditions: (i) High salt-concentration that it is not suitable for irrigation. (ii) Low salt-concentration but with continuous irrigation may gradually build up the salt content in the root zone.

The major dissolved constituents in natural waters are the salts of sodium, calcium and magnesium as their chloride, sulphate, carbonate and bicarbonates. Potassium and nitrate ions are also present sometimes. Irrigation water quality is commonly assessed in terms of soluble salt content *viz*., percentage of sodium, boron and bicarbonates contents. Some other elements also present *viz*., boron, lithium, silicon, iron, manganese, fluorine, aluminum, lead, zinc, vanadium and chromium occur in minor quantities. Occurrence of these elements influences the quality of drinking water.

However, following criteria are described below :

(i) Sodium absorption ratio (SAR), equation (20.1)

(ii) The sodium percentage & is calculated as,

$$Na\% = \frac{Na^+ + k^+}{Ca^2 + M_g^{2+} + Na^+ + k^+} \times 100 \qquad (20.5)$$

(iii) Electrical concluctivity (EC)

(iv) Residual sodium carbonate (RSC)

The recommended classification with above criteria is given Table 20.4.

Table 20.4: Guidelines for evaluation of water quality for irrigation

Water class	Na, %	EC, us/cm	SAR	RS, meg/lit
Excellent	<20	<250	<10	<1.25
Good	20-40	250-750	10-18	1.25 - 2.0
Medium	40-60	750-2250	18-26	2.0 - 2.5
Bad	60-80	2250-4000	>26	2.5-3.0
Very Bad	>80	>4000	>26	73.0

(Source : GWB and PCB, 2000)

Parameters for Water Quality: There is several classification for suitability of irrigation water to crops based on their chemical characteristics. However, no single classification holds true under all conditions. This is mainly due to the large number of factors that affect the suitability of irrigation water.

Residual Sodium carbonate: Presence of carbonate and bicarbonate in irrigation water results in the precipitation of calcium and magnesium present in the soil and thus increases the extent of sodium sodic hazard. For assessing the sodium hazard due to carbonate and bicarbonate of irrigation water. Generally, it is assumed that all the calcium and magnesium would precipitate as carbonates. The residual sodium carbonate (RSC) is expressed as under:

$$RSC = (CO_3^{--} + HCO_3^-) - (Ca^{++} + Mg^{++}) \qquad ....(20.5)$$

where, the concentration of ions are expressed as meq/1. It is evident from experience that waters with more than 2.5 meq/1 residual sodium carbonate are not suitable for irrigation purposes. Waters containing 1.25 to 2.5 meq/1 are marginal, and those containing less than 1.25meq/1 RSC are considered safe (Table 20.4).

According to the U.S. Salinity Laboratory (USSL), irrigation water is classified based on electrical conductivity and sodium adsorption ratio as shown in Fig. 20.1.

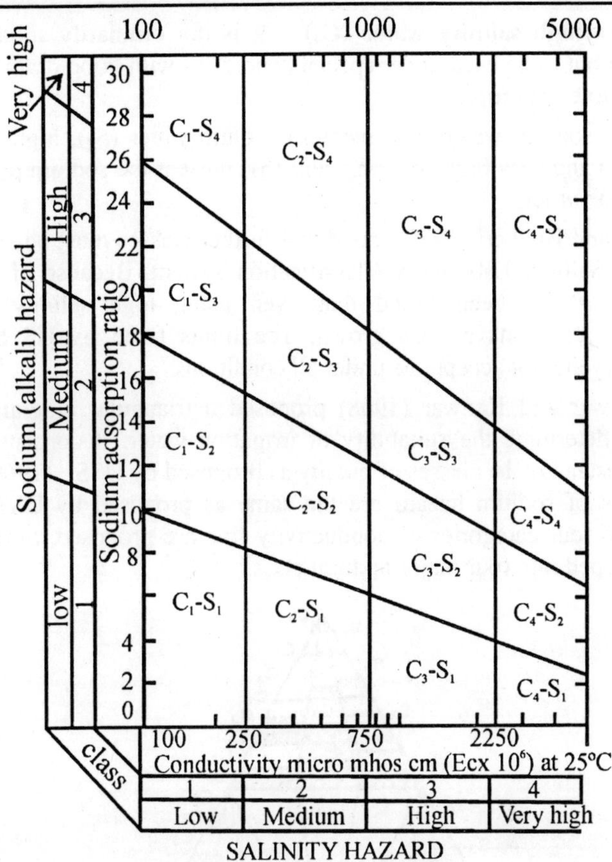

Fig. 20.1: Classification of irrigation water given by USSL staff

It is evident from Fig. 20.1 that C_1, C_2, C_3 and C_4 indicates low salinity, medium salinity, high salinity and very high salinity of water, respectively. While, S_1, S_2, S_3 and S_4 indicates low, medium, high and very high sodium adsorption ratio of the water, respectively. The different of quality of irrigation water used in the following manner:

Low salinity water (C_1) – It can be used for irrigation with most crops and most soils. Some leaching is required.

Medium salinity water (C_2) – It can be used with crops that can tolerate moderate salinity and leaching.

High salinity water (C_3) – It can be used with adequate drainage and salt tolerant crops.

Very high salinity water (C_4) – It is not ordinarily suitable for irrigation but to be used under special conditions with good drainage and with salt tolerant crops.

Low sodium water (S_1), medium sodium water (S_2), high sodium water (S_3) and very high sodium water (S_4) present the sodium problems in the same order.

Pai and Hukkeri (1979) found that further classes must be added to the U.S. Salinity Laboratory Classification system. Because of several instances it has been found that even using high salinity waters successful crops have been grown. The limits fixed by US Salinity Laboratory are not acceptable under all conditions.

Kanwar and Kanwar (1968) proposed a triangular diagram (Fig. 20.2) to determine the suitability of irrigation water. It consists of 20 classes Instead of 16 classes of quality as proposed by USSL (1969). The categories of sodium hazard are the same as proposed by USSL, but instead of four categories of conductivity five are proposed. Soils have been grouped into four major textural class.

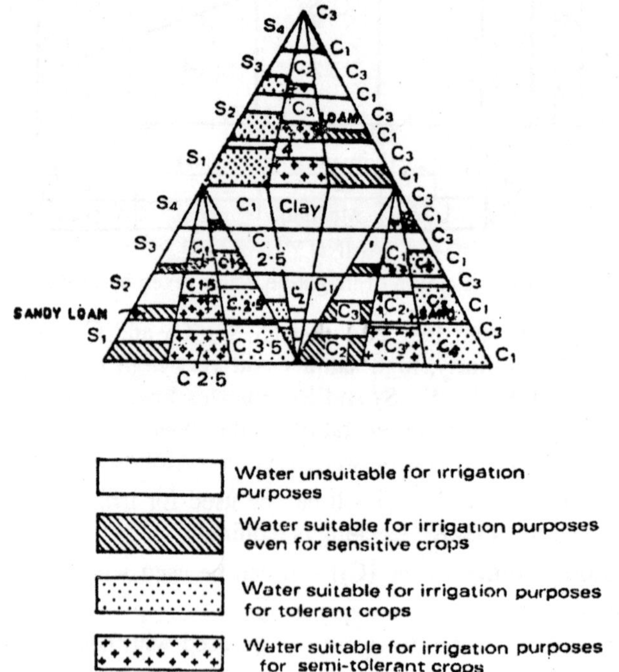

Fig. 20.2: Triangular diagram to determine suitability of irrigation water (Kanwar, 1968)

Based on different studies, a water quality rating is proposed as given in Table 20.5

Table 20.5: Water quality rating as proposed by workers in India

	Nature of soil	Crops to be grown	EC limit (micromhos/cm)
1	Deep black soils and alluvial soils having clay content more than 30 per cent, soils that are fairly to moderately well drained	Semi-tolerant Tolerant	1500 2000
2	Heavy textured soils having a clay content of 20-30 percent, soils that are well drained internally and have a good surface drainage system	Semi-tolerant Tolerant	2000 4000
3	Medium textured soils having a clay content of 10-20 percent, soils that are very well drained internally and have a good surface drainage system	Semi-tolerant Tolerant	4000 6000
4	Light textured soils having a clay content of less than 10 percent, soils that have excellent internal and surface drainage	Semi-tolerant Tolerant	6000 8000

(Source: Pal and Hukkeri, 1979)

Different crops have different tolerance limit for salt is summarized in Table 20.6.

Table 20.6: Salt tolerance of different crops

Sensitive	Semi-tolerant	Tolerant
Apple	Maize	Wheat
Cluster beans	Sorghum	Barley
Cowpeas	Pearl Millet	Oat
Citrus	Rice	Sugar beet
Gram	Safflower	Date Palm
Lentil	Sugarcane	Coconut
Lemon	Cotton	
Mash	Onion	
Peach	Potato	
Pear	Mango	
Peas	Pomegranate	

Boron Concentration: It is needed for normal growth of all plants but the quantity required is very small. When it present above specified limit then it is toxic to plants. The permissible limits of boron to several plant species is given in Table 20.7.

Boron is translocated to the leaves and accumulated in the tip and margin. The symptoms of boron toxicity is the yellowing and burning of tip of the leaves. Then source boron should be findout. If irrigation water has ex corosive boron then it should be pulsed with good quality water of if soil is source of boron then it should be leached out through leaching grass boron solvent crops.

Table 20.7: Boron tolerance of different crops

Tolerant (upto 4 ppm) of boron	Semi-tolerant (upto 2 ppm of boron)	Sensitive (upto 1 ppm of boron)
Date palm	Sunflower	Gripe
Sugatbeet	Potato	Orange
Lucerne	Cotton	Grape fruit
Onion	Radish	Lemon
Cabbage	Barley	
Carrot	Wheat	
	Maize	
	Oat	
	Sweet potato	

The tolerance of crops for total salts, sodium, boron and other toxic constituents would vary not only with the crops variety, but also with the amount of nitrate in the water, the application of fertilizers and manures and agronomic practices *viz.*, seedbed preparation, planting time etc performed.

Example 20.4: An analysis of an irrigation water sample has been done. The following data were recorded:

Constituents Ca Mg Na HCO_3 CO_3 SO_4 Cl mg/1 52.1 38.9 340.4 518.6 18.0 205.7 273.8 pH=7.9 EC at 25°C=2.065 mmhos/cm

Determine SAR & RSC and also comment on the suitability of water for irrigation purpose.

Solution:

Cations			Anions		
Constituent	**mg/1**	**meq/1**	**Constituent**	**mg/1**	**meq/1**
Ca	52.1	2.6	HCO_3	518.6	8-5
Mg	38.9	3.2	CO_3	18.0	0.6
Na	340.4	14.8	Cl	205.7	5.8
			SO_4	273.8	5.7
Total		20.6	Total		20.6

$$SAR = \frac{14.8}{\sqrt{5.8/2}} = 9.18\ (\text{meq}/1)^{1/2}$$

$$RSC = (8.5+0.6) - (2.6+3.2) = 3.3\ \text{meq}/1$$

Comments for suitability for irrigation

(i) According to US Salinity Lab Staff, the classification chart for the irrigation water can be classified as high salinity water (C_3) and low sodium water (S_1). Therefore, this water cannot be used on soil with restricted drainage. Even in soil with adequate drainage special management for salinity control may be required and crops with good salt tolerance to salinity should he sown.

(ii) As proposed by Kanwar (Fig. 20.2), the water can be classified as medium to high salinity (C_3) and moderate sodium (S_2) hazard and accordingly this water is unsuitable for irrigation in clayey soils. However, it can be used in loam and sandy loam soil for growing semi-tolerant crops

(iii) The RSC value of irrigation water is 3.3 meq/1, which is very high according to US. Salinity Lab. Staff and is unsuitable for irrigation. Practical experience in India has shown that this water can be used for irrigation successfully in soils with good drainage for semi-tolerant crops. However, gypsum will have to be applied to soil occasionally.

Influence of rainfall and depth on water quality

The salinity of ground water increases during the summer and is significantly reached during the monsoon due to dilution by rain water the of degree of variation in salinity and its ionic composition depends upon the depth of water table, infiltration capacity of soil and rainfall characteristics. The water quality has been found to deteriorate with depth. In such areas water quality is found superior in dug wells than tube wells. Tube wells should be sealed in such areas to present salinity.

Measures to use poor quality water

These are following measures are recommed for poor quality water:

(1) Addition of gypsum in bad quality water,

(2) Growing of salt tolerent crops,

(3) Use of fertilizers, especially, nitrogeneous fertilizers

(4) Soil management practices, it includes, deep ploughing, green manuring, addition of gypsum and flushing of salts

(5) Irrigation management - amount of water and frequency of irrigation should meet out leaching requirement of soil and the consumptive use of crop gram.

(6) Placement of seed - seed should be placed on raised bed and irrigated with furrow.

Salt balance equation

Salt balance: It is the relationship between the quantity of soluble salts brought into an area by the irrigation water and the quantity of salts removed from the area by the drainage water (leaching) of the area.

Therefore, in an irrigated area, a favourable salt balance – a condition wherein the output of salts equals or exceeds the input must be maintained if irrigated agriculture is to be permanent. This concept may be applied to a given piece of land or to an entire watershed. Maintenance of a favourable salt balance in the soil requires proper water management practices. Hence, it should be ensured that the salt accumulation in the root zone are not beyond permissible limits for crop.

The salt balance equation is expressed as,

$$\text{Salt inputs} = \text{Salt outs} \quad ....(20.7)$$

where,

Salt inputs = Salt in irrigation water (dici) + salt through capillary rise (dg.cg), and

Salt outputs = Salts precipitated in soil (Sp) + salts in crop (Sc) + salts in percolation water (dg.cg)

However, the water inputs and water outputs is consist of as under,

Water inputs = irrigation depth (di) + effective rainfall (dr) + capillary rise (dg)

and

Water output = Percolation (dg) + evapotranspiration (det) + change in moisture (dm)

Hence, salt balance equation may be rewritten as,

$$\text{dici} + \text{dg.cg} = \text{Sp} + \text{Sc} + \text{dp.cp} \qquad \text{....(20.8)}$$

where,

Ci, Cg and Cp represent the concentration expressed as electrical conditivities.

Equation (20.8) may be applied for any time period. Generally, the contribution from capillary rise, plant and soil precipitation losses are negligible. Hence these may be negelected for the evaluation of salt balance.

Thus, Equation (20.7) reduces to dici = dpcp

$$\text{or} \qquad \frac{dp}{di} = \frac{Ci}{Cp} = LR$$

This ratio is known as the leaching requirement (LR). It may be expressed as a fraction or percent.

If all the leaching is to be carried out by the irrigation then it is determined as,

$$\text{di} = (\text{det} - \text{dr}) + \text{dp} \qquad \text{....(20.8)}$$

On putting the value of di in equation (20.3), we obtain,

$$di = \frac{(\det - dr)}{1 - LR} \qquad \text{....(20.10)}$$

Equation (20.9) considers the contribution of rainfall toward crop water requirement but does not consider its leaching contribution.

If the irrigation water contains considerable salts, this care should be taken that the salt accumulations in the rootzone are not beyond permissible limits for crops. The salt balance equation can be written knowing the inputs and outputs of water and salt (Fig. 20.3). Water inputs for a field is expressed as,

irrigation depth (d_i) + Effective rainfall (d_r) + Capi1lary rise (d_g)

Water outputs from a field is expressed as,

Percolation (d_p)+Evapotranspiration (d_{et})+ Change in moisture storage (d_m)

Salt inputs for a field is expressed as,

Salt in irrigation water ($d_i.c_i$) + Salt through capillary rise ($d_g.c_g$)

Salt outputs from a field is expressed as,

Salt precipitated in soil (s_p)+salt in Crop (s_c)+salt in percolation water ($d_p.c_p$)

The salt balance equation is written as,

Salt input = salt output

$$d_i.e_i + d_g.c_g = s_p + s_c + d_p.c_p \qquad ....(20.11)$$

where,

c_i, c_g and c_p = the concentrations expressed as electrical conductivities.

This equation can be applied for any time period. In case where the contribution from capillary rise, plant and soil precipitation losses are negligible. The equation (20.10) can be rewritten as,

$$d_i e_i = d_p c_p$$

$$\text{or } \frac{d_p}{d_i} = \frac{c_i}{c_p} = LR \qquad ....(20.12)$$

These ratios are known as the leaching requirement (LR). It may be expressed as a fraction or percent.

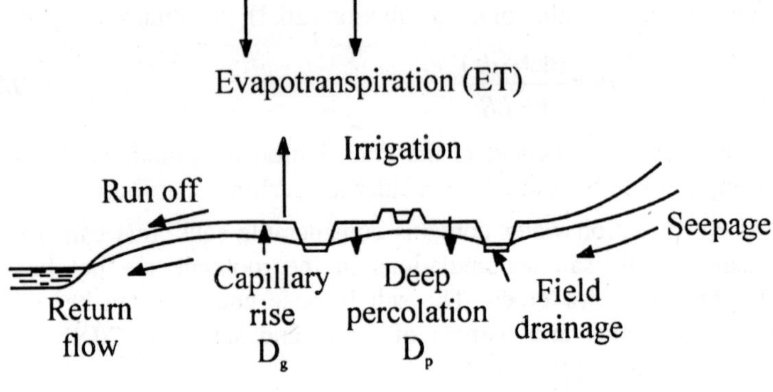

Fig. 20.3: Salt and water inputs and outputs
Source: Withers and Vipond (1974)

If all the leaching is to be carried out by the irrigation waters. Then the amount of irrigation water is the sum of the consumptive use and the drainage water and expressed as,

$$d_i = (d_{et} - d_r) + d_p \qquad ....(20.13)$$

on using Eq. 20.12 and simplification, we obtain,

$$d_i = \frac{(d_{et} - d_r}{1 - LR} \qquad ....(20.14)$$

This equation considers the contribution of rainfall towards crop water requirement but does not consider its leaching contribution. The 100 percent leaching efficiency is not always achievable. While, salt precipitation and removal of salts by crops are significant. Therefore, leaching quantity should be determined from the original salt balance equation.

The above equation is satisfactory on an annual basis. In irrigation practice, it will not be possible to achieve salt balance on a monthly basis, as the water requirements and irrigation water availabilities may differ in each month. Therefore, it is advisable necessary to examine any system on a shorter time basis to prevent high salt contents occurring particularly at crop sensitive periods.

It is evident from Table 20.8 that how salinity can vary during a cropping season with different amounts of irrigation water. It indicated that it needs different management situations as under:

(a) To maintain constant salinity throughout the season, a high rate of water application is necessary.

(b) If a constant amount of irrigation water is applied as in case then deep percolation is reduced and salt build up will result.

(c) When the deep percolation is kept constant as in case some salt build up will take place.

A comparison of the three possible management practices indicate that to maintain constant salinity throughout the season it requires high peak irrigation quantity (320 mm) and also a high drain discharge (160 mm). However, with constant irrigation applications, there is a high peak deep percolation (requiring drainage) and very high salinity level (in 3rd month) which may or may not be acceptable to the particular crop. While, constant deep percolation, results in to a high irrigation requirement and also high salt build up. The above example helps in understanding how the irrigation and drainage can he adjusted to suit the particular site conditions. In the above example 100 percent irrigation application efficiency is assumed. In actual conditions this is seldom achieved and some deep percolation losses occur which at times may satisfy the leaching requirement.

Table 20.8: Example of irrigation, percolation and salinity levels (c_1 = 2 dsm^{-1})

	1st month	2nd month	3rd month	4th month
ET (mm)	**100**	**120**	**160**	**100**
(a) Constant Salinity throughout season				
d_p	100	120	160	100
d_i	200	240	320	200
c_p	6	6	6	6
(b) Constant irrigation throughout season				
d_p	60	40	-	60
d_i	160	160	160	160
c_p	5.3	8	-	5.3
(c) Constant percolation throughout season				
d_p	40	40	40	40
d_i	140	160	-	140
c_p	7	8	10	7

(Source: Withers and Vipond, 1974)

20.5 Reclamation of Salt Affected Soils

The reclamation measures for salt affected soils depend upon the type of the soil *viz.*, (a) saline, (b) alkali or saline-alkali.

(a) *Saline Soils*: The methods of reclamation of saline soils are grouped into three categories as follows:

1. For soils with high concentration of soluble salts into lower depth but with deep water table
   (i) Impounding rain or irrigation water for leaching out injurious salts to a safe limit inside the soil, and
   (ii) Surface drains coupled with flushing to remove salts by surface and partly by subsurface drainage.
2. For soils with high concentration of soluble salts upto great depth but with high water table
   (i) Lowering of the water table either by pumping or subsurface drainage, and
   (ii) Subsurface drainage.

The aim of reclamation of saline soils is to reduce the soluble salt concentration to permissible limits. Leaching is done for this purpose.

For determination of the amount of water needed for leaching with rotation, a leaching curve is developed for particular soil. The following variables are defined for the purpose:

(i) Initial electrical conductivity = (EC_0),

(ii) Final equilibrium electrical conductivity = (EC_{eq}),

(iii) Electrical conductivity of Soil after given depth of leaching water has passed through per unit depth of soil = (EC_e),

(iv) Depth of leaching water = (DW), and

(v) Depth of Soil = (DS).

In addition to this soil samples are also taken before the test and again in the inner cylinder after the water has infiltrated to depths of 50 and 100 cm then salt content of these samples is determined. From the initial moisture content of the soil, the depth of water needed to bring the soil to field capacity is determined and the same is subtracted from the total depth infiltrated in order to find the quantity of leaching water. In this way, soil will tend to reach an equilibrium salt content. The top soil will get to this state during the test and this value (EC eq.) is to be determined. The results are plotted as the leaching curve (Fig. 20.4). This curve will be valid only for the particular soil and the particular quality of the water used for leaching.

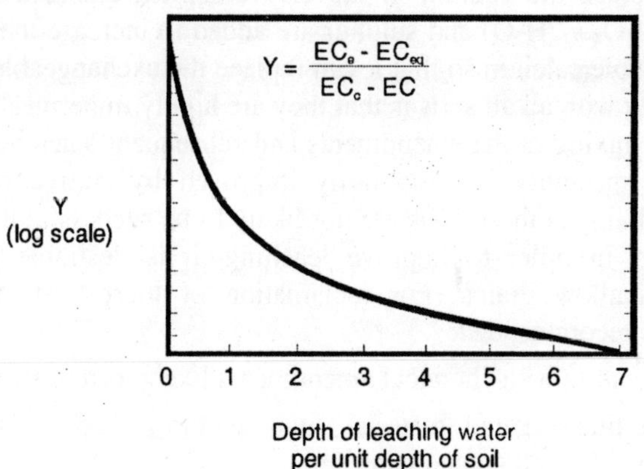

Fig. 20.4: A typical leaching curve

Example 20.6: It is required to reduce the electrical conductivity of the top 60 cm of soil from 20 mmhos/cm to 8 dsm^{-1}. Given EC_{eq}=3 dsm^{-1} evaporation= 1 cm/day and water required to raise 60 cm to field capacity from initial moisture content is 20 cm. Determine the depth of water needed for leaching.

Solution:

Substituting the given values in equation given below:

$$r = \frac{EC_e - EC_{eq}}{EC - EC_{eq}}$$

$$\frac{8-3}{20-3} = \frac{5}{17} = 0.3$$

From the leaching curve DW/DS 0.6cm

36 cm of water is required to leach the top 60 cm of soil and reduce the EC to 8.

With an infiltration late of 2 cm/day, this will take 18 days.

Total water needed = quantity for leaching+evaporation+soil requirement

$$= 36 + 18 + 20$$

$$= 76 \text{ cm}$$

(b) *Alkali Soils*: The aim of reclamation of alkali soils is to reduce the exchangeable sodium percentage and remove the released sodium salts. To replace the sodium in the clay complex, chemicals such as gypsum ($CaSO_4$ - $2H_2O$) and sulphur are added to increase the concentration of soluble calcium so that it can replace the exchangeable sodium. The difficulty with alkali soils is that they are highly impermeable and as such proper mixing of the amendments and subsequent leaching are very difficult. Permeability is temporarily improved by cultivation. These subsurface drains in these soils are not likely to be successful due to low permeability. In order to improve leaching, it is desirable to install temporary shallow drains. For reclamation of these soils following practices are recommended:

1. Treating with chemical amendments like gypsum, sulphur, etc.
2. Adding organic material such as farm yard manure, crop residues and green manuring,
3. Leaching the products of reaction after amendments are added, and
4. Deep ploughing to break any hard pan for improving drainage.

The choice of the chemical amendments depends upon the soil characteristics and on the availability and cost of the material in that particular area. The quantity of gypsum required to reclaim a soil to grow crops depends upon the crops to be grown, nature of soil and its stage of deterioration.

The amount of gypsum needed is determined as illustrated in Table 20.7. Assumed that 0 to 100 cm layer of an alkali soil contains 4 meq. of exchangeable sodium per 100 gm and has a cation-exchange capacity of 10 meq./100 gm. The exchangeable sodium percentage is 40. Suppose it is desired to reduce this to about 10. This will require the replacement of 3 meq. of exchangeable sodium per 100 gm. Assuming quantitative replacement, it will be necessary to apply the amendment at the rate of 3 meq. per 100 gm of soil. Table 20.9 provides the amount of gypsum to be applied for this purpose as 41.8 tonnes per hectare, if I m depth of soil is to be reclaimed.

Table 20.9: Amounts of gypsum required to replace exchangeable sodium

Exchangeable sodium (meq per 100 gm of soil)	Gypsum	2nd month
	Tonnes/ham	Tons/ac.ft.
1	13.9	1.7
2	27.8	3.4
3	41.8	5.2
4	55.7	6.9
5	69.6	8.6
6	83.5	10.3
7	97.4	12.0
8	111.3	13.7
9	125.1	15.5
10	139.2	17.2

(1 m^3 of soil weights approximately 1.5 tonnes, 1 ac.ft. of soil weighs approximately 4 x 10^6 pounds)

Some of the other amendments used in arid ecosystem and their equivalence to gypsum are given in Table 20.10.

Table 20.10: Gypsum equivalent of some amendments

Amendment	Quantity (tones)
Gypsum ($CaSO_4.2H_2O$)	1.00
Calcium chloride ($CaCl_2.2H_2O$)	0.85
Sulphur (S)	0.19
Sulphuric acid (H_2SO_4)	0.57
Iron sulphate ($FeSo_4.7H_20$)	1.62
Iron pyrite (FeS_2, 30 percent sulphur)	0.63
Calcium polysulphide (CaS_5, 24 percent sulphur) (lime sulphur)	0.77

(Source: Abrol and Fireman, 1977)

The USSL Staff (1966) proposed that the values of gypsum requirement given in Table 20.7 should be multiplied by a factor 1.25 to compensate for the lack of quantitative replacement. They also suggest a method known as Schoonover's method for estimating the gypsum requirements. It is evident from the field experience that gypsum requirements determined by the above methods are on higher side of the actual needs. Therefore, it is uneconomically to apply such large quantities of gypsum. It has also been observed that the soil depth to an extent of even 30cm depth need not be reclaimed for crop production.

Abrol *et al.* (1973) found that for alkali soils occurring in the Indo-Gangetic alluvial plains a fair good correlation exists between pH (1:2 soil water suspension) and exchangeable sodium percentage. Therefore, relationship exists between pH and the gypsum requirements of the soil (Fig. 20.5). The relationships given in Fig. 20.5 are approximate and not valid for all other soils. It is recommended that soil with pH lower than 9.0 (1:2 soil water suspension) will seldom need gypsum for growing crops like paddy and barley. But if the pH is above 10.0 no crop is likely to grow unless gypsum is applied.

(c) *Saline Alkali Soils*: These soils are easier to reclaim as compared to alkali soils. In some situations these boils can be reclaimed by leaching alone. Leaching tests should be conducted to ascertain whether additions of amendments are necessary or not. When amendments are needed, it is necessary to add the amendments before leaching the soluble salts or otherwise soil structure may deteriorate. A simple leaching test consists of observing the infiltration rate during a period of 7 to 14 days. If the infiltration rate is similar to the minimum rate during the first few hours then there is a good indication that little damage will result from leaching. If there is deterioration, chemical amendments are necessary.

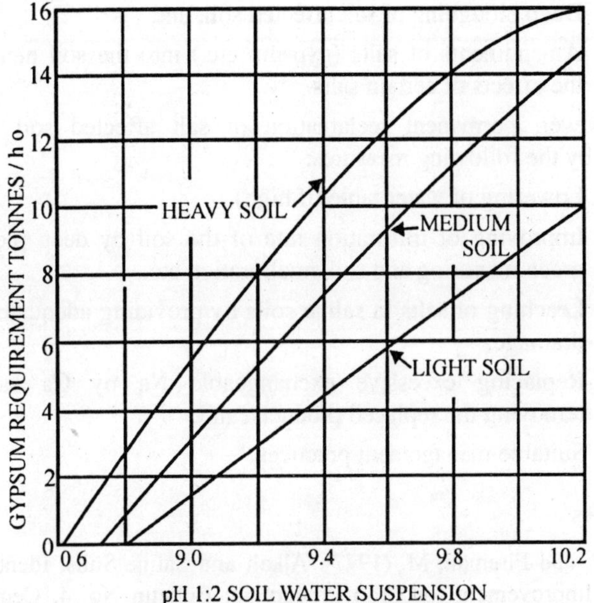

Fig. 20.5: Gypsum requirements

The predation of resalinization of reclaimed soil is essential. This could be achieved by maintaining a salt balance, drainage and controlling the depth of the water table.

The processes and practices involved in bringing saline and alkali soils into productive condition are called reclamation measures.

The reclamation measures for salt affected soils depend upon the proper diagnosis of the problem soil, reclamation procedure depend upon the following factors:

(i) Nature of the soils such as saline, alkali or saline-alkali,

(ii) Degree of salinity or alkalinity in the soil profile,

(iii) Quality of water available for leaching out the salts and the reaction products,

(iv) Drainage characteristics of the soil,

(v) Topography of the land, and

(vi) Presence of any hard pan of lime or clay in the soil strata.

Reclamation procedures: Reclamation of a salt affected soil may be achieved on a temporary basis by the following measures:

(i) Removal of the salt crust from the surface of soil,

(ii) Deep ploughing of salt affected soil, and

(iii) Amendments of salts (gypsum etc.) into the soil neutralizing the effects of certain salts.

However, permanent reclamation of salt affected soil may be achieved by the following measures:

(i) Lowering of water table (if high),

(ii) Improving of infiltration rate of the soil by deep ploughing, green manuring and soil amelioration etc.,

(iii) Leaching of salts in saline soils by providing adequate surface drainage,

(iv) Replacing excessive exchangeable Na by Ca salts and removing the replaced products, and

(v) Suitable management practices.

References

Abrol, I.P. and Fireman, M. (1977). Alkali and Saline Soils, Identification and Improvement for Crop Production, Bulletin No. 4, Central Soil Salinity Research Institute, Karnal.

Abrol, I.P. Dargan, K.S. and Bhumbla, D.R. (1973). Reclaiming Alkali Soils, Bulletin No 2, Central Son Salinity Research Institute. Karnal.

CGWB and CPCB, 2000. Status of ground water quality and pollution, Aspect, NCT Delhi, Jan. 2000.

Kanwar, J.S. and Kanwar, B.S. (1968). Quality of Irrigation Water, 9th International Congress of Sod Sciences Transactions, Vol. 1, Paper 41, 391-403.

Pai, A.A. and Hukkeri, S.B. (1979). Mannual on Irrigation Water Management, Department of Agriculture, New Delhi.

Paliwal, K.V. and Yadav, B.R. (1976). Irrigation water quality and crop management in the Union Territory of Delhi. I.A.R.I. Tech. Bult. 9, Water Technology Centre, I.A.R.I., New Delhi.

Paliwal, K.V., Maliwal, G.L. and Nanawati, G.C. (1975). Effect of bicarbonate rich irrigation water on the growth, nutrient, uptake and synthesis of proteins and carbohydrates in wheat. Plant and Soil 43: pp. 533-536.

Prasad, M. (1973). Effect of different quality irrigation waters on soil properties and crop growth with special reference to magnesium. Ph.D. Thesis, University of Udaipur, Udaipur.

Praving Titovunno (1971). Quality of leachate from Vernal Project soil as affected by certain conditions in the soil and applied wastes. M.Sc. Thesis, Utah State University, Logan, USA.

Ramamoorthy, B. (1970). The scope for living with soil salinity and alkali conditions in India. Proc. Seventh NESA Regional Irrigation Practices Seminar, Lahore, Pakistan, September 16-28, 1968.

Singh, K.N., Bains, S.S. and Agrawal K.N. (1969). Fertilizer use under saline-cum-high water table conditions. *Indian J. Agron*. 14: pp. 26-30.

Singh, S.S. (1963). Studies on some chemical factors determine the quality of irrigation waters. Ph.D. Thesis, I.A.R.I., New Delhi.

United States Department of Agricutlure, Soil Conservation Service (1964). National Engineering Handbook, Section 151, Chapter 1, Soil-Plant-water relationship. U.S.D.A., Washington, D.C.

United States Salinity Laboratory Staff (1969). Diagnosis and Improvement of Saline and Alkali Soil, Agriculture Hand book No. 60, U.S. Department of Agriculture.

United States Salinity Laboratory Staff-Editor Richards, L.A. (1954). Diagnosis and improvement of saline and alkali soils. U.S.D.A. Handbook No. 60.

Wagdi, M.A.H.A. (1968). Some factors influencing the quality of irrigation waters, Ph.D. Thesis, I.A.R.I., New Delhi.

Wilcox, L.V. (1960). Beron injury to crops. U.S. Dept. of Agri. Inform., Bult. 21.

Withers, Bruce and Stanley Vipond (1974). Irrigation: design and practice, B.T. Batsford Ltd. London.

EXERCISES

20.1. Explain the cause of salt build up in arid and semi-arid and regions of India.

20.2. Define the following:

(a) Electrical conductivity

(b) Cation exchangeable capacity

(c) Sodium adsorption ratio

(d) Leaching

(e) Residual sodium carbonate

20.3. Differentiate the following:

(a) Sodium adsorption ratio and exchangeable sodium percentage

(b) Leaching and leading requirement

20.4. Describe the criteria for classification of irrigation water.

20.5. Explain the procedure for reclamation of salt affected soils.

20.6. What is the salt balance and derive the salt balance equation?

20.7 Explain the procedure for reclamation of saline soils.

20.8. Express 2100 ppm and 2200 ppm sodium bicarbonate and calcium chloride concentrations in meq/litre.

Sodium bicarbonate 25 meq/l

Calcium chloride 20 meq/l

20.9. Express the osmotic pressure in atmosphere of a salt solution showing EC values as 3.2, 0.8 and 4.6.

(Ans. 1.152, 0.288, 1.656 atmospheres)

20.10. Express the salt concentration of solutions in ppm showing their EC values as 1.5 dsm^{-1}, 0.2 dsm^{-1} and 1000 micromhos/cm.

(Ans. 960 ppm, 128.000 ppm and 610 ppm)

20.11. Analysis of a 10 gm soil sample showed exchangeable sodium, calcium, magnesium and potassium as 1.46, 1.241, 0.657 and 0.292 meq, respectively. Determine the total cation exchange capacity as well as the degree of saturation of each cation.

(Ans. CES = 36.5; Ex Ha%=40, Ex Ca%=34, Ex Mg%=18 and Ex K%=8)

20.12 A soil sample, weighing 5 g, on shaking with 100 cc of gypsum solution of 31.2 meq/l of calcium concentration showed 26.5 meq/l as Ca+Mg in the filtrate. Determine the gypsum requirement per 100 g soil and also in tonnes/hectare for 15 and 30 cm depth of soil.

20.13. The electrical conductivity of the canal water used for irrigation in a locality is 1 mmhos/cm and its density 1 g/cu cm. The density of salt free loam soil of the area is 1.4 g/cu cm. The saturation per cent is 50. Calculate the volume of water, expressed in depth of application, that will render the soil saline upto a depth of 1 m. **(Ans. 2.8 m)**

■■■■■

Chapter 21

DRAINAGE OF AGRICULTURAL LANDS

21.11. Introduction

Agricultural drainage is defined as removal and disposal of excess water from agricultural lands. Excess water in the root zone hampers the crop production as it restricts soil aeration, affects soil temperatures, hampers tillage operations, etc. Excess water slowly builds up the water table which ultimately rises upto the root zone and even upto the ground level. Waterlogging is defined as the conditions when the water table comes near the surface such that crop growth is influenced. The rising water table may restrict root development of the crops and may also bring up harmful salts to the rootzone.

Drainage problems are widespread in irrigated area due to continued irrigation over a large number of years without adequate drainage facilities. It results into large tracts of irrigated areas unproductive which ultimately reduce productivity. Thus, drainage is essential not only for the removal of excess water but also for removing the salts from rootzone to enhance the productivity.

21.12 Reasons of Waterlogging

Waterlogging is resulted mainly by natural and man-made (artificial) conditions as described below:

(a) Natural reasons: Among natural causes heavy rainfall and floods in basins without adequate outlets result in drainage problems. There are following natural reasons:

(1) Poor natural drainage of the sub-soil under unfavourable geology like existence of hard pan or rock out-crops near the soil surface,

(2) submergence under floods, and

(3) Deep percolation from rainfall.

(b) Artificial reasons: In the man-made reasons, drainage problems are widely caused in the command areas of large irrigation projects consisting of extensive canal networks. A major portion of these canals are unlined and as such continuously add water to the underground reservoir as seepage. Because of lack of adequate control over the irrigation supplies and excessive irrigations contribute to deep percolation losses. While in some situations where canal irrigation has been extended to areas which have low rainfall and no natural drainage channels could be developed. In such areas, water tables have risen fast after introduction of canal irrigation. There are following artificial reasons which build up the water logging:

(1) High intensity of irrigated agriculture irrespective of the soil and subsoil,

(2) Heavy seepage losses from unlined canals, distributaries and farm watercourses,

(3) Enclosing of irrigated fields, with embankments and choking up natural drainage,

(4) Hydraulic pressures from upper saturated areas at higher elevations, and

(5) Non-maintenance of natural drainages, or blocking of natural drainage channels by roads and railways.

21.13 Drainage System

It involves solution to drainage problems as controlling the sources of excess water as well as adequate provision for the removal of the excess water. For achieving the proper drainage, one or a combination of different drainage methods will be needed to solve the drainage problems in the area.

Drainage methods are broadly classified as surface methods and subsurface methods depending upon the surface or subsurface water that is drained.

21.13.1 Surface drainage

It is the safe removal of excess water from the land surface through land shaping and improved or constructed channels. Surface drainage aimed to remove the excess rainfall as well a collection and disposal of excess surface irrigation water wherever it occurs in agricultural lands. Surface drainage is required in flat or nearly flat areas, uneven land surfaces with depressions or ridges preventing natural runoff and in areas without any outlets. Soils having low infiltration rates are susceptible to surface drainage problems. Surface drainage is important for both the rainfed and irrigated areas. Surface drainage should be done without causing soil erosion in sloping area.

Surface drainage is based on the potential energy that exists due to land elevation to provide a hydraulic gradient for the movement of removal of waters. The surface drainage systems consist of three functional parts:

(i) Collection system,

(ii) Disposal system, and

(iii) Outlet.

Water from the individual fields is collected through the collection system and moves through the disposal system to the outlet.

There are four types of surface drainage systems used in flat areas (less than 2 percent slope): (1) random drain system, (2) parallel field drain system, (3) parallel open ditch system, and (4) bedding system.

21.13.1.1 Random drainage system

It is used when small scattered depressions are to be drained over the area. Open drains are designed to connect one depression to another (Fig. 21.1) and water is disposed to an outlet. The drains connecting the depressions may be surface drains (open channels) or subsurface drains. The main drawback of this method is that an open drain occupies land area and is likely to interfere in farming operations.

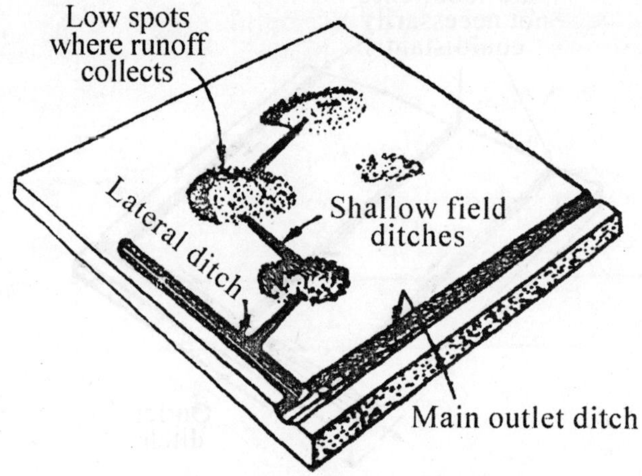

Fig. 21.1: Random drainage system

21.13.1.2 Parallel field drain system

It is the most effective method of surface drainage and is well suited both for irrigated and rainfed areas (Fig. 21.2). In this system, individual fields are properly graded in such way that drains discharge into field drains. The field drains may discharge into field laterals bordering the fields and the laterals in turn lead to the mains. Laterals and mains should be deeper than field drains or ditches to provide free outfall. This is also known to as the field ditch system.

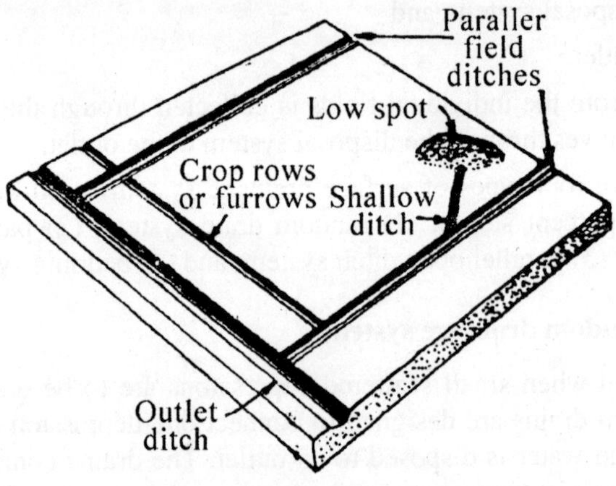

Fig. 21.2: Parallel field drain system

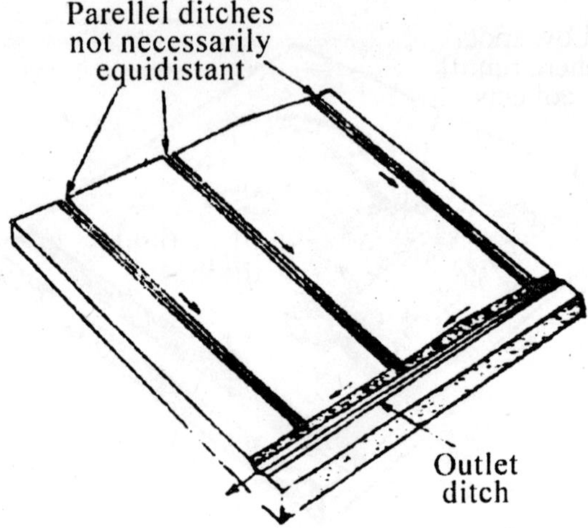

Fig. 21.3: Parallel open ditch system

21.13.1.3 Parallel open ditch system

It is suitable in soils that require both surface and subsurface drainage (Fig. 21.3). It is similar to the parallel field drain system, except that in this system the drains are replaced by open ditches which are comparatively deeper and have steeper side slopes as compared to the field drains. The open ditches cannot be crossed by farm machinery. The spacing of the ditches depends upon the soil and water table conditions and may vary from 60 to 200 m. The open ditches occupy consid-erable land area. Therefore, these should often replaced by subsurface drains. This system is also called as *diversion ditch* system.

24.3.1.4 Bedding drainage system

It is essentially a land forming process. The land is ploughed into beds, separated by dead furrows which run in the direction of prevailing slope (Fig. 21.4). Ploughing should be done parallel to the furrow. All other farming operations can be performed either across the beds or parallel to the furrow. Bedding drainage system is found most efficient on poorly drained soils and on flat lands and also on 1.5 per cent slopes. The bed width depends upon the land use, slope of the field, soil permeability and farming operations. The bed widths recommended for different soils are given in Table 21.1.

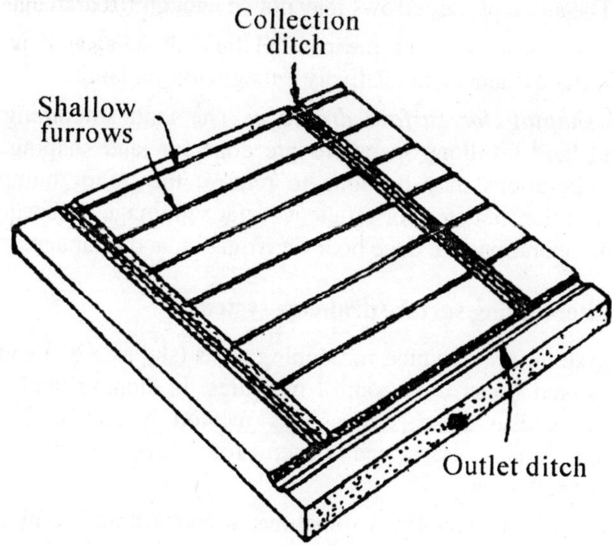

Fig. 21.4: Bedding system

Table 21.1: Recommended bed widths

Permeability	Bed width
Very low (K=0.5 cm/day)	8.12 m
Low (K=5 to 10 cm/day)	15-17 m
Good (K10 to 20 cm/day)	20-30 m

(Source: Murthy, 1985)

The length of the bed depends upon the field conditions and may be vary from 100 to 300 m. The maximum bed height is 20-40 cm. Water from the dead furrows discharge into a field drain which is constructed at the lower end of the field and normal to the dead furrows. The field drain discharge into field laterals and ultimately to the main drains.

Though, the bedding system is successful, however, it has following demerits:

1. The system become unsatisfactory if the crops are grown parallel to the furrows and on ridges,
2. The furrows require regular maintenance,
3. Movement of top soil caused some reduction in yields nearer the furrows, and
4. The slope of the furrows may not be enough for drainage.

Owing to these demerits the parallel field drain system is better suited for surface drainage in relatively flat agricultural lands.

Land shaping for surface drainage: The land smoothing, land grading and land levelling operations are done for land shaping. Land smoothing operations may be done to remove the minor humps and depressions in the first step for efficient surface drainage. Subsequently, land grading operations are done both for irrigation and drainage.

21.13.1.5 Intercepting surface drainage system

This system is well suited for sloping areas (slopes >2%) and same as those adopted for erosion control measures. In sloping lands, water conservation is also an objective. This method is similar to bench terracing and graded bunding and is useful for drainage, erosion control and water conservation.

Cross slope ditch system is sometimes used for drainage of sloping lands. It consists of a series of shallow open drains constructed across the slope with a mild grade (Fig 21.5).

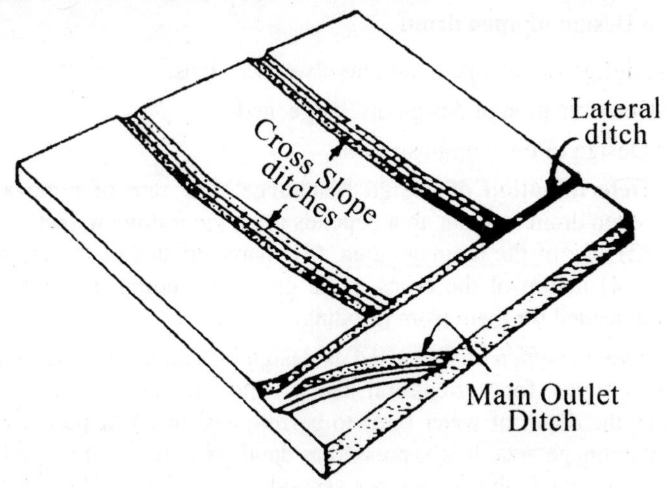

Fig. 21.5: Cross slope ditch system

Open drains: Open drain or ditches are most commonly used for surface drainage because of their convenience in construction. They are also sometimes used for subsurface drainage. Open drain are used as surface drainage channels for individual fields as well as disposal of the drainage water to the outlet.

The open ditches have the following main merits:

(i) Ease in construction either with machines or manual labour,

(ii) Ability to carry large volumes of water and,

(iii) Relatively cheaper compared to a subsurface drainage system.

However, it has the following main demerits:

(i) Loss of land,

(ii) Obstruction in farming operations,

(iii) Higher maintenance costs, and

(iv) Problems encounterd in construction and maintenance under unstable soil conditions.

In view of these demerits, subsurface drains are used for individual fields or a group of fields. Open drains are still to be used to convey the water to the field outlet. However, the problem of maintenance of the open drains is an important one. Weeds come up very quickly in these drains and reduce their capacity to a greater extent.

21.13.1.6 Design of open drain

The design of any open drain involves two steps:

(i) Determination of design discharge, and

(ii) Design of open drain section.

(i) Determination of design discharge: The rate of removal of water by open drain from an area depends upon the following factors: (1) rainfall, (2) size of the drainage area, (3) characteristics of the drainage area, and (4) nature of the crops to be grown including the degree of protection needed for them from flooding.

Drainage coefficient: It is used as design discharge in the design of drainage systems for agricultural lands. The drainage coefficient is defined as the depth of water (cm) to be removed in 24 hr period from the entire drainage area. It is expressed as depth of water (cm) or the flow rate per unit area (cubic metre per second per sq. km.). The drainage coefficient is determined in such way that no appreciable damage is caused to the crops grown in the area. The intensity of rain and its duration are inversely proportional to the time allowed for removal of excess water. It depends up on the kind of crop grown. The past experience with similar soils, climatic conditions and crops are helpful in deciding the drainage coefficient for small agricultural areas. Generally, the value of drainage coefficient varies from 0.6 to 2.5 cm and in extreme cases up to 10 cm.

Example 21.1: A watershed of 1000 ha is discharging through a drain at an average rate of 2.0 cum per second. Determine the drainage coefficient. If the drainage coefficient is 3 cm, what should be discharge through the drain.

Solution:

$$\text{Discharge in 24 hrs.} = 2.0 \times 60 \times 60 \times 24$$

$$\text{Drainage coefficient} = \frac{2.0\times60\times60\times24}{1000\times10} = \frac{0.01768}{10000}$$

$$= 0.01768 \text{ m or } 1.77 \text{ cm.}$$

If the drainage coefficient is 3 cm. then rate of flow through drain

$$= \frac{3}{100}\times\frac{1000\times100000}{24\times60\times60} = 3.47 \text{ cum/sec.}$$

Design discharge or design capacity

The following methods are used for determination of design discharge capacity.

(i) Cypress Creek Formula: It is expressed by the following relationship:

$$Q = CA^{P} \qquad ....(21.1)$$

where,

Q = design discharge (cum/sec.),

C = coefficient depending upon the degree of protection,

p = coefficient (general value 5/6), and

A = area (sq. miles).

For using the formula in a particular region, the value of drainage coefficient should be known for that region. Raddsma and Shulze (1974) proposed the following relationship for determination of drainage coefficient:

$$C = 16.39 + 14.75\,Re \qquad ....(21.2)$$

where, Re is the rainfall excess and expressed in inches and is determined from the maximum 24-hour storm.

(ii) *Boston Society Formula:* The Boston Society of Civil Engineer proposed the following relationship for determination design discharge:

$$Q = C\,A^{½} \qquad ....(21.3)$$

where,

Q = peak discharge (cusecs),

C = coefficient, and

A = peak catchment area (sq mile)

Uppal and Sehgal (1965) proposed the values of C (Table 21.2) for estimation of design discharge (Eq. 21.3). The values obtained mainly confined for the catchments in different parts of Punjab and Haryana states in India.

For economic construction of drainage systems, some times the drains are designed for a discharge varying from 1/4 to 1/12 of the calculated peak discharge (Dhruvanaryana, 1980).

Table 21.2: Values of C in Boston Society Formula

	Values of C		
Catchment area (sq. miles)	**For rainfall less than 20 inches**	**For rainfall between 20 inches and 30 inches**	**For rainfall between 30 inches and 40 inches**
0-25	20	50	100
25-100	400	100	250
100-250	100	200	450
Above 250	200	400	600

(For rainfall above 40 inches a value of C equal to 2500 irrespective of the catchment area is suggested)

Methods for estimation of runoff

Accurate estimation of runoff is difficult as it depends upon several factors. The following methods have been developed after field experience and research.

Rational method: According rational method, the peak rate of runoff is determined with the following relationship:

$$Q = \frac{1}{360} C.I.A. \qquad ....(21.4)$$

where,

Q = peak rate of runoff (cum/sec),

I = Intensity of rainfall (mm/hr or in/hr) for a duration equal to the time of concentration and for the given frequency,

C = runoff coefficient, and

A = area of the catchment (ha).

Runoff coefficient C is defined as the ratio of the peak runoff rate to the rainfall intensity. Gupta *et al.* (1970) worked out the values of C for different slopes and land use conditions (Table 21.3). When the catchment has areas with different values of C, the weighted value of C should be calculated for the whole catchment. If C_1, C_2, C_3, ... are the values of the runoff coefficient for areas a_1, a_2, a_3, ... the weighted value of C is obtained as,

$$C = \frac{C_1 a_1}{A} + \frac{C_2 a_2}{A} + \frac{C_3 a_3}{A} \qquad ....(21.5)$$

where,

A = total area of the catchment.

Equation 21.5 is dimensionally correct in metric units.

The value of the intensity of rainfall (I) to be used in Eq. 21.5 should be determined for the period equal to the time of concentration of the catchment. The time of concentration of a catchment is defined as the time required for water to flow from the most remote point of the catchment to the outlet. If the duration of rainfall equals the time of concentration, then all the parts of the catchment contributes to the discharge at outlet and the discharge will be maximum.

Kirpich (1940) proposed the following empirical formula for determination of the time concentration:

$$t_c = 0.02\, L^{0.77}\, S^{-0.385} \qquad \text{....(21.6)}$$

where,

t_c = time of concentration (minute).

L = length of channel reach (metre), and

S = average slope of channel reach (metre/metre).

According to Hann *et al.* (1982) the above relationship tends to correlate poorly with gauged runoff measurements of the time of concentration for very small (<5 sq. km) catchments. This might be due to over land (shallow) flow conditions dominated in small catchments rather than having a well defined net work of channels. Hann *et al.* (1982) suggested the following equation to overcome this difficulty for estimation the time of concentration:

$$t_c = 0.02\, L^{0.77}\, S^{-0.385} + \left[\frac{2 L_o n}{\sqrt{S_o}}\right]^{0.487} \qquad \text{....(21.7)}$$

where,

L_o = length of overland flow (m),

S_o = slope along the path (m/m), and

n = Manning type roughness coefficient.

The second term accounts for the overland flow in the equation (21.7) the size of watershed increases the maximum length of channel flow L tends to increase with increase in size of the watershed. This indicates that, as watershed size becomes large, the channel flow portion of the time of concentration will dominate the overall value. Thus, the equation (21.7) will be valid for all sizes of watersheds.

The time of concentration can also be determined from the velocities of overland flow and the channel flow. If the overland flow traverses more than one kind of surface, the travel times across them should be added up. Jens and McPherson recommended velocities of 1 m/s for hill slopes of the order of 30 m in length, ranging upto 0.25 m/s for 150 m long areas. The recommended velocities are less than 0.06 m/s for the shorter slopes and 0.08 m/s for the longer slopes, respectively. Bare areas should be intermediate between these values. The area A of the watershed can be known from the topographical map of the area.

The rational method outlined above is simple and widely accepted. However, Jaswant and Singh (1964) discussed and outlined the following limitations:

(i) The method assumes that the rainfall intensity is uniform over the entire watershed during the duration of the storm, a condition rarely satisfied,

(ii) When the runoff process is to starts then the information on losses due to depression storage and initial infiltration must be supplied, and

(iii) There is a retardance to flow caused by effects of surface detention and channel storage.

Loss to consideration, the rational method does not take these factors. The value of the runoff coefficient (C) is assumed to be dependent only on watershed characteristics. This does not seem to be correct as the value of (C) was observed to change with season as well as rainfall characteristics.

Table 21.3: Values of (C) for use in rational formula

Soil Types	Land Use		
	Cultivation	Pasture	Forest
With above average infiltration rate usually sandy or gravelly	0.29	0.15	0.10
With average infiltration rates, no clay pans, barns and similar soils	0.40	0.35	0.30
With below average infiltration rates, heavy clay soils or soils with a clay pan near the surface, shallow soils above impervious rock	0.50	0.45	0.40

Example 21.2: Estimate the peak rate of runoff for a 10 year frequency from a watershed of 25 ha, having 15 ha under cultivation

(C=0.5), 5 ha under forests (C=0.4) and 5 ha under grass cover (C=0.45). There is fall of 5 m in a distance of 700 m. The distance from the remotest point in the watershed to the outlet is 700 m.

Solution

$$\text{Weighted value of C for the entire watershed} = \frac{15\times0.5+5\times0.4+5\times0.45}{25}$$

$$= 0.47$$

For L = 700 m, S = 5/700, time of concentration, using Eq. 21.7.

$$= 21 \text{ minutes}$$

1 hour rainfall intensity for 10 years frequency at the given location from figure in Appendix-D = 100 mm/hour.

Intensity for 21 minutes rainfall, using Fig. 21.6.

$$= 175 \text{ mm/hour}$$

$$\textit{Peak runoff rate } Q = \frac{0.47\times17.5\times25}{360} = 5.7 \text{ cum}$$

Cook's Method. This method takes into consideration the four watershed characteristics such as relief, infiltration, vegetal cover and surface storage. Numerical values are assigned to each one of these factors for the particular watershed. The numerical values along with the watershed conditions are given in Table (21.4).

The sum of these numerical values (SW) is obtained and using this sum and the runoff curves given in Fig. 21.6. The value of peak runoff for a 10-year recurrence interval is obtained. This value has to be modified for the geographic location of the watershed and for the desired recurrence interval. The peak runoff for the given set of condition is determined with the following:

$$Q = PRFS. \qquad \text{....(21.8)}$$

where,

Q = peak runoff .for the specified geographic location,

P = uncorrelated value of runoff (Fig. 21.7),

R = geographic rainfall factor from (Appendix E),

F = recurrence interval factor (Appendix E), and

S = shape factor (Table 21.4).

The accuracy of the method depends upon the accuracy of the numerical values available for a particular location and individual judgement in selecting these values.

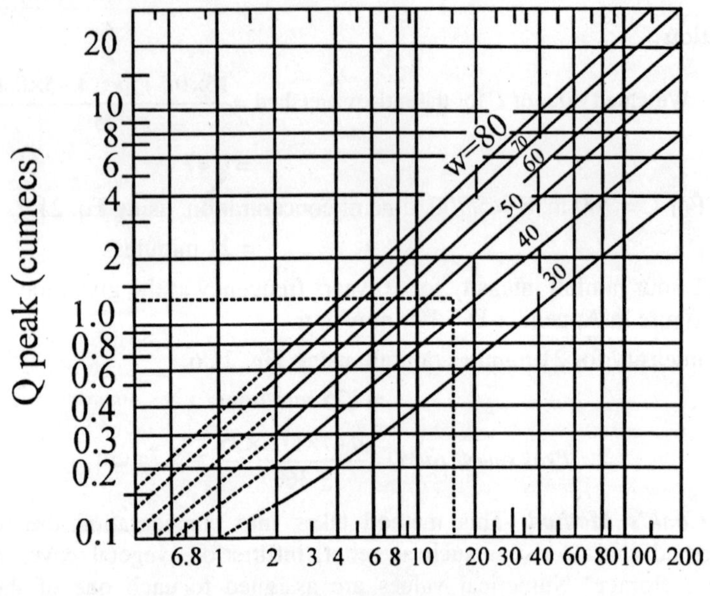

Fig. 21.6: Runoff curves for Cook's method

Table 21.4: Shape factor for Cook's method

Ratio of length to breadth	Watershed area (ha)				
	20	**40**	**80**	**200**	**240**
1	1.00	1.00	1.00	1.00	1.00
1-1 - 1/2	.92	.92	.91	.90	.90
2	.88	.87	.86	.84	.83
2-2 - ½	.85	.84	.82	.80	.78
3	.81	.80	.78	.76	.74
4	.76	.75	.73	.71	.69
5	.74	.72	.70	.68	.66
6	.72	.70	.68	.66	.64
7	.70	.68	.66	.64	.62
	.68	.66	.64	.61	.59

Table 21.5: Values of (W) for Cook's method

Designation of watershed characteristics	Runoff Producing characteristics for the Determination of Summation W			
	Extreme	**High**	**Normal**	**Low**
Relief	(40) to (30) Steep, rugged terrain with average slopes generally about 30%	(30) to (20) Hilly, with average slopes of 10% to 30%	(20) to (10) Rolling, with average slopes of 5% to 10%	(10) to (0) Relatively flat land, with average slopes of 0% to 5%
Soil infiltration	(20) No effective soil cover, either rock or thin soil mantle of negligible infiltration capacity less than 0.25 cm/hr	(15) Slow to take water, clay or other soil of low infiltration capacity 0.25 to 0.75 cm/hour	(10) Normal ; deep, permeable soils 0.75 to 2 cm/hour	(5) High sands, loamy sands, and other loose open soils over 2 cm/hour
Vegetal cover	(20) No effective plant cover, bare or very sparse cover	(15) Poor to fair, clean cultivated crops or poor natural cover, less than 10% of drainage area under good cover	(10) Fair to good, about 50% of drainage area in good grass land or equivalent cover, not more than 50% of area of clean cultivated crops	(5) Good to excellent, about 90% of drainage area in good grass and woodland or equivalent cover
Surface storage	(10) Negligible, surface depressions few and shallow, drainage ways steep and small, no ponds or marshes	(15) Low, well defined system of small drainage ways, no ponds or marches	(10) Normal, considerable surface depression storage, drainage system similar to that of typical prairie lands, lakes, ponds and marshes less than 2% of drainage	(5) High surface depressions storage high, drainage systems not sharply defined, large flood plain storage on a large number of lakes, ponds and marshes

Example 21.3: Determine the peak rate of runoff for 5 years frequency from a watershed comprising of 10 ha cultivated land

(1 percent slope, 0.25 cm/hr infiltration capacity, vegetal cover less than 10 per cent under good cover, low well defined small drainage ways) and 5 ha of grass land (2 per cent slope, infiltration rate of 0.8 cm/hr, vegetal cover excellent, 90 per cent under permanent good cover, low well defined drainage ways). The area is located at 70^0 longitude and 20^0 latitude. The length width ratio is 2.

Solution:

The W values for different watershed characteristics are determined from Table 21.6 as,

	Cultivated land	**Grass land**
Relief	2	4
Soil infiltration	15	11
Vegetal cover	15	5
Surface drainage	15	15
	$\Sigma W = 47$	35

$$\textit{Weighted } \Sigma W \textit{ for the entire watershed} = \frac{47 \times 10 + 35 \times}{15} = 43$$

The peak runoff rate is determined on using the Fig. 21.6 for $\Sigma W = 43$ as,

Peak runoff rate	=	1.3 cum
Rainfall factor (R) for the given location (from Appendix-D)	=	1.2
Frequency factor (from Appendix-D)	=	1.3
Shape factor (from Table 21.6) for a length breadth ratio of 2	=	0.88

Corrected value of runoff is determined by using Equation (21.8),

$$Q = 1.3 \times 1.2 \times 1.3 \times 0.88$$

$$= 1.8 \text{ cum}$$

Soil cover complex (SSC) method: This method is also known as the curve number method and it is based on the recharge capacity of the watershed. The recharge capacity is determined by antecedent moisture conditions and by the physical characteristics of the watershed.

Let I_a be the initial quantity of interception, depression storage and infiltration that must be satisfied by any rainfall before runoff can occur. It is assumed that the ratio of the direct runoff (Q) and the rainfall minus the initial loss (P-I_a) and the storage capacity (S) are determined with the following relationship:

$$\frac{Q}{P-I_a}=\frac{P-Q-I_a}{S} \quad ....(21.9)$$

I_a is assumed to be a faction of S. Let us consider I_a = 0.2S, we obtain,

$$Q=\frac{(P-0.2S)^2}{P+0.8S} \quad ....(21.10)$$

Knowing P and S the value of Q can be determined. Q has the same units of P.

The curve number as defined according to the U.S. Soil Conservation Service (1964), the curve number is expressed as,

$$CN=\frac{1000}{10+S}. \quad ....(21.11)$$

where,

S = recharge capacity of the watershed.

Curve numbers for different land use conditions and hydrologic soil groups are presented in Table 21.6. The values apply to antecedent rainfall condition II which is considered as an average value. Correction factors for other antecedent rainfall conditions are given in Table 21.6.

The recharge capacity S is determined on using Eq. 21.10 and thereafter runoff is determined on using Eq. 21.11. A plot of Eq. 21.10 for different curve numbers is shown in Fig. 21.7.

It should be remembered that value of S in Eq. 21.11 is obtained in inches. If Q is to be obtained in cm, P should be expressed in cm and S is obtained from the following equation.

$$CN=\frac{2540}{25.4+S} \quad ....(21.12)$$

This method involves conversion of rainfall into runoff depth and the peak rate of runoff is determined on using following relationship:

$$Q_{peak}=\frac{0.0208\,A\,Q}{T_p} \quad ....(21.13)$$

Table 21.6: Curve numbers for hydrologic soil cover complexes (for watershed condition II, $I_a = 0.2S$)

Land use or cover	Treatment or practice	Hydrologic condition	Hydrologic soil group			
			A	B	C	D
Fallow-Row crops	Straight row	-	77	86	91	94
	Straight row	Poor	72	81	88	91
	Straight row	Good	67	78	85	89
	Contoured	Poor	70	79	84	88
	Contoured	Good	65	75	82	86
	Contoured and terraced	Poor	66	74	80	82
	Contoured and terraced	Good	62	71	78	81
Small grain	Straight row	Poor	65	76	84	88
	Straight	Good	63	75	83	87
	Contoured	Poor	63	74	82	85
	Contoured	Good	63	75	83	87
	Contoured and terraced	Poor	61	72	79	82
	Contoured and terraced	Good	59	70	78	81
Seeded legumes or rotation meadow	Straight row	Poor	66	77	85	89
	Straight row	Good	58	72	81	85
	Contoured	Poor	64	75	83	85
	Contoured	Good	55	69	78	83
	Contoured and terraced	Poor	63	73	80	83
	Contoured and terraced	Good	61	67	76	80
Pasture or range	-	Poor	68	79	86	89
		Fair	49	69	79	84
		Good	39	61	74	80
	Contoured	Poor	47	67	81	88
	Contoured	Fair	25	59	75	83
	Contoured	Good	6	35	70	79
Meadow (permanent)	-	Good	30	58	71	78
Woodlands (farm woodlots)	-	Poor	45	66	77	83
	-	Fair	36	60	73	79
	-	Good	25	55	70	77
Farmsteads	-	-	59	74	82	86
roads (dirt)			72	82	87	89
roads (hard surface)			74	84	90	92

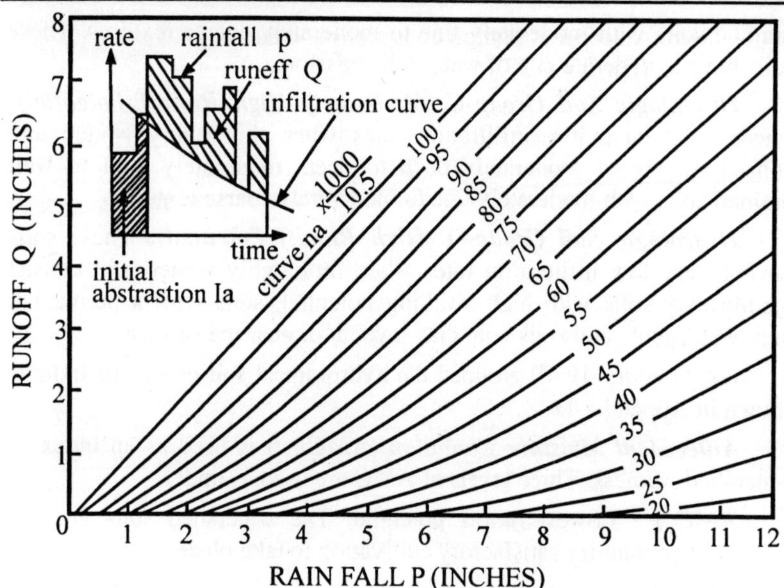

Fig. 21.7: Curve number method

where,

Q_{peak} = peak rate of runoff (cum/sec),

A = area (hectares),

T_p = time of peak (hours),

(given by $T_p = \sqrt{T_e + 0.6T_e}$),

T_e = time of concentration (hours), and

Q = runoff (cm).

When A is expressed in Sq. miles and Q in inches, Q_{peak} in cusecs is expressed by,

$$Q_{peak} = \frac{484\,AQ}{T_p}$$

Hydrologic Soil Group-A (Low Runoff Potential): These soils have high infiltration rates when thoroughly wetted and consist mainly of deep, well to excessively drained sands or gravels.

Hydrologic *soil Group-B (Moderately Low Runoff Potential):* These soils have moderate infiltration rates when thoroughly wetted and consist mainly of moderately deep to deep, moderately well to well drained soils with moderately fine to moderately coarse textures. These soils have a moderate rate of water transmission.

Hydrologic Soil Group-C (Moderately High Runoff Potential): These soils have low infiltration rates when thoroughly wetted and consist mainly of moderately deep to deep, moderately well to well drained soils with moderately line to moderately coarse textures.

***Hydrologic Soil Group-D (High Runoff Potential)*:** These soils having very low infiltration rates when thoroughly wetted and consist mainly clay soils with high swelling potential, soils with a permanent high water-table and soils with clay layer at or near the surface.

Gupta *et al.* (1970) grouped the hydrological soil groups of India is shown in Appendix-D

Antecedent Moisture Condition (AMC): It is used as an index of watershed wetness. Three levels of AMC are used as under,

AMC I : Lowest runoff potential. The watershed soils are dry enough for satisfactory cultivation to take place.

AMC II: Average condition

AMC III : Highest runoff potential. The watershed is practically saturated from antecedent rains.

The AMC group is determined for 5 day antecedent rainfall on using Table 21.7.

Table 21.7: Rainfall limits for estimating antecedent moisture conditions (AMC)

AMC	5 day antecedent rainfall (cm)	
	Dormant season	**Growing season**
I	Less than 1.25	Less than 3.5
II	1.25 to 2.75	3.5 to 5.25
III	Over 2.75	Over 5.25

Example 21.4: Determine the maximum runoff for 50 years, recurrence interval for moisture condition III from a watershed with 50 ha of row crop cultivated good terraced land and 10 ha of good pasture land. The watershed is situated at 30^0 N latitude and 70^0 E longitude. The soil is of hydrologic soil group C. The time of concentration of the watershed is 0.48 hr.

Solution:

Using Table 21.6, the curve numbers for the terraced land is determined as 78 and for the good pasture land is 74.

Weighted curve number for the entire watershed

$$=\frac{78\times50+74\times10}{60}=77.3$$

To convert this curve number to antecedent moisture condition III, the conversion factor of 1.14 is taken from Table 21.8.

Converted Weighted CN = 77.3 x 1.14 = 88.1

6 hour, 50 year frequency rainfall for the given location from figure in Appendix-D = 150 mm

Recharge capacity S is determined on using Eq. 21.11.

$$88.1=\frac{2540}{25.4+8}$$

$$S = 3.46 \text{ cm}$$

or

Runoff from Eq. 24.10

$$Q=\frac{(15-0.2\times3.46)^2}{15+0.8\times3.46} = 11.52 \text{ cm}$$

Empirical Formulae: Empirical formulas have been developed for determination of the peak rate of runoff for larger watersheds.

(1) Dicken's formula: It is expressed with the following relationship:

$$Q = C\,A^{3/4} \qquad ....(21.14)$$

where,

Q = peak rate of runoff (cum/sec),

A = area of the watershed (sq. km),

C = 11.45 for areas with annual rainfall of 610 mm to 1270 mm

=13.88 to 19.42 for areas in Madhya Pradesh

This formula is generally applicable in North and Central India.

(2) Ryve's Formula: It is expressed with the following relationship:

$$Q = C\,A^{2/3} \qquad ....(21.15)$$

Table 21.8: Antecedent rainfall conditions and curve numbers (For $I_a = 0.2S$)

Curve number for condition II	Factor to convert curve number of condition II to	
	Condition I	**Condition II**
10	0.40	2.22
20	0.45	1.85
30	0.50	1.67
40	0.55	1.50
50	0.62	1.40
60	0.67	1.30
70	0.73	1.21
80	0.79	1.14
90	0.87	1.07
100	1.00	1.00

The value of C varies from 6.76 to 40.50 depending upon the location of the catchment. This formula is mainly applicable in South India.

(iv) ***Simplified Hydrological Procedure:*** Raadsma and Schulze (1979) proposed a procedure which consists of analyzing the rainfall data and estimating the number of hours required to remove the excess water using the information about crop tolerance.

The rainfall data is analysed for duration-frequency. A drainage system should be planned taking the crop and the rainfall duration-frequency for the particular crop season into consideration. The rainfall excess is determined and an allowance is made for channel storage. The capacity of the drainage system required is worked out after the drainage coefficient *i.e.* the depth of water to be removed in 24 hr. Example 21.2 illustrates the method. The data assumed in this example is hypothetical.

It is evident from the above example that a storage of 10 mm is uniformly assumed in the surface channels. The information in Table 21.9 along with drainage capacities are plotted in Fig. 21.9. This graph may be used to determine the drainage capacity and the number of hours required to remove the excess surface water. Thus, for example with a drainage capacity of 30 m m/24hrs, a rainfall fall excess occurring once in 5 years can be drained in about 40 hr. This method is useful when a single crop is grown and drainage coefficient is known. When more crops are grown in an area the determination are to be done for each crop separately.

Table 21.9: Example of rainfall data for drainage design

Return period	Rainfall (mm)	Duration of storm period			
		90 Min	24 Hrs	48 Hrs	72 Hrs
1 × 2 years	Annual	40	65	95	105
	Seasonal	30	55	78	93
	Excess (1)	16	30	52	66
	Excess (2)	6	20	42	56
1 × 5 years	Annual	50	86	120	140
	Seasonal	40	72	85	120
	Excess (1)	25	48	58	90
	Excess (2)	15	38	48	80
1 × 10 years	Annual	60	105	140	65
	Seasonal	50	85	120	150
	Excess (1)	35	55	90	120
	Excess (2)	25	45	80	110

(Source: Murthy, 1985)

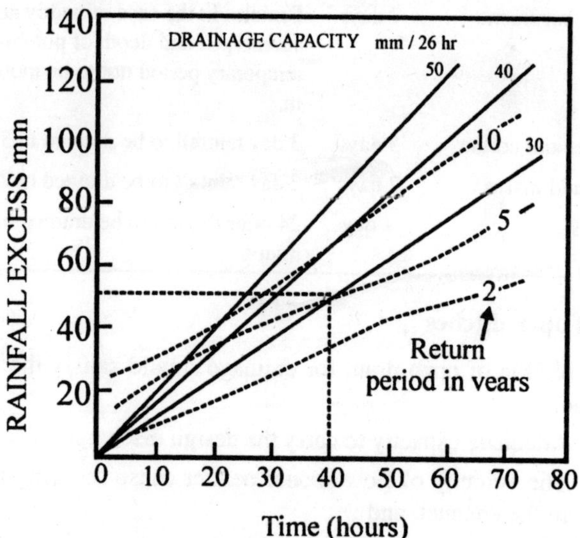

Fig. 21.8: Number of hours required to remove excess surface water

Pai and Hukkeri (1979) proposed that the drains should be designed for 1 day average to 3 day maximum rainfall of 5 year return period with a runoff percentage. These are determined after taking into consideration infiltration rate, evaporation and permissible storage of water at the fields. If no dependable data is available then runoff percentage for different types of soils and intensity of vegetation as given in Table 21.10 may be used.

Table 21.10: Recommended runoff percentages for different soil types

Soil type	Runoff percentage
Sandy soils	20
Loamy soils (largely cultivated)	30
Loamy soils (lightly covered by crops)	40
Gardens and cultivated areas	5 to 25

The period of disposal of excess rainfall depends upon the type of crops and their stage of growth. Pai and Hukkeri (1979) worked out the period of disposal of excess rainfall (Table 21.11).

Table 21.11: Recommended drainage periods for some crops

Crops	Period of Disposal	Remarks
Paddy and sugarcane	7 days	Panicles in the case of paddy not to get submerged and depth of ponding for temporary period not to be more than 0.5 m.
Pearl millet and cotton	3 days	3 day rainfall to be drained in 3 days
Sorghum and maize	2 days	2 day rainfall to be drained in 2 days
Vegetables	1 day	24 hour rainfall to be drained in 24 hours.

Design of open ditches

The section of open drain for drainage should satisfy the following conditions:

1. Adequate capacity to carry the design discharge,
2. The velocity of flow should neither cause scouring nor silting in the channel, and
3. The side slopes should be stable.

Channel cross section: The trapezoid section of open drain is the most convenient cross-section for excavation and maintenance for purpose (Fig. 21.9).

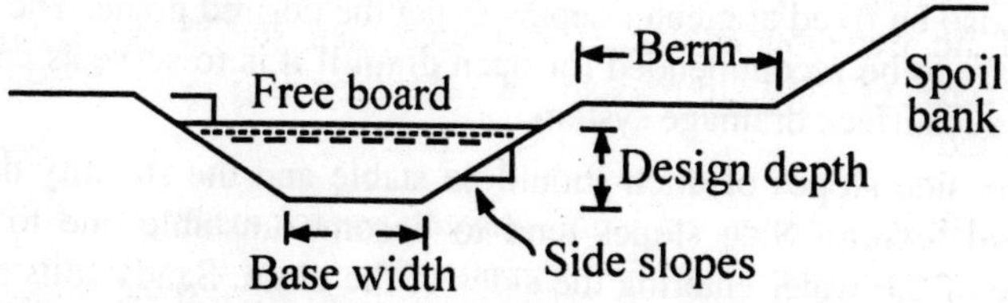

Fig. 21.9: Typical cross-section of an open ditch

Manning's (1890) proposed a formula for determination of the velocity of uniform flow in channels or in pipe expressed as,

$$V = \frac{1}{n} R^{2/3} S^{v2} \qquad \text{....(21.16)}$$

where,

v = velocity of flow, (m/sec.),

R = hydraulic radius, (m),

S = slope of the channel bed, (faction),

n = Manning's retardance coefficient.

For trapezoidal section, the cross-section area is determined as,

$$A = d\ (Bt = Zd) \qquad \text{....(21.17)}$$

where,

A = cross-section area of low, (m)

B = Bottom width of the channel, (m), and

Z = side slope (horizontal: vertical).

The welted perimeter (P) = $B + D\sqrt{z^2 + s}$

The hydraulic radius (R) = $\frac{A}{P}$

The hydraulic radius for a pipe flowing full is determined as,

$$R = \frac{d}{4}$$

where, d diameter of the pipe.

However, the most efficient hydraulic section for drainage ditches is not possible as in irrigation. The main reason is that the side slopes should be stable from the, soil point of view. The depth of the ditch cannot also be fixed at greater depths to get the desired grade. The deeper depths are to be recommended for open drain if it is to serve as an outlet for any subsurface drainage system.

The side slopes of ditch should be stable and the stability depends upon soil texture. Side slopes tend to become unstable due to of the pressure of the water entering the sides of the drain. Sandy soils are less stable as compared to clay soils due to cohesiveness. Stabilization of the side slopes may be achieved by compacting the surface with mechanical means. However, it is costly and also the effect may not last long. The recommended side slopes for different soils is given in Table 21.12.

Table 21.12: Recommended side slopes for open ditches

Soil type	Side slopes
Heavy clay	1: 1
Clay or silt loam	1-½: 1
Sandy loam	2: 1
Loose sand	3: 1

Table 21.13: Permissible velocities in non-vegetated channels

Material excavated	Clear water	Velocity (m/s)	
		Water transporting colloidal silts	Water transporting sand, gravel or fragments
Fine sand, colloidal	0.45	0.75	0.45
Sandy loam, non-colloidal	0.53	0.75	0.80
Silt loam, non-colloidal	0.60	0.90	0.60
Alluvial silts, non-colloidal	0.60	1.05	0.60
Ordinary firm loam	0.75	1.05	0.68
Stiff clay, very colloidal	1.13	1.50	0.90
Fine gravel	0.75	1.50	1.13
Coarse gravel, non-colloidal	1.20	1.80	1.95
Cobbles and shingles	1.80	1.80	1.50

(Source: Fortier and Scobey, 1926)

The grade given to the ditches depends upon the natural topographic conditions. The grade should be such that the velocity of flow in the channel should neither cause silting nor scouring of the bed and the banks. The grade should also permit the outflow of the drainage water into a natural outlet wherever such outlet is available.

Generally velocities of 0.6 to 0.9 m/sec are found to be non-silting. A 20 per cent free board to the designed depth is recommended. This will help as a safety factor for discharge as well as for sediment accumulation.

Example 21.5: Design an open ditch to drain 50 ha of land having a drainage coefficient of 2.5 cm. The soil is silt loam and maximum permissible slope of the channel bed is 0.1 per cent.

Solution:

Quantity of water to be removed in 24 hr

$$=\frac{2.5}{100}\times 500\times 10000\, cu\, m$$

$$\text{Discharge rate (Q)} = \frac{2.5}{100}\times\frac{500\times 10000}{24\times 60\times 60} = 1.5\, m^3/\text{sec}.$$

Side slopes recommended for silt loam = 1.5: 1

$$\tan\theta = 1/1.5 = 0.666 \text{ or } \theta = 17^\circ\, 51'$$

Assuming d =1.2 m, b = 2 d tan θ/2

$$= 2\times 1.2\times .3029 = 0.727 \text{ m}$$

$$\approx 75 \text{ cm}$$

$$\text{Length of side} = \sqrt{(1.5\times 1.2)^3 + (1.2)^2} = 2.16 \text{ m}$$

$$P = 2\times 2.16 + 0.75 = 5.07 \text{ m}$$

$$A = \frac{(3\times 1.2 + 0.75) + 0.75}{2}\times 1.2 = 3.06 \text{ m}^2$$

$$R = \frac{3.06}{5.07} = 0.603\, m$$

$$S = \frac{0.1}{100} = 0.001\, m/m$$

On using Manning's equation

$$V = \frac{(0.603)^{2/2} \times (0.001)^{1/2}}{0.04} = 0.475\,m/\text{sec}.$$

The velocity is within permissible limit.

$$Q = 3.06 \times 0.475 = 1.76 \text{ m}^3/\text{sec}.$$

considering the requirements of free board, the section is suitable.

21.13.1.7 Machineries for construction of open drain

A wide variety of machineries are used for construction of the drains. There are following machineries used for construction of open drains:

(i) Drain ploughs: It is a crawler tractor drawn V-shaped tool. This is used for construction of smaller size field drains.

(ii) Crane shovels: It consists of a lower travel unit which can be either wheel type or crawler type. Crane shovel is mounted on the lower travel assembly and it has an upper revolving frame. Different forms of end attachments are mounted on the top of the upper revolving frame.

(iii) Scrapers: Several types of front end attachments are available for excavation of open drains. However, the most common types of front end attachments are used such as clam shell, bucket, dragline and back hoe. They are described as under:

(a) Clam shell bucket: It has two halves hinged at the top (Fig. 21.10). The buckets can be opened using the hinge and when drawn together a bowl shape is formed. The clam shell bucket is suitable for digging in relatively soft soils and also under high water table conditions. Excavation is performed by dropping the bucket into the soil with the buckets open. The weight of the buckets help in digging into the soil and also soil gets collected into the buckets while closing them using the closing line.

(b) Dragline: This is a common equipment used for excavating open drains (Fig. 21.10). In this, dragging action is used to load the bucket and called as drag line. To excavate the soil, the bucket is thrown away from the machine and using the drag cable the bucket is drawn towards the machine. The excavated soil can he deposited on either side of the ditch or on both sides due to long boom of the machine. The dragline can also be used for excavation of soils under high water table conditions.

(c) Back hoe: It incorporates the similar features as that of dragline (Fig. 21.11). The dipper stick is lowered and using the cable pulled or dragged towards the machine for excavation of the soils. Dipper is raised by the hoist cable after filling it thereafter using the dipper stick the soil is deposited at the required place. The back hoe can be used to dig harder soils as compared to the clam shell and the dragline. More precision in digging is also possible with the back hoe as compared to the dragline.

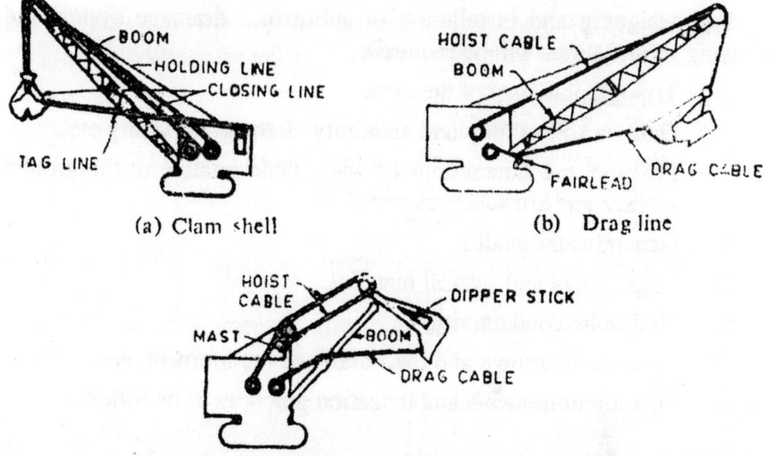

Fig. 21.10: Front end attachments for excavation

21.14 Subsurface Drainage System

It is the removal of excess water present below the ground surface. In general, agricultural lands affected by high water table need subsurface drainage. While surface drainage removes the excess rain *water* before it enters the rootzone consequent by, subsurface drainage lowers the water table and provides a better environment in the rootzone. Whereas surface drainage removes the excess rainwater before it enters the rootzone.

The surface drainage system offers the following advantages:

1. Improve aeration in the rootzone for maximum development of plant roots,

2. Increase opportunity for desirable soil microorganisms through aeration and higher soil temperatures,
3. Improve the field conditions for early cultivation and thus increased crop growth period,
4. Provide effective environment for operation of farm machinery,
5. Removal of undesirable salts from the rootzone, and
6. Enhance storage of rainwater in the rootzone due to low initial water table before the rains.

21.14.1 Investigations for subsurface drainage

For designing and installation of subsurface drainage system the following informations will be required.

1. Topographic map of the area,
2. Data on soil salinity and alkalinity, drainable porosity etc.,
3. Position and fluctuations of water table relative to the ground surface and artesian pressures,
4. Groundwater quality,
5. Logs of soil and subsoil material,
6. Hydraulic conductivity,
7. Crops to be grown and their drainage requirement, and
8. Water requirements and irrigation practices to be followed.

21.14.2 Subsurface drainage system

In subsurface drainage, water moves under the influence of gravity to suitable outlets. This is accomplished using one of the following:

(i) Tile drains including perforated pipes,
(ii) Mole drains,
(iii) Drainage wells,
(iv) Deep open drains, and
(v) Combination of tile and open drains.

Tile drains: It consist of short length pipes (30 to 90 cm) installed at a particular depth from the land surface. The pipes are made of concrete or burnt clay. Water enters the tile drains through the opening available between the pipes. A network of tile lines are laid out with a grade to remove the subsurface water.

Perforated pipes are like tile drains except that they are continuous and water enters the pipes through openings provided on the pipe. PVC pipes are mostly used for drainage purpose. These are also laid out below the soil surface by digging a trench to the required depth then backfilling of soil is done.

Mole drains: These are cylindrical channels formed at a desired depth below the soil surfaces. This is suitable in clay soils. There is no lining material and the inherent stability of the soil at the depth gives stability to the mole drains. Water enters throughout the mole drains and is disposed to the outlet.

It is a semi-permanent method of subsurface drainage. It is similar to tile drainage in layout and operation except permanent tiles, a continuous circular channel is created in the soil below the ground level. This is achieved with the help of a special equipment known as a mole plough and the mole plug (Fig. 21.11).

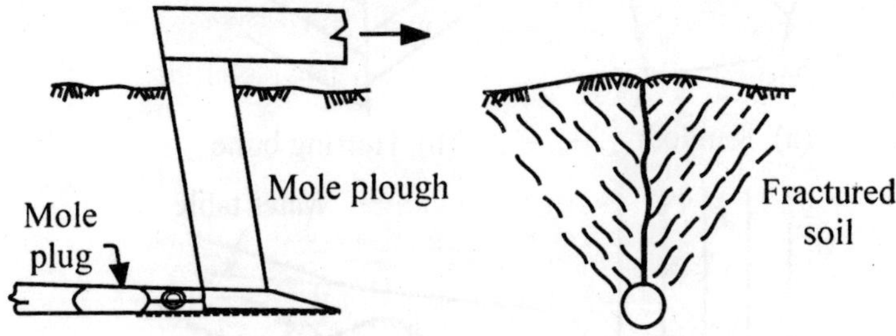

Fig. 21.11: A mole plough and mole drain

Mole drainage is well suited for clay soils as the stability of the channel is most important in this method of drainage. The life of the mole drains is 10 to 15 years depending, upon the structural stability of the subsoil. The other factors that influence the life of the mole drains are as follows:

(i) Amount and intensity of rainfall,

(ii) Temperature variations,

(iii) Method of forming the mole drains, and

(iv) The soil moisture content at the time of formation of the mole drains and the diameter.

The depth of the mole drains vary from 45 cm to 120 cm and it depends upon the moling equipment. The diameter of the mole drains vary from 7.5 to 15 cm. Smaller diameter moles have longer life as compared to larger diameter one. The principles of layout of the mole drains are similar to that of tile drains. However, the mole drains require closer spacing as compared to the tile drain.

21.14.3 Types of tile drainage systems

The tile drainage systems are classified into the following:

(i) Random or natural, (ii) herringbone, (iii) gridiron, and (iv) the interceptor drains. The different types of tile drainage systems are shown in Fig. 21.12. In an area it is possible to use one or a combination of the systems depending upon the site conditions.

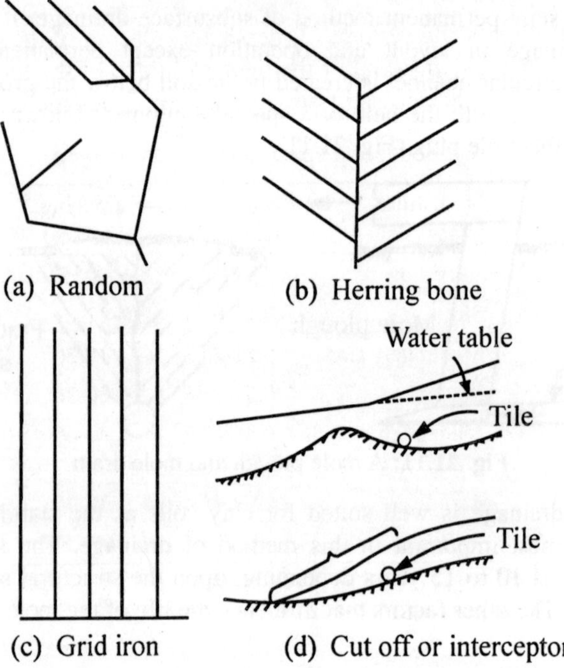

Fig. 21.12: Different types of layouts of tile drainage system

(i) **Random or natural system:** It is used for draining isolated patches and when the entire area does not need drainage. It is economical tile drainage system.

(ii) **Herringbone system:** It is used in areas where a main line or sub-main could be laid in the low area and laterals could be drawn from both sides. The land in this system is well drained.

(iii) **Gridiron system:** It is similar to the herringbone system except that the laterals enter the main only from one side. The system could be more economical than the herringbone system.

(iv) **Interceptor drains system:** The interceptor or cut-off drain is placed at the upper edge of a wet area to intercept the surface seepage. Open drains are also used for the purpose but require considerable land area and maintenance.

Design of tile drain systems

For designing the tile drainage system following parameters will be determined:

(1) selection of the system; (2) depth and spacing of the drains; (3) size of the drains; (4) material of the tiles; (5) grade of the tile lines; (6) envelope materials and accessory structures; and (7) outlets.

Table 21.14: Average depth and spacing of tile drains

Type of soil	Hydraulic Class	Conductivity (cm/day)*	Spacing (metre)	Depth (metre)
Clay	Very slow	0.13	9-15	0.9-1.05
Clay loam	Slow	0.13-0.51	12-21	0.9-1.05
Average loam	Moderately slow	0.51-2.03	18-30	1.05-1.20
Fine sandy loam	Moderate	0.51-6.35	30-36	1.20-1.35
Sandy loam	Moderately rapid	6.35-12.7	30-60	1.20-1.50
Peat and muck	Rapid	12.7-25.4	30-90	1.20-1.50
Irrigated soils	Variable		45-180	1.50-2.40

Hydraulic conductivity as classified O'Neal (1949).
(Source: Shawab *et al.*, 1966)

Depth and spacing

The depth of the tile drains should be such that midway between the drains the water table should be at a satisfactory depth. When an impermeable layer is encountered then the tiles should be placed above such layer, unless when the impermeable layer occurs at very shallow depth. If the tiles are installed below the impermeable layer then backfilling of the trenches should be achieved with permeable soil. The depth and spacing are determined based on different requirements in humid and arid regions. The water table will rise due to heavy rainfall to the near surfaces in humid regions. Thus, the rate of drop of the water table is important in this case. While, in arid regions under irrigation the water table is maintain to a minimum depth for optimum crop growth.

Generally, the deeper the drains will be the wider the spacing and less number of drains is needed. However, deeper drains require more installation costs and may cause outlet problems.

The general depth and spacing that could be recommended under different conditions is summarized in Table 21.14.

Based upon the hydraulics of flow of water in soils different formulas have been proposed for determination of depth and spacing. Out of which few methods are described as under:

Hooghoudt's Equation

This equation gives the relation between the spacing *of* drains and the height of the water table built up between the drains due to a constant rate of irrigation or rainfall. This analysis is applicable for open drains or subsurface drains.

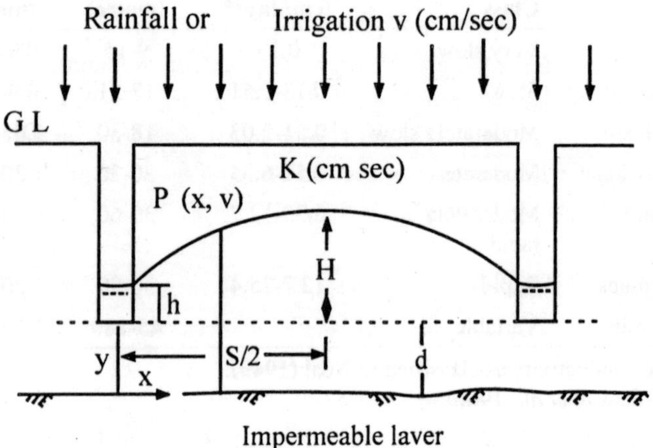

Fig. 21.13: Definition sketch of Hooghoudt's equation

Let us consider the physical situation as shown in Fig. 21.13. The following assumptions are made for deriving the Hooghoudt's equation.

1. The soil is homogeneous,
2. Darcy's law is valid for the flow, (flow should be laminar, Re<1), and
3. The hydraulic gradient at any point is equal to the slope of the water table above that point (i=dy/dx) and the water flows horizontally.

Hooghoudt proposed the following equation for designing of tile drain or open drain:

$$S^2 = \frac{4K(H^2 - h^2 + 2dH - 2dh)}{v} \qquad \text{....(21.18)}$$

where, the terms are same as shown in Fig. 21.13.

This equation gives the spacing of either tile or open drains under the given set of conditions. For all practical purposes the drain is considerable to be empty. On putting h=0, the equation (21.18) may be rewritten as,

$$S^2 = \frac{4KH}{v}(2d + H) \qquad \text{....(21.19)}$$

The equation (21.19) may be sometimes rewritten in the form of an ellipse as follows,

$$\frac{y^2}{S^2 v / 4K} + \frac{x^2}{S^2/4} = 1 \qquad \text{....(21.20)}$$

This is an equation of ellipse having semi-major and semi-minor axes as S/2 and S/2 $\sqrt{v/K}$, respectively.

Application to layered soils: Hooghoudt's equation can be used for layered soils. If K_1 and K_2 are the hydraulic conductivities of the layer above and below the drain line, respectively. The equation (21.20) can be rewritten as,

$$S^2 = \frac{4}{v}(K_1 H^2) + \frac{8}{v}(K_2\, dh) \qquad \text{....(21.21)}$$

For multilayered soil the weighted mean of the horizontal conductivities is determined and used for the same. If the soil above the drain line consists of two layers of conductivity K_1 and K_2 and thickness I_1 and I_2, then the weighted conductivity will be determined as,

$$K_w = \frac{K_1 I_1 + K_2 I_2}{I_1 + I_2} \qquad \text{....(21.22)}$$

In order to account for the convergence losses at the tile drains, Hooghoudt replaced d in Fig. 21.14 by equivalent depth 'd^l', which is smaller than *d*, Fig. 21.9 may be used for determining these values.

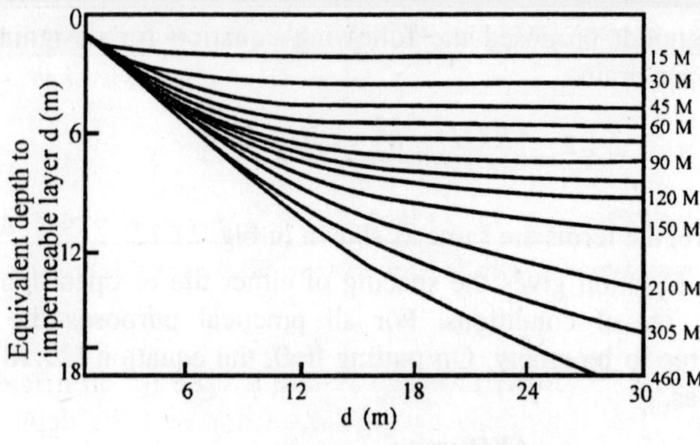

Fig. 21.14: Curves for finding equivalent depth d (Adopted from USBR)

The values of equivalent depths can be determined on using Table 21.15.

Table 21.15: Values for the equivalent depth (*d*) of Hooghoudt (r_o=0.1 m, d and S in m)

S	10	15	20	25	30	35	40	45	50	75	80	85	90	100	150
d															
0.5	0.49	0.49	0.49	0.50	0.50										
1	0.80	0.86	0.89	0.91	0.93	0.94	0.96	0.96	0.96	0.97	0.97	97.00	0.98	0.98	0.99
2	1.08	1.28	1.41	1.50	1.57	1.62	1.66	1.70	1.72	1.80	1.82	1.82	1.83	1.85	1.00
3		1.45	1.67	1.83	1.97	2.08	2.16	2.23	2.29	2.49	2.52	2.54	2.56	2.60	2.72
4			1.81	2.02	2.22	2.37	2.51	2.62	2.71	3.04	3.08	3.12	3.16	3.24	3.46
5			1.88	2.15	2.38	2.58	2.75	2.89	3.02	3.49	3.55	3.61	3.67	3.78	4.12
6					2.48	2.70	2.92	3.09	3.26	3.85	3.93	4.00	4.08	4.23	4.70
7					2.54	2.81	3.03	3.24	3.43	4.14	4.23	4.33	4.42	4.62	5.22
8					2.57	2.85	3.13	3.35	3.56	4.38	4.49	4.61	4.72	4.95	5.68
9						2.90	3.18	3.43	3.66	4.57	4.70	4.82	4.95	5.23	6.09
10							3.23	3.48	3.74	4.74	4.89	5.04	5.18	5.47	6.45
12.5										5.02	5.20	5.38	5.56	5.92	7.20
15										5.20	5.40	5.60	5.80	6.25	7.77
17.5										5.30	5.53	5.76	5.99	6.44	8.20
20											5.62	5.87	6.12	6.60	8.54

S	10	15	20	25	30	35	40	45	50	75	80	85	90	100	150
25											5.74	5.96	6.20	6.79	8.99
30															9.27
35															9.44
∞	1.14	1.53	1.89	2.24	2.58	2.91	3.24	3.56	3.88	5.38	5.76	6.00	6.26	6.82	9.55

(S-spacing of drains, d thickness of aquifer below drain level and r_0-radius of the drains)
(Source: Wessling, 1973)

Example 21.6: A tile drainage system is used for an irrigated area for drainage. The tile drains having 0.2 m dia. and the depth for the drains is 2.5 m below the soil surface. An impermeable layer was found at a depth of 5 m below the soil surface. The hydraulic conductivity of the soil is 1 m/day. The area is irrigated once in 15 days and it is estimated that the losses are 30mm in 15 days so that the average discharge of the drainage system is 2 mm/day. Determine the required spacing of the drains if the water table is to be kept at 1.5 m from the ground level.

Solution:

V = .002 m/day

K = 1 m/day

d = 5 – 2.5 = 2.5 m

H = 2.5 - 1.5 = 1.0 m

Substituting the above values, except d in Eq. 21.19

$$S^2 = \frac{2.5 \times 1.0 \times 1.0(2d + 1.0)}{.002} = 2500d + 1250$$

For the value of d, we should use the equivalent depth. Taking S=80 m, from Table 21.9 (for d = 2.5 m), the equivalent depth is 2.17 m. With this value, we get,

$$S^2 = 2500 \times 2.17 + 500 = 6775$$

which is differing much from 80 x 80 = 6400

Hence considering a spacing of 85 m

$$S^2 = 2500 \times 2.18 + 1250$$

$$= 6700$$

$$S = 85 \text{ m}$$

Hence 85 m spacing is considered.

Example 24.7: Determine the spacing of tile drains using Hoogoudt's equation. If the water table from soil surface is to be lowered by 30 cm in 2 days. The hydraulic of the soil is 4.5 cm/hr, the drains are to be placed at their bottom is 1.0 m below the soil surface while the impermeable layer is 2 m below the soil surface. What is the average flow out of a 80 ha field in the 2 days period.

Solution:

On using Hoogoudt's equation (21.19)

$$S^2 = \frac{4KH(H+2d)}{V}$$

given that,

K = 4.5 cm/hr,

H = 1 – 0.3 = 0.7 m,

d = 2 – 1 = 1 m.

the water table should be lowered by 30 cm in 2 days.

the value of v,

$$V = \frac{30}{24} = 1.25\, cm/hr$$

on substituting these value in Hoogdoudt's equation, we obtain,

$$S^2 = \frac{4 \times 4.5 \times 0.7}{1.25} = 38.88$$

$$S = 6.2 \text{ m}$$

Average out for two days = Area of field × depth of water table to be lowered

$= 80 \times 0.3$

$= 24 \times 10^4 \text{ m}^3$

Example 21.8: How many ha will a 25 cm tile drain with a slope of 0.3 per cent tiles are placed on an area where drainage coefficient can be taken 2 cm?

Solution:

On using Manning's formula

$$V = \frac{1}{n} R^{2/3} S^{v2}$$

For circular drain, R = d/4; for tile drain n = 0.01

Thus,

$$V = \frac{(D/4)^{2/3} S^{v2}}{0.01} = 39.7\ D^{2/3}\ S^{v2}$$

Substituting the values of D and S

$$V = 39.7\ (0.25)^{2/3}\ (0.003)^{1/2}$$
$$= 0.866 \text{ m/sec.}$$

Therefore, the discharge = A × V

$$= \frac{\pi}{4}(0.25)^2 \times 0.866$$

$$= 0.425 \text{ m}^3\text{/sec.}$$
$$= 153 \text{ m}^3\text{/sec.}$$

For a drainage coefficient of 2 cm, drainage,

$$A = \frac{153 \times 24}{0.02 \times 104} = 1835\, ha$$

Example 21.9: The electrical conductivity of a canal irrigation water is 1.2 dsm^{-1} while, the wheat crop can tolerate a maximum conductivity of 6 dsm^{-1}. The crops consumptive is 0.8 cm/day and soils hydraulic conductivity is essential as 0.75 cm/hr.

The area is to be drained by a tile drain system of 20 cm dia laid at a depth of 2.5 m below ground level so as to keep minimum water table depth of 1.5 m. The impermeable layer is located at a depth of 4 m from surface.

(a) Determine the spacing of title drains.

(b) Determine the flow rate out for a 200 ha field.

(c) What will be the size of outlet pipe laid on a gradient of 0.001 m/m.

(d) How much time will be required to drop the water level to 1.2 m below the surface if during rains water table rises to within 1 metre of ground surface.

Solution:

(a) Let *d* be depth of water to be applied per day. Depth of water to be drained

$$= d \times Lr$$

By definition $\quad Lr = \dfrac{Ei}{Ed} = \dfrac{1.2}{6} = 0.2$

Thus, water lost as consumptive use

$= d - d.Lr = 0.8\ d$

Also consumptive use = 0.8 cm/day

Thus d = 2 cm and depth of water to be drained per day,

$$= 0.2 \text{ cm or } \frac{0.2}{24} cm/hr$$

Spacing of drain is determined on using Hoogoudt's equation as,

$$S^2 = \frac{4KH\,(2d+H)}{V}$$

$$= \frac{4\times0.75\times(2.5-1.5)\times(2\times1.5+1)\times24}{0.2} = 1440 \text{ m}^2$$

S = 38 m

(b) Outflow from the field per day

$$= \frac{0.2}{100}\times200\times104$$

$= 4000\ m^2/day$ or $46.3 \times 10^{-3}\ m^3/sec$

(c) Size of outlet pipe is determined on assuming that drain is running full canal Manning's n = 0.01, then,

$$\text{discharge (Q)= av } a\times\frac{1}{n}\ R^{2/3}\ S^{1/2}$$

for pipe flowing full

$$a = \frac{\pi}{4}d^2$$

$$R = \frac{d}{4}$$

and $Q = 46.2 \times 10^{-3}\ m^3/sec$.

Thus, we get

$$Q = \frac{\pi}{4}d^2\,\frac{1}{0.01}\left(\frac{d}{4}\right)^{2/3}(0.001)^{1/2}$$

$\therefore$

On simplification, we get,

d = 44.1 cm say 45 cm

(d) Under the changed conditions of water level, the values used in Hoogoudt's equation will be,

S = 38 m k = 0.75 cm/hr

H = 1.3 m d = 1.5 m

Substituting these values in Hoogoudt's equation, we get,

$$1440 = \frac{4 \times 0.75 \times 1.3}{V}(2 \times 1.5 + 1.3)$$

V = 0.28 cm/day

Adding consumptive use, the total depth of water lost per day

= 0.8 + 0.28

= 1.08 cm

Time required to lower 20 cm water depth

$$= \frac{20}{1.08} = 18.5\, days$$

U.S. Bureau of Reclamation Formula

The Bureau of Reclamation proposed equation for the transient state of the water table or unsteady state condition. In deriving this equation all the assumptions as in case of deriving the Hooghoudt's equation are made.

Let us consider an elemental volume of dx and dz (Fig. 21.15). Taking the surface of the impervious bed to be horizontal, the horizontal component of the flow velocity is given as,

$$V_x = -K.\frac{dy}{dx} \qquad \text{....(21.23)}$$

Where, *y* is the depth of impervious soil layer below around surface. The limit when Δy and Δt and since *x* direction only is considered. The equation written as,

$$f\frac{dy}{dt} = k\frac{d}{dx}\left(\frac{y\,dy}{dx}\right)$$

$$= k\left[\frac{y d^2 y}{dx^2} + \left(\frac{dy}{dx}\right)^2\right] - 24.10$$

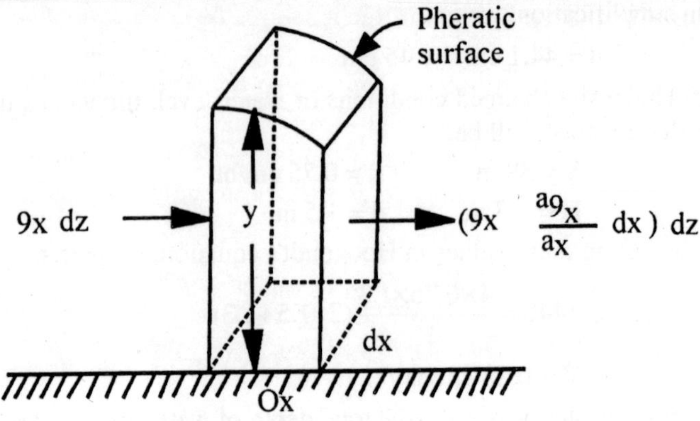

Fig. 21.15: Definition sketch for USBR formula

If *y* is large as compared with changes in *y*, we may consider *y* a constant with the mean value *D* and neglecting second order terms, the equation is rewriteen as,

$$\frac{dy}{dt} = \frac{KD}{f}\frac{d^2y}{dx^2} \qquad \text{....(21.24)}$$

where,

y = height of water table, (m),

K = hydraulic conductivity, (cm/hr),

D = average thickness of the aquifer from the impermeable layer to the water table, (m), and

f = drainable porosity in per cent by volume (equivalent to specific yield).

Equation 21.24 is a second order ordinary differential equation. The solution of this equation with appropriate initial and boundary conditions will give the height of the watertable (y) with respect to time. The solution of equation (21.24) will be obtained as suggested by Carslaw and Jaegar (1959). For this, an initial shape of the water table surface must be assumed. Glover assumed a fourth degree parabola at time t=0. The equation of the parabola is given as,

$$y = \frac{8H}{L^4}(L^3x - 3L^2x^2 + 4Lx^3 - 2x^4 \qquad \text{....(21.25)}$$

where L is the spacing between the drains.

According to the Bureau of Reclamation, the spacing obtained using the fourth degree parabola to represent the initial position of the water table is closer to true situation even though taking initially a flat water table does not cause a large difference.

Considering the water table at the elevation at the water drains, version of the equation obtained for drain spacing on simplication is expressed as;

$$L^2 = \frac{\pi^2 KDt}{f \ln(1.16\frac{y_o}{y_t})} \qquad \text{....(21.26)}$$

where,

y = height of water table after a time interval t, and

y_o= initial height of water table.

Convergence of flow at the drains can be considered by substituting equivalent depth to the impermeable layer for the actual depth as in Hooghoudt's method for the steady state.

An approximation for equivalent depth D is obtained using the formula,

$$D = d + \frac{y_o}{2} \qquad \text{....(21.27)}$$

where d is the equivalent depth from the drain to the barrier as obtained from Hooghoudt's graphs.

Donnan's formula

Donnan developed an equation for design of drain for relief drains based upon certain barrier conditions. According to Donnan, the spacing of the tile drain (S) is given by the following relationship:

$$S = \frac{4k(b^2 - a^2)}{Q_d} \qquad \text{....(21.28)}$$

where,

a = Depth below the centre of tile drain, (m),

b = Depth of water level in between the two drains, (m),

k = Hydraulic conductivity, (cm/sec.) and

Q_d = Drainage coefficient, (cm).

Tile drains should be spaced such that it lowers the ground water table enough for good plant growth within 24 hours after rains. Some desired limits are given in Table 21.16.

Table 21.16: Desired limits for the tile drains

Soil type	Spacing (m)	Depth (m)	Dia. of tile drain (cm)	Min. grade (%)	Max. grade (%)
Clay & clay loam	12-21	0.75-0.90	10	0.1	400
Silt loam	18-30	0.9-1.2	12.5	0.07	600
Sandy loam	30-90	1.0-1.3	15	0.05	900

Example 21.10: Design a subsurface tile drainage using the following data: Soil-sandy loam with K = 2.3×10^{-4} cm/sec. Impermeable barrier at 4 m below land surface as indicated by subsurface borings. Tile drains to be installed at a depth of 1.8 m below land surface. Water table should be held at a minimum depth of 1.2 m below land surface (from farm experience). Depth of irrigation = 10 cm every two weeks. Deep percolation loss from irrigation = 20% (from consumptive use studies). Field ditch loss = 8% of water applied (from ditch flow measurements). There is no significant natural subsurface discharge from the area.

Solution:

Using the Donman's formula, the spacing of the tile drains (S) is determined as,

$$S = \frac{4k(b^2 - a^2)}{Q_d}$$

where, a and b are the depths as shown in Fig. 21.16.

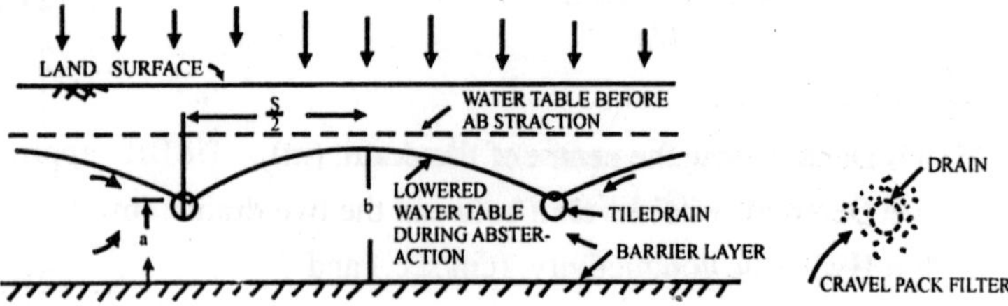

Fig. 21.16: Spacing of tile drains

Q_d = drainage coefficient which is the discharge of an under drainage system expressed as the depth of water to be removed from the area in 24 hr (≈ 1% of a.a.r. in 24 hr).

Water lost through percolation from each irrigation

$$= 10\,(20\% + 8\%) = 2.8 \text{ cm in 14 days}$$

$$\text{Drainage coefficient } Q_4 = \frac{2.8}{14} = 0.2 \text{ cm/day}$$

$$a = 4 - 1.8 = 2.2 \text{ m},\ b = 4 - 1.2 = 2.8 \text{ m}$$

Spacing of the tile drains

$$S = \sqrt{\frac{4k\,(b^2 - a^2)}{Q_d}}$$

$$= \sqrt{\frac{4 \times 2.3 \times 10^{-4}\,(2.8^2 - 2.2^2)}{0.2\,(24 \times 60 \times 60)}} = 34.5\,m$$

The flow in the tile drains and required tile sizes may be computed from the irrigation practice, rainfall intensity, soil type and tile drain spacing. The tile drains are usually of 10-15 cm diameter, laid with open joints with graded filters provided around each joint and the tile drain grades varying from 0.5 to 0.10%.

Capacity of tile drains

The discharge through the tile drains can be determined after determination of drain spacing. Hooghoudt's approach gave the maximum rate of flow per unit length of the drain under steady state condition. Discharge is determined under two conditions namely,

(i) Steady state condition: According to Hooghoudt's approach the maximum rate of flow per unit length of the drain is determined as,

$$q = \frac{S}{2} v \qquad \text{....(21.29)}$$

(ii) Unsteady state condition: According to USBR approach discharge is determined as,

$$q = \frac{2\pi K y D}{L} \qquad \text{....(21.30)}$$

where, q drain discharge per unit length of the drain in unit time.

The design flow of tile drains is determined under humid and arid conditions using different approaches. The main aim is to remove the excess rainfall appearing as sub-surface flow under humid conditions. The concept of drainage coefficient is used for determining the capacity of the tile drains. The values of the drainage coefficient for field crops are summarized in Table 21.17. These values are based on the local soil, crop and surface drainage conditions and some experience is essential for deciding the value of the drainage coefficient.

Table 21.17: Drainage coefficients for tile drains in humid regions

Surface drainage condition	Drainage coefficient (cm per day)	
	Mineral soils	Organic soils
Normal	1.0 – 1.3	1.3 – 1.9
With blind inlets	1.3 – 1.9	1.3 – 2.5
With surface inlets	1.9 – 2.5	1.9 – 3.7

(Source: U.S.S.C.S., 1960)

In determining the capacity of the tile line the area actually drained by that particular line is to be considered. Additional allowance is to be considered where there are surface inlets into the tile line or side hill seepage.

Grades of tile drains

The tile drain should be laid on such a grade that no sedimentation occurs in the tiles. In general, a minimum velocity of 45 cm per second is recommended. A desirable working grade is 0.2 per cent, however, higher grades can be recommended. Higher grades result in higher velocities which may cause un determining of the tiles. These recommendations are made for the laterals. The grades of the main lines depend upon the outlet selected. Grades upto 2 to 3 per cent were found satisfactory for the main lines.

Size of tile drains

The size of the tile drains is determined based on the maximum expected flow and the grade. For determination of size of tile or pipe the Manning formula is used. The value of Manning coefficient (n) is recommended as 0.011 for concrete tiles. If other materials are used a suitable value of "n" can be chosen as given in Table 21.18. The size of the tiles flowing full is determined on using Manning formula as,

$$Q = AV = \frac{\pi}{4}d^2 \frac{\pi}{4}d^{4/3}\ S^{1/2} \quad \text{....(21.31)}$$

For practical purposes the nearest available commercial size is selected. A size slightly greater as compared to the calculated size is selected, to take care of sedimentation and also of any misalignment during installation. In case of concrete or clay tiles the size of 10 cm, 12.5 cm and 15 cm may be used.

Table 21.18: Recommended values of Manning coefficient (n) for various conduit materials

Conduit material	Manning's n
Clay tile Concrete pipe Vitrified clay pipe	0.011
Perforated plastic pipe Corrugated plastic tubing	0.017

(Source: Shawab *et al.*, 1966)

Example 21.11: The laterals were laid out at 50 m apart and the 250 m long and have a grade of 0.25 in a subsurface drainage. (i) if the drainage coefficient of the area is 2 cm, what size tiles will be required? (ii) If the drainage coefficient is increased to 3 cm, what will be the spacing of the laterals?

Solution:

(i) Quantity of water to be drained by the laterals in 24 hrs.

$$= \frac{2 \times 50 \times 250}{100} = 200\, cum$$

$$\text{Rate of flow (q)} = \frac{250}{24 \times 60 \times 60} = 0.002894\, cum/\text{sec.}$$

Consider the rate of flow through a 10 cm tile

$$\text{Wetted perimeter (P)} = \pi \times .1 = 0.3144 \text{ m}$$

$$\text{Hydraulic radius (R)} \ \frac{\frac{\pi}{4} \times .1 \times .1}{\times .1} = 0.25$$

$$V = \frac{R^{2/3}.S^{1/2}}{n} = \frac{(0.25)^{2/3}.\left[\frac{0.25}{100}\right]^{1/2}}{0.0108} = 0.4335 \text{ m/s}$$

Capacity of the tile = (Q) = A.V.

$$= \frac{\pi}{4} \times 1 \times .1 \times 0.4335 = .003405 \text{ cum/sec.}$$

Hence a 10 cm dia tile will be satisfactory.

(ii) Let the laterals be located at a distance of W and length kept at 250 m

Amount of water to be drained in 24 hr

$$= W \times 250 \times \frac{3}{100} \, cu.m.$$

Capacity of the tile line = .003405 x 60 x 60 x 24

$$\therefore \text{Spacing, (W)} = \frac{.003405 \times 60 \times 60 \times 24 \times 100}{250 \times 3} = 49 \text{ or } 50 \text{ m.}$$

Loads on drain pipes

Drain pipes have to bear the load of the soil above them when installed at proper depth. Drains located at shallow depths (less than 1 m) are subjected to both the soil weight and load due to machinery. Drains located at deeper depths are subjected to soil loads only. If the ditch depth is more than 2 or 3 times the outer diameter of the tile then it is considered as a wide trench otherwise a narrow trench.

The following formulas are used for determination of loads on drain pipe.

(i) Ditch conduit formula, and

(ii) Projecting conduit formula

(i) The Ditch Conduit Formula

This formula is used for determination of the load on pipes in narrow trenches and is expressed as,

$$W_o = C_d W B_d^2 \qquad \text{....(21.32)}$$

where,

W = load on the pipe,

C = load coefficient,

W = unit weight of fill material, and

B_d = width of trench at the top of the pipe.

(ii) The projecting conduit formula

It is used for determination of the load on pipes in wide trench and is expressed as,

$$W_c = C_c W B_c^2 \quad ....(21.33)$$

where,

W_c = load on the pipe,

C_c = load coefficient for project conduits,

C_w = unit weight of fill materials, and

B_c = outside diameter of the conduit.

The load coefficients (C_c and C_d) constrains are functions of the frictional coefficients of the soil, the height of the fill (H) and the width of the ditch or the width of the conduit as the case may be. In case of ditch conduits it is assumed that the density of the soil filled is less than that of the original soil. The sides of the ditch offer an upward frictional resistance to the fill material and as such the load on the conduit is less than the weight of the soil directly above it. While, in case of projecting conditions, the load on the conduit is greater than the weight of the soil directly above it. It may be attributary by shearing forces. Due to greater settlement of soil on both sides are downward rather than upward.

Transition widths are those when the loads determined by both the Equations (21.32 & 21.33) are the same on using values of C_d & C_c as given in Fig. 21.17.

Strength test

Drain tiles should be installed in such a way that the loads on them does not exceed their crushing strength.

Testing of tiles

The following techniques are used for testing the strength of concrete or clay tiles:

(i) Three-edge bearing test, and

(ii) Sand bearing test

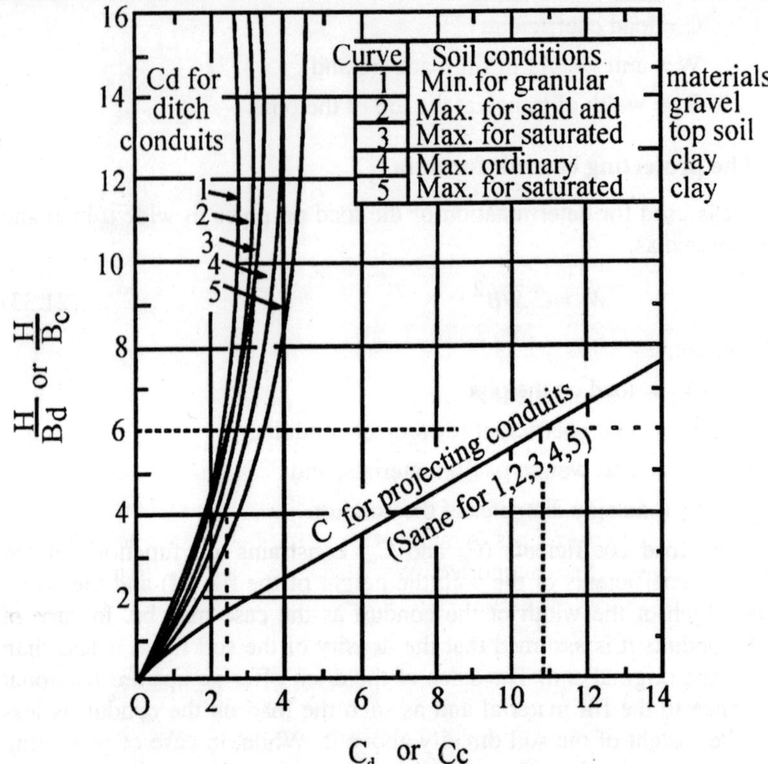

Fig. 21.17: Values of C_d and C_c in ditch conduit formula.
(Source : Spangler, 1947)

(i) Three-edge bearing test: It consists of two wooden blocks on which the tile is placed. In this method, a point load which can be increased is applied on top of the tile (Fig. 21.18). This load is applied either by a mechanical jack or a hydraulic system. The load is gradually increased till the tile breaks and is noted.

(ii) Sand bearing test: In this method, the procedure is the same but the arrangement is different (Fig. 21.18). The bottom of the tile is supported by a bed of sand. This represents the conditions found under normal installation procedures more closely than the three-edge bearing test. The crushing strength found by the sand bearing method is nearly 1.5 times higher as compared to the strength found by the three edge bearing test.

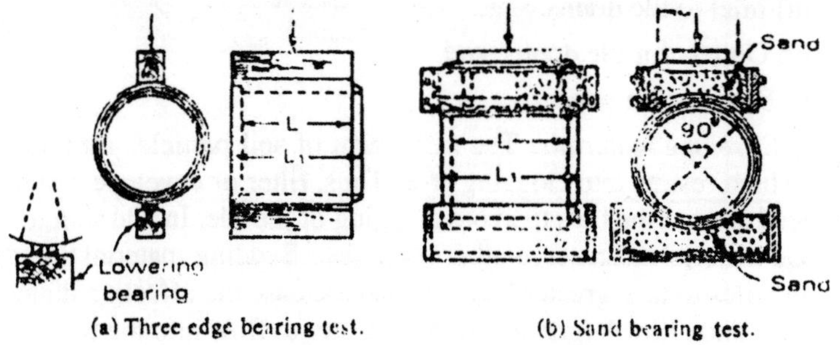

(a) Three edge bearing test. (b) Sand bearing test.

Fig. 21.18: Testing of tiles

Bedding conditions for tile drains

The amount of load that the tile can bear depends upon the bedding conditions of the tiles. The different bedding conditions and the load factors are shown in Fig. 21.19. The crushing strength as determined by the three edge bearing tests is multiplied by the load factor to determine the load that could be supported. While, determining the loads a safety factor is taken into consideration. The determined loads are increased by 50 per cent as a safety factor.

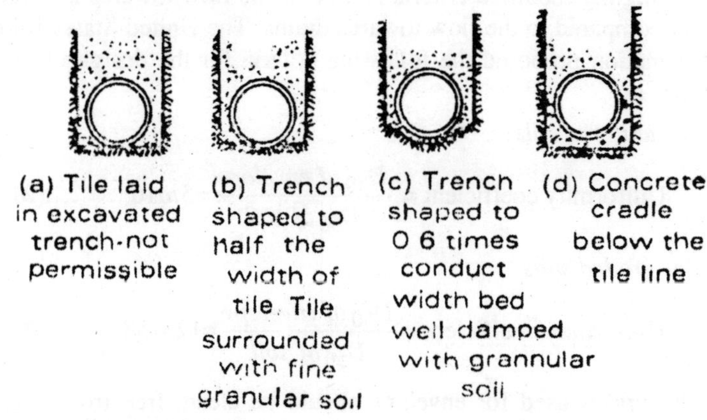

Fig. 21.19: Different bedding conditions of tile drains

Accessories for tile drain systems

The tile drainage systems requires following accessories to function satisfactorily:

(i) Envelope materials,

(ii) Manholes and sedimentation basins,

(iii) Inlet to tile drains,

(iv) Outlets for tile drains, and

(v) Evaporation pond

(i) Envelope materials: The movement of soil particles into the tile drains which results into clogging of it. Thus, filter or envelope materials are placed around the tiles to avoid clogging of the tile. In addition to this envelope materials also serve other purposes. Bedding material and thus allow the tiles to take greater loads. It also increase the effective diameter of the tiles and allow larger quantities of water to flow into the tiles.

A variety of materials are used for envelope in tile drains *viz.*, gravel, coarse sand, organic materials like corn cobs, straw etc. The coir matting has also been used as envelope material for continuous plastic pipe. Among envelope materials, gravel has been the most widely used material because of its efficiency and also of permanent nature.

The gravel envelope provided to avoid the entirely of the soil particles into the tile drains. For this purposes, the gravel envelope should be designed for taking into consideration the nature of the soil around the tile drains. The criteria for the design of envelope are based on the same principles as followed for the design of gravel envelopes for wells. A slightly modified criteria is used as the flow towards the well is faster as compared to the flow towards drains. The United States Bureau of Reclamation proposed the following criteria for the design of gravel envelopes as,

For uniform soils

$$\text{Uniformity coefficient} = \frac{D_{50}\ \textit{of envelope}}{D_{50}\ \textit{of soil}} = 5\ \textit{to}\ 10 \qquad \text{....(21.34)}$$

For graded soils

$$\text{Uniformity coefficient} = \frac{D_{50}\ \textit{of envelope}}{D_{50}\ \textit{of soil}} = 12\ \textit{to}\ 58 \qquad \text{....(21.20)}$$

The gravel used for envelope should be clean, free from organic matter and bentonite clays.

(ii) Manholes and sedimentation basins: These are vertical structures installed at regular intervals along the tile lines to help in inspection of the tile lines and cleaning the accumulated sediment. These are made of either concrete or brick masonry and are of enough dimension for a person to enter it. These structures are installed at junctions of tile lines. There is no general guidelines for their spacing but

decided upon local conditions and experience. The top of the manhole or the sedimentation basin can be kept at ground level or below ground level. Ground level structures may interfere with farming operations and with machineries. It should be located about 60 cm below ground level or below the ground level and covered by a concrete block. Then, it does not obstruct the farming operations.

(iii) Inlet to tile drains: These structures are used to admit surface runoff water into the tile lines. These could also be used for flushing the tile lines. Generally, two types of inlets are used *viz.*, blind inlets (Fig. 21.20) and surface inlets (Fig. 21.21). Blind inlets are cheaper and do not interfere with farming operations. However, these may be likely to get clogged. Surface inlets have arrangement for avoiding entry of the trash and other suspended materials in tile drains.

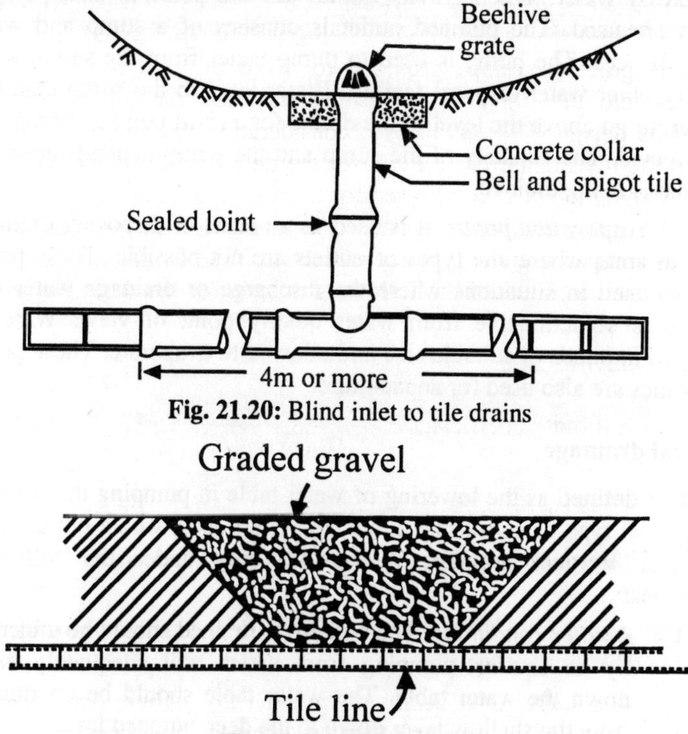

Fig. 21.20: Blind inlet to tile drains

Fig. 21.21: Surface inlet

(iv) Outlets for tile drains: The outlets should satisfy the following requirements for proper functioning of the tile drainage system:

1. Provide free flow of drainage water,

2. Discharge the outflow without any damage to the tile or erosion,
3. Prevent the entrance of flood water,
4. Protect the end of the tile line, and
5. Keep out rodents.

Generally, two types of the outlets are used *viz.*, gravity outlet and the pump outlet. The topography of the land is the main factor which decides the choice of the outlet. In the gravity outlet water flows out of the drainage system by gravity into an open drain, stream, a pond of any other facility that disposes the water. When disposing the water the outlet pipe should be above the water level by about 30 cm so that there is a free fall. It should be ensured that the end of the pipe should not be covered by water. When gravity outlets are not possible then pumped outlets are used. The pumped outlet is consists of a sump and water lifting devices. The pump is used to pump water from the sump into a drain or other water disposal facility. Water level in the sump could be allowed to go above the level of the drains for a short period, if crops are not affected. The capacity of the sump and the pump depends upon the area that is being drained.

(v) Evaporation ponds: It is used as a means of disposing drainage water in areas where any types of outlets are not possible. These ponds are also used in situations where the discharge or drainage water into streams is objectionable from water quality point of view. Water is collected in ponds of suitable size and allowed to evaporate. These ponds sometimes are also used for aquaculture.

Vertical drainage

It is defined as the lowering of water table in pumping the wells. It is a special type of subsurface drainage system used where outlet is not available. Vertical drainage is feasible only under the following conditions:

1. *Aquifer conditions:* The area to be drained should be underlain by an aquifer, pumping from which will adequately lower down the water table. The water table should be continuous from the shallow layer down to the deep pumped layer.
2. *Water quality:* The underground water should be of satisfactory quality. Thus the pumped water could be used for irrigation.

3. *Soil conditions:* The hydraulic conductivity of the soil layers to be drained is important as these soil layers should allow free percolation of water. There should not be any intervening layers of low hydraulic conductivity impeding the movement of water.

Vertical drainage wells are most successful in irrigated areas when the pumped water is of good quality for irrigation purpose. A water balance could be achieved keeping the water table at the desired level.

Types of wells: Pumping of the aquifers through wells lowers the water table levels. The water table or gravity wells are used to lower the water table from the rootzone. Pumping of confined aquifers may not lower the water tables from rootzone immediately. For vertical drainage, a large number of wells are to be installed.

Spacing of the wells: The number and spacing of the wells depend upon the aquifer properties and the extent to which water table is to be lowered. The drainage wells could be designed for mutual interference as the drawdown due to each of the wells are additive. The hydraulics of wells and the hydraulics of interfering wells are useful in designing the well for drainage.

Deep open drains

It can be used for subsurface drainage (Fig. 21.22). Though it requires more land area as well as need constant maintenance. The open drains are used for both as interceptor and relief drains. The use of the deep open drains as interceptor drains is more common as their length required is generally less as compared to the relief drains and it can be used for intercepting both surface and subsurface flows. These drains are similar to diversion drains used in soil conservation practices.

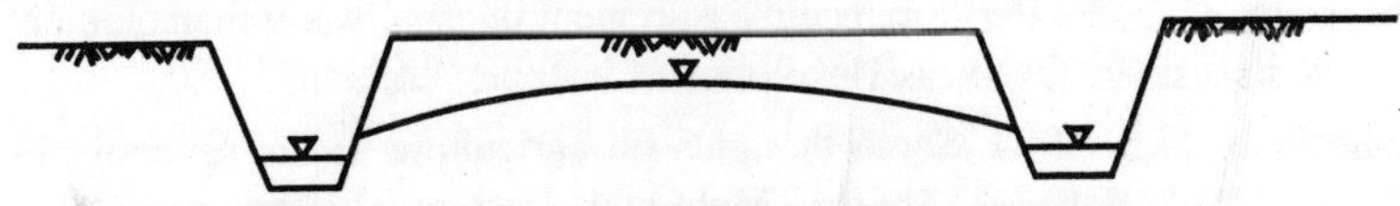

Fig. 21.22: Deep open drains for water table control

The capacity of the interceptor drain can be determined on using Darcy's law when it is intercepting only ground water flow. The flow of water intercepted (Q) is determined as,

$$Q = K\,i\,A \qquad \text{....(21.35)}$$

The cross sectional area intersected is determined as,

$$A = de\,L \qquad \text{....(21.36)}$$

where,

de = Effective depth of drain, and

L = length of drain

On using Equations (21.35) and Equation (21.36), we obtain,

$$Q_i = \frac{K\,i\,de\,L}{24 \times 60 \times 60} \qquad \text{....(21.37)}$$

where,

Qi = discharge (cum/second),

K = hydraulic conductivity, (m/day),

i = hydraulic gradient, (m),

de = effective depth of drain, (m), and

L = length of drain, (m)

This equation is applicable for uniform hydraulic conductivity through-out the soil profile.

Bio-drainage

It is the growing of high water requiring plants to lower the water table and improve the environment for better growth of the plants. When the transpiration is met primarily by drawing groundwater table, the process is knows as bio-drainage. It is suitable where natural outlet is not available. In this system a cluster of plants were planted but major difficulties will be lack of light for photosynthesis. Plant species such as Eucalyptus has been found as very suitable for this purpose.

References

Abernethy, C. 1986. Performance measurement in canal water management - A discussion, Overseas Development Institute, London.

Adiseshiah, M.S. 1985. Some thoughts on agriculture in the seventh plan, 17th Lal Bahadur Shastri Memorial Lecture, Indian Agricultural Research Institute, New Delhi.

Ali, S.H. 1982. Report of the commission for irrigation utilization, Govt. of Andhra Pradesh, Hyderabad.

Anonymous, 1981. Cropping pattern studies for Gomti-Kalyani Doab, Ford Foundation Project, Research Group, WRDTC, University of Roorkee, Roorkee, India.

Anonymous, 1981. Sixth five-year plan 1980-85, Planning Commission, Govt. of India.

Anonymous, 1983. Diagnostic analysis of farm irrigation systems on the Gambhiri irrigation project', Rajasthan, India, Water Management Synthesis Report No. 17, Colorado State University, USA.

Anonymous, 1983. Resource analysis and plan for efficient water management: A Case Study of Mahi Right Bank Canal Command Area, Gujarat, Water Technology Centre, Indian Agricultural Research Centre, New Delhi.

Bottrall, A. 1981. Comparative study of the management and organization of irrigation projects, World Bank Staff Working paper no. 458, Washington.

Carslaw, H.S. and Jaeger, J.C. 1959. *Conduction of heat in solids* (2nd Ed.), Oxford University Press.

Chambers, R. 1988. Managing canal irrigation, Oxford and IBH Publishing Co. Pvt. Ltd., New Delhi.

Chauhan, H.S. 1989. Drip irrigation status and prospects in India, National Seminar on New Perspectives in Water Management held at Indore.

Chawla, A.S. 1989. Conjunctive use of surface and ground waters, National Seminar on New perspectives in Water Management held at Indore.

FAO, 1984. Irrigation and drainage paper Rev.: Irrigation Practices and Water Management, Oxford & IBH Publishing Co. Pvt. Ltd., New Delhi, pp. 63.

Frederick, L.H. 1984. World Bank irrigation experience, Overseas Development Institute, London.

Garg, O.N. and Gupta, D.K. 1988. Economics of canal lining - A Systems Study, Seminar on Canal Lining held at Roorkee.

Ghosh, R. 1989. Waterlogging - problem and remedies, National Seminar on New Perspectives in Water Management held at Indore.

Jayaraman, T.K., *et al.*, 1983. Diagnostic analysis of farm irrigation system on the Mahi-Kadana irrigation project, Gujarat, India, Water Management Synthesis Report 18, Colorado State University, Colorado.

Kirkham, Don. 1946. Proposed method for field measurement of permeability of soil below the water table, *Soil Sci. Soc. Am. Proc.,* 10, 58-68.

Luthin, J.N. 1970. *Drainage Engineering,* Wiley Eastern Private Ltd, New Delhi.

Malhotra, S.P. 1982. The warahandi system and its infrastructure, Publication No. 157, CBIP, New Delhi.

Munot, J.S., Purohit, R.C. and Singh, Jaspal, 1988. Optimal irrigation scheduling for wheat crop of Udaipur Region, *J. of Indian Water Resources Society.*

O' Neal, A.M. 1949. Soil characteristics significant in evaluating permeability, *Soil Science,* 67, 403-409.

Padhi, G.N. 1983. Pricing policy for irrigation water, Ph.D. Thesis, WRDTC, University of Rorkee.

Padhi, G.N. and Suryavanshi, A.R. 1982. Evaluation of water distribution method of left Salawa distributary of upper Ganga canal system, IWRS Workshop on Water Distribution Practices.

Pant, N. (Editor) 1984. Productivity and equity in irrigation systems, Ashish Publishing House New Delhi.

Pant, N. 1981. Some aspects of irrigation administration (A Case Study of Kosi Project), Naya Prakashan, Calcutta.

Sarma, P.B.S., Rao, N.H. and Michael, A.M. 1983. Water resources utilization and management, WTC report on Resource Analysis and Plan for Efficient Water Management: Case Study on Mahi Right Bank Canal Area, Gujarat, Water Technology Centre, Indian Agricultural Research Institute, New Delhi.

Schwab, G.O., Frevert, R.K., Edminister, T.W. and Barnes, K.K. 1966. *Soil and Water Conservation Engineering* (2nd Ed) John Wiley and Sons, New York.

Seckler, D. and Nobe, K. 1983. The critical role of the management factor in developing countries, Paper presented for the International School of Economic Development, Colorado State University.

Singh, Bharat 1989. Agriculture and irrigation, National Seminar on New Perspectives in Water Management held at Indore.

Singh, K.K. (Editor) 1981. Warabandi for irrigated agriculture in India, CBIP Publication No. 146.

Singh, R.V. 1989. Increasing irrigation efficiencies through improved methods of water application, National Seminar on New Perspectives in Water Management held at Indore.

Sivanappan, R.K., Kandaswamy, P. and Selvaraj, K.V. 1984. Water management practices in Parambikulam Aliyar project command area, Tamil Nadu, Proc. of National Symposium on Strategies for Irrigation Development held at Bihar College of Engineering in Patna, Mangalik Prakashan, Saharanpur.

Smiles, D.E. and Youngs, G. 1965. Hydraulic conductivity determination by several field methods in sand tank, *Soil Science,* 99, 83-87.

Spangler, M.G. 1947. Underground conduits-an appraisal of modern research, *Proc. Am. Soc. Civil Engrs.,* 73, 855-884.

Swaminathan, M.S. 1982. Science and integrated rural development, Concept Publishing Company, New Delhi.

U.S. Soil Conservation Service. 1960. *National Eng. Hand Book,* Section 16, Chapter 4, Subsurface drainage in humid areas (lithographed), Washington, D.C.

V.V.N. Murthy. 1985. Land and Water Management Engineering, Kalyani Publishers, pp. 585.

Wessling, J. 1973. Subsurface flow into drains, (in) *Drainage Principles and Applications,* Vol. II. International Institute for Land and Reclamation and Improvement, The Netherlands.

Yoganarsimhan, G.N. 1989. Some musings on people's participation in irrigation water management, *J. of Indian Water Resources Society.*

EXERCISES

21.1. Define the following:

(a) Sodium adsorption ratio

(b) Drainage

(c) Drainage coefficient

(d) Leaching requirement

(e) Hydraulic conductivity

21.2 Differentiate the following:

(i) Surface and sub-surface drainage

(ii) Vertical and bio-drainage

(iii) Parallel and bedding system of surface drainage system

(iv) Herring bone and gridiron system of sub-surface drainage

(v) Interceptor and relief drain

21.3 Write short note on the following:

(i) Mole drainage

(ii) Vertical drainage

(iii) Inlet of tile drain

(iv) Envelope material for tile drain

(v) Manholes and sediment basin

21.4 Explain the causes of water logging.

21.5 Describe surface drainage system.

21.6 How design discharge for open ditches?

21.7 Describe the procedure for design of open ditches.

21.8 How open ditch is constructed?

21.9 Explain the benefit of sub-surface drainage.

21.10 Describe the investigation for sub-surface drainage.

21.11 How hydraulic conductivity is determined?

21.12 Derive the Hooghoudt's equation for sub-surface drainage.

21.13 State Darcy's law.

21.14 Describe the sub-surface drainage system.

21.15 Derive the U.S. Bureau of reclamation formula.

21.16 Describe the materials used for tile drain.

21.17 How load on tile drain is determined?

21.18 How tile drain is tested?

21.19 Discuss the installation of tile drains.

21.20 Describe the mole drainage.

21.21 Discuss the drainage by wells.

21.22 Explain the deep open drain.

21.23 Explain water table contour map.

21.24 Sub-surface drainage system is installed at DFRS, Arjia, Bhilwara. The laterals were laid out at 60 m apart and the 300 m long and have a grade of 0.25.

(i) if the drainage coefficient of the area is 2.3 cm, what size of tiles have to be used?

(ii) If the drainage coefficient is increased to 3.5 cm, what will be the spacing of the laterals?

21.25 A research farm of 150 ha is discharging through a drain at an average rate of 2.5 cum per second. Determine the drainage coefficient, if the drainage coefficient should be 3 cm what should be discharge through the drain.

21.26 Design an open ditch to drain 100 ha of land having a drainage coefficient of 2.5 cm. The soil is silt loam and maximum permissible slope of the channel bed is 0.2 per cent.

21.27 Tile drains have to be installed in an agricultural land which has a soil permissibility of 25,000 lpd/m^2. An impermeable stratum exists at about 3.2 m below the land surface and it is desired to keep the water level at least 1 m below the land surface. What drain spacing would you recommend if the average annual rainfall is 100 cm? (140 m to remove 1% of a.a.r. in 24 hr)

21.28 Irrigation water for growing of wheat has an electrical conductivity of 490 μmhos/cm. If the conductivity tolerance of wheat is 7,000 μmhos/cm, determine the leaching requirement for wheat. (LR = 7%)

■■■■

Chapter 22

WATERSHED MANAGEMENT

22.1 Introduction

The conservation and utilization of natural resources on watershed basis has drawn a widespread interest and is now increasing view as the most appropriate way for the development of highly eroded lands and rainfed areas. About 112.6 million hectare (M ha) are affected by water erosion and 38.7 M ha by wind erosion of which 85 M ha falls under arable land. It is thus imperative that for arresting further land degradation and rehabilitating these lands, a concreted effort is required at all levels.

22.2 Causes of Watershed Deterioration

Uncontrolled, unplanned, unscientific land use and interventions lead to deterioration of the watershed. These activities are summarized as follows:

(1) *Agricultural land:* Cultivation on sloping land without adequate measures, cultivation without agronomic measures to conserve soil and water, cultivation along susceptible nalla banks, cultivation of erosion-permitting crops, overcropping without soil-fertility replenishment, faulty agricultural techniques, etc.

(2) *Grassland:* It includes excessive and uncontrolled grazing, growth of weeds, development of cattle tracks causing damage and compaction of soil resulting in lower infiltration rates, fires, thefts, etc.

(3) *Forest land:* It includes excessive and uncontrolled grazing which inhibits regeneration from seed or coppice stock, clear-felling on steep slope, destruction of forest land by fires and thefts, biotic pressure for fuel, fodder, NTFP and small timber, drastic thinning of plantations along slope, etc.

(4) *Shifting cultivation:* It is also known as shifting cultivation or *jhum* cultivation it is practiced in certain areas of the country has proved very damaging to protective and productive

vegetation. This practice results in damage to the topsoil and inhibits the growth of grasses, shrubs and trees.

(5) *Unscientific mining and construction activities:* These activities damage the vegetation and the landscape. The natural drainage lines are often blocked by debris.

(6) *Fire:* It includes intentional/accidental fires result in loss of vegetation, organic matter and micro-organisms.

(7) *Non-cooperation of the community:* Non-cooperation of the peoples for conserving rainwater and soils, protecting and enriching their ecosystem and CPR is bottleneck problems and it resulted into the most of the ills.

22.3 Concepts of Watershed

Watershed is defined as a geo-hydrological unit with a common drainage point. It is also called catchment or "drainage basis".

Every watershed would have a ridgeline, which forms the boundary and the drainage line. The drainage line may be fed by many streams. A big river may have many small rivers joining it as its tributaries. Thus, each tributary has its own boundary and forms a watershed. There would be first order stream second order stream and so on. If the watershed is bowl shaped, it would form a natural lake. Thus, watershed is the geographical area contributing run-off to a stream/nala/river. Fig. 22.1 shows the division of a watershed into sub watersheds and location of silt observations.

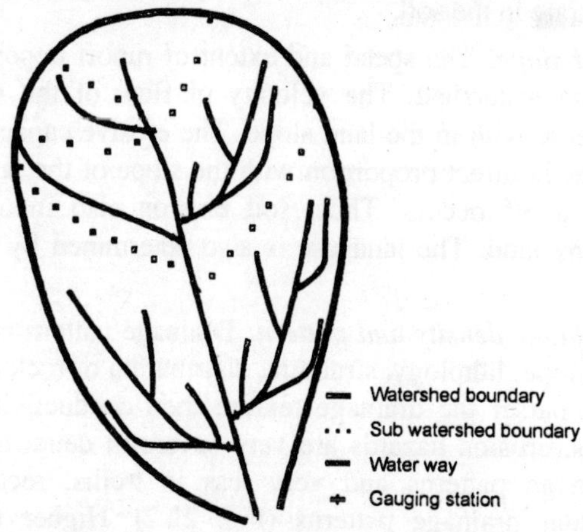

Fig. 22.1: Division of watershed into sub-watersheds for determining priority areas

Delineation of watershed boundaries

Ridgeline: Joins the highest points of watershed.

Drainage line: The stream

In fact, no land is common to two or more watershed. Watershed and sub watershed are the terms used in the watershed nomenclature according to their size and network of streams.

22.4 Watershed Characteristics

There are following watershed characteristics which effects runoff, soil loss and its management.

(i) Physiography, (ii) Edaphic condition, (iii) Vegetation, (iv) Land use, and (v) Rainfall

(i) Physiography – It includes the following:

(a) *Size:* The size or area of watershed is an important characteristics in determining the peak rate of runoff. The runoff rate and volume increases with increase in size of the watershed area.

(b) *Shape:* It affects opportunity time and runoff volume and rate from the watershed. Long and narrow watershed are likely to have longer time of concentration resulting in lower runoff rates than square shaped watershed having the same area. The longer time of concentration, takes more to time to leave the watershed because of increased opportunity for the water to infiltrate in the soil.

(c) *Land slope:* The speed and extent of runoff depends on slope of the watershed. The velocity of flow of the runoff water increase with in the land slope. The erosive capacity of runoff varies in direct proportion with the slope of the land on which the runoff occurs. Thus, soil erosion also increases in the sloppy land. The land use is also determined by the slope of land.

(d) *Drainage density and pattern:* Drainage pattern is affected by the slope, lithology, structure, distribution of rock systems, etc. the coarser the drainage texture then conductivity is higher. Thus, erosion hazards are very severe in dendritic and radial drainage patterns and very less in trellis, rectangular and annular drainage patterns (Fig. 22.2). Higher the drainage density then runoff watershed drains rapidly.

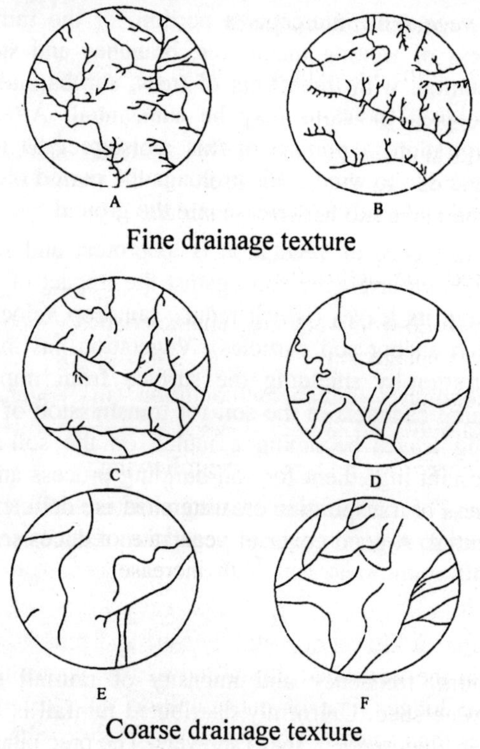

Fig. 22.2 Drainage patterns

(ii) Edaphic conditions

The soil and geology of the watershed determines the infiltration rate of water, percolation of water, runoff and soil erosion.

(iii) Land use

The land in a watershed has to be used for different purposes *viz.*, cultivation, livestock production, housing, water harvesting, etc. The land use affects runoff rates, infiltration and types and quality of vegetation. Suitable land use minimizes the runoff thereby soil erosion.

(iv) Vegetation

The type, quantity and quality of vegetative cover in watershed influences runoff, erosion and sedimentation, rate of evaporation and infiltration rate. There are following effects of vegetation cover.

(a) The vegetation intercepts a portion, of the rain through the process of wetting the leaves, branches and stem. In dense vegetation, with three tiers of trees, shrubs and grasses, this intercepted moisture may be substantial. After the end of precipitation, a portion of this moisture falls to the ground surface due to wind. This prolongs the period of rainfall at the ground and tends to increase into the ground.

(b) The influence of vegetation is to protect and shade the soil. Vegetation arrest the soil against the impact of direct rainfall through its leaves (which reduce raindrop velocity) and roots (which anchor soil particles). Vegetation has major effect on infiltration-by shielding the surface from impact, by roots opening channels in the soil for transmission of water and by falling leaves becoming a mulch on the soil surface. It is important ingredient for soil-building process and through the process of transpiration creating moisture deficiency or storage potential. A good cover of vegetation reduces soil erosion and runoff.

(v) Rainfall

The amount, frequency and intensity of rainfall determine the behaviour of watershed. Uniformly distributed rainfall is less damaging to soil and water than sudden, sharp showers. The precipitation variations effects, runoff, evaporation and percolation in the following manner:

→ Higher precipitation amounts = greater runoff

greater evapotranspiration

greater percolation

→ Higher precipitation intensities = greater runoff

Lower evapotranspiration

Lower percolation

→ Longer precipitation duration = lower runoff

Greater evapotranspiration

Greater percolation

→ Greater areal extent of precipitation = Greater runoff

Greater evapotranspiration

Greater percolation

(vi) Socio-economic factors – It includes the following:

(a) *Demographic profile:* It includes *viz.*, total population of the area, sex age-ethnic group stratification, location and occupation of the population, migration, future demographic trends help to decide the behavioural characteristics of watershed.

(b) *Sociological stratification:* It includes *viz.*, caste, lineage, families, lines of authority, heterogeneity, political organization, groupings, social status and political role of individual members, relations between members, spatial distribution of the community and CPR modify the watershed characteristics.

(c) *Farm structure:* It includes *viz.*, area of farm, number of farm workers, division of labour on farm, collective/individual farming, traditional cropping area, agricultural techniques, land management systems are important factors affecting a watershed.

(d) *Attitudes and behaviour:* It includes *viz.*, attitudes, aptitudes and behavioural patterns, value systems, belief system may also play an important role in the performance of the watershed

(e) *Land tenure systems:* It includes *viz.*, landholding sizes, dominant classes, landholding classes and castes, methods of land acquisition, alienation of land rights, relationship between land rights and social, political or religious systems, land leasing, sharecropping, marketing system of agricultural produce affect watershed functioning.

22.5 Objectives of Watershed Management

There are following objectives of watershed management:

(i) To promote the economic development of the village community which is directly or indirectly dependent on the watershed management which can be achieved through:

  (a) optimum utilization of the available natural resources in the watershed's *viz.*, land, water, vegetation etc., that will mitigate the adverse effects of drought and prevent further ecological degradation, and

(b) generation of employment and development of the human and other economic resources of the village in order to promote savings and other income generation activities.

(ii) To encourage restoration of ecological balance in the village through:

(a) sustained community action for the operation and maintenance of assets created and further development of the potential of the natural resources in the watershed, and

(b) simple easy and affordable technological solutions and institutional arrangements that makes use of, and build upon, local technical knowledge and available materials.

(iii) Special emphasis laid to improve the economic and social conditions of the resource-poor and the disadvantaged sections of the 'watershed community' such as the asset less and the women through the following activities:

(a) more equitable distribution of the benefits of land and water resources' developed and the consequent biomass production; and

(b) greater access to income-generating opportunities and focus on the human resource development *viz.*, self-help group for income and employment generation during lean period.

22.6 Principles of Watershed Management

Watershed is the basis of rainwater management for development of dryland agriculture. Thus, the deterioration of natural resources in an area can be arrested. The total resources can be properly developed only be adopting the watershed approach.

The basic unit of development is a watershed, which is a manageable hydrological unit. In this approach, development is not confined just too agricultural lands alone, but covers all the areas, starting from the highest point of the area (ridge line) to the outlet of the natural stream. The watershed development works mainly covers the following:

(i) Resource development, it covers development and management of resources like soil, water, vegetation and all associated components. By adopting watershed as a unit, different measures are adopted and executed carefully in each of the topo-sequences according to its capability.

(ii) It essentially covers soil and water conservation measures in the watershed which includes proper land use, protection of land against all forms of degradation building and maintaining soil fertility, water conservation and harvesting, proper management of surface and groundwater, flood protection, sediment reduction, increasing productivity from all land uses, etc.

Watershed management is defined as it consist of the process of formulating and carrying out a course of action involving manipulation of natural, agricultural and human resources of a watershed to provide resources that are desired by and suitable to the watershed community, but under the condition that soil and water resources are not adversely affected. Watershed management must consider the social, economic and institutional factors operating within and outside the watershed.

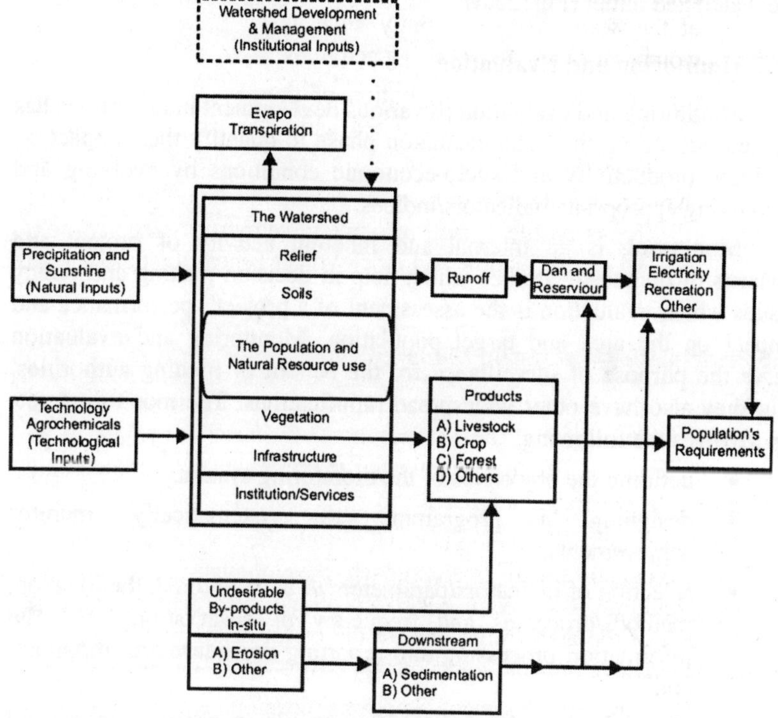

Fig. 22.3 Watershed as a system

Watershed management practices are those which on application changes land use pattern, vegetative cover and other non-structural and structural actions that are taken on a watershed to achieve watershed development objectives or in other words, watershed management are an integrated and interdisciplinary approach. It generally requires land use adjustment measures which contribute to the reduction in soil erosion rates and runoff losses vis-à-vis increased agricultural production, generation of rural employment and balanced growth of the national economy.

Watershed management generally involves the use of the watersheds natural resources, especially the land, water and vegetation by the people of the watershed area with the active participation of their institutions and organizations and in harmony with the ecosystem. It encompasses and incorporates watershed activities into regional or area development plans, taking into account the protection and conservation measures essential and the benefits likely to occur to the people living at the watershed outlet (Fig. 22.3)

22.7 Monitoring and Evaluation

Monitoring and evaluation of various development interventions has carried out during the implementation phase to quantify their impact on ecology, productivity and socio-economic conditions by evolving and employing appropriate indicators/indices.

Monitoring is an internal and inherent activity of project and consists of diagnostic studies which help in decision making and policy issues while, evaluation is the assessment of a project's performance and impact on the area and target population. Monitoring and evaluation serve the purpose of surveillance for the benefit of funding authorities. But they also have other widespread ramifications. The monitoring set-up involves the following:

- defining the objectives of the monitoring system,
- designing a programme to systematically monitor achievements,
- selection of indicators/parameters to be monitored, the location, methods/processes and frequency of observations, and the information processing and reporting procedure are important, and
- organizing training, motivating people to obtain, convey and use the information.

Evaluation is a periodic assessment of the relevance, performance, efficiency and impact of the project in the context of its objectives. It involves comparisons and requiring information from outside the watershed, as well. Evaluation is the process which seeks to analyse and make sense of the data compiled through monitoring. Evaluation organizes and appraises the information gathered by monitoring.

Integrate it with information gathered in other ways and presents the results in an appropriate form and at a place and time that is most beneficial to the planners. Evaluation extracts information from a mass of observational data and presents it in a form that is easy to understand and use. Evaluation may require additional studies or research to obtain information not provided by project monitoring and evaluation analysis. There is a cause and effect relationship between the remedial measures. The benefits may be short, medium and long term. Thus, it compares the monitored data with idealised presumptions.

Monitoring in an integrated watershed development project helps in the following manner:

1. Identification of the targets and goals for the project implementation and the indicators to measure progress against these.
2. Summarise and disseminate the information flowing from various field officials and PIA members.
3. Collection and analyse data from the intended beneficiaries of the project for the purpose of supplementation and complementarily.
4. Identification problems and obstacles encountered by the project and conduct diagnostic and therapeutic studies for the remedial measures.
5. Maintaining various data series over time for the purpose of later evaluation.
6. Preparation of report and indicate range of logical options. Based on the data-base created during the monitoring, evaluation encompasses additional processing supplementary with impact studies and reviewing the information over an extended period to judge achievement.

Evaluation of internal on-going project is continuous and non-stop process and is not an adhoc exercises. It does not involve much data collection. On-going evaluation should be focused on the following:

- What is the response of the project beneficiaries to the inputs, services and other assistance?
- What are the effects of the project on agricultural production, fodder and fuel production in the watershed area?

- Is the project having consequences which were not anticipated?
- Does the intervention model on which the project fs based remain valid in a changing environment?

Monitoring focuses more on implementation issues, while on-going evaluation emphasis on the likely outcomes of the interventions. It provides systematic feedback facilitating the integration of the evaluation process into the management system. The recommendations emerging out of internal on-going evaluation are more comprehensive, practical and relevant. However, it cannot replace or substitute for external evaluation/formal evaluation. Formal evaluation processes are required due to the following reasons:

1. **Decision-making:** Project managers or PIAs need to be able to identify project's strengths, weaknesses, constraints, opportunities, impact and its relevance to the individual, community and nature in order to make sound decisions (realistic and relevant decisions concerning methodology, training, staffing, logistic and other aspects of project management).
2. **Improvement of performance:** Managerial, financial, ecological, beneficiary coverage performance could be improved by the feedback from evaluation results.
3. **Personnel evaluation:** Objective assessment of the performance of project staff (PIA members, WDC secretaries, Project coordinators, community organisers, volunteers, etc.) is required by the project managers.
4. **Allocation of resources:** Prioritisation of resources like time, money, personnel, material, etc., is essential and the result of an evaluation help us to determine how our limited resources could be used most effectively.
5. **Programme justification:** Participants, implementers and funding agency (say DRDA) can make judgements regarding the project direction and impact.
6. **Determination of policy:** Project managers could modify, expand and abandon approaches and modalities tried out in the policy implementation.

22.8 Indicators for Evaluation

The following indicators may be used in project and post-project scenarios to evaluate the overall impact of the watershed development programmes:

(i) Bio-physical indicators
(ii) Economics
(iii) Technical and ecological
(iv) Other indicators

22.8.1 Bio-physical indicators

It indicates the following:

(i) Land Levelling Index

For comparing the efficiency of land levelling activity during the pre-project and post-project periods, an index called as Land Levelling Index can be employed. It is the ratio of recommended land slope to the existing land slope as described by the following relation:

$$\text{Land Levelling Index (LLI)} = \frac{\text{Recommended slope (\%)}}{\text{Existing or treated slope (\%)}} \quad \text{....(22.1)}$$

where, existing slope refers to the individual land slope before the inception of the project and treated slope is the moderated slope resulting from land levelling activities. Higher value of LLI is a measure of better moderation in land slope. LLI can attain a maximum value of 1.0, which refers to a perfectly levelled field.

Crop Diversification Index

Crop Diversification Index (CDI) is used to assess the changes in the cropping patterns/systems due to crop improvement programmes. It is determined by the following relation:

$$\text{Crop Diversification Index (CDI)} = \sum_{i=1}^{n} P_i \log(1/P_i) \quad \text{....(22.2)}$$

where, P_i is proportion of ith crop in comparison with total cropped area and n is the total number of crops in the watershed. The CDI can attain any value >0 and higher value of CDI is an indicator of better crop diversification.

Cultivated Land Utilization Index

Cultivated Land Utilization Index (CLUI) is an indicator of the use of land for intercropping and multiple cropping systems. It indicates the impact of watershed interventions on changes in cultivable land area and duration of crop cultivation in pre-project and post-project periods. It is determined by summing the products of land area planted under each crop, multiplied by actual duration of days of that crop, and dividing the sum by the cultivated land area times 365 days as given below:

$$\text{Cultivated Land Utilization Index (CLU1)} = \frac{\sum_{i=1}^{n} a_i d_i}{A \times 365} \qquad \text{....(22.3)}$$

where, n are the total number of crops; a is the area occupied by i^{th} crop; d_i are the days that the ith crop occupied in the a_i area; and A is cultivated land area. The CLUI can attain a maximum value of 1.0 and higher value of CLUI indicates more usage of available cultivable land area of the watershed and/or duration of crop cultivation period.

Crop Productivity Index

It is used for assessing the overall improvement in crop productivity at the watershed level. Crop Productivity Index (CPI) indicates the extent of crop productivity level in comparison to the normal yield of crops before and after the project. It determined by dividing the crop yield obtained in the watershed by the yield obtained under recommended package of practices and it is expressed as,

$$\text{Crop Productivity Index (CPI)} = \frac{1}{n} \sum_{i=1}^{n} (y_i / Y_i) \qquad \text{....(22.4)}$$

where, n is the total number of crops in the watershed, y_i is the average yield of ith crop cultivated in the watershed, and Y_i is the yield of ith crop with standard package of practices. The CPI can attain any value greater than zero in a given location. Higher value of CPI is indicative of crops' yields closer to the maximum attainable yield under standard package of practices.

Induced Watershed Eco-Index

Induced Watershed Eco-Index (IWEI) is an indicator of additional area brought under vegetative cover due to crops, pasture and grassland development, and horticultural and forestry plantations in a watershed during the project period. It is determined by the following relation:

Induced Watershed Eco-Index (IWEI) =

$$\frac{\text{Additional area vegetatity during the project}}{\text{Total area of the watershed}} \quad ....(22.5)$$

The IWEI can attain a maximum value of 1.0 which indicates that whole of the watershed area has been brought under some form of vegetation.

22.8.2 Participatory and socio-economic indicators

People's Participation Index

It is used for assessing level of participation of the community, in different phases of watershed development programme. Overall Peoples Participation Index (PPI) was used. Overall it is determined by randomly selecting a representative sample of respondents. The responses are recorded in the pre-structured schedule through personal interviews and data are analyzed by employing appropriate statistical methods. Responses of the respondents are recorded on peoples' participation in various interventions of watershed programme and scores were assigned as 1 for "Yes" and 0 for "No" to each statement. PPI is determined by the following relation:

$$PPI = \frac{\text{Mean participation score (P)}}{\text{Maximum participation score (Pi)}} \times 100 \quad ....(22.6a)$$

where,

$$P = \sum_{i=1}^{N} P_i / N \quad ....(22.6b)$$

and,

$$P_i = \sum_{j=1}^{K} (PP_j + PP_j + PM_j + PE_j \quad ....(22.6c)$$

In equations 22.6 (a, b, c), PP_i are the total scores of people's participation in programme planning; PI_j are the total scores of people's participation in programme implementation; PM_j are the total scores of people's participation in programme maintenance; PE_j, are the total scores of people's participation in programmes monitoring and evaluation; K are the total number of statements on which responses of the respondents are recorded; and N is the total number of respondents.

Higher value of PPI is a measure of better participation in the watershed development activities. PPI can attain a maximum value of 100.0, which means participation by the whole watershed community in all the activities.

22.8.3 Economic indicators

The Integrated Watershed Development Programme envisages a change in social and economic conditions of the beneficiaries. The primary objective of a watershed intervention is to catalyse sustained economic growth in project areas and to alleviate poverty. The economic indicators are more specific and change in their values and direction could more easily be adjudged. But, economic indicators are not without ambiguity. There is a widely prevalent difference in opinion about how to define them and how to use them in a project area.

Change in cropped area

Changes in cropped area can be determined by field to field survey. Change in cropping pattern could easily be determined. Comparisons with baseline data would suggest increase in cropped area due to moisture control.

Crop productivity test

Test plots should be selected for various soil types and crops from randomly. Three sub-plots of 10 x 10 m area should be selected in a test plot and crop should be cut and the grains/final produce should be weighed. Findings of the three readings should be averaged.

$$Q = W \times 100 \qquad \text{....(22.7)}$$

where,

Q = Productivity of crop, (kg/ha), and

W = Weight of crop produce in 10x10 m sub-plot.

The average production is determined by the following relation,

$$Qf = \frac{Q1 + Q2 + Q3}{3} \qquad \text{....(22.8)}$$

Revenue department has various prevalent crop-cutting experiment techniques. Almost all *patwaris* (village revenue officers) are well known with such techniques.

Biodiversity index

A survey has to be conducted in a representative manner and in as many compartments as possible. In each compartment quadrates in minimum number and of minimum size have to be laid. Then the number of individual herbs, shurbs and trees have to be counted. The size of the plots can be 2x2 m for herbs, 5x5 m for shrubs, 10x10 m for trees as per the requirement of site. The species-richness and dominance indices can be determined by following relation,

$$D = \text{sigma}\left(\frac{ni}{2}\right)^2 \qquad \text{....(22.9)}$$

where,

D = Diversity index,

ni = Number of individuals of individual species,

n = Total number of individuals, and

High value of D indicates more diversity and less dominance in an area.

Regeneration by root stock

On provident protection to heavily degraded forest areas where live root stock exists, regeneration in the form of coppice shoots emerge. To take stock these 0.1 ha plots are laid on random basis and species-wise counting done. Preferably numerous plots of representative areas are selected. About 1200 coppices per ha can be considered as good regeneration.

Benefit: Cost Ratio

Economic analysis of the projects is carried out separately for arable lands and non-arable lands and also for the watershed as a whole using the discounted measure of the project worth namely Benefit-Cost Ratio (BCR). It is defined as the ratio of present value of gross benefits to the present value of total costs and is expressed as,

$$\text{Benefit - Cost Ratio (BCR)} = \sum_{t=1}^{n} \frac{B_t/(1+i)^t}{C_t/(1+i)^t} \qquad \text{....(22.10)}$$

where, B_t are the Benefits (Rs.) at time t; C_t are the costs (Rs'.) at time t; i is discount rate (%); and t is life of the project. BCR can attain any value equal to or more than zero. Projects with BCR> 1.0 are considered as economically viable and economic soundness of the project increases as the BCR value increases.

Household income

Small landholders are the main groups which would receive maximum inputs and services that are designed and built in the watershed programme to increase production and productivity. To a certain extent landless labourers and rural artisans also benefit. Allied agricultural activities like dairy, fisheries, mushroom cultivation, sericulture, apiculture, etc., would also result in increased household incomes. Non-farm activities like handlooms, handicrafts, small rural manufacturing units etc. would add into the incomes. Non-agricultural works would benefit from the expanded opportunities created in sectors, such as, transportation, processing, marketing, etc. Despite favourable aggregate results, the level of increase in household income would differ and also there would be households with stagnating income levels or fall in their monthly incomes. Evaluation of the economic benefit of particular crops or products could be satisfactorily done. Measurements of total farm or household incomes are extremely difficult. There are definitional problems caused by factors like valuation of labour inputs and consumption of home productions. Similarly, response problems create difficulties in the assessment. The income of each household member is difficult to obtain from a single respondent; incomes from informal activities like small-scale trade or small production units are difficult to cover in an income survey; sporadic marketing may result in erratic income generation. Often objective responses are difficult to elicit in an income survey. Income survey formats should include queries about visible and invisible incomes from primary, secondary and tertiary sectors.

Household expenditure

In the measurement of expenditure, some of the definitional and response problems seen in income evaluation could be curtailed. Expenditure can be assessed accurately if the recall and reference periods are appropriate. Casual daily expenditures such as expenditure on cigarettes/*bidis*, alcohol, fuel-wood, milk, etc., could be appropriately evaluated for a recall period of 2-3 days or at most a week. Items such as expenditures on household provisions, school fees and rent, services and entertainment may need reference period of one month. Irregular major expenditures such as expenditures on health, household repairs, farm activities may have annual reference. Prestige expenditure on social occasions like festivals, marriages, birth/death ceremonies, etc., should also be included in expenditure survey. Changes in expenditure levels, in most cases, reflect changes in income fairly accurately. Expenditure surveys may require multiple visits for an extended period, It may also require multiple interviews in a household. It is advisable to conduct a one to two week survey (with numerous visits within a period) and which could be repeated at certain selected times in the year.

Household savings

Household savings could be easily evaluated. Household savings survey along with expenditure survey could fairly depict the income levels. Savings could be ascertained by a well- prepared survey format. Savings invested in jewellery, precious stones, land, shares, etc., should be included.

Measurement of living standard

The economic change in a household could be evaluated fairly easily by the following:

(a) *Quality housing and related facilities, e.g.*, covered area and number of rooms, type of construction material, nature of the roof, access to potable water, location of bathrooms within the dwelling.

(b) *Acquisition of durable possession, e.g.*, bedding, clothing, footwear, cooking utensils, basic furniture and sundries like watch, radio, television, bicycle, etc.

(c) *Availability of community facilities, e.g.*, access to primary health facilities, all- weather roads, primary schools, markets, etc.

Other economic indicators

1. Number of beneficiaries/families moving above/down the poverty-line.
2. Extent of migration in search of employment (temporary or permanent/within the district or outside the district).
3. Prevailing wage rates in agriculture and non-farm sector.
4. Extent of indebtedness in cash or kind from all sources.
5. Change in Gram Panchayat incomes.
6. Change in incomes from milk, MFP, etc.

7. Agricultural productivity.
8. Prices and incomes from agricultural produce.
9. Change in agricultural and non-agricultural land prices.
10. Diversification of product range and skill upgradation of rural artisan.

22.8.4 Technical and ecological indicators

Numerous technical parameters/indicators are available for evaluating impact of watershed treatment on soil, surface water, groundwater, ecology, etc. A battery of indicators are available for technical and ecological of watershed development programme as described below.

Soil loss

Universal Soil Loss Equation (USLE): It predicts the long-term average soil losses from a specified cropping and management system. The equation used for determination of the losses from sheet and rill erosion is expressed as,

$$A = R \times K \times L \times S \times C \times P \qquad ...(22.11)$$

where,

A = Soil loss (tons/ha/year),

R = The rainfall erosivity factor which is the number of rainfall erosion index units for a particular location,

K = The soil erodibility factor which compares the actual erosion to that erosion from a clean, tilled, fallowed, uniform slope of 9% and length of 22.13 m,

L = The slope length factor is the ratio of soil loss from the field slope length to that from a 22.13 m length under identical conditions,

S = The slope steepness factor is ratio of soil loss from the field slope gradient to that from 9% slope under otherwise identical conditions,

C = The cover and management factor is the ratio of soil loss from an area with specified cover and management to that from an identical area in tilled continuous fallow, and

P = The support practice factor is the ratio of the soil loss with a support practice like contouring, strip cropping, or terracing to that with straight row farming up and down the slope.

Runoff

It is determined by the following methods:

(a) Ogrosky and Mockus formula:

$$Q_{Peak} = \frac{0.0208 \times A \times Qd}{T_p} \quad ....(22.12)$$

where,

Q peak = Peak rate of runoff, (cum/sec),

A = Area (ha),

Qd = Runoff depth (cm), and

Tp = Time to peak, (hr).

(b) Soil Conservation Service formula:

$$Q = \frac{(P - 0.3S)^2}{(P + 0.7S)} \text{ for all soil areas except black soil} \quad ....(22.13)$$

$$Q = \frac{(P - 0.1S)^2}{(P + 0.9S)} \text{ for black soil areas} \quad ....(22.14)$$

where,

Q = Actual runoff,

P = Rainfall, (cm),

S = Potential maximum retention.

(c) Rational method:

$$Q = \frac{C \times 1 \times A}{360} \quad ....(22.15)$$

where,

Q = Peak rate of runoff for the given frequency of rainfall, (cu m/sec),

C = Rational runoff coefficient having values ranging from 0 to 1 depending on watershed conditions,

I = Intensity for design frequency and for duration equal to time of concentration, (mm per hour) and

A = Arca of watershed, (ha).

(d) Central Missouri method:

$$Q = Qt \times L \times I \times T \times S \times V \times C \times P \times F \quad \text{....(22.16)}$$

where,

Q = Peak rate of runoff, (cubic ft/sec),

Qt = Peak rate of runoff from a watershed with a specific set of watershed conditions,

L = Watershed location factor,

I = Topographic factor,

S = Watershed shape factor,

V = Vegetative cover factor,

C = Contour farming factor,

P = Surface storage factor, and

F = Runoff frequency factor.

(e) Cooks method:

$$Q = P \times R \times F \times S \quad \text{....(22.17)}$$

where,

Q = Peak runoff for the specific geographic location,

P = Peak runoff,

R = Geographic rainfall factor,

F = Return period factor, and

S = Shape factor.

Sediment test

This can be determined by using soil turbidity meter. An estimation can be made by collecting sample of running water from a *nalla.* The particles are allowed to settle down. The height of soil/sand particles in the test tube would suggest the extent of sediments in the *nalla* or rivulet. Comparisons could be made with previous years data on comparable rainfall days. This gives rough estimation but could be effectively done by the community.

Waterflow in a stream

(a) Volumetric method: It is the most simple and accurate method and can be used to measure small discharge flowing through a channel or a pipe. Time to fill a known volume of the container is noted with the help of a stop watch. Discharge is obtained by dividing the volume of fill by the required time.

(b) Float method: Any object that floats in water but does not obstruct wind or gets carried away by it can be used as a float. A dry wood of about 1 cu m size can serve the purpose. A straight section of the channel of about 30 m length is selected such that the cross-section is almost uniform through the length. The float is allowed to move from the upstream to the downstream end and the time taken to move 30 m distance is measured. Several measurements are taken and the average time is noted. The distance divided by the time gives the average surface velocity. The surface velocity is always higher than the average stream velocity. Therefore, surface velocity is multiplied by a factor of 0.85 to get the average velocity. Similarly, cross-sectional area of the stream is measured at least in three places and the average discharge is determined as,

$$Q = A \times V \qquad \text{....(22.18)}$$

where,

Q = Discharge, (m^3/s),

A = Cross-sectional area, (m^2), and

V = Average velocity of flow, (m/s).

(c) Current meter method: Current meter can be used to measure speed of water flow in the stream. Two types of current meters are used *viz.*, cup-type and the propeller-type current meter. When the current meter is suspended into a stream, the wheel rotates and the revolutions are recorded on a counter. On a calibration chart the graph gives the relationship between the revolutions per minute and the velocity. The average velocity is determined by taking the average of the velocities at 0.8 m and 0.2 m of the depth from the top. In case of a shallow stream, only one reading may be taken at 0.6 m of the depth. The cross-sectional area should be divided into a number of sub-sections and velocity readings should be taken in each sub-section. If the different areas are A_1, A_2, A_3.... A_n and the corresponding velocities are V_1, V_2, V_3.... V_n, then the average discharge is determined by the following relation,

$$Q = \frac{A_1V_1 + A_2V_2 + A_3 \times V_3 + ... + A_nV_n}{n} \qquad \text{....(22.19)}$$

Measurement of water table

Observation or bench mark wells should be identified before the start of watershed project. Water table readings should be recorded at frequent and regular intervals (say, a fixed date of every month). The simplest method of taking groundwater table is by dipping a measuring tape with chalk marks inside the water in the observation well. The water wets the chalk mark on the tape clearly and the difference of this reading from that at the top gives the water table depth. A suitable weight (say stone) can be tied at the end of the tape for easy lowering (Fig. 22.4). A battery operated automatic water table depth recorder can give more precise measurement.

Yield test

The following equipments are required for conducting a yield test:

- A test pumping unit consisting of a submersible pump (with 3-phase motor) and a diesel generating set.
- Measuring device for time, discharge and water level.

With the help of pumping set, the water from the well is discharged and the volume of water thus drained out is measured into container along with assessment of time taken. Periodic water levels are taken.

$$Y = \frac{A \times d \times 6.25}{T} \qquad \text{....(22.20)}$$

where,

Y = Yield capacity of a well, (gallons/min),

A = Cross-section of the well,

D = Depth of water removed, and

T = Time for refilling the well on itself.

Water table reading should be recorded at frequent and regular intervals from benchmark wells (Fig. 22.5). Simple SWC techniques could increase the water table to astonishing levels.

Productivity of grasses

Randomly 5x5 m plots are selected which are representative of the area. Grasses are cut and weighed. Dry weights may also be taken.

Soil moisture test

Percentage moisture content on weight basis is determined as,

$$Ms = \frac{(Ws - WD) \times 100}{WD} \quad ....(22.21)$$

where,

Ms = % moisture content on weight basis,

Ww = Weight of wet soil, and

WD = Weight of oven dry soil.

Percentage moisture content on volume basis can be determined by multiplying Ms by the bulk density of soil. Soil samples from vertical and horizontal dimensions are collected in air-tight aluminum boxes known as moisture boxes and their lids are closed. The boxes are accurately weighed and after opening their lids they are kept in a thermostatic oven. They are dried at a temperature of 105 degree centigrade for a period of about 24 hours. After drying, the boxes are weighed again. The difference in weight gives the moisture content. This can be expressed either on weight basis or volume basis.

Satellite imagery

Remote sensing images of earmarked areas could be used for comparison and analysis with pre-project images. This would give non-biased impact assessment in the field about changes in cropped area, water bodies, biomass and wastelands.

Other technical indicators

(i) Post project PRA appraisal.

(ii) Soil fertility.

(iii) Stability and effectiveness of structures.

(iv) Movement of sediment.

(v) Fodder and fuel-wood availability etc.

22.8.5 Other indicators

It includes social and behavoural indicators. More specific quantitative indicators may also be used for assessing social, behavioural and organisational impacts of a watershed project as follows:

1. change in literacy rate,
2. change in school enrolment, dropouts etc.,
3. level and extent of prohibition,
4. level and extent of controlled grazing,
5. percentage population linked to watershed activities,

6. level of malnutrition in below one year and under five year age groups,
7. gross birth rate of the village,
8. deliveries in institutions or by trained *dais vis-a-vis* unattended deliveries, and
9. infant mortality rate etc.

Numerous of such quantifiable indicators can be listed and selected.

22.9 A Case Study

22.9.1 Description of the study area on rainwater management based on watershed approach

Hazra (1998) was conducted a case study at Tejpura village watershed. It is located at a distance of 45 km east of Jhansi covering an area of 775.7 ha. A rivulet 6-10 m wide runs through the watershed in an 'L' shaped form for its run of 6.5 km. The soil of the watershed varies from loamy sand to sandy loam on the upper reaches with average slope varying from 3-8 per cent, whereas, the soil of lower reaches are loamy to silty clam loam with gentle slope (1-3%). The average annual rainfall in 940 mm and potential evapotranspiration 1520 mm. The watershed covers 450 farm families of which 195 are below poverty line and 35 are landless.

Table 22.1: Soil and water conservation measures adopted at Tejpura watershed

Works	Achievement
Contour bunding (ha)	23.28
Submergence-cum-field bunding (ha)	558.94
Land levelling (ha)	4.60
Gully plugging (ha)	64.38
Diversion channel (m)	1657
Drainage structure (No.)	70
Underground irrigation channel (No.)	2
Total cost on soil & water conservation (Rs. ha^{-1}) (On the basis of 1985-86)	2342
Total cost on soil & water conservation (Rs. ha^{-1}) (discounted value on 1994-95)	5515

22.9.2 Applied soil and water conservation practices

The entire area was treated with appropriate soil and water conservation measures for efficient rainwater management based on watershed approach. Apart from various soil conservation measures adopted as given in Table 22.1, four check dams were constructed across the *nala* (rivulet) at a vertical interval of 3.5 m for storing runoff water.

Additionally a check dam was also constructed at the highest point by joining two hillocks spaced 350 m apart by earthen bunds. This has helped in bringing back 128.4 ha of wasteland into cultivation. This has also greatly influenced land utilization pattern (Table 22.2). The irrigated area was greatly increased from initial 20.2 ha to 650.8 ha.

Table 22.2: Cultivated area under rainfed and irrigation over the years at Tejpura watershed

Years	**Cultivated area (ha)**			**Degraded and area under miscellaneous uses (ha)**
	Irrigated	**Rainfed**	**Total**	
Before watershed (1982-83)	20.2	505.4	525.6	250.1
After completion (1985-86)	4495	202.0	651 5	124.2
After Three years (1988-89)	541.5	110.5	652.0	123.7
After six years (1991 -92)	603.0	49.5	652.5	123.2
After nine years (1994-95)	650.8	3.2	654.0	121.7

22.9.3 Development of irrigation resources

Initially, a meagre area of 20.2 ha was under irrigation. But with the construction of check dams which not only influenced surface storage of water but also helped in recharging the ground water table to a great extent, which induced in digging out a very large number of wells by the farmer (Table 22.3). This has also reflected in increased irrigation from initial 3.8 per cent to almost 100 per cent at present. The irrigation water availability from the wells also helped in getting about 67 ha under a third crop during summer (Table 22.4). The stored water and other soil and water conservation measures has resulted in rising of water table by 3-6 m over the seasons (Table 22.5)

Table 22.3: Irrigation resources and cropping intensity at productivity of crops Tejpura watershed over period of time

Years	Number of open wells	Water storage in Check dams (ha m)	Irrigated area (%)	Cropping intensity (%)
Before watershed (1982-83)	5	-	3.8	82
After completion (1985-86)	35	12	69.0	156
After Three years (1988-89)	112	13	83.0	188
After six years (1991 -92)	167	18	92.0	206
After nine years (1994-95)	315	23	100.0	209

Table 22.4: Seasonwise cropped area at Tejpura watershed (ha)

Years	*Kharif* (Rainy season)	*Rabi* (Winter season)	Zaid (Summer season)	Fallow (No crop)
Before watershed (1982-83)	116.0	320.0	-	88.8
After completion (1985-86)	365.0	637.3	15.0	-
After Three years (1988-89)	525.0	651.0	48.0	-
After six years (1991 -92)	630.0	651.0	65.0	-
After nine years (1994-95)	650.0	652.0	67.2	-

Initially 436.8 ha was put under crops which was subsequently increased to 525.6 on watershed treatment. An area of 650 ha during *kharif* 652 ha during *rabi* and 67 ha during *zaid* came into cropping increasing thereby the cropping indensity considerably (209 per cent). Various crop production measures coupled with adequate irrigation water availability has resulted in increased crop productivity by 200-400 per

cent in case of' *kharif* crops (Table 22.6) and 217-344 per cent in case of *rabi* crops (Table 22.7). Annual crop productivity was raised to 6.2 Mg ha^{-1} yr^{-1} from a meagre 0.6 Mg ha^{-1} yr^{-1}. This has also reflected in the overall income of the farm families, an increase by 12.4 tunes from the base level (Table 22.8).

Table 22.5: Change in ground water table at Tejpura watershed over the years

Years	Average water table depth (m)			Maximum fluctuation (meter)
	Summer season	Rainy season	Winter season	
Before watershed (1982-83)	14.6	6.0	12.0	8.6
After completion (1985-86)	6.5	2.5	3.5	4.0
After Three years (1988-89)	8.5	4.0	6.0	4.5
After six years (1991 -92)	6.0	3.0	3.5	3.0
After nine years (1994-95)	8.0	4.0	9.0	5.0

± Drought years, received only 560 mm of rainfall against normal rainfall of 940 mm that is 60% of normal rainfall.

Table 22.6: Area and average productivity of *kharif* crops at Tejpura watershed

Crops	Before watershed (1982-83)		After watershed (1991-92)		After nine years of watershed (1994-95)	
	Area (ha)	Productivity (Mg ha^{-1})	Area (ha)	Productivity (Mg ha^{-1})	Area (ha)	Productivity (Mg ha^{-1})
Sorghum	105	0.6	385	1.7	67	1.8
Urad	7	0.3	20	0.8	50	0.8
Paddy	2	1.0	20	4.0	30	4.2
Sesamum	1	0.1	26	0.5	35	0.5
Groundnut	1	0.6	172	1.9	335	2.0
Tomato	-	-	15	30.0	123	26.0

Table 22.7: Area and average productivity of *rabi* crops at Tejpura watershed

Crops	Before watershed (1982-83)		After watershed (1991-92)		After nine years of watershed (1994-95)	
	Area (ha)	Productivity (Mg ha^{-1})	Area (ha)	Productivity (Mg ha^{-1})	Area (ha)	Productivity (Mg ha^{-1})
Wheat	210	0.9	405	4.0	208	4.0
Gram	86	0.6	160	1.9	195	1.9
Mustard	15	0.3	36	1.2	75	1.2
Vegetables	2	7.0	45	17.5	170	26.0

Table 22.8: Crop productivity and annual income of farm families from arable crops at Tejpura watershed

Years	Area Production (Mg ha^{-1} yr^{-1})	Annual income (Rs. ha^{-1} yr^{-1})	Annual income (Rs. Family^{-1} yr^{-1})
1982-83	0.6	1670	2900
1985-86	3.8	10550	18200
1988-89	5.5	14320	24600
1990-91	5.8	15100	26500
1994-95	6.2	24800	36000

22.9.4 Silvipasture development

The total bund area (24 ha) and gully plugs (2 ha) was initially seeded with pigeon pea as long as the soil was loose. This practice gave 900 tonnes of grain yield valued at Rs. 4.25 lakhs (on the basis of 1986 cost structure) which was equivalent to the cost reimbursable from farmers on soil and water conservation measures adopted on individual farmers' field. These areas together with 30.5 ha area of community grazing lands was brought under varied silvipastoral measures coupled with appropriate soil and water conservation practices like staggered contour trenches and contour stone bunding, which greatly helped in improving forage productivity. These vegetational improvement lead to improvement in soil fertility in these earlier degraded sites (Table 22.9). The soil loss was greatly reduced from 38 Mg ha^{-1} to 0.2-0.9 Mg ha^{-1} after seven years. Similarly, the soil fertility was also greatly appreciated in a*rabi*c land (Table 22.10).

Table 22.9: Forage yield, soil physico-chemical properties and soil and water conservation as influenced by silvipastoral development in barren hillocks at Tejpura watershed after seven years

Soil properties	Silvipasture			Past ure	Initial status
	Leucaena Leucocephala	*Acacia nilotica*	*Albizzla lebbeck*		
Organic carbon (%)	1.07	0.80	0.83	0.78	0.22
Available nutrients (kg/ha)					
N	302	263	282	231	87
P	22.5	15.8	16.5	13.9	6.3
K	308	287	283	268	189
S	39	22	24	18	4
Bulk density (Mg rn^{-3})	1.29	1.33	1.33	1.38	1.68
Water stable (>0.25 mm, %)	36	30	31	28	11
Field capacity (%)	16.6	16.0	16.1	15.3	11.2
Runoff water loss (%)	22	26	26	28	78
Soil loss (Mg ha^{-1})	0.2	0.4	0.4	0.9	38
Dry forage yield (Mg ha^{-1})	5.1	5.9	6.2	6.7	0.1

Table 22.10: Physico-chemical properties of the soil of the cropped area before and after nine years of the watershed treatment at Tejpura watershed

Soil Properties	Loamy Sand		Sandy clay loam	
	Before watershed	After nine years	Before watershed	After nine years
Organic carbon (%)	0.17	0.57	0.24	0.69
Available nutrient (kg ha^{-1})				
N	138.0	178.0	143.5	192.0
P	8.9	12.1	11.2	14.4
K	197.0	210.2	195.0	215.0
S	4.7	10.5	10.2	18.0

22.9.5 Animal resources and productivity

The information on changing animal species is presented in Table 22.11, which indicate a great shift in composition with the number being number of cows reduced by 42 per cent and an increase in the number of the buffalos by 82 per cent. This has resulted in doubling the milk production. Productivity per animal was also increased by 78 per cent (Table 22.11). Goat and sheep population was also drastically reduced by 61 and 94 per cent, respectively.

Table 22.11: Animal composition, lactating animals and milk production at Tejpura watershed

Species	Number			Milk (1 day^{-1})		
	Initial	After six years	After nine years	Initial	After six years	After nine years
Cattle						
Cow	139	90	80	0 9	1.5	1.6
Bullock	460	8	550			
Calves	50	40	35			
Buffalo						
She buffalo	51	70	56	1.8	3.0	3.2
Bull	2	3	4			
Calves	15	30	45			
Goat	90	50	35			
Sheep	78	20	5			

22.9.6 Overall productivity

The watershed treatment over the years has resulted in continuous improvement in productivity in field. Spectacular production increase (156 times) was observed in oilseed crops (Table 22.12). Increase in pulses and cereals were of the order of 7 and 4 times, respectively. Availability of forages was improved by 52 per cent. Fish production on check dam reservoirs and other water bodies was observed to be 3 Mg annually. Total annual gross income w s increased by 15 times from the watershed area (Table 22.12).

Table 22.12: Annual production of food crops, forages, fuelwood, milk and fish and gross income from the produce at Tejpura watershed

Commodity	Prior to Watershed (1992-83)		After live years of watershed (1990-91)		After nine years of watershed (1994-95)	
	Prod-uction (t)	Gross income ('000' Rs.)	Prod-uction (t)	Gross income ('000' Rs.)	Prod-uction (t)	Gross income ('000' Rs.)
Cereals	243	656	1900	5110	1079	378
Pulses	60	430	400	2850	410	3690
Oilseeds	5	50	180	1289	778	7002
Vegetables	-	-	650	975	4865	14595
Forages	1080	540	4960	2480	5110	2555
Rielsood	52	21	110	44	280	140
Milk	79	271	126	657	159	1018
Fish	-	-	1	15	3	60
Total	1519	1968	10295	13420	12684	29438

Table 22.13: Employment generation over the years at Tejpura watershed

Years	Area Production (Mg ha^{-1} yr^{-1})	Annual income (Rs. ha^{-1} yr^{-1})
1982-83	41515	30590
1985-86	96520	71120
1988-89	116280	85680
1990-91	127200	92260
1994-95	139900	101480

22.9.7 Employment generation

Most important aspect in overall watershed development is to create employment and income generation opportunity for rural poor and landless families. There were 35 landless families in the watershed constituting 8% of the population and 195 poor families (43% of the population). The employment generated over the years is indicative of the significant impact of the watershed on livelihood of rural poor. Seasonal migration of the entire landless families and few other poor families during crop growing periods have been fully stopped improvement generations for both human and bullock days were 3.4 times than the base year Table 22.13.

26.9.8 Benefit cost of watershed development

Even after ten years (considering this as life period) not a single structure was broken which speaks the quality of work. The overall benefit cost ratio as worked out from individual activities and item-wise cost is presented in Table 22.14. This indicates the benefit: cost of ratio of 2.71 on the basis of cost of 1985-86. Additional expenditure was worked out to be 14.08 lakhs against an additional gain of 52.30 lakhs.

Table 22.14: Overall benefit - cost ratio from direct produce at Tejpura watershed on the basis of 1985-86

Description	Area (ha)	Additional Expenditure		Value of additional produce	
		Rate (Rs. ha^{-1})	Total (Rs.)	Rate (Rs. ha^{-1})	Total (Rs.)
Soil and water conservation work	760	243.24	187978.00	-	-
Improved cultivation practices	650.0	1800.0	1170000.00	7800.00	5070000.00
Pasture grasses	109.4	81.04	8860.00	11344.00	124000.00
Afforestation	36.0	55.04	1980.0	1000.0	36000.00
Establishment	775.0	53.0	42625.00	-	-
Total	-	-	1408443.00	-	5230000.00

Benefit - cost ratio = (52.30-14.08) / 14.08 2.71. ± Calculated on the basis of 10 years life.

Thus, the most important lesson learnt was that close liaisoning with the village communities is a must for successful watershed development activity. The participation of the villagers in the programme implementation and overall management ensures that they have a feeling of sense of belongingness to where they live. Development of land on integrated watershed approach takes care of all round development of village community through their intimate involvement and participation in the programme. The programme of this kind also brings the researchers developmental workers and the farmers to a closer working association and helps in mutual confidence building.

References

Brooks, K. N., Ffolliott, P.F., Gregersen, N.M. and DeBano, L.F. 1997. "Hydrology and the Management of Watersheds". Second Edition. Lowa State University Press. Ames. pp. 502.

Craswell, E.T. and Niamskul 2009. "Watershed management for erosion control on sloppin lahds in Asia", In: Integrated watershed management in the global ecosystem (Ed. R. Lal), CRC Press, Bocaroton, pp. 05-72.

Crosson, P. 1997. "The On-farm Economic Costs of Erosion". In: Methods for Assessment.

Dhuravanarayana, V.V. Shastri, G. and Patnaik, U.S. 1990. Watershed Management "ICAR" Krishi Anusandhan Bhawan, Pusa, New Delhi.

GOI, 1994. Government of India, Ministry of Rural Development. Report of the technical committee on drought development. Report of the technical committee on drought Prone Areas Programme and Desert Development Programme, New Delhi, pp-73.

Grade, R.J. and Ranga Raju, K.G. 1977. Mechanics of Sediment Transportation and Alluvial Stream Problems, Wiley Eastern Ltd., New Delhi.

Hagan, R.M. Rijov, S.N., Aston, M.M. van Bavel, CHM and Raheja, P.C. 1973. Water, plant growth and crop irrigation requirement. In 'Irrigation Drainage and Salinity - An Inter national Source Book' Hutchinson/FAO/UNESCO, Paris, Hutchinson and Co. (Pubis,) Ltd., London, pp. 206-253.

Hanumantha Rao, C.H. 2000. Watershed Development in India- Recent experiences and energying issues. Economic & Political Weekly. 35(45):3943-47.

Hazra, C.R. 1998. Management of rainwater resources on watershed basis for sustainable agricultural production – An experience of Tejpura Watershed (Jhansi) on changing scenario. Proceeding of the National Seminar on Watershed Management for Sustainable Agriculture – Problems and perspective for 21st Century. IARI, New Delhi, April 15-17, 1998, pp. 241-248.

Heinonen, R. 1971. Soil Management and Crop Water Supply, Lankbrk Lanbrakshogskolans kompen dien a mud., 112 pp.

Howard, A. 1924.Crop Production in India Oxford Univ, Press, Bombay, 100 pp.

Indian National Committee fr IHP 1981. Watershed Management in India, Hydcom. Report-I, Council of Scientific and Industrial Research, New Delhi.

Jensen, J., Seth, S., Swaney, T. and Kumar, P. 1996. Watershed Development, Proc. Danida International Workshop on Watershed Development, Water Development Coordination Univ., Danida, New Delhi.

Kung, P. 1971 a. Effect of different intensities and irrigation methods on the economics of crop production. Ph.D. Thesis, *Indian Agric. Res. Inst.*, New Delhi.

Kung, P. 1971b. Irrigation Agronomy in Monsoon Asia. FAQ/Un, Rome, pp. 42-46.

Morris, B.T. and Jhonson, D.C. 1942. Hydraulic Design of Drop Structure for Gully Control. Proc. Am. Soc. Civil Engrs., 68, 17-48.

Parmanik, 1952. Hydrology of Rajasthan desert-Rainfall Humidity and Evaporation. National Inst. Science. India Bulletin 1.

Rajora, R. 2002. Integrated Watershed Management Field manual for equitable Producer and sustainable development. Rawat Publichations, Jaipur, pp 616.

Rajput, R.K. arid Singh, M. 1970. Efficacy of different methods in conserving soil moisture in cotton. *Indian J. Agron.*, 15:41-45.

Rama Rao, M.S.V. 1974. *Soil Conseivation in India,* Indian Council of Agricultural Research, New Delhi.

Rao, K.L. 1975. India's Water Wealth - Its Assessment, Uses and Projection. Orient Longman, New Delhi, 255, pp.

Rhoades, E.E. 2000. Participatory multipurpose watershed Project. Nature's Solution of Suchmagra Nighore in Integrated Watershed Development in the Global Ecosystem (Ed. R.L.A.) CRC Press, Bo Ca Raton, pp. 327-343.

Samra, J.S. 1997. Status of Research on Watershed Management, Dehradun, pp. 44.

Sarkar and Singh 1993, Interactive approach for watershed Management-A Case Study for interaction Tech. Pub. No. WTC/IARI/WSM3-93, IARI, New Delhi.

Schwab, G.O., Prevent, R.K., Edminister, T.W. and Barnes, K.K. 1966. Soil and Water conservation Engineering (2 Ed.) Johan Wiley and sons, New York.

Shmuli., E. 1973. Efficient utilization of water in irrigation In Arid Zone Irrigation' (Eds. Yaron, B., Dunfors, E and Vaadia, Y). Chapman & Hall Ltd., London, Springer Verlag., Berlin. Deidelberg, New York, pp. 411.423.

Singh, 2000. Watershed based research .strategy for sustainable Agriculture WSM Series-I, Indian Agricultural Research Institute, New Delhi.

Singh, H.G. and Singh, B 1970 Preliminary studies on the effect of cycocel on cotton (*G. arborium). Indian. J. Agric. Sci,* 40:562-565.

Singh, V.P. (Ed.) 1995. Computer models of watershed hydrology. Water Resources Publications, U.S.A.

Subba Raju, V. 1969. Design and Construction of Bench Terraces, *Journ of Agril. Engg*., (Indian Soc. Agril. Engrs.), 6(1), 1-7.

Sunderaraman, J. 1957. A design of drop inlet structure. The Harvester, Indian Institute of Technology, Kharagpur, (2), 52-54.

TAC 2000. Integrated water resource management. Background paper No. 4, Technical Advisosy Committee (TAC) of Global Water Partership Stockholm, Sweden.

Tamhane, R.V. 1967. Watershed Protection and Flood Prevention in River Velly Projects, J. Soil and Water Conservation (India), 15(1&2):20-29.

UNCED, 1992. Agra 21: Programme of Action for sustainable development. United Nations Publications, New York.

Wasi Ullah, Gupta, S.K. and Dalal, S.S. 1972. Hydrological measurements for watershed research, Jugal Kishore & Co. Dehradun.

WCED 1987. Our common future (The Brunt land report world commission on environment and development), Oxford University Press.

Weiss, A.O. and Beard, L.R. 1971. A multi basin planning strategy, Water Resources Development. Oxford University Press.

Winkler, P. 1992. Management and challenges the desert', *World Press Review,* Vol. 39(4): 47.

World Resources Institute 1994. 'The 1994 Information Environment Almanac' Houghton Mittin Co., Boston.

EXERCISES

22.1. What is the watershed and cause of watershed deterioration?

22.2. Describe the watershed characteristics which effects runoff and soil loss.

22.3. State the objectives of the watershed management.

22.4. Describe the principles of watershed management.

22.5. Describe how watershed is monitored.

22.6. Discuss the biophysical indicator for watershed evaluation.

22.7. Distinguish the following:

(a) Crop productivity index and crop diversification index
(d) Peoples participation index and cultivated land utilization index
(c) Biodiversity index and crop productivity test

22.8. Explain the different economic indicators for watershed evaluation.

22.9. State the technical and ecological indicators for watershed evaluation.

22.10. Explain how the water flow in a stream in measured.

22.11. Describe the social and behavioral indicators for evaluation of watershed.

22.12. What are the water resources development indicators?

22.13. A small agricultural watershed of 100 ha is developed with different conservation measures. The dry technology interventions take place for crop production and water resources development. The following observations were recorded before and after dryland technology interventions given as under.

S. No.	Observation	Before	After
1	Cultivated land (ha)	500	75.0
2	Crop yield (area)		
(i)	Maize (kg/ha) (ha)	1500 (15)	40 (40)
(ii)	Blackgram (kg/ha) (ha)	4 (5)	12 (15)
(iii)	Groundnut (kg/ha) (ha)	10 (5)	250 (10)
(iv)	Sorghum (kg/ha) (ha)	13 (20)	25 (10)
(v)	Sesame (kg/ha) (ha)	55 (5)	16 (20)
(vi)	Gram (kg/ha) (ha)	8 (10)	10 (20)
(vii)	Mustard (kg/ha) (ha)	5 (15)	10 (20)
(viii)	Wheat (kg/ha) (ha)	-	35 (20)
3.	Water resources (cum)	25000.0	7000.0
4.	Water land in well (m)	10.0	6.5
5.	Runoff %	15.10	8.0
6.	Soil loss (t/ha/year)	6.50	3.0

Determine all the water development indices from above observations?

■■■■

Appendix-A

Table A-1. Extra Terrestrial Radiation, Q_a given in Equivalent Evaporation in mm/day

North Hemisphere

Lat	Jan	Feb	Mar	Apr	May	June	July	Aug	Sept	Oct	Nov	Dec
50^0	3.8	6.1	9.4	12.7	15.8	17.1	16.4	14.1	10.9	7.4	4.5	3.2
48	4.3	6.6	9.8	13.0	15.9	17.2	16.5	14.3	11.2	7.8	5.0	3.7
46	4.9	7.1	10.2	13.3	16.0	17.2	16.6	14.5	11.5	8.3	5.5	4.3
44	5.3	7.6	10.6	13.7	16.1	17.2	16.6	14.7	11.9	8.7	6.0	4.7
42	5.9	8.1	11.0	14.0	16.2	17.3	16.7	15.0	12.2	9.1	6.5	5.2
40	6.4	8.6	11.4	14.3	16.4	17.3	16.7	15.2	12.5	9.6	7.0	5.7
38	6.9	9.0	11.8	14.5	16.4	17.2	16.7	15.3	12.8	10.0	7.5	6.1
36	7.4	9.4	12.1	14.7	16.4	17.2	16.7	15.4	13.1	10.6	8.0	6.6
34	7.9	9.8	12.4	14.8	16.5	17.1	16.8	15.5	13.4	10.8	8.5	7.2
32	8.3	10.2	12.8	15.0	16.5	17.0	16.8	15.6	13.6	11.2	9.0	7.8
30	8.8	10.7	13.1	15.2	16.5	17.0	16.8	15.7	13.9	11.6	9.5	8.4
28	9.3	11.1	13.4	15.3	16.5	16.8	16.7	15.7	14.1	12.0	9.9	8.8
26	9.8	11.5	13.7	15.3	16.4	16.7	16.6	15.7	14.3	12.3	10.3	9.3
24	10.2	11.9	13.9	15.4	16.4	16.6	16.5	15.8	14.5	12.6	10.7	9.7
22	10.7	12.3	14.2	15.5	16.3	16.4	16.4	15.8	14.	13.0	11.1	10.2
20	11.2	12.7	14.4	15.6	16.3	1.4	16.3	15.9	14.8	13.3	11.6	10.7
18	11.6	13.0	14.6	15.6	16.1	16.1	16.1	15.8	14.9	13.6	12.0	11.1
16	12.0	13.0	14.6	15.6	16.1	16.1	16.1	15.8	14.9	13.6	12.0	11.1
14	12.4	13.6	14.9	15.7	15.8	15.7	15.7	15.7	15.1	14.1	12.8	12.0
12	12.8	13.9	15.1	15.7	15.7	15.5	15.5	15.6	15.2	14.4	13.3	12.5

Lat	Jan	Feb	Mar	Apr	May	June	July	Aug	Sept	Oct	Nov	Dec
10	13.2	14.2	15.3	15.7	15.5	15.3	15.3	15.5	15.3	14.7	13.6	12.9
8	13.6	14.5	15.3	15.6	15.3	15.0	15.1	15.4	15.3	14.8	13.9	13.3
6	13.9	14.8	15.4	15.4	15.1	14.7	14.9	15.2	15.3	15.0	14.2	13.7
4	14.3	15.0	15.5	15.5	14.9	14.4	14.6	15.1	15.3	15.1	14.5	14.1
2	14.7	15.3	15.6	15.3	14.6	14.2	14.3	14.9	15.3	15.3	14.8	14.4
0	15.0	15.5	15.7	15.3	14.4	13.9	14.1	14.8	15.3	15.4	15.1	14.8

Southern Hemisphere

Lat	Jan	Feb	Mar	Apr	May	June	July	Aug	Sept	Oct	Nov	Dec
50^0	17.5	14.7	10.9	7.0	4.2	3.1	3.5	5.5	8.9	12.9	16.5	18.2
48	17.6	14.9	11.2	7.5	4.7	3.5	4.0	6.0	9.3	13.2	16.6	18.2
46	17.7	15.1	11.5	7.9	5.2	4.0	4.4	6.5	9.7	13.4	16.7	18.3
44	17.8	15.3	11.9	8.4	5.7	4.4	4.9	6.9	10.2	13.7	16.7	18.3
42	17.8	15.5	12.2	8.8	6.1	4.9	5.4	7.4	10.6	14.0	16.8	18.3
40	17.9	15.7	12.5	9.2	6.6	5.3	5.9	7.9	11.0	14.2	16.9	18.3
38	17.9	15.8	12.8	9.6	7.1	5.8	6.3	8.3	11.4	14.4	17.0	18.3
36	17.9	16.0	13.2	10.1	7.5	6.3	6.8	8.8	11.7	14.6	17.0	18.2
34	17.8	16.1	13.5	10.5	8.0	6.8	7.2	9.2	12.0	14.9	17.1	18.2
32	17.8	16.2	13.8	10.9	8.5	7.3	7.7	9.6	12.4	15.1	17.2	18.1
30	17.8	16.4	14.0	11.3	8.9	7.8	8.1	10.1	12.7	15.3	17.3	18.1
28	17.7	16.4	14.3	11.6	9.3	8.2	8.6	10.4	13.0	15.4	17.2	17.9
26	17.6	16.4	14.4	12.0	9.7	8.7	9.1	10.9	13.2	15.5	17.2	17.8
24	17.5	16.5	14.6	12.3	10.2	9.1	9.5	11.2	13.4	15.6	17.1	17.7
22	17.4	16.5	14.8	12.6	10.6	9.6	10.0	11.6	13.7	15.7	17.0	17.5
20	17.3	16.5	15.0	13.0	11.0	10.0	10.4	12.0	13.9	15.8	17.0	17.4
18	17.1	16.5	15.1	13.2	11.4	10.4	10.8	12.3	14.1	15.8	16.8	17.1
16	16.9	16.4	15.2	13.5	11.7	10.8	11.2	12.6	14.3	15.8	16.7	16.8
14	16.7	16.4	15.3	13.7	12.1	11.2	11.6	12.9	14.5	15.8	16.5	16.6
12	16.6	16.3	15.4	14.0	12.5	11.6	12.0	13.2	14.7	15.8	16.4	16.5

Lat	Jan	Feb	Mar	Apr	May	June	July	Aug	Sept	Oct	Nov	Dec
10	16.4	16.3	15.5	14.2	12.8	12.0	12.4	13.5	14.8	15.9	16.2	16.2
8	16.1	16.1	15.5	14.4	13.1	12.4	12.7	13.7	14.9	15.8	16.0	16.0
6	15.8	16.0	15.6	14.7	13.4	12.8	13.1	14.0	15.0	15.7	15.8	15.7
4	15.5	15.8	15.6	14.9	13.8	13.2	13.4	14.3	15.1	15.6	15.5	15.4
2	15.3	15.7	15.7	15.1	14.1	13.5	13.7	14.5	15.2	15.5	15.3	15.1
0	15.0	15.5	15.7	15.3	14.4	13.9	14.1	14.8	15.3	15.4	15.1	14.8

Table A-2. Mean Daily Duration of Maximum Possible Sunshine (N)

Northern Lats	Jan	Feb	Mar	Apr	May	June	July	Aug	Sept	Oct	Nov	Dec
Southern Lats	July	Aug	Sept	Oct	Nov	Dec	Jan	Feb	Mar	Apr	May	June
50^0	8.5	10.1	11.8	13.8	15.4	16.3	15.9	14.5	12.7	10.8	9.1	8.1
48	8.8	10.2	11.8	13.6	15.2	16	15.6	14.3	12.6	10.9	9.3	8.3
46	9.1	10.4	11.9	13.5	13.5	15.7	15.4	14.2	12.6	10.9	19.5	8.7
44	9.3	10.5	11.9	13.4	14.7	15.4	15.2	14.0	12.6	11.0	9.7	8.9
42	9.4	10.6	11.9	13.4	14.6	15.2	14.9	13.9	12.6	11.1	9.8	9.1
40	9.6	10.7	11.9	13.3	14.4	15	14.7	13.7	12.5	11.2	10.0	9.3
35	10.1	11	11.9	13.1	14.0	14.5	14.3	13.5	12.4	11.3	10.3	9.8
30	10.4	11.1	12.0	12.9	13.6	14.0	13.9	13.2	12.4	11.5	10.6	10.2
25	10.7	11.3	12.0	12.7	13.3	13.7	13.5	13.5	13.0	12.3	11.6	10.6
20	11.0	11.5	12.0	12.6	13.1	13.3	13.2	12.8	12.3	11.7	11.7	10.9
15	11.3	11.6	12.0	12.5	12.8	13.0	12.9	12.6	12.2	11.8	11.4	112
10	11.6	11.8	12.0	12.3	12.6	12.7	12.6	12.4	12.1	11.8	11.6	11.5
5	11.8	11.9	12.0	12.2	12.3	12.4	12.3	12.3	12.1	12.0	11.9	11.8
0	12.0	12.0	12.0	12.0	12.0	12.0	12.0	12.0	12.0	12.0	12.0	12.0

Table A-3. Saturation Vapour Pressure V_{sa} in mb

Temp. °C	0	1	2	3	4	5	6	7	8	9	10	11	12	13
Vsa mbar	6.1	6.6	7.1	7.6	8.1	8.7	9.3	10.0	10.7	11.5	12.3	13.1	14.0	15.0
Temp. °C	14	15	16	17	18	19	20	21	22	23	24	25	26	27
Vsa mbar	16.1	17.0	18.2	19.4	20.6	22.0	23.4	24.9	26.4	28.1	29.8	31.1	33.6	35.7
Temp. °C	28	29	30	31	32	33	34	35	36	37	38	39		
Vsa mbar	37.8	40.1	42.4	44.9	47.6	50.3	53.2	56.2	59.4	62.g	66.3	69.9		

Also actual vapour pressure V_a, can be obtained from this table using available Tdewpoint data. (Example: Tdewpoint is 20°C : V_a is 23.4 mbar)

TABLE A-4 Vapour Pressure, V_a, in mb (Aspirated Psychrometer)

	Depression wet bulb T°C altitude 0-1000 m											
Dry bulb T°C	0	2	4	6	8	10	12	14	16	18	20	22
40	73.8	64.9	56.8	49.2	42.2	35.8	29.8	24.3	19.2	14.4	10.1	6.0
38	66.3	58.1	50.5	43.6	37.1	31.1	25.6	20.5	15.8	11.4	7.3	
36	59.4	51.9	44.9	38.4	32.5	26.9	21.8	17.1	12.7	8.6	4.9	
34	53.2	46.2	39.8	33.8	28.3	23.2	18.4	14.0	10.0	6.2		
32	47.5	41.1	35.1	29.6	24.5	19,8	15.4	I 1.3	7.5	4.0		
30	42.4	36.5	30.9	25.8	21.1	16.7	12.6	8.8	5.3			
28	37.8	3.3	27.2	22.4	18.0	14.0	10.2	6.7	3.4			
26	33.6	28.5	23.8	19.4	15.3	11.5	8.0	4.7	1.6			
24	29.8	25.1	20.7	16.6	12.8	9.3	6.0	2.9				
22	26.4	22.0	18.0	14.2	10.6	7.4	4.3	1.4				
20	23.4	19.3	15.5	12.0	8.7	5.6	2.7					
18	20.6	16.8	13.3	10.0	6.9	4.1	1.4					
16	18.2	14.6	11.4	8.3	5.4	2.7						

14	16.0	12.7	9.6	6.7	4.0	1.5
12	14.0	10.9	8.1	5.3	2.8	
10	12.3	9.4	6.7	4.1	1.7	
8	10.7	8.0	5.5	3.1	0.8	
6	9.3	6.8	4.4	2.1		
4	8.1	5.7	3.4	1.6		
2	7.1	4.8	2.8	0.8		
0	6.1	4.0	2.0			

	Depression wet bulb T^0C altitude 1000-2000 m											
Dry bulb T^0C	0	2	4	6	8	10	12	14	16	18	20	22
40	73.8	65.2	571	49.8	43.0	41.8	31.0	25.6	20.7	16.2	12.0	8.1
38	66.3	58.2	50.9	44.1	37.9	36.7	26.8	21.8	17.3	13.2	9.2	5.7
36	59.4	52.1	45.2	39.0	33.3	32.1	23.0	18.4	14.3	10.4	6.8	3.5
34	53.2	46.4	40.1	34.4	29.1	24.1	19.6	15.4	11.5	8.0	4.6	1.5
32	47.5	41.3	35.5	30.2	25.3	20.7	16.6	12.6	9.1	5.8	2.6	
30	42.4	36.7	31.3	26.4	21.9	17.7	13.8	10.2	6.9	3.8	0.9	
28	37.8	32.5	27.5	23.0	18.9	14.9	11.4	8.0	4.9	2.1		
26	33.6	28.7	24.1	20.0	16.1	12.5	9.2	6.0	3.2	0.5		
24	29.8	25.3	21.1	17.2	13.9	10.3	7.2	4.3	1.6			
22	26.4	22.3	18.3	14.3	11.5	8.3	5.5	2.7	0.2			
20	23.4	19.5	15.9	12.6	9.5	6.6	3.9	1.3				
18	20.6	17.1	13.7	10.6	7.8	5.0	2.5	0.1				
16	18.2	14.9	11.7	8.9	6.2	3.6	1.3					
14	16.0	12.9	10.0	7.3	4.8	2.4	0.3					
12	14.0	11.2	8.4	5.9	3.6	1.4						
10	12.3	9.6	7.0	4.7	2.6	0.4						
8	10.7	8.2	5.8	3.7	1.6							
6	9.3	7.0	4.8	2.7	0.7							

Table A-5 Values of weighting factor at Different Temperature and Altitudes

Temp. ^{0}C2		4	6	8	10	12	14	16	18	20	22	24	26	27	28	30	32	34	36	38	40
	m																				
w	0	.43	.46	.49	.52	.55	.58	.61	.64	.66	.69	.71	.73	.75	.77	.78	.80	.82	.83	.84	.85
at	500	.44	.48	.51	.54	.57	.60	.62	.65	.67	.70	.72	.74	.76	.78	.79	.81	.82	.84	.85	.86
Alt	1000	.46	.49	.52	.55	.58	.61	.64	.66	.69	.71	.73	.75	.77	.79	.80	.82	.83	.85	.86	.87
	2000	.49	.52	.55	.58	.61	.64	.66	.69	.71	.73	.75	.77	.79	.81	.82	.84	.85	.86	.87	.88
	3000	.52	.55	.58	.61	.64	.66	.69	.71	.73	.75	.77	.79	.81	.82	.84	.85	.86	.87	.88	.89
	4000	.54	.58	.61	.64	.66	.69	.71	.73	.75	.77	.79	.81	.82	.84	.85	.86	.87	.89	.90	.90

Table A-6 Correction Factors for Winds Measured at Heights other than 2 m

Measurement height, m	0.50	1.00	1.50	2.0	3.0	4.0	5.0	6.0
Correction factor	1.35	1.15	1.06	1.0	0.93	0.88	0.85	0.83

Table A-7 Albedo for Various Types of Surfaces

Surface	Range of t-values
Close ground crops	0.15 – 0.25
Bare lands	0.05 – 0.45
Water surface	0.05
Snow	0.45 – 0.90

Table A-8 Values of K for Selected Crops for use in Eq. (11.30)

Crop	Value of K	Range of monthly value
Rice	1.10	0.85-1.30
Wheat	0.65	0.50-0.75
Maize	0.65	0.50-0.80
Sugarcane	0.90	0.75-1.00
Cotton	0.65	0.50-0.90
Potatoes	0.70	0.65-0.75

Natural Vegetation:	
(a) Very dense	1.30
(b) Dense	1.20
(c) Medium	1.00
(d) Light	0.80

Table A-9. Monthly Daytime Hours Percentages

North latitude (degree s)	Jan	Feb	Mar	Apr	May	June	July	Aug	Sept	Oct	Nov	Dec
0	8.50	7.66	8.49	8.21	8.50	8.22	8.50	8.49	8.21	8.50	8.22	8.50
10	8.13	7.47	8.45	8.37	8.81	8.60	8.85	8.71	8.25	8.34	7.91	8.10
15	7.94	7.36	8.43	8.44	8.98	8.80	9.05	8.83	8.28	8.26	7.75	7.88
20	7.74	7.25	8.41	8.52	9.15	9.00	9.25	8.96	8.30	8.18	7.58	7.66
25	7.53	7.14	8.39	8.61	9.33	9.23	9.45	9.09	8.32	8.09	7.40	7.42
30	7.30	7.03	8.38	8.72	9.53	9.49	9.57	9.22	8.33	7.99	7.19	7.15
35	7.05	6.88	8.35	8.83	9.76	9.7	9.93	9.37	7.86	6.97	6.97	6.86
40	6.76	6.72	8.33	8.95	10.02	10.08	10.22	9.54	8.39	7.75	6.72	6.52

FREE FLOW ANALYSIS OF CUTTHROAT FLUMES

(*Source:* Skogerboe, *et al.*, 1973)

$$Q = C_1 h_a^{n_1} \qquad \text{.... (B-1)}$$

where,

Q = flow rate, in cubic feet per second,

C_1 = free flow coefficient which is the value of Q when h_a is 1.0 foot; which is the slope of the free flow rating when plotted on logarithmic paper (Fig. B-1).

The values of n_1 can be determined for any flume length between 1.5 feet and 9 feet by simply reading the value from the graphs shown in Fig. C-1.

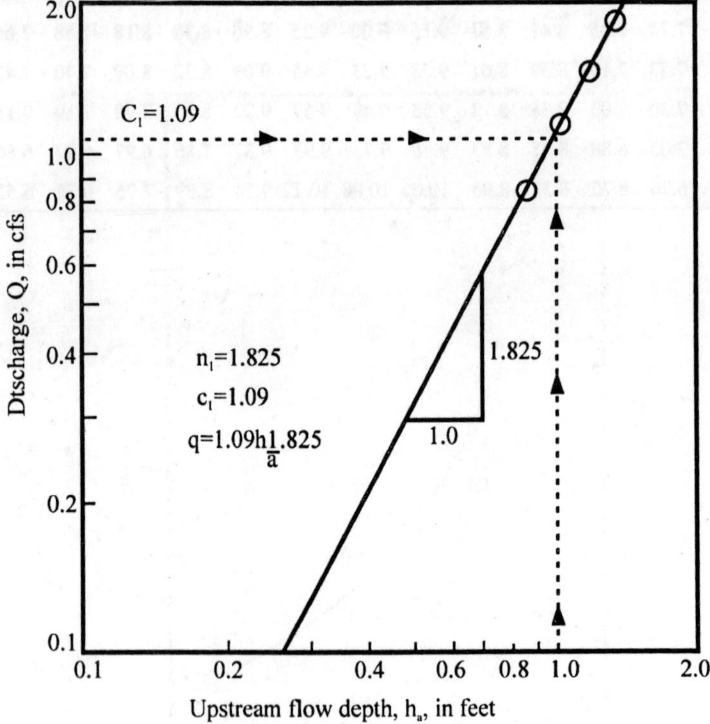

Fig. B-1: Typical free flow rating curve showing actual data points and development of free flow equation

The value of the free flow coefficient is a function of both flume length, *L*, and throat width, *W*. This relationship is:

$$C_1 = K_1 W^{1.025} \qquad \text{.... (B-2)}$$

where,

C_I = is the free flow coefficient

K_I = the flume length coefficient

W = the throat width in feet.

The values of K_I can be obtained from the Fig. B-2.

Having obtained the values of n_I and C_I for the flume being used, the discharge can now be calculated for any ha by using Equation B-1, provided free flow conditions exist in the flume. For accurate discharge measurements, recommended ratio of flow depth to flume length (*ha/L*) should be equal to or less than 0.4, with increasing values of this ratio resulting in greater in accuracies because of higher approach velocities and a amore rapidly changing water surface profile at the flume cross-section where ha is measured.

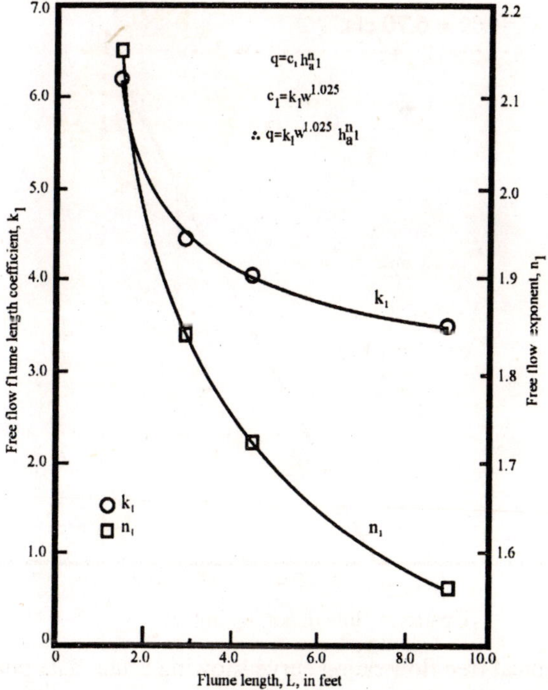

Fig. B-2: Generalized free flow ratings of cutthroat flumes

Example B-1: A free flow rating is needed for a cutthroat flume of length L = 4.0 fee and width W = 1.167 feet. From Fig. B-2 the value of n_l is 1.75 and the value K_l is 4.15. Then using Equation B-2 the value of the free flow coefficient, C_l is calculated. (Source : Michael, 1992)

$$
\begin{aligned}
C_l &= K_l W^{1.025} \\
&= 4.15\,(1.167)^{1.025} \\
&= 4.15\,(1.172) \\
&= 4.86
\end{aligned}
$$

Now, knowing the values of n_l and C_l, the flow rate through the flume can be calculated for any value of ha using Equation B-1. Assuming

$$h_a = 1.20 \text{ feet}$$

$$
\begin{aligned}
Q &= C_1 h_a^{n_1} \\
&= 4.86\,(1.20)^{1.75} \\
&= 4.86\,(1.38) \\
&= 6.70 \text{ cfs.}
\end{aligned}
$$

Appendix-C

RAINFALL FREQUENCY DISTRIBUTION IN INDIA

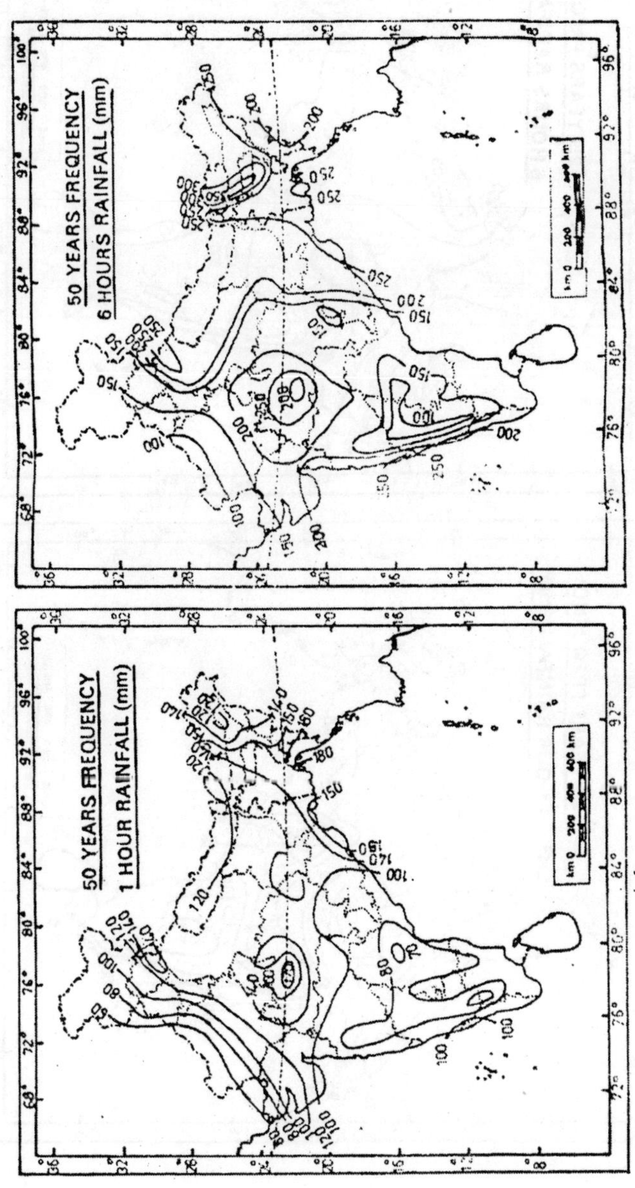

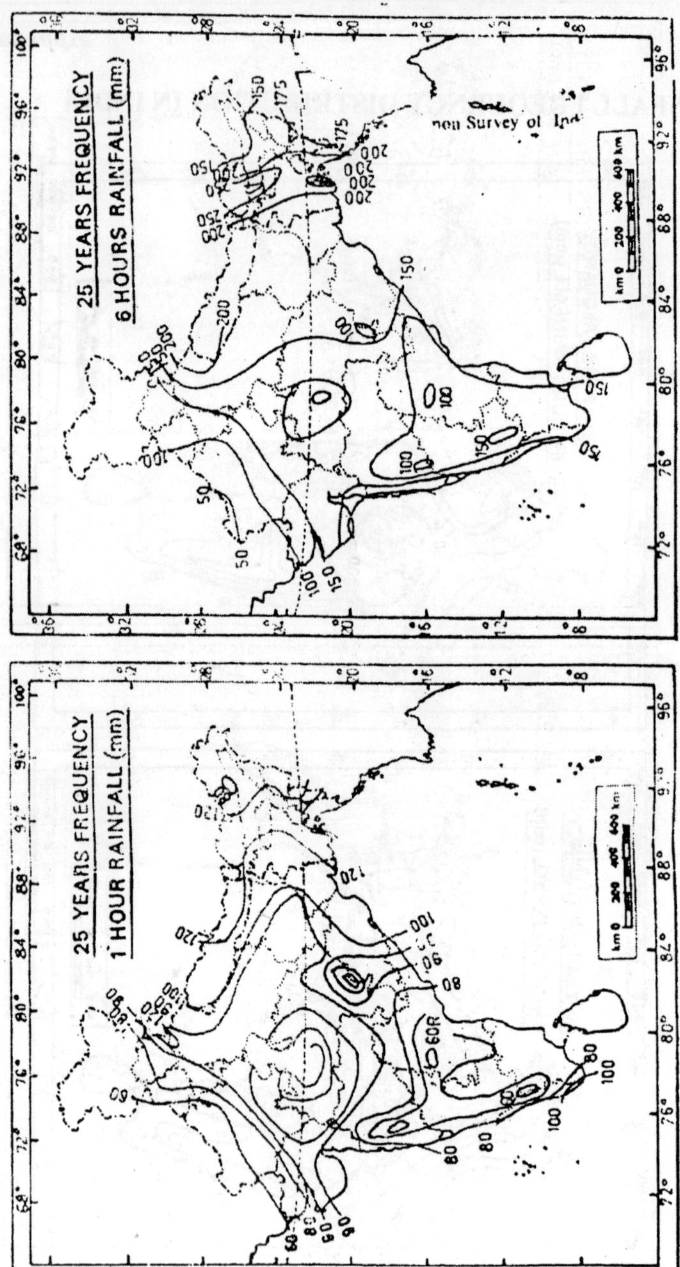
25 YEARS FREQUENCY
6 HOURS RAINFALL (mm)
25 YEARS FREQUENCY
1 HOUR RAINFALL (mm)

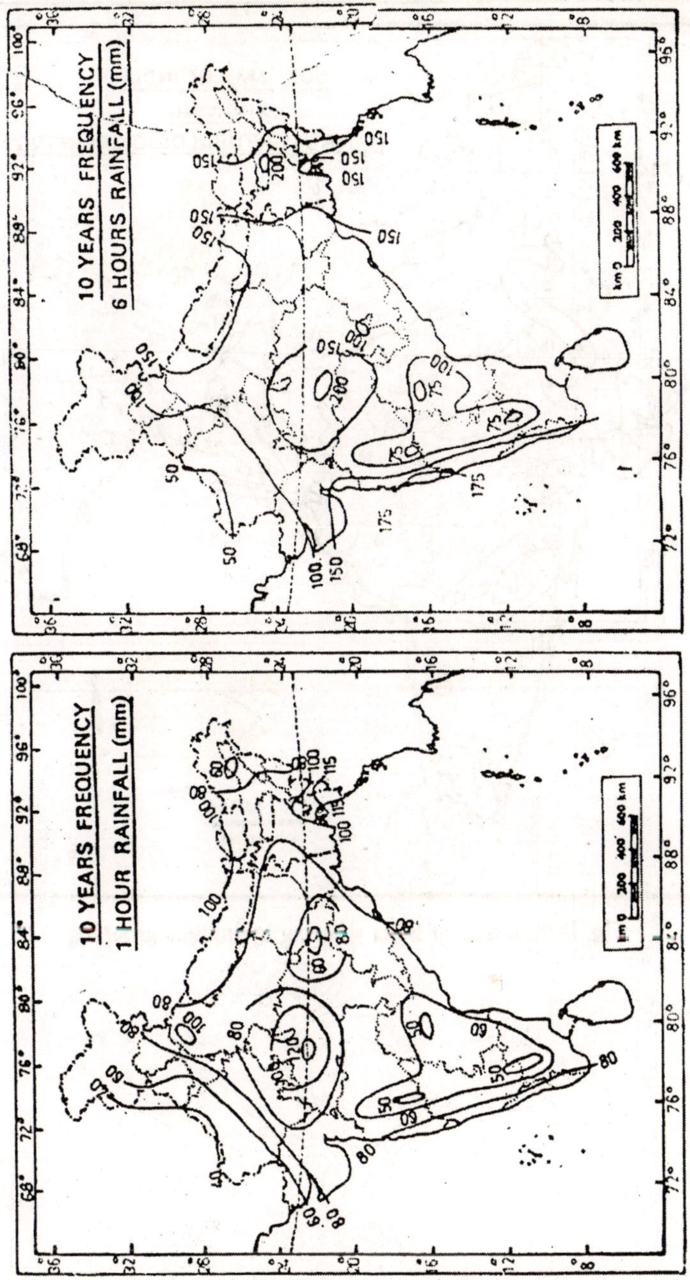

Fig. C. Rainfall frequency distribution in India

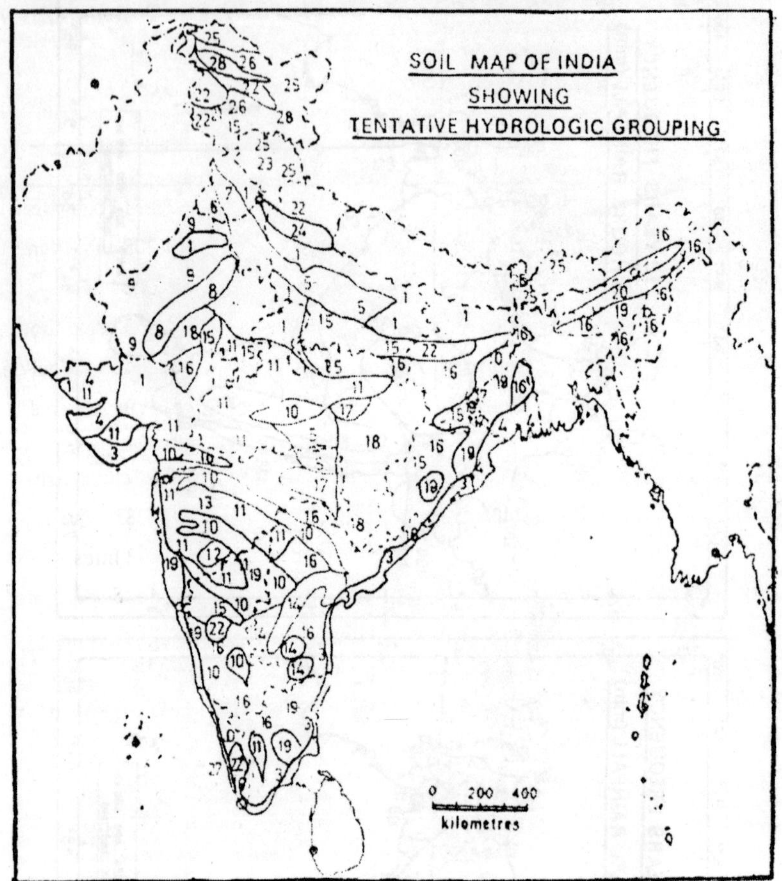

Fig. D. Soil map of India showing hydrologic grouping

Appendix-D

CONVERSION OF FACTORS

Length		Volume	
1 metre	= 3.2808 feet	1 cubic metre	= 35.314 cubic feet
1 metre	= 39.37 inches	1 cubic metre	= 1.308 cubic yards
1 centimetre	= 0.3937 inch	1 cubic metre	= 1,000 litres
1 kilometre	= 0.6214 mile	1 litre	= 0.353 cubic foot
1 foot	= 0.3048 metre	1 litre	= 0.2642 U.S. gallon
1 inch	= 2.54 centimetres	1 litre	= 0.2201 Imperial gallon
1 mile	= 5,280 feet	1 cubic cm	= 0.61 cubic inch
1 mile	= 1.609 kilometres	1 cubic foot	= 0.0283 cubic metre
Area		1 cubic foot	= 23.32 litres
1 square metre	= 10.764 square feet	1 cubic foot	= 7.40 U.S. gallons
1 square cm	= 0.155 square inch	1 cubic foot	= 6.23 Imperial gallons
1 square km	= 100 hectares	1 cubic inch	= 16.39 cubic cm
1 square km	= 0.3861 square mile	1 cubic yard	= 0.7645 cubic metre
1 hectare	= 10,000 square metres	1 U.S. gallon	= 3.7854 litres
1 hectare	= 107.640 square feet	1 U.S. gallon	= 0.833 Imperial gallon
1 hectare	= 2.471 acres	1 Imperial gallon	= 1.201 U.S. gallon
1 square foot	= 0.0929 square metre	1 Imperial gallon	= 4.5436 litres
1 square inch	= 6.452 square cm	1 acre-foot	= 43,560 cubic foot
1 acre	= 43.560 square ft	1 acre-foot	= 1,233.5 cubic metres
1 acre	= 0.4047 hectare	1 acre inch	= 3,630 cubic feet
1 square mile	= 640 acres	1 acre inch	= 102.8 cubic metres
1 square mile	= 258.99 hectares		
1 square mile	= 2.59 square km		
Rates of Flow		**Useful Approximations**	
1 cubic metre per second	= 35.314 cubic feet per second	5 cm	= 2 inches

1 cubic metre per hour	= 0.278 litres per second	10 cm	= 4 inches
1 cubic metre per second	= 4.403 U.S. gallons per minute	30 cm	= 1 feet
1 cubic metre per second	= 3.668 Imperial gallons per minute	10 metres	= 33 feet
1 litre per second	= 0.0353 cubic foot per second	10 km	= 6 miles
1 litre per second	= 15.852 U.S. gallons per minute	16 km	= 10 miles
1 litre per second	= 13.206 Imperial gallons per minute	1 hectare	= 2.5 acres
1 litre per second	= 3.6 cubic metres per second	1 square inch	= 6.5 square cm
1 cubic foot per second (second-foot or cusec)	= 0.0283 cubic metres per second	1 square metre	= 11 square feet
1 cubic foot per second	= 28.32 litres per second	16.4 cubic cm	= 1 cubic inch
1 cubic foot per second	= 448.8 U.S. gallons per minute	35 cubic feet	= 1 cubic metre
1 cubic foot per second	= 373.8 Imperial gallons per minute		
1 cubic foot per second	= 1 acre inch per hour (approx.)		
1 U.S. gallons per minute	= 0.06309 litres per second		
1 Imperial gallons per minute	= 0.07573 litres per second		

English (FPL) Metric Conversion Constants

Measure	FPS		Metric	
	Unit	FPS to Metric	Unit	Metric to FPS
Length	Inch (")	× 25.4	$mm = \frac{1}{1000\,m}$	× 0.03937
	Foot (') = 12"	× 0.3048		× 3.28

	Yard = 3'	× 0.9144	Km = 1,000 m	× 1.09361
	Mile = 1,760 yd.	× 1.6093	Km = 1,000 m	× 0.621
	Nautical mile	× 1.852	cm^2	× 2.540
Area	Square inch	× 6.451	m^2	× 0.1550
	Square feet	× 0.0929	Hectare	× 10.764
	Acres	× 0.4047	cm^3	× 2.471
Volume	Cubic inch	× 16.3871	Litre = $(dm)^3$	× 0.0610
	Imp. gallon (=1.2 American gallon)	× 4.546	M^3	× 0.220
	Imp. gallon (= 10 Ib)	× 0.004546	Litre	× 2.20
	American gallon	× 3.785	hl	× 0.2642
	Petroleum barrel = 42		Litre	× 0.63
	American gallons	× 1.59	M^3	× 0.0353
	Cubic feet	× 28.316	gr	× 35.38
	Cubic feet	× 0.0283		× 15.42
Weight	Grain $= \frac{1}{7000}$	× 0.0648	gr	× 0.035
	Ounce $= \frac{1}{16} lb$	× 28.35	kg	× 2.205
	Pound = 1 lb	× 0.4536	kg	× 0.0197
	Cwt = 112 lb	× 50.802	Tonne	× 1.102
	Short or American ton = 2,000 lb	× 0.907	Tonne	× 0.98421
	Long ton = 2,240 lb	× 1.016		
Pressure and head	lb/sq. in $= \frac{ft\ of\ water}{2.31}$	× 0.0703	kg/cm^2 $= \frac{m\ of\ water}{10}$	× 14.22
	lb/sq. in	× 0.068	atm	× 14.7
	Long ton/sq. in	× 157.5	Kg/cm^2	× 0.00635
	in. mercury $= \frac{ft\ of\ water}{1.13}$	× 345	mm of water	× 0.0029
	In mercury	× 25.4	mm of mercury	× 0.03937

	1 atm (= 34 ft. of water = 14.7 lb/sq. in.)	× 1.033	1 atm. (= 10 m of water or 1 kg/cm^2)	× 0.967
Density	lb/cu. ft.	× 16.02	kg/m^3	× 0.0625
	lb/cu. ft.	× 0.01602	kg/dm^3 or kg/litre	× 62.424
	lb/cu. In.	× 27.6799	kg/cm^3	× 0.0362
Velocity	ft/sec	× 0.3048	m/sec	× 3.28
	ft/min	$\times \frac{5.08}{1000}$	m/sec	× 196.85
Rate of discharge	cu. ft/sec (cusec)	× 0.0283	m^3/sec	× 35.38
	cu. ft/sec (cusec)	× 28.316	lit/sec (= dm^3/sec)	× 0.0353
	cu. ft/min	× 0.472	lit/sec (= dm^3/sec)	× 2.22
	gal/min (gpm)	× 0.0757	lit/sec (= dm^3/sec)	× 13.22
	gal/min (gpm)	$\times \frac{0.757}{10.000}$	m^3/sec	× 13.22

■■■■

SUBJECT INDEX